KV-618-251

FRANK CURTIS LIBRARY
HELLESDON HOSPITAL
NH23145

FUNDAMENTALS OF INFLAMMATION

The acute inflammatory response is the body's first system of alarm signals that are directed toward containment and elimination of microbial invaders. Uncontrolled inflammation has emerged as a pathophysiologic basis for many widely occurring diseases in the general population that were not initially known to be linked to the inflammatory response, including cardiovascular disease, asthma, arthritis, and cancer. To better manage treatment, diagnosis, and prevention of these wide-ranging diseases, multidisciplinary research efforts are under way in both academic and industry settings. The purpose of this book is to provide an introduction to the cell types, chemical mediators, and general mechanisms of the host's first response to invasion. World-class experts from institutions around the world have written chapters for this introductory text. The text is presented as an introductory springboard for graduate students, postdoctoral Fellows, medical scientists, and researchers from other disciplines who wish to gain an appreciation and working knowledge of current cellular and molecular mechanisms fundamental to inflammation.

Charles N. Serhan, PhD, is the Director of the Center for Experimental Therapeutics and Reperfusion Injury at Brigham and Women's Hospital and the Simon Gelman Professor of Anesthesia (Biochemistry and Molecular Pharmacology) at Harvard Medical School, Boston, Massachusetts. He is one of the world's top researchers on the mechanisms and mediators of acute inflammation and its resolution.

Peter A. Ward, MD, is Stobbe Professor of Pathology at the University of Michigan Medical School, Ann Arbor, Michigan. Dr. Ward is past president of the United States and Canadian Academy of Pathology, as well as a number of other scientific societies. He is a world leading authority on sepsis and the impact of inflammation in human disease.

Derek W. Gilroy, PhD, is a Wellcome Trust Senior Fellow in the Centre for Clinical Pharmacology, University College, London, United Kingdom. He has received the Bayer Aspirin Prize and Novartis Prize. He is a world leader in resolution mechanisms in inflammation and the immunopharmacology of anti-inflammatory therapeutics.

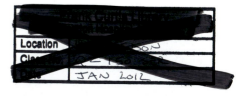

FUNDAMENTALS OF INFLAMMATION

Edited by

CHARLES N. SERHAN

Harvard Medical School, Brigham and Women's Hospital, Boston, Massachusetts

PETER A. WARD

University of Michigan Medical School, Ann Arbor, Michigan

DEREK W. GILROY

University College, London, United Kingdom

CAMBRIDGE UNIVERSITY PRESS
Cambridge, New York, Melbourne, Madrid, Cape Town, Singapore,
São Paulo, Delhi, Dubai, Tokyo

Cambridge University Press
32 Avenue of the Americas, New York, NY 10013-2473, USA

www.cambridge.org
Information on this title: www.cambridge.org/9780521887298

First published 2010

Printed in China by Everbest

A catalog record for this publication is available from the British Library.

Library of Congress Cataloging in Publication data
Fundamentals of inflammation / [edited by] Charles N. Serhan, Peter A. Ward, Derek W. Gilroy.
p. ; cm.
Includes bibliographical references and index.
ISBN 978-0-521-88729-8 (hardback)
1. Inflammation. I. Serhan, Charles N. II. Ward, Peter A., 1934– III. Gilroy, Derek W.
[DNLM: 1. Inflammation – immunology. 2. Immunity, Cellular – physiology.
QW 700 F981 2010]
RB131.F845 2010
616′.0473–dc22 2009050381

ISBN 978-0-521-88729-8 Hardback

CONTENTS

CONTRIBUTORS

Steven B. Abramson
Department of Rheumatology
New York University Langone Medical Center
New York Harbor Veterans Affair Health Care System
New York, New York

Steven J. Ackerman
Department of Biochemistry and Molecular Genetics
Department of Medicine
University of Illinois, Chicago
Chicago, Illinois

Julio Aliberti
Division of Molecular Immunology
Cincinnati Children's Hospital Medical Center
University of Cincinnati College of Medicine
Cincinnati, Ohio

Jonathan Arm
Division of Rheumatology, Immunology and Allergy
Brigham and Women's Hospital
Boston, Massachusetts

Samir S. Ayoub
Biochemical Pharmacology
The William Harvey Research Institute
Barts and the London School of Medicine and Dentistry
Queen Mary College
University of London
London, United Kingdom

Paul L. Beck
Inflammation Research Network and Gastrointestinal Research Group
Faculty of Medicine
University of Calgary
Calgary, Alberta
Canada

M. Blades
William Harvey Research Institute
Barts and the London School of Medicine and Dentistry
Queen Mary College
University of London
Charterhouse Square
London, United Kingdom

Joshua A. Boyce
Harvard Medical School
Allergy and Inflammation Research
Division of Rheumatology, Immunology and Allergy
Brigham and Women's Hospital,
Boston, Massachusetts

Christopher D. Buckley
Rheumatology Research Group
School of Immunity and Infection
Medical Research Council Centre for Immune Regulation
College of Medical and Dental Sciences
University of Birmingham
Edgbaston, Birmingham
United Kingdom

M. Burnet
Synovo GmbH
Germany

Marco A. Cassatella
Department of Pathology
Division of General Pathology
School of Medicine
University of Verona
Verona, Italy

Mary Cavanagh
Imperial College London
Leucocyte Biology Section
National Heart and Lung Institute
London, United Kingdom

Lawrence Chan
Departments of Medicine and Molecular & Cellular Biology
Baylor College of Medicine
Houston, Texas

Sean P. Colgan
Mucosal Inflammation Program
University of Colorado Health Sciences Center
Denver, Colorado

Dianne Cooper
William Harvey Research Institute
Barts and the London School of Medicine and Dentistry
London, United Kingdom

Bruce Cronstein
Department of Medicine
New York University School of Medicine
New York, New York

Constance D'Amato
Department of Pathology
University of Michigan Medical School
Ann Arbor, Michigan

B. Dawson
Bionet Ltd.
Cheviot
Medstead
Alton, Hampshire
United Kingdom

Jeremy S. Duffield
Brigham and Women's Hospital
Harvard Institute of Medicine and Harvard Medical School
Boston, Massachusetts

Neil Dufton
The William Harvey Research Institute
Barts and the London School of Medicine
London, United Kingdom

Sophie Fillon
University of Colorado Denver School of Medicine
Department of Pediatrics
The Children's Hospital, Denver
National Jewish Health Section of Pediatric Gastroenterology, Hepatology and Nutrition
Gastrointestinal Eosinophilic Diseases Program
Aurora, Colorado

János G. Filep
Faculty of Medicine
University of Montreal and Research Center
Masionneuve-Rosemont Hospital
Montréal, Quebec
Canada

Andrew Filer
Rheumatology Research Group
Division of Immunity and Infection
University of Birmingham
Birmingham, United Kingdom
City Hospital, Sandwell and West Birmingham Hospitals NHS Trust
Birmingham, United Kingdom

Emily G. Findlay
Research Fellow
Immunology Unit
Department of Infections and Tropical Diseases
London School of Hygiene and Tropical Medicine
London, United Kingdom

Roderick Flower
Biochemical Pharmacology
The William Harvey Research Institute
Barts and the London School of Medicine and Dentistry
Queen Mary College
University of London
London, United Kingdom

Sarah Fox
Medical Research Council Centre for Inflammation Research
Queen's Medical Research Institute
University of Edinburgh Medical School
Edingburgh, Scotland

Glenn T. Furuta
University of Colorado Denver, School of Medicine
Department of Pediatrics
The Children's Hospital, Denver
National Jewish Health Section of Pediatric Gastroenterology, Hepatology and Nutrition
Gastrointestinal Eosinophilic Diseases Program
Aurora, Colorado

Derek W. Gilroy
Centre for Clinical Pharmacology and Therapeutics
Division of Medicine
University College London
London, United Kingdom

Catherine Godson
Diabetes Research Centre
Conway Institute and School of Medicine and Medical Sciences
University College Dublin
Dublin, Ireland

Caroline Gordon
Division of Immunity and Infection
University of Birmingham
Edgbaston, Birmingham
United Kingdom

Karsten Gronert
University of California, Berkeley
Center for Eye Disease and Development
School of Optometry
Berkeley, California

Jesper Z. Haeggström
Department of Medical Biochemistry and Biophysics
Division of Chemistry II
Karolinska Institute
Stockholm, Sweden

Thorsten Hagemann
Centre for Cancer and Inflammation
Institute of Cancer
Barts and the London School of Medicine and Dentistry
London, United Kingdom

György Haskó
Department of Surgery
University of Medicine and Dentistry of New Jersey – New Jersey Medical School
Newark, New Jersey

Hatice Hasturk
Goldman School of Dental Medicine
Department of Periodontology and Oral Biology
Boston University
Boston, Massachusetts

Joel M. Henderson
Assistant Professor
Department of Pathology and Laboratory Medicine
Boston University School of Medicine
Boston Medical Center
Boston, Massachusetts

F. Humby
William Harvey Research Institute
Barts and the London School of Medicine and Dentistry
Queen Mary College University of London
Charterhouse Square
London, United Kingdom

Tracy Hussell
Imperial College London
London, United Kingdom

Alpdogan Kantarci
Goldman School of Dental Medicine
Department of Periodontology and Oral Biology
Boston University
Boston, Massachusetts

Christopher L. Karp
Division of Molecular Immunology
Cincinnati Children's Hospital Research Foundation
University of Cincinnati College of Medicine
Cincinnati, Ohio

Toby Lawrence
Centre for Cancer and Inflammation
Institute of Cancer
Barts and The London School of Medicine and Dentistry
London, United Kingdom

Giovanna Leoni
William Harvey Research Institute
Barts and the London School of Medicine and Dentistry
London, United Kingdom

Bruce D. Levy
Pulmonary and Critical Care Medicine, Department of Medicine
Brigham and Women's Hospital and Harvard Medical School
Boston, Massachusetts

Andrew P. Lieberman
Department of Pathology
University of Michigan Medical School
Ann Arbor, Michigan

Dennis M. Lindell
Assistant Professor
Department of Immunology
University of Washington
Center for Immunity and Immunotherapies
Seattle Children's Research Institute
Seattle, Washington

Nicholas W. Lukacs
Professor of Pathology
Director, Molecular and Cellular Pathology Graduate Program
Assistant Dean for Research Faculty
Ann Arbor, Michigan

Aksam Merched
Departments of Medicine and Molecular & Cellular Biology
Baylor College of Medicine
Houston, Texas

William M. Nauseef
Department of Internal Medicine – Infectious Diseases
The University of Iowa Health Care
Iowa City, Iowa

Frank O. Nestle
Mary Dunhill Chair of Cutaneous Medicine and Immunotherapy
Division of Genetics and Molecular Medicine
St Johns Institute of Dermatology
King's College London School of Medicine
Guy's Hospital
London, United Kingdom

Lucy V. Norling
William Harvey Research Institute
Barts and the London School of Medicine and Dentistry
London, United Kingdom

Jagdeep Obhrai
Nephrology and Hypertension
Section of Transplant Medicine and Molecular Microbiology and Immunology
Oregon Health and Science University
Portland, Oregon

H.B. Patel
William Harvey Research Institute
Barts and the London School of Medicine and Dentistry
Queen Mary College University of London
Charterhouse Square
London, United Kingdom

Gayathri K. Perera
Clinical Training Research Fellow
Medical Research Council
Cutaneous Medicine and Immunotherapy Unit
St John's Institute of Dermatology
Department of Medical and Molecular Genetics
King's College London School of Medicine
London, United Kingdom

Mauro Perretti
William Harvey Research Institute
Barts and the London School of Medicine and Dentistry
London, United Kingdom

Michael H. Pillinger
Department of Rheumatology
New York University Langone Medical Center
New York Harbor Veterans Affair Health Care System
New York, New York

C. Pitzalis
William Harvey Research Institute
Barts and the London School of Medicine and Dentistry
Queen Mary College University of London
Charterhouse Square
London, United Kingdom

Karim Raza
Division of Immunity and Infection
University of Birmingham
Edgbaston, Birmingham
United Kingdom

Adriano G. Rossi
Medical Research Council Centre for Inflammation Research
Queen's Medical Research Institute
University of Edinburgh Medical School
Edinburgh, Scotland

Aidan Ryan
Diabetes Research Centre
Conway Institute and School of Medicine and Medical Sciences
University College Dublin
Dublin, Ireland

Denise M. Sadlier
Diabetes Research Centre
Conway Institute and School of Medicine and Medical Sciences
University College Dublin
Dublin, Ireland

André L.F. Sampaio
The William Harvey Research Institute
Barts and the London School of Medicine and Dentistry
London, United Kingdom

Jose U. Scher
Department of Rheumatology
New York University Langone Medical Center
New York Harbor Veterans Affair Health Care System
New York, New York

M. Seed
William Harvey Research Institute
Barts and the London School of Medicine and Dentistry
Queen Mary College
University of London
Charterhouse Square
London, United Kingdom

Charles N. Serhan
Director, Center for Experimental Therapeutics and Reperfusion Injury
Department of Anesthesiology, Perioperative and Pain Medicine
Brigham and Women's Hospital
Harvard Medical School
Boston, Massachusetts

David Sloane
Division of Rheumatology, Immunology and Allergy
Brigham and Women's Hospital
Boston, Massachusetts

Thomas E. Van Dyke
Goldman School of Dental Medicine
Department of Periodontology and Oral Biology
Boston University
Boston, Massachusetts

Linda Vong
Farncombe Family Digestive Health Research Institute
McMaster University
Hamilton, Ontario
Canada

John L. Wallace
Director
Farncombe Family Digestive Health Research Institute
McMaster University
Hamilton, Ontario
Canada

Peter A. Ward
The University of Michigan Medical School
Ann Arbor, Michigan

Erika Wissinger
Imperial College London
Leucocyte Biology Section
National Heart and Lung Institute
London, United Kingdom

Kenneth K. Wu
National Health Research Institute
Zhunan, Miaoli
Taiwan
University of Texas Health Science Center
Houston, Texas

Tony L. Yaksh
Department of Anesthesiology
University of California – San Diego
San Diego, California

PREFACE

The acute inflammatory response is the body's first system of alarm signals that are directed toward containment and elimination of microbial invaders. Uncontrolled inflammation has emerged as a pathophysiologic basis to many of the widely occurring diseases in the general population that were not initially known to be linked to events in the inflammatory response. These include cardiovascular diseases and neurodegenerative diseases (including Alzheimer's disease), and it has now become apparent that inflammation is an important component of cancer progression and the persistence of neuropathic pain. These are diseases that cross many disciplines. To better manage treatment, diagnosis, and prevention of diseases, multidisciplinary research efforts are under way in both academic and industry settings. Since knowledge of the acute inflammatory response in itself spans many disciplines, the editors' mission is to provide in this text an introduction to the cell types, chemical mediators, and general mechanisms that are involved in this primordial first response of the host to invasion. It is also now clear that the termination or the resolution of the acute inflammatory response is an active process, which is pivotal and is the outcome of the acute response. As an endogenous programmed response, the terrain of resolution holds many new possibilities for treatment and prevention of uncontrolled inflammation in a wide range of diseases.

World-class experts from many different universities and fields have written the chapters of this introductory textbook. The main sections of this book are focused on the cell types, processes, and molecular events that constitute the acute inflammatory response as we know it today. They cross the biomedical disciplines of hematology, infectious disease, pulmonary medicine, gastroenterology, oral medicine and dentistry, biochemistry, immunology, immunopharmacology, and general pathology. Given the need to gain a more complete understanding of the acute inflammatory response and its resolution, the scope of this text is presented as an introductory springboard intended for graduate students, postdoctoral Fellows, medical scientists, and senior researchers from other disciplines who wish to gain an appreciation and working knowledge of the current cellular and molecular mechanisms of the effector immune system that are fundamental in inflammation.

Part I of this text is devoted to examining acute inflammation, chronic inflammation, wound healing, and resolution, with an emphasis on current concepts in molecular and cellular events and their relevance to health and disease. The first three chapters in Part I thus provide a general view of the terrain and cellular players in inflammation.

Part II of this text brings into focus the individual cell types important in acute and chronic inflammation, their cellular and molecular biology, and, importantly, an introduction of their role in disease processes. Attention is also directed toward the importance of cell–cell interactions in the acute inflammatory response and our current understanding of the key interface between vascular, blood-borne cell types and their relation to interstitial events within inflamed tissues.

Part III stresses the importance of endogenous chemical mediators and local mediators in this process. In this regard, an update is provided on the important role of lipid-derived mediators and protein-derived mediators, including chemokines and cytokines, as well as nucleotide mediators such as adenosine and oxygen-derived reactive oxygen species. The importance of surface adhesion molecules in these processes is also stressed. The role and molecular mechanisms of each of these systems as well as their contributions to host defense is presented in view of their physiology and pathobiology in inflammation.

Since there is considerable interest in understanding the endogenous control mechanisms, as well as new therapeutic approaches to control inflammation in disease, Part IV of this text is devoted to an introduction to immunopharmacology, with a view of current mediators and mechanisms involved in inflammatory pain, currently used nonsteroidal anti-inflammatory drugs, and the importance of cytokines in our current appreciation of the interface between cancer and inflammation.

Part V brings us to one of the unique features of this introductory textbooks. Each of these chapters focuses on the tissue face or histology of inflammation as viewed in human diseases that are characterized by excessive inflammation. The chapters in this part are short and include histology and case reports. This part aims to discuss clinician scientists' and academic pathologists' views about inflammation in relation to widely occurring diseases. The goal is to give readers a picture of inflamed tissues and disease processes that we need to address as researchers to develop better approaches for prevention and treatment via new knowledge and innovative research of these diseases. Part V includes examples from airway inflammation, neural inflammation, sepsis, gastrointestinal diseases, and skin diseases characterized by inflammation, as well as kidney and cardiovascular diseases.

Part VI presents current and widely used animal models that are particularly useful in understanding experimental approaches to study inflammation. This part includes chapters with an emphasis on methodological approaches to address tissue injury and reperfusion of tissues, as these events can be viewed as rapid local acute inflammatory responses in vivo. Chapters are also included that evaluate current asthma, arthritis, ocular, atherosclerosis, and oral inflammation. Chapters in this part include the host's response to pathogens as a classic approach to gain an in-depth appreciation of the cellular and molecular events that have evolved in concert with the microbial world and their dynamic interplay in inflammation.

Each of the chapters is presented as an introduction by experts who are involved in cutting-edge research in their area of expertise. The aim of the editors is to provide a springboard for new investigators and research centers currently devoted to cutting-edge research in these areas. It exposes the reader to the exciting and fascinating cellular and molecular events that are involved in acute inflammation, chronic inflammation, their termination, and our quest for precise pharmacologic control in these life-sparing processes.

Experts worldwide have contributed concise chapters to launch this textbook for students new to this field. The text should be of interest to both students and investigators in academic and industrial settings. The editors trust that the reader will share our enthusiasm and continued excitement for studying the cellular and molecular events in this first response of the human body to invasion, injury, and tissue damage from within the area of inflammation research.

The Editors

1 Acute and Chronic Inflammation

Peter A. Ward

INTRODUCTION

The inflammatory response consists of an innate system of cellular and humoral responses following injury (such as after heat or cold exposure, ischemia/reperfusion, blunt trauma, etc.), in which the body attempts to restore the tissue to its preinjury state. In the *acute inflammatory response*, there is a complex orchestration of events involving leakage of water, salt, and proteins from the vascular compartment; activation of endothelial cells; adhesive interactions between leukocytes and the vascular endothelium; recruitment of leukocytes; activation of tissue macrophages; activation of platelets and their aggregation; activation of the complement; clotting and fibrinolytic systems; and release of proteases and oxidants from phagocytic cells, all of which may assist in coping with the state of injury. Whether due to physical or chemical causes, infectious organisms, or any number of other reasons that damage tissues, the earliest in vivo hallmark of the acute inflammatory response is the adhesion of neutrophils (polymorphonuclear leukocytes, PMNs) to the vascular endothelium ("margination") (Figure 1.1). The *chronic inflammatory response* is defined according to the nature of the inflammatory cells appearing in tissues. The definition of chronic inflammation is *not* related to the duration of the inflammatory response. Reversal or *resolution* of the inflammatory response implies that leukocytes will be removed either via lymphatics or by apoptosis (programmed cell suicide) and that the ongoing acute inflammatory response is terminated. As a consequence, during resolution increased vascular permeability is reversed due to closure of the open tight junctions and PMN emigration from the blood compartment ceases. In both the vascular and extravascular compartments, fibrin deposits are removed by pathways that lead to activation of plasminogen (to plasmin), which degrades fibrin (see section on "Intercommunications between Inflammatory

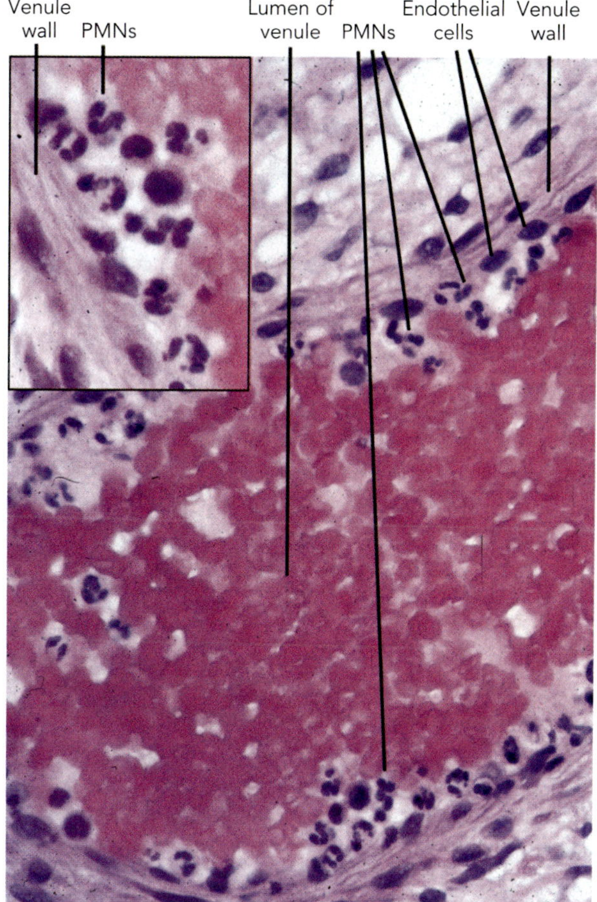

Figure 1.1. Early changes in a human venule involved in an acute inflammatory response. PMNs are "marginating" along the endothelial surface of a venule preparatory to their migrating into the extravascular space due to their adhesion to endothelial surfaces.

Cascades and the Coagulation Cascade"). Cell debris and red blood cells (RBCs) in the extravascular compartment are removed by phagocytosis involving tissue macrophages. There are many situations in which

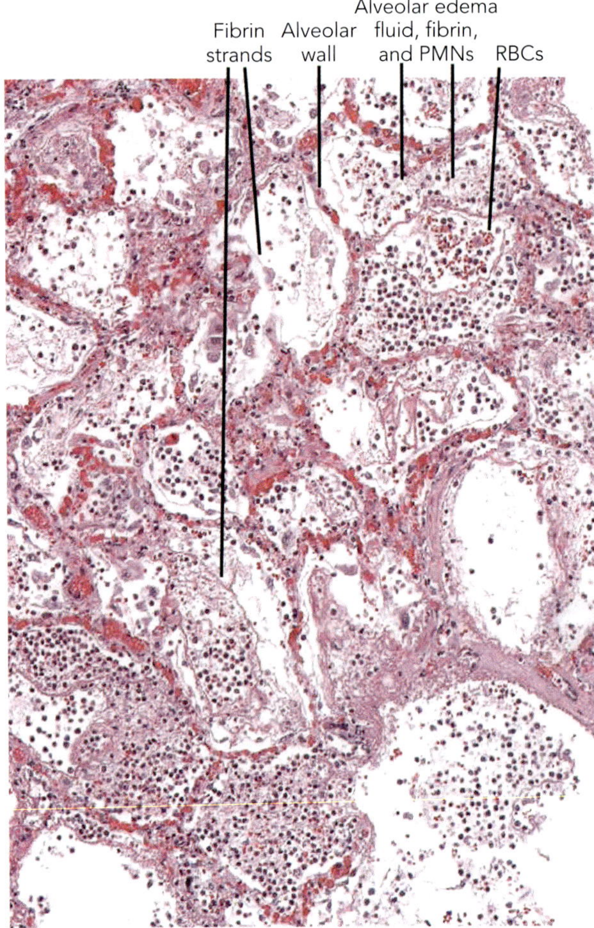

Figure 1.2. Acute bacterial pneumonia (*Streptococcus pneumoniae*) in human lung. Most alveolar spaces are filled with PMNs, fibrin strands, edema fluid, and RBCs, which clinically results in serious interference with gas exchange between the air spaces and vascular compartment.

the acute inflammatory response becomes excessive or prolonged, leading to serious damage of tissues and organs. Examples of unremitting acute inflammatory responses resulting in injury are discussed in the following section. Presented subsequently are concepts regarding acute and chronic inflammation. These are designed to provide the reader with a conceptual framework for an understanding of the inflammatory responses, their causation, and the outcomes.

THE ACUTE INFLAMMATORY RESPONSE

The acute inflammatory response is defined as a series of tissue responses that can occur within the first few hours following injury. In cases of bacterial pneumonia (Figure 1.2), bacterial meningitis (Figure 1.3), or ischemic myocardial injury (Figure 1.4), the inflammatory response is exuberant within the first few hours or days and then gradually declines unless the offending agent (such as bacteria-inducing pneumonia) cannot be cleared by phagocytosis. Resolution of the inflammatory response requires killing of bacteria and removal of their debris. When tissue injury occurs in various organs, the resolution of the inflammatory response may, to an extent, involve regeneration in which an organ can rapidly replace damaged or destroyed cells with an architectural outcome that resembles the original uninjured tissue. A good example of this is acute injury (chemical, viral, etc.) in the liver in which the inflammatory response resolves via regeneration of liver cells (hepatocytes, Kupffer cells, endothelial cells, etc.), reconstituting the damaged or destroyed liver, with the end result being tissue that is virtually identical to that before injury. In other situations, such as in the myocardium, ischemic destruction of cardiomyocytes results in an intense acute inflammatory response with a heavy build-up of PMNs (Figure 1.4). The regenerative abilities of the myocardium are extremely limited, if at all, and the destroyed myocardium must be quickly replaced with a fibrous (collagenous) scar. If scar formation is insufficient, the clinical outcome may be fatal due to cardiac rupture and a filling of the pericardial cavity with blood (hemopericardium). In the central nervous system, the commonly held belief is that vital cells such as neurons are largely unable

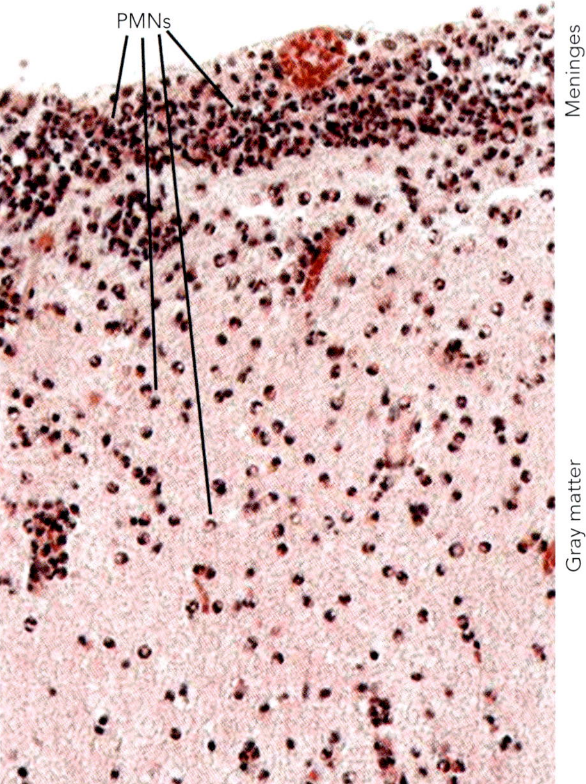

Figure 1.3. Acute bacterial meningitis (*Neisseria meningitidis*) with an intense accumulation of PMNs in the brain covering (meninges) and scattered accumulations of PMNs in the gray matter of the brain.

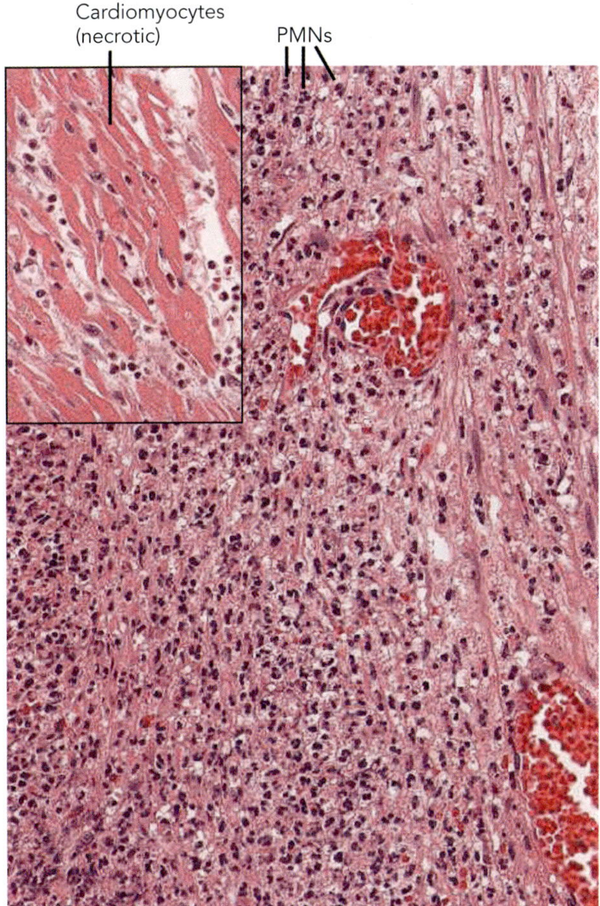

Figure 1.4. An acute inflammatory response in the myocardium 48 h after clinical onset of acute ischemic injury. The main frame shows an intense PMN infiltrate. The inset shows the edge of the infarct area with a necrotic area of myocardium and interstitial accumulation of PMNs.

to regenerate so that, once lost, replacement cannot occur. Most of the theories about cellular regeneration are undergoing reconsideration in the face of extensive stem cell work, which raises the question as to whether locally present or circulating stem cells may become actively involved in replacing cells similar to those that have been destroyed. This is an intensely debated topic that requires much additional investigation.

Figure 1.5 describes changes in tissues related to induction of the acute inflammatory response regardless of what has incited this rejection. There are numerous changes within the vascular compartment that trigger the acute inflammatory response, involving at least six intravascular events. The first change is *activation of endothelial cells*, during which these cells begin to express on their surfaces adhesion molecules for leukocytes (see Chapter 18 for detailed information about these molecules and their "counter-receptors" on leukocytes). Activated endothelial cells also engage in the generation and release of proinflammatory cytokines and chemokines (Chapter 13), which will chemotactically attract and activate PMNs as these cells adhere to the endothelium before their transmigration into the extravascular compartment. Activated endothelial cells often also express tissue factor (TF, aka Factor III) on their luminal surfaces. The complement activation product, C5a, can cause upregulation of TF on endothelial cells. TF is a potent procoagulant that can lead to thrombus formation on the vascular surface. A second event is *reversible opening of endothelial cells tight junctions*, which allows for the leak of protein and fluids from the vascular compartment into the extravascular compartment. When extensive edema develops in closed compartments, such as in the central nervous system (during bacterial meningitis) or in articular joints (after trauma), this can result in increased hydrostatic pressure that can seriously impair organ function. In the lung, extensive edema formation in the alveolar compartment (also known as "alveolar flooding") can seriously compromise air exchange between the alveolar and vascular compartments. Examples of this problem occur in acute high-altitude sickness, during bacterial pneumonia, and during Hanta virus infection of the lung. An inability for adequate gas exchange between

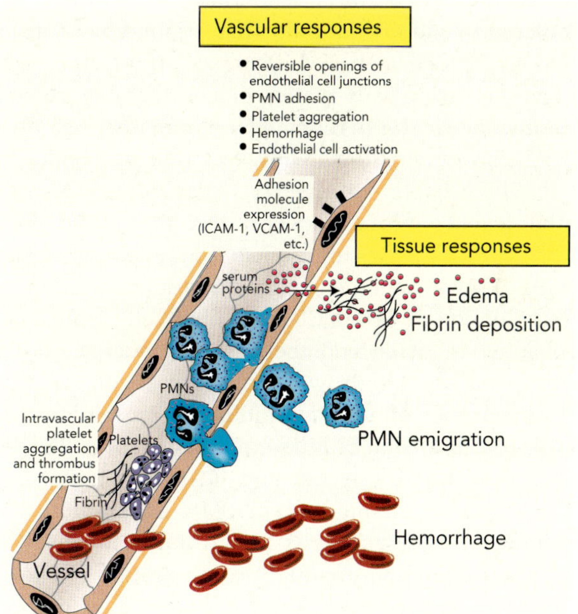

Figure 1.5. Mechanisms of acute inflammatory responses, featuring the initial intravascular events that lead to increased permeability changes, activation of endothelial cells (increased expression of adhesion molecules for neutrophils [PMNs]), adhesion of PMNs to the endothelial surfaces, and platelet activation (resulting in their aggregation and adhesion to one another as well as to endothelial surfaces). Platelets are often present in areas of fibrin deposition. Tissue responses (in extravascular compartment) feature edema, transmigration of PMNs, fibrin deposition, and hemorrhage if the structural integrity of the vascular barrier has been compromised.

TABLE 1.1. Factors affecting vascular integrity

	Events	Factors responsible
Edema	Reversible opening of endothelial tight junctions	Histamine, serotonin, kinins (bradykinin), C3a, C5a, •NO, PAF, prostaglandins
PMN emigration	Movement beyond vascular barrier	Chemoattractants: C5a, cytokines (IL-1β, TNFα), CXC, chemokines: collagen and bacterial peptides, metabolites of arachidonic acid
Hemorrhage	RBCs in extravascular compartment	Physical forces (heat, cold), bacterial products, proteins and oxidants from phagocytes
Platelets	Intravascular aggregation and fibrin formation	PAF, ADP, thrombin activation, etc.

the alveolar and vascular compartments can lead to a life-threatening state of hypoxia requiring intensive resuscitative support. Reversibility of the junctional changes in the vascular endothelium implies that open endothelial junctions can close to contain the amount of edema fluid accumulating in the extravascular compartment. Agents such as histamine are well known to interact with the vascular wall to bring about reversible opening of the tight endothelial junctions (Table 11.1 of Chapter 11). Whether vasopermeability mediators directly affect vascular endothelial cells (resulting in their contractility that opens tight junctions) or whether there are periendothelial cells which are tethered to endothelial cells (that respond by contraction to factors such as histamine, pulling the tight junctions open) is a matter of considerable debate. Another key factor in the acute inflammatory response is adhesive interactions between PMNs and endothelial cells (Chapters 4 and 11). Ordinarily, PMNs and other leukocytes are carried in the center in the blood stream without making contact with endothelial surfaces. PMNs undergo activation responses such as upregulation of CD11b/CD18 on their cell membranes (Chapter 4), while endothelial cells undergo activation most commonly with gene expression, leading to appearance of adhesion molecules on the laminar faces of endothelial cells. Examples of these molecules are P-selectin, E-selectin, and ICAM-1 (Chapter 18). The sequence of intermittent PMN adhesion to the endothelium (described as PMN rolling) followed by tight adhesion and eventual transmigration of PMNs through endothelial cell junctions is described in detail (Chapter 18). It should be pointed out that studies featuring intermittent ("rolling") followed by firm adhesion have been focused on changes in postcapillary venules. In the case of the lung, PMN transmigration occurs in capillaries that would not permit the rolling phenomenon described above because of physical constraints (inadequate space, since PMNs have a diameter equivalent to the diameter of a capillary). Furthermore, as the inflammatory response commences, PMNs become activated and stiff and cannot undergo deformity to adjust to the tight confines of capillaries. Nevertheless, vascular adhesion molecules play an important role in the transmigration of PMNs in the lung, since the absence or blockade of adhesion molecules clearly diminishes the build-up of PMNs in the extravascular space. Another feature of the acute inflammatory response is platelet activation, which is usually associated with the conversion of prothrombin to thrombin. Platelets can also be directly activated by various other agents (Table 1.1). The end result is platelet adhesion to one another (resulting in platelet aggregates) as well as to endothelial cells. This is the forerunner of intravascular thrombosis in which fibrin deposition develops in and around aggregated platelets. Finally, the acute inflammatory response may be associated with hemorrhage because of direct structural damage (reversible or irreversible) to the endothelial barrier. The development of hemorrhage implies that the vasculature has been severely damaged, since RBCs have no intrinsic mobility and are passively carried out of the vasculature if there has been sufficient loss of vascular integrity. Hemorrhage occurs after thermal or cold trauma, in situations of severe platelet dysfunction or platelet deficiency, after infections due to the release of toxins (as from Streptococcal A bacteria), and in patients undergoing excessive anticoagulant therapy, to cite a few examples. As indicated earlier, all of these changes of the acute inflammatory response are reversible. Edema fluid is cleared from the distal airway (alveolar compartment) by uptake of these fluids together with inflammatory cells into the draining lymphatics, with return to the blood compartment. Thus, what comes from the blood compartment often returns to the blood compartment. PMN clearance in tissues may also occur by apoptosis of these cells and their phagocytosis by tissue macrophages. Thrombosis within vessels can be cleared by activation of the fibrinolytic system, involving tissue plasminogen activator (TPA) and other factors that will activate the fibrinolytic enzyme, plasmin (Figure 1.6).

Table 1.1 lists factors that affect the vascular integrity and lead to changes in the endothelium, resulting

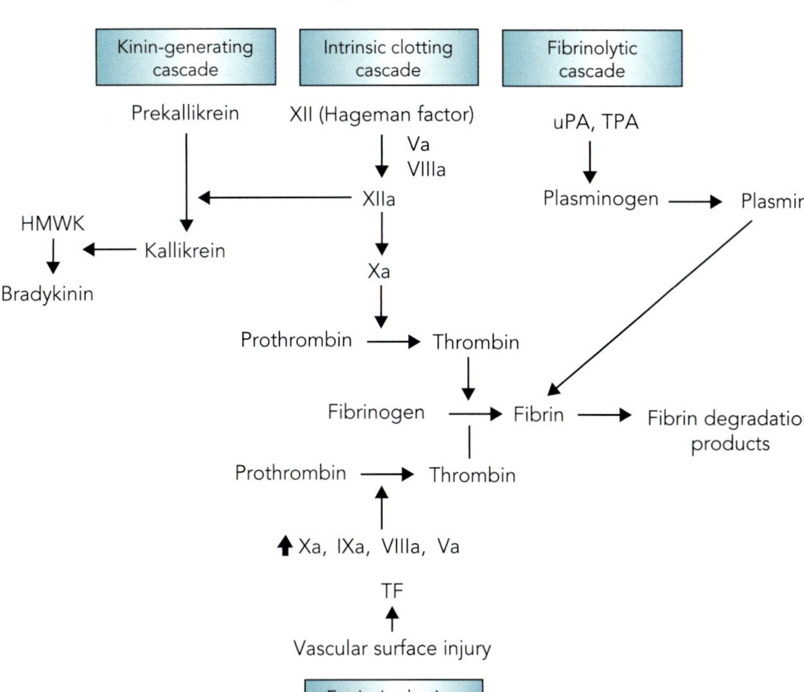

Figure 1.6. The coagulation cascades (intrinsic and extrinsic) and intercommunications with kinin generating and fibrinolytic cascades.

in edema formation, PMN accumulation, platelet activation, and development of hemorrhage. Edema due to reversible openings of endothelial cell tight junctions can be induced by histamine; serotonin; kinins (such as bradykinin); the complement anaphylatoxins (C3a and C5a), which act on mast cells to release histamine; nitric oxide; platelet activating factor (PAF); and certain prostaglandins (Chapter 12). As indicated earlier, these responses resulting in edema fluid accumulation outside the vascular compartment are usually reversible and transient. PMN emigration is preceded by adhesive interactions between these cells and endothelial cells via engagement of adhesion molecules on both cell types (e.g., E- and P-selectins on endothelial cells; CD11b/CD18 on PMNs, etc.) (as described earlier and in Chapter 18). Ordinarily, adhesion molecules are present in low quantities on endothelial and PMN surfaces but activation of either cell type can dramatically and rapidly or slowly increase the levels of adhesion molecules that appear either following fusion of cytosolic granules to the cell membrane in the case of PMNs (the rapid response occurring within minutes) or following transcriptional upregulation (a slow response requiring hours). Adhesion molecules on endothelial cells include ICAM-1 and E-selectin, although in the case of P-selectin, this adhesion molecule can be upregulated either via transcriptionally dependent responses or via transcriptionally independent responses. Rapid expression is due to P-selectin addition to the cell membrane via fusion of cytosolic granules (Weibel–Palade granules). Rapid upregulation of adhesion molecule is found in PMNs, platelets, and endothelial cells. In the case of PMNs, increased expression of cell membrane adhesion molecules (CD11b/CD18) is usually rapid due to fusion of secondary granules in the cytosol (which contain adhesion molecules on their inner surfaces) to the cell membranes of PMNs. Chemoattractants for PMNs responsible for their extravascular migration (emigration) include C5a, cytokines (such as IL-1β and TNFα), CXC chemokines, collagen and bacterial peptides, as well as metabolites of arachidonic acid, all of which are described in Chapters 4, 12, and 13. Proteases and oxidants from activated phagocytic cells cause damage to the endothelial barrier, as well as to cells and connective tissue matrix, resulting in widespread damage of both the vascular and extravascular compartments (Chapter 17). Platelets can be activated by a variety of factors (Chapters 5, 8, 12, and 15), such as PAF, ADP, and thrombin, which is activated when the clotting cascade has been triggered (see section on "Intercommunications between Inflammatory Cascades and the Coagulation Cascade").

As suggested in the earlier comments, the acute inflammatory response is a protective shield against

TABLE 1.2. Regulation of the acute inflammatory response: natural anti-inflammatory factors

	Factors	Targets
Cytokines	IL-4, IL-10, IL-12	• Stabilization of IκB and reduced NF-κB activation
Protease inhibitors	SLPI, TIMP-1, α_1PI, etc.	• Inhibition of serine proteases and nonserine proteases
Antioxidant enzymes	Superoxide dismutase Catalase, Glutathione peroxidase	• Converts •O_2 to H_2O_2 • Destroys H_2O_2 • Catalyzes the breakdown of H_2O_2 to H_2O
Lipoxins		see Chapter 12
Glucocorticoids		• Diverse
Kinases	Hydrolysis of kinins	• Bradykinin, etc.
Phosphatases	Removal of phosphates from proteins	• Transcriptional factors
Transcriptional factors	STAT3, SOCS3	• Blockade of gene activation for proinflammatory mediators

tissue that has been damaged. The purpose of the response is to return the tissue to its predamaged state. In some cases, the response is excessive due to persistence of the damage-causing agent (e.g., bacteria) or the offending trigger (e.g., immune complexes in autoimmune diseases). As will be discussed later, excessive or unregulated inflammatory responses can themselves cause tissue damage. In 1972, Lewis Thomas said "Our arsenals for fighting off bacteria are so powerful, and involve so many different defense mechanisms, that we are more in danger from them than the invaders. We live in the midst of explosive devices; we are mined." (Germs, *N Engl J Med*, 1972, 287:553–555.)

Regulation of the Acute Inflammatory Response

How is the acute inflammatory response kept in check? It is clear that the response is subject to very tight regulation to contain the cascades before they lead to extensive tissue or organ injury. There are numerous, naturally occurring anti-inflammatory factors (Table 1.2). Cytokines such as IL-4, IL-10, and IL-12 in very low concentrations are inducible and are powerful anti-inflammatory factors that contain the acute inflammatory response by stabilizing IκBα, which blocks NF-κB activation. As a result, these regulatory cytokines have greatly diminished the production of proinflammatory mediators and reduced numbers of PMNs accumulating in tissues. There are several protease inhibitors that also contain the response by inhibiting serine proteases, many of which are released from phagocytic cells. The secreted leukocyte protease inhibitor (SLPI) was described as trypsin-like inhibitor that was largely confined to upper airway secretions, produced by and released from nearby epithelial cells of the lung. It is now known that SLPI reduces activation of NF-κB via stabilization of IκBβ. Hydrolysis of the IκB proteins is required for activation of NF-κB. In addition, there are nonserine protease inhibitors such as inhibitors of metalloproteases (MMPs). MMP3 and MMP9 may be the most important MMPs, with targets being elastin, collagen, and altered (denatured) collagens. Tissue inhibitor of MMP2 (TIMP-2) is a common and inducible TIMP and has broad inhibitory activity for MMPs. α1 Protease inhibitor (α_1PI) is abundantly present in plasma and in lung tissue. It is a powerful serine protease inhibitor of trypsin-like enzymes. If α_1PI is absent or present in a functionally defective manner, this is almost always associated with the development of progressive and often fatal pulmonary emphysema in humans. Such individuals may also develop hepatic cirrhosis for reasons that are poorly understood. There are many other naturally occurring protease inhibitors that suppress tissue damaging proteases associated with the induction of acute inflammatory response.

Antioxidant enzymes, such as superoxide dismutase, catalase, and glutathione peroxidase, are abundant in a variety of tissues and can be upregulated in the course of the inflammatory response such as those occurring after hyperoxia, bacterial infection, ischemia–reperfusion, and in various other situations. Upregulation of antioxidant enzymes is especially well documented in the lung, in the case of Gram-negative bacteria (e.g., *Escherichia coli*) lipopolysaccharide (LPS) can rapidly and powerfully upregulate these antioxidant enzymes. Superoxide dismutase converts superoxide anion (•O_2) to H_2O_2, while catalase destroys H_2O_2, reducing it to water and molecular oxygen (Table 1.2). Glutathione peroxidase in the presence of glutathione (GSH) catalyzes the conversion of H_2O_2 to H_2O. If GSH levels are very low in an organ or tissue, it leads to "redox stress" in which the tissue has impaired ability to deal with oxidants

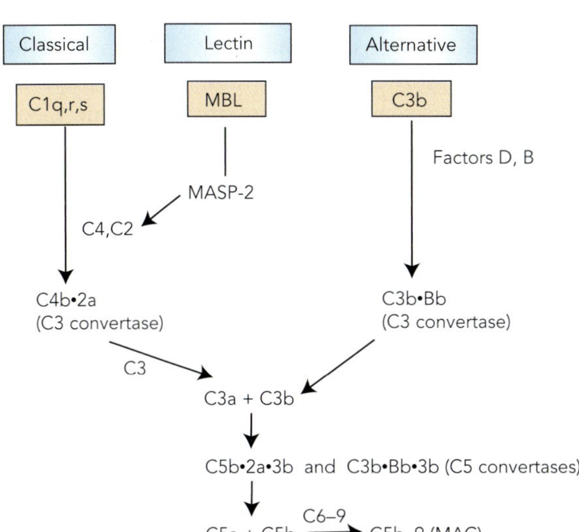

Figure 1.7. The complement cascade, including the three pathways of activation, the C3 and C5 convertases, and the chief complement activation products.

(Chapter 17). Lipoxins represent another source of natural anti-inflammatory factors (Chapter 12). Glucocorticoids are well known to be naturally occurring anti-inflammatory factors. Kininases hydrolyze kinins such as bradykinin and lysyl-bradykinin, leading to their functional inactivation. These enzymes are present in most tissues. The lung vasculature is lined with kininases, such that one pass of blood through the lung can result in complete inactivation of kinins. Phosphatases, such as PTEN, remove phosphates from proteins such as transcriptional and signaling factors, leading to their termination as functionally active moieties. Such regulation can greatly reduce the production of proinflammatory molecules (e.g., adhesion molecules, cytokines, chemokines, etc.). Similar to all other cascades of the inflammatory system, these phosphatases control the production of proinflammatory mediators. Finally, there are several transcriptional regulatory factors such as suppressor of cytokine signaling 3 (SOCS3) and STAT3 that block the activation of proinflammatory genes, resulting in greatly reduced levels of proinflammatory mediators. Obviously, all of these factors are aligned to bring about tightly regulated inflammatory responses before they unleash serious damage to tissues.

THE COMPLEMENT CASCADE

The complement system is an important part of the innate immune system conferring protection especially against invading infectious agents, such as bacteria, viruses, and protozoa. Its role in innate immunity is to generate biologically active products from the pathways of complement activation, resulting in proinflammatory mediators that will call in PMNs as well as the production of opsonic (phagocytosis promoting) and lytic factors for bacteria (C5b-9, membrane attack complex [MAC]) and nucleated cells. The three pathways of complement activation are shown in Figure 1.7. The *classical pathway* is traditionally activated by the presence of IgG or IgM immune complexes. Activation of the first complement component (C1q,r,s) leads to activation of the subunits to active enzymes, with targets being C4 and C2, resulting in fragmentation products (C4a, C4b, C2a, C2b), some of which form the C4b•2a complex, which is a C3 convertase that cleaves C3 into C3a and C3b. C3b can be adducted to the C4b•2a complex to form the complex, C4b•2a•3b, which is a C5 convertase that can convert C5 into C5a and C5b. The second pathway of complement activation is the *lectin pathway* which involves the mannose-binding lectin (MBL), a plasma "collectin," that binds to mannose-related carbohydrates present on surfaces of viruses and bacteria. This leads to the activation of mannose-associated serine protease-2 (MASP-2), a serine protease that has the ability (like C1q,r,s) to interact with C4 and C2 to form the C3 convertase (C4b•2a). The third pathway of activation is the *alternative complement pathway* that can be activated by the presence of C3b which, when interacting with factors B and D, forms a complex, C3b•Bb, which has C3 convertase activity that generates C3a and C3b. Adduction of another molecule of C3b generates the C5 convertase of the alternative pathway, C3b•Bb•3b. The C5a convertases cleave C5 into C5a and C5b. C5b can interact with the terminal complement proteins, C6–9, to form the C5b–9 complex (MAC). In addition to these traditional pathways of complement activation, other serine proteases unrelated to the complement system can interact directly with C3 or C5 to form complement activation products (C3a, C3b, C5a, C5b). For instance, plasmin can interact with C3 to generate C3a and C3b. There are several serine proteases (such as the elastase present in neutrophils and a neutral protease present in macrophages) that will then interact directly with C5 to generate C5a and C5b. In addition, thrombin has the ability to interact with C5 to produce the same activation products. The complement activation pathways are under very rigid and tight control, based on "complement regulatory proteins" (CRPs) that are present both in plasma and on cell surfaces. These CRPs tightly regulate the complement system to either limit the formation of complement activation products or form a protective shield to prevent the activation products from bringing about cell damage. Some complement-mediated human disorders, such as paroxysmal nocturnal hemoglobinemia, result in intensive hemolysis of

RBCs because of a defect in two of the CRPs (decay accelerating factor and CD59).

The complement anaphylatoxins are small peptides (<10 kDa) and consist of C3a, C4a, and C5a. The most abundant of these is C3a since C3 is the complement protein present in highest concentration in plasma. C3a appears to have its major biological activity as induction of histamine release from mast cells, which then leads to greatly increased vasopermeability in the local area. C3b is the major opsonic product generated by the complement system and reacts with receptors on a variety of different cells and microorganisms to bring about greatly enhanced phagocytosis and intracellular killing of microbes. There are relatively few humans with complete C3 deficiency and, as such, they are highly susceptible to life-threatening bacterial infections. The role of C4a is not well understood. C5a is an extremely potent anaphylatoxin which, in very low nanomolar concentrations, can interact with receptors on phagocytic cells, especially neutrophils, either to bring about their priming for enhanced subsequent responses in the presence of a co-stimulus or to bring about direct activation of phagocytic cells by inducing chemotaxis, an intracellular calcium response, generation of reactive oxygen species ($\bullet O_2$, H_2O_2), enzyme release, and a variety of other responses, all of which tend to function as a protective shield in a local setting and bring about accumulation of neutrophils at inflammatory sites. A major function is to contain and kill microorganisms. In some instances, excessive amounts of C5a are generated as in sepsis and in autoimmune diseases (such as rheumatoid arthritis and systemic lupus erythematosus [SLE]). In these cases, major problems can arise such as the signaling paralysis of neutrophils due to excessive generation of C5a and the priming of macrophages for accentuated and excessive inflammatory responses during sepsis. The final product of the complement activation sequence, C5b–9 (MAC), attaches to surfaces of antibody-coated bacteria, leading to their cytolytic destruction. In certain autoimmune disorders in which there are antibodies that can react with epitopes on surfaces of nucleated cells, cell lysis can occur. Soluble C5b–9 has the ability to interact with endothelial cells to bring about their activation with the formation of proinflammatory cytokines and chemokines. Finally, C5b–9 is an important protective factor leading to lysis of Gram-negative bacteria. Details on the biochemistry and functions of the complement system and its role in human diseases are discussed elsewhere.

Intercommunications between Inflammatory Cascades and the Coagulation Cascade

Figure 1.7 demonstrates the intricate intercommunications between three different proinflammatory cascades: the kinin-generating cascade, the clotting cascades, and the fibrinolytic cascade. Central to these intercommunications is the clotting system which involves two activation pathways, the intrinsic and the extrinsic cascades. The intrinsic cascade occurs with the engagement and activation of Hageman factor (Factor XII) which interacts with Factors Va and VIIIa to convert Factor XII to XIIa ("a" signifies the active form of the protein). In turn, this leads to the activation of Factor X, which then directly converts prothrombin to thrombin. Thrombin converts fibrinogen to fibrin, which is the major product involved in in vivo clot formation. Following vascular injury, the extrinsic clotting cascade is activated resulting in the expression of endothelial cells on the surfaces of TF and in the copresence of other clot activating factors (Xa, IXa, VIIIa, Va), there is also conversion of prothrombin to thrombin and generation of fibrin from fibrinogen. The fibrinolytic cascade is activated by urinary plasminogen activator (uPA) and TPA which cause conversion of plasminogen to plasmin. Plasmin directly interacts with fibrin to bring about fibrin degradation products resulting in the breakdown of fibrin clots as they are formed within the intravascular compartment or elsewhere. TPA is used in patients with acute myocardial ischemia to try to bring about lysis of intracoronary arterial clots to allow perfusion to occur. The kinin-generating cascade is also linked to the clotting cascades by the fact that Factor XIIa will convert prekallikrein to kallikrein. Kallikrein interacts with high-molecular-weight kininogen (HMWK) to bring about its hydrolysis and release of bradykinin, which is a powerful vasopermeability agent. Bradykinin also has the ability to slow the heart rate (bradycardia). All of these cascades as well as the complement cascade have, as a common theme, activation of proteins by their limited hydrolysis, after which the split products directly interact with cell receptors to trigger cell responses (e.g., C5a interacting with receptors on PMNs [see earlier]) or the split products can assemble to form an active enzyme (such as C4b•2a, the substrate of which is C3). As with the complement system, the clotting, generating, and fibrolytic systems are each subject to very tight regulation by a series of inhibitors designed to prevent excessive product formation when one of the cascades is activated.

OUTCOMES OF THE ACUTE INFLAMMATORY RESPONSE AND DISORDERED RESPONSES

A clinical example of an acute inflammatory response that is not adequately contained is the acute respiratory distress syndrome (ARDS) in humans where there is a sustained accumulation of PMNs within the distal airway (alveolar) compartment. ARDS occurs in adults and in infants in a variety of clinical

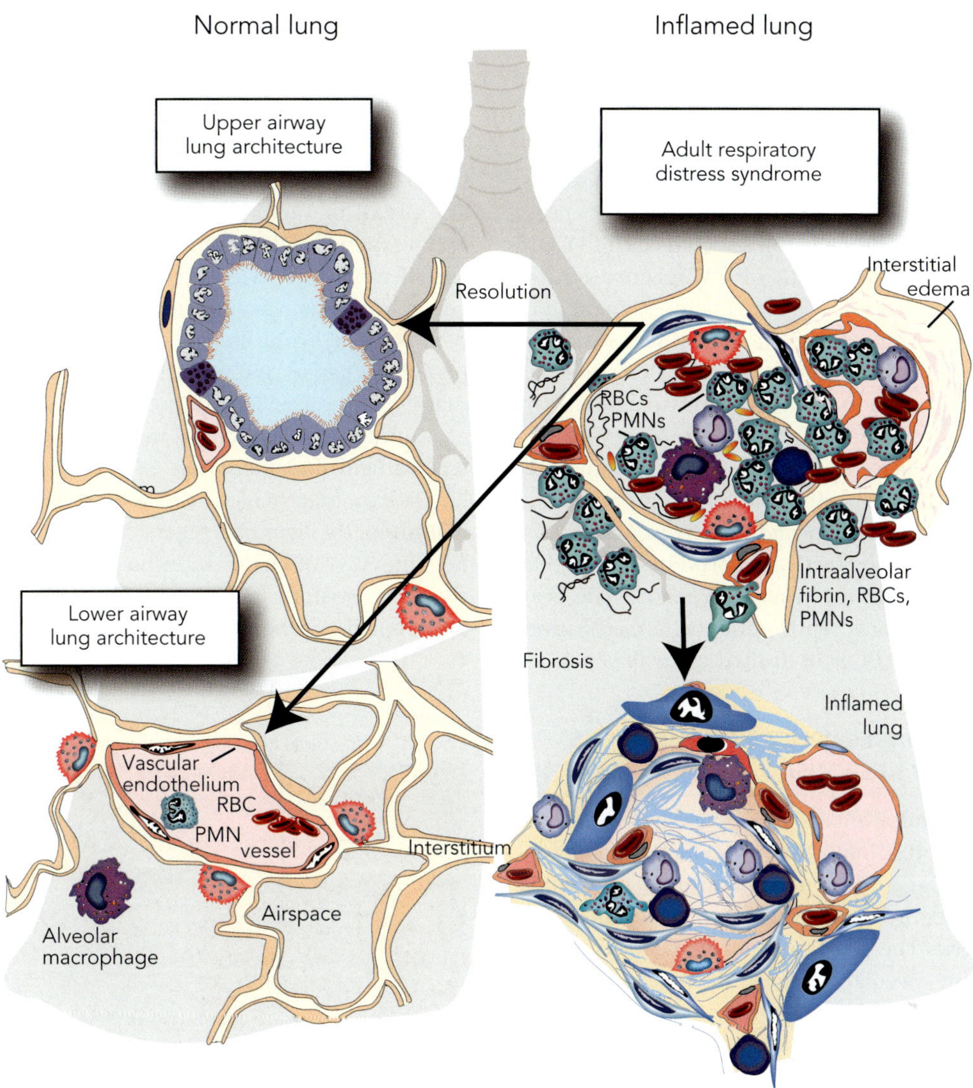

Figure 1.8. Outcomes of the acute inflammatory response in lung in ARDS in humans. In ARDS, edema fluid, fibrin, and PMNs accumulate in the alveolar compartment (upper right area). The inflammatory response can regress (undergo resolution) with return of the lung to its preinflammatory state (lower left area), or it may progress to extensive pulmonary fibrosis (lower right area).

situations, such as premature birth and polytrauma and bacterial pneumonia in adults. The mechanisms responsible for the development of ARDS are very poorly understood. ARDS can be considered to be a sustained and dangerous inflammatory condition in the lung. Bronchoalveolar fluids contain an abundance of PMNs, fibrin split products, and C5a. There is no known specific therapy for ARDS, only supportive treatment (mechanical ventilation, fluid therapy, etc.). ARDS may proceed to resolution or to pulmonary fibrosis that is often fatal (Figure 1.8). The pulmonary infiltrates in ARDS patients often lead to a radiographic "whiteout" in lungs, due to alveolar edema, PMNs accumulation, and fibrin deposition, collectively causing severely compromised gas exchange between the alveolar and vascular compartments, resulting in a high mortality rate. Why some cases of ARDS resolve completely (Figure 1.8) while others progress to pulmonary fibrosis and pulmonary arterial hypertension (Figure 1.8, lower right frame) is entirely obscure. Another clinical example in which the acute inflammatory response is not sufficiently contained is in the setting of sepsis, following bacterial or viral pneumonia, intestinal perforation, or any number of other clinical situations. Sepsis occurs in all age groups and is age related, commonly seen in patients who are immunocompromised (postchemotherapy or age related). The death rate can be as high as 60%. Over the past decade there has been a progressive rise in the incidence of sepsis, which has been linked with bacteria, both Gram-positive (such as *Staphylococcus aureus*) and Gram-negative (such as *E. coli*), viruses, and fungi. Sepsis is clearly a condition in which there has been loss in control of the inflammatory system. For instance, there is a surge of proinflammatory mediators (e.g., TNFα, IL-1β, IL-6, etc.) in the plasma, referred to as the "cytokine storm," although, the cause for high levels of these mediators is poorly understood. The vasculature undergoes upregulation of adhesion molecules for leukocytes. Tissue macrophages are "primed" and show excessive responses (e.g., production of oxygen free radicals, granule enzyme release, etc.). Blood PMNs express high levels of adhesion molecules such as CD11b/CD18 on their surfaces, which indicate PMN activation. In sepsis, there is extensive evidence of complement activation and consumptive coagulopathy (activation of the clotting system with depletion of the clotting factors) based on products measurable in plasma. Why the body has lost control of the inflammatory system during sepsis is entirely unknown. Currently, in addition to traditional supportive therapy (fluid resuscitation, ventilatory support, broad spectrum antibiotics, etc.), only a single drug (activated protein C, APC) has been approved for therapy in sepsis. Because APC is an anticoagulant and most septic patients have consumptive coagulopathy, APC is of limited value in sepsis.

FUNCTIONAL CONSEQUENCES OF THE ACUTE INFLAMMATORY RESPONSE

Once the acute inflammatory response has been triggered in tissues, it is important to understand the consequences. Initially, the acute inflammatory response can be triggered by the presence of infectious agents such as bacteria. The rapid build-up of plasma constituents in the lung alveolar space results in accumulation of antibodies, complement proteins, clotting factors, and other factors that may assist in containment of microorganisms. Traumatic injury triggers an acute inflammatory response in a locale, although in spite of the localized nature of this response, there are often systemic symptoms such as fever, increased numbers of circulating neutrophils (neutrophilia), increased heart rate (tachycardia), and sometimes a feeling of anxiety and apprehension. The local consequences of traumatic injury are well known. The common situation of a sprained ankle results in increased blood flow to the local area which is characterized by redness and increased local temperature. Soon, extensive edema develops, causing soft tissue swelling and pain in the joint area which prevents a full range of motion. Induration (increased thickness of soft tissue) is due to the accumulation of leukocytes and edema fluid. Occasionally, hemorrhage may develop. As described earlier, the acute inflammatory response may resolve with ultimate removal of cell debris, fibrin, red cells, and disappearance of leukocytes. Well-known therapy consists of immobilization of the joint and application of localized cold temperature to reduce blood flow to the area. (The inflammatory response absolutely requires accelerated blood flow to the area.) Sometimes, the acute inflammatory response will not resolve for various reasons, such as the inability to clear an infectious agent whose contained presence will "fan" the flames of the inflammatory response, resulting in persistent and sometimes increasing intensity of inflammation. Ischemia–reperfusion of any organ or region will trigger an acute inflammatory response during the phase of reperfusion when sufficient blood flow is occurring to allow formation of edema and extravascular accumulation of leukocytes, primarily neutrophils. Toxic agents are well known to be able to cause tissue injury and unleash an inflammatory response. In the liver, excessive ingestion of acetaminophen, which is metabolized by liver into free radical intermediates, can result in hepatocellular toxicity (necrosis). If the dose of the ingested drug is limited, the liver can regenerate and replace the lost hepatocytes. Weeks or months later, little evidence of the cell destructive events after drug ingestion may be apparent if regeneration is successful. If the dose of

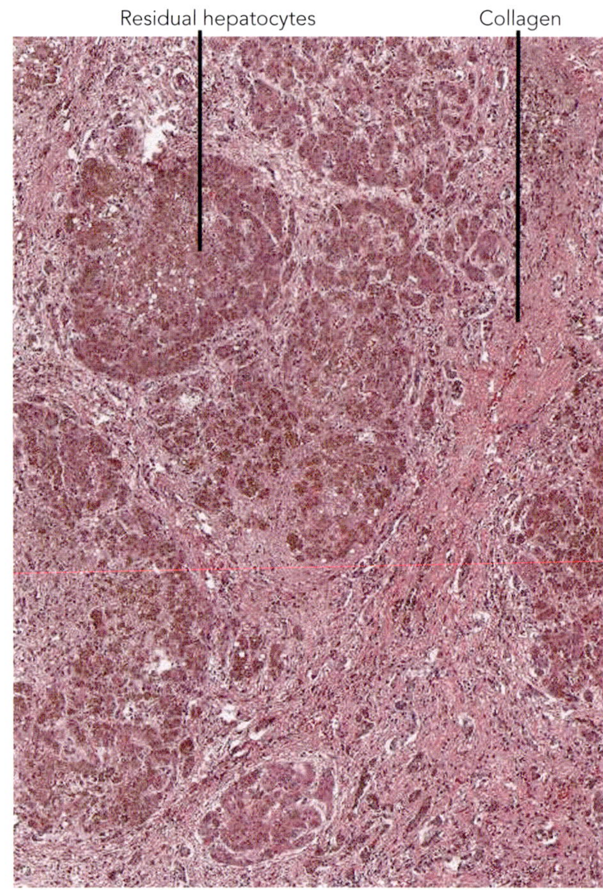

Figure 1.9. Postalcoholic cirrhotic liver. There are residual lobules of hepatocytes, with dense bands of collagenous scar surrounding some lobules and infiltrating others. The presence of dense collagenous scars interferes with the ability of residual hepatocytes to regenerate and restore damaged or destroyed hepatic lobules.

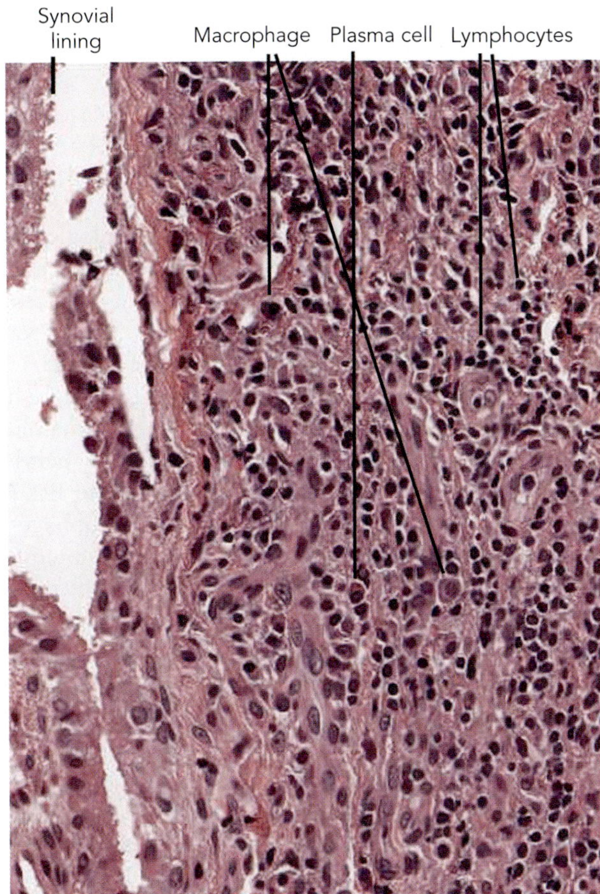

Figure 1.10. Synovium from knee joint of patient with rheumatoid arthritis. The synovium is thickened and contains lymphocytes, macrophages, and plasma cells.

ingested drug is higher, there may be too much damage for regeneration of the hepatic lobules, and bands of collagen may form around hepatic lobules along with chronic inflammatory cells, resulting in a condition referred to as cirrhosis (Figure 1.9). If the dose of ingested drug is very high, it may overwhelm the regenerative abilities of the liver, leading to death of the individual from acute hepatic failure. There are autoimmune diseases, such as rheumatoid arthritis or SLE, in which the predominant targets are joints where an acute inflammatory response persists together with the accumulation of acute and chronic inflammatory cells (Figure 1.10). This ultimately results in fibrosis of the synovial tissues and bone formation, causing fusion of articular bones, creating a "frozen joint." In the case of SLE, the kidney may be continuously bombarded with immune complexes and complement activation products, resulting in neutrophil accumulation, followed by ultimate scarring (hyalinization) of glomeruli and loss of filtration function (Figure 1.11). In other organs such as in the heart, the accumulation of inflammatory cells in the conducting system may lead to damage of cells, resulting in arrhythmias, which are sometimes fatal. Outcomes of the acute inflammatory response may be resolution (as described earlier) or persistence of the inflammatory response, sometimes leading to replacement of PMNs by chronic inflammatory cells (lymphocytes, macrophages).

Granulomatosis Inflammation

Granuloma formation is a special type of inflammatory response, consisting of globular (granular) accumulations of inflammatory and fibrotic nodules, together with chronic inflammatory cells (lymphocytes, macrophages) and giant cells. In tuberculosis, the central part of these granulomas usually undergoes necrosis (caseous necrosis) so that the tissue is largely taken out of functional usefulness (Figure 1.12).

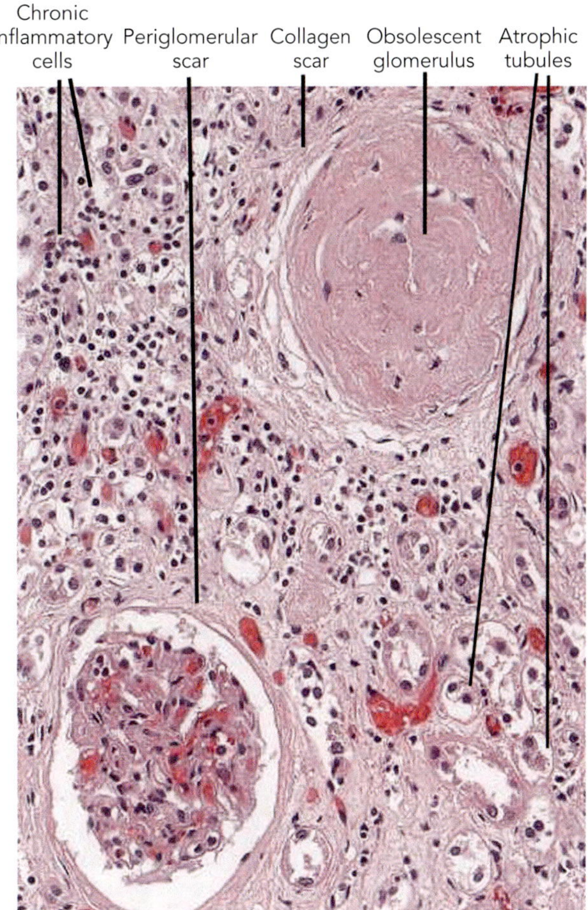

Figure 1.11. Outcome of persistent inflammation in kidney. One glomerulus is "obsolescent" (replaced with hyaline) and nonfunctional, contrasting to an intact glomerulus with thin periglomerular fibrous band (scar). In addition, there is interstitial fibrosis and the presence of chronic inflammatory cells. Extensive tubular atrophy has occurred, featuring greatly reduced numbers of tubular epithelial cells, and a greatly diminished mass of tubules.

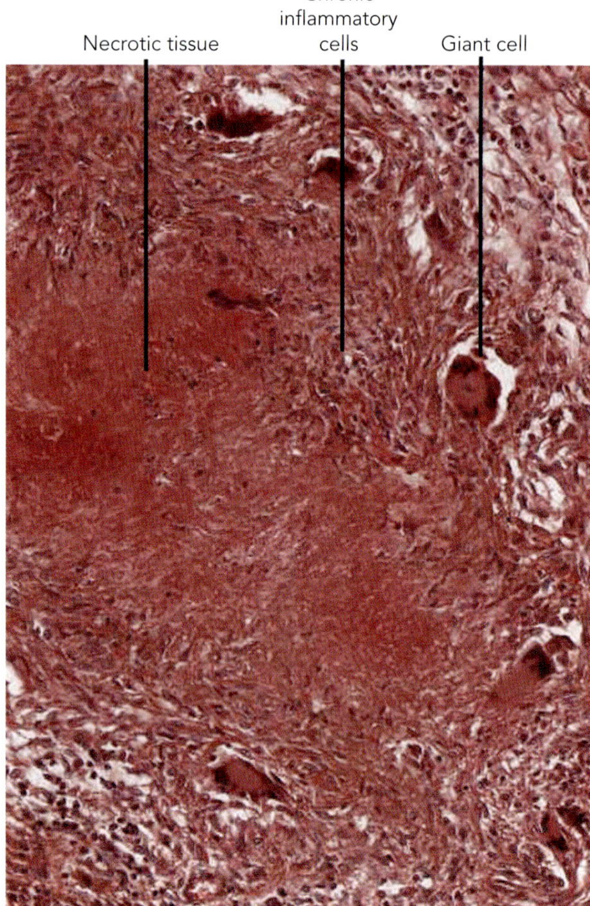

Figure 1.12. Lung mycobacterial granuloma featuring peripheral collagenous connective tissue along with multinucleate giant cells. In the center is necrotic (caseous) tissue, at the edge of which are macrophages and lymphocytes.

These types of inflammatory responses can result in extensive scarring of the affected tissue, with extensive collagenous scars and the development of fibrosis. In addition, tuberculosis granulomas, because of their necrotic centers, may ultimately erode into nearby pulmonary blood vessels, resulting in extensive pulmonary hemorrhage which is sometimes fatal. In granulomas that feature no central necrosis (and these would be the predominant type of granuloma in the United States where tuberculosis is rather infrequent), the causes may be fungi (e.g., histoplasmosis), foreign bodies (surgical suture material), or unknown causes (as in sarcoidosis). In sarcoidosis, there are granuloma nodules in tissues, especially in the lung, associated with the presence of fibroblasts and fibrotic tissue, chronic inflammatory cells (lymphocytes, macrophages, and giant cells) (Figure 1.13). Although necrosis is not seen in diseases like sarcoidosis, there may be extensive pulmonary fibrosis which ultimately leads to death because of interference with gas exchange (Figure 1.14). Under such conditions, there is often development of pulmonary hypertension. The combination of this complication, together with pulmonary fibrosis, may be fatal.

Chronic Inflammation

Chronic inflammation is defined not as the persistence of acute inflammation but is defined morphologically by the presence of lymphocytes, macrophages, and plasma cells in tissues (Figure 1.15). In many cases, the chronic inflammatory response may persist for long periods (months to years). It is considered to be caused by persistent engagement of innate and acquired immune responses, such as in rheumatoid arthritis, in chronic allograft rejection, in berylliosis, and in granulomatous inflammation, to list a few examples. There is evidence that macrophages in these lesions produce a series of proinflammatory mediators that activate fibroblasts to lay down collagen and activate other macrophages and lymphocytes

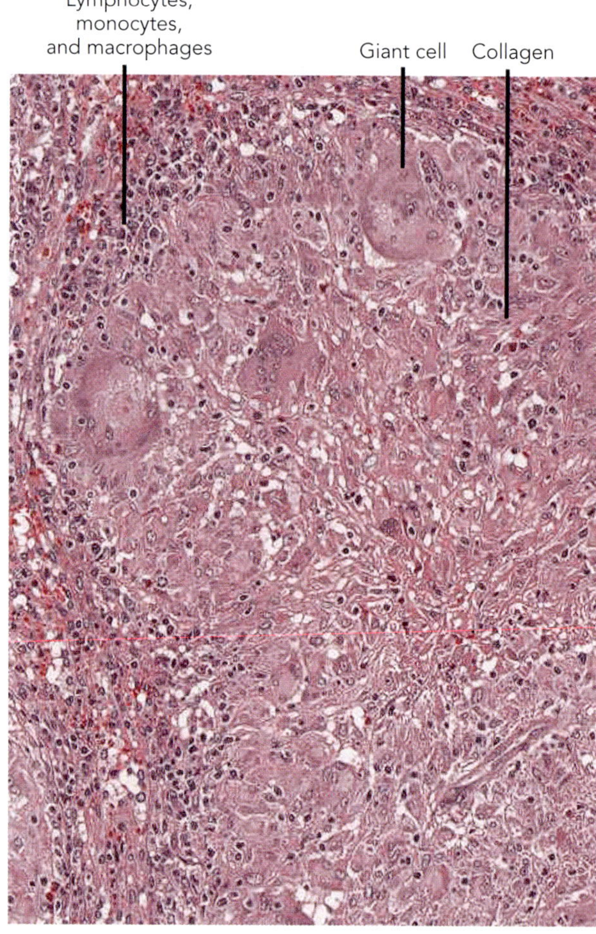

Figure 1.13. Lung biopsy of woman with sarcoidosis. There is a large granuloma surrounded by chronic inflammatory cells (lymphocytes, macrophages, and monocytes). The granuloma contains multinucleate giant cells, macrophages, and deposits of collagen. No necrosis is present.

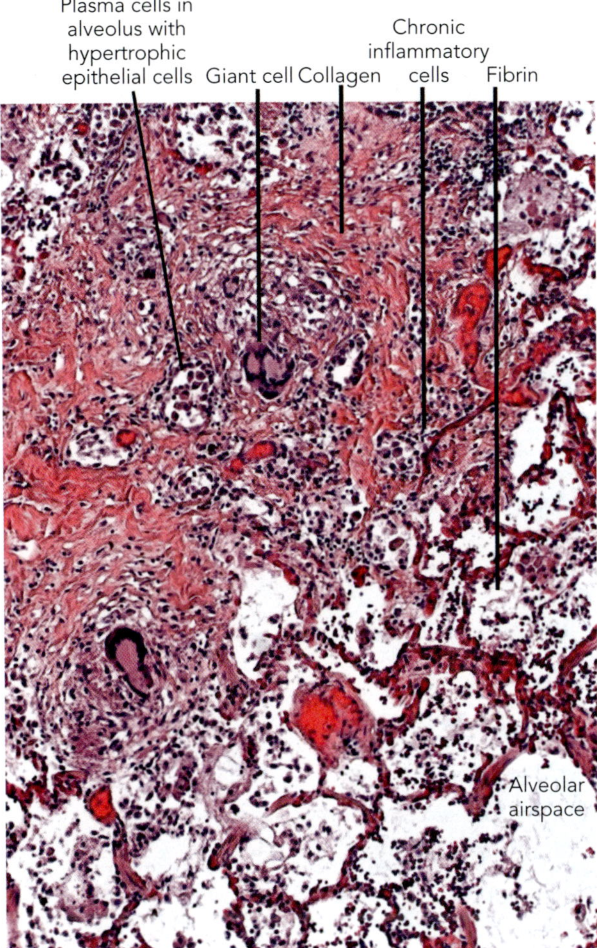

Figure 1.14. Pulmonary sarcoidosis. There are multinucleate giant cells, lymphocytes, and plasma cells replacing interstitial and alveolar spaces. Fibrin is also present. Dense collagenous deposits are starting to build up in lung. No necrosis is present.

to release mediators to perpetuate these inflammatory responses. Why responses to mycobacterial agents result in extensive necrosis of the granulomas is unknown. As shown in Figure 1.16, chronic inflammation is initially triggered by vascular responses that involve the appearance of adhesion molecules on endothelial cell surfaces that will specifically cause adhesion of lymphocytes and mononuclear cells (monocytes), resulting in their subsequent transmigration into the extravascular compartment. In the case of lymphocytes, most are T cells. Like in the acute inflammatory response, lymphocytes and monocytes, as well as endothelial cells, undergo an activation process with lymphocytes and monocytes expressing adhesion molecules that are interactive with vascular (endothelial) adhesion molecules (such as VCAM-1 and other molecules) that promote adhesion and ultimate transmigration of these cells into the extravascular compartment. In any inflammatory response, differences between the types of adhesion molecules expressed on endothelial cells will determine the type of blood leukocytes (e.g., PMNs vs. monocytes vs. lymphocytes) that emigrate (Chapter 18). Since, in general, macrophages are not present in the peripheral blood, the influx of monocytes into the extravascular compartment allows these cells to differentiate into macrophages over a period of several days. In Peyer's patches in the small bowel, endothelial cells constitutively express adhesion molecules that specifically interact with T cells which then transmigrate into the lymphoid follicles (Chapter 11). Lymphocytes and macrophages in extravascular sites secrete factors (e.g., TGFβ) that will activate fibroblasts, resulting in the production of cross-linked collagen, sometimes resulting in extensive collagenous scars (Figure 1.17). In the case of plasma cells, the chief product is antibody which in the case of rheumatoid arthritis results in large amounts of IgGs being present both in the synovial tissue and in the synovial fluids bathing the

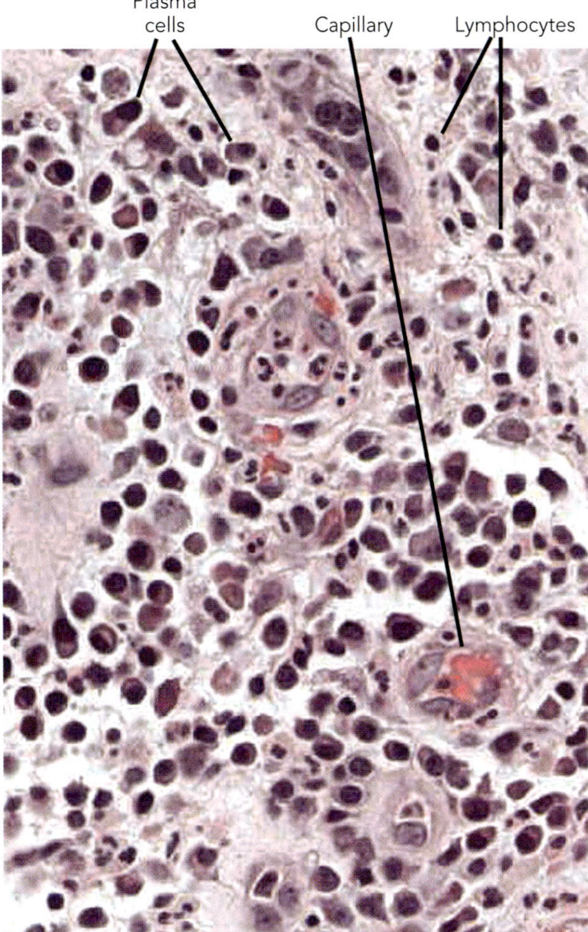

Figure 1.15. Chronic inflammation in soft tissue, featuring large numbers of plasma cells and chronic inflammatory cells (lymphocytes, mononuclear cells, and macrophages) in the interstitium.

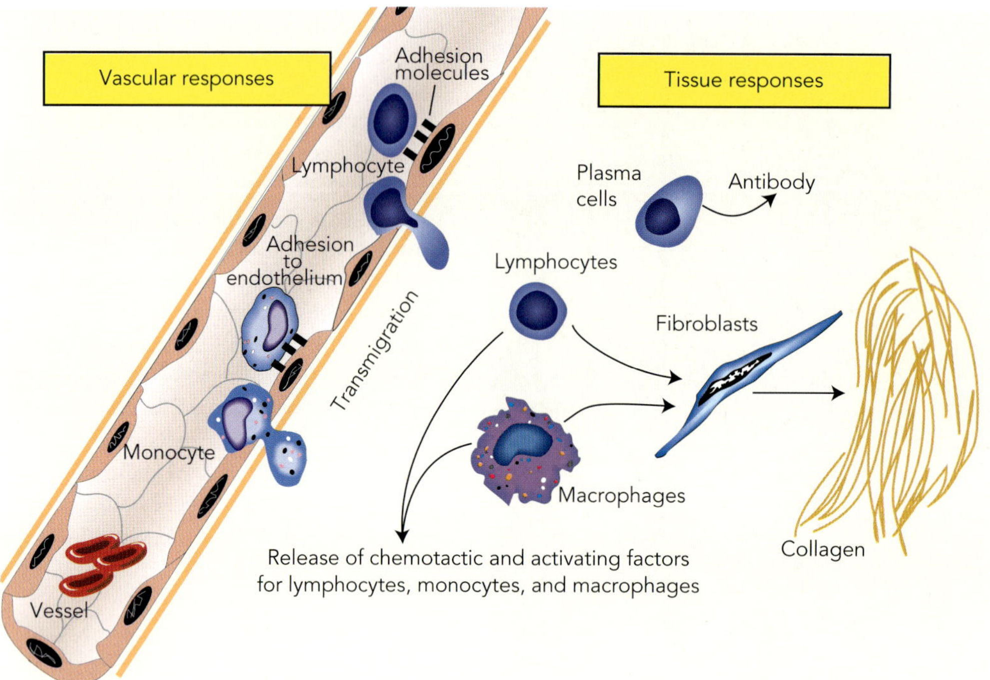

Figure 1.16. Mechanisms of chronic inflammation, with adhesion of lymphocytes and monocytes to the activated endothelium, and the eventual transmigration of these cells into the extravascular space. Activated endothelial cells express adhesion molecules (such as VCAM-1) that facilitate adhesion of lymphocytes and monocytes to endothelial surfaces, followed by their eventual transmigration. In the extracellular compartment, lymphocytes and macrophages secrete factors that stimulate extracellular collagen formation and perpetuate the inflammatory response. Transmigrated monocytes "mature" into macrophages. Plasma cells secrete various subclasses of antibodies.

joint space. The various factors produced by cells that are associated with collagen scar formation are discussed elsewhere (Chapter 13). With reference to chronic inflammation, this response may be short term or long term such as in the response to infectious agents especially such as mycobacterial species, protozoa, and so on. Chronic inflammation may resolve or there may be persistence of the chronic inflammatory response, as described later (Table 1.3). In persistent allograft rejection, recipient T cells infiltrate the transplanted organ and respond to donor histocompatibility antigens, releasing factors (cytokines and chemokines) that damage or even destroy the allograft (Figure 1.18). The purpose of immunosuppressive drugs is to prevent the build-up of T cells in the allografts, thereby preventing loss of function of the allograft or its destruction. There is a special type of rapid graft rejection (hyperacute rejection) which is usually associated with the presence of preformed antibody to graft antigens (often following prior blood transfusions received by the recipient whose immune system generates antibodies to antigens present on lymphocytes of the donor). In hyperacute allograft rejection, the process often starts within hours of vascular anastomosis of the graft and is characterized by endothelial damage in arterioles and capillaries, together with intravascular fibrin clots and PMNs. Immediate immunosuppressive and corticosteroid therapy is initiated. The result of persistent chronic inflammation, as seen in the autoimmune disease, scleroderma, can lead to an intense fibrotic response in organs such as the lung (Figure 1.19) or, as described earlier in the case of a heart allograft (Figure 1.18) to the destruction of cardiac tissue.

THERAPEUTIC INTERVENTIONS IN INFLAMMATION

There are many situations when it is desirable to suppress the inflammatory response if the response is causing an immediate threat (such as abruptly rising intracranial pressure in a patient with bacterial meningitis) or represents a long-term threat (as in rejection of an allotransplanted organ). There are several principles

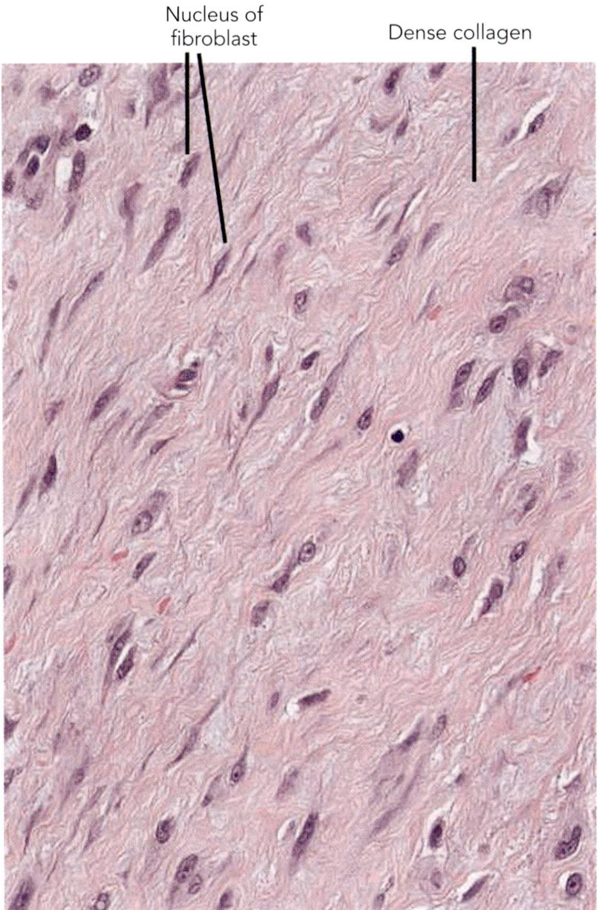

Figure 1.17. Dense collagenous scar with fibroblasts with little, if any, evidence of inflammatory cells. Cylindrical nuclei are in fibroblasts.

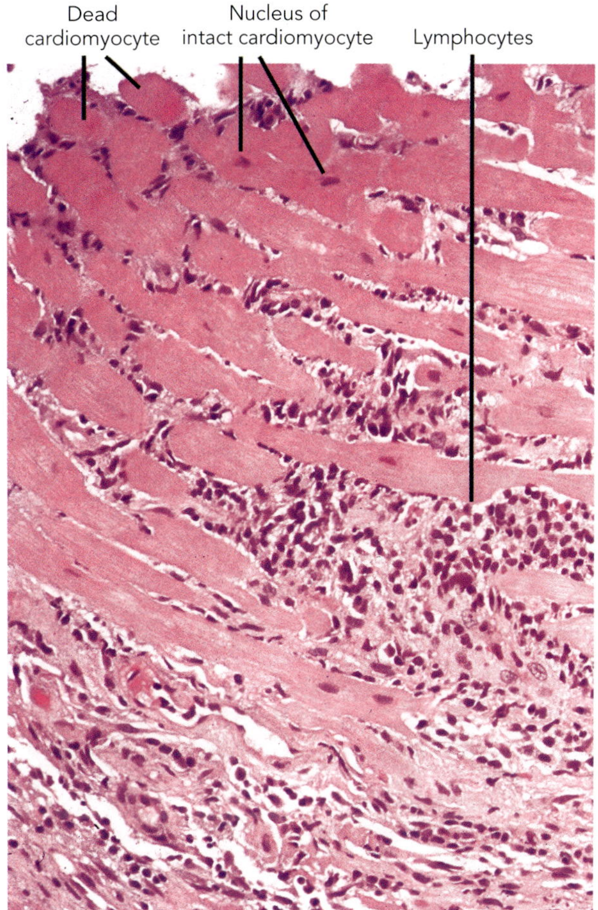

Figure 1.18. Features of chronic rejection of an allotransplanted human heart. Some cardiomyocytes are necrotic as manifested by loss of striations and nuclei and intensified eosin (red) staining. There is an intense accumulation of interstitial lymphocytes, which are known to be of recipient origin and involved in immunological damage (rejection) of the transplanted heart.

TABLE 1.3. Consequences of acute inflammation

Responses	Outcomes
Acute inflammation	• Responses to infectious agents and their containment (e.g., bacteria) • Response to trauma • Response to persistence of trigger (e.g., infectious agents), to toxic agents, or to immunological responses (autoimmune, etc.) • Resolution of persistence of inflammation or progression to chronic inflammation • Outcome may be resolution (clearance of edema fluid, fibrin, RBCs, and leukocytes) – persistence of inflammation, sometimes leading to fibrosis
Chronic inflammation	• Response to persisting inflammatory trigger (infectious agents, autoimmune products, immune complexes) • Inflammatory cells (lymphocytes, macrophages) may persist or be cleared. Giant cells may develop, together with fibrosis (as in sarcoid) and/or necrosis (as in mycobacterial infections)

that need to be considered. The first is the cause of the inflammatory response. In many situations this cannot be clearly determined, as in rheumatoid arthritis and in certain types of glomerulonephritis. If the cause of the inflammatory trigger is known (as in meningitis or pneumonia caused by bacteria), specific therapy (antibiotic drugs) are immediately instituted. If the inflammatory reaction is causing serious functional problems (defective gas exchange in bacterial pneumonia), supportive therapy (mechanical ventilation,

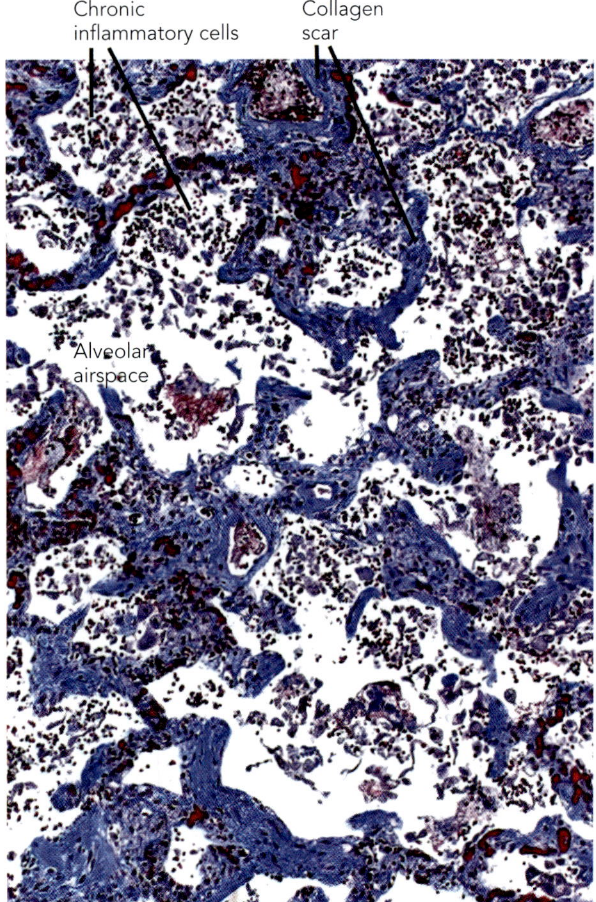

Figure 1.19. Scleroderma lung featuring extensive interstitial fibrosis (blue staining collagen) and alveolar spaces containing lymphocytes and fibrin. (Masson trichrome stain.)

high levels of oxygen, etc.) is instituted. Other interventions can also be employed. Corticosteroid therapy has powerful suppressive effects on all aspects of the inflammatory response. Nonsteroidal drugs (inhibitors of cyclooxygenases I and II) will provide symptomatic relief of pain but these drugs, on balance, have very limited effects on inflammation. That is, they poorly suppress the inflammatory response. If there is evidence that the inflammatory response is being driven by an immune response (as in autoimmune diseases such as SLE or rheumatoid arthritis, or in graft rejection or in many forms of glomerulonephritis), then immunosuppressive drugs are employed and the doses "titrated" up or down depending on clinical symptoms. In some inflammatory diseases, blockade of a single proinflammatory mediator (TNFα) has been found to be effective (rheumatoid arthritis, severe psoriasis, inflammatory bowel disease) using blocking antibody to TNFα or using the soluble TNFα receptor that intercepts TNFα with high affinity before TNFα can react with cell surface receptors to TNFα. Clearly, much more needs to be understood about the inflammatory response before more effective inflammation-blocking strategies are available and ones that, unlike corticosteroids, can be used over the long term without serious side effects.

ACKNOWLEDGMENTS

The generous provision of the photomicrographs from Professor Gerald Abrams (Department of Pathology, University of Michigan Medical School) is gratefully acknowledged. Robin Kunkel (Department of Pathology, University of Michigan Medical School) provided the superb drawings. This study was supported by NIH grants GM29507, GM61656, and HL31963, and DoD grant W81XWH-06-2-0044 (PAW). The excellent secretarial support from Beverly Schumann and Sue Scott is acknowledged.

SUGGESTED READINGS

Medzhitof, R. 2008. Origin and physiological roles of inflammation. *Nature* **454**:428–435.

Rittirsch, D., Flierl, M.A., Nadeau, B.A., et al. 2008. Functional roles for C5a receptors in sepsis. *Nat Med* **14**:551–557.

Rittirsch, D., Flierl, M.A., and Ward, P.A. 2008. Harmful molecular mechanisms in sepsis. *Nat Rev Immunol.* **8**:1–12.

Serhan, C.N. 2007. Resolution phases of inflammation: novel endogenous anti-inflammatory and pro-resolving lipid mediators and pathways. *Annu Rev Immunol* **25**:101–137.

Taraseviciene-Stewart, L., and Voelkl, N.F. 2008. Molecular pathogenesis of emphysema. *J Clin Invest* **118**(2):394–402.

Thelen, M., and Stein, J.V. 2008. How chemokines invite leukocytes to dance. *Nat Immunol* **9**:953–959.

Ward, P.A. 2004. The dark side of C5a in sepsis. *Nat Rev Immunol* **4**:133–142.

Woessner, J.F., Jr. 2002. MMPs and TIMPs – an historical perspective. *Mol Biotechnol* **22**:33–49.

2 Resolution of Acute Inflammation and Wound Healing

Derek W. Gilroy

SUMMARY

It is without doubt that resolution of acute inflammation is under strict checkpoint control by endogenous proresolution factors. It is these factors and mechanisms inherent in resolution that are crucial in preventing excessive tissue injury, autoimmunity, and chronic inflammation. In this chapter, resolution and the factors that control it are detailed to underline its importance in human pathology and highlight new and more effective treatment modalities with fewer side effects for chronic inflammatory diseases.

INFLAMMATION IN HEALTH AND DISEASE

Inflammation is a beneficial host response to foreign challenge or tissue injury that leads ultimately to the restoration of tissue structure and function. It is a reaction of the microcirculation that is characterized by the movement of serum proteins and leukocytes from the blood to the extravascular tissue. This movement is regulated by the sequential release of vasoactive and chemotactic mediators, which contribute to the cardinal signs of inflammation – heat, redness, swelling, pain, and loss of tissue function (Figure 2.1A). Local vasodilation increases regional blood flow to the inflamed area and, together with an increase in microvascular permeability, results in the loss of fluid and plasma proteins into the tissues. Concomitantly, there is an upregulation of adhesion molecule expression on endothelial cells and the release of chemotactic factors from the inflamed site, which facilitate the adherence of circulating cells to the vascular endothelium and their migration into the affected area. These tightly regulated events result in a predominance of polymorphonuclear leukocytes (PMNs, see Glossary) in the inflamed area at the onset of the lesion, which are later gradually replaced by mononuclear cells – mainly monocytes, which then differentiate into macrophages. These phagocytic cells ingest foreign material and cell debris. They also release hydrolytic and proteolytic enzymes, and generate reactive oxygen species that eliminate and digest invading organisms. Finally, the injurious stimulus is cleared and normal tissue structure and function is restored [1].

However, inflammation can cease to be a beneficial event and contribute to the pathogenesis of many disease states. The chronic inflammatory disease rheumatoid arthritis, for instance, is characterized by the accumulation and persistence of inflammatory cells in synovial joints, which results in joint damage. This loss of tissue or organ function as a result of an inappropriate inflammatory response is also seen in various other diseases, such as chronic bronchitis, emphysema, asthma, glomerulonephritis, myocardial infarction, and ischemia reperfusion injury. By contrast, certain inflammatory diseases have an intrinsic capacity for complete resolution without tissue injury – for example, lobar streptococcal pneumonia, which involves the extensive accumulation of PMNs, monocytes, and macrophages in the lungs. Studies of patients who have lobar pneumonia show that most of the lesions resolve without any evident tissue destruction. Experiments in animal models of streptococcal pneumonia show resolution of tissue pathology within days. Therefore, this type of self-limiting inflammatory response is under the strict control of endogenous mechanisms. As continual activation of the adaptive immune system is the driving force behind chronic inflammation, it is crucial to identify the stop signals that are present in self-limiting, self-resolving inflammatory lesions. These signals might be used therapeutically to control the activation of the adaptive immune response and the transition from acute to chronic inflammation, when these signals might be absent or become dysregulated.

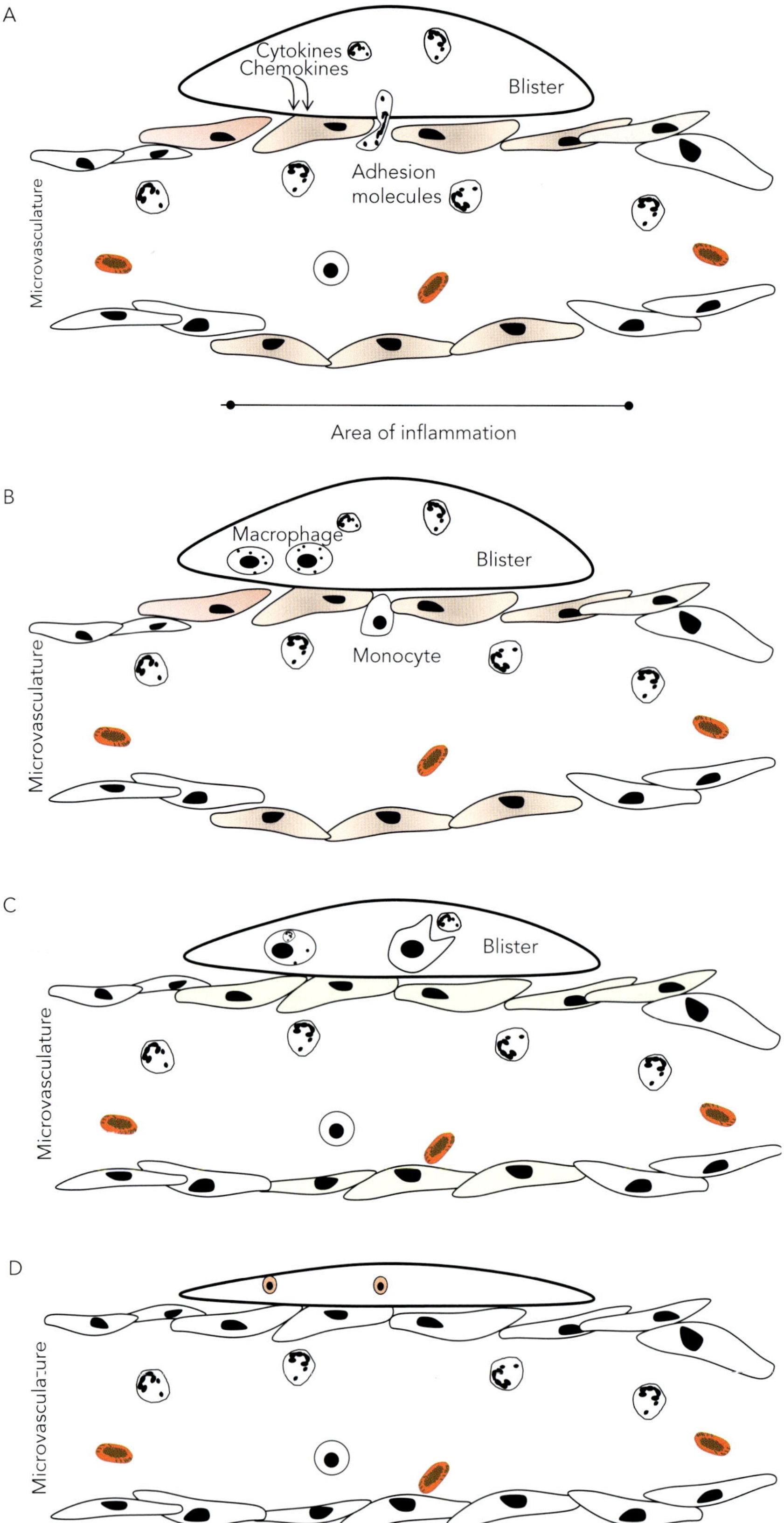

Figure 2.1. Simple depiction of acute inflammation and its resolution in the form of a skin blister. Depending on the nature of the injurious stimuli, the fist sequence of events (A) is communication between the site of injury and the microvasculature that serves the area, whereupon (B) endothelial cells become activated and recruit PMNs. In turn and provided there is no infection, whereupon PMNs will ingest and kill bugs, (C) monocyte migrate from the microvasculature to the extravascular space, differentiate into macrophages, which have a massive capacity for phagocytosing dead and apoptotic leukocytes. (D) In time may either die locally or drain via the lymphatics. Inflammation wanes and homeostasis is restored.

Therefore, in this chapter we will discuss one of the many aspects of the inflammatory response – how acute inflammation resolves. We will also discuss wound healing and the type of resolution associated with substantive tissue injury. In doing so it will be argued that resolution is an active process, whose failure may predispose the host to chronic inflammatory diseases and autoimmunity such as that typified by rheumatoid arthritis and asthma. At the very least it is hoped that this chapter will highlight resolution as a critical facet of the inflammatory response and serve to underline the importance of not altering its normal course of action when developing novel anti-inflammatory drugs. Ultimately, it will be proposed here that resolution is controlled by endogenous pro-resolution factors, which, for the future, may represent a treasure trove for drug discovery in terms of designing drugs that mimic their mode of action or enhance their synthesis [2–4].

WHAT IS INFLAMMATORY RESOLUTION?

To define the fundamental requirements for the successful resolution of either acute innate or acute adaptive immunity, it is becoming increasingly clear that the most simple but absolutely critical determinant for the inflammatory response to switch off is the neutralization and elimination of the injurious agents that initiated it in the first place. Failure to achieve this first step will invariably lead to chronic inflammation with the nature of the agent in question almost certainly dictating the etiology of the developing chronic immune response. For instance, chronic granulomatous disease is characterized by severe, protracted, and often fatal infection, which results from a failure of the phagocytic NADPH oxidase enzyme system to produce superoxide and kill invading infections leading to a predisposition to recurrent bacterial and fungal infections and the development of inflammatory granulomas [5]. Successfully dispensing with the inciting stimulus will signal a cessation to proinflammatory mediator synthesis (eicosanoids, chemokines, cytokines, cell adhesion molecules, etc.) and lead to their catabolism (Figure 2.1B). This would halt further leukocyte recruitment and edema formation. These are probably the very earliest determinants for the resolution of acute inflammation, the outcome of which signals the next stage of cell clearance. The clearance phase of resolution, be it innate PMN or eosinophil driven or adaptive (lymphocyte mediated), also has a number of mutually dependent steps. The clearance routes available to inflammatory leukocytes include systemic recirculation or local death of influxed PMNs, eosinophils, or lymphocytes followed by their phagocytosis by recruited monocyte-derived macrophages (Figure 2.1C). Once phagocytosis is complete, macrophages can leave the inflamed site by lymphatic drainage with evidence that a small population may die locally by apoptosis. If all of these pathways are strictly followed then acute inflammation will resolve without causing excessive tissue damage and give little opportunity for the development of chronic, nonresolving inflammation (Figure 2.1D). In the following, we will discuss the cellular changes that occur throughout acute inflammation that ultimately leads to resolution and the soluble mediators that temper the severity of onset as well as trigger resolution. Finally, we will differentiate between resolution of acute inflammation (leukocyte clearance and inflammatory mediator catabolism) and wound healing, a form of inflammatory resolution associated with parenchymal tissue injury.

SOLUBLE MEDIATORS OF INFLAMMATION AND RESOLUTION

Controlling Onset

It is well known that inflamed tissues generate local proinflammatory stimuli to drive acute inflammation including cytokines, chemokines, and cell adhesion molecules. Indeed, these have acted as targets for drug development in the form of anti-TNFα inhibitors (etanercept, infliximab), for instance. The original targets for the treatment of inflammation-driven disease were the cyclooxygenase (COX)-derived prostaglandins (PGs), with nonsteroidal anti-inflammatory drugs (aspirin, indomethacin, ibuprofen, naproxen) serving as the clinical inhibitor [6,7]. Interestingly, however, the PGs and the arachidonic acid metabolic cascade (COX, lipoxygenase [LOX] pathways) are now recognized to possess potent anti-inflammatory and proresolving properties (see later). Along these lines, there is a range of other systemic and locally produced endogenous mediators that counterbalance the severity of inflammatory onset. Studies in the 1950s and 1960s identified endogenous anti-inflammatory mediators that counteract vascular leakage – namely, adrenaline, noradrenaline, and 5-hydroxytryptamine – and intracellular cyclic AMP, a second messenger induced by several hormones; inflammatory mediators; and cytokines, which dampens leukocyte activation. Elevation of the level of intracellular cAMP – by inhibiting the enzyme system that is responsible for its catabolism (phosphodiesterase) – ameliorates immune and nonimmune inflammation in vivo and suppresses various cellular processes in vitro, including the immunological release of histamine and leukotrienes from mast cells, monocytes, and PMNs; lysosomal enzymes and reactive oxygen species from PMNs; and cytokines and nitric oxide (NO) from macrophages [2]. These data further indicate that cAMP has a central role in the resolution of inflammation. Perhaps the most powerful endogenous anti-inflammatory agents to be described so far are the glucocorticoids [8]. Glucocorticoids and their synthetic mimetics are used for the treatment of several chronic inflammatory diseases, including rheumatoid arthritis, inflammatory bowel disease, asthma, psoriasis, and vasculitis. Most of the actions of glucocorticoids require binding to cytoplasmic steroid hormone receptors that migrate to the nucleus and antagonize proinflammatory gene transcription. However, glucocorticoids also induce the expression of regulatory proteins that have anti-inflammatory actions, of which the peptide annexin 1 (previously known as lipocortin 1) has been well described in vitro and in vivo [9]. Annexin 1 has been shown to inhibit the production of PGs, as well as PMN and monocyte migration, in vivo. The take home point here is that although there are signals that drive the onset of inflammation (PMN trafficking) there are also endogenous factors released by stromal and/or hematopoietic cells that temper the severity of this response, save inflammation becomes too severe or inappropriate in magnitude.

Onset to Resolution

One well-described event in the transition toward resolution is the replacement of PMNs or eosinophils by monocytes and phagocytosing macrophages. However, until recently our understanding of the signals that control this cell profile switch was unclear. Studies addressing this issue of leukocyte infiltration in peritoneal inflammation have suggested that the interaction between interleukin (IL)-6 and its soluble receptor, sIL-6R, forms one of the major determinants of this switch from PMNs to monocytes. It was shown that sIL-6R, produced by the infiltrating PMNs, forms a complex with IL-6 which, in turn, directly modulates CC and CXC chemokine expression. Thus, CXC chemokine synthesis, induced by IL-1 and tumor necrosis factor (TNF)α, was suppressed whereas the CC chemokine monocyte chemoattractant 1 was promoted. This chemokine shift suppresses further PMN recruitment in favor of sustained mononuclear cell influx. In addition to chemokines, the eicosanoids also orchestrate the early transition to resolution in acute inflammation. Transcellular metabolism of arachidonic acid by LOX/LOX interaction pathways gives rise to the lipoxin (LX) family of eicosanoid metabolites. LXs display selective actions on leukocytes that include inhibition of PMN chemotaxis, PMN adhesion to and transmigration through endothelial cells, as well as PMN-mediated increases in vascular leakage (see Glossary). It is unclear at this point whether there is any cross talk between the LXs and IL-6/sIL-6R complex signaling in the control of leukocyte profile switching. Nonetheless, it seems that when acute inflammation needs to resolve the IL-6/sIL-6R, chemokines, and LXs representing some of the earliest signals that control the switch from very early PMNs to monocyte/macrophage. Over and above this, there is the need for proinflammatory mediator catabolism, specifically the removal of cytokines and chemokines that drive inflammation. To do this, D6, a scavenger receptor expressed on lymphatic endothelial cells binds and neutralizes inflammatory members of the b-chemokine family but not constitutive b-chemokines or members of the other chemokine subfamilies such that its absence predisposes to failed resolution. Indeed, CCR5 expression on apoptotic PMNs and apoptotic T cells also sequester and effectively clear CCL3 and CCL5 from sites of inflammation, with scavenging CCR5 expression being inhibited by proinflammatory TNFα, for instance, but upregulated by proresolving lipids (LXs and resolvins). Thus, there is the coordinated interaction with factors that drive inflammation (PMN influx, cytokines, chemokines, vasoactive amines) counterbalanced by endogenous factors that temper the severity of inflammation (PGD2, LXs, cAMP) overshadowed by factors that ensure the smooth transition to resolution.

Resolving Inflammation

Studies on the resolution of acute inflammation have already revealed novel mediators with potent anti-inflammatory properties. Determining their basic structure and function might help in the development of unique anti-inflammatory therapeutics. So far, these proresolving mediators have been shown to exert powerful anti-inflammatory effects in various experimental models of inflammatory diseases. Of these, lipid-mediator derivatives of the COX-2 and LOX/LOX interaction pathways of arachidonic acid, eicosapentaenoic acid (EPA), and docosahexaenoic acid (DHA) metabolism will be described [10].

PGD_2 has emerged recently as an eicosanoid with both pro- and anti-inflammatory properties. PGD_2 undergoes dehydration in vivo and in vitro to yield biologically active PGs of the J_2 series, including PGJ_2, $\Delta^{12,14}$-PGJ_2, and 15-deoxy-$\Delta^{12,14}$-PGJ_2 (15d-PGJ_2). In addition to being a high-affinity natural ligand for anti-inflammatory peroxisome proliferator-activated receptor gamma (PPARγ) (PPAR is explained in the Glossary), 15d-PGJ_2 also exerts its effects through PPARγ-dependent and -independent mechanisms to suppress proinflammatory signaling pathways and the expression of genes that drive the inflammatory response. 15d-PGJ_2 also preferentially inhibits monocyte rather than PMN trafficking through the differential regulation of cell adhesion molecule and chemokine expression. We have shown that COX-2–derived PGD_2 metabolites contribute to the resolution of acute inflammation (pleuritis) through the preferential synthesis of PGD_2 and 15d-PGJ_2, which, along with the alternative DNA-binding p50–p50 homodimers complexes of nuclear factor kappa B (NF-κB) (see Glossary), bring about resolution by inducing leukocyte apoptosis. Indeed, there is an increasing body of evidence detailing the differential effects of PGD_2 metabolites on leukocyte apoptosis as well as the signaling pathways involved. In addition to the well-known eicosanoids, there is a new generation of lipid mediators showing promise as endogenous anti-inflammatories. Resolvins and docosatrienes are fatty-acid metabolites of the COX/LOX pathways, where the omega-3 fatty-acid constituents of fish oils (DHA and EHA) are the substrates and not arachidonic acid. Thus, transcellular metabolism of arachidonic acid by LOX/LOX interaction pathways gives rise to the LX family of eicosanoid metabolites. LXs display selective actions on leukocytes that include inhibition of PMN chemotaxis, PMN adhesion to and transmigration through endothelial cells, as well as PMN-mediated increases in vascular permeability. In contrast to their effects on PMN and eosinophils, LXs are potent stimuli for peripheral blood monocytes, stimulating monocyte chemotaxis and adherence without causing degranulation or release of reactive oxygen species. In fact, LXs and their stable analogues accelerate the resolution of allergic pleural edema and enhance phagocytosis of apoptotic PMNs by monocyte-derived macrophages in a nonphlogistic fashion (see Glossary), paving the way for a return to tissue normality. LXA4 and aspirin-triggered 15-epi-LXA4, as well as their stable analogues, act with high affinity at a G-protein–coupled receptor, LXA4 receptor (ALXR; also referred to as formyl peptide receptor-like 1 or FPRL1). FPRL1 is a member of the family of seven transmembrane G-coupled receptors, which has at least two other members – FPRL2 and the formyl-Met-Leu-Phe receptor (FPR). By contrast, LXB4 does not bind the ALXR and, although functional studies have indicated the existence of a receptor that is activated by LXB4, this receptor has not been cloned. As with the cyPGs, the LXs have also been identified as being expressed during and being crucially important for the resolution of acute inflammation. In a model of rat allergic edema, for instance, LXA4 was identified along with PGE_2 as being present during the clearance of edema in this model. Inhibition of their synthesis prolonged edema clearance, which was rescued using stable analogues of these eicosanoids. A recent analysis of eicosanoid synthesis in a murine dorsal air pouch of acute inflammation elicited by TNFα has revealed a switch in lipid class metabolism reminiscent of that found in the rat carrageenin-induced pleurisy. In response to TNFα, levels of leukotriene B4 increased rapidly, followed by PMN infiltration, which coincided with a rise in inflammatory exudate PGE_2. Concomitant with the eventual reduction in PMN numbers and PGE_2 was an increase in LXA4. It was concluded that PGE_2 induced a switch in lipid mediator synthesis from predominantly 5-LOX-generated leukotriene B4 to 15-LOX-elicited proresolving LXA4. Along with our findings in the rat carrageenin-induced pleurisy in terms of PG metabolism, this work indicates that, in acute inflammation, lipid-mediator biosynthesis is biphasic, with a role for eicosanoids in the initiation as well as termination of the inflammatory response [2].

Arachidonic acid is not the only fatty-acid substrate that can be transformed by COXs and LOXs to bioactive mediators with roles in anti-inflammation and resolution. DHA and EPA – omega-3 fatty-acid constituents of fish oils – were shown recently to be metabolized during the resolving phase of an aspirin-treated TNFα-induced inflammation to potent anti-inflammatory products, named resolvins. For instance, endothelial cells expressing COX-2 and treated with aspirin convert EPA to 18*R*-hydroxyeicosapentaenoic acid (HEPE) and 15*R*-HEPE. Both are subsequently used by PMNs to generate separate classes of novel trihydroxy-containing mediators that potently inhibit human PMN transendothelial migration. Similarly, aspirin-acetylated COX-2 converts DHA to 17*R*-HDHA, which is subsequently transformed by PMNs into two sets of novel di- and trihydroxy products that can inhibit microglial cell cytokine expression and ameliorate experimental models of dermal inflammation and leukocyte accumulation in peritonitis at nanogram doses. Even in the absence of aspirin, human whole blood converts DHA to 17*S* series resolvins as well as novel dihydroxy-containing docosanoids. DHA-loaded glial cells stimulated with zymosan also release docosanoids, with these novel resolvins possessing such potent anti-inflammatory effects as inhibiting leukocyte trafficking in vivo and proinflammatory cytokine release by stimulated human glial cells. Collectively, the LXs and resolvins represent novel classes of anti-inflammatory agents that are tightly associated with the resolution of acute inflammation and shown to be implicated in the pathogenesis of disease processes, including atherosclerosis, periodontitis, chronic liver disease, and asthma. Moreover, LXs and their analogues are proving to be highly effective therapeutics in a range of experimental disease models, including immune-mediated glomerulonephritis and renal ischemia-reperfusion injury, a range of skin inflammation-like diseases and gastritis. Lipid mediators of this sort are not only natural and essential components of acute inflammatory resolution, but show that when applied to inflammatory disease processes are highly effective, thereby providing the rationale for the development of compliant and stable mimetics that target key aspects of chronic inflammation, either ongoing or recurrent, forcing them down a revolving pathway and into remission [11].

GETTING RID OF LEUKOCYTES

One of the hallmarks of acute inflammation is white blood cell accumulation (PMNs and eosinophils, for instance), designed to neutralize and eliminate the injurious agents. Once the PMNs and eosinophils have done their job and their help is no longer needed, what happens next? At this juncture it must be borne in mind that these are a formidable cell type and if left unchecked could do untold damage to an already inflamed site. After all, these cells are designed to combat infection by releasing hydrolytic and proteolytic enzymes as well as generating reactive oxygen species. Therefore, PMNs and eosinophils must be disposed of in a controlled and effective manner. To oversee this, nature has come up with an ingenious way of defusing such potentially explosive cells called programmed cell death or apoptosis. Apoptosis of inflammatory cells is a physiological process for the nonphlogistic removal of cells. During apoptosis, cells maintain an intact membrane and, therefore, do not release their potentially histotoxic agents. Necrosis of inflammatory leukocytes, on the other hand, involves a loss of membrane integrity leading to the release of potentially toxic intracellular contents [12,13]. Moreover, apoptotic cells express a repertoire of surface molecules that allow their recognition and phagocytosis by macrophages. In fact, the way these cells die helps the resolution process enormously. Recognition of these apoptotic cells by macrophages does not liberate proinflammatory agents from the macrophages themselves but can release anti-inflammatory signals such as IL-10 and TGFβ (endogenous immunosuppressive agents). Thus, not only is apoptosis a noninflammatory way of disposing of cells, but this method has the added advantage of conferring upon macrophages an anti-inflammatory phenotype conducive to resolution and curtailment of ensuing adaptive immune responses. It is important to note that if not recognized and disposed of, apoptotic cells will eventually undergo secondary necrosis releasing damaging intracellular contents and amplifying the inflammatory response. Therefore, increasing the rate of apoptosis, as a potentially anti-inflammatory strategy, must be matched by a mechanism that upregulates macrophage phagocytic clearance capacity. Thus, the removal process might also be susceptible to selective modulation by pharmacological agents for therapeutic gain.

Enhanced undesirable apoptosis occurs in many neurological diseases, such as Alzheimer's disease, Parkinson's disease, Huntington's disease, and multiple sclerosis. Furthermore, inappropriate inflammatory responses or dysfunctional vascular effects leading to tissue damage with increased apoptosis have been observed. So, there is good evidence of cell or tissue apoptosis during myocardial infarction, stroke, or sepsis. Consequently, a therapeutic strategy to delay or inhibit apoptosis would seem a viable option assuming that cell specificity can be achieved. On the other hand, there is much evidence indicating that reduced apoptosis occurs in most cancers. Essentially, uncontrolled cell division or proliferation, there is an apparent failure of cancerous cells to undergo apoptosis. It has also been proposed that in many inflammatory diseases (e.g., rheumatoid arthritis, atopic dermatitis, Crohn's disease, asthma, and chronic obstructive pulmonary disease) there might be delayed apoptosis of key inflammatory cells, thereby prolonging the functional responsiveness of these potential histotoxic cells. A strategy to specifically promote death of cancer cells or tissue-damaging inflammatory cells is therefore likely to be therapeutically beneficial. However, as stated, any attempts to induce cell, especially inflammatory cell, apoptosis must be matched by effective noninflammatory clearance by phagocytic cells (e.g., macrophages). Failure to remove these apoptotic cells may lead to the cells becoming necrotic, thereby increasing the potential for tissue damage. Great progress has been made in recent years in the elucidation of the complex mechanisms that are involved in recognition of apoptotic cells (or apoptotic bodies) by phagocytes (at least 10 recognition mechanisms have been identified so far). Furthermore, phagocytosis of apoptotic cells has been shown to be a highly regulatable process and therefore likely to be amenable to pharmacological manipulation. It has been shown, for example, that elevation of cAMP by PGs can downregulate macrophage capacity to ingest apoptotic cells, whereas treatment of phagocytes with glucocorticoids, LXs (arachidonic acid metabolites via the COX/LOX or LOX/LOX interaction pathways), or even certain cytokines can markedly increase macrophage clearance of apoptotic cells. Interestingly, the environment in which phagocytes are likely to reside can also upregulate apoptotic cell clearance. For example, interactions with extracellular matrix components, such as fibronectin, and ligation of macrophage CD44 with cross-linking antibodies can augment the capacity of phagocytes to engulf apoptotic cells. So, clearance of apoptotic cells in a noninflammatory manner by phagocytes is a therapeutic possibility. Indeed, there is already evidence in animal models that CD44 has an important role in resolving lung inflammation and that glucocorticoids might exert some of their therapeutic beneficial anti-inflammatory effects in patients with asthma by influencing apoptosis and apoptotic cell clearance. Some of these novel developments have led to the design of drugs that have even gone into clinical trials. On this note, it is interesting to speculate upon how many clinically used anti-inflammatory drugs trigger hither unknown proresolution pathways in addition to their classic role of dampening conventional proinflammatory events.

Currently, there are no drugs in the clinic that are purposefully based on the elicitation of proresolving pathways with the exception of those drugs that target apoptosis. One extremely active area of drug development that targets apoptosis is the identification of small-molecule caspase inhibitors. Caspases are a family of cysteinyl aspartate–specific proteases that are of fundamental importance in the initiation and execution of apoptosis, ultimately being responsible for the dismantling of the cell during apoptosis. A number of small-molecule inhibitors of caspases (e.g., ZVAD-fmk) have already been tested in animal models of human disease with remarkable success. For example, specific caspase inhibitors have been shown to be effective in preventing or reducing the effects of ischemia (e.g., organ failure and death) in various animal models. These inhibitors of apoptosis are now being tested preclinically or have reached clinical trials for hepatic disease, acute myocardial infarction, and sepsis. A good example of a novel and specific broad-spectrum caspase inhibitor that has been successfully used preclinically is PF-03491390 (formerly named IDN-6556). This compound, administered by a number of routes, was shown to be potent and efficient in reducing signs of liver damage in in vivo rodent models of liver disease. It was shown that the compound seemed to exert its activity by effectively inhibiting caspase activity. In a recent study, the induction of PMN apoptosis during acute inflammation using R-roscovitine (Seliciclib or CYC202), a cyclin-dependent kinase inhibitor (cyclin-dependent kinases override antiapoptotic survival signals from survival factors such as GM-CSF), resulted in enhanced PMN apoptosis and early inflammatory resolution. There are a number of other drugs that are in development for inhibiting apoptosis, but so far the most promising therapeutic development is in the induction of apoptosis, especially for the treatment of a number of cancers. The strategy for inducing apoptosis has been to block powerful survival pathways, for example, by inhibiting BCL2-mediated survival using antisense oligonucleotides, and interference of survival pathways that are mediated by NF-κB, phosphoinositol-3-kinase, and tyrosine kinase activation. Another strategy is to directly induce apoptosis by engaging death receptor pathways (such as FASR, TNFR, and TRAILR) or by other less well-defined mechanisms.

MACROPHAGES AND RESOLUTION – CELLULAR PLAYERS AND HOMEOSTASIS

From our text book reading the role of the lymphatic system in localized acute inflammation, it is clear that it plays a role in draining inflammatory mediators and effete leukocyte away from the inflamed site. We have already discussed the importance of PMN clearance to the resolution of acute inflammation, but it is equally important that phagocytosing inflammatory macrophages are cleared away from the inflamed site to prevent local macrophage-induced tissue damage and potential granuloma tissue damage and the development of chronic inflammation. However, despite the need to understand the endogenous control of macrophage clearance during acute inflammatory resolution, little is known about this field. There is increasing evidence that macrophage clearance from an inflamed site is a highly regulated event. Using an experimental model of acute resolving peritonitis, it was shown that macrophages adhere specifically to mesothelium overlying draining lymphatics and that their emigration rate is regulated by the state of macrophage activation providing the first evidence that macrophage emigration from the inflamed site is controlled by adhesion molecule regulation of macrophage–mesothelial interactions. This report highlights the importance of adhesion molecules controlling clearance of inflammatory macrophages into the draining lymphatic circulation, thus underscoring new pathways in the resolution of acute inflammation.

Despite the need to clear macrophage from sites of injury, there is emerging evidence that such cells play an important role in eliciting the final phase of inflammation – triggering homeostasis and immune recovery. Macrophages are generally classified as either classically (M1) or alternatively (M2) activated [14]. While this nomenclature is based on the phenotype macrophages acquire in response to defined stimuli in vitro, inflammatory characteristics of macrophages at sites of inflammation in vivo are less well studied. In particular, the phenotype of macrophages found during resolving inflammation is little unknown. Despite this, we have made some advances in understanding this by characterizing the inflammatory nature of macrophages found at the site of resolving peritonitis (Table 2.1). Interestingly, these so-called resolution-phase macrophages, in the context of resolving peritonitis at least, are neither classically nor alternatively activated but are a hybrid of both canonical definitions while they express mannose receptor, synthesize IL-10, and arginase 1 but also express

TABLE 2.1. Phenotype of resolution-phase macrophages

	Resolving (rM)	Nonresolving (M1)
Proinflammatory cytokines/chemokines	0	+++
Anti-inflammatory cytokines/chemokines	+++	0
Bactericidal	+	+++
Mannose receptor	+++	+
HMGB-1	0	+++
iNOS	+++	+
COX-2	+++	+
cAMP	+++	+

M1 macrophage markers, that is, COX-2 and iNOS (Figure 2.2). And while elucidating their precise role in resolution is in its infancy, we are finding that resolution-phase macrophages play a critical role in signaling the influx of innate-type lymphocytes to sites of resolving inflammation. It transpires that as inflammation resolves lymphocytes repopulate the cavity comprising B1, NK, gamma/delta T, CD4+/CD25+, and B2 cells. In particular, repopulating lymphocytes do not bring about resolution but are critical for modulating inflammatory responses to secondary infection and associated mortality (Figure 2.3). While the repopulation of innate-type lymphocytes has been found in several experimental models including mouse and man, the signals that control their postinflammation repopulation and the

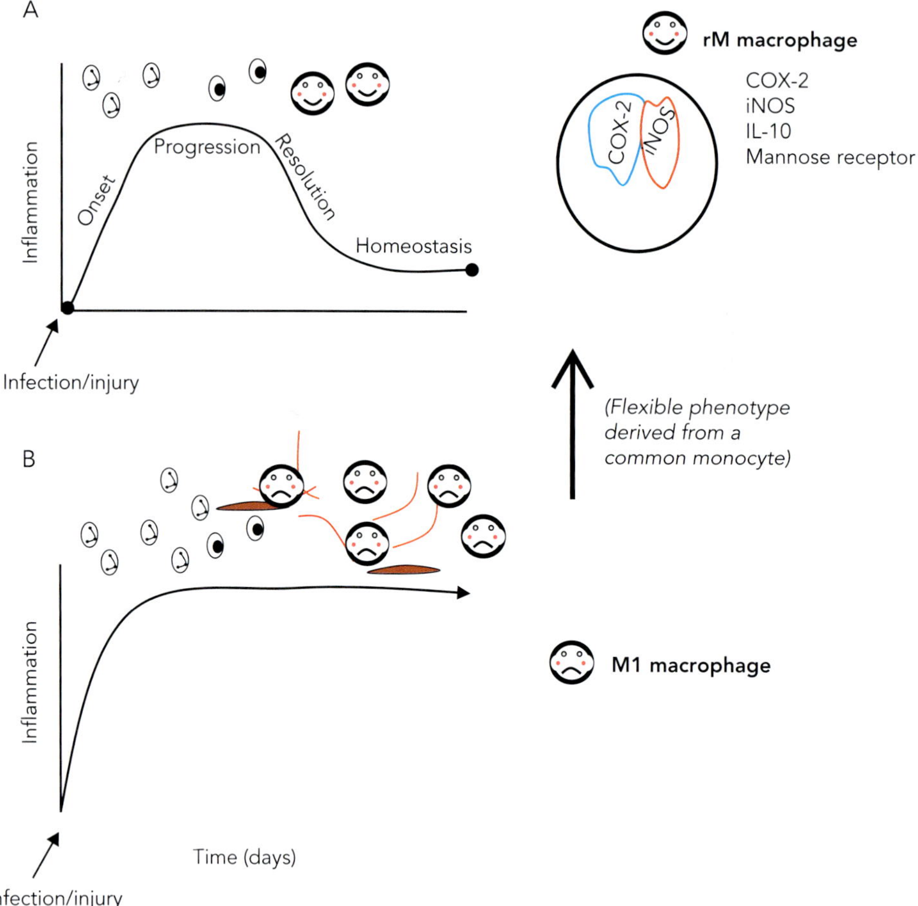

Figure 2.2. Inflammatory cell profile and resolution-phase leukocyte phenotype. In response to injury or infection leukocyte migrate to sites of injury. (A) Provided the injurious agent is neutralized inflammation will resolve leading to the injured tissue regaining its prior physiological function/state which we have recently found is characterized by a population of novel macrophages that possess a unique and distinct phenotype, termed rM cells for resolution-phase macrophages. However, (B) if the injurious agent is not cleared or there is failure of proresolution pathways, inflammation will persist characterized by M1 or classically activated proinflammatory macrophages that propagate the response and cause tissue injury. The objective, therefore, is for inflammation to resolve with stromal cells and leukocytes attaining an immunosuppressive "resolution" phenotype.

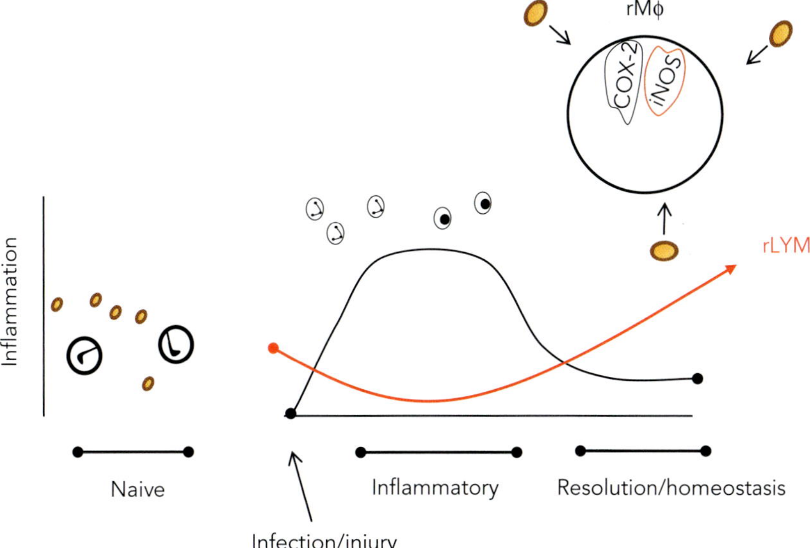

Figure 2.3. Resolution leading to homeostasis. One of the objectives of resolution is for inflamed tissues to resolve with limited tissue injury and to reacquire the physiological state it enjoyed before injury. To emphasize this, we need to appreciate that naive tissue have a complement of innate immune cells that control the severity of inflammatory responses. The peritoneal or pleural cavities, for instance, have a defined population of innate-type lymphocytes which control the severity of acute responses as well as tissue-resident macrophages, in a state of immune suppression. Once inflammation ensues, these protective cells transiently disappear and remain absent for the duration of the response only to repopulate tissue postinflammation/resolution. The objective therefore is to achieve the state the tissue enjoyed before injury to restore homeostasis and immune function.

subtypes that confer protection and signal homeostasis are unknown at this stage. Thus, macrophages are important for resolution and restoration of homeostasis after inflammation with the phenotype (M1, M2, or rM) dictating whether inflammation abates or progresses to wound healing. This will depend on the degree of inflammation and associated tissue injury. We certainly know, for instance, that the phenotype of macrophages involved in the progression and resolution of liver injury or chronic renal damage have distinct inflammatory characteristics and that the true phenotype of such protective cells will most likely depend on the tissues involved and may even be species specific. The important point is that while we have gained a great deal of insight into macrophage function based on in vitro cell studies and in response to defined inflammatory stimuli, there is now a greater need to define the inflammatory characteristic of macrophages taken directly from resolving tissues. Taking this further, I would suggest that in addition to defining resolution as the active clearance of leukocytes, we need to include macrophage phenotype and innate-type lymphocyte repopulation. This will become important when developing drugs with the objective of bringing about resolution. The indices for resolution, therefore, should not only include proinflammatory mediator catabolism and leukocyte clearance but also consider including macrophage-resolution phenotype and innate-type lymphocyte repopulation.

RESOLUTION OF ADAPTIVE IMMUNE RESPONSES

In the previous sections we have discussed how resolution of innate inflammation resolves, discussing the fate of inflammatory PMNs, eosinophils, and macrophages. However, adaptive immunity, specifically, a Type III hypersensitivity (Arthus reaction) (discussed in the Glossary) or a Type IV delayed type hypersensitivity (DTH), are also acute in nature and resolve in a matter of hours or days. For instance, in a purified protein derivative–induced DTH response, it was shown that the induction and resolution of this response may depend on the expression of cytokines, such as IL-2 and IL-15, which regulate both proliferation and apoptosis in T cells. Failure to control either of these phases of the reaction may contribute to the chronicity of T lymphocyte–mediated inflammatory reactions. In another important series of studies, the endogenous factors that control the longevity of granulomatous autoimmune thyroiditis revealed that the ratio of CD4+/CD8+ T cells are critical determinants of its resolution. In this disease process, CD4+ T cells outnumber CD8+ T cells when lesions progress to fibrosis, while CD8+ T cells outnumbered CD4+ T cells in thyroids that resolved.

WOUND HEALING

Importantly, successful resolution will limit excessive tissue injury and give little opportunity for the development of chronic, immune-mediated inflammation. However, if the host is unable to neutralize the injurious agent and/or there is a failure of endogenous proresolving mediators to invoke resolution, then acute inflammation might perpetuate, resulting in varying degrees of tissue injury. If tissue injury is mild, necrotic parenchymal cells will be replaced by new cells of the same type in a process known as regeneration. If, however, tissue damage is extensive, or when fibrin is not rapidly cleared after acute inflammation, the process of healing is by repair. This involves the in-growth from the surrounding connective tissue of an initially vascular tissue containing capillary loops, fibroblasts, and leukocytes, and is known as granulation tissue. With time, the fibroblasts lay down collagen and the capillaries disappear, leaving an avascular area of fibrosis or scar. Repair by granulation and fibrosis occurs in many parts of the body where a deposit of clot, exudate, or dead tissue occurs and is given the general name of organization. Undoubtedly, this is also a form of inflammatory resolution associated with tissue damage. As in rheumatoid arthritis and asthma, for example, there will be continuous or repeated bouts of acute inflammation, resulting in ongoing tissue damage. Attempts at wound healing would result in granulomatous tissue formation, angiogenesis, fibrosis, and scar formation, all occurring concurrently. This is chronic inflammation and might be defined as a continuous inflammatory disease state that could be driven by the development of an immune response to an endogenous antigen (autoimmunity).

CONCLUSION

The take-home point from all these studies is that regardless of the etiology of the inflammatory response, if it is resolving in nature then this resolving event is highly controlled and regulated by endogenous factors and mechanisms that, if interfered with, may lead to chronic inflammation and autoimmunity. Therefore, this emerging concept of resolution presents with drug discovery opportunities. For instance, we can develop drugs that mimic endogenous proresolution factors or enhance their synthesis. By analogy, just like IL-1β and TNFα, which are expressed during the early onset phase of acute inflammation, can be used to initiate acute inflammation experimentally, then factors expressed during resolution may be useful in switching off an ongoing inflammatory response. Alternatively, we can develop drugs that enhance the synthesis of endogenous proresolution factors. Equally, we can develop drugs or refine existing ones (e.g., glucocorticoids) to bring about selective leukocyte apoptosis (eosinophil) while at the same time enhance their phagocytosis. If these aims seem too unrealistic at present, then for the foreseeable future perhaps we should bear in mind that when developing novel anti-inflammatories such drugs should not interfere with the synthesis or mode of action of proresolution mediators and/or the mechanisms that are critical for resolution. Finally, from a basic concept perspective this chapter aimed at underscoring the complexity of acute inflammation and the fact that while there are endogenous factor that drive the response, there are also factors produced by inflammatory cells and injured tissues that counterbalance the degree of inflammation and prevent the response from causing unnecessary tissue injury. After all, inflammation is a good thing but it has the potential to do great harm. Therefore, not only the magnitude of the initial response to injury/infection must be tightly controlled, but also its longevity and the factors that switch it off.

GLOSSARY OF TERMS

Polymorphonuclear leukocyte Polymorphonuclear leukocytes (PMNs) were discovered by Paul Ehrlich who used contemporary fixing and staining techniques to identify the lobulated nucleus and the granules that typify cells that we now classify as eosinophils, basophils, and PMNs.

Vascular leakage The process of the escape of plasma and plasma proteins along with white blood cells from the vessel is known as exudation. This inflammatory exudate accounts for an increase in the volume of interstitial fluid (edema) and tissue swelling at the local site of injury.

Peroxisome proliferator-activated receptor Peroxisome proliferator-activated receptors (PPARs) are members of the nuclear receptor family that regulate the transcription of genes involved in lipid and lipoprotein metabolism, glucose and energy homeostasis, as well as cellular differentiation and consist of three isotypes, alpha (NR1C1), gamma (NR1C3), and beta/delta (NRC1C2) with a differential tissue distribution.

NF-κB Nuclear factor kappa B (NF-κB) is a group of sequence-specific transcription factors that are best known as a key regulator of the innate and adaptive inflammatory responses, cell survival, and oncogenesis. In mammals, NF-κB consists of five structurally related and functionally conserved proteins, RelA (p65), RelB, c-Rel, NF-κB1 (p105 and p50), and NF-κB2 (p100 and p52).

Nonphlogistic Noninflammatory. This term is used to describe the clearance of leukocytes in a manner that does not elicit an inflammatory response.

Hypersensitivity reactions A delayed type hypersensitivity or Type IV hypersensitivity is mediated by T lymphocytes and not by antibody–antigen complexes (Arthus or Type III hypersensitivity). Typically, this response occurs 24–72 hours after the sensitized host is exposed to the offending antigen. For example, a DTH reaction may be set up experimentally by sensitizing to methylated bovine serum albumin in Freund's complete adjuvant and challenging 12 days later with the same antigen.

REFERENCES

1. Majno, G., and Joris, I. 2004. *Cells, Tissues and Disease. Principles of General Pathology.* New York: Oxford University Press.
2. Gilroy, D.W., Lawrence, T., Perretti, M., and Rossi, A.G. 2004. Inflammatory resolution: new opportunities for drug discovery. *Nat Rev Drug Discov* **3**:401–416.
3. Lawrence, T., Willoughby, D.A., and Gilroy, D.W. 2002. Anti-inflammatory lipid mediators and insights into the resolution of inflammation. *Nat Rev Immunol* **2**:787–795.
4. Serhan, C.N., and Savill, J. 2005. Resolution of inflammation: the beginning programs the end. *Nat Immunol* **6**:1191–1197.
5. Segal, A.W., Geisow, M., Garcia, R., Harper, A., and Miller, R. 1981. The respiratory burst of phagocytic cells is associated with a rise in vacuolar pH. *Nature* **290**:406–409.
6. Vane, J.R. 1971. Inhibition of prostaglandin synthesis as a mechanism of action for aspirin-like drugs. *Nat New Biol* **231**:232–235.
7. Vane, J.R. 1987. Anti-inflammatory drugs and the many mediators of inflammation. *Int J Tissue React* **9**:1–14.
8. Hench, P.S., Kendall, E.C., Slocumb, C.H., and Polley, H.E. 1949. The effect of a hormone of the adrenal cortex (17-hydroxy-11-dehydrocorticosterone: compound E) and of pituitary adrenocorticotropic hormone on rheumatoid arthritis: preliminary report. *Proc Staff Meet Mayo Clin* **24**:181–197.
9. Flower, R.J. 1988. Eleventh Gaddum memorial lecture. Lipocortin and the mechanism of action of the glucocorticoids. *Br J Pharmacol* **94**:987–1015.
10. Serhan, C.N. 2004. A search for endogenous mechanisms of anti-inflammation uncovers novel chemical mediators: missing links to resolution. *Histochem Cell Biol* **122**:305–321.
11. Serhan, C.N., and Chiang, N. 2008. Endogenous pro-resolving and anti-inflammatory lipid mediators: a new pharmacologic genus. *Br J Pharmacol* **153**(Suppl 1): S200–S215.
12. Savill, J., and Fadok, V. 2000. Corpse clearance defines the meaning of cell death. *Nature* **407**:784–788.
13. Savill, J., Gregory, C., and Haslett, C. 2003. Cell biology. Eat me or die. *Science* **302**:1516–1517.
14. Gordon, S., and Taylor, P.R. 2005. Monocyte and macrophage heterogeneity. *Nat Rev Immunol* **5**:953–964.

SUGGESTED READINGS

Flower, R.J. 1988. Eleventh Gaddum memorial lecture. Lipocortin and the mechanism of action of the glucocorticoids. *Br J Pharmacol* **94**:987–1015.

Lawrence, T., Willoughby, D.A., and Gilroy, D.W. 2002. Anti-inflammatory lipid mediators and insights into the resolution of inflammation. *Nat Rev Immunol* **2**:787–795.

Serhan, C.N., and Savill, J. 2005. Resolution of inflammation: the beginning programs the end. *Nat Immunol* **6**:1191–1197.

Serhan, C.N., Brain, S.D., Buckley, C.D., et al. 2007. Resolution of inflammation: state of the art, definitions and terms. *FASEB J* **21**(2):325–332. Commentary.

3 Links between Innate and Adaptive Immunity

Christopher L. Karp

THE INNATE AND ADAPTIVE IMMUNE SYSTEMS: DEFINITIONS, CONTEXT, AND CONTRASTS

Standard accounts of the immune system emphasize the antigen-specific immunity and memory afforded by the adaptive immune system, contrasting it with the "nonspecific" defenses provided by the phylogenetically more ancient innate immune system. While study of the innate immune system has undergone a recent renaissance, most immunology textbooks still present innate immunity as an initial stopgap defense that holds the line until the "real" (efficient, effective, sophisticated) adaptive immune system can take over. There are obvious flaws in such formulations. First, while adaptive immunity may usefully be seen as a single system – with its cells (B and T lymphocytes) and antigen receptors (immunoglobulins [Ig], T-cell receptors) depending directly on the evolution of the recombination-activating gene (*RAG*) in jawed vertebrates – innate immunity, present in all metazoans, is a congeries of pathways. "Innate immune systems" is a much better term. Second, the innate immune systems are in no way less sophisticated than the adaptive immune system, having been under evolutionary pressure for far longer. Third, the innate immune systems are not of secondary importance; the adaptive immune system is directly dependent on the former for efficient and appropriate activation. Fourth, innate immune effector mechanisms are not less effective than adaptive immune effector mechanisms. (As an example, chemotherapy-induced ablation of neutrophils leads to a high risk of fatal infection with otherwise harmless commensal flora in a telescoped time frame compared with the similar risk of infectious mortality attendant on lymphocyte dysfunction, e.g., in infants with severe combined immunodeficiency.) Fifth, the standard superficial view of the kinetics of innate and adaptive immune responses, the former handing off to the latter, is incorrect; the innate immune systems do not become quiescent when the adaptive immune system is activated.

The divide between those studying innate (previously "cellular") and adaptive (previously "humoral") immunity dates back to the earliest days of immunology as a science. The ascendancy of those studying adaptive immunity in the field over the past several decades is something of a sociological/historical accident, albeit fueled in part by the compelling nature of the problems being investigated (e.g., the generation of lymphocyte receptor diversity, the nature of tolerance to self, the biology of functional polarization of effector and regulatory cell types) and by the mechanistic insights thereby obtained. The recent molecular identification of key innate immune receptors that signal the presence of microbial products has revitalized the study of innate immunity; the pendulum is swinging back. But, in many ways, the distinction between innate and adaptive immunity is an artificial one. The two are inextricably linked in vertebrates – something that forms the focus of this chapter.

Prior to outlining these links (an outline that will, perforce, be illustrative as opposed to comprehensive), it is useful to sketch out and contrast innate and adaptive immunity. The essential function of these immune systems includes protection against the microbial universe and injury (broadly defined). These functions can usefully be broken down into (a) discrimination of microbes or injury (immune recognition/activation), (b) containment or elimination of microbes or injury (immune effector responses), and (c) control of immune response vigor and duration (counter-regulation, resolution) – all of which needs to be done without harming the host itself. It should also be noted that, apart from this, the immune systems also play critical roles in development and homeostasis.

Immune Recognition

Adaptive immune recognition is a function of receptors that are clonally distributed on individual T cells (T-cell receptors) and B cells (B-cell receptors: surface Ig), or secreted by the latter (Ig). The receptor specificities of T and B cells are essentially infinite, this repertoire being generated through somatic rearrangement of gene segments and somatic mutation. Specific receptor–bearing lymphocytes are selected for (or against) during the development of each individual organism. The ability of the adaptive immune system to generate receptors that can recognize any molecular pattern has consequences for self-recognition. Deleterious recognition of self by the adaptive immune system is largely avoided by diverse mechanisms collectively referred to as "immunological tolerance," including developmental selection against lymphocytes bearing receptors with inappropriate affinity for self-antigens; alteration of the function of lymphocytes that encounter (self-) antigens in the absence of cues from the innate immune system (*vide infra*); and active suppression of (perniciously) autoreactive lymphocytes by specialized populations of regulatory lymphocytes.

Immune recognition in the innate immune system is quite different. For one, the receptors are germ line–encoded and non-rearranging (with the exception of isoform generation). This means, perforce, that innate immune receptors are relatively few in number, and that receptor repertoire selection reflects evolutionary processes. In 1989, Charles Janeway wrote a landmark theoretical paper outlining an elegant framework for understanding these constraints on innate immune recognition, conceived as the discrimination of noninfectious self from infectious nonself. Given the enormous molecular variability and high mutation rate of microorganisms, he postulated that (a) the molecular structures recognized must be shared by large groups of pathogens, (b) such structures must be tightly constrained from mutational variation by being essential to microbial survival, and (c) such structures must be completely distinct from host structures. In this schema, recognition of microbial nonself occurs through pattern recognition receptors (PRRs) that bind to pathogen-associated molecular patterns (PAMPs; something of a misnomer: the structures recognized by PRRs are not unique to pathogens). The subsequent discovery in 1998 of the Toll-like receptor (TLR) family of membrane-bound PRRs that signal in response to conserved microbial products (Chapter 13) not only appeared to fit this schema beautifully, but, more generally, revitalized study of the innate immune system. This led, quite rapidly, to discovery and/or delineation of other PRR families, including the NOD-like receptors (NLRs) and RIG-I-like helicase receptors (RLRs). A competing theoretical framework for understanding innate immune recognition, the danger hypothesis, was put forward by Polly Matzinger in the early 1990s. In this model, the primary driving force for the immune system and immune recognition is not self/nonself discrimination but protection against, and hence detection of, danger. Obviously, the presence of microbial products in normally sterile sites represents danger. But nonmicrobial (e.g., trauma, tissue ischemia and infarction, crystal deposition) danger abounds, and also leads to innate immune activation, suggesting that there must be receptors for structures induced, upregulated, or solubilized by cellular injury. The fact that sterile inflammation often mirrors microbe-driven inflammation suggests likely overlap between PAMPs and receptors for altered or injured self damage-associated molecular patterns (DAMPs), something that has been borne out experimentally. Finally, it should be noted that, despite the desire for theoretical simplicity, evolution works in an ad hoc fashion. It is thus not surprising that there is at least one other mode of innate immune recognition that does not fit easily into either the microbial nonself or the danger rubric (theoretical constructs that are, in any case, not mutually exclusive) – recognition of missing self. In this mode, seen with both natural killer (NK) cells (Chapter 8) and the alternative pathway of complement activation, the engagement of molecular structures only expressed by normal host cells (major histocompatibility complex [MHC] class I and regulators of complement activation family members, respectively) inhibits activation. An overview of innate immune receptors is provided in Table 3.1.

Immune Effectors

B cells and α/β T cells (including CD8+ and CD4+ T cells) are the effector cells of the adaptive immune system. In terms of the effector pathways mediated by such cells, B cells produce Ig, the binding of which to antigens, microbes, and/or host-derived structures can facilitate phagocytosis ("opsonization"), activate complement, activate or inhibit the activation of cells bearing receptors for the Fc portion of Ig, and/or alter the function of the structures so bound (e.g., neutralization of viruses and microbial toxins). Like all immune cells, B cells also produce a variety of secreted protein mediators (cytokines) that act in autocrine, paracrine, and systemic fashion to alter the function of other cells, regulating the activity of other immune effector cells and controlling inflammatory responses. The effector mechanisms of CD4+ and CD8+ T cells include diverse pathways of inducing apoptosis ("cytolysis") as well as the regulation of inflammation and immune responses via cytokine production and cell surface molecules.

TABLE 3.1. Pattern recognition receptors of the innate immune systems

Receptor class/receptor	Location	Ligands	Comments
Toll-like receptors (TLRs) (TLR1–13)	TLR1,2,4,5,6,10–13: plasma membrane; TLR3,7,8,9: endosomal compartments	Diverse PAMPs and DAMPs (Chapter 13)	Activation of NF-κB, MAP kinase, and IRF pathways
NOD-like receptors (NLRs) NALP1–14 NOD1–5 NAIP, IPAF, CIITA	Cytoplasmic	Diverse PAMPs and DAMPs	Activation of NF-κB; activation of caspase 1 leading to IL-1β cleavage or cell death
RIG-I-like helicase receptors (RLRs) RIG-I MDA5	Cytoplasmic	Viral RNA: RIG-I: 5′ triphosphate ssRNA; MDA5: dsRNA	Activation of NF-κB and IRF pathways
DAI	Cytoplasmic	dsDNA from pathogens, damaged host cells	Activation of NF-κB and IRF pathways
PKR	Cytoplasmic	dsRNA	Serine/threonine kinase: phosphorylation of eIF2α, blocking protein synthesis; activation of NF-κB and MAP kinase pathways
2′-5′-Oligoadenylate synthase	Cytoplasmic	dsRNA	Activation of RNaseL: RNA degradation
Dectin-1	Plasma membrane	β-Glucans from fungi and other microbes	C-type lectin; cooperation with TLR2; activation of NF-κB and MAP kinase pathways; phagocytosis
Phagocytic receptors numerous, including Scavenger receptors (class 1–6) MΦ mannose receptor DEC-205 DC-SIGN Langerin CR3, CR4	Plasma membranes of macrophages and/or dendritic cells (DCs)	Diverse microbial and altered host components	Multiple protein families; expressed on the surface of professional phagocytes: macrophages, DCs, neutrophils; those on DCs are important for the delivery of antigen to processing compartments
Secreted PRRs numerous, including			
Lipid transferases			
BPI		Lipid A	Killing/opsonization of Gram-bacteria transfer of LPS to CD14
LBP		Lipid A	
Collectins			
MBL		Microbial sugars	Complement activation, opsonization
SP-A		Diverse pathogens	Opsonization
SP-D		Diverse pathogens	Opsonization
Pentraxins			
C-reactive protein			Opsonization, complement activation
Serum amyloid P			Opsonization, complement activation
PTX3			Opsonization, complement activation
PGRPs		Peptidoglycan	Bacterial killing

Receptor class/receptor	Location	Ligands	Comments
fMLR	Plasma membrane, neutrophils	N'-formylated peptides	G-protein–coupled receptor; chemotaxis, activation
Activating NK cell receptors numerous, including Activating KIRs NKG2D NKp46, 44, 30	Plasma membrane	Diverse viral, and stress-inducible host ligands	

Note: Both this and the following table, like the chapter as a whole, are illustrative as opposed to comprehensive. BPI, bactericidal permeability-increasing protein; CR, complement receptor; LBP, LPS-binding protein; MBL, mannose-binding lectin; PGRP, peptidoglycan receptor protein; SP, surfactant protein.

The innate immune system consists of an array of specialized immune cells: monocytic cells (including monocytes, macrophages, and most dendritic cells [DCs]), other DC subtypes (such as plasmacytoid DCs, whose origin remain somewhat unclear), granulocytes (including neutrophils, eosinophils, mast cells, and basophils), and innate lymphocytes (including NK cells, NKT cells, γ/δ T cells, and "innate" B cells). Innate immunity comprises more than just specialized immune cells, however. Epithelial cells lining body surfaces are central to innate immunity: acting as barriers, producing antimicrobial effectors, secreting substances that prevent microbial attachment and entry, and secreting mediators that regulate inflammation. Endothelial cells are similarly central to regulation of both innate and adaptive immunity, regulating immune cell trafficking into tissues (Chapter 11). Hepatocytes play essential roles through the secretion of humoral receptors and effectors. More broadly, the presence of intracellular sensors and effectors that respond to viral nucleic acids in all nucleated cells suggests that, at a fundamental level, all cells are part of the innate immune systems. An overview of innate immune effector pathways is provided in Table 3.2. It will be noted that, in addition to cell-associated effector pathways, a variety of humoral effector pathways (including proteolytic cascades such as the complement and coagulation systems, and secreted PRRs) are part of the innate immune systems. Finally, the induction and regulation of local and systemic inflammation is, in large part, a function of the production of mediators by innate immune cells.

Immune Counter-Regulation

While inflammatory responses are critical for protection against the microbial universe (and danger), all inflammatory responses have the potential for harming the host. Indeed, it has become clear that dysregulated inflammation is central to the pathogenesis of a wide spectrum of diseases, including infectious (e.g., sepsis), autoimmune (e.g., multiple sclerosis), autoinflammatory (e.g., juvenile rheumatoid arthritis), allergic (e.g., allergic asthma), vascular (e.g., atherosclerosis), neurodegenerative (e.g., Alzheimer's disease), metabolic (e.g., type II diabetes and other metabolic sequelae of obesity), and monogenic (e.g., cystic fibrosis) diseases. As a result, immune responses must be tightly regulated in space, time, amplitude, and character. While immunologists have traditionally focused on the molecular mechanisms underlying immune activation and class specification, control of the amplitude and resolution of immune responses is just as important (Chapter 2). In recent years, numerous, often overlapping and redundant immune counter-regulatory mechanisms that control innate and/or adaptive immune responses have been defined, including specialized cells (e.g., Foxp3+ regulatory T cells), cytokines (e.g., IL-10), enzymes (e.g., indolamine 2,3-dioxygenase [IDO]), cell surface receptors (e.g., CTLA-4), intracellular signaling modulators (e.g., SOCS proteins), lipid mediators (e.g., lipoxins), and cellular reprogramming (e.g., endotoxin tolerance).

Relationship between Acute versus Chronic Inflammation and Innate versus Adaptive Immunity

What is the relationship between acute and chronic inflammation, defined and described in Chapter 1, and the innate and adaptive immune systems? Classic, acute (neutrophilic) inflammation – as seen early in pyogenic infection and with tissue infarction – represents an innate immune response (albeit likely under the control of counter-regulatory mechanisms deriving from both the innate and adaptive immune systems). Chronic (lymphocytic, monocytic) inflammation – as seen in delayed type hypersensitivity and granulomatous responses – represents coordinate activation of both the innate and adaptive immune systems. It should be noted, however, that classical "acute" and "chronic" inflammation represent only small portions of the spectrum of inflammatory responses mounted by the human (and murine) innate and adaptive immune systems. For example, many of the disease-associated

TABLE 3.2. Effector mechanisms of the innate immune systems

System or mechanism	Functions
Epithelia	Barrier function Secretion of antimicrobial peptides and proteins Secretion of iron-binding proteins Secretion of mucins (mucosae; inhibition of microbial attachment) Mucociliary clearance (airway epithelia) Secretion of opsonins (including those activating complement) Secretion of chemotactic cytokines and lipid mediators Secretion of proinflammatory cytokines Secretion of complement components
Phagocytosis (macrophages, DCs, neutrophils)	Intracellular killing via Antimicrobial peptides Hydrolytic enzymes Reactive oxygen species Reactive nitrogen species Nutrient competition Antigen processing and presentation
Complement system	Opsonization Direct antimicrobial activity Regulation of inflammation, immunoregulation
Cytolysis	Induction of apoptosis in virally infected and stressed cells by NK cells (via perforin/granzyme, via surface-expressed TNF family members) and by IFN-α/β-inducible gene products
Noncytolytic antiviral effectors	Induction of apoptosis (above), inhibition of viral transcription, replication, assembly (Mx proteins, GBP proteins); blockade of protein synthesis (PKR); RNA degradation (OAS) Inhibition of viral gene expression and replication by IFN-γ, TNFα
Extracellular killing	Neutrophils (via extracellular "nets") Basophils, eosinophils, mast cells (defense against helminths)
Hepatocytes	Proinflammatory cytokine-driven production of "acute phase proteins": opsonins, complement components
Coagulation system	Walling off of infected tissues, regulation of inflammation

inflammatory responses noted earlier – such as allergic inflammation (involving innate and adaptive immunity); the chronic, low-grade inflammatory state associated with obesity (apparently, largely innate); the neurotoxic inflammation associated with Alzheimer's disease (also largely innate) – fit poorly, if at all, into these categories.

REGULATION OF ADAPTIVE IMMUNITY BY THE INNATE IMMUNE SYSTEM

At a fundamental level, efficient activation of the adaptive immune system is dependent on innate immunity – while the adaptive immune system has the ability to respond to essentially any molecular structure, it relies on the innate immune system to discriminate what structures should be responded to.

Activation of Adaptive Immune Responses: T Cells

It has been known for 40 years that effective lymphocytic responses to antigens do not occur in the absence of mononuclear phagocytic cells. In contemporary terms, the generation of effector T cells from naive CD4+ and CD8+ T cells is dependent on antigen presentation by a class of innate immune cells: DCs. DC presentation of peptides on MHC class I and II molecules to T-cell receptors (TCRs) on naive CD4+ and CD8+ T cells, respectively, with appropriate affinity for the specific peptide/MHC complex ("signal 1") is necessary, but not sufficient, for T-cell activation, clonal expansion, and the generation of effector and memory cells. In the absence of costimulation, the provision of signals ("signal 2") through cell surface–expressed molecules on the APC (e.g., CD80 and CD86, CD40) to cognate receptors on the T cell (CD28, CD40L), the result of such antigen presentation is deletion or suppression. DCs (which come in a variety of subtypes, with somewhat different functions) are prodigious samplers of their antigenic environment, both external and internal. In part, external sampling is a function of phagocytosis induced by innate immune PRRs (e.g., complement receptors and scavenger receptors); in part, this is due to fluid phase sampling via macropinocytosis. Effective antigen presentation depends on

DC maturation (antigen processing, upregulation and loading of MHC class I and II complexes, upregulation of costimulatory molecule expression) and migration to the T-cell areas of lymph nodes. In turn, such maturation is a result of signaling through innate immune PRRs.

This necessity for DC maturation provides a mechanistic explanation for adjuvanticity – what Janeway called the "immunologists' dirty little secret." Purified, soluble proteins on their own fail to drive adaptive immune responses. The ability of bacterial products to act as adjuvants has long been appreciated. Lipopolysaccharide (LPS), acting through TLR4, is a paradigmatic case in point. TLRs are clearly not the only PRRs that drive such DC activation and maturation; both microbial and endogenous danger signals can suffice. For example, aluminum hydroxide (alum), the adjuvant most widely used clinically, now appears to act through induction of uric acid, a DAMP that signals through the NLR, Nalp3. While the key molecular details of PRR-driven DC maturation remain somewhat unclear, the type I IFNs (IFN-α[s] and -β) appear to be of special importance. A variety of other stimuli are also able to drive DC maturation, including interactions with other innate immune cells (NK cells and NKT cells, whose production of IFN-γ after PRR engagement can drive DC maturation in a paracrine fashion), neutrophils, and mucosal epithelia. Cells rendered apoptotic by NK cells can drive robust adaptive immune responses, a phenomenon dependent on signaling pathways downstream of the TLRs. An overview of innate immune regulation of T-cell activation is presented in Figure 3.1.

While the generation of effector T cells from naive T cells appears to depend largely on antigen presentation by DCs in lymph nodes, the *reactivation* of memory cells in tissues can occur efficiently via other antigen-presenting cells (APCs), among which cells of the macrophage lineage appear to play a dominant role. Further, macrophages (along with other innate immune cell types and epithelial cells) act as critical, tissue-specific sensors of infection and injury, secreting chemotactic cytokines that drive and regulate not only innate immune inflammation, but also the recruitment and activation of DCs and lymphocyte populations.

Activation of Adaptive Immune Responses: B Cells

Efficient activation of most B-cell responses is also dependent, either directly or indirectly, on the innate

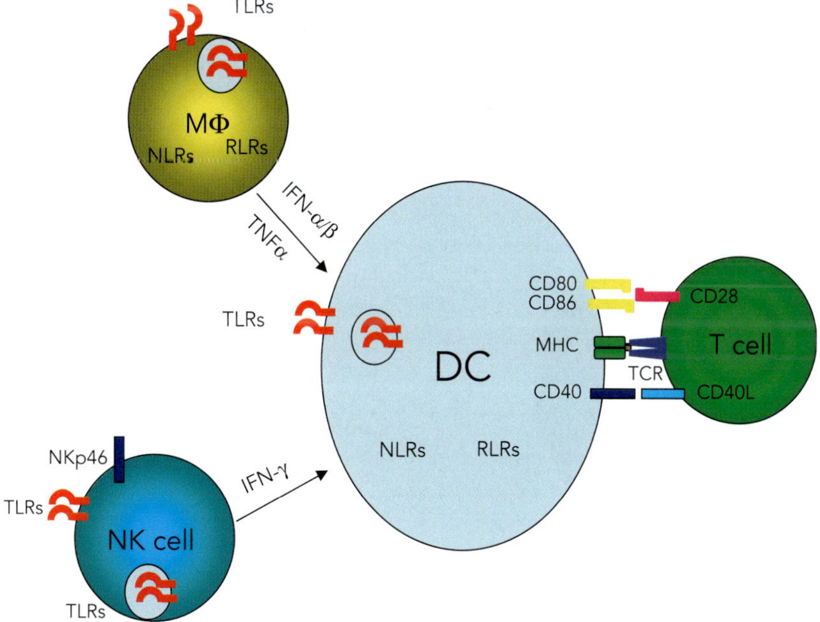

Figure 3.1. Innate immune regulation of T-cell activation. DC presentation of peptides on MHC molecules to T-cell receptors (TCRs) on naive T cells ("signal 1") is necessary, but not sufficient, for T-cell activation, clonal expansion and the generation of effector and memory cells. In the absence of costimulation, the provision of signals ("signal 2") through cell surface–expressed molecules on the APC (e.g., CD80 and CD86, CD40) to cognate receptors on the T cell (CD28, CD40L), the result of antigen presentation is deletion or suppression. "Effective" antigen presentation to T cells is the result of DC maturation: upregulating antigen processing, MHC class I and II expression and loading, and costimulatory molecule expression. DC maturation is a result of PAMP and DAMP signaling through innate immune PRRs (e.g., TLRs, NLRs, RLRs) on DCs. Maturation can also occur through interactions with other innate immune cells. IFN-γ secretion by NK cells (pictured; activated through widely expressed PRRs such as TLRs, or through NK activating receptors such as NKp46) or other innate lymphocyte populations can drive DC maturation; as can TNFα and type I IFN secretion by macrophages activated through PRRs. MΦ, macrophage; MHC, major histocompatibility complex; NLR, NOD-like receptor; RLR, RIG-I-like helicase receptor; TCR, T-cell receptor.

immune systems. Cognate antigen-induced clustering of the B-cell receptor (BCR, surface Ig) triggers B-cell activation. The presence of complement activation fragments (C3d) on the antigen lowers the threshold for B-cell activation by 1000-fold, through cross-linking of complement receptor 2 (CR2) and associated molecules (CD19, CD81) to the BCR and its associated signaling molecules (Igα and Igβ) (Figure 3.2). Beyond the role of complement, B-cell responses to protein antigens require CD4+ T cell help – the provision of membrane-bound (CD40L) and paracrine (cytokines) signals by antigen-specific CD4+ T cells that stimulate B-cell proliferation (clonal expansion) as well as Ig isotype switching, synthesis, and secretion. As noted earlier, the generation of such helper T cells is itself dependent on antigen presentation by DCs, cells that can also directly present antigen and provide stimulation to B cells through CD40/CD40L interactions.

Immune Class Specification

In addition to activation and clonal expansion, T cells that avoid deletion in the aftermath of antigen presentation by DCs undergo variable degrees of differentiation down pathways of functional polarization, leading to the generation of cells with diverse effector and/or

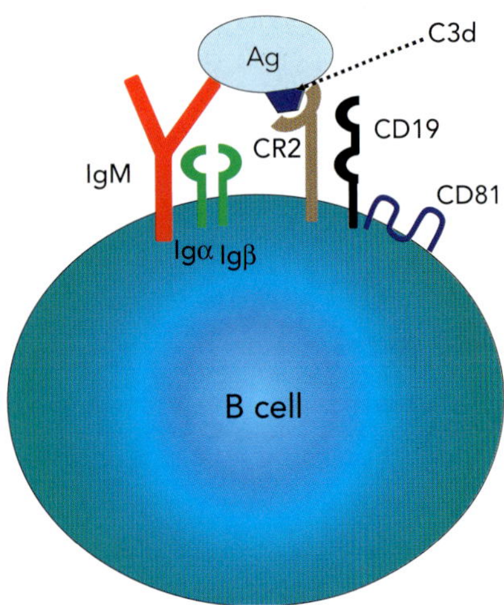

Figure 3.2. Innate immune coactivation of B cells. Antigen-induced clustering of the B-cell receptor (BCR, surface Ig) triggers B-cell activation. Decoration of antigen with the complement activation fragment, C3d, lowers the threshold for B-cell activation by a factor of 1,000 via cross-linking of complement receptor 2 (CR2) and its associated molecules, CD19 and CD81, with the BCR and its associated molecules (Igα and Igβ) – leading to the coordinate activation of signaling cascades from both the BCR and the CR2 complexes. In addition to the effects of complement, the innate immune systems play a critical role in B-cell responses to protein antigens. Such responses demand help from antigen-specific CD4+ T cells, the activation of which is dependent on DCs (cells that can also directly present antigen, and provide stimulation to B cells through CD40/CD40L interactions).

regulatory properties. The molecular details are best understood for CD4+ T cells, which can differentiate down a variety of pathways that promote specific classes of immune responses. Such polarized CD4+ T-cell populations include **Th1 cells** (producing IFN-γ, among other cytokines; facilitating systemic immunity to intracellular pathogens through activation of the microbial effector pathways of macrophages, generation of complement-fixing Ig isotypes, and activation of CD8+ cytolytic T cells and innate lymphocyte populations; implicated as well in organ-specific autoimmune diseases); **Th2 cells** (producing cytokines such as IL-4, IL-13, IL-5, and IL-9; facilitating cellular and humoral responses to helminths; also implicated in allergic disease and pathological fibrotic responses); **Th3 cells** (producing transforming growth factor [TGF-β], among other cytokines; facilitating mucosal immunity, providing immune counter-regulation); **Th17 cells** (producing IL-17, among other cytokines; facilitating the mobilization of neutrophils to epithelial barriers; also implicated in organ-specific autoimmune diseases); and various populations of regulatory cells, such as **induced Foxp3+ regulatory T cells** ("Treg"; producing cytokines such as IL-10 and TGF-β; suppressing both innate and adaptive immune responses via diverse mechanisms). Control of the functional polarization of CD4+ (and CD8+) T cells is under the instructive guidance of cytokines ("signal 3") produced by different DC subsets (activated under different conditions) as well as by other innate immune cells – the production of which is driven by innate immune PRRs. Innate immune regulation of CD4+ T-cell differentiation and polarization is schematized in Figure 3.3.

In turn, CD4+ T-cell polarization leads to polarization of B-cell Ig isotype production. Help from Th1 cells drives the synthesis of complement-fixing Ig isotypes; that from Th2 cells driving IgE synthesis; and that from Th3 cells driving IgA synthesis.

Counter-regulation of Adaptive Immune Responses

In addition to activation and class specification, the innate immune systems play a central role in regulation of the amplitude and resolution of adaptive immune responses. Numerous pathways have been implicated. Among counter-regulatory cytokines, IL-10 appears to play a uniquely important role, restraining the vigor of inflammatory responses: local and systemic, innate and adaptive, and polarized along diverse axes of effector response. IL-10 is produced by a plethora of cell types including diverse innate (myeloid cells, innate lymphocytes, epithelial cells) and adaptive immune cells. It appears likely that the more polarized a B or T cell is, the more likely it is to produce IL-10 for immune counter-regulation. As noted earlier, the generation of such polarization (including the generation of specific classes of counter-regulatory

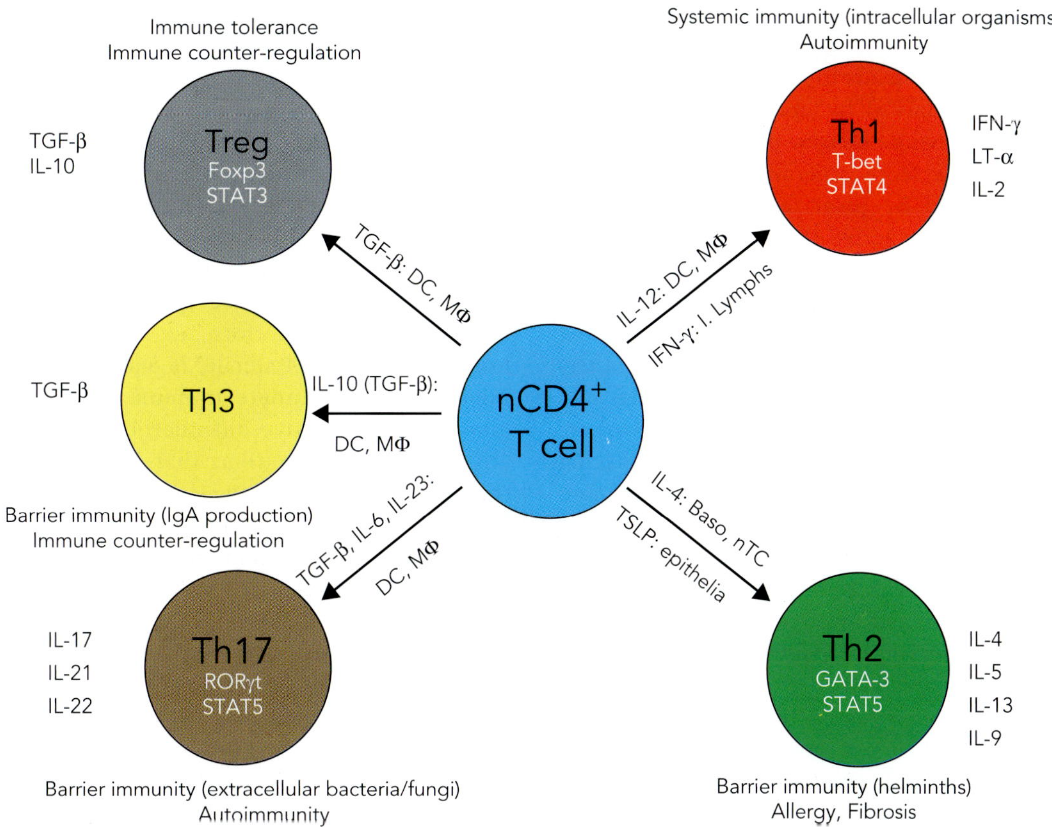

Figure 3.3. Regulation of CD4+ T-cell polarization by the innate immune systems. Specification of the effector (or regulatory) class of immune responses by CD4+ T cells is largely the result of functional polarization of naive CD4+ T cells in the periphery into differentiated populations that produce unique patterns of cytokines and have distinctive biological properties. Such polarized CD4+ T-cell populations include Th1 cells, Th2 cells, Th3 cells, Th17 cells, and induced regulatory T cells (Treg; *N.B.*, natural Treg are already differentiated upon leaving the thymus). Control of the functional polarization of CD4+ (and CD8+) T cells is under the instructive guidance of cytokines ("signal 3"), largely derived from DCs, as well as other innate (as well as adaptive) immune cells. The relevant "signal 3," its sources, and the transcription factors (the latter in white, in the cells) critical for the differentiation of these polarized populations, along with their signature cytokine profiles and biological activities, is schematized here. It appears that the more polarized a T cell, the more likely it is to express IL-10 and have counter-regulatory properties. Baso, basophils; I. Lymphs, innate lymphocytes; LT, lymphotoxin; MΦ, macrophages; nTC, naive T cells; TSLP, thymic stromal lymphopoietin.

cells such as Tregs) is largely specified by the innate immune systems. In addition to other inhibitory cytokines (e.g., TGF-β), the innate immune systems provide counter-regulation to adaptive immune responses through inhibitory enzymes (e.g., inducible nitric oxide synthase and IDO), lipid mediators (e.g., lipoxins and resolvins), cell surface molecules (e.g., CTLA-4 and membrane-bound TGF-β); antigen destruction and clearance (e.g., by phagocytic or cytolytic destruction of microbial pathogens, and by complement receptor–mediated clearance of immune complexes); and killing of APCs (e.g., by NK cell–mediated killing of DCs).

REGULATION OF INNATE IMMUNITY BY THE ADAPTIVE IMMUNE SYSTEM

If the innate immune systems are essential for adaptive immune function, the reverse is no less true. T cells regulate inflammation, in part, through regulation of innate immune cell differentiation state, recruitment, activation, and homeostasis. Innate immune activation is also under counter-regulatory control, both direct and indirect, by the adaptive immune system.

Regulation of Innate Immune Cell Function

The generation of specific classes of inflammatory responses by the adaptive immune system involves the selective marshaling of innate immune cell functions (along with adaptive immune functions). Again, this has been best worked out for CD4+ T cell–driven immunoregulation, illustrated by the effects of cytokines produced by polarized CD4+ effector T cells on macrophage populations. Th1 cell production of IFN-γ drives antigen-specific, "classical activation" of macrophages, marked by upregulation of microbicidal effector functions (e.g., production of reactive oxygen species and nitric oxide) and secretion of proinflammatory cytokines. On the other hand, Th2 cell production of IL-4 and IL-13 leads to "alternative activation" of macrophages, marked by upregulation of repair functions (e.g., driving collagen production and cellular proliferation). More generally, these and other cytokines produced by polarized CD4+ effector T cells have wide effects across the innate immune systems. For example, IFN-γ affects neutrophil function (facilitating recruitment, inhibiting apoptosis, priming for cytokine and chemokine production, and the generation of reactive oxygen species), epithelia (altering chemokine production by airway epithelia) and endothelia (regulating adhesion molecule expression); IL-4 and IL-13 alter the properties of epithelia (driving mucous metaplasia, altering chemokine production) and endothelia (regulating adhesion molecule expression); other Th2-derived cytokines regulate the differentiation, recruitment, apoptosis, and/or priming of eosinophils (IL-5) and mast cells (IL-9); and IL-17 plays an important role in neutrophil homeostasis, recruitment, and activation.

These same cytokines alter the function of innate immune APCs (e.g., DCs), regulating the subsequent generation of polarized populations of effector lymphocytes. Polarized effector lymphocytes also regulate DC function via cell surface ligands that up- or downregulate antigen presentation (e.g., via CD40L and CTLA-4, respectively).

Counter-Regulation of Innate Immune Responses

As might be expected, the adaptive immune system plays a role in counter-regulation of innate immune responses. Again this occurs through multiple, often overlapping pathways, including specialized cell types (e.g., Tregs, Th3 cells, regulatory B cells), inhibitory cytokines (e.g., IL-10, TGF-β, IL-4), and inhibitory enzymes (e.g., inducible nitric oxide and IDO – both upregulated by IFN-γ).

TISSUE-SPECIFIC IMMUNITY: ORGAN-SPECIFIC REGULATION OF IMMUNE RESPONSES

As the increasingly convoluted descriptions of immune activation and regulation found above should make clear, distinctions between innate and adaptive immunity are, in many ways, artificial. The immune system acts in a coordinated fashion to respond to insults. Reductive distinctions between various facets of the immune system have theoretical and experimental utility – but, in the end, such facets can only be understood in the wider context of the immune system as a whole.

Another critical, contextual piece to be kept in mind is that immune responses occur in specific tissues. The biologically appropriate class of immune response varies not only by type of insult, but also by tissue. For example, the lung represents a very large (approximately a tennis court in surface area), very thin (two cells thick) interface with the external environment that is specialized for gas exchange. Constantly exposed to danger in the form of inhaled and aspirated microbes and microbial products, the normal function of the lung can rapidly be compromised by the destructive power of inflammatory responses. It is therefore not surprising that the lung, normally kept sterile, is equipped with a prodigious array of innate immune mechanisms that allow for constitutive antimicrobial activity in the absence of inflammation; that immune activation, the class of immune activation and the amplitude of immune activation is tightly controlled in the lung; and that the lung is marked by particularly strong immune counter-regulation. On the other hand, the gut mucosa is bathed by commensal flora in numbers that exceed the total number of cells in the host organism. Failing to respond to commensal flora, responding appropriately to pathogenic microbes, and doing the latter without functional derangement of gut function are obvious, important problems for the gut mucosal immune system. Again, gut immune responses (often marked by IL-10 and TGF-β expression and secretory IgA production) are tightly controlled in terms of activation, class, and amplitude. Similar biological considerations come to the fore in sites such as the eye, brain, testis, and placenta, where functional and structural constraints dictate the biological utility of specific classes of immune response. Matzinger has pointed out that immunologists have traditionally talked about such sites as being "immunologically privileged," because the sorts of immune responses that immunologists were used to measuring were hard to generate in these sites. But, in fact, this represents not a failure of immune activation, but immune class deviation; these sites mount quite effective, biologically appropriate immune responses.

Our knowledge of how specific organs and tissues "predispose" to certain classes of immune response remains somewhat spotty. In part, this is likely a function of tissue-specific parenchymal recruitment and "education" of DC subtypes (master regulators of immune class specification), tissue macrophage populations (long known to have dramatic, organ-specific functional heterogeneity), and innate lymphocyte populations (including NKT cells in the liver, innate B cells in the peritoneum, and γ/δ T cells in the gut). In turn, organ-specific innate immune cell repertoires (including, quite importantly, parenchymal cells such as airway epithelia) are likely to regulate the induction (and resolution) of organ- (and class-) specific immune responses, both innate and adaptive.

SUGGESTED READINGS

Beutler, B. 2004. Inferences, questions and possibilities in Toll-like receptor signalling. *Nature* **430**:257–263.

Janeway, C.A., Jr. 1989. Approaching the asymptote? Evolution and revolution in immunology. *Cold Spring Harb Symp Quant Biol* **54(**Pt 1):1–13.

Matzinger, P. 2002. The danger model: a renewed sense of self. *Science* **296**:301–305.

Metchnikoff, E. 1908. On the present state of the question of immunity in infectious diseases. *Nobel Lecture*. http://nobelprize.org/nobel_prizes/medicine/laureates/1908/mechnikov-lecture.html. Accessed September 3, 2009.

Meylan, E., Tschopp, J., and Karin, M. 2006. Intracellular pattern recognition receptors in the host response. *Nature* **442**:39–44.

Sakaguchi, S. 2004. Naturally arising CD4+ regulatory T cells for immunologic self-tolerance and negative control of immune responses. *Annu Rev Immunol* **22**:531–562.

Steinman, R.M., and Banchereau, J. 2007. Taking dendritic cells into medicine. *Nature* **449**:419–426.

Zhu, J., and Paul, W.E. 2008. CD4 T cells: fates, functions, and faults. *Blood* **112**:1557–1569.

4A Neutrophils I

Jose U. Scher, Steven B. Abramson, and Michael H. Pillinger

INTRODUCTION

Polymorphonuclear leukocytes or granulocytes are hematopoietically derived phagocytes characterized by multilobed nuclei and the presence of multiple, distinct granules within their cytoplasm. Three different polymorphonuclear leukocytes are distinguished according to their granular staining properties: neutrophils (polymorphonuclear neutrophils or PMNs), basophils, and eosinophils. Neutrophil granules stain preferentially with neutral dyes, whereas basophil granules stain with basic dyes, and eosinophilic granules stain with acidic colorants such as eosin. These three types of leukocytes differ not only in their tinctorial properties, but also in their functions and roles during the inflammatory process. They constitute key effector cells in innate immunity, and the frontline of host defense in response to foreign antigens and microorganisms. In this chapter, we will focus on the biology and role of neutrophils. The other polymorphonuclear leukocytes are discussed elsewhere.

NEUTROPHIL HOMEOSTASIS: MYELOPOIESIS AND DESTRUCTION

Neutrophil myelopoiesis is a closely regulated process that begins with the differentiation of pluripotent stem cells into primitive myeloid progenitors, which in turn differentiate into specific myeloid precursors. Contact with specific adhesion molecules, hematopoietic growth factors, and cytokines promotes the progression of myeloblasts along unique pathways to mature as neutrophils, eosinophils, and basophils, as well as monocytes. The sequence that leads to neutrophil formation begins with the neutrophilic promyelocyte and progresses through several maturation steps (neutrophilic myelocyte, metamyelocyte, band cell, and mature neutrophil) (Figure 4A.1). This process of neutrophil development takes approximately 14 days; cell division is terminated at the metamyelocyte stage, while granule development continues. Eventually, terminally differentiated neutrophils are released from the bone marrow. At that point they are unable to divide, and their synthetic machinery has become partially inactivated. During steady-state granulopoiesis, roughly 10^{11} neutrophils are released into the bloodstream daily. Neutrophil half-life in blood is 6–8 hours and, under normal circumstances, neutrophils constitute roughly 60% of all leukocytes in human peripheral blood [1].

The coordinated expression of a number of myeloid transcription factors (including PU.1, CCAAT enhancer binding proteins α and ε [C/EBPα and C/EBPε], and GFI 1) is necessary for the regulation of neutrophil development [2]. Among the extracellular factors that direct pluripotent stem cells to differentiate into neutrophils, granulocyte colony-stimulating factor (G-CSF) plays an essential role [3]. G-CSF has been shown to induce myeloid differentiation [4], stimulate proliferation of granulocytic precursors, and provoke neutrophil release from the bone marrow [5]. The biological effects of G-CSF are mediated via its receptor (G-CSFR or CD114), a member of the hematopoietic cytokine receptor family [6]. Other hematopoietic cytokines contributing to neutrophil development in vivo include granulocyte-macrophage colony-stimulating factor (GM-CSF), interleukin-6 (IL-6), and IL-3 [7–9].

In the absence of inflammatory stimuli, neutrophil populations are maintained within a relatively narrow range. Although incompletely understood, the mechanisms controlling neutrophil homeostasis regulate both neutrophil production and clearance. Recent observations implicate the SDF-1/CXCR4 signaling system in the process of neutrophil clearance. SDF-1 (CXCL12) is a CXC chemokine that is secreted from bone marrow and attracts neutrophils by engaging the CXCR4 receptor. Senescent neutrophils upregulate CXCR4 and acquire the ability to migrate

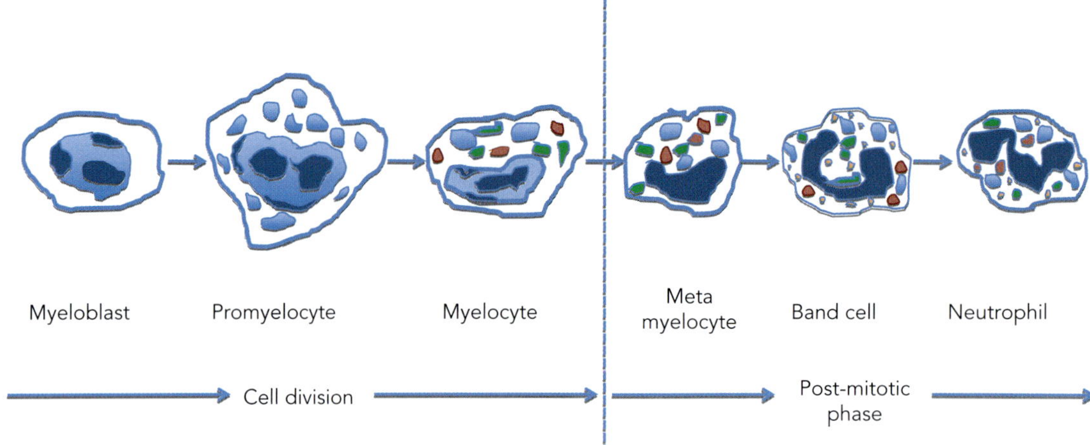

Figure 4A.1. Neutrophil development.

toward SDF-1, leading to homing to the bone marrow and clearance of aging cells from the blood [10,11]. In the bone marrow, as well as in the spleen and liver, damaged and aging neutrophils are cleared by tissue macrophages. Reduction of neutrophil populations is also effected through induction of apoptosis. For example, aged neutrophils undergo programmed cell death when exposed to tumor necrosis factor alpha (TNFα) or Fas ligand (CD95) [12,13].

A feedback loop downregulating neutrophil production has recently been identified. Following phagocytosis of apoptotic neutrophils by tissue macrophages, the latter shut down their secretion of IL-23. In consequence, IL-17 production by TH17 cells is reduced. Lack of IL-17 then results in decreased levels of G-CSF and, consequently, reduced neutrophil development and release. In the setting of increased neutrophil populations (and hence increased populations of apoptotic neutrophils), this system is poised to maintain neutrophil populations within a constant range [14].

NEUTROPHIL ANATOMY: MORPHOLOGY AND GRANULES

The development and formation of neutrophil granules occurs in a sequential process during myeloid cell differentiation (Figure 4A.1). Granule formation begins in the early promyelocyte stage, a period during which the majority of nascent granules are rich in myeloperoxidase (MPO), an enzyme that catalyzes the formation of hypochlorous acid. These *primary* granules are also referred to as *azurophilic*, owing to their affinity for the basic dye azure A [15]. They are functionally similar, although not identical, to lysosomes of other cells. Azurophilic granule maturation is largely complete at the myelocyte stage of neutrophil development, at which point peroxidase-negative granules begin to form. On the basis of their time of appearance and content, the latter are subdivided into secondary (or specific) and tertiary (or gelatinase) granules. Specific granules develop in myelocytes and metamyelocytes and are rich in lactoferrin, whereas gelatinase granules form later (neutrophil band stage) and lack lactoferrin [16]. A fourth category of granules, secretory vesicles, are smaller than the others and appear during the late stage of nuclear neutrophil segmentation [17].

In addition to MPO, azurophilic granules contain a variety of acidic hydrolases and serine proteinases (e.g., elastase, proteinase-3, and cathepsins), as well as enzymes directed at nucleic acids and sugars (Table 4A.1). Like lysosomes, azurophilic granules contain granulophysin (CD63) in their membrane [18]. In contrast to lysosomes, however, primary granules are deficient in lysosome-associated membrane proteins 1 and 2 (LAMP-1 and LAMP-2). Similarly, the mannose-6-phosphate receptor system – essential for lysosomal enzyme targeting [19] – is not utilized by neutrophil azurophilic granules [20]. Thus, some authors consider that azurophilic granules are best described as regulated secretory granules, rather than as specialized lysosomes [21].

In addition to soluble vesicular contents, specific granules possess an extensive array of membrane-associated proteins including cytochromes, signaling molecules, and receptors (Table 4A.1). Specific granules thus form a storage depot of proteins that can be directed to the surfaces of phagocytic vacuoles as well as the plasma membrane. One particularly important family of soluble proteinases found in neutrophil-specific granules are the matrix metalloproteinases (MMPs), including neutrophil collagenase-2 (MMP-8), gelatinase-B (MMP-9), and leukolysin (MMP-25). Many of these are stored as inactive proenzymes. They undergo proteolytic activation following granule

TABLE 4A.1. Content of neutrophil granules

Azurophil granules	Specific granules	Gelatinase granules	Secretory vesicles
CD63 (granulophysin)	fMLP receptor	fMLP receptor	fMLP receptor
CD68	CD11b/CD18	CD11b/CD18	CD11b/CD18
Stomatin	CD66	Cytochrome b_{558}	Cytochrome b_{558}
Presenilin 1	CD67	SCAMP	Leukolysin (MMP-25)
	CD15	SNAP-23, -25	CD10
	Cytochrome b_{558}	Leukolysin (MMP-25)	CD13
	TNF receptor		CD14
	Fibronectin receptor		CD16
	Laminin receptor		CD45
	Thrombospondin receptor		CR1
	Vitronectin receptor		C1q receptor
	SCAMP		DAF
	SNAP-23, -25		
	Leukolysin (MMP-25)		
	NB1 antigen		
Myeloperoxidase	Lysozyme	Gelatinase (high concentrations)	Plasma proteins
Acid mucopolysaccharide	Lactoferrin	Lysozyme	
α-1 Antitrypsin	Gelatinase (MMP-9)	Acetyltransferase	
α-Mannosidase	Collagenase (MMP-8)	β2-Microglobulin	
β-Glycerophosphatases	β2-Microglobulin		
β-Glucuronidase	Histaminase		
Cathepsins	Heparanase		
Defensins	Sialidase		
Elastase	hCAP-18		
Lysozyme			
Proteinase-3			
BPI			

fusion and interaction with azurophilic granule contents, and are capable of disrupting major structural components of bacteria and/or the extracellular membrane. Neutrophil MMPs appear to be crucial not only for bacterial killing, but also for neutrophil extravasation and migration [22].

The remaining two classes of vesicles are less well studied. Gelatinase granules resemble specific granules in size and density, and share some proteins in common with them. Gelatinase granules differ from specific granules, however, in that they contain high concentrations of gelatinase, a latent enzyme with the capacity for tissue destruction. Secretory vesicles are smaller and lighter than the other classes. They lack proteolytic enzymes but are rich in membrane-associated proteins, including receptors otherwise identified with the plasma membrane. Moreover, they are preferentially directed not to phagosomes, but to the plasma membrane. Thus, secretory vesicles appear to serve mainly as a reservoir of neutrophil plasma membrane proteins (Table 4A.1).

NEUTROPHIL PHYSIOLOGY: ACTIVATION AND FUNCTION

In order to carry out their role in acute inflammation, bloodstream neutrophils must first sense chemical mediators (chemoattractants) via cell membrane receptors. Next, neutrophils must attach to the activated endothelium through several interactions involving adhesion molecules and their receptors (rolling and adhesion). After passing through postcapillary venules (diapedesis), neutrophils migrate toward sites of inflammation where they recognize their target, engulf it (phagocytosis), and finally destroy it. Activated neutrophils also play an important role in further upregulating the inflammatory process. Eventually, neutrophils participating in acute inflammation must be cleared from the inflammatory site as the inflammation resolves.

Stimuli and Receptors

Chemoattractants playing a role in neutrophil activation include lipid mediators (e.g., leukotriene B_4 [LTB_4] and platelet-activating factor), as well as proteins and peptides such as bacterial formylated peptides and the complement split product C5a. One important group of neutrophil chemoattractants are the CXC chemokines, characterized by the presence of paired cysteines ("C"), separated by any other amino acid ("X") at the C-terminus. CXC chemokines that appear to be most critical for neutrophil recruitment include IL-8 (CXCL8), MIP-2 (CXCL2), and KC (CXCL1). At sites of inflammation, chemoattractants are either produced by inflammatory cells (e.g., LTB_4 or IL-8), liberated from previously synthesized proteins (e.g., C5a), or derived from bacteria themselves (e.g., fMLP). Although all neutrophil chemoattractants can stimulate multiple aspects of neutrophil activation, they differ in both their relative potencies and the kinetics of their effects [23]. For example, neutrophil mobilization from the bone marrow by G-CSF takes 4–6 hours, whereas IL-8-driven neutrophil mobilization into peripheral tissues takes only minutes [24,25].

Neutrophil responses to chemoattractants and other stimuli depend on the presence of specific surface receptors. On the basis of microarray analysis, receptors for various inflammatory stimuli are significantly expressed only in terminally differentiated neutrophils [26]. Important examples include receptors for interferon (IFN) α and γ; IL receptors IL-1R, IL4R, IL-6R, IL-10R, and IL-17R; tumor necrosis factor (TNF) receptors 1 and 2; and CXC and CC chemokine receptors such as IL-8R-α and β, CXCR-4 and CCR-1, 2, and 3. Whereas growth factors and cytokines signal through various classes of protein tyrosine kinase–based receptors, chemoattractants signal mostly via G-protein–coupled receptors (GPCRs).

Once a GPCR is activated by its ligand, the receptor undergoes a conformational change, causing activation of an associated heterotrimeric GTP-binding protein (G protein). In neutrophils, the predominant G proteins are of the G_i family [27]. All G proteins function as "molecular switches," alternating between an inactive guanosine diphosphate (GDP)- and active guanosine triphosphate (GTP)-bound state, ultimately going on to regulate downstream cell processes.

Intracellular Signaling and Second Messengers

Multiple intracellular systems propagate signals from the surface of neutrophils to their interiors. One particularly important system is constituted by low molecular weight (20–25 kDa) GTP-binding proteins (LMW-GBPs), also known as Ras-related or Ras superfamily proteins. In contrast to heterotrimeric G proteins, these proteins are monomeric. However, they share with G proteins the ability to bind and hydrolyze GTP and serve as molecular switches. Four LMW-GBP subfamilies play important roles in neutrophils: the Ras subfamily itself, whose members regulate cell growth and division; the Rho subfamily, functioning in cytoskeletal rearrangements and superoxide generation [28]; and the Rab and Arf subfamilies, critical for vesicular and endomembrane trafficking [29].

Second Messengers

Neutrophils also employ second messenger systems, in which small molecules generated at the plasma membrane diffuse through the cell to affect targets. Among the best-studied of the systems in neutrophils is phospholipase C (PLC). At the cell membrane, activated PLC cleaves phosphoinositol bisphosphate (PIP_2) into diacyl glycerol (DAG) and inositol trisphosphate (IP_3), which in turn increases calcium influx to induce protein kinase C activation. A second critical phospholipase in neutrophils is $cPLA_2$, which hydrolyzes the sn-2 position of membrane glycerophospholipids to liberate arachidonic acid (AA), a precursor of eicosanoids including prostaglandins (PGs) and leukotrienes (LTs) [30,31].

Nonlipid second messengers regulating neutrophil activation include cAMP, whose intracellular concentrations can rise rapidly upon neutrophil stimulation. It is thought that cAMP provides a mostly negative regulatory (off) signal, as direct exposure to membrane-permeable analogs of cAMP inhibits many neutrophil responses, probably via activation of protein kinase A [32,33]. cGMP can also be generated by neutrophils, and appears to have a modest enhancing effect on some neutrophil responses.

Low levels of nitric oxide (NO) are also expressed in neutrophils following stimulation [34]. In addition to phagocytic and tumoricidal activity, NO in neutrophil appears to play an important role in signal transduction. Exogenously added NO exerts a variety of effects on neutrophils, including inhibition of the NADPH oxidase, actin polymerization, and chemotaxis [35]. Excessive NO production is seen during the course of a variety of rheumatic and inflammatory diseases, attesting to the possible importance of its effects on neutrophils [36]. See Chapters 14 through 16 for further discussion on second messengers and NO functions.

Kinases and Kinase Cascades

Intracellular kinases and kinase cascades also participate in neutrophil regulation. Protein kinase C (PKC) is a family composed of three subsets (PKC-α, β, and γ) with overlapping but distinct functions. PKCs are activated in response to several chemoattractants, and studies support a role for PKC in neutrophil survival and function. Activation of specific isotypes of PKC is known to be involved in membrane alteration and motility, oxidative phosphorylation, and apoptosis modulation of neutrophils. Exogenous activators of PKCs, including phorbol myristate acetate (PMA) and palmitoyl-2-linoleoylglycerol hydroperoxide (PLG-OOH), stimulate neutrophil responses such as adhesion and O_2-generation [37,38]. Moreover, inhibitors of PKC (including chelerythrine chloride and staurosporine) block stimulation of neutrophil functions [39].

Mitogen-activated protein kinases (MAPK) are a family of serine/threonine-specific protein kinases that respond to extracellular stimuli (mitogens) and regulate cellular activities including gene expression, mitosis, differentiation, and cell survival/apoptosis. MAPKs include the Erk, p38, and Jun kinase (Jnk) subfamilies. In neutrophils, chemoattractants, colony stimulating factors, and other stimuli such as LPS are capable of activating all three classes of MAPK [40–42]. Current evidence support a role for Erk in neutrophil activation [43,44], including adhesion and phagocytosis processes [32,45].

Phosphatidylinositol 3-kinases (PI 3-Ks) phosphorylate lipid (phosphotidylinositol) targets. The main active product of class I PI 3-Ks appears to be phosphatidylinositol trisphosphate (PIP_3). Chemoattractants such as *N*-formyl-methionyl-leucyl-phenylalanine (FMLP) rapidly activate PI 3-K in neutrophils, regulating O_2^- generation, adhesion, chemotaxis [46], and degranulation [47]. A role for PI 3-K in the regulation of neutrophil survival and apoptosis has also been established [48].

THE CONSEQUENCES OF NEUTROPHIL ACTIVATION: ADHESION, MIGRATION, AND BACTERIAL KILLING

Neutrophil Adhesion and Migration

The ability of bloodstream neutrophils to adhere to endothelium is an essential step required to concentrate them and direct them to sites of inflammation. Membranes on both neutrophils and endothelial cells have the capacity to express multiple families of interacting adhesion molecules. These include selectins, integrins, and intercellular adhesion molecules (ICAMs), as well as sialylated glycoproteins. Under noninflammatory conditions, neutrophils and other leukocytes travel primarily through the center of the blood vessel lumen, where flow is fastest. In response to proinflammatory signals, both the neutrophils and the blood vessels undergo a series of changes. As a consequence of vascular dilatation blood flow slows, facilitating leukocyte interactions with endothelial cells. A process of rolling adhesion ensues, in which P- and L-selectins on neutrophils, and E-selectins on endothelia, interact with sialyl-Lewis^x moieties (s-Le^x) on their respective partners. These interactions are reversible and transient, and prepare the cells for a tighter binding step to follow. In response to CXCL8 (released by tissue macrophages and transported actively through the venular endothelial cells [49]), complement or formyl peptide fragments generated at the inflammatory site, integrins (predominantly LFA-1 and CR3 [Mac-1]) already expressed on neutrophil surfaces are stimulated to undergo conformational changes that render them adhesive for their cognate receptors. Concurrently, macrophage-derived TNFα and/or IL-1β induce endothelial cell expression of the integrins' cognate ligands, ICAM-1 and ICAM-2. The result is tight binding, and arrest of neutrophil motion [50].

Movement of neutrophils out of the circulatory system, or diapedesis, requires first the interaction between molecules of CD31 (expressed by PMNs and also by the intercellular junctions of endothelial cells) allowing the cell to pass through the endothelium [51]. The secretion of a broad range of MMPs degrades the basement membrane and permits the neutrophil passage through the acellular matrix. By mechanisms not fully established, the basement membrane appears to reseal behind the exited neutrophil. Once in the interstitial compartment, neutrophils migrate along the chemotactic gradient toward the site of injury or infection [52]. The mechanism of chemotaxis involves, in part, directed localization of chemoattractant receptors to the leading edge of the neutrophil ("headlight phenomenon") and cytoskeletal rearrangements to permit unidirectional motion.

Phagocytosis and Bacterial Killing

Once recruited to the inflammatory site, neutrophils directly recognize, phagocytose, and destroy foreign pathogens. Phagocytosis – a specialized form of endocytosis – is the cellular process of engulfing particles by the cell membrane to form an internal phagosome. Initiation of phagocytosis can be achieved by neutrophil recognition of pathogen-associated molecular patterns, or PAMPs, small repetitive molecular motifs that are found on bacteria and/or viruses but not on mammalian cells. PAMPs are recognized by Toll-like receptors (TLRs) [53,54] and other pattern recognition receptors (PRRs). Indeed, it has been demonstrated

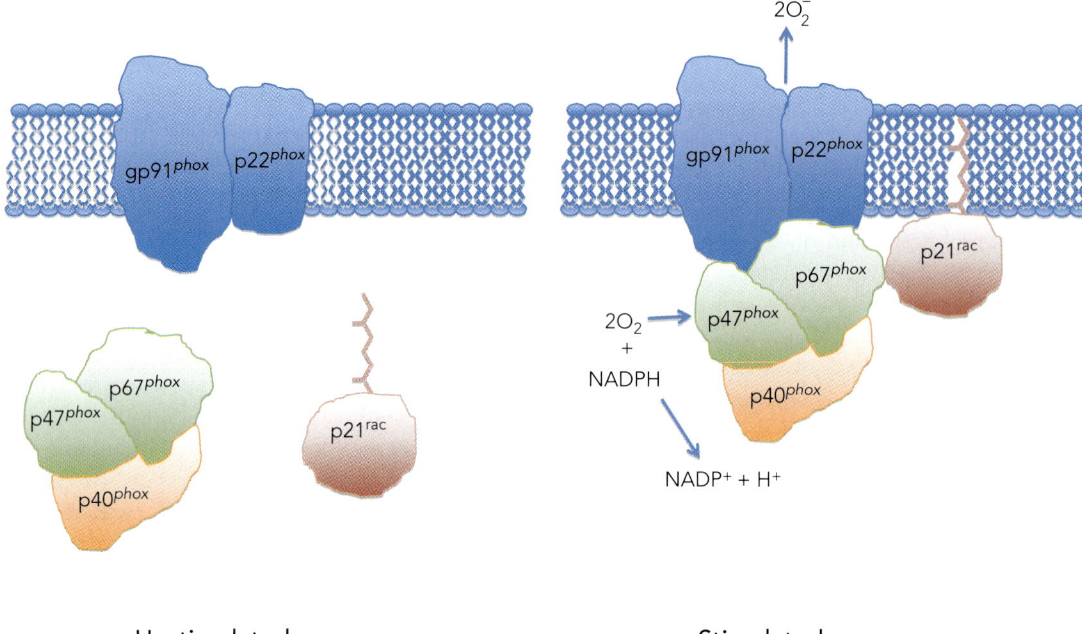

Figure 4A.2. Assembly and activation of the phagocyte oxidase system, resulting in the production of superoxide anions.

that human neutrophils are able to express all TLRs except for TLR3. Upon TLR stimulation, neutrophil shape is altered and phagocytosis is increased [55].

Neutrophil phagocytosis is greatly facilitated by opsonization, or the coating of bacterial or other targets with immunoglobulins and complement C3b fragments. Antibody-coated pathogens are recognized by both complement and immunoglobulin Fc receptors on the neutrophil surface [56,57]. Neutrophils possess receptors to the IgG isotypes IgG_1 and IgG_3 (FcγRI [CD64], FcγRII [CD32], and FcγRIII [CD16]). Opsonization of bacteria by immunoglobulins plays a particularly important role in the response of neutrophils to bacteria with polysaccharide capsules, since their capsules help them evade direct phagocytosis [58].

Neutrophil mechanisms of pathogen destruction are multiple, and involve granule fusion, toxic oxygen radical production, activation of latent proteolytic enzymes, and the activity of antibacterial proteins. The phagosome undergoes fusion with neutrophil granules to form a phagolysosome, the protected space in which pathogen degradation occurs.

Oxygen-dependent degradation of pathogens requires a NADPH oxidase system and a *respiratory burst*: a rapid production and release of reactive oxygen species (ROS) such as NO, superoxide anion, and hydrogen peroxide, all of which are highly toxic to bacteria. The NADPH oxidase enzyme is a membrane-bound multimeric complex that assembles from membrane and cytoplasmic proteins upon cell stimulation (Figure 4A.2). The NADPH oxidase system converts oxygen molecules (O_2) into superoxide anions (O_2^-). A second enzyme, superoxide dismutase (SOD), then converts O_2^- into hydrogen (H_2O_2), which can kill microorganisms. H_2O_2 can be further converted by myeloperoxidase into hypochlorous acid (HOCl, chlorine bleach), which also has potent antibacterial activity. Whereas the membrane components of the NADPH oxidase are localized to specific granules, myeloperoxidase is localized to primary granules; thus, the production of HOCl cannot occur until both granule classes fuse into the phagolysosome.

Oxygen-independent degradation also depends on the fusion of granules into the phagolysosome. As detailed at the beginning of this chapter, primary, specific, and gelatinase granules contain proteolytic enzymes such as lysozyme and metalloproteinases, as well as defensins and cationic proteins (e.g., bactericidal permeability inducing protein, BPI) with intrinsic antimicrobial properties required for the destruction of invading microorganisms [59].

Most recently, a unique, nonphagocytic mechanism of bacterial killing has been identified in neutrophils. Neutrophil extracellular traps (NETs) are extracellular neutrophil structures composed of chromatin and granule proteins that bind and kill microorganisms [60]. In settings of extreme stimulation, and in the presence of stress from ROS, neutrophils may undergo a novel form of cell death characterized by cell membrane breakdown and the release of NETs. These NETs can bind and neutralize extracellular pathogens. Because this process requires the sacrifice of the neutrophil, the term "beneficial suicide" has been proposed [61].

NEUTROPHIL PROPAGATION OF INFLAMMATION: PRODUCTION OF INFLAMMATORY MEDIATORS

In addition to engulfing and destroying foreign invaders, activated neutrophils enhance the inflammatory response by producing mediators that promote additional inflammation and facilitate host defense.

Arachidonic Acid Metabolites

Upon stimulation, neutrophils release AA from membrane phospholipids. While AA can act as a neutrophil chemoattractant [45,62], its main role is as a precursor for the generation of other, more potent lipid compounds, most prominently the prostaglandins and leukotrienes.

Prostaglandins are generated by the action of cyclooxygenases (COXs, endoperoxide synthases) on AA. COXs are heterobifunctional enzymes that first reduce AA to PGG_2, then convert PGG_2 into PGH_2. PGH_2 serves as a substrate for generating multiple prostanoids. Among the latter, the most directly relevant to inflammation is PGE_2. Proinflammatory effects of PGE_2 include vasodilation, increased vascular permeability, and pain. COX activity in resting neutrophils is negligible, but activated neutrophils upregulate COX-2 to produce PGE_2.

Neutrophils generate LTs through the action of 5-lipoxygenase (5-LO), like COX a heterobifunctional enzyme. 5-LO first converts AA to 5-hydroxyeicosatetraenoic acid (5-HETE), then converts 5-HETE into LTA_4. Both 5-HETE [45] and LTA_4 possess intrinsic biological activity, but mainly serve as substrates for additional reactions. Neutrophils convert LTA_4 into LTB_4 via the action of LTA_4 hydrolase; LTB_4 is itself a potent chemoattractant and activator of neutrophils.

Interestingly, some neutrophil lipid products have anti-inflammatory effects. Lipoxins are generated by the sequential actions of neutrophil 5-lipoxygenase and a related enzyme (either 12- or 15-lipoxygenase) in another cell type, either platelets or endothelial cells [63,64]. The requirement for two cells, passing substrates between them, suggests that lipoxins may not be produced until an inflammatory response is well underway, when they may serve to promote the resolution of inflammation [65]. Some prostaglandins – notably 15d-PGJ_2 and possibly $PGF_2\alpha$ – also characteristically have anti-inflammatory effects [66]. Even PGE_2 may, in some cases, have anti-inflammatory effects, underlining the importance of context on the determination of inflammatory outcomes.

Cytokine Production

Although the neutrophil protein synthetic apparatus is partly disabled, neutrophils remain capable of limited protein synthesis, including the production of cytokines. Cytokines released by neutrophils include MIP-1-α and β (CCL3 and CCL4), transforming growth factor-β (TGF-β), as well as IL-8 (CXC8), growth-related oncogene-α (Gro-α/CXCL1), and the granulocyte chemotactic protein-2 (CXCL6) [67]. IL-1α and β, TNFα, and IL-6 may also be produced by neutrophils, although this remains a matter of discrepancy [68]. The production of each chemokine requires a relatively selective combination of stimulants [69]. For example, in the presence of LPS and TNFα, only IL-8, Gro-α, and MIP-1-α are produced, while LPS and IFNγ additionally induce the synthesis of CXCL 9 and 10. Neutrophils under conditions of inflammation may upregulate the synthesis not only of the cytokines, but also of their cognate receptors, leading to autocrine-stimulation loops that regulate cell trafficking and immune responses [70].

B-lymphocyte stimulator (BLyS), a member of the TNF superfamily of ligands, and TNF-related apoptosis-inducing ligand (TRAIL) have been reported to be secreted by G-CSF and IL-8–timulated neutrophils, respectively [71,72]. Whereas BlyS supports the survival and maturation of B cells, TRAIL acts to stimulate the antitumor and antiviral actions of T cells. Through these and other products, neutrophil activation may not only provide direct attack on pathogens, but also support the transition from innate to adaptive immunity.

Just as neutrophils have the capacity to make anti-inflammatory lipids, they may also produce anti-inflammatory proteins. TGF-β has a dual role. On the one hand, it is a potent chemoattractant for human neutrophils [73], and enhances neutrophil cytokine production, including that of TGF-β itself [74]. On the other hand, TGF-β also has potent antiinflammatory effects [75] that may contribute to the resolution of the inflammatory state. Activated neutrophils also produce IL-1 receptor antagonist (IL-1Ra), an endogenous inhibitor of IL-1β signaling [76]. The value of recombinant IL-1Ra (anakinra) in the treatment of autoinflammatory conditions such as Still's disease [77] highlights the biological importance of endogenous IL-1Ra.

REFERENCES

1. Bainton, D.F., Ullyot, J.L., and Farquhar, M.G. 1971. The development of neutrophilic polymorphonuclear leukocytes in human bone marrow. *J Exp Med* **134**(4):907–934.

2. Rosmarin, A.G., Yang, Z., and Resendes, K.K. 2005. Transcriptional regulation in myelopoiesis: hematopoietic fate choice, myeloid differentiation, and leukemogenesis. *Exp Hematol* **33**(2):131–143.
3. Gallin, J.I., and Snyderman, R. 1999. *Inflammation: Basic Principles and Clinical Correlates*. Philadelphia, PA: Lippincott Williams & Wilkins, 3rd ed., p. xxiii, 1335.
4. Richards, M.K., Liu, F., Iwasaki, H., Akashi, K., and Link, D.C. 2003. Pivotal role of granulocyte colony-stimulating factor in the development of progenitors in the common myeloid pathway. *Blood* **102**(10):3562–3568.
5. Lord, B.I., Bronchud, M.H., Owens, S., et al. 1989. The kinetics of human granulopoiesis following treatment with granulocyte colony-stimulating factor in vivo. *Proc Natl Acad Sci USA* **86**(23):9499–9503.
6. Fukunaga, R., Seto, Y., Mizushima, S., and Nagata, S. 1990. Three different mRNAs encoding human granulocyte colony-stimulating factor receptor. *Proc Natl Acad Sci USA* **87**(22):8702–8706.
7. Pojda, Z., and Tsuboi, A. 1990. In vivo effects of human recombinant interleukin 6 on hemopoietic stem and progenitor cells and circulating blood cells in normal mice. *Exp Hematol* **18**(9):1034–1037.
8. Metcalf, D., Begley, C.G., Johnson, G.R., Nicola, N.A., Lopez, A.F., and Williamson, D.J. 1987. Quantitative responsiveness of murine hemopoietic populations in vitro and in vivo to recombinant multi-CSF (IL-3). *Exp Hematol* **15**(3):288–295.
9. Metcalf, D., Begley, C.G., Johnson, G.R., Nicola, N.A., Lopez, A.F., and Williamson, D.J. 1986. Effects of purified bacterially synthesized murine multi-CSF (IL-3) on hematopoiesis in normal adult mice. *Blood* **68**(1):46–57.
10. Nagase, H., Miyamasu, M., Yamaguchi, M., et al. 2002. Cytokine-mediated regulation of CXCR4 expression in human neutrophils. *J Leukoc Biol* **71**(4):711–717.
11. Martin, C., Burdon, P.C.E., Bridger, G., Gutierrez-Ramos, J-C., Williams, T.J., and Rankin, S.M. 2003. Chemokines acting via CXCR2 and CXCR4 control the release of neutrophils from the bone marrow and their return following senescence. *Immunity* **19**(4):583–593.
12. Murray, J., Barbara, J.A., Dunkley, S.A., et al. 1997. Regulation of neutrophil apoptosis by tumor necrosis factor-alpha: requirement for TNFR55 and TNFR75 for induction of apoptosis in vitro. *Blood* **90**(7):2772–2783.
13. Tortorella, C., Piazzolla, G., Spaccavento, F., Pece, S., Jirillo, E., and Antonaci, S. 1998. Spontaneous and Fas-induced apoptotic cell death in aged neutrophils. *J Clin Immunol* **18**(5):321–329.
14. Stark, M.A., Huo, Y., Burcin, T.L., Morris, M.A., Olson, T.S., and Ley, K. 2005. Phagocytosis of apoptotic neutrophils regulates granulopoiesis via IL-23 and IL-17. *Immunity* **22**(3):285–294.
15. Borregaard, N., Lollike, K., Kjeldsen, L., et al. 1993. Human neutrophil granules and secretory vesicles. *Eur J Haematol* **51**(4):187–198.
16. Borregaard, N., Sehested, M., Nielsen, B.S., Sengeløv, H., and Kjeldsen, L. 1995. Biosynthesis of granule proteins in normal human bone marrow cells. Gelatinase is a marker of terminal neutrophil differentiation. *Blood* **85**(3):812–817.
17. Cowland, J.B., and Borregaard, N. 1999. The individual regulation of granule protein mRNA levels during neutrophil maturation explains the heterogeneity of neutrophil granules. *J Leukoc Biol* **66**(6):989–995.
18. Cham, B.P., Gerrard, J.M., and Bainton, D.F. 1994. Granulophysin is located in the membrane of azurophilic granules in human neutrophils and mobilizes to the plasma membrane following cell stimulation. *Am J Pathol* **144**(6):1369–1380.
19. Dahms, N.M., Lobel, P., and Kornfeld, S. 1989. Mannose 6-phosphate receptors and lysosomal enzyme targeting. *J Biol Chem* **264**(21):12115–12118.
20. Nauseef, W.M., McCormick, S., and Yi, H. 1992. Roles of heme insertion and the mannose-6-phosphate receptor in processing of the human myeloid lysosomal enzyme, myeloperoxidase. *Blood* **80**(10):2622–2633.
21. Cieutat, A.M., Lobel, P., August, J.T., et al. 1998. Azurophilic granules of human neutrophilic leukocytes are deficient in lysosome-associated membrane proteins but retain the mannose 6-phosphate recognition marker. *Blood* **91**(3):1044–1058.
22. Owen, C.A., and Campbell, E.J. 1999. The cell biology of leukocyte-mediated proteolysis. *J Leukoc Biol* **65**(2):137–150.
23. Kobayashi, Y. 2006. Neutrophil infiltration and chemokines. *Crit Rev Immunol* **26**(4):307–316.
24. Cohen, A.M., Zsebo, K.M., Inoue, H., et al. 1987. In vivo stimulation of granulopoiesis by recombinant human granulocyte colony-stimulating factor. *Proc Natl Acad Sci USA* **84**(8):2484–2488.
25. Hechtman, D.H., Cybulsky, M.I., Fuchs, H.J., Baker, J.B., and Gimbrone, M.A., Jr. 1991. Intravascular IL-8 inhibitor of polymorphonuclear leukocyte accumulation at sites of acute inflammation. *J Immunol* **147**(3):883–892.
26. Theilgaard-Monch, K., Porse, B.T., and Borregaard, N. 2006. Systems biology of neutrophil differentiation and immune response. *Curr Opin Immunol* **18**(1):54–60.
27. Bokoch, G.M. 1995. Chemoattractant signaling and leukocyte activation. *Blood* **86**(5):1649–1660.
28. Hall, A. 2005. Rho GTPases and the control of cell behaviour. *Biochem Soc Trans* **33**(Pt 5):891–895.
29. Kawasaki, M., Nakayama, K., and Wakatsuki, S. 2005. Membrane recruitment of effector proteins by Arf and Rab GTPases. *Curr Opin Struct Biol* **15**(6):681–689.
30. Morgan, C.P., Sengelov, H., Whatmore, J., Borregaard, N., and Cockcroft, S. 1997. ADP-ribosylation-factor-regulated phospholipase D activity localizes to secretory vesicles and mobilizes to the plasma membrane following N-formylmethionyl-leucyl-phenylalanine stimulation of human neutrophils. *Biochem J* **325**(Pt 3):581–585.
31. Murakami, M., and Kudo, I. 2002. Phospholipase A2. *J Biochem* **131**(3):285–292.
32. Pillinger, M.H., Feoktistov, A.S., Capodici, C., et al. 1996. Mitogen-activated protein kinase in neutrophils and enucleate neutrophil cytoplasts: evidence for regulation of cell-cell adhesion. *J Biol Chem* **271**(20):12049–12056.
33. Smolen, J.E., Stoehr, S.J., and Kuczynski, B. 1991. Cyclic AMP inhibits secretion from electroporated human neutrophils. *J Leukoc Biol* **49**(2):172–179.
34. Amin, A.R., Attur, M., Vyas, P., et al. 1995. Expression of nitric oxide synthase in human peripheral blood mononuclear cells and neutrophils. *J Inflamm* **47**(4): 190–205.
35. Scher, J.U., Pillinger, M.H., and Abramson, S.B. 2007. Nitric oxide synthases and osteoarthritis. *Curr Rheumatol Rep* **9**(1):9–15.
36. Clancy, R.M., Amin, A.R., and Abramson, S.B. 1998. The role of nitric oxide in inflammation and immunity. *Arthritis Rheum* **41**(7):1141–1151.
37. Nauseef, W.M., Volpp, B.D., McCormick, S., Leidal, K.G., and Clark, R.A. 1991. Assembly of the neutrophil respiratory burst oxidase. Protein kinase C promotes cytoskeletal and membrane association of cytosolic oxidase components. *J Biol Chem* **266**(9):5911–5917.

38. Kambayashi, Y., Takekoshi, S., Tanino, Y., et al. 2007. Various molecular species of diacylglycerol hydroperoxide activate human neutrophils via PKC activation. *J Clin Biochem Nutr* **41**(1):68–75.
39. Yamamoto, Y., Kambayashi, Y., Ito, T., Watanabe, K., and Nakano, M. 1997. 1,2-Diacylglycerol hydroperoxides induce the generation and release of superoxide anion from human polymorphonuclear leukocytes. *FEBS Lett* **412**(3):461–464.
40. Suzuki, K., Hino, M., Hato, F., Tatsumi, N., and Kitagawa, S. 1999. Cytokine-specific activation of distinct mitogen-activated protein kinase subtype cascades in human neutrophils stimulated by granulocyte colony-stimulating factor, granulocyte-macrophage colony-stimulating factor, and tumor necrosis factor-alpha. *Blood* **93**(1):341–349.
41. Nick, J.A., Avdi, N.J., Young, S.K., et al. 1999. Selective activation and functional significance of p38alpha mitogen-activated protein kinase in lipopolysaccharide-stimulated neutrophils. *J Clin Invest* **103**(6):851–858.
42. Avdi, N.J., Nick, J.A., Whitlock, B.B., et al. 2001. Tumor necrosis factor-alpha activation of the c-Jun N-terminal kinase pathway in human neutrophils. Integrin involvement in a pathway leading from cytoplasmic tyrosine kinases apoptosis. *J Biol Chem* **276**(3):2189–2199.
43. Chang, L.C., and Wang, J.P. 1999. Examination of the signal transduction pathways leading to activation of extracellular signal-regulated kinase by formyl-methionyl-leucyl-phenylalanine in rat neutrophils. *FEBS Lett* **454**(1–2):165–168.
44. Chen, L.W., Lin, M.W., and Hsu, C.M. 2005. Different pathways leading to activation of extracellular signal-regulated kinase and p38 MAP kinase by formyl-methionyl-leucyl-phenylalanine or platelet activating factor in human neutrophils. *J Biomed Sci* **12**(2):311–319.
45. Capodici, C., Pillinger, M.H., Han, G., Philips, M.R., and Weissmann, G. 1998. Integrin-dependent homotypic adhesion of neutrophils. Arachidonic acid activates Raf-1/Mek/Erk via a 5-lipoxygenase-dependent pathway. *J Clin Invest* **102**(1):165–175.
46. Thelen, M., Uguccioni, M., and Bosiger, J. 1995. PI 3-kinase-dependent and independent chemotaxis of human neutrophil leukocytes. *Biochem Biophys Res Commun* **217**(3):1255–1262.
47. Capodici, C., Hanft, S., Feoktistov, M., and Pillinger, M.H. 1998. Phosphatidylinositol 3-kinase mediates chemoattractant-stimulated, CD11b/CD18-dependent cell-cell adhesion of human neutrophils: evidence for an ERK-independent pathway. *J Immunol* **160**(4):1901–1909.
48. Kilpatrick, L.E., Lee, J.Y., Haines, K.M., Campbell, D.E., Sullivan, K.E., and Korchak, H.M. 2002. A role for PKC-delta and PI 3-kinase in TNF-alpha-mediated antiapoptotic signaling in the human neutrophil. *Am J Physiol Cell Physiol* **283**(1):C48–C57.
49. Middleton, J., Neil, S., Wintle, J., et al. 1997. Transcytosis and surface presentation of IL-8 by venular endothelial cells. *Cell* **91**(3):385–395.
50. Diacovo, T.G., Roth, S.J., Buccola, J.M., Bainton, D.F., and Springer, T.A. 1996. Neutrophil rolling, arrest, and transmigration across activated, surface-adherent platelets via sequential action of P-selectin and the beta 2-integrin CD11b/CD18. *Blood* **88**(1):146–157.
51. Muller, W.A., Weigl, S.A., Deng, X., and Phillips, D.M. 1993. PECAM-1 is required for transendothelial migration of leukocytes. *J Exp Med* **178**(2):449–460.
52. Baggiolini, M., and Clark-Lewis, I. 1992. Interleukin-8, a chemotactic and inflammatory cytokine. *FEBS Lett* **307**(1):97–101.
53. Underhill, D.M., and Ozinsky, A. 2002. Toll-like receptors: key mediators of microbe detection. *Curr Opin Immunol* **14**(1):103–110.
54. Ozinsky, A., Underhill, D.M., Fontenot, J.D., et al. 2000. The repertoire for pattern recognition of pathogens by the innate immune system is defined by cooperation between toll-like receptors. *Proc Natl Acad Sci USA* **97**(25):13766–13771.
55. Hayashi, F., Means, T.K., and Luster, A.D. 2003. Toll-like receptors stimulate human neutrophil function. *Blood* **102**(7):2660–2669.
56. Indik, Z.K., Park, J.G., Hunter, S., and Schreiber, A.D. 1995. The molecular dissection of Fc gamma receptor mediated phagocytosis. *Blood* **86**(12):4389–4399.
57. Ravetch, J.V., and Kinet, J.P. 1991. Fc receptors. *Annu Rev Immunol* **9**:457–492.
58. Karakawa, W.W., Sutton, A., Schneerson, R., Karpas, A., and Vann, W.F. 1988. Capsular antibodies induce type-specific phagocytosis of capsulated *Staphylococcus aureus* by human polymorphonuclear leukocytes. *Infect Immun* **56**(5):1090–1095.
59. Elsbach, P., and Weiss, J. 1998. Role of the bactericidal/permeability-increasing protein in host defence. *Curr Opin Immunol* **10**(1):45–49.
60. Brinkmann, V., Reichard, U., Goosmann, C., et al. 2004. Neutrophil extracellular traps kill bacteria. *Science* **303**(5663):1532–1535.
61. Brinkmann, V., and Zychlinsky, A. 2007. Beneficial suicide: why neutrophils die to make NETs. *Nat Rev Microbiol* **5**(8):577–582.
62. Abramson, S.B., Leszczynska-Piziak, J., and Weissmann, G. 1991. Arachidonic acid as a second messenger. Interactions with a GTP-binding protein of human neutrophils. *J Immunol* **147**(1):231–236.
63. Samuelsson, B., Dahlen, S.E., Lindgren, J.A., Rouzer, C.A., and Serhan, C.N. 1987. Leukotrienes and lipoxins: structures, biosynthesis, and biological effects. *Science* **237**(4819):1171–1176.
64. Chiang, N., Arita, M., and Serhan, C.N. 2005. Anti-inflammatory circuitry: lipoxin, aspirin-triggered lipoxins and their receptor ALX. *Prost Leukot Essent Fatty Acids* **73**(3–4):163–177.
65. Serhan, C.N., Hong, S., Gronert, K., et al. 2002. Resolvins: a family of bioactive products of omega-3 fatty acid transformation circuits initiated by aspirin treatment that counter proinflammation signals. *J Exp Med* **196**(8):1025–1037.
66. Scher, J.U., and Pillinger, M.H. 2005. 15d-PGJ2: the anti-inflammatory prostaglandin? *Clin Immunol* **114**(2):100–109.
67. Kasama, T., Miwa, Y., Isozaki, T., Odai, T., Adachi, M., and Kunkel, S.L. 2005. Neutrophil-derived cytokines: potential therapeutic targets in inflammation. *Curr Drug Targets Inflamm Allergy* **4**(3):273–279.
68. Schroder, A.K., von der Ohe, M., Kolling, U., et al. 2006. Polymorphonuclear leucocytes selectively produce anti-inflammatory interleukin-1 receptor antagonist and chemokines, but fail to produce pro-inflammatory mediators. *Immunology* **119**(3):317–327.
69. Scapini, P., Lapinet-Vera, J.A., Gasperini, S., Calzetti, F., Bazzoni, F., and Cassatella, M.A. 2000. The neutrophil as a cellular source of chemokines. *Immunol Rev* **177**:195–203.
70. Theilgaard-Monch, K., Jacobsen, L.C., Borup, R., et al. 2005. The transcriptional program of terminal granulocytic differentiation. *Blood* **105**(4):1785–1796.
71. Scapini, P., Carletto, A., Nardelli, B., et al. 2005. Proinflammatory mediators elicit secretion of the

intracellular B-lymphocyte stimulator pool (BLyS) that is stored in activated neutrophils: implications for inflammatory diseases. *Blood* **105**(2):830–837.

72. Cassatella, M.A., Huber, V., Calzetti, F., et al. 2006. Interferon-activated neutrophils store a TNF-related apoptosis-inducing ligand (TRAIL/Apo-2 ligand) intracellular pool that is readily mobilizable following exposure to proinflammatory mediators. *J Leukoc Biol* **79**(1):123–132.
73. Reibman, J., Meixler, S., Lee, T.C., et al. 1991. Transforming growth factor beta 1, a potent chemoattractant for human neutrophils, bypasses classic signal-transduction pathways. *Proc Natl Acad Sci USA* **88**(15):6805–6809.
74. Fava, R.A., Olsen, N.J., Postlethwaite, A.E., et al. 1991. Transforming growth factor beta 1 (TGF-beta 1) induced neutrophil recruitment to synovial tissues: implications for TGF-beta-driven synovial inflammation and hyperplasia. *J Exp Med* **173**(5):1121–1132.
75. Huynh, M.L., Fadok, V.A., and Henson, P.M. 2002. Phosphatidylserine-dependent ingestion of apoptotic cells promotes TGF-beta1 secretion and the resolution of inflammation. *J Clin Invest* **109**(1):41–50.
76. McColl, S.R., Paquin, R., Menard, C., and Beaulieu, A.D. 1992. Human neutrophils produce high levels of the interleukin 1 receptor antagonist in response to granulocyte/macrophage colony-stimulating factor and tumor necrosis factor alpha. *J Exp Med* **176**(2):593–598.
77. Fitzgerald, A.A., LeClercq, S.A., and Yan, A. 2005. Rapid responses to anakinra in patients with refractory adult-onset Still's disease. *Arthritis Rheum* **52**(6):1794–1803.

SUGGESTED READINGS

Burg, N.D., and Pillinger, M.H. 2001. The neutrophil: function and regulation in innate and humoral immunity. *Clin Immunol* **99**(1):7–17.

Kobayashi, Y. 2006. Neutrophil infiltration and chemokines. *Crit Rev Immunol* **26**(4):307–316.

Theilgaard-Monch, K., Porse, B.T., and Borregaard, N. 2006. Systems biology of neutrophil differentiation and immune response. *Curr Opin Immunol* **18**(1):54–60.

4B Neutrophils II

Marco A. Cassatella

INTRODUCTION

Polymorphonuclear neutrophil leukocytes (PMN) are "professional" phagocytic cells of the innate immune system that act as the first line of defense against invading pathogens, principally bacteria and fungi but also viruses. Because of their powerful microbicidal equipment, they have a major role in inflammatory responses other than anti-infectious defenses. In fact, after agonist challenge, neutrophils have the capacity to generate reactive oxygen species (ROS) and release lytic enzymes with potent antimicrobial activity, which are all essential for pathogen killing. Conversely, PMN-derived ROS and proteases may also damage the surrounding tissues if released in an uncontrolled manner, as observed in inflammatory diseases dominated by neutrophils. Nevertheless, it is now clear that the role of neutrophils goes far beyond phagocytosis and pathogen killing, as uncovered in the past two decades (Figure 4B.1). For instance, it has been documented that neutrophils have the capacity to migrate toward the lymph nodes and to express major histocompatibility complex class II (MHC II) molecules, once appropriately activated. An additional and fascinating aspect that has gradually come to light is the ability of neutrophils to newly express a number of genes, whose products lie at the core of inflammatory and immune responses. Not only neutrophils synthesize numerous proteins involved in their effector functions, including some complement components and Fc receptors, but they also produce a variety of cytokines and chemokines. Since the latter molecules exert a broad spectrum of biological activities, it is reasonable to assume that neutrophils, by producing and releasing cytokines, may significantly contribute to the regulation of many different processes and, in turn, act as key regulators of cross talks among immune, endothelial, stromal, and parenchymal cells, as well as important players in autoimmune diseases and antitumoral responses. Finally, the application to neutrophils of the new paradigm of apoptosis, defined as the process of regulated cell death that prevents release of their cytotoxic and proteolytic contents, is now used to explain how some processes lead to resolution of inflammation or infection without tissue damage, simply by allowing neutrophil apoptosis to occur.

NEUTROPHIL DIFFERENTIATION, MATURATION, AND OTHER GENERALITIES

Polymorphonuclear neutrophils are the most abundant circulating leukocyte type in humans, normally present at 2.5–7.5 × 10^9 cells/L in the blood. PMN are members of the granulocyte family of leukocytes, which also comprises eosinophils and basophils and that were thus named because they are not well stained either by eosin, a red acidic stain, or by methylene blue, a basic or alkaline stain. All three types of granulocytes contain heterogeneous cytoplasmic granules (500–1,500/neutrophil), which are storage pools for mostly cell-specific intracellular enzymes, cationic protein, receptors, and discrete proteins. A morphological peculiarity of the neutrophil concerns their nucleus, which is polymorphous and usually consists of three to five sausage-shaped masses of chromatin connected by fine threads.

Mature neutrophils are terminally differentiated, nondividing cells, with a rich supply of cytoplasmic glycogen. They develop and maturate in the bone marrow from pluripotent CD34+ stem cells (over a 7–14 day period, at a rate of approximately 5–10 × 10^{10} cells on a daily basis), under the regulatory effects of several colony-stimulating factors (CSFs), including interleukin-3 (IL-3), granulocyte-macrophage CSF (GM-CSF) and granulocyte CSF (G-CSF). Transcriptional profiling studies have suggested that, whereas macrophages represent the default myeloid cells that mature,

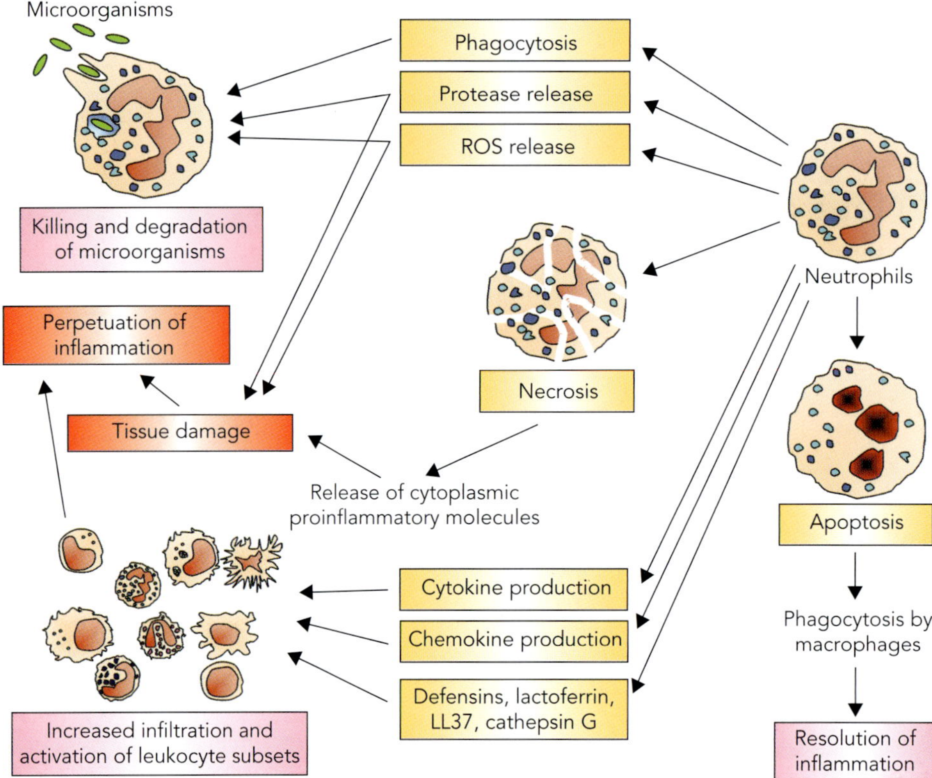

Figure 4B.1. Potential effector functions of neutrophils at sites of infection and inflammation.

granulocytes arise through the selective expression of a subset of transcription factors (Egr1, HoxB7, STAT3) during differentiation. Colony-stimulating factors exert their growth and differentiation effects on progressively more committed stages of cell maturation, starting from the myeloblast, the promyelocyte, the myelocyte, the metamyelocyte, the nonsegmented granulocyte (band neutrophils), and, finally, the mature neutrophil with segmented nuclei. Targeted genetic disruption experiments in mice have clearly shown that G-CSF and its receptor are essential for the maintenance of normal levels of blood neutrophils. Importantly, both GM-CSF and G-CSF can also positively influence the survival and functions of mature neutrophils.

Within the circulation, PMNs are 10–20 nm in diameter and exist in two pools in a dynamic equilibrium: a circulating pool and a "marginated" pool, the latter believed to be sequestered within the microvasculature of many organs. Cell-labeling experiments have shown that the lifespan of neutrophils in the circulation is short, with a half-life of approximately 7–12 hours. In the resting uninfected host, the production and elimination of neutrophils are balanced, resulting in fairly constant concentration in peripheral blood. Neutrophil turnover must therefore be under strict control, as demonstrated by the spontaneous apoptosis that neutrophils undergo before their removal by macrophage in the lung, spleen, and liver. During pathological conditions (for instance, during a bacterial infection), the number of circulating neutrophils dramatically increases (even up to 10-fold, a condition called neutrophilia), as a result of either an accelerated release of neutrophils from the bone marrow combined with a stimulated maturation of immature neutrophils by CSFs, or a demargination from the lungs or the spleen. Under these conditions, the generation of specific chemotactic agents triggers the migration of neutrophils to the site of infection where their phagocytic activities and defensive functions are exerted. In the absence of an inflammatory stimulus, one mechanism shown to control the release and reuptake of neutrophils in the mouse bone marrow involves the interaction of stromal cell–derived factor-1α/CXCL12 and its cognate receptor on the neutrophil, CXCR4. Expression of CXCR4 increases, in fact, as neutrophils age, and consequently is thought to result in cells homing to murine bone marrow. The importance of CXCR4 is illustrated by the findings in patients with the myelokathexis syndrome, also referred to as WHIM (warts, hypogammaglobulinemia, infections, and myelokathexis) syndrome, which is now specifically attributable to a defect in CXCR4 expression. Patients who have this disorder have severe leukopenia and neutropenia, with accumulation of neutrophils in the bone marrow and alterations in the trafficking of lymphocytes and hematopoietic progenitor cells as well.

NEUTROPHIL MICROBICIDAL MECHANISMS

Neutrophils have been described as mobile arsenals that seek and destroy a variety of targets, mainly following phagocytosis. Two primary processes are utilized by neutrophils to eliminate invading pathogens. One of them involves the generation of ROS, which include O_2^- (superoxide anion), H_2O_2 (hydrogen peroxide), singlet oxygen, and other products derived from the metabolism of H_2O_2. The other process involves the release of their intracellular granules, which contain lytic enzymes and antimicrobial polypeptides. The former mechanism is referred to as the "respiratory burst," which is defined as an increase in the oxidative metabolism following the uptake of particles, or in response to soluble inflammatory stimuli, leading to the generation of O_2^- through the activation of an enzymatic system that is unique to phagocytic cells (neutrophils, eosinophils, monocytes, and macrophages), the NADPH oxidase. This latter enzyme consists of a multiprotein complex which is dissociated and thus dormant in unstimulated PMN, formed by a flavocytochrome-b_{558}, which is an heterodimer of $gp91^{phox}$ and $p22^{phox}$ (*phox* standing for phagocyte oxidase) chains, four cytoplasmic components, namely $p40^{phox}$, $p47^{phox}$, $p67^{phox}$, and the small GTP-binding regulatory protein, Rac1/2. Upon cell stimulation, the cytosolic components become phosphorylated and assemble together with the cytochrome and Rac1/2 on the plasma membrane, thus forming the active enzyme. NADPH oxidase then produces the superoxide free radical by catalyzing the transfer of electrons from NADPH to molecular oxygen. Once discharged into the phagosome, O_2^- is converted to hydrogen peroxide (by superoxide dismutase), which, in the presence of neutrophil myeloperoxidase and the abundant Cl^- ions taken up from extracellular fluids, is metabolized into hypochlorous acid (HOCl). The latter, in addition to numerous other microbicidal oxidants, synergizes with granule proteins to kill microbes in the neutrophil phagosome. While O_2^- and H_2O_2 are moderately bactericidal, the production of HOCl is one of the neutrophil's major weapons against microbes. Furthermore, O_2^- can act as a substrate for the reactions with other cellular radicals, for example, O_2^- reacts with nitric oxide to form the very reactive peroxynitrite ($OONO^-$) molecule, a potent cytotoxic oxidant. In the phagosome, the high concentrations of ROS are very toxic, and, ultimately, contribute to the killing of the phagocytosed microbe. The same oxidants can however damage "innocent bystanders" and provoke the tissue destruction. Interestingly, neutrophils contain large reserves of endogenous antioxidants, including glutathione and ascorbate, which serve to protect them from the oxidative stress.

The critical role of NADPH oxidase and its products in host defense is best illustrated by chronic granulomatous disease (CGD), which is a rare condition with an incidence of about 1/250,000 individuals. CGD is the most common clinically significant inherited disorder of neutrophil function, in which mutations in any of the NADPH oxidase complex subunits (except $p40^{phox}$) can lead to a severe immunodeficiency characterized by defective killing of phagocytosed pathogens by phagocytic leukocytes because of their inability to generate ROS. This condition leads to recurrent life-threatening infections that are difficult to treat and which can lead to early death. These infections typically involve microorganisms for which oxidant-mediated killing is particularly critical for effective host defense, including *Staphylococcus aureus*, *Aspergillus* species, *Nocardia*, and a variety of Gram-negative enteric bacilli. Additional and distinctive hallmarks of CGD patients are aberrant inflammatory responses and the formation of inflammatory tissue granulomas, which can obstruct the gastric outlet or urinary tract or cause granulomatous colitis and the symptoms of inflammatory bowel disease. Such chronic inflammatory complications are thought to reflect an important role for superoxide in downregulating the inflammatory response, even though a definitive confirm is required.

The second mechanism used by neutrophils for the elimination of pathogens relies on the release and activity into the phagosome of potent degradative enzymes contained in their granules (Figure 4B.2). Granule proteins are cytotoxic and immune regulatory molecules synthesized at defined stages of neutrophil maturation in the bone marrow, with no further de novo expression occurring once the PMN has entered the bloodstream. Traditionally, neutrophil granules are subdivided into peroxidase-positive granules (based on the presence of myeloperoxidase), also called primary or azurophil granules, and peroxidase-negative granules that include the specific (secondary) granules, the gelatinase (tertiary) granules, and the secretory vesicles. Distinctions between granule classes can be made upon analysis of marker proteins, even if their strict compartmentalization is not a dogma (Figure 4B.2). For instance, proteinase 3, a serine protease described in azurophil granules is also localized in the membrane of secretory vesicles, the most mobilizable compartment of neutrophils, and in the plasma membrane as well.

The *azurophilic* or primary granules, so named because they are the first to appear during hemopoiesis in the promyelocyte stage, constitute the densest

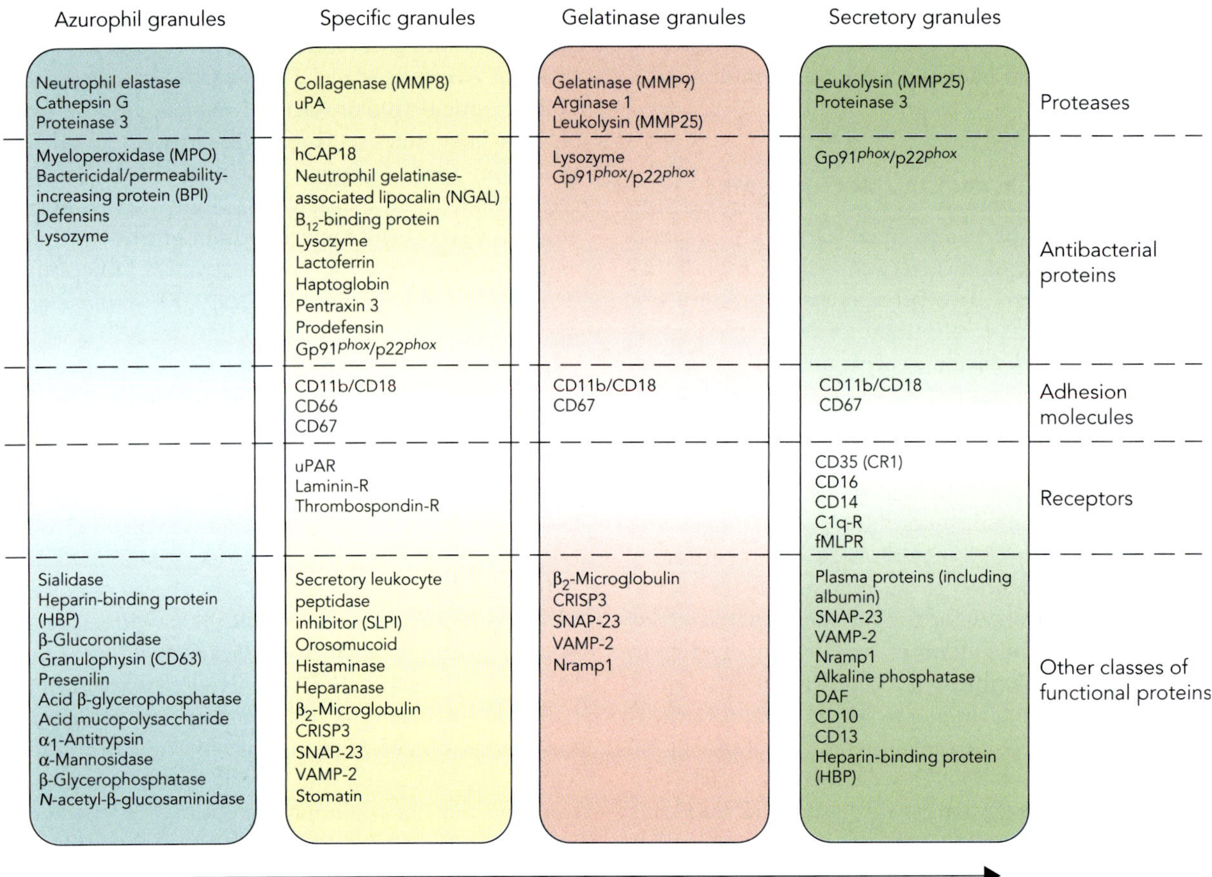

Figure 4B.2. Main constituents of neutrophil granules.

granule populations and undergo limited exocytosis in response to stimulation. Azurophil granules are packaged with acidic hydrolases and antimicrobial proteins and they are believed to contribute primarily to the killing and degradation of engulfed microorganisms that take place in the phagolysosome. They contain myeloperoxidase (MPO, which is responsible for the characteristic greenish color of pus and is often used as specific marker for *azurophilic* granules), hydrolases, lysozyme, matrix metalloproteinases, and three structurally related serprocidins (serine proteases with microbicidal activity): proteinase-3, cathepsin G, and elastase, which display proteolytic activity against a variety of extracellular matrix components, such as elastin, fibronectin, laminin, type IV collagen, and vitronectin. Azurophil granules also contain antimicrobial molecules such as bactericidal/permeability-increasing protein/BPI (which is important for killing Gram-negative bacteria) and α-defensins, a family of small cysteine-rich antibiotic peptides with broad antimicrobial activity against bacteria, fungi, and certain enveloped viruses.

The second major type of granule is termed specific or secondary. Specific granules are formed in the myelocytic stage, are less dense and smaller than the azurophil ones, and stain heavily for glycoproteins. They are known to contain relatively unique constituents, including collagenase, haptoglobin, and vitamin B_{12}-binding protein. Secondary granules mainly participate in the antimicrobial activities of the PMN by mobilization of their arsenal of antimicrobial substances, for example, lactoferrin, NGAL (neutrophil gelatinase-associated lipocalin), lysozyme, hCAP18 (cathelicidin human cationic antimicrobial protein of 18 kDa, the proform of LL-37), and pentraxin-3, either to the phagosome or the exterior of the cell, their mobilization being very important in modulating inflammation. In general, the mechanisms of the inhibitory effect of antimicrobial proteins include depriving ions essential for microbial survival, degrading structural components of microorganisms (e.g., peptidoglycan), and disrupting the integrity of target cell membrane by punching pores in the membrane or by perturbing membrane integrity.

Tertiary granules (whose primary constituent is gelatinase, a metallo-enzyme) are produced at the metamyelocyte stage during differentiation and more easily exocytosed by mature neutrophils than secondary granules. These characteristics reflect the fact that tertiary granules are important primarily as a reservoir of matrix-degrading enzymes and membrane receptors needed during PMN extravasation and diapedesis. Thus, whereas the larger specific granules are rich in antibiotic substances, the smaller gelatinase granules are not. Finally, the secretory vesicles (the fourth type of granules) are created by endocytosis during terminal neutrophil maturation in the bone marrow and are the most readily mobilizable. The propensity for exocytosis is reflected in the density of vesicle-associated membrane protein (VAMP-2), a fusogenic protein associated with the granule membrane. Secretory vesicles contain plasma proteins like albumin and thus do not release toxic substances when they fuse with the plasma membrane and empty their content. Their importance lies in their membrane, which is the main store of receptors (including β2-integrins, the complement receptor 1 [CR1], receptors for formylated bacterial peptides (formylmethionyl-leucyl-phenylalanine [fMLP]-receptors), CD14, the FcγIII receptor CD16, and the metalloprotease leukolysin), which, when translocated to the cell surface, allow the neutrophil to interact with other ligands and to respond to soluble mediators. Thereby, the relatively inactive, on the endothelium rolling PMN, is transformed into a cell that is capable of interacting with the endothelium, monocytes, and dendritic cells (DC) and to better receive inflammatory signals from the environment. An additional protein described to be stored in secretory vesicles is heparin-binding protein (HBP, also known as CAP37 and azurocidin), whose release at the initial stage of PMN extravasation is thought to be of essential importance in the PMN-induced increase in vascular permeability. Finally, peroxidase-negative granules also contain three metalloproteases with great physiological and pathophysiological significance, namely collagenase, matrix metalloproteinase-8 (MMP-8), gelatinase (MMP-9), and leukolysin (MMP-25) (Figure 4B.1).

It is now established that PMN granules are not just emptied like bags, but are rather released upon specific signals that the PMN senses in the local environment. The mechanisms underlying the secretion of the four morphologically distinct populations of granules may be under separate control, as reflected by the order of exocytosis observed after ionophore-induced progressive elevation of cytosolic calcium: secretory vesicles, gelatinase granules, specific granules, and, lastly, azurophilic granules. It is thus plausible that during its journey from the bloodstream to the inflammatory site, the PMN releases its granules in a hierarchical order because not all stored proteins are needed at the same time. Primary neutrophil defects causing disorders of degranulation are very rare. Two such genetic disorders affecting neutrophil granules have been described, the Chediak–Higashi syndrome and the neutrophil-specific granule deficiency, both of which are associated with increased frequency of bacterial infections. Another common disorder of granules is the MPO deficiency, the most common inherited disorder of phagocytes, in which deficiency in MPO prevents the formation of hypochlorous acid from chloride and hydrogen peroxide. There is a remarkable lack of clinical symptoms in most individuals with MPO deficiency, despite in vitro defects in the neutrophil ability to kill *Candida albicans* and *Aspergillus fumigatus* hydra.

NEUTROPHIL RECEPTORS

Since their interactions with the external milieu are crucial for innate responses, neutrophils have developed a recognition apparatus that is able to specifically bind a wide range of extracellular ligands. Following ligation of one or more types of surface receptors, neutrophils generate a number of activation steps, via the generation of a cascade of intracellular "second messengers." These steps are biochemical events which mediate the transmission of biological information between membrane receptors and various intracellular components involved in specific effector functions. Consequently, neutrophil receptors are able to regulate a wide range of functions, including adhesion, migration, phagocytosis, survival, cell activation, gene expression induction, proinflammatory mediator release, and target cell killing. The transduction machinery is mainly, but not exclusively, located in the plasma membrane and is composed of a series of enzymes (i.e., kinases, phosphatases, adenylate cyclase, phospholipases, and other enzymes involved in lipid metabolism) or regulatory proteins (G-protein subunits, channel proteins, anchoring and adaptor proteins) which, in turn, generate, or whose activities are regulated by, several other second messengers (calcium ions, inositol phosphates, diacygIycerol, phosphatidate, cAMP, and so forth). A nonexhaustive list of neutrophil receptors includes (i) receptors for proinflammatory mediators (i.e., the anaphylotoxin complement component C5a, leukotriene B_4 [LTB_4], platelet-activating factor [PAF], substance P, and bacterial formylated peptides typified by fMLP); (ii) receptors for cytokines such as interferon-γ (IFNγ), IL-1, IL-4, IL-6, IL-10, IL-13, IL-15, IL-18, tumor necrosis factor-α (TNFα), G-CSF, GM-CSF, and many others; (iii) receptors for chemokines, including those for IL-8/CXCL8 and GROα/CXCL1, called CXCR1 and CXCR2; (iv) receptors/adhesion molecules for the endothelium (see later); (v) receptors for tissue matrix proteins and for lectins; and (vi) opsonin receptors, such as those for the Fc portion of γ-immunoglobulins (FcγRs) and those for the major cleavage fragments of the complement system (CRs). Neutrophils constitutively express the low-affinity FcγRs (FcγRIIA/CD32A and FcγRIIIA/CD16A), and, when exposed to IFNγ or G-CSF, the high-affinity FcγR (FcγRI/CD64) as well. CR expressed by neutrophils are CR1 (also known as CD35), which binds to complement components C1q, C4b, C3b, and Mannan-binding lectin (MBL); CR3 ($\alpha_M\beta_2$ integrin, CD11b/CD18, or MAC-1), which binds to iC3b, ICAM-1, and some microbes; CR4 ($\alpha_X\beta_2$ integrin, CD11c/CD18), which binds to iC3b. By expressing these latter receptors, neutrophils are able to recognize and bind, in a cooperative manner, IgG-opsonized particles and/or complement-opsonized microbes, and then activate their phagocytosis.

Neutrophils also express a variety of pattern recognition receptors (PRRs). The latter represent an emerging class of sensors that function by recognizing the so-called pathogen-associated molecular patterns (PAMPs). The latter ligands constitute a limited set of conserved molecular patterns that are unique to microbes and that induce signaling cascades that ultimately activate a plethora of effector mechanisms by immune and nonimmune cells, which all serve to clear the invading pathogens. PRRs are involved in the initiation of the inflammatory response upon infection, activating multiple pathways of host defense, cytokine production, cell survival, and ROS production, without, by themselves, triggering phagocytosis. Examples among PRRs include the Toll-like receptors (TLRs) and nucleotide-binding oligomerization domain (NOD) proteins, two families of innate immune recognition receptors which are required for the detection of a broad range of microbial products including, for instance, lipopolysaccharide (LPS, recognized by TLR4), peptidoglycan and bacterial lipopeptides (recognized by TLR2), foreign DNA (recognized by TLR9), viral RNA (recognized by TLR3, 7, and 8), and muramyldipeptide (recognized by NOD1). Neutrophils express all TLRs (with the exception of TLR3), whose activation has been shown to influence many functional responses. For instance, engagement of TLR can (a) modulate neutrophil expression of adhesion molecules; (b) regulate neutrophil recruitment directly through effects on chemokine receptor expression and indirectly through effects on CXCL8 generation by neutrophils themselves; (c) "prime" (see later) neutrophils for enhanced ROS production; (d) prolong neutrophil survival both directly and indirectly (via monocyte activation). Crucial in the regulation of many neutrophil effector functions is TLR4. Being the specific receptor for LPS, great efforts have been made to characterize the molecular mechanisms whereby TLR4 regulates the expression of target genes in myeloid cells. From studies mainly performed in macrophages or DCs of gene-targeted mice, it has emerged that LPS triggers two classes of genes via TLR4, whose regulatory signals are transduced by discrete intracellular adaptor molecules with TIR (Toll-IL-1 receptor) domains that bind to the same TLR4. One class is defined as "MyD88 (myeloid differentiation factor-88)-dependent," because it is not induced in $MyD88^{-/-}$ mice, and that mostly includes proinflammatory mediators such as TNFα, IL-1, IL-12p40, and CXCL8. In this pathway, MyD88 and TIRAP/MAL (TIR domain containing adapter protein/myeloid differentiation factor-88 adapter-like protein) mediate a rapid and early activation of the transcription factor NF-κB, whose nuclear translocation is essential for transcriptional induction of the above-mentioned proinflammatory genes. The other class is defined as "MyD88-independent" because it relies on a more delayed activation of both NF-κB and IRF (IFN-regulatory factor)-3 transcription factors in $MyD88^{-/-}$ mice, which ultimately lead to the expression of IFNβ and subsequently to an IFNβ-dependent STAT1 activation. Genes that belong to this last class are thus regulated by endogenous IFNβ and activated in a more delayed manner compared to those of the "MyD88-dependent" pathway. They include, for instance, a number of antiviral genes, Th1-activating chemokines such as MIG (monokine induced by IFNγ)/CXCL9, IP-10 (IFN-inducible protein-10)/CXCL10 and I-TAC (IFN-inducible T-cell alpha chemoattractant)/CXCL11, anticancer molecules such as TRAIL (TNF-related apoptosis-inducing ligand), and iNOS (inducible NO synthase). Along the "MyD88-independent" pathway, both the TRIF (TIR domain-containing adapter inducing IFNβ) and TRAM (TIR domain-containing adapter-inducing IFNβ-related adapter molecule) adaptor proteins transduce the activation of redundant protein kinases, TBK1 (TRAF family-associated NF-κB binding kinase [TANK]-binding kinase-1) and IKK (IκB kinase, IKK)-ε, which phosphorylate IRF3 on Ser/Thr residues. As a result, IRF3 dimerizes, translocates to the nucleus, associates with other coactivators (for instance, CBP/p300), and ultimately contributes to activate IFNβ gene transcription.

Recent studies, aimed at clarifying the molecular mechanisms whereby the LPS–TLR4 complex transduces its signals in human neutrophils, have uncovered that LPS does not mobilize the so-called MyD88-independent pathway for its inability to activate TBK1 and, in turn, IRF3 phosphorylation (Figure 4B.3). Consequently, neither IFNβ nor all the genes classically grouped as "IFNβ-dependent" is inducible in LPS-treated neutrophils. Curiously, lack of TBK1 activation by TLR4 ligands occurs despite the constitutive expression of all the intracellular signaling and regulatory components involved along

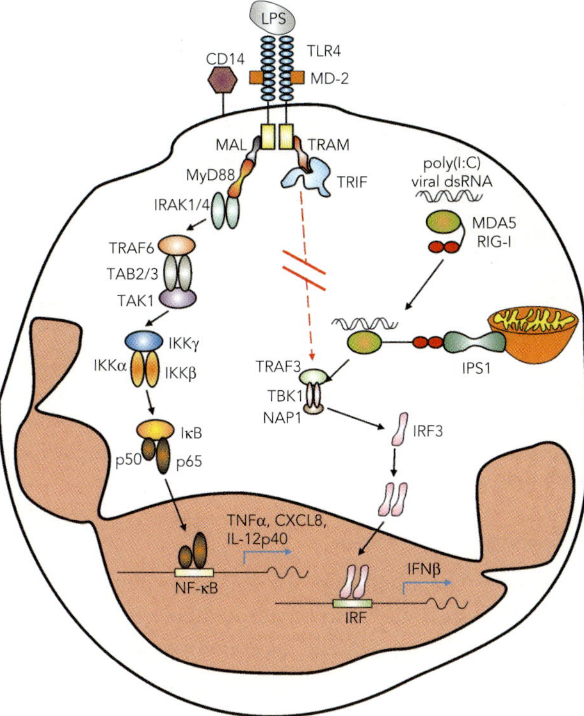

Figure 4B.3. TLR4- and helicase-mediated signal transduction pathways operating in human neutrophils.

the MyD88-independent cascade, including TRAM, TRIF, NAP1, TRAF3, TBK1, IKKε, and IRF3 in neutrophils. In view of the fact that the other known TLR-dependent transduction pathway ultimately signaling through TBK1 and IRF3, namely the "TLR3/TRIF-dependent pathway," is also nonfunctional in neutrophils (for their lack of TLR3 expression) it is intriguing that human neutrophils should constitutively express TBK1 and IRF3. This puzzle has been recently clarified, as revealed by capacity of human neutrophils to trigger an immunoregulatory and antiviral gene expression program, in response to intracellularly administered polyinosinic:polycytidylic acids [poly(I:C)], specifically via TBK1 and IRF3 activation (Figure 4B.3). Poly(I:C) is a synthetic mimetic of viral double-stranded RNA (dsRNA), which is known to interact either with endosomal TLR3 or with cytoplasmic RNA helicases (involved in viral RNA recognition) such as MDA5 (melanoma differentiation-associated gene 5) and RIG-I (retinoic acid–inducible gene I). Accordingly, human neutrophils constitutively express significant levels of both MDA5 and RIG-I. The biological relevance of these findings were strengthened by the observations that murine neutrophils infected with picornaviruses such as encephalomyocarditis virus (EMCV), whose dsRNA recognition specifically requires MDA5 (Figure 4B.3), were found to produce large amounts of either IFNβ or TNFα. Taken together, these data have revealed that neutrophils are able to activate antiviral responses via helicase recognition and thus contribute to local or systemic IFNβ production.

THE ROLE OF NEUTROPHILS IN ACUTE INFLAMMATION

When pathogenic agents enter into (or localize within) a tissue, neutrophils are rapidly recruited to the infected site by a sequence of events that begins with the elaboration of mediators able to promote their migration from the intravascular compartment. Such inflammatory mediators include not only cytokines such as TNFα, IL-1β, or IL-17A (this latter being one of the most abundant product of the Th17 subset of proinflammatory T lymphocytes), which orchestrate neutrophil recruitment through their capacity to trigger the production of chemotactic factors, but also various products derived from numerous sources, including activated plasma components and tissue macrophages or endothelial cells. Classical chemotactic factors include C5a, LTB_4, fMLP, and chemokines of the CXC family (CXCL8/IL-8, CXCL1/GROα, CXCL5/epithelial cell neutrophil-activating protein-78, etc.), the latter molecules representing the main and more specific chemoattractant involved in neutrophil recruitment, according to many in vivo (gene-targeted) models.

When neutrophils need to be recruited to an inflammatory site, they adhere to endothelial cells, transmigrate by diapedesis (the process of squeezing between endothelial lining cells), and crawl toward a corresponding chemotactic gradient (a process called chemotaxis) to the injury site. The detection of chemotactic stimuli by neutrophils leads, in fact, to the activation of signaling pathways, cytoskeletal rearrangement, and changes in cell surface molecules that coordinate to facilitate migration. Recently, chemotaxis has been shown to be dependent on phosphoinositide 3-kinase (PI3K)-γ, which belongs to a family of class I PI3Ks. Adhesion of blood neutrophils (and of other leukocytes as well) to the endothelium during inflammation is a multistep and highly complex phenomenon, which requires specific leukocyte–endothelial interactions involving different families of adhesion molecules, which may vary according to the precise stimulatory conditions. Adhesion molecules include members of the selectin family and their cognate carbohydrate and glycoprotein ligands, which mediate leukocyte deceleration along the vessel wall (a process called "rolling"), as well as the integrins and their cognate immunoglobulin superfamily ligands, which mediate high-affinity adhesion of leukocytes to venules. Neutrophil integrins consist of a family of at least three different glycoproteins, each one composed of an identical β-subunit (β_2, known as CD18) noncovalently linked to different α-subunits (known as CD11a/CD18, lymphocyte function antigen [LFA-1], CD11b/CD18 [Mac-1], CD11c/CD18 [p150,95], and CD11d/CD18). Upon encountering chemoattractants, probably displayed bound to glycosaminoglycans on the vessel wall, integrins are converted from an inactive to an active conformation. Activated integrins then interact with counter-receptors (e.g., ICAM-1 [intercellular adhesion molecule-1] and ICAM-2) on the surface of the endothelium, to lead to the strong adhesion and arrest of the leukocyte. The neutrophils then flatten and migrate between and right through the endothelial cells of postcapillary venules to the surrounding tissue, following the gradient of specific chemotactic factor(s). The former process involves homophilic interaction of CD31 and JAM-A on neutrophils and endothelial cells where CD31 and JAM-A act sequentially to mediate neutrophil migration through the venular walls. The process of transmigration under the influence of adhesion molecules or chemotactic factors, as well as the exposure to other stimuli in local tissues, often result in a marked release by neutrophils not only of preformed proinflammatory mediators, such as those contained in the granules, but also in the rapid production of ROS or the activation of cytokine gene expression.

The fundamental importance of adhesive glycoproteins for the neutrophil functions in vivo is illustrated by those individuals, genetically lacking the leukocyte adhesive proteins, who display an abnormally high susceptibility to bacterial infections. Several types of leukocyte adhesion deficiency (LAD) disease have been identified. LAD-I is an autosomal recessive disorder in which mutations of the β2 integrin family typically eliminate expression of all three integrin complexes. Because of the multiple defects in adhesion-related functions, LAD-I patients develop recurrent bacterial and fungal infections, typically with *S. aureus* or Gram-negative enteric microbes. Characteristic clinical features include frequent skin and periodontal infections, delayed separation of the umbilical cord and omphalitis, and deep tissue abscesses. Neutrophilia with paucity of neutrophils at inflamed or infected sites is also characteristic. LAD-II, caused by mutations in the membrane transporter for fucose, is associated with loss of expression of fucosylated glycans on the cell surface. Fucosylated proteins such as sialyl-Lewis X (CD15s) are ligands for endothelial selectins and are important for the early phases of leukocyte adhesion to endothelial cells. Patients with LAD-II also have leukocytosis and form pus poorly, although infections tend to be less severe in LAD-II patients compared to LAD-I patients. Finally, LAD-III has been described in four patients and appears to be an autosomal recessive disorder. LAD-III is characterized by defects in cell signaling that interfere with activation of multiple classes of integrins downstream of G-protein–coupled receptors. The phenotype of this disorder is similar to LAD-I, but it can also be associated with defects in integrin-mediated platelet aggregation and excessive bleeding.

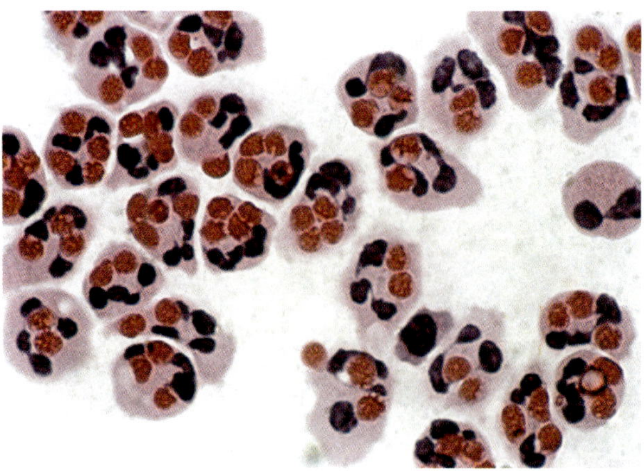

Figure 4B.4. Smear of neutrophils that have phagocytosed opsonized red cells.

After the process of emigration, during which neutrophils undergo shape changes and mobilize their granules to the cell surface, cells arrive to infection foci, where they adhere to extracellular matrix components such as laminin and fibronectin, and begin to react with the etiopathogenetic agent. When neutrophils are exposed to inflammatory mediators such as cytokines (IFNγ, GM-CSF, G-CSF) or LPS, they become "primed" or hyperresponsive to subsequent activating stimuli, in this way acquiring the capacity to exhibit maximal degranulation and respiratory burst responses and enhanced cytokine and lipid mediator release. Accordingly, agents that "prime" neutrophils usually also extend their lifespan by inhibiting apoptosis. In addition, activated neutrophils at inflammatory sites also synthesize large quantities of important inflammatory lipid mediators, including LTB_4 and PAF, which together facilitate the extravasation and recruitment of additional waves of granulocytes to inflamed tissues.

Although neutrophil granulocytes do not apparently show any particular specificity for antigens, they nevertheless play an important role in host protection against microorganisms, by eliminating them through phagocytosis (the process of eating particles). In fact, many cells in our body are capable of phagocytosis, but only neutrophils and macrophages do it to an extent sufficient to consider them as "professional phagocytes" (up to 10–12 particles [e.g., bacteria] can be engulfed by a single neutrophil [see Figure 4B.4]). Phagocytosis is triggered upon the binding of opsonized microorganisms through, for instance, FcγRs and CRs, or through nonspecific glycosylated receptors that recognize certain lectins on target microorganisms. During the phagocytic process, the foreign particle is internalized, initially through membrane recruitment to the site of particle contact, and then via membrane extensions outward to surround the particle. Subsequently, particle engulfment within cytoplasmic phagosomes occurs, which eventually fuse with granules, thereby forming phagolysosomes. This, in turn, results in the subsequent release of proteolytic enzymes and other bactericidal components into the phagolysosome. At the same time, NADPH oxidase assembles on the phagosome membrane after phagocytosis and starts to generate ROS into the phagolysosome to kill bacteria by oxidizing microbial proteins and lipids. The activity of NADPH oxidase also leads to the acidification of the phagosome, which enhances the effectiveness of pH-sensitive antimicrobial compounds. In addition, activation of gene transcription and cytokine production occur during phagocytosis, representing another critical feature of neutrophils for the development of an effective immune response. For instance, recruited neutrophils that encounter a pathogen respond by producing further chemokines, in particular CXCL8, to amplify the innate immune response.

Beside oxygen-dependent and independent killing mechanisms, neutrophils also use powerful structures to capture and kill microbes in the extracellular space, the so-called NETs (neutrophil extracellular traps), thus fulfilling their antimicrobial function even beyond their life span. NETs consist of threads of nuclear chromatin, with a diameter of approximately 15 nm, which are decorated with antimicrobial peptides and enzymes (e.g., BPI, neutrophil elastase, and cathepsin G), but lack membranes and cytosolic markers. Typically during NET formation, the nuclear envelope and the granule membranes gradually dissolve, two phenomena that allow mixing of the cytoplasmic components. Pathogens trapped by NETs are killed by the high local concentration of antimicrobial peptides and enzymes, similar to the situation in the phagolysosome. NETs entrap not only various Gram-positive and Gram-negative bacteria, such as *S. aureus*, *Salmonella typhimurium*, *Streptococcus pneumoniae*, and group A streptococci, but also pathogenic fungi, such as *C. albicans*. Current data suggest that the formation of NETs involves a novel cell death mechanism, which is neither apoptotic nor necrotic, and that does not seem to be dependent on the activation of caspases. NADPH oxidase–dependent ROS are instead formed, and subsequent signaling ultimately leads to the disintegration of nuclear and cellular membranes and to the release of DNA and effector molecules that mix to form extracellular traps. Accordingly, NETs are now thought to be important in CGD, as neutrophils from these individuals do not form NETs. This raises the possibility that the absence of NETs might be another major explanation for the pathology seen in CGD.

While all of the above-mentioned events are necessary and essential for neutrophil-mediated microbicidal activity and digestion of engulfed particles, the same effector mechanisms can also be involved in other effects of PMN, such as amplification of the inflammatory process, preparation of the tissue-healing processes, tumoricidal activity and cytotoxicity, and tissue matrix injury if ROS and proteases are massively released or leaked into the external milieu.

NEUTROPHIL-DERIVED CYTOKINES

Cytokines orchestrate the complex networks mediating the myriads of cellular interactions that ultimately regulate effector cell functions. T cells, DC and natural killer (NK) cells, monocytes, and macrophages are considered to be major producers of cytokines during natural and adaptive immunity. However, numerous in vitro and in vivo studies have made it clear that PMNs also have the ability to synthesize and release immunoregulatory cytokines. Such findings, while challenging the traditional concept of neutrophils viewed as "terminally differentiated cells, devoid of transcriptional and protein synthesis activity," have made it clear that the contribution of neutrophils to host defence and natural immunity extends well beyond their traditional role as professional phagocytes. It is important to mention that, at least in vitro, neutrophils usually (but not always) make fewer molecules of a given cytokine than do monocytes/macrophages or lymphocytes on a per-cell basis. In vivo, however, neutrophils constitute the majority of infiltrating cells in inflamed tissues and often outnumber mononuclear leukocytes by one to two orders of magnitude. Thus, the fact that neutrophils clearly predominate over other cell types under various in vivo conditions suggests that, under those circumstances, the contribution of neutrophil-derived cytokines can be of foremost importance. Interestingly, at the molecular level, studies addressing cytokine release by neutrophils have revealed that the induction of cytokine production in neutrophils is usually preceded by an increased accumulation of the related mRNA transcripts. Similarly to other cell types, cytokine expression in neutrophils can be regulated at the transcriptional, posttranscriptional, translational, and posttranslational levels. Elegant studies have also shown that the inflammatory cytokines produced by neutrophils are generally encoded by immediate-early response genes, which in turn depend on the activation of transcription factors such as those belonging to the NFκB, STAT (signal transducers and activators of transcription), ETS, CCAAT/enhancer-binding proteins (C/EBP), and AP-1 families of transcription factors. Moreover, a wide range

TABLE 4B.1. Cytokines expressed in resting or activated human neutrophils

C-X-C chemokines	*Proinflammatory cytokines*	*Colony-stimulating factors*
IL-8/CXCL8	**TNFα**	G-CSF
GROα, GROβ, GROγ	**IL-1α, IL-1β**	M-CSF(?)
ENA-78/CXCL5	IL-16(?), **IL-17**(?)	GM-CSF(?)
CINC-1*, CINC-2α	IL-18	IL-3(?)
CINC-3/MIP-2/CXCL1	MIF	SCF*(?)
PF4/CXCL4	**IL-6**(?), IL-7, IL-9	
GCP-2/CXCL6		*Angiogenic and fibrogenic factors*
IP-10/CXCL10	*Anti-inflammatory cytokines*	**VEGF**
MIG/CXCL9	**IL-1ra**	HB-EGF
I-TAC/CXCL11	**TGFβ$_1$, TGFβ$_2$**	**FGF-2**
	IL-4(?), **IL-10**(?)	**TGFα**
C-C chemokines		**HGF**
MCP-1/CCL2	*Immunoregulatory cytokines*	
MIP-1α, CCL3	IFNβ*, **IFNβ**, **IFNγ**(?)	*TNF superfamily members*
MIP-1β, CCL4	**IL-12**	**FasL**
MIP-3α, CCL20	IL-23(?)	CD30L
MIP-3β, CCL19		TRAIL
PARC/CCL18	*Other cytokines*	**LIGHT***
MDC/CCL22*	**Oncostatin M**	**Lymphotoxin-β***
	GDF (?)	APRIL, **BAFF/BLyS**
	NGF*, BDNF*, NT4*	RANKL

Cytokines in bold refer to neutrophil studies in animal models confirming human findings. *: mRNA only; (?): it requires definitive corroboration.

of stimuli that are able to induce cytokine synthesis in PMN has been already identified. It is clear, for instance, that cytokines themselves, chemotactic factors (fMLP, LTB$_4$, PAF, C5a, and CXCL8), phagocytic particles, microorganisms (such as fungi, viruses, and bacteria), and PRR ligands can all induce the synthesis and release of cytokines by neutrophils. Not only do the magnitudes and kinetics of cytokine release vary substantially depending on the stimulus used, but also the pattern of production is influenced to a great extent by the stimulus used, each one inducing a characteristic signature. Thus, since neutrophils might usually represent the first cell type encountering, and interacting with, the etiologic agent in an inflammatory context, a stimulus-specific response of neutrophils in terms of cytokine production might help in predicting the evolution of certain types of inflammatory and immune reactions.

Table 4B.1 lists all the cytokines that, to date, have been shown to be released by neutrophils in vitro, either constitutively or following appropriate stimulation, or in vivo. Numerous in vivo observations, in fact, not only have confirmed and reproduced the in vitro findings, but often have clarified their biological meanings and implications. As outlined in Table 4B.1, neutrophils can produce proinflammatory, anti-inflammatory, immunoregulatory, angiogenic and fibrogenic cytokines, chemokines, and ligands belonging to the TNF superfamily. At first sight, an analysis of the table immediately suggests that the ability of neutrophils to produce such a variety of cytokines enables them to significantly influence many processes, not only the multiple aspects of the inflammatory and immune responses, but also hematopoiesis, wound healing, and antiviral defense. For instance, other than typical proinflammatory cytokines such as TNFα and IL-1β, being expected to be produced by inflammatory cells, neutrophils have been shown to also secrete BLyS/BAFF (B-lymphocyte stimulator/B-cell activating factor) and APRIL (a Proliferation-Inducing Ligand). Since BAFF and APRIL are well known to be essential for B-lymphocyte homeostasis and related pathologies, it is plausible to assume a role of neutrophils not only in sustaining B and plasma cell antibody production and survival, but also in promoting B-cell-dependent autoimmune diseases and tumors, as already elegantly demonstrated in the case of B-cell lymphoma. Along the same line, the recent findings that neutrophils exposed to type I and II IFN express and produce another TNF superfamily member selectively involved in tumor cell killing and autoimmunity, namely TRAIL, have opened an additional perspective to exploit neutrophils for novel roles in anticancer responses, whose importance has been recently highlighted in the BCG immunotherapy for bladder cancer. Experimental studies of tumor cure and prevention have, in fact, suggested that, at least in some models, engagement of neutrophil functions can be crucial for the establishment of an effective antitumoral immune response and immune memory reaction. Cytotoxic mediators of tumor and endothelial cell killing produced by neutrophils include TNFα, defensins, proteases (such as elastase and cathepsin G), ROS, nitric oxide, and angiostatic chemokines (CXCL9/MIG, CXCL10/IP-10, and CXCL11/I-TAC). In addition, it has been observed that neutrophils may indirectly generate massive amounts of bioactive, angiostatin-like fragment, inhibiting basic FGF (fibroblast growth factor) plus VEGF (vascular endothelial growth factor)-induced endothelial cell proliferation. However, as described for macrophages, neutrophils can also favor malignant growth and progression, in relation to the type of tumor environment in which they reside: they may do so, for instance, via the production of proangiogenic cytokines such as VEGF and CXCL8, or acting as a distinct myeloid-derived suppressor cell subpopulation releasing, at least in mice, arginase 1–mediated urea and ornithine.

NEUTROPHIL CROSS TALK WITH OTHER CELL TYPES: ROLE OF NEUTROPHIL-DERIVED CHEMOKINES

Amongst the cytokines produced by neutrophils, chemokines are particularly relevant because of their ability to selectively recruit discrete cell populations into sites of injury and thereby effectively regulate leukocyte trafficking. In addition, chemokines play fundamental roles in coordinating the immune system responses, in regulating B- and T-cell development and in modulating angiogenesis. As displayed in Figure 4B.5, activated neutrophils may potentially produce several chemokines, including IL-8/CXCL8, GROα/CXCL1, monocyte chemotactic protein-1 (MCP-1/CCL2), macrophage inflammatory protein-1α (MIP-1α/CCL3), and MIP-1β/CCL4 and MIG/CXCL9, IP-10/CXCL10, and I-TAC/CXCL11. As examples, LPS or TNFα-treated neutrophils produce CXCL8, CXCL1, CCL3, and CCL4; fMLP-treated neutrophils produce CXCL8 only; and IFNγ plus LPS- or IFNγ plus TNFα-treated neutrophils specifically produce, among other chemokines, CXCL9, CXCL10, and CXCL11. Because the chemokines produced by neutrophils are primarily chemotactic for neutrophils, monocytes, DCs, NK, and T-helper type 1 and type 17 cells, a potential role for neutrophils in orchestrating the sequential recruitment to, and activation of, distinct leukocyte types in the inflamed tissue is plausible. Broadly speaking, a general scenario that has already been validated in various experimental models is that, after migration to a focus of infection, neutrophils are stimulated by the etiological agent to initially produce the chemokines that are chemoattractive for neutrophils themselves, including CXCL8. This would

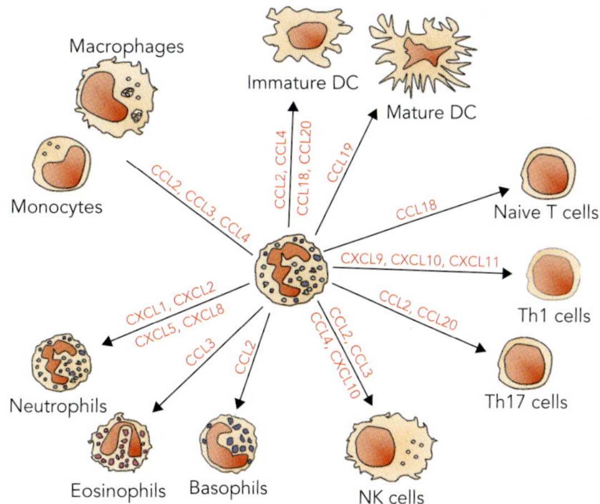

Figure 4B.5. Chemokines produced by neutrophils and their target cells.

activate a positive-feedback loop to induce accumulation of a larger number of neutrophils. Then, specific CC-chemokines would be produced by neutrophils to serve an instrumental role in recruiting the required leukocyte types necessary for the subsequent phases of infection.

Remarkably, neutrophils have also been shown to produce biologically active MIP-3α/CCL20 and MIP-3β/CCL19, two structurally related CC-chemokines that have been suggested to play a fundamental role in trafficking of, respectively, immature and mature DC to mucosal surfaces and lymphoid organs. DCs are professional antigen-presenting cells that are pivotal in the induction of T-cell responses to combat infection. Immature DCs are scattered throughout the peripheral tissues, where they sample antigens and process these into peptides that are loaded onto MHC II molecules for presentation to T cells. Such a process occurs during maturation of DC, which is paralleled by upregulation of MHC II molecules, co-stimulatory molecules and proinflammatory cytokines that enable DC to stimulate pathogen-specific T cells efficiently. DC can produce IL-12 in response to intracellular bacteria and instruct Th1 polarization to produce IFNγ and trigger effective cellular immune responses. By contrast, DC instruct Th2 polarization in response to extracellular parasites, to produce IL-4, IL-5, and IL-13 and to induce appropriate anti-parasite immunity. Thus, DC tailor immune responses to the type of pathogen for rapid and complete clearance of infection. It has been postulated that neutrophil-derived MIP-3α/CCL20 and MIP-3β/CCL19 might contribute to the recruitment of DC, at various stages of maturation, to sites of inflammation and disease for cross talks with immunocompetent T lymphocytes and regulation of the immune response. Likewise, several antimicrobial compounds released by neutrophils, for instance, α-defensins, lactoferrin, LL-37, and cathepsin G (Figure 4B.1) have been found to act as chemoattractants for immature DC and T cells. In addition, neutrophils proteolytically activate prochemerin to generate chemerin, one of the few chemokines that attracts both immature DC and plasmacytoid DC. On the other hand, immature DC are also known to produce CXCL8/IL-8 early upon stimulation, thereby attracting more neutrophils for a more proximal colocalization. This ensures, for instance, that neutrophils and DC are present at the same place at the same time during pathogenic challenge to enable cross talk between these cells. Current evidence now indicates that neutrophils can indirectly play a role in adaptive immunity, not only by recruitment of immune cells, such as DC or specific T lymphocyte subsets, for instance Th17 or Th1, but also by instructing directly DC and inducing adaptive immune responses. Upon coculture of immature DC with activated neutrophils, in fact, DC maturate, as demonstrated by upregulation of HLA-DR, the maturation marker CD83, the co-stimulatory molecules CD86 and CD40, and IL-12-production, and, in turn, trigger T-cell proliferation and strong Th1 cell responses. Such neutrophil-induced maturation of DC is in part mediated by neutrophil-derived TNFα, as well as by cellular contacts regulated by a defined set of receptors expressed by both neutrophils and DC. This is also reflected in vivo, for instance during Crohn's disease, which is characterized by inflammation of the colon mucosa and Th1 responses, in which DC and neutrophils can be found in close proximity.

Recent reports suggest that neutrophils might travel to the lymph nodes during infections and express MHC II and co-stimulatory molecules, in the latter case when appropriately activated by IFNγ, TNFα, and GM-CSF. However, whether neutrophils transmit signals to naive T cells remains still puzzling. It is possible that within the lymph nodes neutrophils undergo apoptosis and are taken up by DC. As a consequence, DC can present PMN-derived antigens to T cells. Other evidence indicates that neutrophils have also a role in directing T-cell polarization, for instance through their capacity to produce the Th1-inducing cytokine, IL-12. The latter has been clearly demonstrated in mice, in which strong Th1-dependent T-cell responses that result in pathogen clearance are elicited upon infection with *C. albicans*, *Helicobacter pylori*, or *Legionella pneumophila*: strikingly, depletion of neutrophils reverses the Th1 responses into a predominant Th2-response, therefore making the mice susceptible to infection. Another mechanism whereby neutrophils may interact with Th1 cells is through the production of chemokines such as CCL3/CCL4, CXCL9, CXCL10, and CXCL11, which act on CCR5 and CXCR3 of Th1 cells but not on the chemokine receptors expressed by Th2 cells. Indeed, secretion of these chemokines has been shown to enable neutrophils to augment Th1 cell responses by preferential attraction of Th1 cells to the sites of infection. Therefore, it is tempting to speculate that the production of cytokines and chemokines by neutrophils not only influences the development and control of Th1 responses, but may also be involved in the cross talk with Th17 cells. Finally, it has also been proposed that mature postmitotic neutrophils can also "transdifferentiate" into much-longer-lived cells with macrophage- or DC-like characteristics, which might constitute a further way for neutrophils to act as regulatory cells of the adaptive immune response.

NEUTROPHIL APOPTOSIS

As previously mentioned, neutrophils are short-lived cells. In culture, neutrophils undergo spontaneous death, with a half-life of around 12–16 hours. It is believed that this behavior reflects a phenomenon that has adapted to ensure a strict control on neutrophil turnover and maintain the number of neutrophils relatively stable in healthy individuals. Deregulation of the neutrophil death rate may in fact cause unwanted and exaggerated inflammation, autoimmunity, or cancer, instead of providing a perfect balance between their immune functions and their safe clearance. An accelerated neutrophil death may lead to a decrease of neutrophil counts (neutropenia), which augments the chance of contracting bacterial or fungal infections and impairing the resolution of such infections, while a delayed neutrophil death elevates neutrophil counts (neutrophilia). The importance of appropriately controlling the turnover of neutrophils is clearly manifest at inflammatory foci, where it contributes to limit neutrophil activities and minimize the risk of externalization of excessive amounts of cytotoxic molecules (i.e., from the mobilization of neutrophil cytoplasmic granules). Conversely, the removal of apoptotic neutrophils is thought to be another important step in preventing the release of granule contents and cytoplasmic danger signals into the extracellular fluid, thereby halting further injury during resolution of inflammation. One can, in fact, imagine that suppression or delay of the basal apoptotic rate of PMN by an agent would be deleterious for the host during inflammation, because PMN might remain activated and thus would perpetuate tissue damage by releasing their toxic products. A paradigmatic example for complete resolution of inflammation, without relevant subsequent lung injury, is pneumococcal pneumonia, which is characterized by an extensive neutrophil accumulation that is followed by neutrophil apoptosis and timely removal of apoptosing cells by lung macrophages.

Neutrophil spontaneous death in culture shares many features of classical apoptosis, such as cell body shrinkage, cellular crenation, exteriorization of phosphatidylserine (PS) from the inner to the outer leaflet of the plasma membrane, exposition of "eat-me" signals for phagocytosis by scavengers, vacuolated cytoplasm, mitochondria depolarization, nuclear condensation, and internucleosomal DNA fragmentation. Neutrophil apoptosis proceeds, in vitro, through an execution phase during which cell dismantling is initiated, with or without fragmentation into apoptotic bodies but with maintenance of a near-intact cytoplasmic membrane. This cascade is ultimately followed by a transition to a necrotic cell elimination traditionally called "secondary necrosis." Secondary necrosis involves an activation of self-hydrolytic enzymes and swelling of the cell or of the apoptotic bodies, as well as a generalized and irreparable damage to the cytoplasmic membrane culminating with cell disruption. For instance, elimination of alveolar macrophages in a model of pneumococcal pneumonia was shown to affect the clearance of accumulated neutrophils and to result in extensive apoptotic secondary necrosis of neutrophils followed by an exaggerated lung inflammation and increased lethality.

Neutrophil apoptosis is also associated with the loss of receptor expression (for instance, CD16) and greatly reduced responsiveness to external stimuli, as revealed by impaired chemotaxis, respiratory burst, degranulation, phagocytosis, and cytokine synthesis. It is thus plausible that, within an inflammatory context, neutrophils become functionally isolated from their environment. Another form whereby neutrophils can die, for instance following a dramatic insult by an exterior aggression damaging enough to produce irreversible alterations, is primary necrosis. In this mode of cell death, there is an immediate, extensive, and irreparable damage to the cytoplasmic membrane associated with loss of plasma membrane integrity, so that release of harmful neutrophil contents is not prevented. Interestingly, some of the features of classical apoptosis, including those observed in constitutive neutrophil death, are also shared by autophagic cell death, another recently characterized physiological cell death process. Autophagy is a caspase-independent, cell death program, featured by the degradation of cellular components in autophagic vacuoles within the dying cell and little chromatin condensation. The autophagosomes fuse with endosomes or lysosomes to form amphisomes or autolysosomes, respectively. Lysosomal proteases ultimately degrade the luminal content of the autophagic vacuole of the autolysosome. In vivo, it has been demonstrated that cells dying through autophagy are phagocytosed by surrounding cells. In primary human neutrophils, this form of death has been shown to be induced by Siglec-9 (sialic acid binding immunoglobulin-like lectin-9) ligation, with the concurrent stimulation with certain proinflammatory cytokines. While there is almost nothing known about the role and the mechanisms triggering autophagy in neutrophils, one critical question is whether the autophagic-like neutrophil death shares the anti-inflammatory properties of apoptosis or whether it induces additional inflammation.

Whatever the case is, the key role of apoptosis in resolution of neutrophilic inflammation has gained increasing attention over the past few years and is appreciated as crucial to allow safe clearance of potentially dangerous neutrophils from tissues. In view of its relevance, studying and discovering molecules that modulate the PMN apoptotic rate is of great importance in biology and medicine. Many of the molecular events involved in the apoptosis pathway have been already identified, with their exquisite regulation becoming better appreciated. For instance, evidence now suggests that neutrophil death is mediated by a complex network of intracellular death/survival signaling pathways, which can be modulated by a variety of extracellular stimuli. Signals such as adhesion, transmigration, hypoxia, prosurvival factors, and cytokines (LPS, G-CSF, GM-CSF, and IFNγ), can all delay neutrophil apoptosis and extend their lifespan, thereby enhancing the ability of neutrophils to function during inflammatory challenge. Even glucocorticoids have been shown to delay neutrophil apoptosis in vitro, a finding which might have potential clinical implications. For instance, while in severe asthma, eosinophils and neutrophils are often found together, neutrophils may gradually replace eosinophils in proportion to the severity and/or duration of the disease, perhaps reflecting the ability of corticosteroids to induce eosinophil apoptosis while inhibiting this process in neutrophils. By contrast, death signals (for instance, "death" receptor engagement, cytokine depletion, or ingestion of bacterial pathogens such as *Escherichia coli* or *S. aureus*) intervene to trigger apoptosis once neutrophil function is complete, and prepare their phagocytic removal. At the molecular level, neutrophil apoptosis is a highly and exquisitely regulated process involving many molecules and proteolytic cascades such as FasL, TNFα, TRAIL, ROS, caspases, IAP (inhibitor of apoptosis protein), survivin, and Bcl-2 family members such as Mcl-1 and A1. Both intrinsic (via mitochondria) and extrinsic (via death receptor signaling) apoptotic pathways have been described in neutrophils. For instance, neutrophils often undergo apoptosis after phagocytosis, in a ROS-dependent fashion and following the cleavage and activation of caspases 8 and 3. A novel, recently identified but unexpected player in the regulation of PMN apoptosis is CDK (cyclin-dependent protein kinase), which has been shown to induce PMN apoptosis and overrides the effect of potent survival mediators, such as LPS and GM-CSF, by a mechanism involving caspase activation and reduction of Mcl-1. The possible use of synthetic CDK inhibitors in the nonproliferating neutrophils is exciting and opens new opportunities of therapeutic strategies, for instance in models of pulmonary inflammatory diseases.

Although it is of great importance to discover and explain the mode of action of apoptotic neutrophil modulators, it is also of major interest to fully understand the mechanisms involved in the recognition of apoptotic PMNs. First of all, apoptotic cells release lysophosphatidylcholine to attract monocytes and macrophages. To allow macrophages to ingest apoptotic PMN, and differentiate them from viable cells, neutrophil membrane is altered in the way that they express specific molecules and receptors. Recognition and engulfment is, for instance, facilitated by the loss of phospholipids asymmetry and phosphatidylserine (PS) exposure ("eat-me signal") to the outer leaflet of the apoptotic cell surface. However, this phenomenon does not seem to be sufficient for the recognition of apoptotic cells. Other proteins, such as calreticulin or annexin-1, the latter being externalized on apoptotic cell membrane colocalized with PS, might participate in the removal of neutrophils by macrophages. Several other receptors have been reported to play a role in the ingestion of apoptotic cells by macrophages: some of them involved in tethering to the phagocyte and others in signaling pathways leading to actin cytoskeleton rearrangement and engulfment during membrane extension and fusion. These receptors on phagocytes include the controversial PSR, the vitronectin receptor, scavenger receptors like CD36 and CD14, lectins, Mer, integrins ($\alpha_v\beta_5$ and $\alpha_v\beta_3$), CR2 (complement receptor-2), CR3, β2-GPI receptor, and CD91. They can interact directly with apoptotic cells or indirectly through bridging proteins, like C1q, β_2-GPI, Gas6, and MFG-E8. In addition, components of the innate immune system play a role in regulating apoptotic cell clearance since the opsonization of apoptotic cells increases their removal by macrophages. Interestingly, although the presence of such "eat me" signals has been largely documented, the presence of "don't eat me" signals on the surface of viable nonapoptotic cells has also been recognized. In this respect, CD47 or CD31 have been identified as "don't eat me" molecule and the proposed model is that loss or inactivation of CD47 effects accompanies apoptosis and allows the apoptotic cells to be recognized and cleared. On the other hand, it appears that this phenomenon of discrimination between apoptotic and viable cells involves redistribution of eat-me signals into patches (to increase their avidity toward receptors) as well. Engulfment of apoptotic cells by macrophages (also called "efferocytosis") is influenced by cytokines such as IL-1 and TNFα, glucocorticoid hormones, prostaglandins, and also interaction of surface adhesion molecules with neighboring cells and extracellular matrix components. Lipoxins also stimulate macrophages to phagocytose apoptotic PMN. Expression of 15-lipoxygenase on nonphlogistic macrophages, which had phagocytosed apoptotic PMN, induces lipoxin A4 release, thus resulting in a decreased PMN recruitment and in an increase in apoptotic PMN removal. After apoptotic cell ingestion, macrophages phenotypes switch to induce tissue repair or emigrate in the lymphatic system, suggesting that phagocytosis of apoptotic cells is involved in the negative regulation of macrophage activation and that PMN apoptosis could be considered as endogenous "active" anti-inflammatory process. Furthermore, phagocytosis of apoptotic neutrophils by human macrophages in vitro suppresses the release of macrophage-derived proinflammatory mediators such as TNFα, IL-1β, or GM-CSF and it induces the release of anti-inflammatory mediators, like TGF-β1, IL-10, PGE_2, leukocyte protease inhibitor, and reparative growth factor. This process thus confers an anti-inflammatory phenotype to macrophages that helps to lead to resolution of inflammation, but that could predispose to development of autoimmunity if not efficient.

NEUTROPHILS IN HUMAN DISEASES

Previous sections have already summarized some of the most common inherited disorders concerning the functions of neutrophils that impair critical responses for host defense. Typically, patients with defects in neutrophil function present, in infancy or childhood, with recurrent and/or difficult to treat bacterial infections. The microorganisms causing these infections are often unusual or opportunistic pathogens and typically infect the skin, mucosa, gums, lung, or draining lymph nodes, or cause deep tissue abscesses. However, many of these disorders have characteristic clinical and microbiological features that are related to the specific nature of the defect in neutrophil function.

Other than in inherited impairments of neutrophil functions, dramatic clinical consequences may be observed also during acquired neutropenia. For instance, susceptibility to infectious diseases increases sharply when neutrophil counts fall below 1,000 cells/μL. When the absolute neutrophil count falls to < 500 cells/μL, control of endogenous microbial flora (e.g., mouth, gut) is critically impaired; when it is < 200/μL, the inflammatory process is absent. Neutropenia can be due to depressed production (i.e., hereditary neutropenias), increased peripheral destruction, or excessive peripheral pooling. The most common neutropenias are iatrogenic, resulting from the use of cytotoxic or immunosuppressive therapies for malignancy or control of autoimmune disorders. Interestingly, acute neutropenia, such as that caused by cancer chemotherapy, is more likely to be associated with increased risk of infection than neutropenia of long duration. Recombinant human G-CSF usually reverses this form of neutropenia. A neutrophilia occurs instead when the number of circulating neutrophils dramatically increase (even up to 10-fold), for instance from increased neutrophil production, increased marrow release, or defective margination. The most important acute cause of neutrophilia is infection, which is caused by both increased production and increased marrow release of neutrophils. Increased marrow release and mobilization of the marginated leukocyte pool are also induced by glucocorticoids. Persistent neutrophilia with cell counts of 30,000–50,000/μL is called a leukemoid reaction, a term often used to distinguish this degree of neutrophilia from leukemia. In a leukemoid reaction, the circulating neutrophils are usually mature and not clonally derived.

Apart from the disorders associated with genetic dysfunctions or quantitative alterations of neutrophils, there are certain pathological situations in which neutrophils themselves become the predominant contributors to tissue injury, especially when the mechanisms supposed to control and inactivate their hypothetical beneficial and protective effector functions are deregulated. For instance, the latter may occur when the acute insult cannot be resolved, when the shut off of the neutrophil influx (that may involve inactivation of proinflammatory mediators and/or temporal change in the pattern of chemokines produced) is impaired, or when the safe clearance of dying neutrophils from the inflammatory site is altered. An uncontrolled production and release of oxidants and proteases by neutrophils, acting either individually or in concert, appear to be responsible for much (but not all) of the tissue injury observed in such cases. Examples of such pathologies, tentatively classified according to the major neutrophil-activating event (some of them covered in other chapters), are (i) diseases caused by ischemia-reperfusion injury (i.e., myocardial infarction, transplantation); (ii) bacterial infections (adult respiratory distress syndrome, endotoxic shock, osteomyelitis, endocarditis, acute and chronic pyelonephritis); (iii) cytokine-mediated diseases (rheumatoid arthritis, inflammatory bowel diseases); (iv) diseases caused by crystal deposition (gout, articular chondrocalcinosis); (v) immune complex–mediated diseases (vasculitis, lupus, Goodpasture's syndrome); (vi) anti-neutrophil cytoplasmic antibody (ANCA)-associated vasculitis (Wegener's granulomatosis, pauci-immune necrotizing crescentic glomerulonephritis); (vii) airway diseases in which considerable observational and experimental data support an association between neutrophils and the severity and progression (chronic obstructive pulmonary disease, bronchiectasis, bronchiolitis, cystic fibrosis, and even certain forms of asthma are characterized by neutrophil infiltration of the airway wall). One of the key challenges in neutrophil-dominated conditions is how to manipulate neutrophil function to abolish their destructive potential in a way that does not compromise their antibacterial and antifungal capacity. This has been difficult to achieve, as successful treatments in animal models have frequently proven ineffective or limited by side effects when used in human inflammatory diseases.

CONCLUSIONS

Contrary to what is traditionally thought, the neutrophil is now recognized as a highly versatile and sophisticated cell with significant synthetic capacity and an important role in linking the innate and adaptive arms of the immune response. Even though the ability of neutrophils to transcribe many genes is no longer a matter of debate, the research conducted in the recent years has brought forward exciting discoveries that have greatly broadened our knowledge on the functional role of this cell type and uncovered novel links involving neutrophils to unsuspected physiopathologic processes. For instance, evidence on their capacity to change phenotype under specific circumstances; or on their active involvement in the resolution of inflammation (other than in its regulation); or also on their unquestionable regulatory role in angiogenesis and tumor fate; or else on their response to, and release of, a wide variety of cytokines and chemotactic molecules have made it clear that the obsolete concept of the neutrophil as a "terminally differentiated, synthetically inert cell" found in most biomedical textbooks is clearly no longer tenable. Time is now mature for pathologists, cell biologists, physicians, and, why not, the same immunologists to definitively change such an outdated view and start thinking the neutrophil biology in a more modern way.

SUGGESTED READINGS

Borregaard, N., Sørensen, O.E., and Theilgaard-Mönch, K. 2007. Neutrophil granules: a library of innate immunity proteins. *Trends Immunol* **28**(8):340–345.

Cassatella, M.A., ed. 2003. The neutrophil: an emerging regulator of inflammatory and immune response. *Chem Immunol Allergy* **83**:1–225.

Luo, H.R., and Loison, F. 2008. Constitutive neutrophil apoptosis: mechanisms and regulation. *Am J Hematol* **83**(4):288–295.

Nathan, C. 2006. Neutrophils and immunity: challenges and opportunities. *Nat Rev Immunol* **6**(3):173–182.

van Gisbergen, K.P., Geijtenbeek, T.B., and van Kooyk, Y. 2005. Close encounters of neutrophils and DCs. *Trends Immunol* **26**(12):626–631.

Witko-Sarsat, V., Rieu, P., Descamps-Latscha, B., Lesavre, P., and Halbwachs-Mecarelli, L. 2000. Neutrophils: molecules, functions and pathophysiological aspects. *Lab Invest* **80**(5):617–653.

5 Mast Cells as Sentinels of Inflammation

Joshua A. Boyce

INTRODUCTION

Mast cells are tissue-dwelling hematopoietic effector cells that are endowed with a range of potent inflammatory effector molecules. They are implicated in both allergic and nonallergic diseases, as well as in innate and adaptive immunity to infectious agents based on animal studies. Mast cells constitutively reside in a perivascular distribution in connective tissues and at mucosal surfaces (Figure 5.1). They are especially abundant in tissues that form interfaces with the external environment (skin, conjunctivae, intestinal, and airway mucosa), suggesting a strategic placement so as to function in a first line of host defense. Mast cells are best known for their role as effectors of classic type 1 hypersensitivity (allergic) reactions. In such reactions, mast cell activation is initiated by the binding of multivalent allergen to membrane-bound IgE that is coupled with the tetrameric high-affinity Fc receptor for IgE (FcεRI) on mast cells. IgE-dependent activation of mast cells results in their release of preformed inflammatory mediators that are stored in their secretory granules, including histamine, neutral proteases, preformed cytokines, and proteoglycans (Figure 5.2). In addition, activated mast cells secrete newly synthesized lipid mediators that are the products of endogenous arachidonic acid metabolism, such as prostaglandin (PG)D_2, leukotriene (LT)B_4, and LTC_4, the parent molecule of the cysteinyl leukotrienes (cys-LTs). Finally, activated mast cells synthesize and secrete a host of proinflammatory cytokines over a period of hours. Thus, the net result of IgE-dependent tissue mast cell activation includes the rapid development of plasma extravasation, tissue edema, bronchoconstriction, leukocyte recruitment, and persistent inflammation. These events contribute to the pathogenesis of anaphylaxis, urticaria, angioedema, and acute exacerbations of asthma, each of which are diseases in which mast cells play a role.

In the past decade, experimental models have linked mast cells and their products in a broad range of inflammatory diseases, as well as host defense. Moreover, numerous activating stimuli for mast cells have been identified that do not require signaling through FcεRI. These findings suggest several contexts in which mast cells contribute to protective immunity, and disease states in which IgE-independent mechanisms of mast cell activation are functionally relevant. It therefore seems likely that mast cell–targeted treatment strategies have potential efficacy in the context of nonallergic inflammatory diseases, as well as in their traditionally accepted place in allergy.

CHARACTERISTICS OF MAST CELLS

Mature mast cells can be found in all vascularized organs. Morphologically, mast cells in tissues range between 7 and 20 μm in diameter and appear as round, spindle-shaped, or spider-like cells in tissues with round or oval nuclei. They have thin 1–2 μm processes (microplicae) emanating from their plasma membranes (Figure 5.3). Ultrastructurally, mast cells have multiple electron-dense granules (Figure 5.3). In mice infected with helminthic worms, mast cells appear in the mucosal epithelium of the intestine with strikingly altered, stellate granule morphology (Figure 5.3). Mast cells are readily identified in tissues using cationic dyes that bind their unique secretory granules. These dyes (toluidine blue, methylene blue) impart a blue-to-purple change in color known as "metachromasia" (Figure 5.1). Of all hematopoietic cells, only basophils share this staining feature. The metachromatic staining of mast cell granules reflects their content of highly sulfated heparin and chondroitin sulfate glyosaminoglycans bound to a protein core, termed "serglycin" due to the presence of alternating serine and glycine residues in its amino acid structure. The sulfated glycosaminoglycans of the mast cell granule

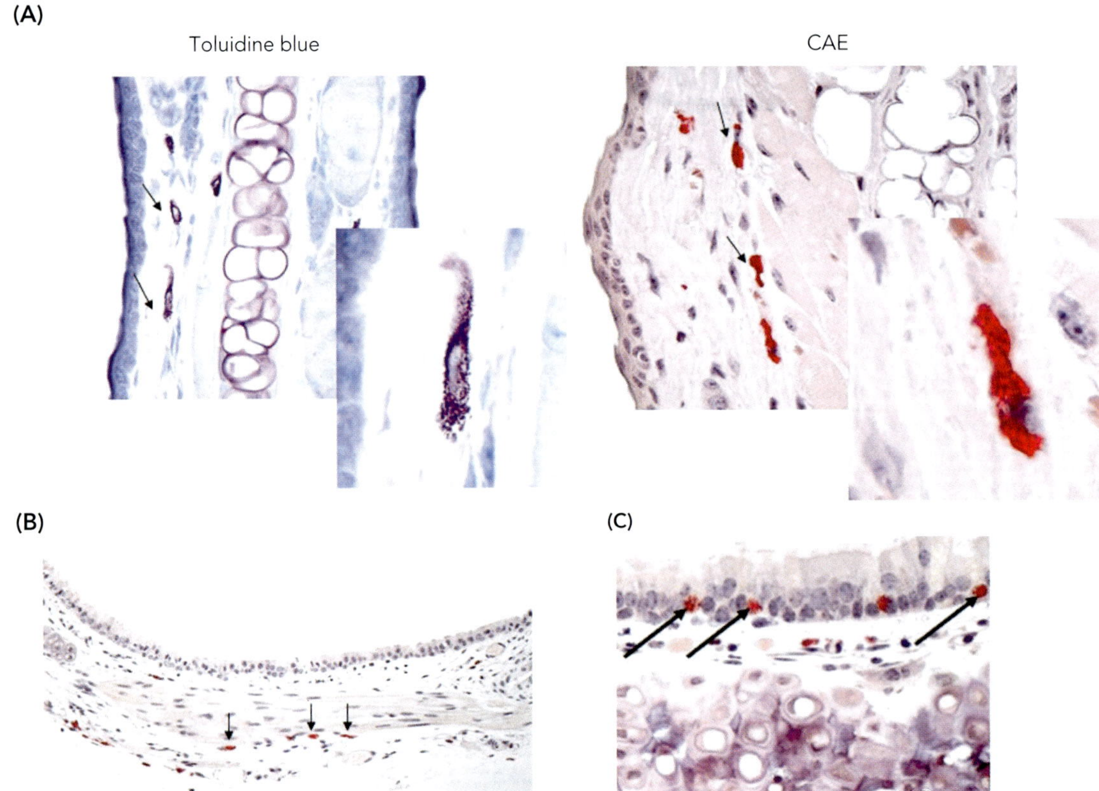

Figure 5.1. Staining features of mast cells. (A) Sections of a mouse ear showing toluidine blue (left) and chloroacetate esterase (CAE) (right) stains. Mast cells are shown using arrows. Higher magnification images (insets) are shown of typical spindle-shaped mast cells in the connective tissue of the ear. (B) CAE stains of a cross-section of a naive mouse trachea showing mast cells (arrow) in proximity to smooth muscle bundles in the submucosa. (C) Trachea of an antigen sensitized and challenged mouse showing CAE-stained intraepithelial mucosal mast cells (arrows). (Courtesy of Wei Xing, M.D., Ph.D., Harvard Medical School.)

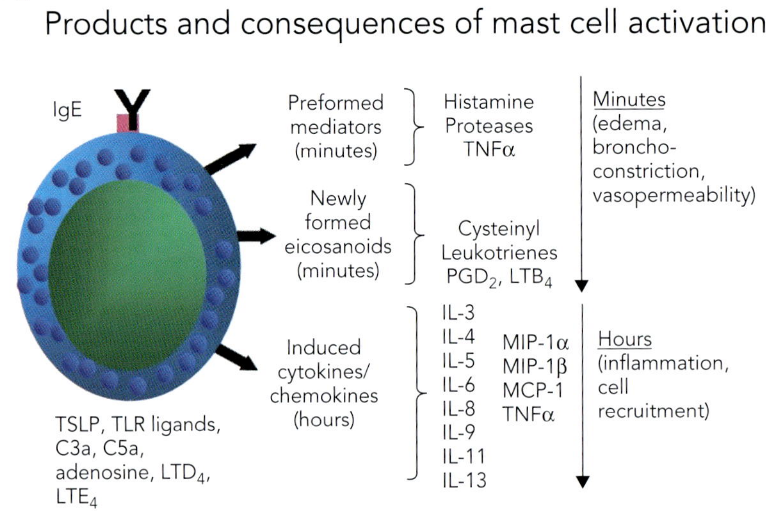

Figure 5.2. Mast cell activation products and bioactive effects. Cross-linkage of FcεRI with multivalent antigen induces exocytosis and eicosanoid formation within minutes, with attendant effects on the vascular and airway smooth muscle. Later, cytokines and chemokines are generated de novo and secreted. Several additional receptor-dependent mechanisms can induce mast cell activation independently of IgE.

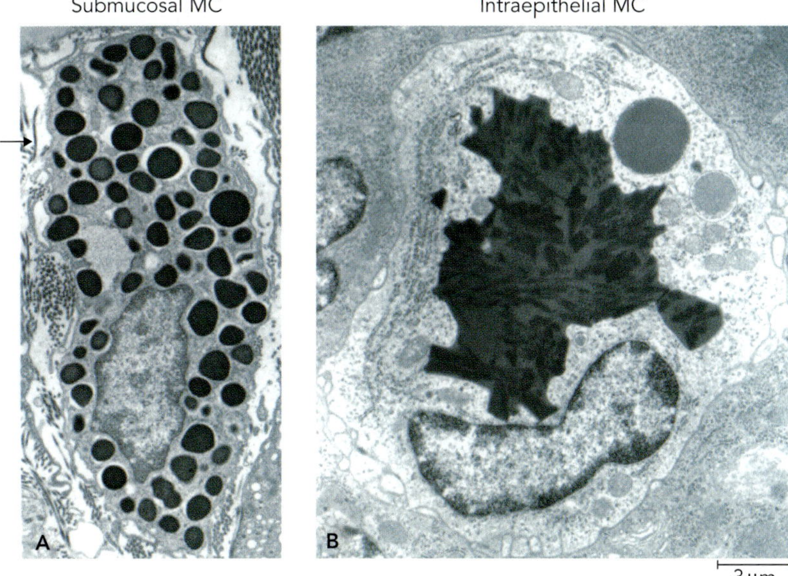

Figure 5.3. Ultrastructure of submucosa (left) and intraepithelial mast cells (right) from the jejunum of a mouse infected with *Trichinella spiralis.* Note microplicae (arrow) and dense cytoplasmic granules in the submucosal mast cell, and the large, stellate granules of the intraepithelial mast cell. (Photomicrograph courtesy of Daniel Friend, M.D., Harvard Medical School.)

core are essential for the normal storage of proteases (tryptases, chymases) and amines (histamine and serotonin). Accordingly, since proteases account for a major portion of the mast cell's granule by weight, mast cells can also be identified by immunostains or biochemical reagents that detect their proteases, such as chloroacetate esterase activity (an indicator of chymotryptic proteases) (Figure 5.1) or antibodies against tryptase or chymase in the human.

Although all known mast cells stain metachromatically with toluidine blue dye, they vary in their protease content, which is the best-established index of mast cell heterogeneity. Subsets of human tissue mast cells that contain tryptases, chymase, cathepsin G, and carboxypeptidase A (CPA) are found in perivascular connective tissues, skin, and muscularis of the uterus and intestine. In contrast, mast cells containing tryptases but lacking chymase, cathepsin G, and CPA are the dominant subtype found in lung, as well as the mucosal surfaces of the gut and nose. Although the mouse genome encodes an array of mast cell–restricted inducible chymases that do not exist in the human, it is noteworthy that two of these proteases (mouse mast cell protease [mMCP]-1 and mMCP-2) are selectively expressed by mast cells arising in the mucosal epithelium of the gut during helminth-induced intestinal inflammation. Humans with deficient T-cell function selectively lack tryptase-positive mast cells in the gut mucosa, whereas mice without functional T cells cannot develop mMCP-1- and mMCP-2-positive mast cells in the intestinal mucosa. Thus mast cell development in mucosal epithelial surfaces is T-cell-dependent, whereas mast cell development in other locations is independent of T cells. Moreover, the profile of protease expression is a marker of the subsets (see the following section). Rodent and human studies indicate that regional differences in protease expression are accompanied by heterogeneity in other effector properties, such as proteoglycan content, cytokine generation, and eicosanoid production. It seems plausible that mast cells can alter their phenotypic properties in response to factors derived from the adaptive immune system so as to fine tune their capabilities in accord with the context. It is important to note that numerous additional members of the tryptase family have been identified in recent years that are also present in mast cell granules. On the basis of precedent in rodents, it seems very likely that human mast cells will prove to be even more heterogeneous than currently thought, again reflecting "fine tuning" of their functions by the microenvironment.

MAST CELL GROWTH AND DEVELOPMENT

Role of the Kit Receptor Tyrosine Kinase and Stem Cell Factor

While mast cells originate from the bone marrow, the fact that they are found only in tissues implies that their terminal development occurs in their ultimate tissue destinations. The molecular basis of mast cell development has been studied primarily in rodents.

The receptor protein-tyrosine kinase Kit (CD117) is encoded by the proto-oncogene c-*kit*. Two mast cell–deficient strains of spontaneously emerging laboratory mice proved to have loss-of-function mutations at the W locus at which c-*kit* resides (termed W/W^v and W^{sh}/W^{sh} mice, respectively). Another strain of mast cell–deficient mouse (Sl/Sl^d) has a loss-of-function mutation at the Steel (Sl) locus that encodes the membrane-bound form of the tissue-associated growth factor, stem cell factor (SCF). Since SCF is the ligand for Kit, the molecular defects of these three mouse strains define the fundamental requirements for mast cell development. Although no mast cell deficiency is reported in humans, a gain-of-function point mutation in Kit is associated with systemic mastocytosis, a disease of increased mast cell burden. This confirms the importance of the SCF-Kit axis in mast cell development across species. SCF is constitutively expressed as a membrane-bound protein by fibroblasts, endothelial cells, and other cell types that juxtapose to mast cells in peripheral tissues. In vitro studies using mouse and human mast cells confirm that soluble SCF protects mast cells from apoptosis, induces proliferation, can cause migration, and can induce some degree of activation and secretion independent of FcεRI and IgE. The mutant mice lacking normal Kit or SCF function are often used for diseases models to implicate functional roles for mast cells.

The identity of the Kit-positive cell population giving rise to mast cells in the tissue became evident over the past three decades as a result of several independent experimental observations. Schrader and colleagues reported that the intestinal mucosa of mice contained cells with lymphocyte-like morphology that were Thy-1 negative and capable of mast cell differentiation in vitro when treated with a splenocyte-conditioned medium. Rodewald and colleagues identified a population of cells isolated from fetal mouse blood that expressed high levels of c-Kit and low levels of Thy-1, which gave rise to pure colonies of FcεRI-positive mast cells when cultured with a combination of recombinant SCF and IL-3. These cells, termed "promastocytes," lacked developmental potential for other hematopoietic lineages. Intraperitoneal injection of purified promastocytes reconstituted the population of peritoneal mast cells in genetically mast cell–deficient W/W^v mice. More recently, fully committed mast cell progenitors, marked by their expression of CD34, FcεRI, and β_7 integrin, were identified in the intestine and in the allergen-challenged lungs of adult mice.

Human mast cell progenitors in the bone marrow and circulation have not been definitively identified but are contained within the CD34+/Kit+ cell subpopulation and are clearly distinct from the basophil lineage. Cells isolated from human blood that are CD34+/Kit+ and also positive for CD13 (aminopeptidase N), give rise to three colony types: mast cell colonies, macrophage colonies, and colonies containing mixtures of both cell types. Thus fully committed human mast cell progenitors are likely a subpopulation of the CD34+Kit+CD13+ population of cells in peripheral blood and in tissues.

Comparatively little is known about the mechanisms that control the trafficking of mast cell progenitors to tissue, their movement within tissue, and their incremental recruitment with inflammation. In mice, the β_7 integrin is instrumental to the homing of mast cell–committed progenitors to the intestine, and is also involved in the de novo recruitment of these cells to the lung during experimentally induced allergic pulmonary inflammation. Subsequent studies have implicated potential roles for the chemokine receptor, CXCR2, and for the inflammatory leukotriene LTB_4 in the recruitment of mast cell progenitors to sites of inflammation. A range of chemokines are active on human and mouse mast cells in vitro, but to date, no mast cell–selective chemokine has been identified.

Accessory Mast Cell Growth Factors

Allergic mucosal inflammation is characteristic of asthma, rhinitis, and other human diseases associated with atopy. Each of these diseases involves increases in the numbers of mast cells in the involved mucosa surface. The likely mechanisms responsible for these increases have been identified largely through studies of helminth infection in rodents. As is the case for human allergic disease, antihelminthic intestinal immune responses typically induce eosinophilia, IgE production, and Th2 cytokine generation, along with striking increases in the numbers of mast cells arising in the intestinal epithelial surface. This mucosal mast cell hyperplasia is essential for the elimination of some helminths. Not surprisingly, mast cell hyperplasia does not occur with normal kinetics or magnitude in mice lacking SCF and Kit. However, mast cell hyperplasia is also severely blunted or absent in T-cell–deficient mice, as well as mice with targeted deletions in the genes encoding IL-3 or IL-4. Antibody neutralization of either IL-3 or IL-4 also decreases the magnitude of helminth-induced mast cell hyperplasia. In contrast to the mucosal surface, the losses of IL-3 or IL-4 do not change the number of mast cells in the intestinal submucosa or in other connective tissue locations. Furthermore, mice with tissue-specific overexpression of an IL-9 transgene display a hyperplasia of mast cells in the mucosal compartments of both the intestine and lung. Thus in vivo studies implicate at least three T-cell–derived cytokines in mast cell hyperplasia that are generated during Th2 immunity in mice.

In vitro studies indicate that IL-3, IL-4, and IL-9 can control the rate of SCF-driven mast cell proliferation (comitogenesis), rate of apoptosis, or expression of certain key genes. IL-3 can substitute for SCF for inducing mast cell growth in vitro from mouse bone marrow (although it lacks this activity in cultures of human bone marrow or cord blood). Although human IL-3 fails to promote mast cell growth by itself, the IL-3 receptor is constitutively expressed by human mast cell progenitors, is inducible on mature human mast cells, and mediates an SCF-driven comitogenic effect on human mast cells and their progenitors in vitro. In the absence of SCF, IL-3 also sustains the survival of human mast cells derived from cord blood or isolated from surgically resected intestinal tissue. IL-9 induces the expression of the mucosal proteases mMCP-1 and mMCP-2 by mouse mast cells in vitro, and upregulates the expression of several mRNA transcripts involved in the control of cell survival and cell cycle progression in human mast cells. IL-10 shares the feature of inducing mMCP-1 and mMCP-2 expression by mouse mast cells with IL-9, and also strongly stimulates histamine synthesis and granule development.

Of the T-cell–derived cytokines, IL-4 may be especially pivotal regulating functional responses of mast cells in mucosal inflammation. First, IL-4 is important for Th2 cell development and the production of IgE. Second, IL-4 has several direct effects on the development and/or function of mast cells. In vitro, IL-4 amplifies SCF-driven proliferation of both mouse and human mast cells in vitro, and strongly upregulates the expression of leukotriene C_4 synthase (LTC_4S), the terminal enzyme needed for the production of LTC_4, the parent of the cys-LTs, in the mast cells of both species. Surprisingly, the induction of LTC_4S is necessary for a normal proliferative response to IL-4, which is markedly blunted in human or mouse mast cells in which cys-LT generation is blocked by either pharmacologic or genetic means. Extracellular cys-LTs can induce mast cell proliferation in vitro by acting at the type 1 receptor for cys-LTs ($CysLT_1R$) and inducing transactivation of Kit. These findings may explain a profound deficiency in intraepithelial mast cells identified in LTC_4S knockout mice subjected to a model of allergen-induced pulmonary inflammation. Collectively, these studies confirm the dynamic nature of mast cell phenotype and its dependency on tissue-specific determinants such as local cytokines, and implicate a major potential role for cys-LTs as signaling molecules in amplifying mast cell development with mucosal inflammation.

Transcription Factors and Signaling Molecules

The microphthalmia transcription factor (MITF) is a basic helix–loop–helix leucine zipper-type protein that is central to mast cell and melanocyte development. Mice with a loss-of-function mutation in the *mitf* gene have reduced the number of tissue mast cells, albeit not to the same extent as mice bearing mutations in the *c-kit* or *scf* genes. MITF controls c-*kit* transcription in mast cells, and regulates the expression of several proteases. MITF is also important in controlling the expression of hematopoietic prostaglandin D_2 synthase (H-PGDS), the terminal enzyme required for the synthesis of prostaglandin D_2 (PGD_2), the most abundant cyclooxygenase product of mast cells. The guanine nucleotide exchange factor RasGRP4, which is highly mast cell restricted, is also required for the normal expression of H-PGDS, suggesting that RasGRP4 and MITF might cooperate in a coordinated transcriptional system. To date, the parallel roles of MITF and RasGRP4 have not been defined in human mast cells.

The GATA family of zinc finger transcription factors are also involved in mast cell development. GATA-1 is involved in the differentiation and survival rates of mast cell precursors, and promotes expression of the FcεRI α- and β-chains. GATA-2 is important for the differentiation of mast cells from yolk sac precursors. The GATA consensus sequence is found in the promoter region of genes expressed by mast cells, including CPA, several proteases, and FcεRIα. PU.1, an Ets family transcription factor, may also play a role in the regulation of mast cell development, possibly in cooperation with GATA-2.

MAST CELL ACTIVATING RECEPTORS

FcεRI

FcεRI-dependent mast cell activation is a primary mechanism by which allergens induce effector responses in allergic disease. All tissue mast cells express the variant of FcεRI composed of an Fc-binding α subunit, a β subunit, and two γ subunits. Occupied by monomeric allergen-specific Ig, FcεRI is cross-linked by multivalent allergens, resulting in signal transduction events that result in immediate granule fusion and exocytosis, arachidonic acid metabolism, and induction of cytokine and chemokine gene transcription. The importance of these processes to the pathophysiology of allergic responses is undisputed. Early and late-phase allergic responses to allergen challenge in the nose and lung of atopic individuals are blocked by the administration of cromolyn, a drug that inhibits mast cell activation, and can also be blocked by the administration of a therapeutic anti-IgE antibody. IgE-driven early and late-phase reactions induced experimentally in mice are also mast cell dependent.

While the role of FcεRI in allergic disease is well established, its role in the normal function of the immune system is less clear. Mice genetically deficient in FcεRI develop normally and have normal T- and B-cell functions, and can eliminate *Schistosoma mansoni* from the intestine with normal kinetics after experimental infection. Nonetheless, they do develop increased hepatic granulomas and hepatic fibrosis. Deficiencies in mast cells or IgE also impair the ability of mice to eliminate *Haemaphysalis longicornis* ticks, suggesting a role for IgE-mediated hypersensitivity for resistance against this parasite. The elimination of adult *T. spiralis* worms from the intestine is delayed in an IgE-null strain of mice, and the number of viable larvae encysting in the skeletal muscles doubled in the IgE-null mice compared to infected wild-type controls. Thus, IgE may serve an amplifying function in the containment and/or elimination of certain multicellular parasites. However, there is no infectious disease recognized to date that absolutely requires IgE or FcεRI for protective immunity.

FcεRI expression on the surface of mast cells and basophils is regulated at several levels. FcεRI is stabilized at the cell surface on encounter with IgE. Peritoneal mast cells from IgE-null mice show strikingly lower levels of surface FcεRI than do wild-type littermates, which increase with the intravenous administration of IgE. Circulating basophils from atopic individuals show levels of surface FcεRI that correlate with circulating levels of IgE. Moreover, depletion of IgE in humans using a therapeutic antibody sharply decreases FcεRI surface expression by circulating basophils and mast cells in both the lungs and the skin. Surface expression of FcεRI on mast cells ex vivo are increased, in a dose-dependent manner, by the addition of IgE. Incubation of in vitro bone marrow–derived mast cells with IgE enhanced not only their FcεRI expression but also signal transduction and mediator release in response to FcεRI cross-linking. Certain clones of IgE can also induce cytokine secretion and sustain the survival of mouse mast cells independently of antigen in vitro. It is not clear whether this mechanism operates in vivo or in humans.

Cytokines can also regulate FcεRI expression and signaling. Human mast cells derived in vitro respond to the addition of exogenous IL-4 with dose- and time-dependent enhancement of their surface FcεRI expression. Mast cells from human intestine have similar responses to IL-4. In both instances, the effect of IL-4 is to drive transcription of FcεRIα, providing more subunits for surface expression. Not surprisingly, this mechanism is synergistic with the stabilization of surface receptors by the addition of exogenous IgE. Both mechanisms amplify mediator release in response to cross-linkage of FcεRI.

Fcγ Receptors

In rodents, anaphylaxis can occur through IgG-dependent activation of the low-affinity receptor for IgG, FcγRIII. This receptor shares β and γ subunits with the FcεRI, but has not been identified on human mast cells. Human mast cells exposed ex vivo to IFNγ inducibly express high-affinity FcγRI receptors, and generate the same mediator profile in response to cross-linkage of this receptor as they do via FcεRI. The importance of mast cell activation by these IgG-dependent pathways in human health and disease is unknown.

Pattern Recognition Receptors

In recent years, several studies have linked the activities of mast cells to the innate immune response, reflecting their capacity for activation in response to pathogen-derived molecules and endogenously generated mediators. On the basis of studies of *c-kit* mutant mice, mast cells are required for protective neutrophilia occurring in response to experimentally induced Gram-negative septic peritonitis and pneumonia. The activation of mouse mast cells in response to virulent strains of *Klebsiella pneumoniae* was linked to CD48, a glycosylphosphatidylinositol-anchored protein. Mouse mast cells also express a number of Toll-like receptors (TLRs). In Gram-negative septic peritonitis, expression of TLR4 by mast cells is required to induce normal protective neutrophilia for microbial clearance. Skin mast cells are activated by bacterial peptidoglycan through TLR2 to mediate the influx of granulocytes to the site of intracutanous challenge with staphylococcal peptidoglycan.

G-Protein–Coupled Receptors

The complement fragments C3a and C5a ("anaphylotoxins") are established agonists for mast cell activation that can be generated as by-products of both adaptive and innate immune responses. Each induces signaling through a specific GPCR. Human mast cells stimulated by exogenous cys-LTs secrete cytokines and chemokines but do not release histamine. LTD_4, the initial extracellular metabolite of LTC_4, triggers mast cell activation through the type 1 receptor for cys-LTs ($CysLT_1R$). LTE_4, the terminal product of the cys-LTs, also activates human mast cells ex vivo to generate PGD_2 and chemokines, but does so through a pathway involving an uncharacterized GPCR that induces a secondary signal through peroxisome proliferator-activated receptor (PPAR)γ, a lipid-activated transcription factor. Mast cell activation through the adenosine A3 receptor can also induce mediator release that potentiates airway reactivity in mice. It is likely that these GPCR-dependent mechanisms for mast cell activation can elicit or modify mediator release in nonallergic diseases and host defense responses that involve contributions from mast cells.

Effector Cytokine Receptors

Recently, tissue-derived effector cytokines have been identified as potentially major inducers of mast cell activation. IL-33, which signals through T1/ST2, and thymic stromal lymphopoietin (TSLP), which signals through a heterodimeric receptor sharing a subunit with the IL-7 receptor, are both generated by injured or inflamed epithelium. Mast cells express the receptors for both cytokines, and both can potently induce mast cells to generate cytokines and chemokines. Thus, IL-33 and TSLP provide potential mechanisms whereby mucosal injury can induce a direct response from mast cells without a requirement for antibody or FcεRI.

INHIBITORY RECEPTORS

Immunoglobulin-like and Fc Receptors

Mast cells express several receptors that inhibit their activation, potentially contributing to homeostasis in mast cell–dependent inflammatory responses. Mouse mast cells possess at least four inhibitory receptors that contain an immunoreceptor tyrosine-liked inhibitory motif (ITIM). These include the closely related low-affinity Fc receptors for IgG, FcγRIIb1 and FcγRIIb2, leukocyte immunoglobulin-like receptor B4 (LILRB4, previously called GP49B1), and CD300a. Each of these receptors inhibits exocytosis and eicosanoid generation when colligated with FcεRI. Mice with targeted disruption of FcγRIIβ or LILRB4 show exaggerated experimentally induced IgE-dependent anaphylactic responses in vivo. RNA transcripts for LILRB4 have been detected in human mast cells, but their function has not been confirmed. Unlike FcγRII, the natural counterligands for LILRB4 and CD300a are not known.

G-Protein–Coupled Receptors

Mast cell activation can be inhibited by GPCRs that couple to G_s proteins and adenylyl cyclase (AC). Stimulation of mast cells through the β_2 adrenergic receptor, the A2B receptor for adenosine, and the EP_2 receptor for PGE_2 all block IgE-dependent mast cell activation responses. Inhalation of PGE_2 by subjects with asthma prevents both early and late asthmatic responses. A recent study revealed that the A2B receptor for adenosine plays a major role in regulating endogenous levels of cyclic AMP in mouse mast cells, and A2B knockout mice display exaggerated IgE-dependent anaphylactic responses. Since GPCRs are good targets for pharmacologic compounds, these G-coupled receptors all represent promising therapeutic targets, and agonists of the β_2 adrenergic receptor are stables of asthma treatment.

MAST CELL PATHOBIOLOGY

Anaphylaxis

IgE-dependent activation of mast cells (and possibly of basophils) is thought to underlie the pathophysiology of all systemic allergic reactions. The panel of mediators derived from mast cells include those that induce urticaria (histamine), vascular leak (histamine, LTD_4, PGD_2), hypotension (histamine), bronchoconstriction (histamine, LTC_4, PGD_2), flushing (PGD_2), and intestinal peristalsis (LTC_4), all of which are associated signs or symptoms of anaphylaxis. The serum levels of mature β-tryptase, a protease stored in abundance in the granules of mast cells, are elevated during the first 2–4 hours after many anaphylactic events. Because of its relatively long half-life when compared with histamine, β-tryptase is a more useful marker of mast cell activation in vivo than is histamine. Interestingly, "pseudoallergic" or "anaphylacticoid" systemic reactions to cyclooxygenase inhibitors and radiocontrast media, which are not thought to be mediated by IgE, can also be associated with elevated levels of β-tryptase, indicating the likely involvement of mast cells in these syndromes after their activation by idiosyncratic mechanisms. Surprisingly, β-tryptase levels are often not elevated in anaphylaxis caused by food ingestion. Although this may indicate a principal role for basophils, the lack of basophil-specific markers of activation precludes a definitive explanation at this time.

Asthma

Most evidence from human studies supports a role for mast cells and their mediators as effectors in asthma. Bronchial biopsies from patients with asthma contain increased numbers of mast cells relative to biopsies obtained from nonasthmatic control subjects, irrespective of whether the asthmatic patients are atopic or nonatopic. The increased numbers of mast cells reflect distributions in the mucosal epithelium and in the bronchial smooth muscle layer, the latter being a region in which mast cells are rarely found in nonasthmatic individuals or patients with eosinophilic bronchitis (a syndrome of bronchial mucosal inflammation without AHR or airflow obstruction). In contrast, substantial numbers of mast cells are found at this tissue site in asthmatic patients. The increase in the numbers of mast cells in the mucosal epithelial surface of biopsy specimens from asthmatic patients likely reflects the T-cell–driven pathway that amplifies mast cell development. The numbers of mast cells in bronchial biopsy specimens from asthmatic patients predict therapeutic failure during weaning from glucocorticoids.

Mast cell degranulation is prominent in postmortem bronchial tissues from patients who die from asthma and is also observed in bronchial biopsy specimens from living asthmatics. The levels of tryptase and other mast cell mediators often are markedly increased in the bronchoalveolar lavage (BAL) fluid of asthmatic patients relative to their levels in specimens obtained from nonasthmatic controls, even without allergen provocation. Thus, both mast cell hyperplasia and ongoing mast cell activation are characteristic features of asthma. Therapeutic anti-IgE is an efficacious treatment for atopic asthma. However, it is unlikely that FcεRI-dependent mechanisms account for the entirety of mast cell contributions to asthma, particularly since mast cell numbers (and indices of local mast cell activation) are increased in both atopic and nonatopic individuals.

Mastocytosis

Mastocytosis refers to a group of disorders that are characterized by increased mast cell burden in the skin and/or internal organs. When mastocytosis is limited to the skin (cutaneous mastocytosis), it may be isolated to one lesion (mastocytoma), or more commonly is manifested by multiple brown-to-tan, flat lesions (urticaria pigmentosa) that swell with gentle pressure (Darier's sign). The lesions of urticaria pigmentosa are usually present in cases of systemic mastocytosis as well, defined by mast cell infiltration of internal organs (lymph nodes, spleen, bone marrow). As many as 80% of patients with systemic mastocytosis bear a somatic mutation of the *c-kit* gene resulting in a single amino acid substitution (D816V) that renders the Kit receptor constitutively active. There are several subtypes of systemic mastocytosis (indolent, aggressive, and others) that vary in clinical presentation and prognosis, all of which are associated with this single mutation, raising the likelihood of additional genetic or environmental factors that modify the clinical course. Unfortunately, the D816V variant of Kit resists inhibition by currently available tyrosine kinase inhibitors. Most of the signs and symptoms of systemic mastocytosis (flushing, pruritis, headache, diarrhea and abdominal pain, and hypotension) are attributable to nonspecific mast cell mediator release. Total pro-β-tryptase levels (but not mature β-tryptase levels) are generally elevated in systemic, but not cutaneous, mastocytosis, as a reflection of total body mast cell burden.

Other Allergic Conditions

All atopic diseases (i.e., associated with allergen-specific IgE) involve mast cell activation. These include allergic rhinitis, rhinosinusitis, allergic conjunctivitis, oral allergy syndrome, and atopic dermatitis. The comparative success of H1 histamine receptor blockade for the treatment of these conditions provides strong support for the pathogenetic contributions of mast cells. In addition, diseases in which allergen-specific IgE is not demonstrable, such as chronic urticaria, urticaria, or anaphylaxis provoked by physical stimuli (such as exercise, cold exposure, or heat exposure) also involve contributions from mast cells. In most of these circumstances, the mechanism of activation remains obscure but may involve autoantibodies or neural circuits.

Atherosclerosis

Degranulated mast cells characteristically present in atherosclerotic plaques and sites of plaque rupture. The serum of patients presenting with acute myocardial infarction show elevations of both histamine and mature β-tryptase, suggesting mast cell activation occurs concomitantly with ischemia. Mast cell granule constituents (heparin, proteases) can contribute to foam cell formation and cholesterol accumulation through effects on lipoproteins, and plaque instability through effects on the extracellular matrix and inhibition of smooth muscle proliferation. Mast cell chymase also facilitates angiotensin-II formation and endothelin metabolism. Mast cell–derived eicosanoids can serve as vasoconstrictors and can promote neutrophil recruitment. Thus, there are several potential mechanisms by which mast cells could contribute to the pathogenesis of acute vascular events.

Roles Implicated by Animal Models

The increasing use of *c-kit* mutant mice in experimental models has led to a steadily increasing number of diseases involving mast cells. As noted earlier, mast cells are essential for the initiation of protective innate immunity to bacteria in certain experimental models. This function in part reflects their ability to store and rapidly generate and release TNFα. More recently, a critical role for the tryptase mMCP-6 in septic peritonitis was inferred from studies of mice lacking this protease. Other studies have linked the protective neutrophilia to the production of LTB_4, a powerful neutrophil chemotactic factor. Models of contact dermatitis, inflammatory arthritis, aortic aneurisms, bullous pemphigoid, and multiple sclerosis all implicate potential roles for mast cells as effectors. Although evidence for mast cell contributions to these diseases in humans is inferential at present, it seems possible that mast cells may prove to be therapeutic targets in at least some of these disorders, provided the responsible mediators and activation mechanisms become evident. Since animal models also support prominent roles for mast cells and their mediators in protective immunity to helminths, bacteria, and a range of other pathogens, the challenge for therapeutic development is to maintain these benefits without sacrificing efficacy.

SUGGESTED READINGS

Bischoff, S. C., Sellge, G., Lorentz, A., et al. 1999. IL-4 enhances proliferation and mediator release in mature human mast cells. *Proc Natl Acad Sci USA* **96**(14):8080–8085.

Enerback, L. 1966. Mast cells in rat gastrointestinal mucosa. I. Effects of fixation. *Acta Pathol Microbiol Scand* **66**(3):289–302.

Godfraind, C., Louahed, J., Faulkner, H., et al. 1998. Intraepithelial infiltration by mast cells with both connective tissue-type and mucosal-type characteristics in gut, trachea, and kidneys of IL-9 transgenic mice. *J Immunol* **160**(8):3989–3996.

Hsieh, F. H., Lam, B. K., Penrose, J. F., et al. 2001. T helper cell type 2 cytokines coordinately regulate immunoglobulin E-dependent cysteinyl leukotriene production by human cord blood-derived mast cells: profound induction of leukotriene C(4) synthase expression by interleukin 4. *J Exp Med* **193**(1):123–133.

Jiang, Y., Feng, C., Backskai, B., and Boyce, J.A. 2007. CysLT2 receptors interact with CysLT1 receptors and down-modulate cysteinyl leukotriene-dependent mitogenic responses of mast cells. *Blood* **100**:3263–3270.

Jiang, Y., Kanaoka, Y., Feng, C.L., Nocka, K., Rao, S., and Boyce, J.A. 2006. Cutting edge: interleukin 4-dependent mast cell proliferation requires autocrine/intracrine cysteinyl leukotriene-dependent signaling. *J Immunol* **177**:2755–2759.

Kambe, N., Hiramatsu, H., Shimonaka, M., et al. 2004. Development of both human connective tissue-type and mucosal-type mast cells in mice from hematopoietic stem cells with identical distribution pattern to human body. *Blood* **103**(3):860–867.

Kirshenbaum, A.S., Goff, J.P., Semere, T., et al. 1999. Demonstration that human mast cells arise from a progenitor cell population that is CD34(+), c-kit(+), and expresses aminopeptidase N (CD13). *Blood* **94**(7):2333–2342.

Kitamura, Y., and Go, S. 1979. Decreased production of mast cells in S1/S1d anemic mice. *Blood* **53**(3):492–497.

Kitamura, Y., Go, S., and Hatanaka, K. 1978. Decrease of mast cells in W/Wv mice and their increase by bone marrow transplantation. *Blood* **52**(2):447–452.

Li, L., Yang, Y., and Stevens, R.L. 2003. RasGRP4 regulates the expression of prostaglandin D2 in human and rat mast cell lines. *J Biol Chem* **278**(7):4725–4729.

Madden, K.B., Urban, J.F., Jr., Ziltener, H.J., et al. 1991. Antibodies to IL-3 and IL-4 suppress helminth-induced intestinal mastocytosis. *J Immunol* **147**(4):1387–1391.

Morii, E., and Oboki, K. 2004. MITF is necessary for generation of prostaglandin D2 in mouse mast cells. *J Biol Chem* **279**(47):48923–48929.

Paruchuri, S., Jiang, Y., Feng, C., Francis, S.A., Plutzky, J., and Boyce, J.A. 2008. Leukotriene E4 activates peroxisome proliferator activated receptor gamma and induces prostaglandin D2 generation by human mast cells. *J Biol Chem* **283**:16477–16487.

Reynolds, D.S., Stevens, R.L., Lane, W.S., et al. 1990. Different mouse mast cell populations express various combinations of at least six distinct mast cell serine proteases. *Proc Natl Acad Sci USA* **87**(8):3230–3234.

Rodewald, H.R., Dessing, M., Dvorak, A.M., et al. 1996. Identification of a committed precursor for the mast cell lineage. *Science* **271**(5250):818–822.

Ruitenberg, E.J., and Elgersma, A. 1976. Absence of intestinal mast cell response in congenitally athymic mice during *Trichinella spiralis* infection. *Nature* **264**(5583): 258–260.

Schrader, J.W., Scollay, R., and Battye, F. 1983. Intramucosal lymphocytes of the gut: Lyt-2 and thy-1 phenotype of the granulated cells and evidence for the presence of both T cells and mast cell precursors. *J Immunol* **130**(2):558–564.

Schwartz, L.B., Metcalfe, D.D., Miller, J.S., et al. 1987. Tryptase levels as an indicator of mast-cell activation in systemic anaphylaxis and mastocytosis. *N Engl J Med* **316**(26):1622–1626.

Toru, H., Ra, C., Nonoyama, S., et al. 1996. Induction of the high-affinity IgE receptor (Fc epsilon RI) on human mast cells by IL-4. *Int Immunol* **8**(9):1367–1373.

Vyas, H., and Krishnaswamy, G. 2006. Paul Ehrlich's 'Mastzellen' from aniline dyes to DNA chip arrays: a historical review of developments in mast cell research. *Methods Mol Biol* **315**:3–11.

Weidner, N., and Austen, K.F. 1991. Ultrastructural and immunohistochemical characterization of normal mast cells at multiple body sites. *J Invest Dermatol* **96**(3 Suppl):26S–31S.

6 Basophils

Jonathan Arm and David Sloane

INTRODUCTION

Named after their affinity for alkaline stains and possessed of numerous potent inflammatory substances, basophils are the rarest of the granulocytes, typically constituting 0.5%–1.5% of peripheral blood leukocytes. Studied both for their suspected effector roles in parasitic and allergic diseases and for their similarity to mast cells, these peripheral blood cells are increasingly hypothesized to have important immunoregulatory functions. This chapter focuses primarily on the human basophil, but makes reference to the basophils of other species to point out phenotypic differences from human basophils, to describe phenomena that have been elucidated best in nonhuman basophils, and to describe murine and other mammalian models of disease in which the role of the basophil has been explored.

IDENTIFICATION

Historical Discovery

Paul Ehrlich, the German Nobel prize winner who made foundational contributions to immunology, is credited with the first description of the human basophil in 1879 in a paper in which he also proposed the name *Mastzellen* for connective tissue cells that also took up aniline dyes in their numerous cytoplasmic granules. Because of their infrequency among peripheral blood leukocytes, basophils were given scant attention until their functional similarity to mast cells in terms of IgE-mediated histamine release was appreciated more than eight decades later.

Development

Like other leukocytes, basophils derive from totipotent hematopoietic stem cells, and are believed to arise from the common myeloid precursor by means of the common granulocyte precursor. However, the penultimate and ultimate steps in basophil development are presently unclear. There are two models of basophil ontogeny, both of which are supported by experimental data. The first posits that basophils and mast cells share a common parent cell distinct from that which produces eosinophils. This hypothesis is most strongly supported by a murine model in which a bipotent cell could differentiate into either mast cells or basophils in the mouse spleen. The discovery of a cell with ultrastructural features of mast cells and basophils in humans with chronic myelogenous leukemia agrees with this model. However, a genetic analysis using the D816V mutation in the c-kit receptor in patients with the disease systemic mastocytosis as a trackable marker did not suggest a close lineage relationship between mast cells and basophils, as basophils were found to carry the mutation in a minority of patients (5 of 33) and only when other lineages (monocytes, neutrophils) also harbored D816V. If a common basophil mast cell precursor cell exists, it would be expected to carry the D816V mutation more frequently in mastocytosis and would occur at least occasionally in the absence of other lineage involvement.

The second, competing hypothesis is that there is a basophil–eosinophil precursor distinct from a cell that gives rise to mast cells. This hypothesis is supported by the finding that peripheral blood cells from atopic donors can, when cultured in methylcellulose and with conditioned media, produce colonies from single cells that are either pure basophils or mixtures of basophils and eosinophils, suggesting the presence of a circulating basophil–eosinophil precursor. Quantification of such precursors demonstrates that atopic patients have fluctuating numbers of these bipotent progenitors, with fewer observed during steroid treatment and more detected after allergen challenge. Similar cells have been obtained from the bone marrow and peripheral blood of patients with myeloid leukemias, and hybrid cells with histochemical and ultrastructural features of both basophils and eosinophils can be derived from normal human cord blood precursors.

Drawing firm conclusions from these studies is limited by the potentially confounding effects of pathological conditions in human donors (allergic diseases, mastocytosis, malignancy), medications donors take to treat these diseases, and the lack of details and variation in culture conditions (cytokines, growth factors, and conditioned media) among studies. The possibility that there may be two or more basophil subpopulations, one more closely related to mast cells, another more akin to eosinophils, has not been explored, but would resolve the apparently conflicting data and would suggest that basophils resemble dendritic cells in their developmental heterogeneity. An alternative reconciling hypothesis is that basophils and/or eosinophils may, in some contexts, (de)differentiate into hybrid cells, but definitive studies documenting this are presently lacking.

The current dogma posits that basophils exit the bone marrow as fully differentiated cells, as they are identified in peripheral blood relatively easily, and once isolated are functional for mediator release. However, basophils in inflamed tissues have distinct biochemical properties and may survive significantly longer than those in the circulation, suggesting that blood basophils may be partially functional cells that only fully mature after diapedesis or that they have a more plastic phenotype than currently appreciated.

Phenotypic Characteristics

Appearance

As at the time of their discovery, basophils are still identified routinely by their microscopic appearance. In addition, their identity is assessed by analysis of their pattern of expression cell surface proteins.

Light Microscopy Image

Basophils from peripheral blood avidly take up alkaline stains, such as alcian blue and trypan blue. Importantly, the granules of basophils (and mast cells) acquire a purple, not blue (as seen in other granulated cells such as monocytes and neutrophils), color upon interaction with these dyes; hence, the staining is termed "metachromatic" (Figure 6.1A). Basophils have numerous, moderately sized, metachromatic cytoplasmic granules and a bean-shaped or bilobed nucleus. Immunohistochemically, peripheral blood basophils from normal donors generally have a very low content of tryptase and do not contain chymase, classical markers of the mast cell.

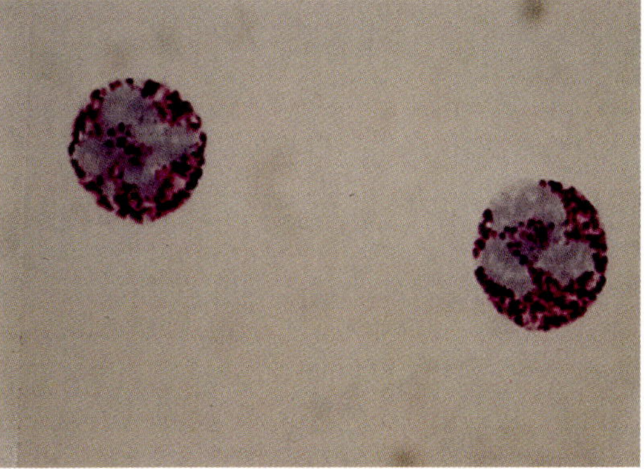

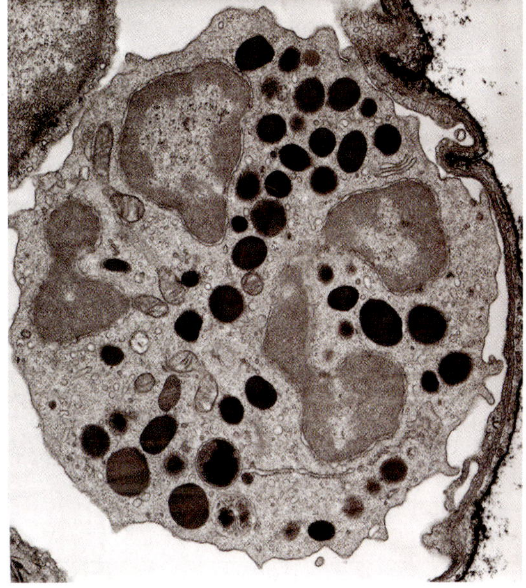

Figure 6.1. (A) Light micrographs of human peripheral blood basophils stained with Wright's Giemsa. Note the numerous dark purple, metachromatic granules, and the bilobed or trilobed (close up) nucleus. (B) Electron micrograph of a human basophil in a capillary.

Electron Microscopy Image

Viewed by electron microscopy, human basophils have a diameter of 5–7 microns, numerous amorphous round granules, and "stubby" cell surface processes (Figure 6.1B). Compared to basophils, tissue mast cells are larger; have more numerous, smaller, and more electron dense granules that have either a lattice or scroll substructure; and possess distinctly longer and thinner plasma membrane processes. Basophils undergo dramatic ultrastructural changes with activation, but the pattern and degree of alteration vary depending on the particular stimulus used. Thus, for example, "anaphylactic degranulation," with granule-to-granule fusion or granule-to-plasma membrane fusion was seen best with stimuli that engage the high-affinity Fc receptor for IgE (FcεRI), while recombinant histamine releasing factor was most potent in inducing uropod formation, and the more gradual process of piecemeal degranulation was observed after stimulation by monocyte chemotactic protein-1.

Cell Surface Markers

Basophils are often identified by their pattern of plasma membrane proteins. Like mast cells, basophils have high cell surface binding of IgE by means of their expression of FcεRI. CD123, the α chain of the IL-3 receptor, is also easily detected on these cells, as are CD11b and CD13. Often, the positive expression of these two proteins is used in conjunction with the absence of expression of a panel of lineage markers (such as CD2, CD14, CD16, CD19, and MHC class II) to identify basophils on flow cytometry (Figure 6.2).

Over a decade ago, the murine monoclonal antibody 2D7 was found to bind to the secretory granules of basophils (but not those of eosinophils, neutrophils, or mast cells). Likewise, the antibody BB1+ has been reported to be basophil specific. Antibody J175–7D4 stains basophils based on its binding to promajor basic protein, which is not expressed by mature eosinophils.

Cell surface protein expression is highly dynamic. With basophil activation through cross-linking of FcεRI, there is increased expression of a number of basophil cell surface markers, including CD13, CD63, CD107a, CD164, and CD203C. This observation is the basis of various *in vitro* basophil activation assays that have been used to identify circulating allergen/antigen-specific IgE. Expression of specific cell surface receptors through which basophils may be activated is described later.

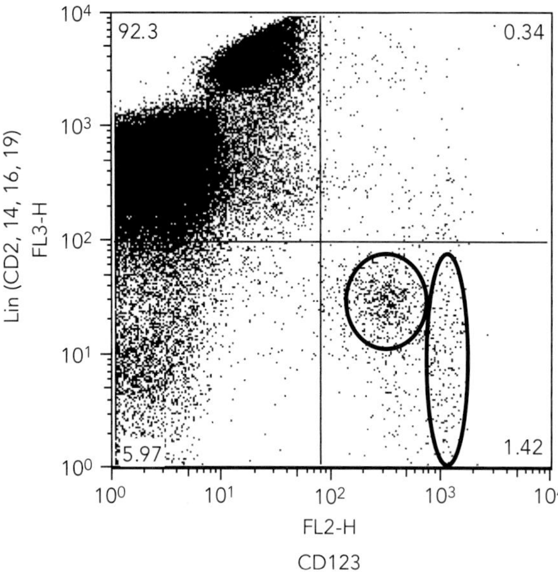

Figure 6.2. Flow cytometric identification of peripheral blood basophils. Basophils (right lower quadrant, within the bold circle) are typically positive for CD123 (the IL-3 receptor α chain, x-axis) and negative for the lineage markers IL-2, -14, -16, and -19 (y-axis). The cells within the ellipse are plasmacytoid dendritic cells, which may be distinguished from basophils by their expression of other cell surface markers such as blood dendritic cell antigen-2 (BDCA-2, CD303), and BDCA-4 (neuropilin-1, CD304), not shown here.

Products of Basophils

Histamine

Histamine, the best known mediator released by basophils, is present preformed in cytoplasmic granules, and is rapidly released in response to stimuli that induce degranulation. While the platelets of some species (e.g., rabbits) also contain histamine, basophils seem to be the exclusive source of this mediator in human peripheral blood. The histamine content of basophils is typically on the order of 1.0 pg per cell. When quantifying the release of histamine from basophils, results are often reported as a percent of histamine release, with the total histamine content being determined by complete lysis of the cells using hypotonic and/or acidic media. Thus, for example, extensive cross-linking of FcεRI on a population of basosphils with allergen, anti-IgE, or anti-FcεRI may result in 50%–90% histamine release, with exocytosis beginning within 2–5 minutes of activation and completed by 30–45 minutes.

Lipid Mediators

Unlike histamine, the major lipid mediators of basophils, leukotriene C4 (LTC_4), and platelet activating factor (PAF) are newly synthesized upon cell activation. Although not preformed, the kinetics of their release after basophil activation with physiological and nonphysiological stimuli rivals that of histamine, initially being detected 1–5 minutes after activation and approaching maximal release by 15 minutes. When purified basophils are activated, the released LTC_4 (<5–80 ng/million basophils) remains stable in solution. However, when basophils are selectively stimulated in the presence of other leukocytes, such as the mononuclear cells (monocytes, T cells, B cells, NK cells, and dendritic cells) with which they copurify during density gradient centrifugation, there is further metabolism of LTC_4 to the most stable of the cysteinyl leukotrienes, LTE_4. Unlike mast cells, basophils lack the enzyme hematopoietic prostaglandin (PG)D_2 synthase, and are thus incapable of generating PGD_2.

Cytokines and Chemokines

The discovery that basophils release significant amounts of important immunoregulatory proteins is arguably the seminal finding that rescued the cell from being relegated to the status as a "mere" innate immune effector cell or surrogate mast cell. It also justified the exploration of what roles basophils might play in the initiation, maintenance, and resolution of inflammation, especially that of allergic diseases.

Chief among the cytokines produced by basophils is interkeukin-4 (IL-4), which is released by basophils in response to stimulation both through FcεRI (by means of specific allergen, anti-IgE, or anti-FcεRI – see the discussion of this receptor later) and via IgE-independent stimuli such as the complement fragment C5a. Basophils possess mRNA encoding IL-4 at baseline, but cell activation augments this by more than threefold. The mass of IL-4 released varies among donors, ranging from less than 20 to more than 600 pg/million basophils. The kinetics of secretion are slower than that of histamine and the lipid mediators, with a half time to maximal release of approximately 1.5 hours and maximal release generally seen by 4–8 hours. The inhibition of IL-4 release by basophils in the presence of cyclohexamide indicates that new protein synthesis is required.

Basophils also release IL-13 (<20–2,000 pg/million cells) in response to IgE cross-linking and to non-IgE-mediated stimulation. As with IL-4, new protein synthesis is required, but maximal release may not be achieved until 16–24 hours, depending on the agonist. One study of normal and asthmatic subjects revealed that during the first 6 hours after specific allergen challenge *in vitro*, basophils outnumbered T cells by fourfold as sources of IL-4 and by twofold as sources of IL-13.

Basophils release macrophage inflammatory protein-1α (MIP-1α) in amounts and according to kinetics similar to those of IL-13 secretion, and are also sources of the cytokine GM CSF chemokines produced by basophils in response to IgE cross-linking include IL-8, MIP-5, and eotaxin.

Others

Although best known as the products of eosinophils, major basic protein and Charcot–Leyden crystals containing lysophospholipase are produced by basophils. Likewise, basophils contain small amounts of mast cell tryptase, and release this mediator in response to anti-IgE. Recently, the cytotoxic serine protease granzyme B has been detected from activated basophils, and basophils were able to directly kill target cells at least in part by means of this mediator, in a caspase-dependent mechanism and to a degree rivaling natural killer (NK) cells.

ISOLATION TECHNIQUES

Positive Selection

Basophils comprise 0.5%–1.5% of circulating leukocytes. They can be enriched modestly by density gradient centrifugation over Ficoll or Percoll, during which they co-sediment with the mononuclear cells with rather efficient removal of contaminating polymorphonuclear neutrophil (PMN) and eosinophils. This is sufficient for many purposes, in which selective basophil activation can be effected by stimulation through FcεRI and monitored by measuring histamine release. Employing multistep density gradient centrifugation, peripheral blood mononuclear cells can be enriched for basophils from their baseline frequency 10–20 fold. Thereafter, basophils can be purified further by countercurrent elutriation. Relying on the rarity of circulating mast cell precursors, the absence of FcεRI expression by resting eosinophils, and the relative paucity of FcεRI expression by plasmacytoid dendritic cells, early techniques positively selected basophils by means of magnetic microbeads coated with anti-FcεRI. Once bound, the basophils are separated from the non-basophils by application to a column placed in a magnetic field that restrains the microbeads while all unbound cells flow through the column.

Negative Selection

Because positive selection techniques that depend on ligation of cell surface receptors such as FcεRI risk activating the cells, they have largely been replaced by the inverse method – the binding and removal of all non-basophils in a sample to purify unbound basophils by negative selection. Such an approach depends on a panel of cell surface markers strongly and consistently expressed by eosinophils, neutrophils, monocytes, NK cells, T cells, B cells, and various subtypes of dendritic cells. To succeed in purifying basophils, an antibody against at least one marker on each non-basophil must be included. In this case, magnetic beads bind to non-basophils and the cell mixture is again subjected to column filtration. The unlabeled basophils flow through the magnetic field and are collected while other cells are trapped. Basophils of >95% purity and >50% recovery can be achieved using negative selection techniques.

STIMULI FOR SECRETION

FcεRI and ITAM-Linked Receptors

FcεRI has been the most extensively studied cell surface receptor on the basophil. As in mast cells, this receptor is composed of an α chain whose extracellular domain binds to the CH3 region of the IgE heavy chain, a β chain, and two copies of a γ chain shared with numerous other receptors such as those for IgG. Although other cells, including eosinophils and plasmacytoid dendritic cells, express the α and γ chains, and thus bind IgE and react to its cross-linking, they lack the β chain. The intracellular domains of the β and γ chains contain immunoreceptor tyrosine-based activating motifs (ITAMs). Cross-linking of IgE results in the phosphorylation of critical tyrosine residues in these motifs, providing docking sites for proteins such as the tyrosine kinase Syk and the induction of a complex intracellular signaling network that culminates in mediator release. In particular, it was first demonstrated in basophils (and later confirmed in mast cells) that the expression of FcεRI is dynamic and directly proportional to the ambient concentration of IgE. This has been exploited clinically with the development of therapy for allergic diseases such as asthma and food allergy with monoclonal antibodies directed against FcεRI. These antibodies bind to circulating IgE and thereby prevent its interaction with FcεRI. Not only does this lead to reduced binding of specific IgE to FcεRI, but also reduced expression of FcεRI, which together lead to blunted IgE-dependent responses. The presence of FcεRI is often detected by means of flow cytometry using a monoclonal antibody directed against the extracellular portion of the α chain, such as 22E7 (which binds to the α chain whether occupied by IgE or not) and 15A5 (which only binds unoccupied α chain), or by binding of labeled IgE. Because of the dynamic expression of FcεRI, basophils may have anywhere from a few thousand to hundreds of thousands of these receptors on their surface. The cross-linking threshold for mediator release from a given basophil may likewise range broadly from a few hundred to tens of thousands of FcεRI receptors.

Basophils also consistently express the Ig superfamily member leukocyte immunoglobulin-like receptor A2 (LILRA2, also known in alternative nomenclature systems as LIR7, ILT1, and CD85h). The ligands for this receptor are as yet unidentified, but cross-linking by means of a monoclonal antibody results in mediator release from all three basophil "compartments" – preformed histamine from granules, newly synthesized LTC_4, and IL-4. As with FcεRI, Syk tyrosine kinase is involved in signal transduction from LILRA2, as predicted for a receptor that also employs the common γ chain.

GPCR

Formyl Peptide Receptors

The tripeptide formylated methionine-leucine-phenylalanine (fMLP) is a bacterial product capable of inducing histamine release and LTC_4 generation by basophils. However, as a direct agonist, it is unable to elicit significant release of the cytokines IL-4 and IL-13 from these cells. Basophils consistently express at least three formyl peptide receptors: FPR, FPRL1, and FPRL2.

C5a

After FcεRI, the most potent activating basophil receptor is that of the complement fragment C5a. Agonization of C5aR leads not only to mediator release from all three basophil compartments when basophils are primed or stimulated in the presence of IL-3, but to chemotaxis as well. Basophils also express C3aR, the receptor for a second anaphylotoxin, but this molecule is a less potent basophil secretogogue when compared to C5a.

Chemokines

The impressive (and likely as yet incompletely determined) variety of chemokine receptors expressed on the surface of basophils allows them to respond to a broad gamut of factors that influence cell trafficking (see Table 6.1). Many ligands for basophil surface receptors have both chemotactic or chemokinetic and variable prosecretory activities, making impossible a sharp

TABLE 6.1. Basophil GPCR receptors and their counterligands

Receptor	Ligands
Platelet activating factor receptor	PAF
C5aR	C5a
FPR, FRPL1, FPRL2	fMLP, urokinase plasminogen activator (uPA)
CCR1	MIP-1α, MCP-3
CCR2	MCP-1 (CCL2), MCP-3
CCR3	Eotaxin (CCL11), RANTES, MCP-3
CXCR1	IL-8
CXCR2	Nap-2
CXCR4	Stromal cell–derived factor 1 (SFD1, CXCL12)
Oxoeicosanoid receptor (OXE)	5-oxo-ETE

distinction among these components of cell activation. Thus, for example, C5a and fMLP elicit both mediator release from and chemotaxis of basophils, while IL-3 augments basophil chemokinesis but not chemotaxis. Basophils express the stem cell factor (SCF) receptor c-kit, allowing SCF, which is not directly chemotactic, to augment basophil movement in response to C5a, MIP-1α, MCP-1, and eotaxin. Some chemotactic factors promote only selected mediator release, as exemplified by the abilities of MIP-1α, MCP-1, and eotaxin, which all direct basophil movement, to elicit LTC_4 generation but not IL-4 or IL-13 production.

One study of the transendothelial migration of basophils using human vascular endothelial cells (HUVEC) documented a tenebrous array of molecules that initiate or influence basophil movement, often with complex additive or synergistic effects and distinct (sometimes partially overlapping) time courses of maximal receptor expression and ligand potencies.

Basophils express $\alpha_4\beta_1$ (CD29/CD49d) and β_2 integrins (specifically CD11a/CD18 and CD11b/CD18), both of which contribute to diapedesis by allowing basophils to bind VCAM-1 and ICAM-1 and adhere to endothelial cells. As noted earlier, basophils produce a number of chemotactic factors, many of which may bind to their receptors contributing to an autocrine and/or a positive feedback system in which basophils promote their own recruitment. The results of numerous studies suggest a convoluted system of receptors and ligands that orchestrate the movement, modulate the activation, and promote the longevity of basophils in inflamed tissues. Although labyrinthine in its details, this network resembles that regulating eosinophils, a fact that has been emphasized by those who assert a close onotologic relationship between these two granulocytes.

CRTH2

Among the most prominent biochemical differences between mast cells and basophils is the ability of the former and the inability of the latter to produce the lipid mediator PGD_2. However, basophils express the CRTH2 receptor (chemotactic receptor homologous molecule expressed on TH2 cells), which is the second receptor for PGD_2, DP2. Thus, basophils can respond to this chemotactic factor, suggesting that activation of mast cells (e.g., through IgE by allergen or parasite antigen) may lead to the recruitment of basophils to inflamed tissues. Agonization of CRTH2/DP2 also leads to Ca^{2+} influx, augmentation of degranulation induced by anti-IgE and by fMLP, and increased expression of CD11b.

Histamine and Leukotriene Receptors

Interestingly, basophils possess cell surface receptors for the inflammatory mediators they produce at least one for histamine (H_2) and two for LTC_4 (CysLTR1 and CysLTR2). While this suggests autocrine actions of histamine and cysteinyl leukotrienes, the implications of this have not been fully elucidated. In addition, expression of BLT1 allows LTB4 to induce degranulation of basophils.

Cytokines

IL-3 is the best studied cytokine influencing basophils, and the α chain of the IL-3 receptor (CD123) is so strongly and consistently expressed on the cell surface that it is often used as a marker to help identify basophils. Exposure of basophils to IL-3 has profound effects on both cell viability and mediator release in response to activation. IL-3 at a concentration of 10 ng/mL retards basophil apoptosis and augments the release of histamine, LTC4, and IL-4 in response to IgE-dependent and -independent stimuli by "priming" (e.g., a 15-minute exposure before activation). At this and higher concentrations, IL-3 can elicit some mediator release on its own. *In vitro* experiments suggest that IL-3 is necessary (and possibly sufficient) to drive precursor cells toward the basophil phenotype.

Basophils also express receptors for IL-2 (CD25), IL-4, IL-5, and IL-8. The influence of these cytokines as well as of GM-CSF and SCF on basophil development and the longevity and function of mature basophils are less well established than those of IL-3. IL-5, and GM-CSF are able to prime basophils for LTC_4 synthesis and the release of IL-4 and IL-13 in response to C5a, but with less potency than IL-3. Unlike IL-3, these cytokines do not elicit mediator release directly. Nerve growth factor (NGF) augments mediator release by basophils in response to C5a.

Toll-Like Receptors

Basophils have been shown to express both mRNA and protein for Toll-like receptors (TLRs), TLR2 and TLR4. In importance, extensive experience demonstrates little reproducible effects of the TLR4 ligand lipopolysaccharide (LPS) on basophils, possibly because these cells do not express CD14, the LPS co-receptor present on LPS responsive cells such as macrophages. In contrast, the TLR2 ligand peptidoglycan directly induces IL-4 and IL-13 secretion from basophils. This activation appears to be cytokine selective, as it occurs in the absence of any histamine or LTC_4 release, though peptidoglycan augments their release in response to anti-IgE.

Histamine Releasing Factors

Originally named for their ability to induce degranulation, these enigmatic mediators also induce IL-4 release from basophils and augment cytokine release in response to IgE-mediated stimulation of these cells. Early reports suggested the existence of at least two histamine releasing factors (HRFs), one that depends on cell surface IgE and at least one that is IgE-independent. Some HRF activity was subsequently attributed to the action of chemokines. HRFs are produced by a variety of cells, such as endothelial cells, platelets, monocytes, and B lymphocytes. Much about HRFs remains to be explored. The receptors for HRF are currently unknown, and basophils from some donors do not respond to HRF. One particular HRF, human recombinant HFR (HrHRF) has been purified, sequenced, and studied in detail. This 23 kDa protein induces histamine release from the basophils of a subpopulation of atopic individuals. Responsiveness to HrHRF depends on IgE present on the basophils and in the serum of these donors, termed IgE+. This IgE+ can confer HrHRF responsiveness to any basophil, as opposed to the IgE from HrHRF unresponsive donors, termed IgE–, which cannot. While HrHRF depends on IgE+, it appears to have its own receptor, as it also primes unresponsive basophils for histamine release in response to anti-IgE and fMLP. In addition to inducing histamine release from the basophils of responders, it elicits IL-4 secretion from these cells (also in an IgE+ dependent manner) and primes even nonresponder basophils for IL-4 and IL-13 secretion in response to anti-IgE. The signal transduction mechanisms of HRF-mediated activation involves the tyrosine kinase Syk, but other critical details, such as the identity of the HrHRF receptor, are lacking.

Inhibitory Receptors

Appreciation of the importance of inhibitory receptors on the functions of basophils (as with those of other inflammatory cells) has increased dramatically during the past decade. Basophils consistently express the leukocyte immunoglobulin-like receptor B3 (LILRB3), a member of the inhibitory subgroup of LILRs. Although the related inhibitory receptors LILRB1 and 2, which are expressed on the surface of basophils from a minority of donors, are known to bind nonpolymorphic regions of MHC class I molecules, the ligands for LILRB3 are currently undetermined. Co-ligation of LILRB3 to an activating receptor such as FcεRI or LILRA2 results in the attenuation of mediator release from all three basophil compartments. LILRB3 possesses a relatively long cytoplasmic tail that contains immunoreceptor tyrosine-based inhibitory motifs (ITIMs). When phosphorylated, the tyrosine residues in these regions allow the docking and activation of phosphatases that antagonize the kinases involved in the propagation of activating signal transduction networks. It is tempting to hypothesize that evolutionarily related receptors such as LILRB3 and LILRA2 (whose genes are found along with those encoding the other LILRs in the leukocyte receptor complex on chromosome 19q13.42) are paired immunoregulatory switches, differentially expressed on various leukocytes and potentially binding related ligands with distinct affinities and kinetics to exert fine control over the functions of basophils and other inflammatory cells.

Basophils also express the inhibitory IgG receptor FcγRIIB (CD32), which possesses a single ITIM. Costimulation of basophils through both FcεRI and FcγRIIB delays and limits Ca^{2+} influx and attenuates degranulation and IL-4 release. These effects appear to be mediated by phosphorylation of the ITIM tyrosine residues, providing a docking site for the SH2 domain-containing phosphatase SHP-1, which then dephosphorylates key activating signal transduction components such as Syk. These observations form the basis of two novel approaches to antagonizing basophil and mast cell contributions to allergic inflammation. The first is a bifunctional Fcγ–Fcε fusion protein composed of part of the Fc portion of IgG that binds to FcγRIIB linked to a part of the Fc portion of IgE that binds to FcεRI. This protein effectively co-ligates FcγRIIB to FcεRI, constraining all IgE-mediated activation of basophils and mast cells. The second is a fusion protein composed of the same part of the Fc portion of IgG that binds to FcγRIIB linked to the major cat allergen Fel d1. By co-ligating FcγRIIB to Fel d1 specific IgE, this protein is an effective allergen-specific antagonist of basophil and mast cell activation appropriate for immunotherapy.

CD200R is a further example of an inhibitory receptor expressed by basophils. Intriguingly, a number of pathogenic viruses, including the human herpes viruses HHV-6, -7, and -8 express homologues of the ligand CD200, and are capable of downregulating activation of basophils.

FUNCTIONAL IMPLICATIONS OF BASOPHIL ACTIVATION

Tissue Effects of Mediator Release

The widespread activation of circulating basophils, with the release in minutes of preformed granule contents and the rapid *de novo* synthesis of lipid mediators, is predicted to result in systemic vasodilation and thus a decreased blood pressure. Indeed, since histamine, LTC_4, and PAF possess such vasoactive properties, this is a model of anaphylaxis especially attractive to explaining the rapid systemic allergic reactions to specific allergens administered intravenously (e.g., those to IV antibiotics) and to nonspecific secretogogues such as radiocontrast media and hyper- and hypotonic solutions. However, unlike mature, tissue-bound mast cells, circulating basophils must be recruited to tissues to participate in more typical allergic or antiparasitic inflammatory reactions, and this process may take hours to days. In the skin, the manifestations of basophil activation due to the release of these mediators include edema, erythema (wheal and flare reactions), and pruritus. In the gut, mucosal changes such as edema and the transmucosal loss of fluid, accompanied by increased peristalsis result in nausea, vomiting, diarrhea, and cramping abdominal pain (colic) as seen in food allergies. In the upper respiratory tract, similar changes in the nasal mucosa appear as congestion and rhinorrhea, while bronchoconstriction and wheezing typify the effects of LTC_4 and other mediators in the lower respiratory tract. The relative contribution of mast cells and basophils to IgE-dependent allergic responses has been the subject of debate. However, the observation that histamine but not PGD_2 is released into the nasal secretions during late-phase allergic reactions are argued for involvement of the basophil.

Recruitment and Modulation of Other Immune Cells

Their expression of numerous cytokines and chemokines suggests that over a longer time course (hours to days), basophils may recruit other immune cells to inflammatory lesions, and kinetic studies support this by demonstrating that basophils arrive in inflammatory tissues concomitant with or before other cells such as eosinophils and lymphocytes. Once there, basophils may modulate the function of these cells by means of extracellular mediators and by cell surface receptors recognizing their counterligands. The details of such cell–cell interactions await complete elucidation, but likely vary among different disease models.

The most intriguing advance in the appreciation of the potential immunoregulatory role of the basophil is the observation that activation of this innate immune cell can induce ε isotype switching and IgE production by human B cells when the two cell types are cocultured *in vitro*. Originally observed in response to nonphysiological stimulation of peripheral blood basophils (with PMA and ionomycin), this effect on B cells is also seen with activation of KU812 cells (a basophil-derived cell line) and with FcεRI-mediated activation of human cord blood–derived basophils. In all these systems, B-cell isotype switching is accompanied by and dependent on the production of IL-4 and IL-13 by the basophil (or KU812 cell) and expression of the costimulatory molecule CD40L (CD154). This phenomenon has traditionally been interpreted as representing the ability of basophils to *reinforce* previous Th2 skewing: once IgE production is established by Th2 cells, the IgE sensitizes basophils, and IgE-dependent stimulation of these basophils leads to their elaborating additional pro-IgE production signals, further encouraging B cells to make IgE. Importantly, however, recent data from *in vivo* murine models suggests that basophils may be an early, innate influence on B cells that *initiates* IgE production. This is not a direct effect of basophils on B cells, but rather is mediated by basophils inducing Th2 skewing, and it is the Th2 cells that directly induce B cells to undergo ε isotype switching. Thus, the basophil may be an answer to the mystery of the origins of Th2 skewing as well as a player in a potential positive feedback loop of IgE production for those antigens that can activate basophils for IL-4 production and CD40L expression by an IgE-independent mechanism.

EVIDENCE FOR ROLES IN DISEASES

Asthma and Rhinitis

Determining the precise contribution of the basophil to human diseases has been challenging. Basophils are considered to participate in allergic and antiparasitic inflammatory reactions, based on their biochemical and morphological resemblances to mast cells and eosinophils.

Their role in asthma was suggested by early studies that documented increased numbers of peripheral blood basophils and histamine concentrations in symptomatic asthmatics. Numerous basophils are present in sputum samples from patients with untreated asthma exacerbations, while 95% of the IgE-positive cells in bronchoalveolar lavage (BAL) fluid from asthmatic subjects have the morphologic and flow cytometric characteristics of basophils. Allergen inhalation by subjects with atopic asthma induces increases in sputum basophils at both early (7 hours after inhalation) and late phase (24 hours after inhalation) time points. The magnitude of this increase correlated with airway hyperresponsiveness (AHR) in response to methacholine in some studies, but this finding is controversial as other studies do not support or confirm this result.

Using the basophil-specific antibody 2D7, significantly more basophils were detected in postmortem lung section of patients with fatal asthma than in control patients who had asthma but died of other causes. The basophils were observed in both large and small airways and were present in the epithelium, submucosa, and alveolar wall. A subsequent study using 2D7 to identify basophils in biopsy specimens after allergen-induced late asthmatic responses showed that basophils accounted for a significant percentage of cells positive for IL-4 mRNA and the majority (72%)

of cells positive for IL-4 protein. Employing the BB1 antibody that recognizes basophil granules and defining basophils and tryptase negative, increased numbers of basophils were seen at baseline in the lungs of atopic asthmatics compared to atopic subjects without asthma and control patients with nonatopic asthma. Basophil numbers increased further after antigen inhalation, but these cells were outnumbered tenfold by eosinophils and mast cells, suggesting that the basophil may play a predominantly regulatory role rather than being an effector cell.

Using flow cytometry to detect intracellular cytokine production in peripheral blood mononuclear cells from asthmatic and nonasthmatic, nonatopic subjects, one study determined that, as expected, specific antigen elicited IL-4 and IL-13 secretion only from the basophils of the asthmatics. Of note, under such circumstances, basophils outnumbered T cells as IL-4 producing cells by fourfold. Two hours after antigen stimulation basophils were the exclusive cellular source of IL-13, suggesting that in the peripheral blood basophils are the dominant cellular source of Th2 cytokines in the first 6 hours after stimulation with allergen.

Like asthma, allergic rhinitis is an allergen-driven mucosal inflammatory condition with an immediate and a late phase (6–8 hours later). Among such biphasic responders, there is an increase in the histamine content of nasal secretions at both time points. However, elevated concentrations of PGD_2 are observed only in the immediate phase. As, in contrast to mast cells, basophils lack the ability to produce PGD_2, these findings suggest that the immediate phase is dominated by the activation of mast cells and that mediator release in the late phase is derived from newly recruited basophils. Data in support of this model were obtained in a study of patients with nasal allergy to grass pollen. At baseline, biopsies of the nasal mucosa demonstrated no basophils (as assessed by staining with the BB1 antibody). Within 1 hour after local allergen challenge, biopsies demonstrated a significant influx of basophils into the epithelium and lamina propria, and basophils remained detectable in the lamina propria at 24 hours and 1 week after challenge. Basophil numbers increase in the circulation of patients with ragweed allergic patients during ragweed season, suggesting that an antigen induced signal reaches the bone marrow and augments basophil production and/or attenuates basophil apoptosis.

Thus, the role of basophils in asthma and rhinitis is suggested by their recruitment to the airways in acute asthma and following antigen challenge; by the pattern of mediator production in the late phase allergic response; and by the increased numbers in the circulation of basophils producing Th2 cytokines after antigen challenge.

Immunologic Skin Disease

Cutaneous allergic inflammation takes a number of forms, one of the most common of which is urticaria (hives) with or without angioedema (swelling). Analogous to studies of the airways, the involvement of basophils in antigen-induced late phase skin reactions (LPR), a model of cutaneous allergy, has been assessed by staining skin biopsies with the BB1 antibody. After local antigen challenge, basophils were detected in the dermis (especially around blood vessels) at 6 hours, peaked at 24 hours, and started to resolve at 48 hours. Maximal basophil numbers reached 40% of peak eosinophil numbers, and approximately one-third of the basophils had an appearance consistent with partial degranulation. A similar study employing the 2D7 antibody also demonstrated a dramatic influx of basophils into the dermis around blood vessels 6 and 24 hours after allergen injection. Importantly, these 2D7-positive dermal basophils were negative on stating for metachromatic granules with toluidine blue, suggesting that degranulation had occurred and indicating that light microscopy might underestimate the presence of basophils.

A comparison of biopsies of allergen-induced cutaneous LPR with those from patients with chronic idiopathic urticaria (CIU) and those from normal controls demonstrated increased numbers of BB1 positive basophils in the LPR and CIU tissues compared with nonatopic controls. Approximately 35%–50% of patients with CIU have IgG autoantibodies directed against IgE or FcεRI, but there were no statistically significant differences in the numbers of dermal basophils between those with and without such antibodies. In situ hybridization demonstrated mRNA for IL-4 in both LPR and CIU biopsies, but it was unclear what fraction of IL-4 mRNA positive cells were basophils (as opposed to Th2 lymphocytes). A number of studies have used peripheral blood basophils in assays for histamine release to detect anti-IgE or anti-FcεRI autoantibodies in the serum of patients with CIU, and found that basophil histamine is dependent on or augmented by the complement fragment C5a.

Basophils have been detected infiltrating the lesions of bullous pemphigoid, erythema multiforme, incontinentia pigmenti, and rejection of first set skin allographs. Thus, basophils may be involved in the pathobiology of many immunologic skin diseases.

Anaphylaxis

Anaphylaxis is a severe, rapidly progressive, life threatening systemic allergic reaction typically triggered by allergens such as foods (e.g., peanut, shellfish), drugs (e.g., penicillin, aspirin), insect venom, and latex. Typical routes of allergen entry include ingestion (foods and some drugs), injection (intradermal for venom, intravenous for some drugs), and inhalation (latex). Anaphylaxis is a clinical diagnosis, dependent on the appearance of common symptoms and signs in specific organ systems (e.g., pruritus, urticaria, and angioedema; asthma; and abdominal pain, nausea, vomiting, and diarrhea) in the context of allergen exposure. With the exception of food-induced anaphylaxis, an elevation in the serum concentration of mast cell tryptase is so frequently seen that its absence casts some doubt upon the diagnosis, though it does not categorically rule out anaphylaxis. This in conjunction with decades of histological and biochemical data have established that mast cells are central to most forms of anaphylaxis.

Whether basophils are activated in one or more subtypes of anaphylaxis, depending perhaps on the triggering allergen and the route of exposure is the subject of investigation. The absence of elevations in serum tryptase in food induced anaphylaxis has led some to hypothesize that basophils may play a dominant role in this subtype of the disease. Individuals who develop anaphylaxis to the venom of stinging insects can be effectively treated with incremental subcutaneous injection of increasing doses of the responsible antigen over several hours or, more commonly, several weeks. This treatment is called immunotherapy. One recent study of peripheral blood basophils from patients with venom-induced anaphylaxis and those from patients with only large local sting reactions indicated that CD63 expression was higher at baseline (i.e., prior to sting challenge) among patients with systemic reactions (anaphylaxis) who were receiving venom immunotherapy compared to patients with large local reactions who were receiving immunotherapy. Basophils increased their expression of CD69 and CD203C in response to *in vivo* sting challenge. The authors suggested that basophils may thus be useful biomarkers of venom-induced anaphylaxis, but whether basophil activation is either necessary or sufficient for human anaphylaxis is unclear.

Parasitic Infections

Parasitic infections typically elicit elevated serum concentrations of IgE and eosinophilia. Evidence for basophils participating in immune responses against parasites is largely extrapolated from animal models (see later). While basophilia (increased numbers of circulating basophils) is commonly observed in nonhuman models, it is very rarely observed in human infections. One study documented that basophilia occurred in only 4 out of 668 patients with confirmed parasitic infection. However, data from other studies suggest that human basophils are functional in these diseases. Basophils from patients with hookworm infection (*Necator americanus*) degranulate in response to hookworm antigen, and the proteases secreted by this parasite have been shown to induce cytokine release from purified human basophils. This latter finding suggests that hookworm proteases may be innate (i.e., IgE-independent) activators of basophils important in early antiparasitic responses that later include IgE production (see earlier section for a discussion of basophil induction of B-cell ε isotype switching). Likewise, peripheral blood basophils from patients harboring active infection with the filarial worm *Brugia malayi* produced significant amounts of IL-4 in response to *B. malayi* antigen. Importantly, this IL-4 response was quantitatively on a par with that of parasite-specific T cells and was triggered by lower concentrations of antigen. Together, these studies suggest that at least for some infections, parasite antigens may elicit IL-4 release from basophils in an IgE-independent manner and that basophils may therefore be influential in establishing Th2 immune antiparasitic immune responses.

Other Human Diseases

Using the BB1 antibody, basophils have been identified in the gastric mucosa of patients with moderate to severe gastritis due to *Helicobacter pylori* infection. Hp(2–20), a peptide derived from the gut pathogen, is a potent chemoattractant for basophils, likely through the FPR-like 1 and 2 receptors (Table 6.1). While Hp(2–20) could induce Ca^{2+} influx into basophils, this stimulus did not elicit histamine release or cytokine (IL-4 or IL-13) secretion from these cells *in vitro*. The nature, mechanisms, and effects of basophil activation, if it occurs in the context of *H. pylori* infection, remain to be fully elucidated.

Peripheral blood basophils from normal individuals, unlike their mast cells, express a poorly characterized 45 kDa cell surface protein that binds the IgM mAb Bsp-1. As noted earlier, these cells have trace amounts or no detectable mast cell tryptase, chymase, or carboxypeptidase in their granules, and little cell surface expression of *c-kit*. However, in the peripheral blood of patients with asthma, atopy (allergic rhinitis, atopic dermatitis, urticaria), and drug allergy there are significant numbers of metachromatic, Bsp-1 positive cells with segmented nuclei (i.e., with the morphology of basophils) that possess tryptase, chymase, and/or carboxypeptidase in their granules. In addition, these cells expressed detectable amounts of *c-kit* on their surface. These data suggest that basophils in allergic disease may have a plastic phenotype and/or that basophils and mast cells are developmentally related, though this is controversial (see the discussion of basophil ontogeny earlier).

Guinea Pig and Mouse Models of Disease

Decades ago, studies of basophils from other mammalian species were frequently performed in guinea pig models and occasionally in rats and mice. Recently, the murine model has come to dominance, but advances have been limited by two difficulties. The first is in definitively identifying mouse basophils, as murine basophils appear to differ morphologically and in terms of cell surface receptor expression from their human counterparts. The second is the lack of a genetic basophil "knockout" model corresponding to mast cell deficient mice strains.

Guinea pig and murine models of parasitic infections, such as ectoparasites (ticks) in the former and *Nippostrongylus brasiliensis* in the latter, have been used to explore both the contribution of basophils to host defense and the ontogeny of basophils, as both species (unlike humans) typically develop basophilia in response to parasite invasion. These studies established basophils as important sources of cytokines such as IL-4 and IL-13.

Recently, a murine model of response to an allergen (the protease papain) indicated a critical immunoregulatory role for basophils. After immunization with papain, basophils entered draining lymph nodes and initiated Th2 differentiation by releasing IL-4 and the cytokine thymic stromal lymphopoietin (TSLP), leading to papain-specific IgE production. Depletion of basophils using the MAR-1 antibody revealed the dependence of Th2 development on basophils (by means of IL-4 and TSLP) but not on mast cells. Other recent studies have investigated the roles of murine basophils in a model of IgG– (as opposed to IgE–) mediated anaphylaxis and in antibody memory responses.

PHARMACOLOGICAL REGULATION OF BASOPHIL RESPONSES

Several drugs that are commonly used to treat asthma and allergic diseases have been shown to have an effect on basophil numbers and function. The relevance of these findings to the therapeutic efficacy of these medications depends on the putative role of basophils in these conditions. As outlined earlier, this remains somewhat controversial, and is the subject of ongoing investigation.

Steroids

Administration of intravenous hydrocortisone and inhaled beclomethasone was shown three decades ago to decrease the number of peripheral blood basophils concomitant with their therapeutic effects in asthmatics. Indeed, steroids appear to augment basophil apoptosis (programmed cell death) just as they do in eosinophils.

LT Receptor Antagonist

Basophils both produce and respond to cyLTs. Antagonism of the cysLT1 receptor by montelukast in patients with asthma had no effect on basophil degranulation in response to anti-IgE, but did result in modest inhibition of LTC_4 release from these cells, indicating differential regulation of these distinct mediator compartments and a possible cysLT feedback mechanism.

Anti-IgE

As noted earlier, the expression of FcεRI by basophils is directly proportional to the ambient concentration of free IgE. Humanized monoclonal antibodies (e.g., omalizumab) that bind to the C_H3 domain of the IgE heavy chain compete with the binding site of the FcεRI α chain, thus preventing IgE from adhering to the surface of basophils by means of the high-affinity receptor. This leads to a dramatic and nearly complete decline in FcεRI expression on the surface of basophils. This greatly impairs FcεRI-mediated activation of basophils, including histamine release, LT secretion, and cytokine production. Over the course of weeks after cessation of treatment, free IgE concentrations, FcεRI expression, and basophil mediator responses return to baseline.

DIAGNOSTIC TESTS USING BASOPHILS

A number of tests using basophils have been developed to assist in the diagnosis of allergic diseases by means of an *in vitro* assay that is ideally easily performed, minimally invasive (requiring a peripheral blood sample as opposed to skin testing or allergen challenge), rapid, reproducible, and widely applicable. Generally, these tests involve exposing basophils, either in a peripheral blood sample or purified to varying degrees, to a suspected allergen, such as environmental substances (e.g., plant pollens, animal dander, dust mite antigens), drugs (e.g., NSAIDs, antibiotics, chemotherapeutic agents), foods, and Hymenoptera venom. Because basophils are believed to be the exclusive source of histamine in human peripheral blood, measuring histamine release 30–60 minutes after exposing the patient's blood cells to the antigen determines whether basophil activation occurred. Other tests rely on changes in the expression of certain basophil cell surface markers, such as CD63, CD69, and CD203C as assayed by flow cytometry. Some assays combine measurements of mediator release with flow cytometry.

Some potential pitfalls with these tests include their inability to determine whether basophil activation is IgE dependent or independent; the assumption that an absence of basophil activation correlates with nonresponsiveness of mast cells to a given antigen; and the possibility that an allergen might elicit the release of one or more important but less cell-specific mediators (e.g., LTC_4, PAF) but not histamine, resulting in a false negative test. It is important also to note that the basophils of some individuals do not release histamine or leukotrienes in response to FcεRI cross-linking. This so-called nonreleasing phenotype has been attributed to a delayed and blunted Ca^{2+} influx in response to FcεRI cross-linking, possibly due to impaired expression and/or activity of the signaling molecule, Lyn kinase.

CONCLUSIONS

Human basophils are the rarest of peripheral blood granulocytes. Their expression of the high-affinity Fc receptor for IgE, the range of mediators they produce, and their presence in tissues and circulation following allergen challenge suggest that they are involved in asthma, allergies, and selected infections. While their precise roles in health and disease are still incompletely understood, recent data suggest that they may be important immunoregulators, especially in allergic diseases and parasitic infections. Their role as immunoregulatory cells, particularly a role in "priming" Th2-dependent allergic reactions, as opposed to effector cells of allergic disease would be consistent with their presence in low numbers in the resting state. This is the subject of intense investigation, the conclusions of which will depend in part on the relevance of animal models to human disease and our ability to directly interrogate basophil function in humans. There is still much to explore in terms of the basic mechanisms and intercellular interactions of basophils.

SUGGESTED READINGS

Devouassoux, G., Foster, B., Scott, L.M., Metcalf, D.D., and Prussin, C. 1999. Frequency and characterization of antigen-specific IL-4 and IL-12-producing basophils and T cells in peripheral blood of healthy and asthmatic subjects. *J Allergy Clin Immunol* **104**:811–819.

Falcone, F.H., Haas, H., and Gibbs, B.F. 2000. The human basophil: a new appreciation of its role in immune responses. *Blood* **96**(13):4028–4038.

Gauchat, J.-F., Henchoz, S., Mazzei, G., et al. 1993. Induction of human IgE synthesis in B cells by mast cells and basophils. *Nature* **365**:340–343.

Lee, J.J., and McGarry, M.P. 2007. When is a mouse basophil not a basophil? *Blood* **109**(3):859–861.

MacGlashan, D.W., Jr., Bochner, B.S., Adelman, D.C., et al. 1997. Down-regulation of FcεRI expression on human basophils during in vivo treatment of atopic patients with anti-IgE antibody. *J Immunol* **158**:1438–1445.

Marone, G., Triggiani, M., and de Paulis, A. 2005. Mast cells and basophils: friends as well as foes in bronchial asthma? *Trends Immunol* **26**(1):25–31.

Ochensberger, B., Tassera, L., Bifrare, D., Rihs, S., and Dahinden, C.A. 1999. Regulation of cytokine expression and leukotriene formation in human basophils by growth factors, chemokines, and chemotactic agonists. *Eur J Immunol* **29**:11–22.

Sokol, C.L., Barton, G.M., Farr, A.G., and Medzhitov, R. 2008. A mechanism for the initiation of allergen-induced T helper type 2 responses. *Nat Immunol* **9**(3):310–318.

7 Eosinophils

Sophie Fillon, Steven J. Ackerman, and Glenn T. Furuta

INTRODUCTION

Eosinophilic granulocytes, commonly referred to as eosinophils, or less commonly as acidophils, were first identified by Paul Ehrlich in 1879. He named these bilobed nucleated cells as eosinophils because of their intense staining with the acidic dye eosin. Ehrlich completed his original studies that defined the eosinophil's morphological features during a time when he was committed to a clinical practice as well as basic research, an early example of a physician scientist. Combining his clinical expertise with his laboratory discoveries, he proposed a variety of functions for eosinophils that included phagocytosis, granule secretion, and chemotaxis.

Eosinophils are normally present as a minority of the peripheral blood leukocyte pool and primarily reside within tissues containing mucosal surfaces such as the uterus and gastrointestinal tract. While their exact function is not certain, eosinophils are thought to possess both beneficial and deleterious properties as participants in both innate and adaptive host immune responses. As an arm of the innate defense system, eosinophils are strongly proposed to be responsible in part for combating parasitic infections, particularly with helminths. When recruited in excess, eosinophils are currently thought to participate in mediating the pathogenesis of allergic diseases such as asthma, atopic dermatitis, gastrointestinal diseases, and hypereosinophilic syndromes.

EOSINOPHIL MORPHOLOGY

Eosinophils are approximately 12–17 μm in diameter, and represent 1%–6% of the total blood leukocyte population (Figure 7.1). The human eosinophil's nucleus is characteristically bilobed, and usually circular or ovoid [1]. The cytoplasm of the eosinophil is filled with a number of secretory organelles that include primary and secondary (crystalloid/specific) granules, small granules, and secretory vesicles. Primary and secondary granules together number approximately 200 per cell and contain an electron-dense matrix containing the eosinophil's characteristic cationic proteins including eosinophil peroxidase, the eosinophil-derived neurotoxin, and the eosinophil cationic protein (the eosinophil's two ribonucleases RNase2 and RNase3, respectively) (Figure 7.2). In addition, the secondary granules contain an electron-dense crystalloid core composed of two related cationic proteins, major basic protein-1, and major basic protein-2. The small, amorphous granules contain arylsulfatase and acid phosphatase [2]. The Charcot–Leyden crystal (CLC) protein is present within the primary granule population (which represent less than 5% of total granules) and in the agranular spaces beneath the eosinophil's plasma membrane. When released by eosinophils that undergo cytolytic degranulation in tissues such as the lung and gastrointestinal (GI) tract, the CLC protein may form the characteristic hexagonal bipyramidal structures known as CLCs, a hallmark of eosinophil involvement in host allergic and other immune responses. These crystals can be observed in the stool or sputum of patients with nonspecific eosinophilic inflammation associated with inflammatory bowel diseases, allergic intestinal disease and asthma, respectively (Figure 7.3). The CLC are slender, hexagonal bipyramidal crystals that stain purplish-red in tissue sections stained with hematoxylin and eosin (H&E). Eosinophils also contain lipid bodies, nonmembrane-bound lipid-rich organelles often confused with granules and present in many types of leukocytes and other cells [3]. Eosinophils contain greater numbers of lipid bodies than do neutrophils, and the numbers of lipid bodies increase during eosinophil activation in vitro and engagement in inflammatory reactions in vivo [4]. The functions of lipid bodies include incorporation of fatty acids such as arachidonate. They likely serve as intracellular depots for its storage and metabolism,

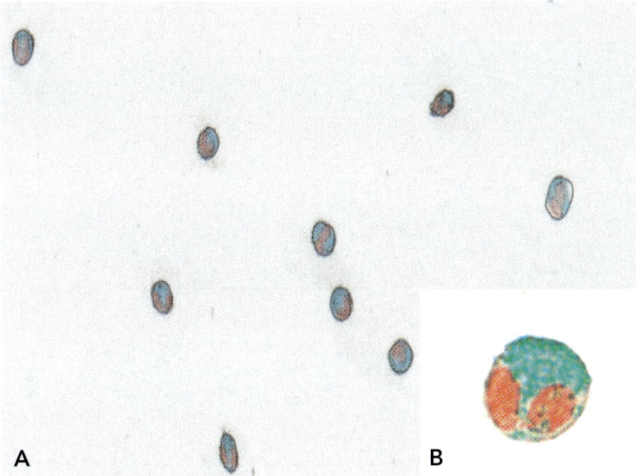

Figure 7.1. Eosinophils in peripheral blood smear. Fast green staining of eosinophils in human peripheral blood with a neutral red counterstain. Cytoplasmic granules appear brilliant turquoise green and the bilobed nucleus is stained red (Magnification A: 20×, B: 100×).

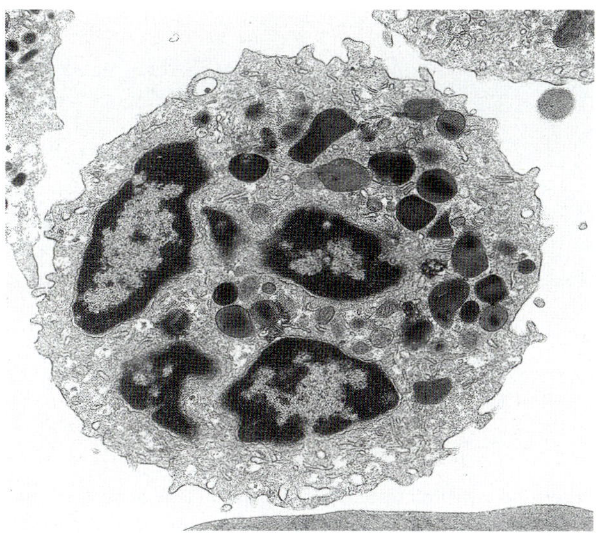

Figure 7.2. Electron microscopic view of the human eosinophil. Electron micrograph of a human eosinophil with granules showing electron dense cores (2500×, Courtesy of Dr. Ann Dvorak, Beth Israel Deaconess Medical Center, Boston, MA).

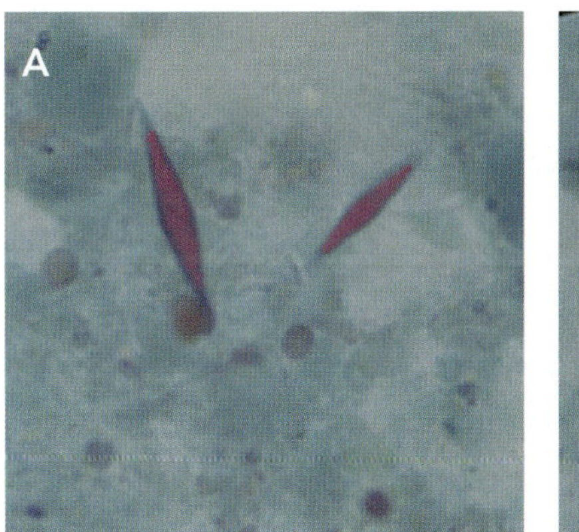

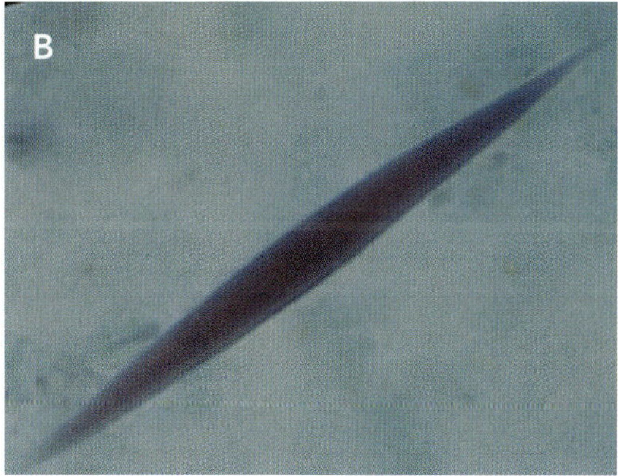

Figure 7.3. Charcot–Leyden crystals from bronchial lavage fluid. Photomicrographs of birefringent Charcot–Leyden crystals, a hallmark of degranulated eosinophils (A: 40×, B: 100×).

as suggested by the localization of many of the eosinophil's eicosanoid-forming enzymes including 5-lipoxygenase, leukotriene C4 synthase, and cyclooxygenase in the lipid body [5]. These four organelles, along with vesiculotubular structures and small vesicles involved in transport and secretion in the activated cell, serve as the major subcellular sites for the eosinophils' armamentarium of preformed cytotoxic and inflammatory mediators.

EOSINOPHIL MEDIATORS

Granule Proteins

Studies of the biochemistry, biologic activities, and, in particular, the localization in tissues of the distinctive enzymatic and nonenzymatic cationic protein constituents of the eosinophil's secondary (specific) granule provided the first convincing evidence for the role of the eosinophil in the pathogenesis of inflammation and tissue damage in eosinophil-associated diseases [6]. In addition, identification of eosinophil-specific granule constituents in tissues by a variety of sensitive and specific immunochemical methods provided additional evidence for the participation of the eosinophil in diseases not normally associated with blood or tissue eosinophilia, for example, certain skin diseases [7]. Studies in the past 15–20 years of the biochemistry, functions, and localization in tissues of the unique enzymatic and nonenzymatic cationic proteins present in eosinophil granules provide compelling evidence supporting a pathologic proinflammatory and effector

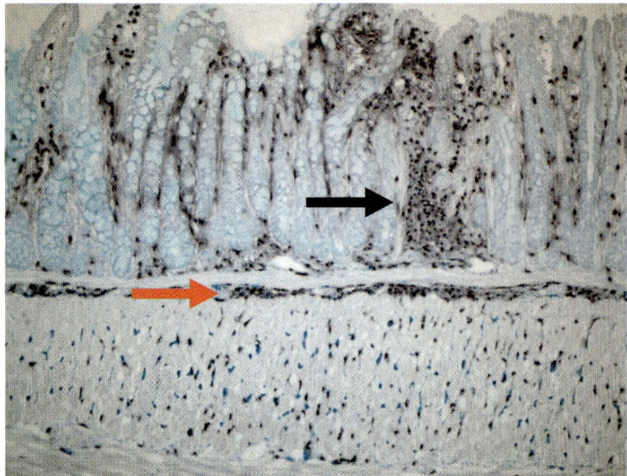

Figure 7.4. Gastrointestinal eosinophilic inflammation. Immunoperoxidase detection of major basic protein-1 in a section of inflamed murine small intestine. Black arrow shows eosinophilic infiltration in epithelial surface whereas orange arrow shows inflammation in the muscularis (20×).

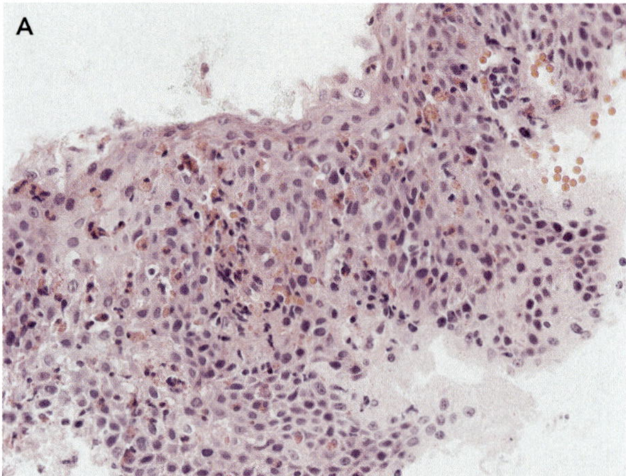

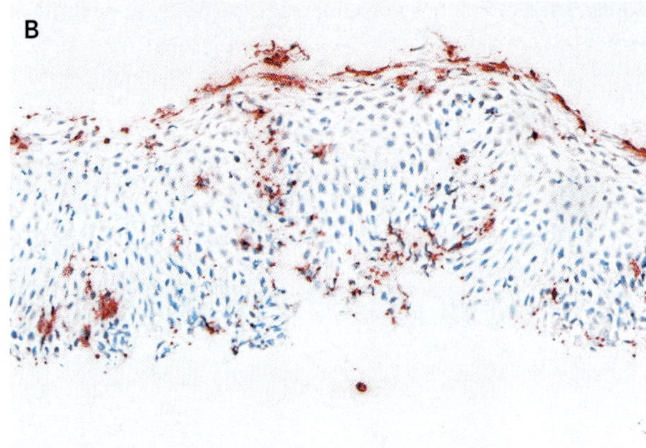

Figure 7.5. Eosinophilic esophagitis inflammation. (A) H&E of the esophageal epithelium from a patient with eosinophilic esophagitis (20×). (B) Eosinophil peroxidase staining of the esophageal epithelium from a patient with eosinophilic esophagitis (20×). (Courtesy of Cheryl Protheroe and James Lee, Mayo Clinic Scottsdale, AZ.)

role for the eosinophil in directly inducing tissue damage. The distinctive cationic granule protein constituents of the eosinophil include the two major basic proteins (MBP-1, MBP-2) (Figure 7.4), the eosinophil peroxidase (EPO) (Figure 7.5), and the ribonucleases eosinophil cationic protein (ECP), and eosinophil-derived neurotoxin (EDN). Eosinophils also have the capacity to express a variety of toxic oxidative intermediates [8], eicosanoid and other lipid mediators of inflammation, and most recently, inflammatory and hematopoietic cytokines integral to the normal and pathologic roles of this granulocyte in health and disease. Studies of the mechanisms by which the eosinophil granule cationic proteins are secreted or released into tissues have demonstrated that they can be selectively secreted, depending on the activation stimulus, by a number of different secretory pathways ranging from classical granule fusion and exocytosis (e.g., in killing parasitic helminths), piecemeal degranulation (PMD; vesicular transport from the granules in the absence of classical exocytosis), and cytolytic degranulation (release of intact membrane-bound granules directly into the tissue upon eosinophil apoptosis/cell death) [9].

Cytokines

Eosinophils are a rich source of both Th1 and Th2 cytokines that may participate in immune regulation, leukocyte chemotaxis, and cell growth (Table 7.1) [10]. While other cell types such as lymphocytes can produce larger quantities of a variety of cytokines localized within inflammatory sites, eosinophils may act in a different role. For instance eosinophils, like mast cells, have the capacity to synthesize cytokines, and then store them in their cytoplasmic granules. Thus, rather than a lag period between stimulation and release of cytokine, eosinophils are able to provide immediate and focused cytokine release at sites of inflammation [11].

Traditionally, eosinophils are associated with the generation of Th2 immune responses and are increased in immune environments rich with Th2 cytokines such as IL-4, IL-5, and IL-13 as seen in asthma. IL-5 mediates eosinophil expansion and differentiation from progenitors in the bone marrow, recruitment, priming, and activation, and prolongs tissue survival in response to allergic stimuli [12]. Activated eosinophils also express granulocyte-macrophage colony-stimulating factor (GM-CSF) which functions in an autocrine fashion to prevent apoptosis and prolong their survival in tissues such as the lung in asthma. Interestingly, basal levels of eosinophil development can occur in the absence of IL-5, and antigen-induced tissue eosinophilia can occur independent of IL-5 as

TABLE 7.1. Eosinophils release interleukins, chemokines, and growth factors
Interleukins
IL-1, IL-2, IL-3, IL-4, IL-5, IL-6, IL-9, IL-10, IL-11, IL-12, IL-13, IL-16, IL-17
Chemokines
Epithelial cell–derived neutrophil activating peptide (CXCL5), eotaxin (CCL11), growth-related oncogene (CXCL1), interleukin-8 (CXCL8), IFN-γ-inducible protein (CXCL10), IFN-inducible T-cell alpha chemoattractant (CXCL11), macrophage inflammatory protein 1 (MIP-1α), monocyte chemoattractant protein 1 (CCL3), monokine induced by IFN-γ (CXCL9), MCP-3 (CCL7), MCP-4 (CCL13), RANTES (CCL5)
Growth factors
Heparin-binding epidermal growth factor-like binding protein (HB-EGF-LBP), nerve growth factor (NGF), transforming growth factor (TGF-α), transforming growth factor-beta (TGF-β1)
Others
Interferon (IFN-γ), tumor necrosis factor (TNF-α), granulocyte-macrophage colony-stimulating factor (GM-CSF)

suggested by the presence of tissue eosinophils in asthmatic patients treated with anti-IL-5 [13]; these observations and others suggest a redundancy in eosinophil growth factors capable of driving eosinophil terminal differentiation and prolonging eosinophil survival in tissues, factors that include IL-3 and GM-CSF. In contrast, numerous studies have shown the importance of IL-5 during allergic responses, particularly for the amplification of blood and tissue eosinophilia. Employing experimental models of asthma, investigators have demonstrated that IL-5 neutralization can block various aspects of the asthmatic response [14,15]. Animal models have also shown that inhibition of IL-5 suppresses pulmonary eosinophilia in response to antigen inhalation [16]. Moreover, eosinophil trafficking to the allergic lung is significantly reduced in IL-5–deficient (knockout) mice and in animals treated with anti-IL-5 antibodies [17–19]. Finally, overexpression of IL-5 in transgenic mice is sufficient for the development of eosinophilia [20–22]. IL-4 activates human vascular endothelial and respiratory epithelial cells to produce eosinophil chemoattractant cytokines [23]. IL-4 and IL-13 produced by Th2 cells recruits and activates IgE-producing B cells, and enhances IgE-mediated responses [24]. IL-13 appears to have its own distinct role in allergic inflammation, acting as a key regulator of allergen-induced airway inflammation, and goblet cell metaplasia [25]. Both molecules induce the expression of eotaxins by a STAT6-dependent pathway [26]. IL-13 alone is also an important eosinophil chemoattractant [12].

Alternatively, eosinophils themselves can produce Th1 cytokines such as IL-2, IL-6, IL-12, TNF-α, and IFN-γ [27–29]. Th1 cells are recruited by CXCL9 (monokine induced by IFN-γ) and CXCL10 (IFN-γ-inducible protein-10) [30]. IL-6 has been shown to inhibit Th1 and increase Th2 differentiation of eosinophils through induction of naive T-cell IL-4 production [31]. In some Th1 models of inflammation, eosinophil-derived IFN-γ is increased [10]. In addition TNF-α is essential for IFN-γ induced secretions of Th1 chemokines [28]. Eosinophils secrete chemokines such as the eotaxins, RANTES, and macrophage inflammatory protein (MIP)-1α. Eotaxins (1, 2, and 3), secreted primarily by activated epithelial cells and T cells, are highly selective eosinophil chemoattractants that bind their cognate CCR3 receptor on the eosinophil's surface. Eotaxin secretion by eosinophils strongly suggests their ability to initiate and perpetuate an allergic inflammatory response. RANTES can stimulate eosinophil and neutrophil influx while MIP-1α is primarily involved in neutrophil trafficking [32].

Eosinophils also may participate in other biological processes. IL-4-deficient mice have reduced transplant-associated eosinophilia. Impaired Th2 inflammation at the site of incompatible allograft transplants was also observed in these mice suggesting that eosinophils may modulate immune responses leading to transplant rejection [33]. Moreover, IL-4 and IL-13 are strongly associated with tumor cell death in several types of cancer [34]. In addition, the use of eosinophil-specific antibodies has demonstrated that almost all human or mouse cancers are associated with an important eosinophil infiltrate at some point in tumor growth [35,36]. Finally, eosinophils are associated with tissue remodeling (e.g., epithelial cell hyperplasia, angiogenesis) and fibrosis (fibroblast activation, transdifferentiation into myofibroblasts increased deposition of extracellular matrix) in many eosinophil-associated diseases, and are a rich source of TGF-β [37].

Lipid Mediators

Eosinophils express cysteinyl leukotriene receptors (CysLT1R and CysLT2R), prostaglandin (PG)D2 type 2 receptor, and platelet activating factor (PAF) receptor [38–40]. Eosinophil recruitment is induced by leukotrienes (LTB4, LTD4, LTE4), PAF, and 5-oxo-6,8,11,14-eicosatetraenoic acid, suggesting that these lipid mediators may play a role in eosinophil transmigration [41–43]. Eosinophils have the capacity to secrete lipid mediators including PAF and leukotrienes such as LTC_4. The release of these lipids increase leukocyte trafficking, endothelial adhesion, smooth muscle contraction, vascular permeability and mucus secretion and are considered one of the proinflammatory activities of the eosinophil [44].

EOSINOPHIL DEVELOPMENT

Life Cycle

Eosinophils are derived from hematopoietic stem cells and myeloid-committed progenitors that arise within the bone marrow. The human eosinophil progenitor is derived from the common myeloid progenitor pool, expresses a number of unique cell surface markers including IL-5Rα, CD34, CD38, IL-3Rα, is CD45RA negative, and is distinct from the human granulocyte-macrophage progenitor (GMP) population that gives rise to neutrophils and macrophages [45]. Two groups of transcription factors, the C/EBP family and GATA-1, appear to synergistically participate in eosinophil development [46]. The importance of GATA-1 and the C/EBP factors (C/EBPα and C/EBPε) is emphasized by the fact that murine models with targeted deletions (knockouts) of these three factors are all eosinophil-deficient [47]. However, GATA-1 appears to be the most important transcription factor for eosinophil lineage specification, since transgenic deletion of the high-affinity GATA-binding site in the HS-2 region of the GATA-1 promoter itself leads to a selective loss of eosinophils in the mouse [48]. Both GATA-1-specific expression and activity in eosinophils appears to be mediated by this high-affinity palindromic GATA site. This double GATA site is present in the GATA-1 promoter and in the regulatory regions of many eosinophil specific genes (e.g., eotaxin receptor CCR3, CLC/Gal-10, MBP-1, and IL-5Rα) [48–50].

Eosinophil differentiation and expansion from IL-5Rα+ common myeloid progenitor cells occurs in response to three important cytokines, IL-3, IL-5, and GM-CSF [51–53]. These three cytokines provide permissive proliferative and differentiation signals through transcription factors that include GATA-1, PU.1, C/EBPα, and C/EBPε. IL-5 is most specific to the eosinophil lineage of these cytokines and is responsible for selective differentiation of eosinophils [54] and stimulates eosinophil release from the bone marrow into the peripheral circulation [55]. At these sites, eosinophils are primed by Type 2 cytokines released from Th2 helper cells. IL-5, GM-CSF, and IL-3 are important as well for eosinophil terminal differentiation, activation, and are antiapoptotic, promoting long-term eosinophil survival in vitro and in tissues.

Eosinophils produce and store secondary granule proteins before their exit from the bone marrow. PU.1 plays a critical role in eosinophil granule protein production as evidenced by the fact that PU.1 null mice show attenuated MBP and EDN expression and addition of PU.1 induces MBP expression. A synergistic role for PU.1 in amplifying GATA-1-mediated gene transcription has been demonstrated in the eosinophil lineage [49]. Direct PU.1 protein–protein interaction alters GATA-1 conformation and/or binding to high-affinity double GATA sites present in many eosinophil-specific gene promoters, increasing its activity and synergizing to strongly upregulate transcription of eosinophil lineage-specific genes such as MBP in developing eosinophil progenitors [49].

Eosinophil Migration

After maturation, eosinophils migrate through the vascular space and traffic to mucosal surfaces including the gastrointestinal tract (Figures 7.4 and 7.5) in response to a variety of different chemoattractants produced by cells present in the mucosa, particularly epithelial cells. These chemoattractants or chemokines (chemotactic cytokines) including CCL11 (eotaxin-1), CCL24 (eotaxin-2), CCL26 (eotaxin-3), and CCL5 (RANTES) participate in the recruitment of eosinophils to mucosal surfaces and sites of allergic inflammation. Arachidonic acid metabolites, especially certain leukotrienes such as (LTB_4) are synthesized in leukocytes from arachidonic acid by 5-lipoxygenase. Leukotrienes are thought to play a role in allergic and asthmatic reactions by stimulating bronchoconstriction and increasing vascular permeability [56]. Eotaxin production can lead to tissue eosinophilia. For instance, experimental induction of cutaneous and pulmonary late-phase responses in humans leads to synthesis of eotaxins by resident cells as well as infiltrative cells such as eosinophils [26].

Transmigration of eosinophils through the vascular endothelium into mucosal surfaces involves rolling, tethering, adhesion, and transendothelial migration [57,58]. Eosinophils express the transmembrane glycoproteins, selectins, that regulate rolling on endothelia [59]. Adhesion and transmigration of eosinophils across the vascular endothelia to tissue is a tightly regulated process. Of the cytokines implicated in modulation of leukocytes recruitment, only IL-5 and the eotaxins have been shown to selectively regulate eosinophil trafficking [60]. Eosinophil chemoattractants alter the expression of vascular and leukocyte adhesion molecules and receptors such as selectins, integrins, and integrin receptors. Mucosal addressin cell adhesion molecule (MAdCAM)-1, vascular cell adhesion molecule (VCAM)-1, and intercellular adhesion molecule (ICAM)-1 present on endothelial cells can be differentially expressed during transmigration [61]. VLA-4 of the p-integrin family is expressed on the surface of eosinophils [62] where it can interact selectively with the activated endothelial ligand, VCAM-1 [63]. Endothelial cell products and adhesion molecules (β1-, β2-, and β7-intergrins) [62] also participate in eosinophil trafficking.

TABLE 7.2. Eosinophils' chemoattractants

Platelet activating factor (PAF), complement factor C5a, leukotriene B_4 (LTB_4), LTC_4, *N*-formyl-methyionyl-leucyl-phenylalanine (fMLP), 5-oxo-15-HETE, 5,15-diHETE
Cytokines
IL-2, IL-3, IL-5, IL-16, GM-CSF, Ecalectin
Chemokines
Eotaxins 1–3, MCP-2, MCP-3, MCP-4, RANTES, MIP-1α, IL-8, MDC

Eosinophils express the chemokine receptors CCR3 and CCR1 [64–66]. Ligation of these receptors by macrophage inflammatory protein (MIP)-1α, RANTES, macrophage chemotactic protein (MCP)-2, MCP-3, MCP-4, eotaxin-1, -2, and -3, and mucosa-associated epithelial chemokine (MEC) leads to increased expression of these receptors. Eosinophils also express chemokine receptor such as CXCR3, CXCR4, CCR5, CCR6, and CCR8 after IL-5 activation [67–69]. RANTES as well as the acidic mammalian chitinase participates in eosinophil chemoattraction [26,70] (Table 7.2). Trafficking of eosinophils into inflammatory sites involves several cytokines such as IL-4, IL-5, and IL-13. These cytokines, particularly IL-5, modulate leukocyte recruitment. IL-4 and IL-13 act as potent inducers of the eotaxins [71,72].

ACTIVATION OF EOSINOPHIL DEGRANULATION AND MEDIATOR RELEASE

For eosinophils to participate in the pathogenesis of local tissue inflammation, damage, remodeling, and fibrosis, more must occur besides their selective recruitment and accumulation. In fact, the state of activation of recruited eosinophils is now considered to be a critical aspect of their participation in disease pathophysiology. For example, a number of strains of IL-5 transgenic mice have massively increased numbers of eosinophils in their blood, spleen, and other tissues and organs, but without a second signal, these mice show little evidence of eosinophil-mediated pathology and are relatively healthy. One major pathway by which eosinophils are activated is through the cross-linking of surface immunoglobulin receptors, especially those involving IgA, and to a lesser extent IgG [73]. Although there is general agreement that murine eosinophils do not express the high-affinity FcεRI IgE receptor, its expression remains controversial for human eosinophils, which express very low levels, if any, FcεRI on their surface, and if present, clearly lack the β chain; so significant, direct activation of eosinophils via IgE remains highly unlikely [74–76]. Activation of eosinophils by a variety of cytokine and lipid stimuli leads to eosinophil granule protein release (e.g., ECP, EDN, EPO, MBP), superoxide generation, and the synthesis of LTC_4 [73,77], although secretion of the granule cationic proteins is difficult to induce with traditional eosinophil-activating stimuli. Eosinophils also elaborate the lipid mediator PAF and a wide range of cytokines and chemokines, not the least of which include IL-1β and TGF-β, two key players in the eosinophil-mediated tissue remodeling and fibrosis seen in many eosinophil associated diseases [78–81]. Although the quantities of cytokines and chemokines released by the eosinophil compared to other leukocytes vary widely, GM-CSF is among the cytokines produced in greatest quantities by eosinophils [73], and as noted earlier, functions in part in an autocrine fashion to prevent apoptosis and prolong eosinophil survival once they are recruited into sites of tissue inflammation. Eosinophil priming for increased secretory and oxidative responses, for example, by IL-5 and the eotaxins, and subsequent activation in vitro by a number of physiologically relevant agonists that include IL-5, IFN-γ, sIgA, and others, has been shown to induce secretion of the granule cationic proteins (EPO, MBP1, EDN, ECP) and eosinophil-expressed cytokines (e.g., RANTES and IL-4) in a unique process termed PMD [82,83] that involves the differential mobilization followed by vesicular transport of these proteins to the cell surface and into the extracellular space [84–86]. This is in marked contrast to more classical secondary granule fusion and exocytosis, events rarely observed for eosinophils at sites of inflammation in tissues [87] but frequently observed in eosinophil-mediated killing of the larval stages of parasites (helminths), for example, schistosomula of *Schistosoma mansoni*. Once secreted, the eosinophil granule cationic proteins are reported to have multiple proinflammatory activities that have been well-defined both in vitro and in vivo, including membrane, cell, and tissue-damaging cytotoxicity [88,89], the ability to selectively activate other inflammatory cells, such as mast cells and basophils to release inflammatory mediators (e.g., histamine) [90], and oxidative burst in neutrophils, potent blocking activity for inhibitory M2 muscarinic receptors in the airways in allergic asthma models [91], as well as the ability to augment TGF-β primed fibroblast elaboration of the inflammatory and profibrotic IL-6 family of cytokines including IL-6 and IL-11 [92], to name just a few. Thus, eosinophils (1) come fully armed with preformed mediators of inflammation, tissue damage, remodeling, and fibrogenesis that are secreted at sites of eosinophilic inflammation in tissues in many eosinophil-associated allergic diseases such as asthma and hypereosinophilic syndromes such as eosinophilic esophagitis and (2) have the capacity to generate newly formed protein (cytokines, chemokines) and lipid mediators (LTC_4, PAF) of inflammation when primed

TABLE 7.3. Eosinophils' activators
IL-3, IL-5, GM-CSF, IFN, PAF, LTB4, eotaxins, sIgA

and further activated in the process of their recruitment from the bone marrow into the tissue in response to allergic (e.g., mast cell, basophil) and other inflammatory and/or Th2 T-cell stimuli [93] (Table 7.3).

SIGNAL TRANSDUCTION

The process in which eosinophils convert external stimuli into biochemical reactions, similar to many aspects of eosinophil biology, is understudied. A number of factors are known to rapidly activate eosinophils ranging from growth factors, extracellular matrix proteins, cytokines, and neural factors. For instance, fibronectin can modify cell signaling and subsequent migratory capacity of eosinophils [94]. This is of particular interest since fibronectin levels from bronchoalveolar lavage fluids of asthmatic patients increase after antigen challenge, and correlates with eosinophilic inflammation [95,96]. The increase in airway fibronectin may contribute to eosinophil persistence by modulating the survival, recruitment, and retention of lung eosinophils [96,97].

While a number of factors lead to cell surface activation of signal transduction pathways in eosinophils, a new body of evidence suggests that eosinophil granules may also undergo activation with release of preformed proteins and cytokines. Granule surfaces contain IFN-γ receptors and G-protein–coupled receptors for eotaxin providing a means for rapid release [98].

Other studies have investigated the role of eosinophils on activating other resident mucosal cells such as nerves. For instance, the adhesion of eosinophils to cholinergic nerves led to a rapid and sustained activation of the nuclear transcription factors such as nuclear factor (NF)-κB and activator protein (AP)-1. Eosinophil binding to neuronal ICAM-1 contributes to a rapid activation of ERK1/2 whereas eosinophil adhesion to VCAM-1 results in AP-1 activation, mediated in part by rapid activation of the p38 mitogen-activated protein kinase signaling pathway [99].

SUMMARY

Eosinophils are multifunctional leukocytes that likely participate in the initiation and propagation of inflammatory responses. Much is yet to be learned about eosinophil biology. To date, their function is not certain but their presence at mucosal surfaces, such as the gastrointestinal tract, in the healthy host suggests a protective role. Their increasing recognition in increased numbers in disease states supports a role in the pathogenesis of allergic, inflammatory, and neoplastic diseases. As more knowledge accumulates regarding the mechanisms of mediator production and release, chemotaxis, and life cycle, we will begin to understand the role of this enigmatic cell in health and disease.

KEY POINTS

Eosinophils can synthesize a number of biologically active mediators and contain a number of preformed proteins both of which can be released upon appropriate stimulation.

A variety of mediators including eotaxins, PAF, LTB_4, fMLP are potent eosinophil chemoattractants.

While eosinophils occupy 1%–6% of the peripheral white blood cell count, they are normally present in larger numbers within mucosal surfaces and increase during disease states.

ACKNOWLEDGMENTS

We are grateful to Joanna Grenawalt and Joanne Masterson for their expertise in obtaining human and murine micrographs.

REFERENCES

1. Sparrevohn, S., and Wulff, H.R. 1967. The nuclear segmentation of eosinophils under normal and pathological conditions. *Acta Haematol* **37:**120.
2. Afshar, K., Vucinic, V., and Sharma, O.P. 2007. Eosinophil cell: pray tell us what you do! *Curr Opin Pulmon Med* **13:**414.
3. Weller, P.F., Bozza, P.T., Yu, W., et al. 1999. Cytoplasmic lipid bodies in eosinophils: central roles in eicosanoid generation. *Int Arch Allergy Immunol* **118:**450.
4. Weller, P.F., Monahan-Earley, R.A., Dvorak, H.F., et al. 1991. Cytoplasmic lipid bodies of human eosinophils. Subcellular isolation and analysis of arachidonate incorporation. *Am J Pathol* **138:**141.
5. Bozza, P.T., Yu, W., Penrose, J.F., et al. 1997. Eosinophil lipid bodies: specific, inducible intracellular sites for enhanced eicosanoid formation. *J Exp Med* **186:**909.
6. Gleich, G.J., Adolphson, C.R., and Leiferman, K.M. 1993. The biology of the eosinophilic leukocyte. *Annu Rev Med* **44:**85.
7. Leiferman, K.M. 1991. A current perspective on the role of eosinophils in dermatologic diseases. *J Am Acad Dermatol* **24:**1101.
8. Weiss, S.J., Test, S.T., Eckmann, C.M., et al. 1986. Brominating oxidants generated by human eosinophils. *Science* **234:**200.
9. Moqbel, R., and Lacy, P. 2000. New concepts in effector functions of eosinophil cytokines. *Clin Exp Allergy* **30:**1667.
10. Spencer, L.A., Szela, C.T., Perez, S.A., et al. 2009. Human eosinophils constitutively express multiple Th1, Th2, and immunoregulatory cytokines that are secreted rapidly and differentially. *J Leukoc Biol* **85:**117.
11. Lacy, P., and Moqbel, R. 2000. Eosinophil cytokines. *Chem Immunol* **76:**134.
12. Rosenberg, H.F., Phipps, S., and Foster, P.S. 2007. Eosinophil trafficking in allergy and asthma. *J Allergy Clin Immunol* **119:**1303.

13. Flood-Page, P.T., Menzies-Gow, A.N., Kay, A.B., et al. 2003.Eosinophil's role remains uncertain as anti-interleukin-5 only partially depletes numbers in asthmatic airway. *Am J Respir Crit Care Med* **167:**199.
14. Foster, P.S., Hogan, S.P., Ramsay, A.J., et al. 1996. Interleukin 5 deficiency abolishes eosinophilia, airways hyperreactivity, and lung damage in a mouse asthma model. *J Exp Med* **183:**195.
15. Hamelmann, E., and Gelfand, E.W. 2001. IL-5-induced airway eosinophilia – the key to asthma? *Immunol Rev* **179:**182.
16. Boyce, J.A., and Austen, K.F. 2005. No audible wheezing: nuggets and conundrums from mouse asthma models. *J Exp Med* **201:**1869.
17. Foster, P.S., Mould, A.W., Yang, M., et al. 2001. Elemental signals regulating eosinophil accumulation in the lung. *Immunol Rev* **179:**173.
18. Hogan, S.P., Koskinen, A., Matthaei, K.I., et al. 1998. Interleukin-5-producing CD4+ T cells play a pivotal role in aeroallergen-induced eosinophilia, bronchial hyperreactivity, and lung damage in mice. *Am J Respir Crit Care Med* **157:**210.
19. Mattes, J., Yang, M., Mahalingam, S., et al. 2002. Intrinsic defect in T cell production of interleukin (IL)-13 in the absence of both IL-5 and eotaxin precludes the development of eosinophilia and airways hyperreactivity in experimental asthma. *J Exp Med* **195:**1433.
20. Lee, J.J., McGarry, M.P., Farmer, S.C., et al. 1997. Interleukin-5 expression in the lung epithelium of transgenic mice leads to pulmonary changes pathognomonic of asthma. *J Exp Med* **185:**2143.
21. Lee, N.A., McGarry, M.P., Larson, K.A., et al. 1997. Expression of IL-5 in thymocytes/T cells leads to the development of a massive eosinophilia, extramedullary eosinophilopoiesis, and unique histopathologies. *J Immunol* **158:**1332.
22. Mishra, A., Hogan, S.P., Brandt, E.B., et al. 2002. Enterocyte expression of the eotaxin and interleukin-5 transgenes induces compartmentalized dysregulation of eosinophil trafficking. *J Biol Chem* **277:**4406.
23. Terada, N., Hamano, N., Nomura, T., et al. 2000. Interleukin-13 and tumour necrosis factor-alpha synergistically induce eotaxin production in human nasal fibroblasts. *Clin Exp Allergy* **30:**348.
24. LaPorte, S.L., Juo, Z.S., Vaclavikova, J., et al. 2008. Molecular and structural basis of cytokine receptor pleiotropy in the interleukin-4/13 system. *Cell* **132:**259.
25. Wills-Karp, M. 2004. Interleukin-13 in asthma pathogenesis. *Immunol Rev* **202:**175.
26. Zimmermann, N., Hershey, G.K., Foster, P.S., et al. 2003. Chemokines in asthma: cooperative interaction between chemokines and IL-13. *J Allergy Clin Immunol* **111:**227.
27. Kita, H. 1996. The eosinophil: a cytokine-producing cell? *J Allergy Clin Immunol* **97:**889.
28. Liu, L.Y., Bates, M.E., Jarjour, N.N., et al. 2007. Generation of Th1 and Th2 chemokines by human eosinophils: evidence for a critical role of TNF-alpha. *J Immunol* **179:**4840.
29. Straumann, A., and Simon, H.U. 2004. The physiological and pathophysiological roles of eosinophils in the gastrointestinal tract. *Allergy* **59:**15.
30. Bisset, L.R., and Schmid-Grendelmeier, P. 2005. Chemokines and their receptors in the pathogenesis of allergic asthma: progress and perspective. *Curr Opin Pulmon Med* **11:**35.
31. Rincon, M., Anguita, J., Nakamura, T., et al. 1997. Interleukin (IL)-6 directs the differentiation of IL-4-producing CD4+ T cells. *J Exp Med* **185:**461.
32. Ottonello, L., Montecucco, F., Bertolotto, M., et al. 2005. CCL3 (MIP-1alpha) induces in vitro migration of GM-CSF-primed human neutrophils via CCR5-dependent activation of ERK 1/2. *Cell Signal* **17:**355.
33. Surquin, M., Le Moine, A., Flamand, V., et al. 2002. Skin graft rejection elicited by beta 2-microglobulin as a minor transplantation antigen involves multiple effector pathways: role of Fas-Fas ligand interactions and Th2-dependent graft eosinophil infiltrates. *J Immunol* **169:**500.
34. Debinski, W., Obiri, N.I., Pastan, I., et al. 1995. A novel chimeric protein composed of interleukin 13 and Pseudomonas exotoxin is highly cytotoxic to human carcinoma cells expressing receptors for interleukin 13 and interleukin 4. *J Biol Chem* **270:**16775.
35. Cormier, S.A., Taranova, A.G., Bedient, C., et al. 2006. Pivotal advance: eosinophil infiltration of solid tumors is an early and persistent inflammatory host response. *J Leukoc Biol* **79:**1131.
36. Jacobsen, E.A., Taranova, A.G., Lee, N.A., et al. 2007. Eosinophils: singularly destructive effector cells or purveyors of immunoregulation? *J Allergy Clin Immunol* **119:**1313.
37. Aceves, S.S., Newbury, R.O., Dohil, R., et al. 2007. Esophageal remodeling in pediatric eosinophilic esophagitis. *J Allergy Clin Immunol* **119:**206.
38. Wang, H., Tan, X., Chang, H., et al. 1999. Platelet-activating factor receptor mRNA is localized in eosinophils and epithelial cells in rat small intestine: regulation by dexamethasone and gut flora. *Immunology* **97:**447.
39. Fujii, M., Tanaka, H., and Abe, S. 2005. Interferon-gamma up-regulates expression of cysteinyl leukotriene type 2 receptors on eosinophils in asthmatic patients. *Chest* **128:**3148.
40. Zinchuk, O., Fukushima, A., Zinchuk, V., et al. 2005. Direct action of platelet activating factor (PAF) induces eosinophil accumulation and enhances expression of PAF receptors in conjunctivitis. *Mol Vis* **11:**114.
41. Bandeira-Melo, C., Bozza, P.T., Diaz, B.L., et al. 2000. Cutting edge: lipoxin (LX) A4 and aspirin-triggered 15-epi-LXA4 block allergen-induced eosinophil trafficking. *J Immunol* **164:**2267.
42. Ohshima, N., Nagase, H., Koshino, T., et al. 2002. A functional study on CysLT(1) receptors in human eosinophils. *Int Arch Allergy Immunol* **129:**67.
43. Powell, W.S., Chung, D., and Gravel, S. 1995. 5-Oxo-6, 8,11,14-eicosatetraenoic acid is a potent stimulator of human eosinophil migration. *J Immunol* **154:**4123.
44. Rothenberg, M.E., and Hogan, S.P. 2006. The eosinophil. *Annu Rev Immunol* **24:**147.
45. Mori, Y., Iwasaki, H., Kohno, K., et al. 2008. Identification of the human eosinophil lineage-committed progenitor: revision of phenotypic definition of the human common myeloid progenitor. *J Exp Med*.206:183–193.
46. Bedi, R, Du, J., Sharma, A.K., et al. 2009. Human C/EBP-epsilon activator and repressor isoforms differentially reprogram myeloid lineage commitment and differentiation. *Blood* **113:**317.
47. McNagny, K., and Graf, T. 2002. Making eosinophils through subtle shifts in transcription factor expression. *J Exp Med* **195:**F43.
48. Yu, C., Cantor, A.B., Yang, H., et al. 2002. Targeted deletion of a high-affinity GATA-binding site in the GATA-1 promoter leads to selective loss of the eosinophil lineage in vivo. *J Exp Med* **195:**1387.
49. Du, J., Stankiewicz, M.J., Liu, Y., et al. 2002. Novel combinatorial interactions of GATA-1, PU.1, and

C/EBPepsilon isoforms regulate transcription of the gene encoding eosinophil granule major basic protein. *J Biol Chem* **277:**43481.

50. Zimmermann, N., Daugherty, B.L., Kavanaugh, J.L., et al. 2000. Analysis of the CC chemokine receptor 3 gene reveals a complex 5' exon organization, a functional role for untranslated exon 1, and a broadly active promoter with eosinophil-selective elements. *Blood* **96:**2346.
51. Metcalf, D., Begley, C.G., Nicola, N.A., et al. 1987. Quantitative responsiveness of murine hemopoietic populations in vitro and in vivo to recombinant multi-CSF (IL-3). *Exp Hematol* **15:**288.
52. Metcalf, D., Burgess, A.W., Johnson, G.R., et al. 1986. In vitro actions on hemopoietic cells of recombinant murine GM-CSF purified after production in *Escherichia coli*: comparison with purified native GM-CSF. *J Cell Physiol* **128:**421.
53. Yamaguchi, Y., Suda, T., Suda, J., et al. 1988. Purified interleukin 5 supports the terminal differentiation and proliferation of murine eosinophilic precursors. *J Exp Med* **167:**43.
54. Sanderson, C.J. 1992. Interleukin-5, eosinophils, and disease. *Blood* **79:**3101.
55. Collins, P.D., Marleau, S., Griffiths-Johnson, D.A., et al. 1995. Cooperation between interleukin-5 and the chemokine eotaxin to induce eosinophil accumulation in vivo. *J Exp Med* **182:**1169.
56. Capra, V., and Rovati, G.E. 2004. Leukotriene modifiers in asthma management. *IDrugs* **7:**659.
57. Wardlaw, A.J., Walsh, G.M., and Symon, F.A. 1994. Mechanisms of eosinophil and basophil migration. *Allergy* **49:**797.
58. Wardlaw, A.J. 2000. The role of adhesion in eosinophil function. *Chem Immunol* **78:**93.
59. Sriramarao, P., von Andrian, U.H., Butcher, E.C., et al. 1994. L-selectin and very late antigen-4 integrin promote eosinophil rolling at physiological shear rates in vivo. *J Immunol* **153:**4238.
60. Rankin, S.M., Conroy, D.M., Williams, T.J. 2000. Eotaxin and eosinophil recruitment: implications for human disease. *Mol Med Today* **6:**20.
61. Hogan, S.P., Rothenberg, M.E., Forbes, E., et al. 2004. Chemokines in eosinophil-associated gastrointestinal disorders. *Curr Allergy Asthma Rep* **4:**74.
62. Bochner, B.S., and Schleimer, R.P. 1994. The role of adhesion molecules in human eosinophil and basophil recruitment. *J Allergy Clin Immunol* **94:**427.
63. Elices, M.J., Osborn, L., Takada, Y., et al. 1990. VCAM-1 on activated endothelium interacts with the leukocyte integrin VLA-4 at a site distinct from the VLA-4/fibronectin binding site. *Cell* **60:**577.
64. Ponath, P.D., Qin, S., Post, T.W., et al. 1996. Molecular cloning and characterization of a human eotaxin receptor expressed selectively on eosinophils. *J Exp Med* **183:**2437.
65. Phillips, R.M., Stubbs, V.E., Henson, M.R., et al. 2003. Variations in eosinophil chemokine responses: an investigation of CCR1 and CCR3 function, expression in atopy, and identification of a functional CCR1 promoter. *J Immunol* **170:**6190.
66. Elsner, J., Dulkys, Y., Gupta, S., et al. 2005. Differential pattern of CCR1 internalization in human eosinophils: prolonged internalization by CCL5 in contrast to CCL3. *Allergy* **60:**1386.
67. Sullivan, S.K., McGrath, D.A., Liao, F., et al. 1999. MIP-3alpha induces human eosinophil migration and activation of the mitogen-activated protein kinases (p42/p44 MAPK). *J Leukoc Biol* **66:**674.
68. Nagase, H., Miyamasu, M., Yamaguchi, M., et al. 2000. Glucocorticoids preferentially upregulate functional CXCR4 expression in eosinophils. *J Allergy Clin Immunol* **106:**1132.
69. Oliveira, S.H., Lira, S., Martinez, A.C., et al. 2002. Increased responsiveness of murine eosinophils to MIP-1beta (CCL4) and TCA-3 (CCL1) is mediated by their specific receptors, CCR5 and CCR8. *J Leukoc Biol* **71:**1019.
70. Zhu, Z., Zheng, T., Homer, R.J., et al. 2004. Acidic mammalian chitinase in asthmatic Th2 inflammation and IL-13 pathway activation. *Science* **304:**1678.
71. Moser, R., Fehr, J., and Bruijnzeel, P.L. 1992. IL-4 controls the selective endothelium-driven transmigration of eosinophils from allergic individuals. *J Immunol* **149:**1432.
72. Sher, A., Coffman, R.L., Hieny, S., et al. 1990. Ablation of eosinophil and IgE responses with anti-IL-5 or anti-IL-4 antibodies fails to affect immunity against Schistosoma mansoni in the mouse. *J Immunol* **145:**3911.
73. Kita, H., Adolphson, C.R., and Gleich, G.J. 2003. Biology of eosinophils. In *Allergy Principles and Practice*, Adkinson, J., Yunginger, J.W., Busse, W.W., et al. (eds.), 6th Ed., p. 305 Philadelphia: Mosby.
74. Kayaba, H., Dombrowicz, D., Woerly, G., et al. 2001. Human eosinophils and human high affinity IgE receptor transgenic mouse eosinophils express low levels of high affinity IgE receptor, but release IL-10 upon receptor activation. *J Immunol* **167:**995.
75. Kita, H., Kaneko, M., Bartemes, K.R., et al. 1999. Does IgE bind to and activate eosinophils from patients with allergy? *J. Immunol* **162:**6901.
76. Sihra, B.S., Kon, O.M., Grant, J.A., et al. 1997. Expression of high-affinity IgE receptors (FcεRI) on peripheral blood basophils, monocytes, and eosinophils in atopic and nonatopic subjects: relationship to total serum IgE concentrations. *J Allergy Clin Immunol* **99:**699.
77. Bandeira-Melo, C., and Weller, P.F. 2003. Eosinophils and cysteinyl leukotrienes. *Prostaglandins Leukot Essent Fatty Acids* **69:**135.
78. Gharaee-Kermani, M., and Phan, S.H. 1998. The role of eosinophils in pulmonary fibrosis (Review). *Int J Mol Med* **1:**43.
79. Gomes, I., Mathur, S.K., Espenshade, B.M., et al. 2005. Eosinophil-fibroblast interactions induce fibroblast IL-6 secretion and extracellular matrix gene expression: implications in fibrogenesis. *J Allergy Clin Immunol* **116:**796.
80. Levi-Schaffer, F., Garbuzenko, E., Rubin, A., et al. 1999. Human eosinophils regulate human lung- and skin-derived fibroblast properties in vitro: a role for transforming growth factor beta (TGF-beta). *Proc Natl Acad Sci USA* **96:**9660.
81. Spry, C.J. 1989. The pathogenesis of endomyocardial fibrosis: the role of the eosinophil. *Springer Semin Immunopathol* **11:**471.
82. Dvorak, A.M., Ackerman, S.J., Furitsu, T., et al. 1992. Mature eosinophils stimulated to develop in human-cord blood mononuclear cell cultures supplemented with recombinant human interleukin-5. II. Vesicular transport of specific granule matrix peroxidase, a mechanism for effecting piecemeal degranulation. *Am J Pathol* **140:**795.

83. Melo, R.C., Perez, S.A., Spencer, L.A., et al. 2005. Intragranular vesiculotubular compartments are involved in piecemeal degranulation by activated human eosinophils. *Traffic* **6:**866.
84. Logan, M.R., Odemuyiwa, S.O., and Moqbel, R. 2003. Understanding exocytosis in immune and inflammatory cells: the molecular basis of mediator secretion. *J Allergy Clin Immunol* **111:**923.
85. Moqbel, R., and Coughlin, J.J. 2006. Differential secretion of cytokines. *Sci STKE* **2006:**pe26.
86. Moqbel, R., and Lacy, P. 1999. Exocytotic events in eosinophils and mast cells. *Clin Exp Allergy* **29:**1017.
87. Dvorak, A.M., Ackerman, S.J., and Weller, P.F. 1990. Subcellular morphology and biochemistry of eosinophils. In *Blood Cell Biochemistry: Megakaryocytes, Platelets, Macrophages and Eosinophils*. Harris, J.R. (ed.), Vol 2, p.237. London: Plenum Publishing Corporation.
88. Gleich, G.J. 2000. Mechanisms of eosinophil-associated inflammation. *J Allergy Clin Immunol* **105:**651.
89. Martin, L.B., Kita, H., Leiferman, K.M., et al. 1996. Eosinophils in allergy: role in disease, degranulation, and cytokines. *Int Arch Allergy Immunol* **109:**207.
90. Thomas, L.L., and Page, S.M. 2000. Inflammatory cell activation by eosinophil granule proteins. *Chem Immunol* **76:**99.
91. Jacoby, D.B., Costello, R.M., and Fryer, A.D. 2001. Eosinophil recruitment to the airway nerves. *J Allergy Clin Immunol* **107:**211.
92. Rochester, C.L., Ackerman, S.J., Zheng, T., et al. 1996. Eosinophil-fibroblast interactions. Granule major basic protein interacts with IL-1 and transforming growth factor-beta in the stimulation of lung fibroblast IL-6-type cytokine production. *J Immunol* **156:**4449.
93. Sedgwick, J.B., Calhoun, W.J., Vrtis, R.F., et al. 1992. Comparison of airway and blood eosinophil function after in vivo antigen challenge. *J Immunol* **149:**3710.
94. Holub, A., Byrnes, J., Anderson, S., et al. 2003. Ligand density modulates eosinophil signaling and migration. *J Leukoc Biol* **73:**657.
95. Meerschaert, J., Kelly, E.A., Mosher, D.F., et al. 1999. Segmental antigen challenge increases fibronectin in bronchoalveolar lavage fluid. *Am J Respir Crit Care Med* **159:**619.
96. Meerschaert, J., Vrtis, R.F., Shikama, Y., et al. 1999. Engagement of alpha4beta7 integrins by monoclonal antibodies or ligands enhances survival of human eosinophils in vitro. *J Immunol* **163:**6217.
97. Anwar, A.R., Moqbel, R., Walsh, G.M., et al. 1993. Adhesion to fibronectin prolongs eosinophil survival. *J Exp Med* **177:**839.
98. Neves, J.S., Perez, S.A., Spencer, L.A., et al. 2008. Eosinophil granules function extracellularly as receptor-mediated secretory organelles. *Proc Natl Acad Sci USA* **105:**18478.
99. Walsh, M.T., Curran, D.R., Kingham, P.J., et al. 2004. Effect of eosinophil adhesion on intracellular signaling in cholinergic nerve cells. *Am J Respir Cell Mol Biol* **30:**333.

SUGGESTED READINGS

Fillon, S., Robinson, Z.D., Colgan, S.P., and Furuta, G.T. 2009. Epithelial function in eosinophilic gastrointestinal diseases. *Immunol Allergy Clin North Am* **29**(1)**:**171–178.

Gleich, G.J., and Leiferman, K.M. 2009. The hypereosinophilic syndromes: current concepts and treatments. *Br J Haematol* **145**(3)**:**271–285.

Hogan, S.P., Rosenberg, H.F., Moqbel, R., et al. 2008. Eosinophils: biological properties and role in health and disease. *Clin Exp Allergy* **38**(5)**:**709–750.

8 Macrophages

Sarah Fox and Adriano G. Rossi

INTRODUCTION

Macrophages are a major leukocyte involved in orchestrating inflammatory responses. Their name is derived from the Greek term "big eaters" (*makros*, large; and *phagein*, eat). This gives some insight into the primary function of this cell in clearance of invading pathogens, cell debris, and apoptotic cells by a process of engulfment and digestion called "phagocytosis." However, the role of the macrophage goes beyond that of what its name suggests. They are endowed with the ability to rapidly react to and secrete a plethora of biological agents and mediators that can influence the initiation, progression, and resolution of an inflammatory response and coordinate processes to establish acquired immunity against specific pathogens. This chapter is an overview of the basics of macrophage biology and function, with particular insights into the involvement of macrophages in disease pathogenesis as well as pharmacological modulation of macrophage responses as targets for treatment of disease.

IDENTIFICATION

Historical Discovery

Elie Metchnikoff first used the term "macrophage" to describe large mononuclear phagocytic cells he observed in tissues over 100 years ago. Macrophages are now recognized as the major phagocytic cell type with diverse characteristics and localities around the body where they are important for both innate and acquired immune responses as well as maintenance of tissue homeostasis and regulation of various processes subsidiary to the immune defense such as hematopoiesis.

Hematopoietic Lineage

In man, macrophages are derived from bone marrow hematopoietic cells and constitute the terminally differentiated cell type from the mononuclear phagocytic cell lineage. Monocytes are the immediate precursor cell types of macrophages and these cells have important functions in inflammatory responses on their own. It is monocytes that are liberated from bone marrow to circulate in the bloodstream where they are recruited to sites of inflammation. Tissue macrophages are present constitutively in the absence of inflammation where they are thought to contribute to the maintenance of many processes (see heterogeneity). During an inflammatory response, monocytes differentiate into macrophages within the tissue where they remain until their useful function is completed. It is thought that they either die by apoptosis *in situ* or are cleared by drainage to the lymphatic system.

Phenotypic Characteristics

Macrophages are characterized by their large size (approx 5–50 µm in diameter) and irregular shape. They generally contain a large kidney-shaped nuclear lobe (mononuclear). The cytoplasm is abundant and contains many large and small granules, which contain the chemical cocktail needed to kill and digest invading pathogens.

Cell Surface Markers, Isolation of Macrophages

Macrophages are endorsed with an impressive repertoire of surface receptors, which allow these cells to have functionally diverse actions. Many of these receptors are not exclusive to macrophage cell populations and are shared by other cell types such as granulocytes and lymphocytes. There are macrophage-specific cell surface markers, which are used to identify and isolate these cells from unpurified cell populations. The F4/80 antibody was developed as one of the first macrophage-specific markers and remains one of the most common markers used for this purpose. The F4/80 antibody recognizes human CD97, a glycoprotein 7-transmembrane domain protein receptor resembling the G-protein-coupled peptide hormone receptor family. It is expressed highly in tissue macrophage populations, dendritic cells (DCs), Langerhans cells, and microglia. The receptor is also expressed in small to moderate levels in blood monocytes and eosinophillic granulocytes. No clear function of CD97 is apparent although the high degree of posttranslational modification of this protein could be linked to diverse functional characteristics of macrophages including cell–cell recognition and adhesion. Fluorescent markers conjugated to antibodies against the F4/80 antigen can be used to distinguish macrophages from other cell types. In this context, a cell sorter can distinguish those cells with higher fluorescent profiles, and due to the relatively specific binding capacity of the antibody for the macrophage F4/80 antigen, a purified macrophage population can then be singled out for analysis.

Cell Surface Receptors; Pathogen Recognition Receptors

The ability to discriminate between self and foreign particles by cells involves the recognition motifs expressed on prokaryotic cells. These motifs are highly conserved within the pathogenic organisms and are subsequently not subjected to the high rates of mutation occurring in other aspects of prokaryotic biology. Pathogen recognition receptors (PRRs) are receptors located both on the cell surface and in the intracellular environment that can recognize an array of pathogen-associated molecular patterns (PAMPs). On the cell surface, the Toll-like receptors (TLRs) form the major class of PRRs found on the surface of macrophages. There are 10 known TLRs in humans and 13 in mice, and most of these TLRs are located on the cell membrane although some are found in intracellular membranes such as those found on endosomes or lysosomes. These receptors endow the cell with the ability to recognize a range of PAMPs including lipid and carbohydrate residues from Gram-positive bacteria (TLR1, TLR2, and TLR6), LPS from Gram-negative bacteria (TLR4), fungal pathogens (TLR2), bacterial proteins such as flagellin and profilin (TLR5 and TLR11 respectively), nucleic acid and its derivatives (TLR3, TLR7, TLR8, and TLR9), double-stranded RNA (TLR3), synthetic antiviral compounds and guanine nucleotide analogues (murine TLR7 and human TLR8), and DNA containing unmethylated CpG-motifs abundant in bacterial DNA (TLR9). The TLRs can also form heterodimers with other subtypes of TLRs or form complexes with accessory molecules (LPS-binding protein and CD14 in the case of TLR4). TLR-independent recognition of pathogens is important particularly for intracellular recognition of pathogens and can occur via the nucleotide-binding oligomerization domain (NOD)-like receptors (NLRs) or retinoic acid–inducible gene I (RIG-I)-like helicases (RLHs). Macrophages use these receptors to form robust responses to various types of pathogenic infection, which stimulate an array of second messenger pathways to mount diverse functional immune responses.

Heterogeneity

All members of the mononuclear phagocytic system (MPS) cell lineage share homology in morphological and ultrastructural features, expression of specific enzymes and surface receptors, and the ability to engulf various particles. However, there are many different morphologically and functionally diverse subsets of macrophages that have adapted to perform tasks ideally suited to the surrounding environment.

Bone Marrow Macrophages

Cells of the monocytes-macrophage line are recruited into the bloodstream where they circulate and migrate into tissues where they differentiate into specific tissue macrophages or, depending on the inflammatory status of the tissue environment, they will form activated inflammatory macrophages. However, mature macrophages are also found within the bone marrow where they play an important role in support and regulation of hematopoiesis. Release of granulocyte-macrophage colony-stimulating factor (GM-CSF) can stimulate maturation of granulocyte macrophage and eosinophil colony formation in vitro.

Alveolar Macrophages

Alveolar macrophages are a distinct population of cells found within the lung where they provide protection against pathogens and particles inhaled from the air. They are the only type of macrophage that are exposed to environmental air and, as such, have developed characteristics appropriate for the task of clearing the abundance of potentially harmful bodies entering the body through this route. They characteristically contain a large number of membrane-bound cytoplasmic inclusions, indicative of their ability to rapidly release a barrage of proteolytic enzymes and reactive oxygen species to kill and clear foreign bodies. Damage to these cells is an important factor associated with increased host susceptibility to airborne infection and environmental toxins.

Splenic Macrophages

Resident macrophages in the spleen are thought to function to trap and process foreign antigens, coordinate with T and B cells in the lymphoid areas, as well as the engulfment and digestion of red blood cells in the red pulp. The specific phenotype of macrophages in the spleen varies between different locations and likely reflects the degree of functional specialization of these cells.

Kupffer Cells

The resident macrophages in the liver called Kupffer cells are responsible for clearance of particulate and soluble substances from the portal circulation primarily by phagocytosis. They are responsible for the detoxification of endotoxin from the portal circulation and can present antigens to initiate T-cell lymphocyte responses. Ingestion of microorganisms by Kupffer cells triggers the release of a deluge of inflammatory mediators that rapidly recruits monocytes to the liver to aid in clearance of invading pathogens. Thus, these cells are considered as an important front in the defense against infection but also contribute to acute liver damage sustained during liver toxicity (e.g., alcohol consumption) and cancer.

Microglial Cells

Microglial cells are the first and main line of defense against pathogens of the central nervous system (CNS). They differentiate from myeloid progenitor cells normally found within the bone marrow and differ from macrophages in their responses to various stimuli and their turnover rate. Microglial cells do not need to be replaced as often as other macrophages and have adapted a range of quiescent and activated states to maintain their function over longer periods. They are termed nonprofessional phagocytes due to their need to be activated before functions such as phagocytosis and antigen presentation are performed. They can rapidly respond to CNS challenges due to the ability of these cells to use potassium channels and changes in potassium levels as sensors of injury or infection.

Langerhans Cells

Langerhans cells are a type of DCs found predominantly and in large numbers in the epidermis where they function to recognize and capture antigens. DCs are the main cells responsible for antigen processing and presentation. They are derived from activated monocyte precursors and found in areas likely to come into contact with antigens such as the skin, nose, lungs, stomach, and intestines. They differ from tissue macrophages in their morphology and primary function that in the case of the latter is phagocytosis. During an immature stage, they extend large cytoplasmic projections called dendrites (not the same as neuronal dendrites but from which they get their name) that increase their surface area and increase contact area with surrounding antigens. The surface of DCs are endowed with PRRs such as the TLRs that recognize and uptake various bacterial and viral antigens for processing via the major histocompatibility complex (MHC) molecule. Upregulation of the surface chemotactic receptor (CCR7) encourages the transport of the DCs to lymph nodes where phenotypic changes occur to facilitate the presentation of the processed antigens on their surface with naive T cells, thus orchestrating the adaptive immune response.

Mesangial Cells

Mesangial cells are specialized phagocytic cells located in the glomerular lobules of the kidney. They are thought to contribute to the support of these kidney structures and to aid in filtration and regulation of blood flow to the glomerular capillaries through contractile responses, similar to smooth muscle cells. In fact, these cells are rare examples of phagocytic cells derived from smooth muscle and not monocytes. Mesangial cells have many other characteristics that differ from that of the smooth muscle, they can respond to and secrete a number of vasoactive and immunomodulatory substances as well as antigen presentation making these cells important in the local response to injury in the kidney.

Other

Macrophages are present in many other areas of the body particularly where you would expect exposure to or processing of foreign pathogens such as the gut and lymphoid tissues. Interestingly, macrophages can fuse to form multinucleated cells often termed giant cells. The giant cells tend to form two main phenotypes (Figure 8.1): the Langhans cells (Figure 8.1A) where the nuclei are arranged in a horseshoe or ring shape around the outer periphery of the cell and the so-called foreign body giant cell (Figure 8.1B) where the nuclei are more disorganized but often form a group of nuclei in the center of the cell. The precise functions of multinucleated giant cells, however, are unknown although they appear to be found in granulomatous disease and the origin of multinucleated osteoclasts.

MACROPHAGE ACTIVATING STIMULI

The surface of the macrophage is equipped with many receptors that stimulate macrophage function such as adhesion, chemotaxis, secretion, and phagocytosis. Activating stimuli are often found in the inflammatory milieu and are commonly produced by macrophages themselves, other inflammatory cell types, or the invading organisms. Thus, the inflammatory environment

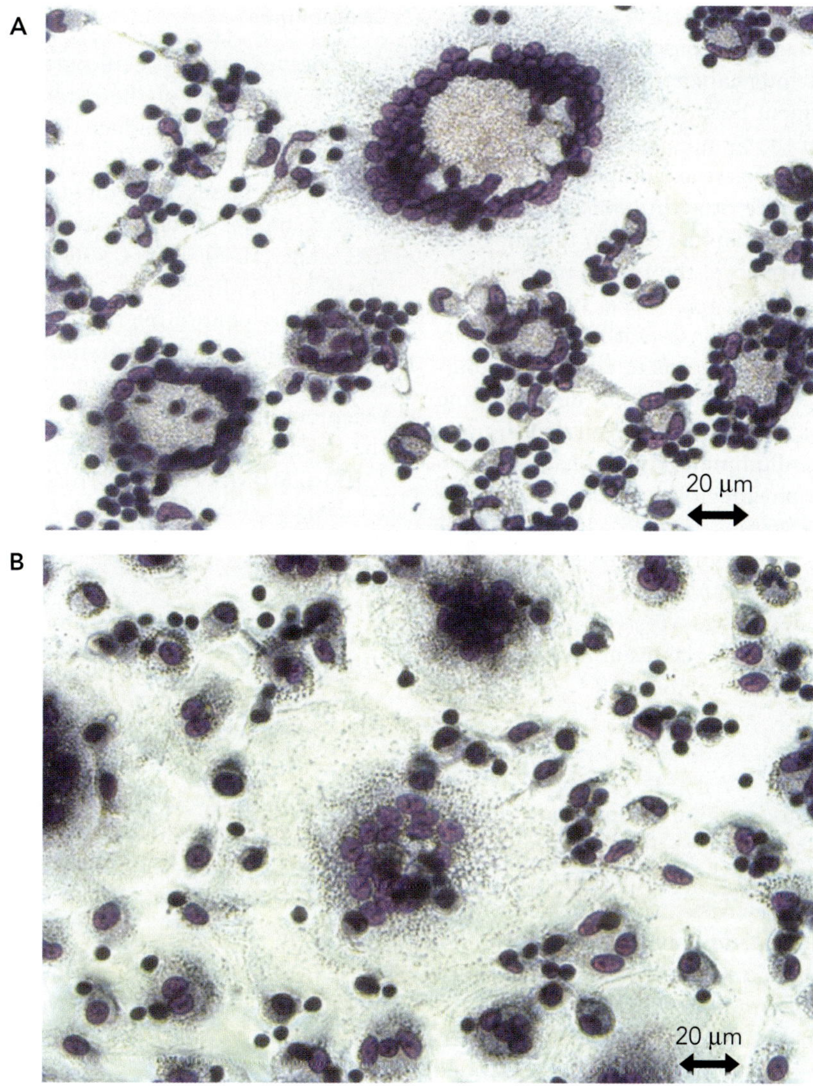

Figure 8.1. Multinucleated giant cells of human macrophages differentiated from blood-derived monocytes. (A) Represents typical images of Langhans cells and (B) depicts examples of foreign body cells. (Images derived from work by Martin F. Lister, Charlotte M. Hamilton, and Adriano G. Rossi.)

is instrumental in determining the type and intensity of the inflammatory response by the macrophage. It is imperative for macrophages to be able to distinguish between cells that are "self" and those that are "non-self" during the inflammatory process. Surface receptors expression by both macrophages and surrounding cells are a way in which macrophages are able to do this.

Fc and Complement

Immunoglobulin (Ig) antibody complexes and complement proteins are found in abundance in the serum. They bind to the outside of pathogenic material to facilitate the recognition and clearance of these particles by phagocytes. This process is called opsonization and, thus, Ig antibodies and complement are important particle opsonins. The Fc portion of the Ig complex is prominent upon opsonization of a particle and it is this portion that is readily recognized by Fc receptors (FcRs) on the surface of the macrophage. The FcRs involved in the internalization of Ig complexes are the FcγRI, FcγRIIA, and the FcγRIII. Particles opsonized by complement proteins are recognized and facilitate phagocytosis through the stimulation of the complement receptors (CRs), also located on the surface of the macrophage. Several CRs exist on macrophages, namely CR1, CR3, and CR4, which recognize several complement proteins. For example, the CR1 binds C3b, C4b, and

C3bi. The CRs can also function to facilitate different aspects of the phagocytic process such as particle binding (CR1) and internalization (CR3 and CR4). The Fc and complement pathways of particle uptake are different in that the Fc-mediated process is constitutively active whereas the complement system can require the presence of additional stimuli, such as tumor necrosis factor alpha (TNF-α) for phagocytosis to occur. Furthermore, the initial macrophage surface changes differ between these two receptor-mediated phagocytic response with Fc receptors producing membrane projections to envelop the particle, whereas the complement system does not and the particle appears to sink into the cell. Differences also arise in the inflammatory response of macrophages following phagocytosis mediated by either receptor; the FcR process is coupled with the release of proinflammatory mediators such as reactive oxygen species (ROS) and arachidonic acid metabolites whereas CR-stimulated phagocytosis does not induce the release of either of these classes of molecules and is considered less proinflammatory.

Cytokines

During inflammation, cytokine release from inflammatory cells is greatly increased. Many different cytokines exist that orchestrate a number of important immunomodulatory roles. Macrophages can be either classically activated or alternatively activated depending on the cytokine profile of the surrounding inflammatory environment. Classically activated macrophages are driven by cytokine activators interferon gamma (IFN-γ), TNF-α, and also by microbial products such as LPS. Classically activated macrophages are highly proinflammatory and microbicidal cells that produce exaggerated levels of ROS and cytokines responsible for immune defense. In contrast, alternatively activated macrophages are a consequence of predominant IL-4 and IL-13 production from Th2 cells dominant during parasitic infection. They induce macrophage production of specific chemokines, macrophage-derived chemokine (MDC) and thymus and activation-regulated chemokine (TARC), and upregulate the expression of receptors responsible for endocytosis and antigen presentation and enzymes responsible for arginase metabolism and granuloma formation. The alternatively activated macrophage does not produce a major respiratory burst and ROS products are considered less proinflammatory but nonetheless are critical for clearing of parasitic infections. Other cytokines are also important for macrophage deactivation such as IL-10 and transforming growth factor-beta (TGF-β), which promotes the resolution of the macrophage inflammatory response.

Chemokines

Chemokines are a specific group of peptide inflammatory mediators of the cytokine family that act as chemoattractants to induce the recruitment of inflammatory cells to the site of injury or infection. There are various subclasses of chemokines that are described as CC (e.g., monocyte chemoattractant protein-1; MCP-1), CXC (e.g., IL-8), and CX3C (fractalkine). They are classed depending on the molecular structure with CC chemokines containing two cysteine residues together and CXC chemokines containing two cysteine residues separated by another amino acid. A number of CC receptors and CXC receptors have been identified on the macrophage that can bind chemokines including MIP-1α, MIP-1β, RANTES, MCP-1, and IL-8.

Lipoproteins

Macrophages express receptors that allow them to recognize, uptake, and digest cholesterol-containing lipoproteins such as low-density lipoproteins (LDLs). Macrophages are also involved in oxidation of lipoproteins by macrophage produced ROS, a process linked with atherosclerotic plaque progression. The uptake of LDLs occurs through macrophage recognition receptors for apoplipoprotein E and apoplioprotein B-100 expressed on the LDL molecule. Scavenger receptors have also been implicated in the uptake of modified lipoproteins.

Integrins

Integrin molecules are expressed on the surface of many cell types including neutrophils, endothelial cells, and macrophages. They are important for cell adhesive interactions and are commonly upregulated on the surface of cells during inflammation. Cell adhesion during inflammation is particularly important during macrophage migration and cell adhesion during phagocytosis. Integrins are recognized by macrophage receptors to facilitate adhesion and to initiate macrophage behaviors that allow cell movement including actin rearrangement.

Oligosaccharides

Oligosaccharide molecules terminating in fucose, mannose, galactose, and sialic acid are recognized by specific macrophage receptors. The discrimination of self and nonself particles is partly due to expression of cell specific oligosaccharide molecules. Bacterial and fungal membranes characteristically contain terminal monosaccharide residues that are recognized by the macrophage mannose receptor to facilitate phagocytic clearance of these particles.

TABLE 8.1. Major macrophage secretory products

Complement proteins	**Cytokines**
C1	Interleukin (IL)-1α and β
C2	IL-6
C3	MIP-2
C3b	IFN-α and γ
C3b inactivator	TGF-β
C4	TNF-α
	IL-8
Enzymes	Colony stimulating factors (G-CSF, M-CSF, and GM-CSF)
Lysozyme	
Plasminogen activator	
Collagenase	**Reactive oxygen species**
Elastase	Superoxide anions
Angiotensin converting enzyme	Hydrogen peroxide
Cysteine proteinase	Hydroxyl radical
Proteases	Nitric oxide
Lipases	
Deoxyribonucleases	**Antioxidants**
Phosphatases	Glutathione
Glycosidases	
Sulfatases	**Coagulation factors**
Arginase	Factor V, VII, IX, and X
	1,25-dihydroxyvitamin D3
Proteins	Thromboplastin
Alpha-proteinase inhibitor	Prothrombin
Alpha-macroglobulin	Prothrombinase
Plasminogen activator inhibitor	Tissue factor
IL-1 inhibitor	Plasminogen activator
Neutrophil migration inhibitor	
Inhibitor of fibroblast growth	**Bioactive lipids**
Lipomodulin	Prostaglandin E, $F_2\alpha$
Glycoproteins	Cyclooxygenase
Fibronectin	Thromboxane A_2
Transferrin	Leukotriene B_4, C_4, D_4, E
Transcobalamin II	Platelet-activating factor
Apolipoprotein A and E	

Others

The above description of macrophage activating agents is by no means an exhaustive list as there are many other equally important mediators that have been demonstrated to activate or modulate macrophage responsiveness. For example, there are many lipid mediators including prostaglandins, leukotrienes, lipoxins, and ether lipids (e.g., platelet-activating factor) which can influence macrophage function. Other agents known to affect macrophage function include adenosine, ATP, neuropeptides, nitric oxide, and so on, which are covered in other chapters.

MACROPHAGE SECRETORY PRODUCTS

Much has been discovered about macrophages over the past few decades that go beyond the phagocytic function from which they were originally described. These cells release many agents that influence and orchestrate many processes including enhancement and resolution of the inflammatory response, microbial killing, antigen presentation, and wound healing to name a few (Table 8.1). The specific profile of agents released is dynamic and can switch from proinflammatory to antiinflammatory phenotypes, further demonstrating the multi-faceted roles of the macrophage in inflammation.

Enzymes

Various enzymes are released from macrophages both constitutively and in response to Fc or complement receptor engagement, cytokines, or bacterial products. They are generally released to aid digestion of cell debris and pathogens by breaking down components of cell membranes, connective tissue, and nuclear material (DNA). Lysozyme is secreted constitutively and in large amounts by macrophages where it functions to digest bacterial cell walls. Enzymes that are released upon macrophage stimulation include proteases, lipases, deoxyribonucleases, phosphatases,

glycosidases, collagenase, elastase, and sulfatases. Enzyme secretion by macrophages also serves to activate or inhibit various pathways including the complement pathway, renin-angiotensin system, and the kinin-kallikrein system.

Cytokines

Cytokines, many of which can themselves influence macrophage behavior (see above) are described as soluble peptide factors that can influence cell growth, function, and differentiation of many target cell types. Cytokine secretion by macrophages plays a pivotal role in orchestration of many stages of the immune response. Most cytokines are released during an inflammatory response when concentrations of substances such as TNF-α and IL-6 can be significantly increased in the inflammatory environment. Although TNF-α was one of the first cytokines discovered that is secreted by macrophages, these cells release many different cytokines although many share a common target and often work in synergy to achieve their function. Furthermore, the inflammatory milieu can influence the function of secreted cytokines. The cell priming phenomena is an example of how the inflammatory environment can affect macrophage responses. Stimuli that do not necessarily evoke an inflammatory response on their own, when presented with another inflammatory stimulus can cause a reaction that is massively increased. Such processes help explain how macrophages can adapt their responses in a manner appropriate to the inflammatory insult. The most well known macrophage priming agent is IFN-γ.

Complement Components

Macrophages can secrete many components of both the classical and alternative complement systems. These include C1, C2, C3, C4, C5, factor B, factor D, properdin, C3b inactivator, and α-IH. In addition, proteases secreted by macrophages can generate active fragments of the complement system such as C3a, C3b, C5a, and Bb. The complement system is involved in many immunoregulatory functions including lysing of foreign cells by forming C5–9 complex or by acting as opsonins to promote recognition and uptake of particles by phagocytosis. Complement factors can also act directly to promote or suppress the inflammatory response by binding directly to complement receptors on the macrophage surface. Factors C3a and C5a are well-known macrophage chemoattractants whereas factor Bb, a product of the alternative complement pathway, suppresses macrophage migration.

Coagulation Factors

Macrophages secrete coagulation factors VII, IX, X, and V. The coagulation cascade is a critical pathway stimulated in response to injury or trauma where it promotes blood clotting and through recruitment and activation of platelets and wound healing through deposition of fibrin.

Reactive Oxygen Intermediates

Small molecular weight molecules such as O_2^- and NO are released by activated macrophages. The rapid increase in macrophage oxygen consumption and reactive oxygen intermediate (ROI) production occurs in response to various inflammatory stimuli. This process is called the respiratory burst. These highly reactive molecules are the product of macrophage NADPH and iNOS enzyme activity. They are directly micobicidal but can also react with surrounding molecules to form other cytotoxic species including H_2O_2, $ONOO^-$, and OH^- that also contribute to defense against infection. The importance of these substances during infection is shown through the use of mouse models deficient in macrophage-derived ROS, which show markedly increased susceptibility to infection by various organisms.

Arachidonic Acid Intermediates

Macrophages are known to be important sources of products of the arachidonic acid pathway such as prostacyclin, thromboxane, PGE_2, and LTB_4. Macrophages contain the necessary enzyme components involved in synthesis, control, and metabolism of these products such as PLA_2 that forms arachidonic acid by cleaving phospholipid precursors and the cyclooxygenase and lipoxygenase enzymes that control the metabolism of arachidonic acid into active prostacyclin, prostaglandins, thromboxanes, and leukotrienes. The release of metabolites of arachidonic acid has multiple functions during inflammation including effects on macrophage functions. Lipoxin A4 production increases the phagocytosis of neutrophils by macrophages and is thus considered important in the resolution phase of the inflammatory response. Macrophages are a significant source of prostaglandin E that acts to control body temperature through its action on the temperature regulatory organ, the hypothalamus. The elevation of body temperature is considered an important defense mechanism that develops to create an environment less suitable for pathogens to thrive in while simultaneously facilitating immune functions.

FUNCTIONAL RESPONSES OF MACROPHAGES

Phagocytosis of Cell Debris and Invading Organisms

As already mentioned, the process of engulfing and digesting particles (phagocytosis) is the characteristic most associated with, although not exclusive to, macrophages. Most cells have some phagocytic capacity, such as some epithelial cell types, but these cells are not equipped with the complex repertoire of receptors specific for recognition of a large particle range and modulation of phagocytic rate. The process of phagocytosis is complex and involves a large number of recognition receptors and signaling events to initiate this process. The fundamental steps surrounding phagocytosis are as follows: recognition of particles requiring to be phagocytosed, rearrangement of macrophage cytoskeleton to allow it to extend "arms" (pseudopodia) and surround the particle and, lastly, to digest the particle. Different particles can stimulate an array of different receptors and signaling cascades and these processes are capable of cross talk and synergy adding further complexity to the regulation of phagocytosis. Thus, phagocytosis is a very dynamic process that allows for a diverse reactionary repertoire appropriate to the inflammatory insult. For example, the ingestion of Gram-negative microbial particles involves receptors for LPS that trigger unique intracellular signaling pathways to induce phagocytosis but which also modulate transcriptional activity of the macrophage. In particular, expression of the inflammatory mediator TNF-α is upregulated and released into the inflammatory environment, where its presence can recruit and activate other inflammatory cells, thus allowing for an intense inflammatory response to clear the infection. In contrast to this, the ingestion of innate senescent cells, such as apoptotic neutrophils, occurs in a noninflammatory manner using distinct recognition mechanisms that utilize surface molecules on the surface of apoptotic cells (e.g., phosphatidylserine) and specific receptors of the surface of the phagocytic macrophage. Interestingly, when macrophages phagocytose apoptotic cells (a process termed efferocytosis) their phenotype is changed in that the macrophage starts to reduce the amount of proinflammatory mediator being synthesized and liberated. Indeed, there is now strong evidence that these macrophages start to express mediators that can have anti-inflammatory potential (e.g., TGF-β and IL-10) thereby aiding the resolution of the inflammatory response. Phagocytosis in general is a highly regulatable process that can be modulated by endogenous mediators and by pharmacological intervention. The clearance by macrophages of apoptotic cells can be modulated by complement components, CD44 ligation, certain prostaglandins, lipoxins, and glucocorticoids, to name a few. Enhancement of macrophage clearance of apoptotic cells as an anti-inflammatory strategy is a realistic target for the pharmaceutical industry as the process is likely to be important in the resolution of inflammation. Indeed, convincing evidence indicates that failed clearance of apoptotic cells contributes to the pathogenesis of diseases such as systemic lupus erythematosus.

Chemotaxis

Chemotaxis is described as the movement of cells in response to a chemical stimulus (chemokine). These stimuli are often released by foreign pathogens or inflammatory cells (e.g., neutrophils) allowing cells such as macrophages to accurately migrate to the site of infection or injury. Chemotaxis is initiated by the stimulation of receptors on the macrophage surface, which in turn facilitate the rearrangement of cytoskeletal structures, resulting in movement of the cell in the direction of the stimulus. Rearrangement of cytoskeletal structures also allows the migration of cells through blood vessel cell layers into tissues in a process called transmigration. The most common chemotactic stimuli are bacterial products including formylated peptides and LPS although other inflammatory cells release factors such as IL-8, MCP-1, RANTES, and GM-CSF which can directly stimulate or enhance the chemotactic response of macrophages.

Antigen Processing and Presentation

In addition to the crucial role macrophages play in innate immune defenses, these cells are also important in orchestrating acquired immune responses. Antigen processing and presentation by macrophages to T- and B-cell lymphocytes allows for heightened recognition and clearance of pathogen-infected cells and antibody production against the pathogen, which allows for rapid clearance of the pathogen upon subsequent exposure. DCs are the major cell type responsible for antigen processing and presentation although macrophages are also known to perform such duties and are classed as professional antigen presenting cells (APCs) along with antibody expressing B cells. These cells process pathogenic proteins in the cytoplasm following phagocytosis of virus or bacteria initiated through stimulation of PRRs on the surface of the APCs. The processing of antigens is dependent on the MHC molecules of which there are various classes (e.g., MHC I and MHC II). These complexes are brought into contact with foreign proteins within the endoplasmic reticulum (MHC I) where they are processed into denatured or peptide fragments (antigens) by catabolism. The MHC associated with the antigen is then transported to the APC cell surface where it is recognized by

CD4+ and CD8+ lymphocytes that can initiate a rapid cytotoxic response by initiating apoptosis of APCs that are infected with viral or bacterial pathogens. APCs also promote recruitment and clonal proliferation of T cells endowed with the specific recognition motifs for the MHC:antigen complex. Memory B cells expressing the antibody complex that recognizes the specific pathogenic antigen remain in circulation following resolution of the infection where they can rapidly activate T-cell–mediated responses and clearance of the pathogen should the host be infected again. This method of immunity is exploited in vaccine development where processed or fragments of bacterial or viral antigens are given to a host along with an adjuvant to enhance the response by activating APCs. Antibody expressing memory B cells persist and protect the host from infection without ever being exposed to the full brunt of the disease producing pathogen.

Secretion

Macrophages actively produce and secrete a large number of factors that have important immunomodulatory effects. Macrophages generate and secrete (a) cytokines and chemokines usually via transcriptional activation (e.g., NF-kB and AP-1) and protein synthesis dependent mechanisms, (b) de novo generation of production of lipids and reactive oxygen species, and (c) preformed enzymes (see earlier). These factors can directly cause microbial killing, enhance or dampen an immune reaction, and facilitate the wound healing process.

Tumor Cell Control

Tumor-associated macrophages (TAMs) have recently been shown to be important in tumor biology as they secrete factors that can actively suppress cell proliferation or cause lysis of tumor cells. There is a role for cell–cell contact in macrophage-mediated tumor control, which can specifically recognize and lyse neoplastic cells in a nonphagocytic process. Release of cytostatic and cytotoxic agents includes prostaglandins, IL-1, TNF-α, ROS, and proteases.

IMPLICATIONS OF MACROPHAGE DYSREGULATION

The role of the macrophage in innate immunity has evolved to protect the host against injury and infection; however, macrophage function must be very tightly regulated to ensure minimal toxicity to host tissue and exacerbation of the inflammatory response. Dysfunctions in macrophage biology have been implicated in many disease pathologies including arthritis and atherosclerosis. Once the macrophage performs its duties, it must be decommissioned by processes such as apoptosis *in situ* or by drainage by the lymphatics to the lymph nodes. Macrophages can respond to their environment through the vast array of receptors on their surface, which influences the behavior and toxicity of the macrophage. However, the complex biology of macrophages leaves plenty of scope for things going awry. Dysfunction in maturity, migration, and phagocytic responses of macrophages ultimately result in a compromised ability to clear infection. Also, the inability to control macrophage responses can result in host tissue damage due to increased release of toxic substances.

EVIDENCE FOR ROLES IN DISEASES

Arthritis

Patients with arthritis have elevated levels of various macrophage-derived products that are associated with increased cartilage and bone degradation leading to compromised joint function. The chronic inflammatory response in the joints of patients with arthritis is intimately associated with prolonged and continued cytokine production by an abundant inflammatory cell infiltrate, which includes large numbers of macrophages. TNF-α, IL-1, and IL-6 have been identified as targets for anticytokine therapies for arthritis.

Atherosclerosis

The uptake and oxidation of cholesterol containing lipoproteins is associated with the production of foam cells; modified lipid-laden macrophages. These cells contribute to the progression and instability of atherosclerotic plaques and can promote inflammatory environments that lead to plaque rupture and thrombus formation, the leading cause of myocardial infarction and stroke.

IBD

Irritable bowel disease (IBD) includes Crohn's disease and ulcerative colitis, which are chronic inflammatory diseases of the gastrointestinal tract. These diseases are caused by inappropriate or hypersensitive inflammatory reactions to normal gut flora, a reduced capacity to resolve gut inflammation, and increased inflammatory cell burden. Increased levels of inflammatory cell cytokines are found in patients with IBD, including macrophage-produced cytokines such as TNF-α and IL-6, which can contribute to increased tissue damage (gut ulceration) and enhanced inflammatory cell recruitment, further exacerbating the inflammatory response. Patients with IBD have increased risk of developing malignancies within the gut.

Lung Fibrosis

Lung fibrosis has a poor prognosis once developed. Idiopathic pulmonary fibrosis, asbestosis, silicosis, and interstitial lung disease all have hallmark scarring and thickening of the lung tissue which leads to impaired gas exchange in the alveolar spaces of the lungs. Fibroblast proliferation, collagen deposition, and interstitial tissue remodeling lead to lung fibrosis, which is usually the result of a chronic inflammatory reaction. Macrophages are the driving force initiating and maintaining the fibrotic response by producing profibrotic cytokines such as TGF-β and proinflammatory cytokines such as IL-6, IL-1β, and TNF-α.

Tumor Biology

Macrophages produce various factors important in the regulation of cell growth, proliferation, and survival. Macrophage products can initiate neoplastic cell death and are important in preventing cancers. In contrast, chronic inflammatory reactions can drive macrophage responses into "repair" mode where they switch functions and can promote tumor growth and progression. Tumor-associated macrophages (TAMs) (see earlier) are found at high density in some tumors where they support tumor growth through the production of angiogenic factors, promotion of tissue remodeling and repair mechanisms, and release of growth factors. The major growth factor released by macrophages is vascular endothelial growth factor (VEGF) that promotes angiogenesis and can also attract monocytes to the tumor site. Many cancers are associated with chronic inflammatory responses such as lung cancer in asbestosis patients and stomach cancer associated with *Helicobacter pylori* infection indicating a detrimental role for macrophages in cancer biology. In addition, the inflammatory microenvironment is rich is mutagenic factors such as ROS that can stimulate neighboring cells into a neoplastic phenotype.

Infection

Many pathogens have evolved defense mechanisms to avoid eradication by the immune response when invading the body. In some instances, pathogens use macrophages and their products to facilitate infection. The human immunodeficiency virus (HIV) enters the macrophage through receptors on its surface where it can then replicate. HIV also relies on macrophages to sustain reservoirs of virus that are resistant to antiretroviral therapies, thereby forming a continuing cycle of infection and disease progression. Increases in levels of M-CSF production by macrophages increases macrophage numbers and longevity associated with the maintenance of tissue reservoirs of HIV. M-CSF also promotes virus replication, underpinning the importance of macrophages in the pathogenesis and persistence of HIV. Some species of *Leishmania* parasites can target macrophages by expressing phosphatidylserine (PS) residues to encourage phagocytosis (via mechanisms evolved to ingest apoptotic cells) of the organism where they then establish themselves as intracellular parasites. In addition, the PS-mediated Phagocytosis stimulates the macrophage into an anti-inflammatory mode by increasing release of IL-10 and TGF-β and by decreasing NO production, which aid in the survival of extracellular parasites and increase the susceptibility of the host to infection.

PHARMACOLOGICAL REGULATION OF MACROPHAGE RESPONSES

Macrophages play a pivotal role in immune defense and resolution of inflammation. Thus, pharmacological manipulation of macrophage functions such as enhancement of phagocytosis have become of increasing interest in conditions of chronic inflammation. In addition, we have already touched upon how the macrophage can contribute to disease pathogenesis, highlighting a possible route by which drugs can dampen the effects of macrophages in these conditions. Proinflammatory cytokines released by macrophages, such as TNF-α, can be significantly elevated in various diseases causing tissue damage and exacerbation and perpetuation of the inflammatory response. Although many anti-inflammatory and immunomodulatory drugs can influence macrophage responsiveness two major examples of such agents are highlighted in the subsequent text.

Steroids

Pharmacological intervention in inflammatory conditions is spearheaded by the use of steroidal anti-inflammatory drugs. Glucocorticoids (GCs) are commonly used to treat inflammation such as rheumatoid arthritis and asthma but have significant drawbacks due to side effects. The primary mechanism of action of GCs is the modulation of transcription at the DNA level by activation or repression of GC response elements or by inhibition of transcription factors such as NF-κB and AP-1. In addition, the effects of GC treatment on macrophage phagocytic responses have provided further insight into the anti-inflammatory effects of these drugs. Macrophages treated with GCs such as dexamethasone were shown to markedly enhance the phagocytic capacity of macrophages for apoptotic cells. This function of GCs has obvious advantages when considering the promotion of resolution of inflammation, particularly those that are associated with large numbers of infiltrating neutrophils that require clearance following apoptosis to avoid release of histotoxic contents and inflammatory exacerbation.

Anti-TNF Therapy

A major inflammatory mediator associated with numerous diseases including RA and Crohn's disease is TNF-α. This has sparked research targeting this pathway as a novel treatment for chronic inflammatory conditions. Monoclonal antibodies directed against TNF and soluble TNF receptors have produced promising results in this arena. They work by inhibiting the effector functions of TNF-α by competitively binding the molecule in the extracellular environment, thus reducing the binding of TNF-α to its receptors. The rationale for targeting this cytokine came from the discovery that inhibiting TNF-α could also downregulate the production of other proinflammatory mediators including IL-1, IL-6, and GM-CSF. A potential drawback to the use of anti-TNF-α antibodies is the increased susceptibility to infection, a side effect found in TNF knockout mouse models.

CONCLUSIONS

Knowledge about macrophage biology has been greatly expanded over the past few decades. These cells are now considered integral cells in the innate immune system, involved in orchestrating many immune functions including phagocytic clearance of foreign materials or cell debris, secretion of a plethora of immunomodulatory molecules that can heighten or resolve the immune response and promote wound healing. Furthermore, these cells are also now known to have a role in initiation of the adaptive immune response. In chronic inflammatory conditions, the contribution of macrophages to the pathology of many diseases is now widely recognized and has highlighted many areas through which pharmacological manipulation of macrophage responses can be used to treat these diseases, which include anticytokine therapies and promotion of resolution of inflammation by stimulating the phagocytic capacity of macrophages. There is a continued research effort focused around these cells including using these cells as vectors for gene or drug therapy that home to areas of inflammation or injury. The complex signaling and functional responses of these cells leaves much to be understood, especially surrounding the changing inflammatory profiles of macrophages throughout an inflammatory response; the so-called alternatively or classically activated macrophages. Many new agents are being researched that can directly or indirectly enhance the anti-inflammatory potential of these cells by promoting anti-inflammatory cytokine production or by driving apoptosis of neutrophils to enhance phagocytic clearance using inhibitors of cyclin-dependent kinases (CDKs). Such therapies hold significant promise in the treatment of a number of conditions where current therapeutic strategies are limited, such as idiopathic lung fibrosis.

SUGGESTED READINGS

Aderem, A., and Underhill, D.M. 1999. Mechanisms of phagocytosis in macrophages. *Annu Rev Immunol* **17:**593–623.

Burke, B., and Lewis, C.E., eds. 2002. *The Macrophage*, 2nd ed. Oxford: Oxford University Press, ISBN 0192631977.

Fujiwara, N., and Kobayashi, K. 2005. Macrophages in inflammation. *Curr Drug Targets Inflamm Allergy* **4**(3)**:**281–286.

Gilroy, D.W., Lawrence, T., Perretti, M., and Rossi, A.G. 2004. Inflammatory resolution: new opportunities for drug discovery. *Nat Rev Drug Discov* **3**(5)**:**401–416.

Gordon, S. 2007. The macrophage: past, present and future. *Eur J Immunol* **37**(Suppl 1)**:**S9–S17.

Gordon, S., and Taylor, P.R. 2005. Monocyte and macrophage heterogeneity. *Nat Rev Immunol* **5**(12)**:**953–964.

Gregory, C.D., and Devitt, A. 2004. The macrophage and the apoptotic cell: an innate immune interaction viewed simplistically? *Immunology* **113**(1)**:**1–14.

Hallett, J.M., Leitch, A.E., Riley, N.A., Duffin, R., Haslett, C., and Rossi, A.G. 2008. Novel pharmacological strategies for driving inflammatory cell apoptosis and enhancing the resolution of inflammation. *Trends Pharmacol Sci* **29**(5)**:**250–257.

Hart, S.P., Dransfield, I., and Rossi, A.G. 2008. Phagocytosis of apoptotic cells. *Methods* **44**(3)**:**280–285.

McColl, A., Michlewska, S., Dransfield, I., and Rossi, A.G. 2007. Effects of glucocorticoids on apoptosis and clearance of apoptotic cells. *Sci World J* **7:**1165–1181.

Plüddemann, A., Mukhopadhyay, S., and Gordon, S. 2006. The interaction of macrophage receptors with bacterial ligands. *Expert Rev Mol Med* **8**(28)**:**1–25.

Serhan, C.N., Brain, S.D., Buckley, C.D., et al. 2007. Resolution of inflammation: state of the art, definitions and terms. *FASEB J* **21**(2)**:**325–332.

Sica, A., Rubino, L., Mancino, A., et al. 2007. Targeting tumour-associated macrophages. *Expert Opin Ther Targets* **11**(9)**:**1219–1229.

Szekanecz, Z., and Koch, A.E. 2007. Macrophages and their products in rheumatoid arthritis. *Curr Opin Rheumatol* **19**(3)**:**289–295.

Taylor, P.R., Martinez-Pomares, L., Stacey, M., Lin, H.H., Brown, G.D., and Gordon, S. 2005. Macrophage receptors and immune recognition. *Annu Rev Immunol* **23:**901–944.

9 Lymphocytes

Tracy Hussell, Mary Cavanagh, Erika Wissinger, and Emily G. Findlay

Lymphocytes are a type of white blood cell or leukocyte and are critical for the defense of the body against invaders such as infectious pathogens and foreign materials. There are approximately 10^{12} lymphocytes in the average human body, though this fluctuates considerably during illness. Indeed, fluctuation from a normal range is used as an indicator of disease. For example, immune deficiency leads to reduced lymphocyte numbers in the blood whereas infection or allergy leads to an increase. Several different types of lymphocyte exist, and though they display highly specialized and diverse functions, they all derive from a common hematopoietic stem cell in the bone marrow (Figure 9.1). Mature lymphocytes are found throughout the body: in the blood and lymphatic system; concentrated in immune specialized regions (lymph nodes, thymus, spleen, and Peyer's patches); and scattered in most other tissues.

Lymphocytes can be divided on the basis of size and granularity. Natural killer (NK) cells are large and highly granular, and there are two main types of small lymphocyte: B lymphocytes (or B cells) and T lymphocytes (or T cells). As we will see in this chapter, these are further subdivided based on their expression of different surface molecules, their secretion of different products important for immunity, and their ultimate effector function. B cells (so-called because they were originally identified in birds in the Bursa of Fabricius) are not only produced in the bone marrow but also mature there. However, the precursors of T cells leave the bone marrow and mature in the thymus (which explains the notation "T"). In general, B cells secrete proteins called antibodies that recognize and bind foreign particles and toxins found free within the body interior or protruding from an infected cell. T cells assist other immune cells such as B cells (via secreted proteins called cytokines) or attack body cells when they have been infected by viruses or have become cancerous. T and B cells differ from most other white blood cells in that they express a variable receptor on their surface that specifically detects foreign molecules. T cells express the T-cell receptor (TCR) and B cells express the B-cell receptor (BCR) which is identical to their secreted antibody molecules. The method of producing these receptors leads to enormous diversity. As a result, B cells and T cells are highly specific for pathogenic molecules but cells of each specificity are relatively rare. B and T lymphocytes are therefore referred to as components of acquired or adaptive immunity. Unlike "innate" immune cells, which are more numerous but have less specific receptors, they "acquire" their specificity with time and are adaptable once differentiating. The following chapter details the function of these lymphocyte subsets, what signals cause them to become activated and factors that assist their development into further subtypes. This adaptability is extremely important as it determines their ultimate function and enables the human body to eliminate a hugely diverse range of potentially dangerous organisms and foreign particles.

ANTIGEN-PRESENTING CELLS

Though part of the "adaptive" immune system, lymphocytes must interact with the "innate" immune system and one of these interactions is antigen presentation – the key event in pathogen recognition by adaptive immune cells. To recognize pathogenic molecules, B cells use the BCR which binds pathogens directly. In contrast, T cells rely on other, more innate, antigen-presenting cells (APCs) to break down pathogenic proteins into peptides, which are then presented to the TCR by the major histocompatibility complex (MHC). Innate APCs tend to be large, phagocytic cells that patrol the body and constantly ingest particles, dead cells, and other debris from tissues. Once ingested, these particles, typically proteins, are broken down within the cell and transferred onto carrier molecules

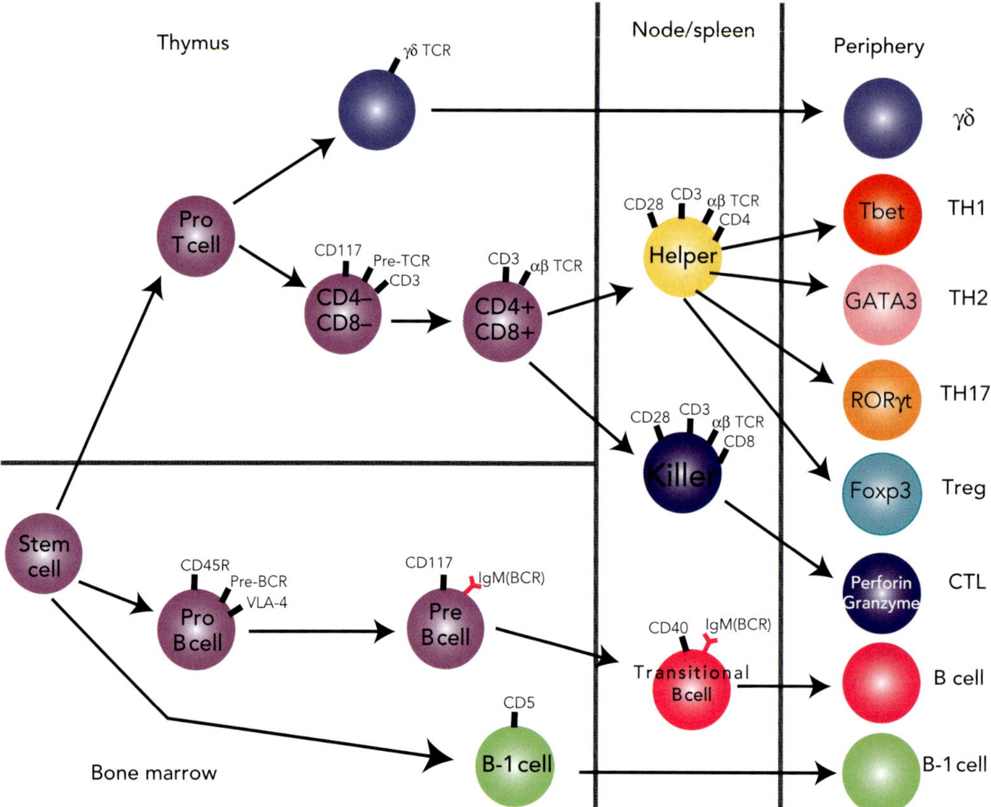

Figure 9.1. T-cell and B-cell development. T and B cells derive from a common stem cell in the bone marrow. The precursors of T cells leave the bone marrow and mature in the thymus. Here, they express different molecules which aid the maturation process. Some of these are shown in the figure. Pro B cells express VLA-4 which binds VCAM-1 on bone marrow stromal cells. This interaction induces the expression of CD117 (also known as Kit) on the B cell, which is the receptor for stem cell growth factor (SCF). In the bone marrow, BCR rearrangement occurs. This process is mirrored in the thymus where pro T cells begin by expressing Kit and rearrange their TCR to produce a highly diverse population. In the thymus, γδ T cells diverge from the αβ T-cell lineage. Once the TCR and BCR have been selected, T and B cells migrate from the thymus and bone marrow to the periphery. Now the cells have reached maturity but are naive. They then circulate in the body until they encounter specific antigen, become activated and differentiate further. The different subsets of mature T cells are shown on the right along with internal transcription factors and other internal molecules used to identify them.

known as the major histocompatibility complexes or MHC. The ways in which APCs process and present antigen are illustrated in Figure 9.2.

The two main types or classes of MHC are known as MHC class I and MHC class II. MHC class I molecules are found on the surface of all nucleated cells and present peptides derived from the cell itself and from intracellular pathogens such as viruses. MHC I molecules continuously cycle from the endoplasmic reticulum (where they pick up antigen) to the cell surface and back. If the cell is uninfected, the antigens presented on the cell surface will not be recognized by cytotoxic T cells. However, if the presenting cell is infected or cancerous, the T cell will target it for death as described later.

MHC class II molecules are generally found on the surface of professional APCs and typically present antigen from extracellular pathogens. The pathogens are ingested and broken down in endosomes and, in contrast to MHC class I, MHC class II molecules are loaded with antigen within these endosomes after passing through the endoplasmic reticulum. Once on the cell surface, MHC class II molecules present antigen to CD4+ T helper cells, which become activated and provide help to other immune cells as detailed subsequently.

For many years, it was thought that these systems of presentation were mutually exclusive but it is now known that MHC class II–destined peptides can access the MHC class I pathway by a process called cross-presentation. This process is thought to provide a way of presenting antigen from viruses that have evolved mechanisms to subvert the classical MHC class I presentation pathway.

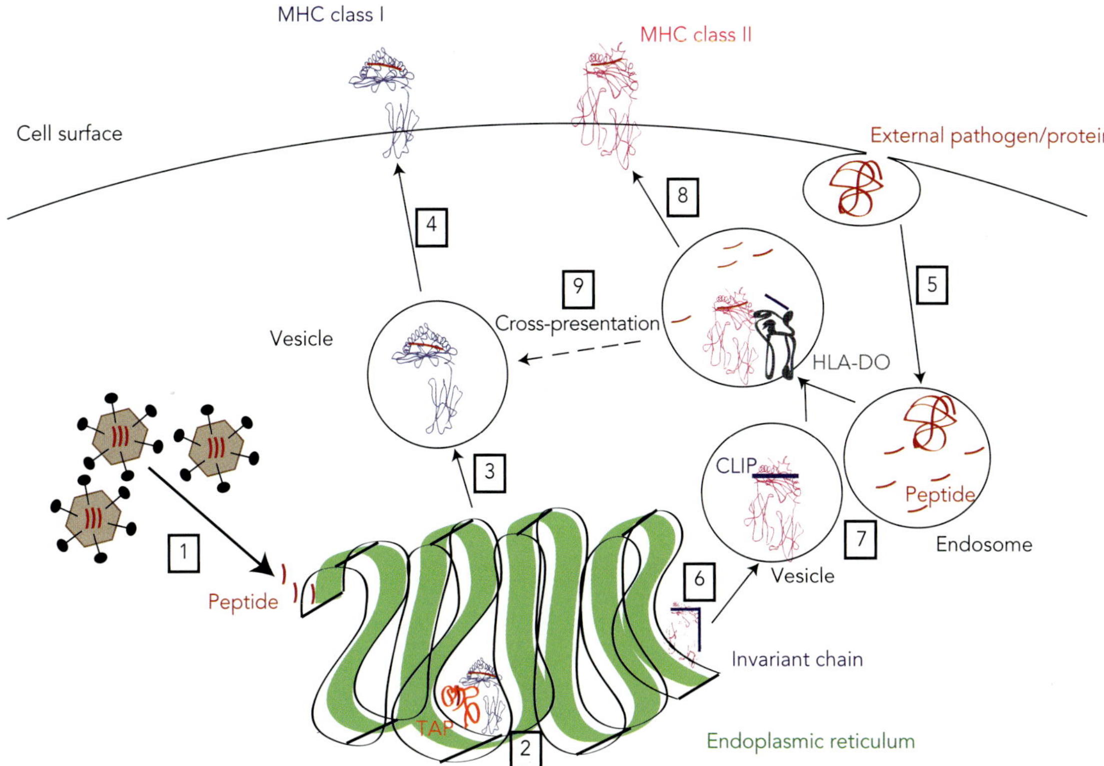

Figure 9.2. MHC processing and presentation. MHC class I molecules typically present peptides derived from intracellular sources whereas MHC class II molecules present those from extracellular sources.

During viral replication, some proteins are released from the proteosome (1) and are taken into the endoplasmic reticulum (2). Here, they encounter the membrane-associated transporter protein TAP which loads the peptide onto the MHC class I binding domain. The peptide-loaded MHC I molecule then leaves the ER in a vesicle (3) and is transported to the cell surface where it displays the peptide to passing CD8+ T cells (4).

Innate APCs can ingest proteins from the extracellular space (5). These are internalized and degraded in endosomes. Though MHC class II molecules are processed in the ER (6), they leave this compartment in vesicles before being loaded with peptide. The peptide-binding site of MHC II is protected by the CLIP fragment of the invariant chain which is only completely removed once the MHC-containing endosome has fused with the antigen-containing endosome (7). CLIP is removed by a related MHC class II protein (HLA-DM or HLA-DO), allowing the foreign peptide to bind, and the peptide:MHC complex is then transported to the cell surface where it can present antigen to CD4+ T helper cells (8).

On rare occasions, externally derived peptides can be presented by MHC class I in a process known as cross-presentation (9).

T CELLS

T-Cell Development

T cells play multiple roles in immunity and disease depending on their ultimate phenotype after differentiation. Unlike the BCR (antibody), the TCR does not see foreign particles or pathogens directly but requires them to be digested and presented by MHC molecules. The outcome of this recognition then depends on the T cell in question and will be discussed later. The TCR is a heterodimer of two chains but also requires a collection of other membrane-proximal chains for signaling (see Figure 9.3); these additional chains are known collectively as CD3. All T cells, regardless of their end phenotype, express CD3. However, broadly speaking there are two different types of CD3-expressing T cells. Helper T lymphocytes also express CD4 and cytotoxic T lymphocytes (CTLs) express CD8. As their names suggest, CD4+ T cells secrete products to help other cells of the immune system and CD8+ CTLs cause death of the cell with which they interact. As we will see later, CD4+ and CD8+ cells can be further subdivided based on the products they secrete or additional receptors they express.

Both main T-cell subsets mature in the thymus where rearrangement of the genes encoding the two variable chains (α and β) of the TCR takes place. As this rearrangement is a random process the thymus is also a sorting factory and ensures, through a process of selection, that T cells do not exit the thymus capable of recognizing our own proteins. If they did, then autoimmunity would occur. Progenitor T cells (ProT) from the bone marrow move into the outer cortex of the thymus and proliferate. At this stage, they do not express CD4 or CD8 but express Kit (the receptor for

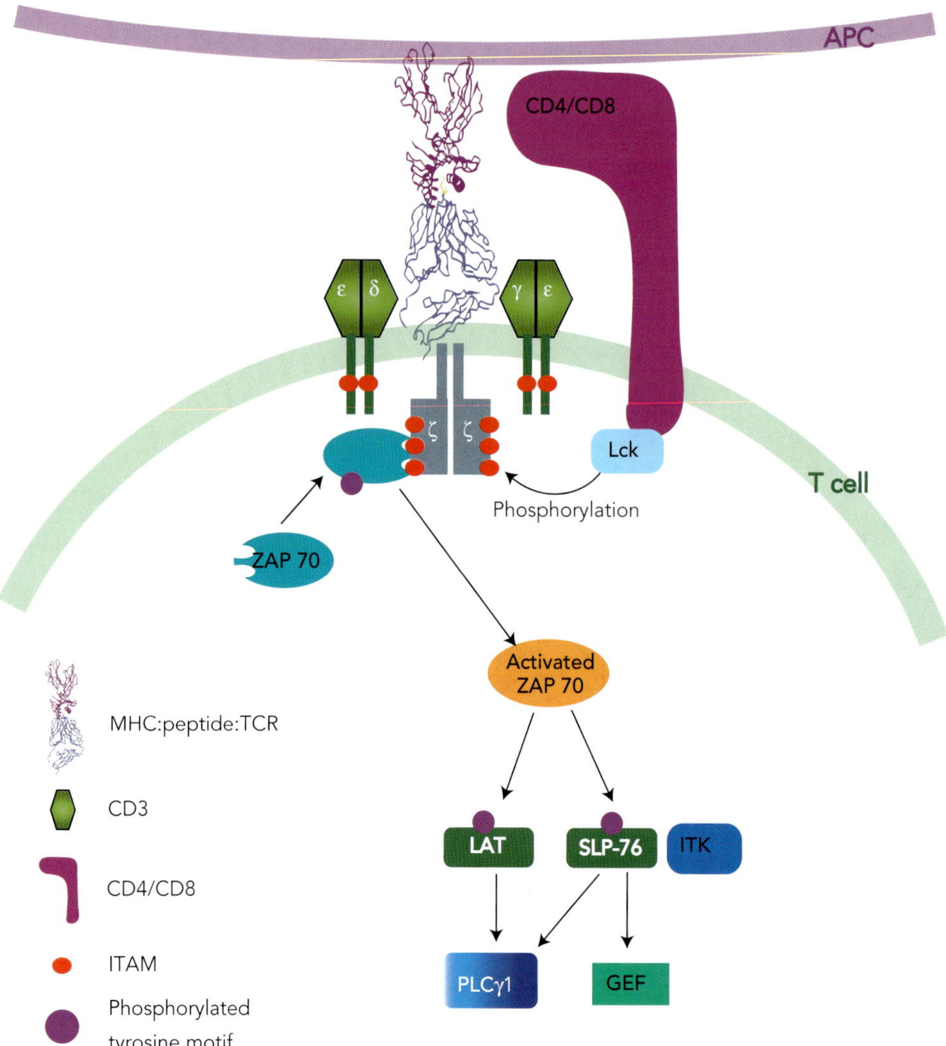

Figure 9.3. TCR signaling. A T cell is activated when its TCR binds to antigenic peptide presented by an antigen-presenting cell (APC). This peptide is bound to MHC, and a co-receptor (CD4 or CD8) stabilizes the interaction by binding to nonpolymorphic regions of the MHC molecule. Protein tyrosine kinases, such as Lck, quickly act to phosphorylate the immunoreceptor tyrosine–based activation motifs (ITAMs) located on the intracellular portions of CD3 and the ζ chains of the TCR. The ζ chain–associated protein ZAP 70 then binds and phosphorylates the membrane-bound adapter proteins, linker of activated T cells (LAT), and SH2 domain-containing leukocyte protein of 76 kDa (SLP-76). LAT, SLP-76 and another kinase, ITK, then activate phospholipase C (PLC)γ1 and recruit guanine nucleotide exchange factors (GEFs) to continue the signaling cascade.

stem cell growth factor [SCF]), CD44 and CD25 (part of the receptor for the T cell growth factor interleukin [IL]-2). Following a number of rounds of division, CD25 and CD44 are lost from the surface and the TCR gene rearrangement takes place.

The β chain genes of the TCR are first rearranged through the actions of the recombination-activating genes RAG-1 and -2, and combine with a pre-α chain to generate what is known as a pre-TCR. This combination determines whether the β chain rearrangement is successful. Successful signaling through the pre-TCR allows the T cell to progress to the next stage. Cells that pass the test move on to a CD4+CD8+ (double positive) stage, which is marked by rapid proliferation and rearrangement of the α chain. With a functional α and β chain, the TCR of the T cell is ready to be checked for function via processes called positive and negative selection. Remember, T cells require antigen to be presented to them via MHC molecules. Cells that only weakly recognize self-antigen in MHC molecules die by neglect (they do not receive sufficient stimulation to survive); those which recognize it too strongly are deemed dangerous and are induced to die by a process known as apoptosis. Only those with moderate affinity for MHC plus antigen survive (only approximately 5% of potential T cells). Once past this selection stage RAG-1 and -2 are downregulated and the cells become single positive for either CD4 or CD8.

The process by which cells become helper (CD4+) or cytotoxic (CD8+) has not been fully determined. Two different types of MHC molecules present antigen to CD4 or CD8 T cells (MHC class II and class I, respectively) and it is possible that the CD4 or CD8 phenotype is determined by which MHC the functional TCR encounters first; interaction with MHC II may induce CD4+ T cells whereas MHC I may promote CD8+ T cells. However, the final distinction may arise as a result of the strength of action of TCR signaling components. For example, MHC II binding the TCR enhances signaling to Lck compared to MHC I, and artificially attenuating TCR signaling allows MHC II–specific CD4+ T cells to become CD8+ over the course of the long differentiation process. Longer interactions and stronger TCR signals therefore favor CD4+ (helper) T-cell development over CD8+ (cytotoxic) T cells. Other signaling pathways, particularly involving the Notch proteins, are also important for lineage development (for an in-depth discussion of this, see Yang 2006).

T-Cell Receptor Signaling

The TCR is composed of several subunits that colocalize within the cell membrane, drawn together by virtue of their positive and negative charges. Eight subunits are required to form a complete TCR. The α and β chains form the foreign peptide:MHC recognition site but lack a signaling domain, which is provided by two intracellular ζ chains. This central $\alpha\beta\zeta\zeta$ structure is flanked by two other complexes each composed of a heterodimer of either one γ and one ε or one δ and one ε chain (collectively known as CD3). The intracellular domains of the γ, δ, ε, and ζ chains provide the substrate for the first step in TCR signal transduction. Contained within them are immunoreceptor tyrosine-based activation motifs (ITAMs) which contain tyrosine residues that, once phosphorylated by the Src family kinases (such as Lck), recruit the 70 kDa ζ chain–associated protein ZAP-70. On binding to one or more ITAMs, ZAP-70 is activated setting off a series of phosphorylation events that culminate in the transcription of a number of genes. These pathways are summarized in Figures 9.3 and 9.4.

The Requirement for Co-Receptors

As described earlier, T cells differentiate during development into CD4+ and CD8+ subsets. These subsets are defined by their expression of the CD4 or CD8 co-receptor on the cell surface. CD4 and CD8 colocalize with the TCR and recognize invariant

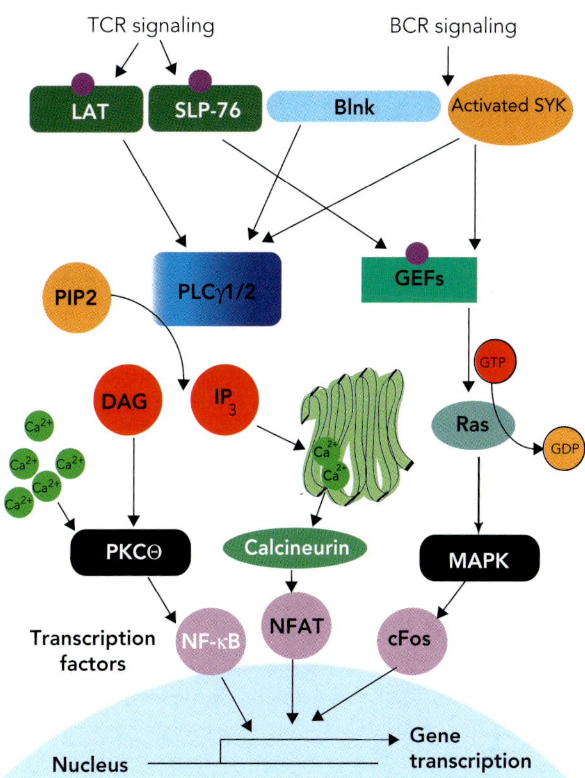

Figure 9.4. Downstream signaling following TCR or BCR activation. Signaling through SLP-76, downstream of the TCR, and through Syk, downstream of the BCR, results in the activation of guanine nucleotide exchange factors (GEFs) which promote the activation of Ras-GTPases via the dissociation of GDP allowing Ras to bind fresh cytosolic GTP. This process acts as a molecular switch and leads to the activation, via other factors, of MAP kinase and the eventual activation and nuclear translocation of the transcription factor cFos.

LAT signaling in the T cell results in the activation of the enzyme phospholipase C (PLC)γ1. Syk in the B cell interacts with the B cell linker Blnk and Tec kinases to activate phospholipase C (PLC)γ2. These enzymes convert inactive phosphatidylinositol bisphosphate (PIP2) into diacylglycerol (DAG) and inositol triphosphate (IP_3). IP_3 promotes the release of calcium ions from the endoplasmic reticulum and thus the activation of the phosphatase calcineurin. Calcineurin dephosphorylates the transcription factor nuclear factor of activated T cells (NFAT) which then translocates to the nucleus and activates the transcription of genes involved in the synthesis of interleukin (IL)-2. DAG also acts with calcium ions to activate protein kinase C (PKC)θ at the plasma membrane. This kinase also activates a transcription factor, this time NF-κB, which translocates to the nucleus to initiate the transcription of a wide variety of genes involved in T-cell activation, proliferation, and survival.

motifs on the MHC molecule. CD4 recognizes MHC class II molecules, which are mainly expressed on professionalAPCs; CD8 recognizes MHC class I molecules, which are expressed on the surface of all nucleated cells. Both receptors provide intracellular docking sites for the kinase Lck which, as described, initiates the phosphorylation cascade following TCR engagement. However, CD4 and CD8 vary in their downstream effects on the T cell. CD4+ T cells receive signals encouraging the production of cytokines and chemokines and the expression of molecules such as CD40L, which interacts with CD40 on B cells to encourage B-cell maturation and antibody class switching. Because CD4+ cells act, for the most part, in this indirect manner, they are known as T helper (T_h) cells. In contrast, CD8+ T cells receive signals encouraging the release of cytotoxic compounds such as perforin and granzyme. Their direct, lytic activity means that CD8+ cells are known as CTLs.

Costimulatory Signaling

We have discussed the αβ TCR that recognizes foreign particles bound to MHC, the associated intracellular chains that make up the CD3 complex and the co-receptors CD4 and CD8 that bind to MHC molecules and stabilize the interaction. Collectively this "signal 1" is still not enough for full T-cell activation. In fact, T cells that only receive signal 1 lose the ability to respond to antigen and acquire a quiescent, nonproliferative phenotype. This state of "anergy" is central to the development of peripheral tolerance and ensures that harmless antigens, encountered in the absence of other "danger" signals, do not trigger a potentially damaging autoimmune or allergic response. Full activation requires "signal 2," which is delivered by the cell that presents antigen to the T cell via MHC molecules: the APC. The additional molecules required to provide signal 2 are called costimulatory molecules. Evolutionarily, it is beneficial to require this second signal as it means that only "professional" APCs (macrophages, dendritic cells, and B cells) can fully activate a T cell. Without the need for costimulation, T-cell activation could occur widely leading to bystander tissue damage. The constitutively T-cell–expressed molecule CD28 provides the first, and strongest, of these costimulatory signals. CD28 on T cells binds to B7 molecules (CD80 and CD86) on professional APCs and results in T-cell survival and proliferation. CD28's association with lipid rafts leads to the close association of these structures with the TCR and an amplification of the protein tyrosine phosphorylation associated with TCR signaling. Downstream signals also increase the amount of IL-2 produced by the T-cell. IL-2 is the main cytokine involved in T-cell proliferation, increasing the expression of survival factors and pushing T cells to enter the cell cycle.

Late Costimulation

Though CD28 provides strong survival signals, further T cell–APC interactions are required for full T-cell effector function and to prevent T-cell death. Several other costimulatory molecules are induced on the surface of T cells following their initial recognition of peptide: MHC complexes. These induced costimulatory molecules can be divided into three families based on their structure: the Ig-like superfamily, the integrins, and the TNF/TNF receptor superfamily.

CD28 is itself a member of the Ig-like superfamily and interacts with CD80 and CD86 (also known as B7.1 and B7.2) on the surface of APCs. This interaction results in the signaling cascades described earlier. However, if this interaction continues, the T cell will remain in a proinflammatory state and may be subject to activation-induced cell death (AICD). To counteract this, CD28 signaling results in the production of a related molecule: CTL antigen (CTLA)-4. CTLA-4 is very similar in structure to CD28 and displaces it, but instead of sending a positive signal via ITAMs, it delivers inhibitory signals to the T cell that arrest the activation cascade and "turn off" the inflammatory response. In most circumstances, however, we would require the T cell to continue its effector function. How then do we override CTLA-4? If the threat persists and stimulates T cells strongly, other "late costimulators" are induced to bypass CTLA-4. The inducible costimulator (ICOS) on T cells is one such molecule and another member of the Ig-like superfamily. Integrins, such as leukocyte function-associated antigen (LFA)-1, also provide T cells with activation signals following interaction with their ligands on APCs, but are reliant on the presence of other costimulators, in particular CD28, to exert this effect. The TNF receptor superfamily (TNFRSF) includes several costimulatory molecules, which are typically upregulated late in T-cell activation, such as 4–1BB and OX40. These bind to equally inducible ligands on APCs. Without these additional late costimulators, T-cell activation would not persist and certain pathogens might escape elimination. Furthermore, these late costimulators are vital for the development of memory T-cell populations. Without them, T-cell activation and progression is prevented too early to seed the memory T-cell pool. Development of memory is vital to thwart a similar infection on re-encounter. Memory cells are more abundant and more easily activated than naive cells so can clear a second infection with the same pathogen more quickly and efficiently. A basic sequence of expression of costimulatory molecules on the T-cell surface and their ligands on the APC is illustrated in Figure 9.5.

Further Differentiation of T Cells

So far, we have discussed the T-cell subcategories of CD4+ helper and CD8+ cytotoxic T cells. CD4+ helper T cells (and recently CD8+ T cells) are further subcategorized based on the cytokines they secrete. Most work has focused on helper CD4+ T cells and will be discussed here. So far, the subcategories are called

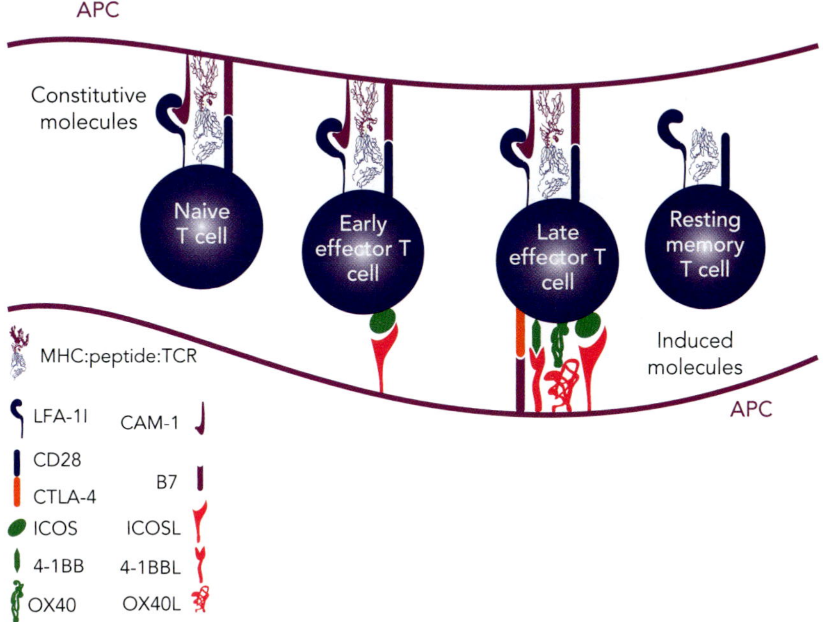

Figure 9.5. Upregulated quickly after TCR stimulation, ICOS acts to boost the signals provided by CD28. As activation continues, late costimulatory molecules begin to appear along with the coinhibitory molecule CTLA-4. These molecules result in a cessation of proinflammatory cytokine production and an increase in the expression of prosurvival factors such as members of the Bcl family of antiapoptotic genes. On receiving these signals, the T cell stops proliferating, but avoids AICD and progresses into the memory pool. Memory T cells are able to survive long term and do not continue to express costimulatory molecules. However, the kinetics of costimulatory molecule reexpression are much faster.

T_h1, T_h2, T regulatory (Treg), and T_h17 cells. T_h1 cells secrete the cytokines interferon (IFN)-γ, TNF, and IL-2; T_h2 cells secrete IL-4, IL-5, and IL-13; T_h17 cells secrete IL-17 and TNF-α; and regulatory T cells secrete IL-10 and/or TGF-β (although IL-10 is also produced by T_h2 cells). Researchers have attempted to find surface molecules that distinguish these different helper T-cell subsets for many years. However, those described so far remain equivocal and clear distinctions between the subsets, in terms of surface molecule expression, remain elusive.

The cytokine profiles of these cell types allow them to induce discrete immune responses according to the nature of the threat. T_h1 cytokines enable a cell-mediated immune response to target intracellular pathogens by activating CTLs, macrophages, and NK cells, whereas the T_h2 cytokine response induces a humoral antibody response targeting extracellular pathogens. T_h2 cell cytokines activate mast cells and basophils through IL-4 and eosinophils through IL-5. T_h subset polarization also dictates the final type of antibody produced by B lymphocytes (see the subsequent text).

It is still unclear how these different helper T-cell subtypes develop. The nature of the pathogen and its dose contribute, as do the level of costimulation, signals from the tissue microenvironment and transcription factors. At present however, there is no overriding consensus, as it seems to vary with experimental condition and/or the pathogen under investigation. Longer contact of MHC with the TCR is reported to produce a stable T_h2 response; it is also suggested that a higher degree of stimulation is required for T_h1 polarization. However, a number of parameters dictating cytokine profile outcome is standard. For example, differentiation of T_h1 responses involves activation of the *IFN-γ* gene and concomitant silencing of the IL-4 gene, while the opposite is seen on T_h2 differentiation. Type I and II interferon signaling causes the STAT1-dependent formation of the IL-12 receptor; ligation of this IL-12R leads to activation of T-bet, a transcription factor critical for the development of T_h1 responses, and STAT4-dependent potentiation of IFN-γ production. In T_h2 responses, IL-4 signaling to a naive cell activates STAT6 and the master transcription factor GATA-3. GATA-3 suppresses STAT4 and the IL-12R whilst inducing positive epigenetic changes in the T_h2 cytokine cluster of Il4, Il5, and Il13 genes (see Figure 9.6 for a summary).

For two decades, the T_h1/T_h2 polarization model was largely unaltered, until the recent description of

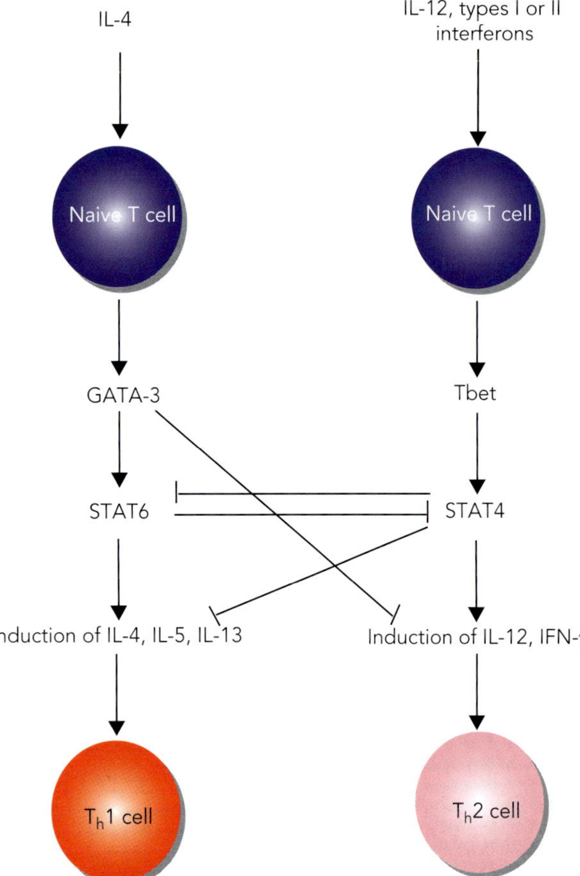

Figure 9.6. Following signals from inflammatory cytokines, naive T cells begin to polarize into T_h1 or T_h2 cells. Transcription factors, induced following cytokine signaling, direct the changes in cell phenotype. The T_h1 master transcription factor, Tbet, is induced after IL-12R ligation and STAT4 signaling. This induces further IL-12 production and expression of IFN-γ, and inhibits the activation of GATA-3: the T_h2 master regulator. Similarly, GATA-3 activation in T_h2 cells inhibits T-bet and STAT4 and allows the transcription of IL-4, IL-5, and IL-13 by inducing epigenetic changes such as induction of DNase I hypersensitive sites.

the T_h17 subset of cells. These are characterized by IL-17, IL-6, IL-22, and TNF-α production. T_h17 cells are entirely separate from the T_h1 and T_h2 lineages; cytokines from T_h1 cells antagonize T_h17 generation, and it is likely that they have evolved to deal with different pathogens to those efficiently cleared by T_h1 and T_h2 responses. IL-23 polarizes naive cells into T_h17 cells in an ICOS- and CD28-dependent manner (T_h1 cells are dependent on CD28 but not ICOS). Generation of T_h17 cells is independent of the STAT1, STAT4, and STAT6 signaling proteins but does require STAT3. In APC-free culture conditions, T_h17 cells can be generated with the addition of only TGF-β and IL-6; these cytokines induce differentiation by activation of the nuclear receptor retinoic acid receptor-related orphan receptorγ (RORγ). The specific isoform RORγt and the receptor RORα are regulated by STAT3 and are highly expressed in T_h17 cells.

Effector Function of T Helper Cell Subsets

By virtue of their secreted cytokines (IFN-γ, TNF, and IL-2), T_h1 effector cells are very efficient at promoting cell-mediated immunity; the cytokines they produce help to activate macrophages and NK cells, and promote survival of CTLs. These functions make T_h1 cells important for the clearance of intracellular pathogens such as viruses. T_h1 cytokines also induce antibody class (or isotype)-switching in B cells, leading to the production of IgG2a, as discussed later. However, an excessive T_h1 response is frequently associated with immunopathology, or immune-mediated damage to the host, and has been implicated in some autoimmune diseases and inflammatory disorders. T_h2 cells have a very different function; by producing cytokines such as IL-4, IL-5, and IL-13, they help to activate eosinophils and mast cells, and are important for the clearance of extracellular pathogens, especially parasites. T_h2 cytokines can also induce class switching in B cells, but promote the secretion of IgE, IgA, and IgG1. However, in some individuals, a disproportionate T_h2 response is mounted which is a hallmark of allergic reactions and atopic disease such as asthma. Activated T_h17 cells making IL-17, IL-21, and IL-22 are also important in host defense to infection, particularly for extracellular bacteria, as T_h17 cytokines efficiently recruit neutrophils to the site of inflammation. Once again, T_h17 cells are highly inflammatory and, like T_h1 cells, play a role in the pathogenesis of autoimmune disease. Figure 9.7 summarizes the main roles of T_h1 and T_h2 cells in health, disease, and autoimmunity, and also highlights the complications that can arise if an inappropriate response is initiated during infection.

Regulatory T cells (Treg) comprise another important T-cell subset that suppresses the activation of other cells in a contact-dependent or -independent manner. Treg interact with and suppress both T_h1 and T_h2 effector T cells. They can achieve this indirectly, by affecting APCs, or directly through T–T interactions. There are four main Treg subsets; natural Treg constitutively express CD4 and CD25 as well as the transcription factor Foxp3. Their generation is dependent on TGF-β. Tr1 cells, by contrast, are induced Treg, which do not express Foxp3 but express high levels of CTLA-4 and require IL-10 for their differentiation and suppression. In addition to these Treg subsets, anergic CD4+ and CD8+ cells are capable of suppression by competing for antigen and IL-2; they have a very high-stimulation threshold and do not therefore become activated except in situations of very high antigen exposure.

The main functions of each T_h subset and the cytokines they produce are summarized in Figure 9.8.

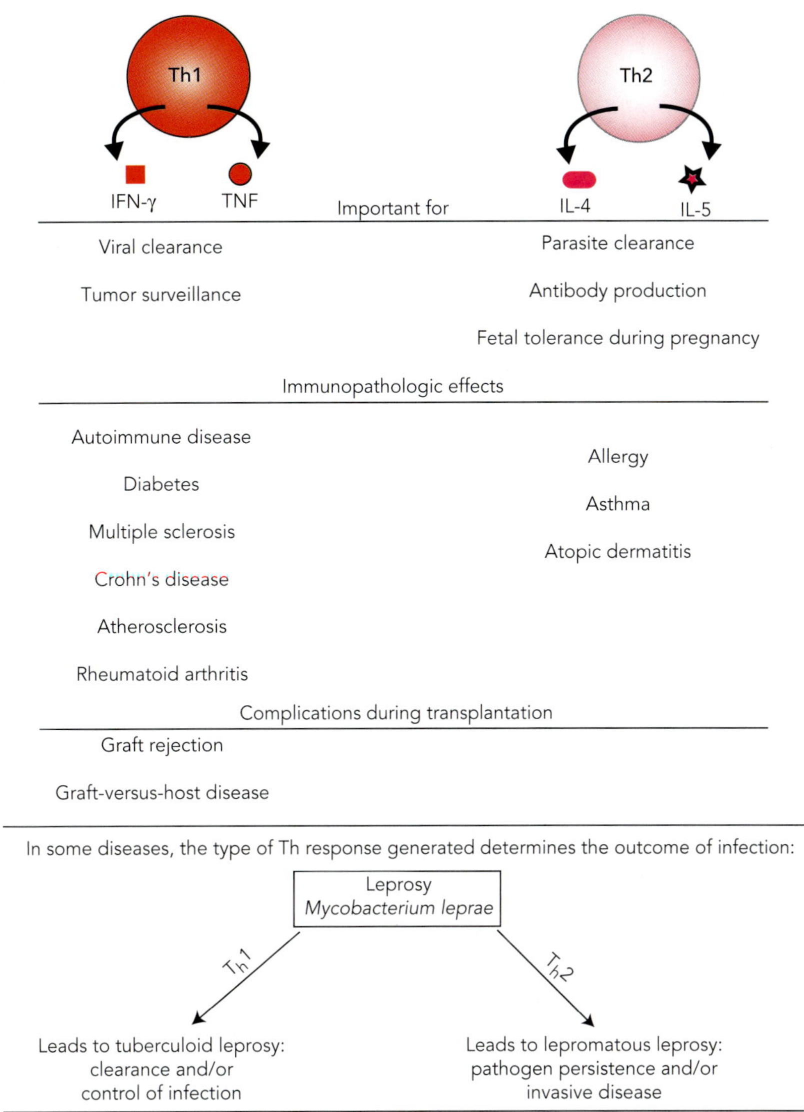

Figure 9.7. The role of T helper cell responses in pathogen clearance and immunopathology. Th1 and Th2 responses are important for pathogen clearance. However, both responses can exacerbate autoimmunity and in some diseases, the type of Th response generated can determine the outcome of infection as in the case of leprosy. The Th1/Th2 paradigm has been at the forefront of research for many years but it is now clear that Th17 cells take part in many Th1-based responses and that these cells are also involved in many autoimmune diseases. Similarly, regulatory T cells can influence T cell responses, play a role in foetal tolerance and can inhibit tumor immunity.

Intraepithelial T Cells

As with every biological system, there are a few exceptions to the rules. We cannot discuss lymphocytes without highlighting a subset of T cells that resides among the epithelial cells that line our mucosal surfaces (i.e., gut, lung, genital tract, and skin); the intraepithelial lymphocytes (IELs). There can be as many as 20 IELs per 100 epithelial cells. Within this population a high proportion (50% in humans and 90% in mice) have an alternate TCR heterodimer of γδ (do not confuse these with the γ and δ chains within the CD3 cluster). IELs are mostly CD8+ but some have CD3 but no CD4 or CD8. Their origin and development is unclear but they exist in athymic mice, albeit in reduced numbers. Unlike conventional T cells, γδ IELs do not recognize antigen presented by classical MHC molecules. Instead, it is believed that they are primed by the surrounding epithelial cells possibly via nonclassical MHC such as CD1. Despite their mucosal localization, and therefore the vast array of antigens that they must encounter, the γδ TCR is much less diverse than the αβ TCR. Rather than recognizing highly specific pathogen-derived sequences, they respond to molecules such as heat

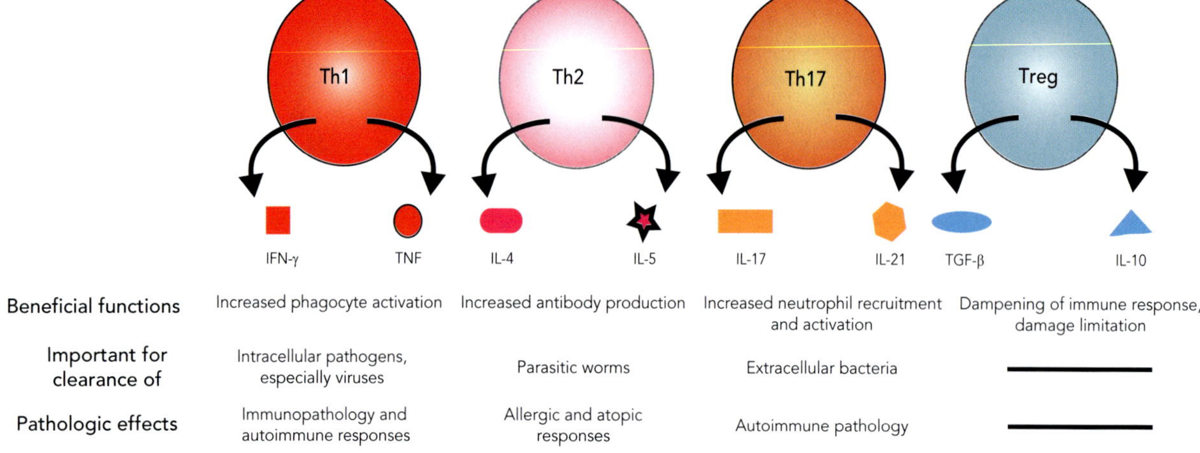

Figure 9.8. T helper cell subsets. CD4+ T helper cells can be subcategorized according to the cytokines they secrete. These cytokines are primarily responsible for the effector function of each subset and these functions are listed. If the immune response is uncontrolled or responds inappropriately to harmless antigens, for example in allergy or autoimmunity, T cells can have detrimental effects and can contribute to pathology.

shock proteins that are upregulated on the epithelial cells following infection or transformation. The suggestion therefore is that IEL behave more like innate immune cells than adaptive lymphocytes.

Effector Function of Cytotoxic (Killer) T Lymphocytes

As described previously, when an antigen-specific CD8+ T cell encounters the appropriate cognate peptide antigen bound to MHC class I on a target cell, the T cell becomes activated. For CTLs, this activation results in two major effector functions: cytokine secretion and targeted killing. Similar to effector CD4+ cells, there are subsets of effector CD8+ T cells, based on the profile of cytokines they secrete. T_c1 cells produce large quantities of IFN-γ and TNF, and T_c2 cells secrete IL-4 and IL-5. However, unlike their CD4+ counterparts, T_c2 cells are rare in vivo, and as yet no equivalent population of Treg cells has been identified for CD8+ T cells.

As well as producing cytokines, CTLs perform a vital function by killing infected cells and tumors. Following the recognition step, during which the CTL binds the antigen-bearing target cell, the TCRs, and associated co-receptors on the T cell come together, forming a large cluster called the supramolecular activation cluster (SMAC). This aggregation triggers a reorganization of the actin cytoskeleton within the CD8+ T cell, which allows it to correctly position its lytic granules to target the site of contact; a mechanism which limits nonspecific damage to surrounding cells. The highly specialized lytic granules found in CD8+ CTLs are also found in natural killer (NK) cells and contain several proteins important for triggering programmed cell death. Perforin, first, polymerizes to form pores in the target cell membrane allowing granzymes to enter. Granzymes are a family of enzymes with serine protease activity which, after entering the target cell, degrade its internal proteins. A specific member of this family, granzyme B, cleaves and activates caspase-3, which initiates the caspase cascade ultimately culminating in the degradation of cellular DNA, one of the hallmarks of cell death by apoptosis. Perforin and granzyme are both required for efficient killing of target cells, although some cytotoxic activity can be mediated by perforin alone. A further component of CTL lytic granules is granulysin, another protein, which can induce apoptosis and has antimicrobial properties. In killing a target cell, the CTL itself is unharmed and is able to kill several infected targets in rapid succession using internal stores of preformed lytic granules. The action of inducing cell death in the target cell via perforin/granzyme can be programmed in as little as 5 minutes, although the process of apoptosis may take several hours. Controlled cell death is required to contain infection because if cells die by necrosis (uncontrolled death), they can release infectious intracellular pathogens in the process. CD8+ T-cell killing can be measured using the method outlined in Figure 9.9.

Another major mechanism by which a CTL can kill an infected cell involves interactions of cell surface receptors rather than stores of toxic granules. Fas ligand (FasL), a member of the TNF family, is expressed on the surface of activated CTLs as a homotrimer. It binds to Fas, a member of the TNFR superfamily on the target cell, and in doing so, produces a conformational change in Fas allowing the recruitment of death-domain containing adaptor proteins such as FADD. This clustering of adaptor proteins ultimately results in activation of the caspase cascade, as described earlier for killing via perforin and granzyme, but is independent of these

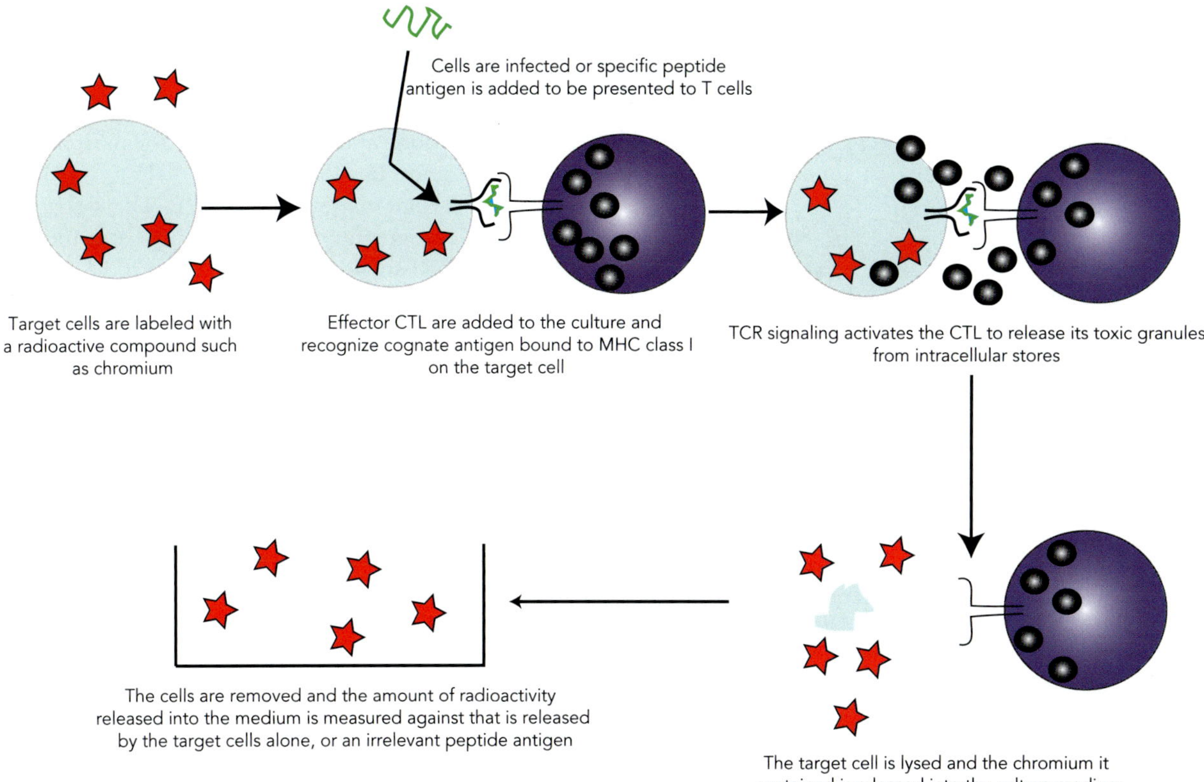

Cells are infected or specific peptide antigen is added to be presented to T cells
Target cells are labeled with a radioactive compound such as chromium
Effector CTL are added to the culture and recognize cognate antigen bound to MHC class I on the target cell
TCR signaling activates the CTL to release its toxic granules from intracellular stores
The cells are removed and the amount of radioactivity released into the medium is measured against that is released by the target cells alone, or an irrelevant peptide antigen
The target cell is lysed and the chromium it contained is released into the culture medium

Figure 9.9. Measuring CTL activity in vitro. Activated cytotoxic T lymphocytes (CTLs) are programmed to kill cells that have been infected by viruses or that have become cancerous. However, not all CTLs are able to kill with equal efficiency and not all stimuli induce equally strong CTL responses. The assay outlined above can be used to quantify CTL killing in vitro and thus give an indication of the strength of the CTL response in vivo.

enzymes. Since activated effector T cells can express both Fas and FasL, this is one of the major mechanisms by which CTLs can kill off other CTLs (a process known as fratricide) at the end of a robust immune response to reduce cell numbers back to homeostatic levels (known as the contraction phase).

B CELLS

B-Cell Development

B cells develop in the bone marrow in adults (the liver in foeti) and are continually produced throughout life. The stromal cells present in the marrow are vital for B-cell differentiation owing to their production of adhesion molecules, which bind to and activate the progenitor cells, and to their secretion of prodifferentiation mediators such as SCF, IL-7, IL-11, and chemokines such as CXCL12. Development of a B cell in the bone marrow is marked by the expression of molecules which interact with these stromal cells, and in the periphery by the expression of immunoglobulin genes. ProB (progenitor-B) cells are distinguished from the hematopoietic stem cells from which they are derived by the upregulation of B220 (CD45R) and the integrin VLA-4. These allow the B cell to bind VCAM-1 on the stroma, which induces the expression of Kit on the B cell. Kit (CD117) is able to bind SCF, which is a proliferation factor.

The BCR is composed of two identical heavy chains (pink in Figure 9.10), of which there are five different alternatives or isotypes; and two identical light chains (green in Figure 9.10), of which there are two types (κ or λ). The two heavy and two light chains are held together by disulfide bonds. Different regions of the antibody have different functions in host defense. The two Fab portions (derived from fragment antigen binding) are shown at the top of the antibody molecule and recognize foreign antigen. Unlike the TCR, the Fab portion of the BCR can do this directly without the need for presentation. The Fc portion (derived from fragment crystalizable) binds to Fc receptors that are present on a wide variety of other immune cells including NK cells, macrophages, and dendritic cells (discussed later). This binding of soluble antibody to Fc receptors has significant consequences for the Fc receptor–bearing cells and different heavy chain isotypes have different effects. Though antibody has significant functions as a secreted molecule, B cells also retain it on their cell surface where it functions as the BCR. As with the TCR, the development of the BCR is a random process leading to a diverse and extensive

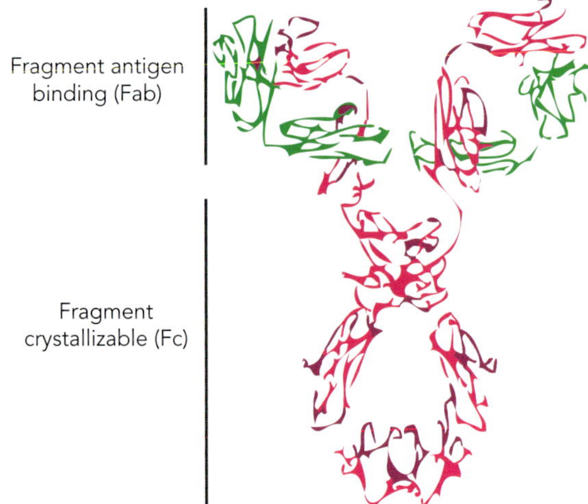

Figure 9.10. Antibody structure. The BCR is composed of an antibody molecule which itself has four chains. Two of these are the identical heavy chains and two are the identical light chains. Light and heavy chains are composed of variable and constant regions. The variable regions colocalize in the two Fab fragments at the top of the molecule and recognize antigen. The lower Fc fragment is composed of the constant regions of the heavy chains and varies between antibody isotypes allowing each to perform a different function. When tethered in the B-cell membrane as the BCR, the Fc fragment loses this function and instead associates with CD79 to transduce signals to the B cell.

repertoire capable of directly binding to a vast array of foreign proteins. Each B cell selectively expresses only one type of antibody and therefore recognizes one antigen. The specific nature of foreign protein binding by the B cell means it also functions as a highly sensitive and specific antigen-presenting cell.

During development, immunoglobulin expression depends on functional gene rearrangement, which is dependent, like TCR production, on the recombinase genes *RAG-1* and *RAG-2*. ProB cells rearrange the μ heavy chain gene to form an early BCR (preBCR). This surrogate allows testing of the ability of heavy chains to associate with light chains and other molecules important for BCR signaling such as CD79 α/β. If rearrangement of this initial heavy chain is unsuccessful or results in autoreactivity, secondary rearrangements are induced. During development in the bone marrow, signaling through the surrogate preBCR lowers the threshold for IL-7 responses, meaning cells which have productively rearranged the BCR are able to proliferate and outnumber cells which have not. All immature B cells express IgM, and strong signals received through self-antigen binding to IgM in the bone marrow deletes self-reactive B cells in the same way that self-reactive T cells are deleted in the thymus. Some of these cells can be rescued by "receptor editing," which allows the light chain recombination to continue until the self-reactive light chain has been replaced. Once the B cell has passed the test in the bone marrow, the *RAG* genes are no longer expressed and the light chain present at that time remains. These cells are now committed to the B-cell lineage and mature further in the bone marrow until the need for interaction with stromal cells is lost; the cells can then move into the center of the bone and finally out into the periphery as mature but naive B cells.

The functional maturation of the B cell, so that it is capable of recognizing antigen, begins from the earliest pro–B-cell stage and continues once the cell has left the bone marrow and moved out into the periphery. This maturation involves swapping to different heavy chain isotypes and the expression of membrane immunoglobulins. The expression of different membrane and surface immunoglobulins is the key difference between subsets of B cells, which do not diverge and mature separately as do T cells. However, the site at which B cells settle does affect their role and the molecules they express. Once they have left the bone marrow, B cells can either circulate the body in the lymphatics, or they can settle in a defined site such as the peritoneum, the pleural cavity, the lymph nodes, or the spleen. Cells in these sites are long-lived and have varied characteristics depending on the role they are required to play. Like the IEL T-cell population, a separate subset of B cells exists that expresses the CD5 molecule. CD5+ B cells (also known as B1-cells) are found in the peritoneal and pleural cavities of mice, and the tonsils and blood in humans. Like IEL they are of a separate lineage to all other B cells, are dominant early in life, and gradually decline with age. Also like IEL, they have limited receptor diversity (in this case Ig). They respond to polysaccharide, rather than protein, antigens and produce IgM very early (within 48 hours) during inflammation. A key role of CD5+ B cells is to produce "natural" antibodies; these polyreactive low-affinity antibodies bind a number of antigens including DNA and α1–3 dextran, and may thus provide a first line of defense against bacterial infection.

B-Cell Receptor Signaling

As mentioned earlier, the BCR consists of membrane-tethered immunoglobulin (mIg): a tetramer consisting of a heavy (H) chain homodimer and two κ or two λ light (L) chains. Like the TCR, the BCR requires additional molecules to facilitate signal transduction via immunoglobulin. In the case of a B cell, this is a heterodimer of Igα/Igβ (CD79 α/β) subunits, which is noncovalently bound to surface immunoglobulin. Antigen binding to the Fab portion of mIg causes rapid phosphorylation of the ITAMs in the intracellular domains of CD79 α/β by Src-family kinases, such as Lyn, leading to the activation of Syk (Figure 9.11). Syk, in turn, activates B-cell linker (Blnk) which phosphorylates and activates phospholipase Cγ2 (PLCγ2) and leads to the recruitment of guanine nucleotide exchange factors (GEFs). These then utilize the same

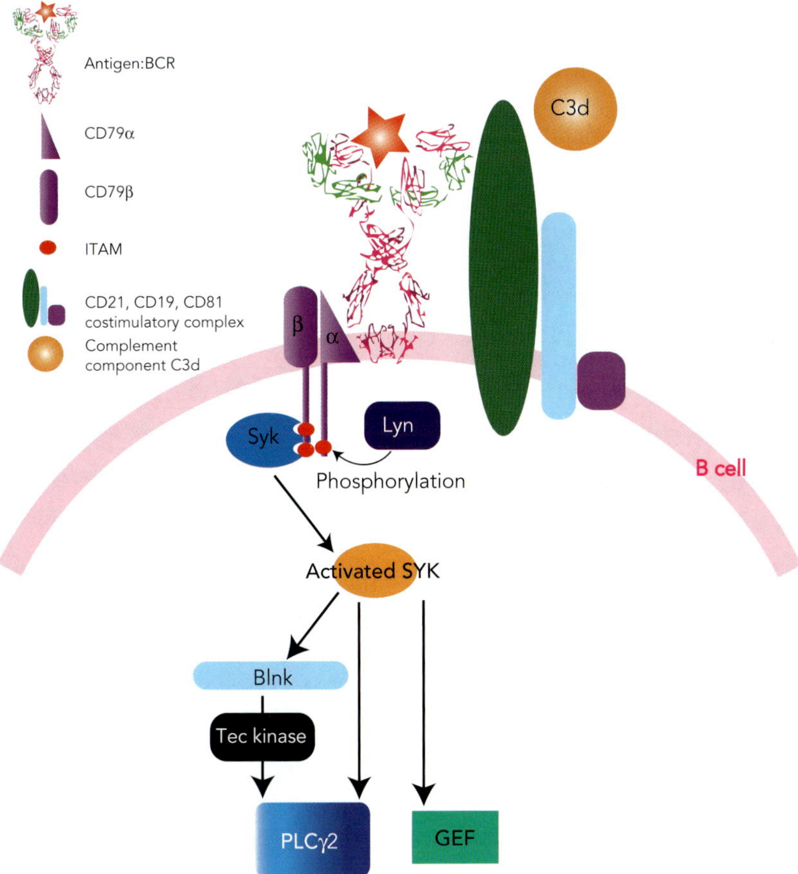

Figure 9.11. BCR signaling. Like the T cell, the B cell needs a signal through its BCR to become activated. However, unlike the T cell, the B cell does not need an APC to present processed peptides. B cells can recognize and respond to native foreign antigens. Like the TCR, the BCR lacks any significant intracellular signaling domains and must associate with accessory molecules in order to signal. In this case, signals are transmitted by CD79, the α and β chains of which associate with the BCR and also contain ITAMs. When the BCR encounters antigen, Src family kinases, such as Lyn, quickly act to phosphorylate the ITAMs located on the intracellular portions of CD79. Another kinase, Syk, is then able to bind and becomes activated going on to phosphorylate the B cell linker (Blnk) and recruiting Tec kinases and guanine nucleotide exchange factors (GEFs) to continue the signaling cascade.

signaling pathways as PLCγ1 and the GEFs involved in T-cell signaling (see Figure 9.4) to induce a variety of cellular processes, including regulation of gene expression, reorganization of the cytoskeleton, and immunoglobulin-mediated internalization of antigen complexes. Antigen internalized through the BCR is then processed in endosomal compartments and presented by MHC class II molecules to specific T cells.

The Requirement for Co-receptors

As with TCR signaling, simple stimulation through membrane immunoglobulin and its associated co-receptors does not provide the full picture. Both mIg and CD79 interact with elements of the cytoskeleton causing an accumulation of immunoglobulin–antigen complexes in a central SMAC (cSMAC). In established activation (but not required for immediate early signaling) this is encircled by a ring of integrins forming the peripheral SMAC (pSMAC).

CD19 is an essential co-receptor on the surface of the B-cell membrane and its cytoplasmic domains contain a number of binding sites for intracellular signaling and adaptor molecules. It colocalizes with CD21 and CD81 to function as an adaptor-like protein that facilitates the recruitment and activation of signaling molecules such as Vav and PI3K, to the BCR (for a review of BCR signaling, see Hasler and Zouali, 2001).

B-Cell Effector Function

All naive B cells express IgM and IgD. However, different antigens require different responses, and to fulfill these, the surface immunoglobulin isotype is changed in a process known as class switching. This involves a heavy chain gene rearrangement in the mature B cell resulting in a different immunoglobulin class becoming expressed.

It is one of the functions of activated T helper cells to promote immunoglobulin production and class switching in B cells. One of the most important of these interactions is binding of CD40L (CD154) on the activated T cell to CD40 on the B cell. This ligation promotes B-cell proliferation, and drives the B cell to terminally differentiate into a plasma cell that secretes large quantities of immunoglobulin or antibody. Furthermore, different T-cell subsets will cause the expression of different immunoglobulin isotypes in response to the cytokines the T-cell secrete (see Table 9.1).

TABLE 9.1. The influence of different cytokines on the immunoglobulin expressed and secreted by B cells

	Cytokines		
Ig isotype	**IFN-γ**	**IL-4**	**TGF-β**
IgM	↓*	↓	↓
IgG1	↓	↑	
IgG2a	↑	↓	
IgG2b			↑
IgG3	↑	↓	↓
IgE	↓	↑	
IgA			↑

*↓ indicates an inhibitory effect on that immunoglobulin isotype, ↑ a stimulatory effect, and no arrow indicates no known effect.

The primary functions of antibodies are neutralization and opsonization of pathogens; sensitization of effector cell types, such as NK cells, phagocytes, and mast cells; and activation of complement. Opsonization is the coating of foreign particles by antibodies to facilitate their uptake via Fc receptors (see Figure 9.12). Similarly, neutralization involves the coating of pathogens or the toxic compounds they release rendering them harmless to the host (see Figure 9.13). Each isotype varies in its efficiency for each of these functions, largely determined by the anatomical location of the antibody class. For example, IgM, the only isotype of antibody that can be produced without immunoglobulin class switching, is found primarily in the blood. Its large pentameric structure makes it difficult for the IgM molecule to diffuse into tissues, but means it is very effective at activating complement proteins, which are plentiful in the blood. In contrast, IgG and IgE are always found as monomers, and diffuse easily into tissues. The method for detecting soluble levels of secreted antibody is shown in Figure 9.14.

IgG is the most abundant immunoglobulin isotype in plasma, largely as a result of its long half-life in the circulation, and is important for neutralization, opsonization, and complement fixation. IgA can be found as a monomer or dimer, and is the most abundant isotype found at mucosal epithelia, such as the gastrointestinal and respiratory tracts – sites of frequent pathogen exposure. IgA molecules are produced in very large quantities (measured in grams each day in humans) and are actively transported through epithelial cells of the mucosa (as are pentameric IgM molecules). Because of its location at the front line of immune defense, IgA is highly efficient at neutralizing invading organisms. Very little IgD is produced in the body, and it has a negligible role in humoral responses. Only small quantities of IgE are produced and yet it fulfills an important biological role. It is found at very low concentrations in the blood or tissues because once secreted it is tightly bound to the surface of mast cells, via their high-affinity Fc receptors (discussed later).

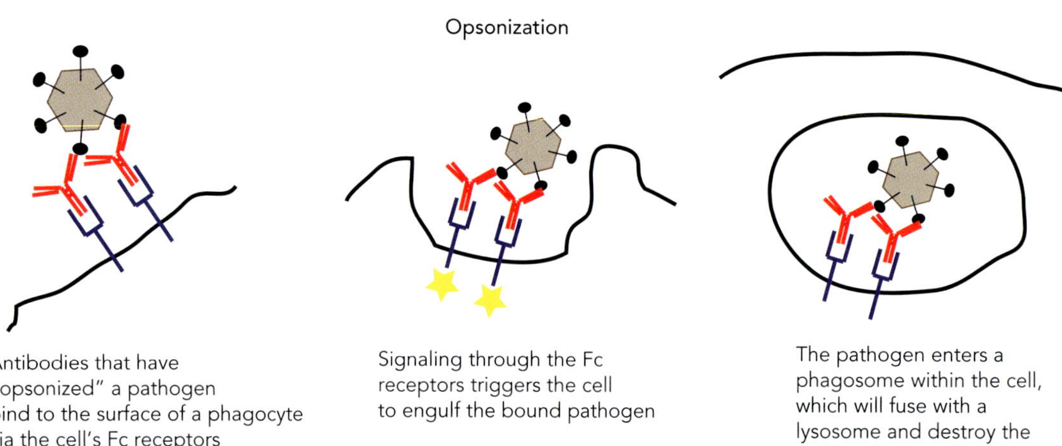

Opsonization
Antibodies that have "opsonized" a pathogen bind to the surface of a phagocyte via the cell's Fc receptors
Signaling through the Fc receptors triggers the cell to engulf the bound pathogen
The pathogen enters a phagosome within the cell, which will fuse with a lysosome and destroy the pathogen

Figure 9.12. Opsonization of pathogens by antibody. Another major role of antibody is to 'opsonize' foreign particles. The variable chains of antibodies can bind specific foreign peptides on the surface of pathogens. Once the variable regions have bound, the constant regions of the antibodies are recognized by Fc receptors on phagocytes, which engulf the antibodies along with the pathogen to which they have bound. The antibody/pathogen complex is then degraded in a phagolysosome within the phagocyte.

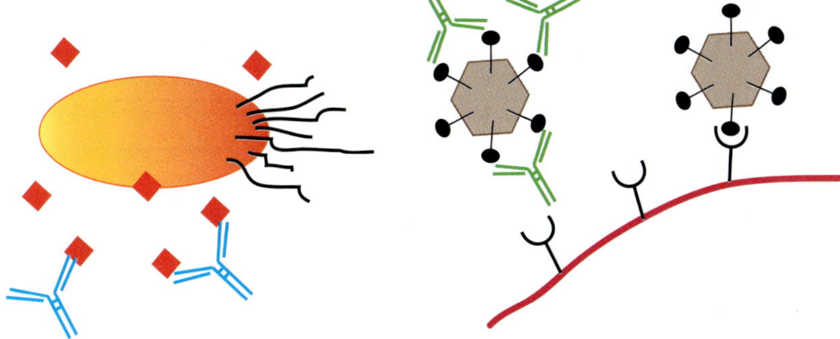

Antibodies can bind and neutralize harmful substances, preventing damage to host cells (e.g., secreted bacterial toxins)

Antibodies bound to pathogens can block attachment to a host cell (e.g., Ab against many viruses can prevent receptor-mediated endocytosis)

Figure 9.13. Neutralization/immune exclusion by antibody.

For soluble cytokine

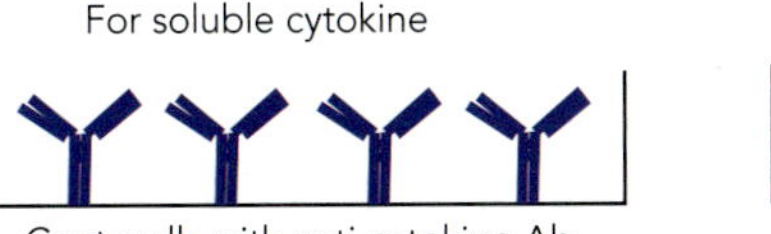

Coat wells with anti-cytokine Ab, for example, anti-IFN-γ; this is called the capture antibody.

For antibodies

Coat wells with target antigen of interest, for example, influenza virus.

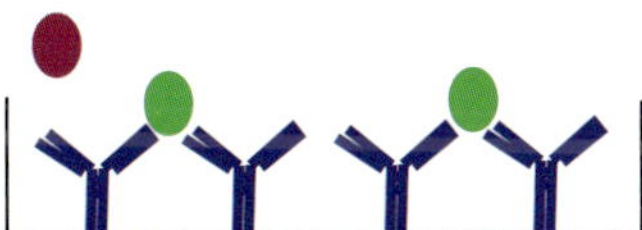

Add samples containing cytokine of interest, for example, cell culture supernatants, serum, etc. Antibodies bind only the IFN-γ in the culture, no other cytokines.

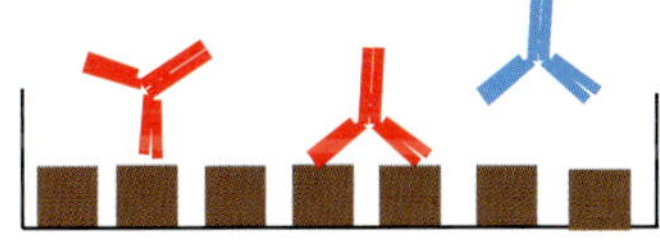

Add samples containing antidodies of interest, for example, serum from flu-infected animals. Only flu-specific antibodies bind to the plate.

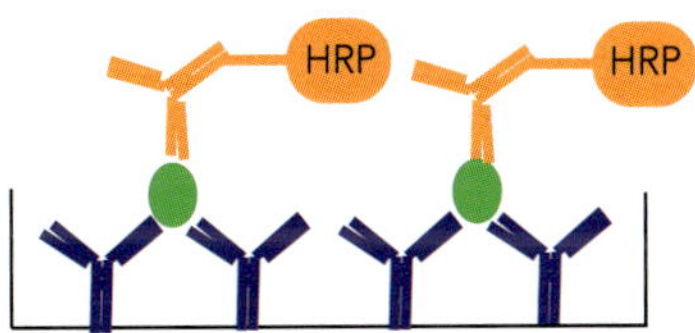

Add anti-IFN-γ antibody of a different specificity (binds a different epitope on IFN-γ protein). This is called the detection antibody and often has an enzyme coupled to it, such as horseradish peroxidase (HRP).

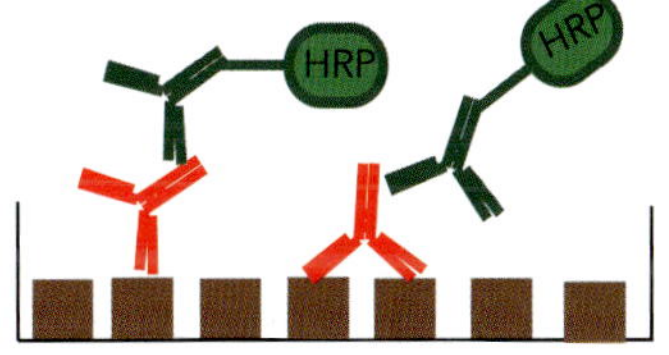

Add an enzyme-coupled detection antibody specific to the species or isotype of the target antibody, for example, rat anti-mouse IgG1.

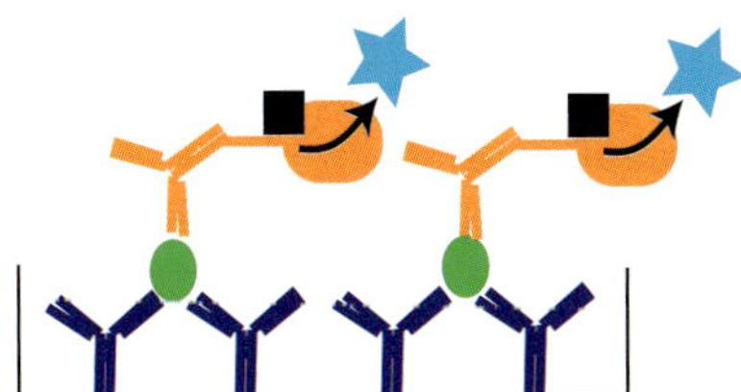

A substrate is added, which when converted by the enzyme on the detection antibody, produces a colored solution. The absorbance of the solution can be analyzed and is proportional to the amount of cytokine or target antibody of interest found in the sample.

Figure 9.14. Enzyme-linked immunosorbent assay.

TABLE 9.2. The dominant effector function (denoted in red) of secreted immunoglobulin isotypes

Ig isotype	Neutralization	Opsonization	Complement activation	NK cell activation (ADCC)	Mast cell activation
IgM					
IgD					
IgG					
IgE					
IgA					

IgE is therefore the primary activator of mast cells, which are important for T_h2-driven immune responses and feature in allergic reactions. (See Table 9.2 for a summary of antibody isotype functions.)

In addition to direct neutralization, activation of complement (leading to pore formation and lysis of the pathogen), and immune exclusion of pathogens (in which antibodies bind to and inactivate the micro-organism, or block its binding to the host cell), antibodies can also exert protective effects via binding to Fc receptors (FcR) on the surface of immune cells. At least eight different FcR have been identified, including several splice variants of a single receptor. FcR are named based on the isotype of immunoglobulin they bind, so FcεR binds IgE, FcγR binds IgG, and so on. FcR vary in their distribution on cells and their binding affinity, but at least one type of FcR is expressed on DCs, macrophages, NK cells, neutrophils, eosinophils, basophils, mast cells, and B cells.

In most cases, binding of an immunoglobulin to an FcR via the Fc region results in cross-linking and signaling that activates the FcR-bearing cell. In the case of macrophages, this binding via the FcR stimulates phagocytosis and increases the antimicrobial properties of the macrophage, allowing destruction of any engulfed pathogens. FcR and complement receptors are both found on the surface of many phagocytic cells such as macrophages and neutrophils, and signaling through both of these pathways has a synergistic effect on uptake and killing by the phagocyte. This process of coating a pathogen (with antibody and/or complement) to increase its uptake by phagocytic cells is called opsonization. Cross-linking of FcR on other cell types, such as mast cells and NK cells, can trigger exocytosis of toxic mediators stored in granules within the cell. Parasitic worms are frequently too large to be engulfed by phagocytes, but when coated with IgE, become targets for activated eosinophils, which bind via their FcεR. This binding causes release of granules containing destructive enzymes and toxic proteins, helping to clear the pathogen. NK cells can be triggered to kill an infected cell by a similar mechanism. During a viral infection, viral peptides can be displayed on the surface of the infected cell and recognized by circulating immunoglobulin. The exposed Fc regions are recognized by FcRs on NK cells, triggering them to release toxic granules that result in apoptosis of the target cell, using the same mechanism as that employed by CTL discussed earlier. This NK cell activation through FcR via immunoglobulin bound to the surface of an infected cell is called antibody-dependent cell-mediated cytotoxicity (ADCC) and is shown in Figure 9.15.

NATURAL KILLER CELLS

NK cells, so called because it was initially thought that they did not require activation in order to kill cancerous cells, are a type of lymphocyte involved in innate immunity against pathogens and tumors. They arise from a common lymphocyte progenitor but are larger and more granular than T cells and B cells. NK cells account for up to 15% of peripheral blood lymphocytes and are located in several organs including the liver, peritoneal cavity, and placenta in homeostasis. However, numbers increase during disease because they are recruited and activated in response to interferons and other macrophage-derived cytokines produced during infection. By inducing apoptosis in infected cells, NK cells act early during the immune response and are able to limit viral spread until the adaptive immune response has developed. The importance of this role is apparent in patients with NK cell deficiency who are more susceptible to viral infection.

NK Cell Function

NK cells were originally defined by their ability to kill tumor cells in vitro. They sit between the innate and adaptive arms of the immune response and detect the presence of "altered self" by scrutinizing cells for the presence of MHC class I: the "missing self hypothesis." In health, all nucleated cells express MHC class I. As discussed earlier, MHC class I molecules can present viral proteins to CTLs which then recognize the cell as infected and induce apoptosis. To escape recognition by CTLs, a number of viruses and tumors downregulate MHC class I. It is thought that NK cells evolved

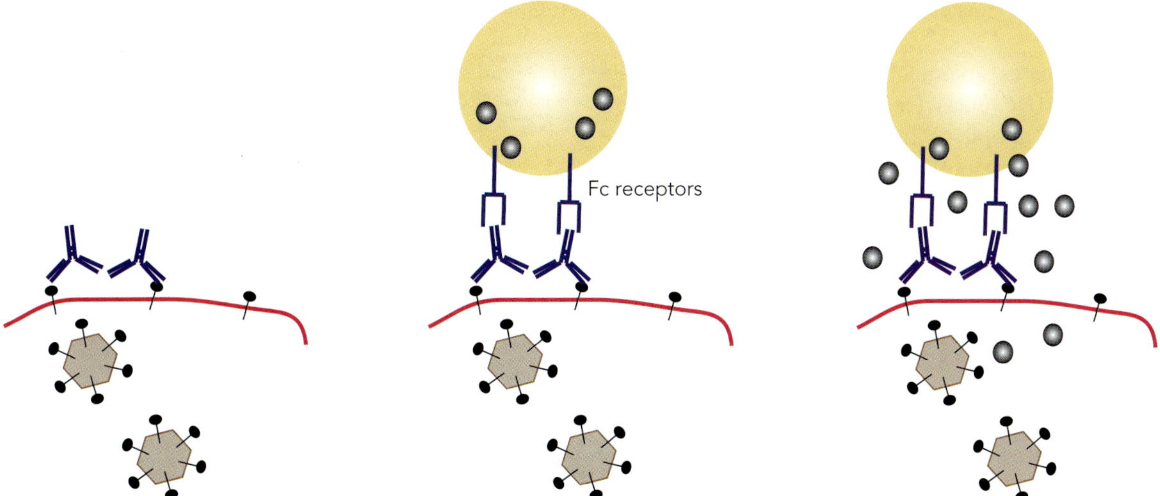

Fc receptors
Virally infected cells express some viral proteins at the cell surface that can be recognized and bound by specific antibodies
An NK cell bearing Fc receptors can recognize Ab bound to the virus-infected cell, and this interaction activates the NK cell
The activated NK cell releases its cytotoxic granules, inducing apoptosis in the infected cell and limiting viral spread

Figure 9.15. Antibody-dependent cell-mediated cytotoxicity (ADCC).

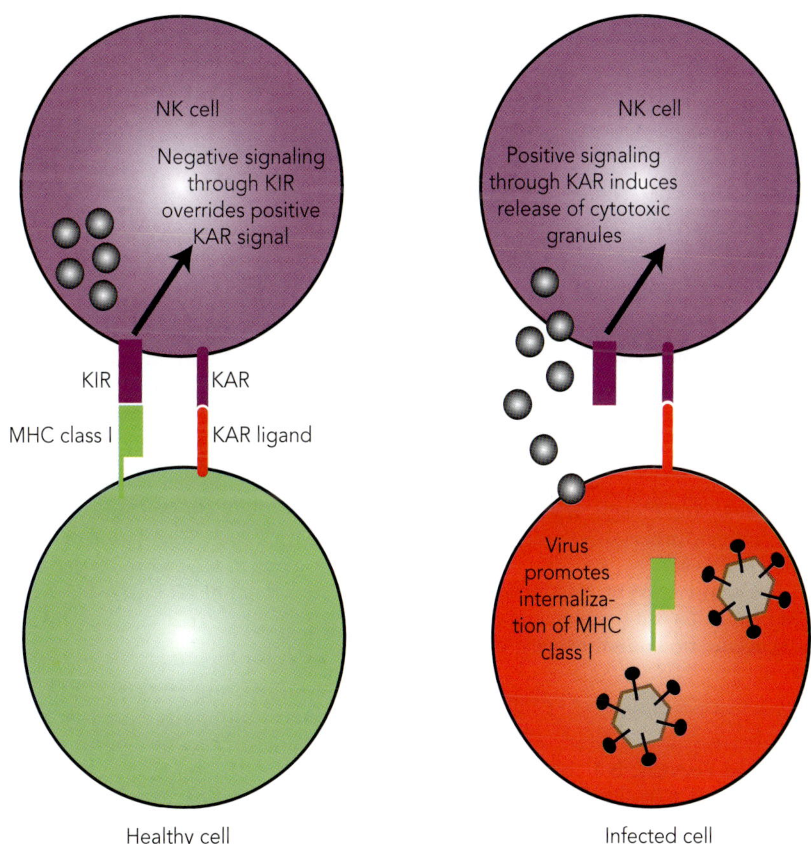

NK cell
Negative signaling through KIR overrides positive KAR signal
KIR
KAR
MHC class I
KAR ligand
Healthy cell
NK cell
Positive signaling through KAR induces release of cytotoxic granules
Virus promotes internaliza-tion of MHC class I
Infected cell

Figure 9.16. NK cell killing. NK cells have evolved to recognize MHC class I molecules. When a NK cell interacts with a healthy cell, it receives signals through both killer activating receptors (KARs) and killer inhibitory receptors (KIRs). The balance of these signals suppresses the killing response and the healthy cell is released. If, however, the cell is infected with a virus or becomes cancerous, it loses MHC class I from its surface. The NK cell no longer receives an inhibitory signal through its KIRs so the KAR signal promotes the release of cytotoxic granules from the NK cell which induces apoptosis in the infected cell.

to counter this escape mechanism by recognizing the absence of MHC class I as a hallmark of infection and inducing apoptosis in the manner of CTLs (see Figure 9.16). In response, a number of viruses have now evolved to produce proteins that mimic MHC class I on the surface of the infected cell and thus avoid NK cell–mediated killing.

To control their cytotoxic activity, NK cells possess two types of surface receptor: killer activating receptors (KARs) and killer inhibitory receptors (KIRs). On encountering a healthy cell, KIRs recognize and bind to MHC class I molecules on the cell surface. KIRs contain immunoreceptor tyrosine-based inhibitory motifs (ITIMs) in their cytoplasmic domains that recruit intracellular tyrosine phosphatases and inhibit NK cell activation. If MHC class I is absent, then positive signals through a KAR drive the NK cell to perform its cytotoxic duties and secrete inflammatory cytokines; principally IFNγ. NK cell killing is mediated by the same factors involved in CTL killing: perforin, which polymerizes to form pores in the infected cell; granzymes, which enter the target cell and activate the caspase cascade; and the Fas–FasL interaction. In the case of KARs, the ligands inducing the killing response are still unclear though one KAR, NKG2D, has been shown to recognize a stress-induced protein with similar structure to MHC class I.

Some NK cells express the TCR. These are known as NKT cells and can recognize antigen presented by APCs. However, the TCR they express is only semivariant and recognizes a limited number of antigens on the nonpolymorphic CD1 molecule. As a result, they are also more "innate" than other T cells.

SUGGESTED READINGS

Antigen presentation

Vyas, J.M., Van der Veen, A.G., and Ploegh, H.L. 2008. The known unknowns of antigen processing and presentation. *Nat Rev Immunol* **8:**607–618.

Zinkernagel, R.M., and Doherty, P.C. 1974. Restriction of *in vitro* T cell-mediated cytotoxicity in lymphocytic choriomenigitis within a syngenic or semi-allogeneic system. *Nature* **248:**701–702.

T-cell development

Laky, K., Fleischacker, C., and Fowlkes, B.J. 2006. TCR and Notch signalling in CD4 and CD8 T cell development. *Immunol Rev* **209:**274–283.

Moroy, T., and Karsunky, H. 2000. Regulation of pre-T cell development. *Cell Mol Life Sci* **57**(6)**:**957–975.

T-cell receptor signaling

Nel, A.E. 2002. T-cell activation through the antigen receptor. Part 1: Signaling components, signaling pathways, and signal integration at the T-cell antigen receptor synapse. *J Allergy Clin Immunol* **109**(5)**:**758–770.

Nel, A.E., and Slaughter, N. 2002. T-cell activation through the antigen receptor. Part 2: Role of signaling cascades in T-cell differentiation, anergy, immune senescence, and development of immunotherapy. *J Allergy Clin Immunol* **109**(6)**:**901–915.

Costimulatory signaling and late costimulators

Alegre, M-L., Frauwirth, K.A., and Thompson, C.B. 2001. T-cell regulation by CD28 and CTLA-4. *Nat Rev Immunol* **1**(3)**:**220–228.

Croft, M. 2003. Co-stimulatory members of the TNFR family: keys to effective T-cell immunity? *Nat Rev Immunol* **3**(8)**:**609–620.

Effector function of T helper cell subsets

Reiner, S.L. 2007. Development in motion: Helper T cells at work. *Cell* **129**(1)**:**33–36.

Intraepithelial T cells

Hayday, A., Theodoridis, E., Ramsburg, E., and Shires, J. 2001. Intraepithelial lymphocytes: exploring the Third Way in immunology. *Nat Immunol* **2**(11)**:**997–1003.

Effector function of cytotoxic (killer) T lymphocytes

Anderson, M.H., Schrama, D., thor Straten, P., and Becker, J.C. 2006. Cytotoxic T cells. *J Invest Dermatol* **126**(1)**:**32–41.

B-cell development

Milne, C.D., Fleming, H.E., Zhang, Y., and Paige, C.J. 2004. Mechanisms of selection mediated by interleukin-7, the preBCR, and hemokinin-1 during B-cell development. *Immunol Rev* **197:**75–88.

B-cell receptor signaling

Hasler, P., and Zouali, M. 2001. B cell receptor signaling and autoimmunity. *FASEB J* **15**(12)**:**2085–2098.

Pierce, S.K. 2002. Lipid rafts and B-cell activation. *Nat Rev Immunol* **2**(2)**:**96–105.

The requirement for co-receptors

Rickert, R.C. 2005. Regulation of B lymphocyte activation by complement C3 and the B cell coreceptor complex. *Curr Opin Immunol* **17**(3)**:**237–243.

B-cell effector function

Casadevall, A., and Pirofski, L.A. 2006. A reappraisal of humoral immunity based on mechanisms of antibody-mediated protection against intracellular pathogens. *Adv Immunol* **91:**1–44.

Martin, F., and Kearney, J.F. 2000. B-cell subsets and the mature preimmune repertoire. Marginal zone and B1 B cells as part of a "natural immune memory." *Immunol Rev* **175:**70–79.

NK cell function

Cerwenka, A., and Lanier, L.L. 2001. Natural killer cells, viruses and cancer. *Nat Rev Immunol* **1**(1)**:**41–49.

Nimmerjahn, F., and Ravetch, J.V. 2008. Fcγ receptors as regulators of immune responses. *Nat Rev Immunol* **8**:34–47.

Sigal, L.H. 2007. Basic science for the clinician 47: Fcgamma receptors. *J Clin Rheumatol* **13**(6):355–358.

References for protein structures

MHC class I – Protein databank accession number 2vaa. Primary citation

Fremont, D.H., Matsumura, M., Stura, E.A., Peterson, P.A., and Wilson, I.A. 1992. Crystal structures of two viral peptides in complex with murine MHC class I H-2Kb. *Science* **257**:919–927.

MHC class II – Protein databank accession number 1iao. Primary citation

Scott, C.A., Peterson, P.A., Teyton, L., and Wilson, I.A. 1998. Crystal structures of two I-Ad-peptide complexes reveal that high affinity can be achieved without large anchor residues. *Immunity* **8**:319–329.

B cell receptor – Protein databank accession number 2RCJ. Primary citation

Perkins, S.J., Nealis, A.S., Sutton, B.J., and Feinstein, A. 1991. Solution structure of human and mouse immunoglobulin IgM by synchrotron X-ray scattering and molecular graphics modelling: a possible mechanism for complement activation. *J Mol Biol* **221**:1345–1366.

T cell receptor:peptide:MHC complex – Protein databank accession number 1ao7. Primary citation

Garboczi, D.N., Ghosh, P., Utz, U., Fan, Q.R., Biddison, W.E., and Wiley, D.C. 1996. Structure of the complex between human T-cell receptor, viral peptide and HLA-A2. *Nature* **384**:134–141.

10 Fibroblasts and Stromal Cells

Andrew Filer and Christopher D. Buckley

WHAT IS A STROMAL CELL?

The architecture of organs and tissues is closely adapted to their function to provide microenvironments in which specialized functions may be carried out efficiently. The nature and character of such microenvironments are primarily defined by the stromal cells that reside within the tissues. The most abundant cell types of the stroma are fibroblasts, which are responsible for the synthesis and remodeling of extracellular matrix (ECM) components. In addition, their ability to produce and respond to growth factors and cytokines allows reciprocal interactions with adjacent epithelial and endothelial structures and with infiltrating leukocytes. As a consequence, fibroblasts play a critical role during tissue development and homeostasis and have often been described as having a "landscaping" function. In this chapter, we concentrate on fibroblasts as the prototype stromal cell. However, the stroma also consists of blood and lymphatic vessels, and a wider definition of stromal cells might include endothelial cells, specialized cells such as pericytes (blood vessel supporting cells), and even tissue resident macrophages.

Stromal Cell Identity and Microenvironments

Tissue resident macrophages in the liver (Kupffer cells) and lung (alveolar macrophages) perform very different functions compared to macrophages in the brain (glial cells) or skin (Langerhans cells), yet they are all members of the monocyte/macrophage family. Until recently, fibroblasts had been thought of as ubiquitous, generic cells with a common phenotype even within different tissues. However, we now know that fibroblasts from different organs are more like their macrophage counterparts, with unique morphology and repertoires of ECM proteins, cytokines, costimulatory molecules, and chemokines specialized to the different microenvironments in which they are found (Figure 10.1). Furthermore, when their transcriptional profiles are examined using microarray techniques, fibroblasts hold a strong memory of their position and function in the body. Early studies demonstrated that fibroblast transcriptomes (the global picture of transcribed genes measured using microarrays) could be clustered into peripheral (synovial joint or skin fibroblasts) versus lymphoid (tonsil or lymph node) groups according to their organ of origin, with the potential to shift their transcriptional profiles by treatment with inflammatory mediators such as tumor necrosis factor alpha (TNF-α), interleukin-4 (IL-4), or interferon-γ. More extensive analysis of expression profiles from primary human fibroblasts by Rinn et al. has shown large-scale differences related to three broad anatomical divisions: anterior–posterior, proximal–distal, and dermal–nondermal. Genes involved in pattern forming, cell-signaling, and matrix remodeling were found to predominantly account for these divisions. The gene expression profile of adult fibroblasts may therefore play a significant role in assigning positional identity within an organism. This in turn provides evidence for the concept of a stromal address code, analogous to the endothelial area post code. According to this hypothesis, chemokines, cytokines, and adhesion molecules produced by stromal cells control the accumulation, retention, survival, and differentiation of leukocytes in a site or organ-dependent manner, providing a plausible explanation for the well described but as yet poorly understood clinical finding that relapses in chronic inflammatory diseases are often tissue and site specific. For instance, despite some similarities at the level of leukocyte infiltrates rheumatoid arthritis (RA) affects predominantly the joints, whereas multiple sclerosis affects the central nervous system. Support for this hypothesis also comes from the cancer field, where defective or altered stromal cell

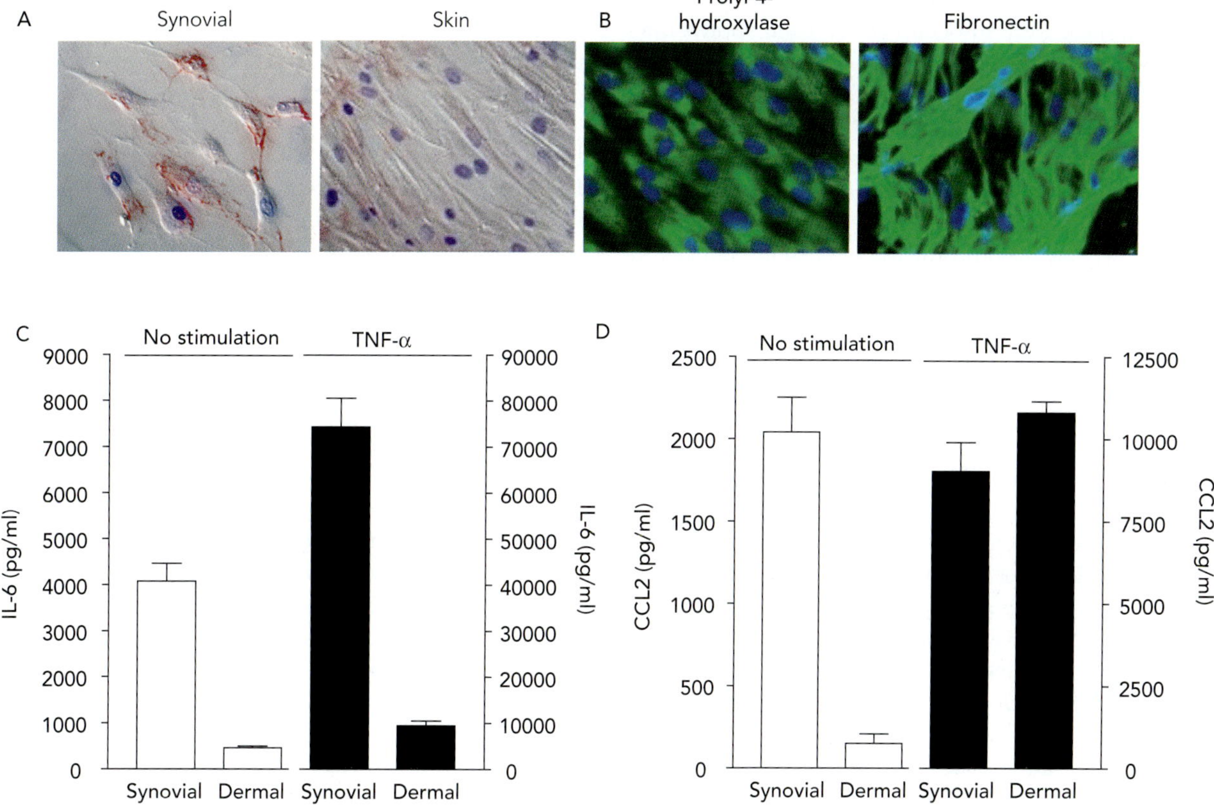

Figure 10.1. Fibroblast phenotype. (A) staining of live fibroblast cells in culture illustrating typical morphology and marked differences between synovial fibroblasts of the rheumatoid arthritis joint and skin fibroblasts. Red stain fibronectin demonstrates matrix production. (Blue stain = nuclear stain.) (B) Stromal cell status is confirmed by fluorescence microscopy of cells showing collagen synthetic enzymes (prolyl-4-hydroxylase) and matrix production (fibronectin) in skin fibroblasts. (C) Production of IL-6 measured by ELISA in unstimulated and TNF-α-stimulated fibroblasts is constitutively higher in synovial than skin cells. (D) A similar high basal CCL2 (MCP-1) chemokine production is seen in synovial fibroblasts; this is obliterated by TNF-α stimulation.

function has been strongly associated with tumorigenesis and progression of tumors to metastasis.

Stromal Cells and Inflammation

Inflammatory reactions proceed against the backdrop of specialized stromal microenvironments. The response to tissue damage involves a carefully choreographed series of interactions between diverse cellular, humoral, and connective tissue elements. For an inflammatory lesion to resolve, dead or redundant cells that were recruited and expanded during the active phases of the response must be removed. In addition, resident stromal cells, largely defined by fibroblasts, attempt to repair damaged tissue. It is becoming increasingly clear that fibroblasts are not just passive players in immune responses, but that they play active roles in determining the switches that occur governing progression from acute to chronic inflammation, and also those governing resolution or the progression to chronic, persistent inflammation. The "switch to resolution" is an important signal that permits tissue repair to take place and enables immune cells to return to draining lymphoid tissues (lymph nodes) for immunological memory to become established. However, in immune-mediated inflammatory diseases, stromal cells contribute to the inappropriate recruitment and retention of leukocytes, leading to chronic persistent inflammation. Research on the roles of stromal cells in inflammation and cancer medicine and specifically the parts played by fibroblasts in physiological and pathological inflammation will be examined later.

Stromal Cells: In Vitro and In Vivo

By virtue of their role in defining the geography of specialized tissues, fibroblasts and other stromal cells exist in living organisms within 3-dimensional environments, whereas the majority of experiments performed using fibroblasts in the laboratory are conducted within 2-dimensional environments. Furthermore, fibroblasts are frequently grown in nonphysiological stimuli such as serum, to which fibroblasts would not normally be exposed unless tissue damage were to occur. It has been shown that the behavior of fibroblasts is different when cultured in artificial 3-dimensional environments. It is therefore all the more remarkable that fibroblasts cultured using conventional 2-dimensional techniques retain characteristics such as positional memory and unique cytokine profiles. This suggests a degree of epigenetic imprinting may account for the persistent changes in gene expression seen in these cells.

EMBRYOLOGICAL ORIGINS OF FIBROBLASTS

As mentioned earlier, even within a single tissue there is growing evidence that fibroblasts are not a homogeneous population, but exist as subsets of cells, much like tissue macrophages and dendritic cells (DCs). It is likely that connective tissue contains a mixture of distinct fibroblast lineages with mature fibroblasts existing side by side with more immature fibroblasts that are capable of differentiating into other connective tissue cells. Stromal cells have been defined in terms of their embryological origins and lineage relationships, and are generally considered to be mesenchymal in origin. However, the embryological mesenchyme, from which fibroblasts are derived, is not in itself a germ layer (usually defined as ecto-, endo-, and mesoderm), but is variably considered to be either wholly composed of embryonic mesoderm, or a combination of the mesoderm and ecto- or endoderm layers. For instance, in the head and neck some mesenchyme is derived from neural crest cells (and hence from ectoderm). Moreover, cell populations have now been identified which appear to blur the distinction between hemopoietic and nonhemopoietic populations. In addition, other unexpected shifts in lineage have been reported, including differentiation from neural stem cells into myeloid and lymphoid hemopoietic lineage. Classification by such lineages that are no longer as restrictive as previously thought is therefore becoming increasingly awkward.

The problem of distinguishing fibroblasts of differing origin or maturity has historically been very difficult due to a lack of cell surface specific markers. Whereas the cluster of differentiation (CD) markers have revolutionized the isolation and study of leukocyte subsets, there have been relatively few, poor quality discriminatory markers allowing the identification of fibroblast subpopulations. Fibroblasts have frequently been identified by their spindle-shaped morphology (Figure 10.1), elaboration of ECM, and lack of positive markers for endothelium, epithelial, and hemopoietic cells. However, recent studies have begun to identify novel markers which demarcate distinct subpopulations of stromal cells during development and which have the potential to act as markers for different subpopulations of fibroblasts each with different roles. Such markers include smooth muscle actin, which marks out a population of secretory, activated cells termed myofibroblasts, and more recently discovered markers such as CD248 and GP38 (podoplanin) (Table 10.1). The question of the origin of fibroblasts is an important one. Both inflammation and wound healing are characterized by the formation of new

TABLE 10.1. Comparison of phenotypic expression of blood-borne mesenchymal precursors and RA synovial fibroblasts

Marker	MSC/MPC	RA synovial fibroblasts	Tissue myofibroblasts	Fibrocytes
Collagen I	+	+	+	+
Collagen III	+	+	+	+
Fibronectin	+	+	+	+
Vimentin	+	+	+	+
HLA-DR	+	+*	–	+
CD44	+	+	+	+
CD45	–	–	–	+
CD34	–	–	–	+
VCAM-1	+	+	–	–
Smooth muscle actin	+	+/–	+	–**
Bone morphogenic protein receptors	+	+	–	Not known
DAF (CD55)	+	+	–	Not known
BST-1	+	+	–	Not known

*, Rapidly lost in culture; **, fibrocytes differentiate into SMA; –, positive fibroblasts.

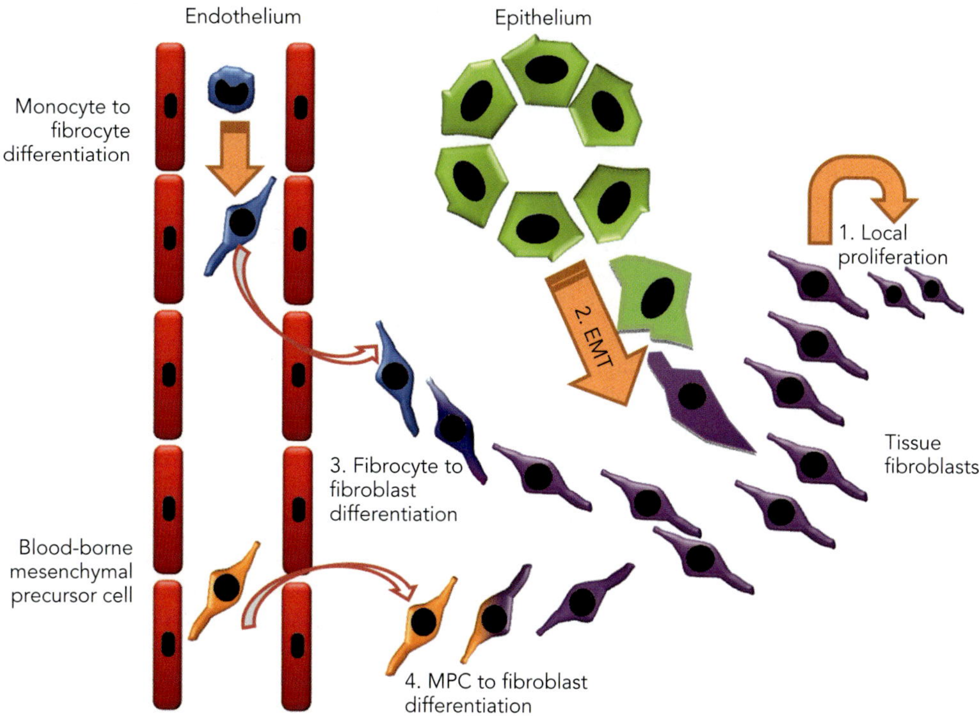

Figure 10.2. Routes of differentiation to tissue fibroblasts. In response to wounding or inflammation, increased numbers of fibroblasts are produced within tissue. (1) Fibroblasts can proliferate locally to generate new fibroblasts. (2) The transition of epithelial to stromal cell populations has been shown to occur in diseases of the lung and kidney. (3) Fibrocytes arise from the monocyte population in blood, then differentiate toward fibroblasts in tissue. (4) Blood borne mesenchymal precursor cells may be recruited to tissues and undergo local differentiation to tissue fibroblasts.

tissue. However, recent findings have suggested that the new cells that form the remodeled tissues are not, as was hitherto assumed, necessarily derived from the proliferation of cells that are resident in the adjacent noninjured tissue.

Sources of Tissue Fibroblasts

It is widely accepted that the principle origin for fibroblasts is from primary mesenchymal cells and that upon appropriate stimulation fibroblasts can proliferate locally to generate new fibroblasts. Though an increase in fibroblast numbers caused by local proliferation does occur, fibroblasts may arise from other sources (Figure 10.2). The first of these is local epithelial to mesenchymal transition (EMT). This is an essential, physiologically important developmental mechanism for diversifying cells in the formation of complex tissues. However, fibroblasts also appear to be derived by this process in adult tissue following epithelial stress such as inflammation or tissue injury. EMT disaggregates epithelial cells and reshapes epithelial cells for movement. The epithelium loses polarity as defined by the loss of adherens junctions, tight junctions, desmosomes, and cytokeratin intermediate filaments. Epithelial cells also rearrange their F-actin stress fibers and express filopodia and lamellopodia. A combination of cytokines and matrix metalloproteinases (MMPs) associated with digestion of the basement membrane is believed to be secreted and important in the process. The transition of epithelial to mesenchymal cell populations has been shown to occur in diseases of the lung and kidney where the process has been implicated in fibrotic disease (see Kalluri et al.).

An alternative explanation for the accumulation of stromal cells in chronic inflammatory conditions such as RA lies in the possibility of blood borne precursors. In the mid 1990s it was shown that vascular precursors (angioblasts) could be found circulating in the blood of normal individuals, and that they could be recruited to sites of vasculogenesis in a rabbit ischemic hind limb model. This demonstrated that circulating mesenchymal precursors exist outside the hemopoietic system. Subsequent work has confirmed the presence of circulating cells of a mesenchymal phenotype in human subjects. These cells bear a remarkable resemblance to the synovial fibroblasts found in the joints of patients with RA, which accumulate in enormous quantities in the joint lining despite little evidence of proliferation. Interestingly, Marinova-Mutafchieva et al. showed that an influx of such cells preceded inflammation in a mouse collagen-induced arthritis model, suggesting that there may be a role for blood-borne stromal cell precursors in the initiation of inflammatory diseases.

Another circulating precursor cell that could account for the accumulation of fibroblasts in some diseases is a cell called the fibrocyte. Fibrocytes appear to comprise 0.1%–0.5% of nonerythrocytic cells in peripheral blood, and have been shown to rapidly enter sites of tissue injury and contribute to tissue remodeling in models of inflammatory lung disease. They are adherent cells with a spindle-shaped morphology which express MHC class II as well as type I collagen and which arise from within the CD14-positive (monocyte) fraction of peripheral blood. Fibrocytes are capable of matrix elaboration, and have been proposed to differentiate along a fibroblast lineage under the influence of cytokines, particularly TGF-β. The mere fact that a cell type apparently arising from within the monocyte lineage may become a "mesenchymal" stromal cell such as a fibroblast implies a further degree of plasticity and blurring of the apparently clear line between hemopoietic and nonhemopoietic lineages.

Fibroblasts versus Mesenchymal Precursor Cells

The potential role of circulating mesenchymal cell precursors (variously termed mesenchymal stem cells [MSC], mesenchymal stromal cells or mesenchymal precursor cells [MPC]) as sources of tissue fibroblasts is highlighted by the remarkable capacity of these cells to differentiate into other members of the connective tissue family including cartilage, bone, adipocyte, and smooth muscle cells. This ability was initially demonstrated in bone marrow stromal cells, RAL synovial fibroblasts, and circulating mesenchymal cells, and was therefore suggested to define a characteristic mesenchymal phenotype, based on the hypothesis that the rheumatoid synovium could become populated by a large proportion of circulating mesenchymal precursor cells exported from the bone marrow. However, the property of trilineage differentiation ("pluripotentiality") has now been shown to be a property of many adult tissue fibroblasts, though varying somewhat between fibroblasts from different tissues, implying a hitherto unsuspected degree of plasticity in the body's stromal cell populations. The two previously separate fields of mesenchymal precursor cell biology and largely disease-centered fibroblast biology have therefore rapidly converged.

STROMAL CELLS IN IMMUNE RESPONSES

Physiological inflammation is not a stable state. Such inflammation either resolves, and the tissue reverts to normal, or develops into chronic persistent inflammation. Once established, chronic inflammation is hard to "cure" as evidenced by the fact that despite increasingly focused approaches to suppressing active inflammation within tissues, a cure for chronic immune-mediated inflammatory diseases such as RA remains elusive, since on the withdrawal of effective anti-inflammatory therapies the disease inevitably relapses.

As an inflammatory process reaches its conclusion, the resolution of inflammatory leukocyte infiltrates within a microenvironment is governed by a number of dynamic factors: first, the balance between cell recruitment and emigration; second, the balance between cell death and proliferation; and third, the coordinated release of proresolution factors such as resolvins and adenosine (which will not be discussed further in this chapter). Recent evidence suggests that tissue stromal cells are able to determine the type and duration of leukocyte infiltrates in an inflammatory response. At the resolution of such responses, stromal cells contribute to the withdrawal of survival signals and normalization of chemokine gradients, allowing infiltrating cells to undergo apoptosis or leave via draining lymphatics (Figure 10.3). Subversion of these pathways results in a switch to persistent inflammation that remains remarkably stable over time. We shall use the stromal microenvironment of the rheumatoid joint as a model to illustrate the role played in persistence of inflammation by stromal cells.

Stromal Cells as Innate Immune Sentinels

Classically, macrophages have been studied as sources of inflammatory cytokines and chemokines in response to innate immune stimuli, and portrayed as immune sentinel cells accordingly. However, when activated by substances released during tissue injury or the products of invading microorganisms, fibroblasts are capable of elaborating a broad repertoire of inflammatory mediators which fully justifies their classification as immune sentinel cells. Through expression of toll-like receptors, fibroblasts respond to bacterial products such as LPS by generating chemokines capable of recruiting inflammatory cells. Furthermore, fibroblasts are capable of bridging the innate and adaptive immune responses through expression of the molecule CD40. This molecule was initially assumed to be restricted in its expression to antigen-presenting cells such as macrophages and DCs. However, it is widely expressed by fibroblasts within discrete tissues. Engagement of CD40 by its ligand CD40L expressed on a restricted population of immune cells including activated T lymphocytes (Figure 10.4) is critical for the further induction of proinflammatory cytokines and chemokines during an immune response, as well

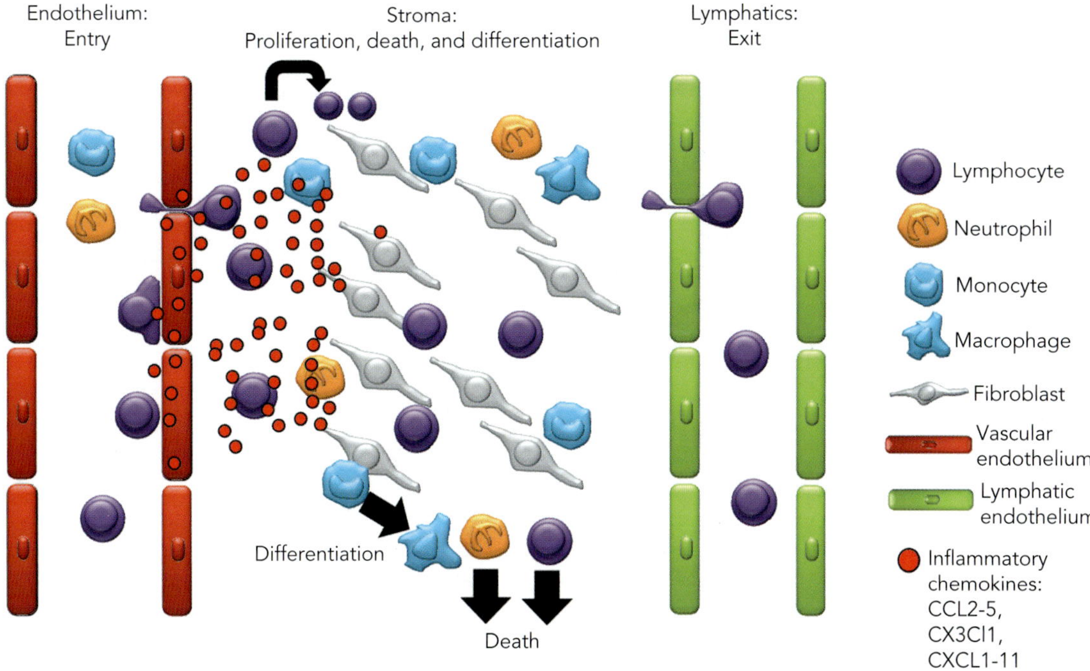

Figure 10.3. Regulation of inflammatory responses by the stroma. Evolution of an inflammatory response is characterized by recruitment of immune system cells via production of inflammatory chemokines. Recruited cells may undergo proliferation and/or differentiation. As an inflammatory process reaches its conclusion, the resolution of inflammatory leukocyte infiltrates within a microenvironment is governed by a number of dynamic factors including the balance between cell recruitment and emigration and the balance between cell death and proliferation. Recent evidence suggests that tissue stromal cells are able to determine the type and duration of leukocyte infiltrates in an inflammatory response. At the resolution of such responses, stromal cells contribute to the withdrawal of survival signals and normalization of chemokine gradients, allowing infiltrating cells to undergo apoptosis or leave via draining lymphatics.

as for antibody production by CD40 expressing B lymphocytes.

In some pathological states, activation of fibroblasts becomes persistent, leading to the hypersecretion of cytokines, chemokines, and matrix, which contribute to disease. The rheumatoid synovial fibroblast is a prime example of such persistent activation of fibroblasts. In RA, the fibroblasts take on a characteristic activated phenotype that remains stable even after culturing in vitro for many months. These cells play a direct role in tissue damage through secretion of multiple MMPs and cathepsins that degrade cartilage and bone tissues in the joint (Figure 10.4). Attachment to and direct, autonomous invasion of fibronectin-rich matrix such as cartilage has been demonstrated in the SCID mouse model of arthritis, in which cultured rheumatoid, but not osteoarthritis synovial fibroblasts invade and destroy co-implanted human cartilage. In vitro models such as the matrigel invasion assay produce corresponding results, in which the degree of invasion correlates with the degree of damage seen in the joints of patients from whose samples the fibroblast were initially grown. These models demonstrate the remarkably stable and disease-specific phenotype of cultured RA synovial fibroblasts which includes high basal and stimulated expression of signature cytokines such as IL-6, and chemokines (discussed later). RA synovial fibroblasts also express characteristic adhesion and immune modulating molecules such as VCAM-1, galectin-3, and a specific repertoire of Toll-like receptors, which initiate innate immune cellular responses. A satisfactory molecular explanation for the stable phenotype of RA synovial fibroblasts has until recently evaded the field. However, epigenetic changes including DNA methylation, histone modifications such as acetylation, and microRNA expression have now been suggested to underlie the observed persistent changes in fibroblast gene transcription and posttranscriptional repression.

Signal Transduction Processes

Both innate and adaptive immune responses triggered in synovial fibroblasts initiate characteristic signal transduction events. These result from stimulation such as Toll-like receptor engagement, or exposure to a proinflammatory cytokine network within a persistent chronic inflammatory infiltrate, as is seen in the case of RA. Many of the genes activated by such stimuli belong to the classical proinflammatory signaling

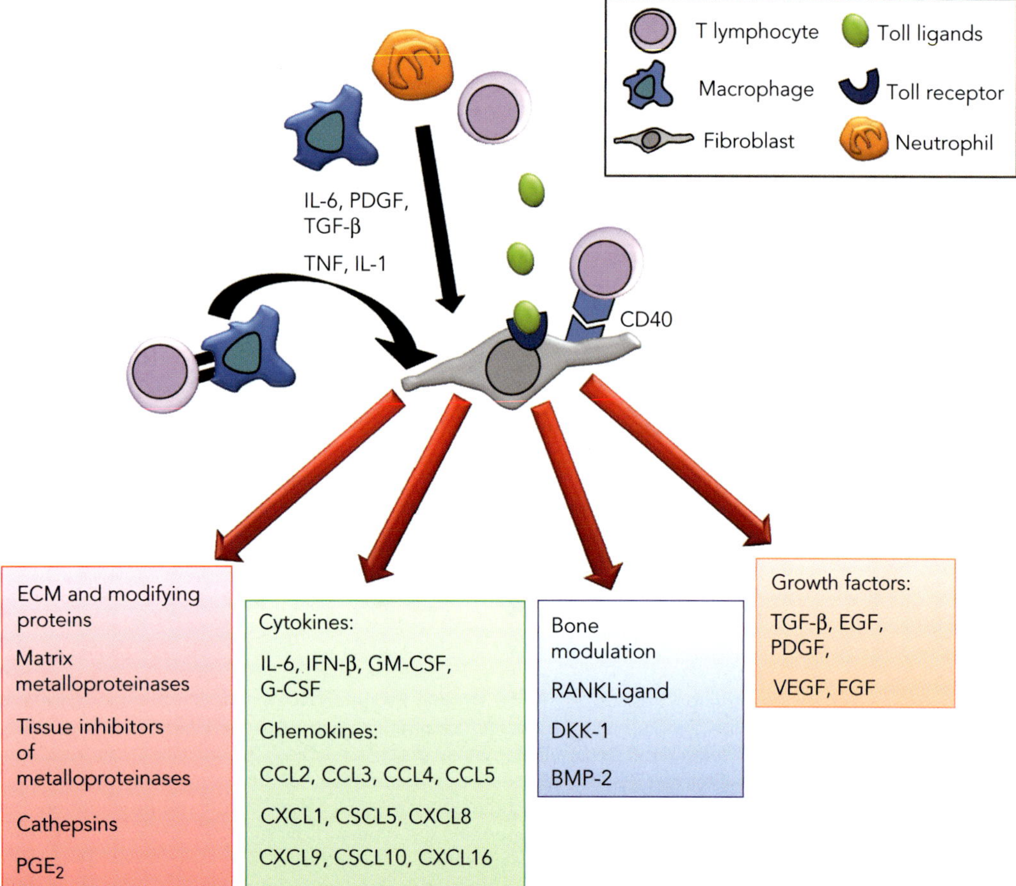

Figure 10.4. Fibroblast roles in inflammation. Fibroblasts may be subjected to various inflammatory stimuli. Bacterial cell products activate toll receptors, proinflammatory cytokines, such as TNF, IL-1, and IL-6 are most abundantly secreted by infiltrating macrophages and neutrophils. Infiltrating T lymphocytes may activate macrophages by direct contact to produce further inflammatory cytokines, and also activate fibroblasts by CD40Ligand–CD40 interactions. In response, fibroblasts break down and elaborate new extracellular matrix (ECM) by means of metalloproteinases and prostaglandin E2 (PGE_2). A wide range of cytokines including IL-6 and GM-CSF and a broad range of chemokines may be elaborated. In conditions such as rheumatoid arthritis, interaction with bone metabolism is important: RANKLigand promotes osteoclast differentiation and activation, while DKK-1 (Dickkopf) inhibits osteoblast activation. Alternatively, bone formation may be stimulated by bone morphogenic proteins (BMP) and vascular endothelial growth factor (VEGF), which also has a primary role in promoting vasculogenesis. Fibroblasts are a rich source of other growth factors which modulate stromal cell survival such as platelet-derived growth factor (PDGF) and epidermal and fibroblast growth factors (EGF, FGF).

pathways such as NF-κB and the mitogen-activated protein kinases (MAPK) with increasing recognition of the importance of the phosphatidylinositol 3-kinase (PI3K) signaling pathways, particularly of the delta and gamma subtypes.

All three subpathways within the MAPK system (JNK, ERK-p42/p44, and p38) are heavily implicated in inflammatory signaling cascades of fibroblasts in the rheumatoid synovium. Furthermore, MAPK activation frequently overlaps with proinflammatory NF-κB pathways. Activation of these pathways leads to further induction of proinflammatory cytokines and chemokines, as well as regulating the survival of stromal cells by inhibiting apoptosis. Work using knockout mice has demonstrated that JNK MAPK is crucial to the elaboration of MMPs such as the collagenases, while P38 MAPK has also been implicated in regulation of other MMPs by synovial fibroblasts. The importance of these signaling pathways has led to the suggestion that MAPK should be targets for rational drug design. PI3K is frequently implicated in production of mononuclear cell attracting chemokines such as CCL2 and CCL5 in stromal cells. For instance, PI3K and JNK are involved in IL-18 regulation of CCL2 production by synovial fibroblasts. PI3K signaling frequently lies upstream of NF-κB activation, which plays an important role in regulating the cellular response to proinflammatory stimuli. Furthermore, signaling due to IL-6 binding results in activation of the JAK-STAT pathways (Janus kinases [JAKs] and signal transducers and activators of transcription [STATs], particularly STAT-3). These pathways govern cell proliferation, differentiation, and apoptosis.

Fibroblasts also express chemokine receptors and respond to chemokine ligands. All functioning chemokine receptors appear to signal through pertussis toxin–sensitive heterotrimeric G proteins. Shape changes, firm adhesion, and chemotaxis, which involve restructuring of the cytoskeleton, utilize the Rho GTPase and phosphatidylinositol-3-OH kinase (PI3K) pathways. Other pathways downstream from PI3K include the modulation of gene transcription via the MAPK. Activation of pertussis toxin insensitive pathways such as the JAK-STAT pathway has been proposed to result from either hetero- or homo-oligomerization of chemokine receptors, leading to a range of functional outcomes such as cellular adhesion. A further emerging signaling pathway of note is PLC-γ (phospholipase-C-γ). This phospholipid hydrolase phospholipase plays a major role in regulation of cell proliferation, development, and cell motility. It has been implicated in the control of stromal cell apoptosis and production of MMPs, which are important both in progression of cancer mediated by stromal cells and in damage to structures within the rheumatoid joint. PLC-γ downstream effects are usually mediated through the ERK MAPKinases.

STROMAL CELL–LEUKOCYTE INTERACTIONS AND PERSISTENT INFLAMMATION

The maintenance of a persistent leukocyte infiltrate at sites of chronic inflammation reflects a distorted balance between those factors that enhance cellularity (leukocyte recruitment, proliferation, and retention) and those that decrease cellularity (cell death and emigration) (Figure 10.3). While the mechanisms responsible for the recruitment of leucocytes into and their proliferation within tissues have been well studied, those responsible for their survival, retention, and emigration within tissues have attracted much less attention.

Recruitment of Inflammatory Infiltrates into the Joint

Stromal elements such as synovial fibroblasts are subject to a proinflammatory cytokine network within the inflamed synovium. Direct contact interactions with other infiltrating cells such as T lymphocytes leads to high levels of expression of many inflammatory chemokines (Figure 10.3). Neutrophil-attracting chemokines are expressed at high levels by stimulated fibroblasts and include CXCL8 (IL-8), CXCL5 (ENA-78, epithelial-cell–derived neutrophil attractant 78), and CXCL1 (GROalpha, growth-related oncogene alpha). Monocytes and T cells are recruited by a range of chemokines found at high levels in the synovium; CXCL10 (IP-10) and CXCL9 (Mig) are highly expressed in synovial tissue and fluid. CXCL16 is also highly expressed in the RA synovium and acts as a potent chemoattractant for T cells. CCL2 (MCP-1) is found in synovial fluid and known to be produced by synovial fibroblasts; it is considered to be a pivotal chemokine for the recruitment of monocytes. CCL3 (Mip-1alpha), CCL4 (Mip-1beta), and CCL5 (RANTES) are chemotactic for monocytes and lymphocytes, and are known products of synovial fibroblasts. CCL20 (Mip-3alpha) is also overexpressed in the synovium, and has a similar chemoattractant profile via its specific receptor, CCR6. CX3CL1 (Fractalkine) is also widely expressed in the rheumatoid synovium. A number of chemokine receptors have been shown to differ between peripheral blood and synovial leucocytes, suggesting that they are enriched in the synovium either though their selective recruitment by endothelial expressed chemokines, or following upregulation by the microenvironment after their recruitment.

Stromal Support for Leukocyte Survival

Stromal cell support for the survival of leukocyte populations fulfills a physiological role in certain organs within the body. The selective recruitment and support of hemopoietic subsets is an essential physiological function of stromal cells in specific microenvironments. For instance, immature B lymphocytes are completely dependent on factors such as IL-6 produced by bone marrow stromal cells. Though the bone marrow niche plays a critical role in the early development of all hemopoietic leukocyte populations, it also acts as an active reservoir for terminally differentiated leukocyte subpopulations, including CD4 and CD8 T cells and neutrophils. The bone marrow stromal microenvironment therefore maintains not only the selective survival, differentiation, and proliferation of all lineages of immature hemopoietic cells, but, in some cases, also the survival of their mature counterparts. The stromal microenvironment plays a crucial role in the maintenance of such survival niches, which are not generic, but highly specific to certain organs and tissues, resulting in site-specific differences in the ability of different stromal cells to support the differential accumulation of leukocyte subsets.

In the case of an inflammatory response, successful resolution requires the removal of the vast majority of immune cells that were recruited and expanded during the active phase of the inflammation. A

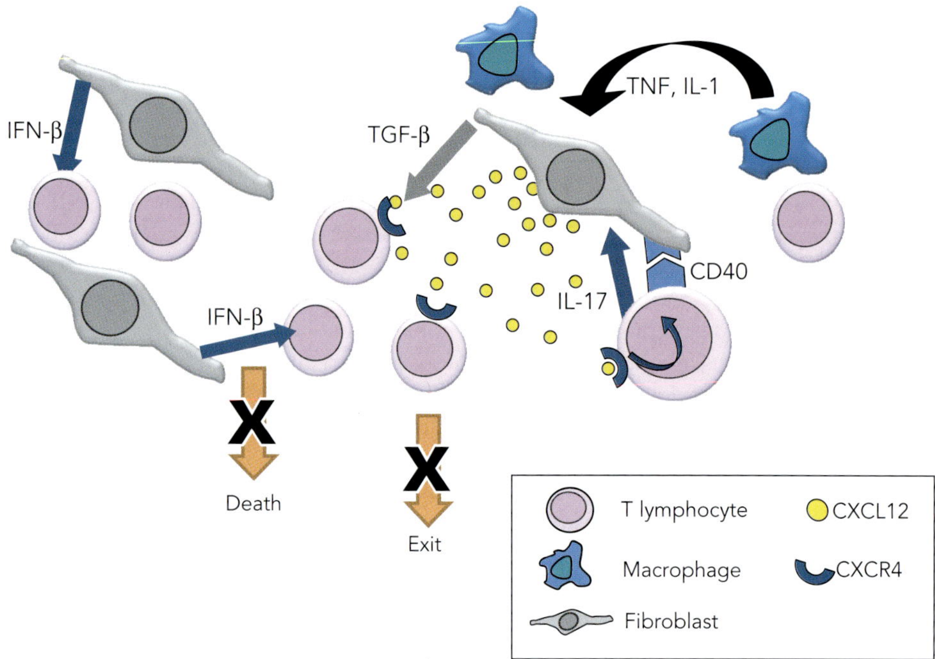

Figure 10.5. Inappropriate survival and retention of T lymphocytes mediated by synovial fibroblasts. Interferon-β, produced by synovial fibroblasts, has been identified as one of the principal factors responsible for prolonged T-cell survival in the rheumatoid joint. Similar mechanisms exist for stromal-induced survival of B cells. It is likely that these mechanisms of stromal cell–induced leukocyte survival occur in many chronic inflammatory conditions in which lymphocytes accumulate. In addition, aberrant ectopic expression of constitutive chemokines such as CXCL12 by synovial stromal cells contributes to the retention of T cells within the RA synovium. CXCL12 (SDF-1) and its receptor CXCR4 plays an important role, both in the constitutive traffic of lymphocytes and in the recruitment and retention of hemopoietic cells within the bone marrow. In the rheumatoid joint, fibroblast-derived TGF-β induces high levels of CXCR4 receptors in T cells, leading to retention of cells. CXCL12 is highly expressed on endothelial cells and fibroblasts at sites of T-cell accumulation. Positive feedback loops enhance CXCL12 production by fibroblasts via CD40–CD40 ligand interactions.

number of studies have shown that during the resolution phase of viral infections, the initial increase in T cell numbers in peripheral blood that is seen within the first few days is followed by a wave of apoptosis occurring in the activated T cells. This situation is mirrored within tissues, where apoptosis induced by the molecule Fas occurs at the peak of the inflammatory response and may be responsible for limiting the extent of the immune response. In contrast, the resolution phase appears to be principally triggered by cytokine-deprivation–induced apoptosis, during which leukocytes compete for a shrinking pool of survival factors provided by the microenvironment, leading to programmed death of those cells that are surplus to requirements.

In RA, the resolution phase of inflammation becomes disordered. Recent studies have shown that a failure of synovial T cells to undergo apoptosis or programmed cell death contributes to the persistence of the inflammatory infiltrate. The T-cell survival pathway shares all the essential hallmarks of a stromal cell, cytokine-mediated mechanism (high Bcl-X_L, low Bcl-2, and lack of cell proliferation). Type I interferons (interferons alpha and beta), produced by synovial fibroblasts and macrophages, have been identified as one of the principal factors responsible for prolonged T-cell survival in the rheumatoid joint (Figure 10.5). Interestingly while type I interferon has been shown to be beneficial in multiple sclerosis (a disease in which tissue scarring and low levels of T-cell infiltrates are observed) these results suggest that type I interferon is not likely to be a successful therapy for RA patients, a prediction which has been borne out in clinical trials. It is likely that this mechanism of stromal cell–induced leukocyte survival occurs in many chronic inflammatory conditions in which T cells accumulate. Not surprisingly, other leukocyte subpopulations have been shown to derive support from stromal cells. While fibroblast support for T-cell and B-cell survival exhibits site-specific properties, neutrophil survival is dependent on prior cytokine activation of fibroblasts, and shows no differences between fibroblasts taken from different anatomical sites.

Stromal Cell Retention of Leukocytes in Persistent Inflammatory Disorders

While the inhibition of T-cell death by stromal cells at sites of chronic inflammation contributes to T-cell accumulation, it is unlikely to be the only mechanism, because lymphocytes should be able to leave the inflamed tissue during the resolution of inflammation, even if their death is inhibited. A number of studies have recently reported that the synovial microenvironment contributes directly to the inappropriate retention of T cells within the joint, by an active chemokine-dependent process. The presence of high levels of inflammatory chemokines, produced by stromal cells, is a characteristic of environments such as the rheumatoid synovium. However, recent data suggest that paradoxically constitutive chemokines, which are involved in the recruitment of lymphocytes to secondary lymphoid tissues, are ectopically expressed in immune-mediated inflammatory diseases. The constitutive chemokine CXCL12 (SDF-1) and its receptor CXCR4 emerged as unexpected but crucial players in the accumulation of T lymphocytes within the rheumatoid synovial microenvironment (Figure 10.5). This chemokine receptor pair plays an important role, both in the constitutive traffic of lymphocytes and in the recruitment and retention of hemopoietic cells within the bone marrow. Unexpectedly, CD45RO+ T lymphocytes in the rheumatoid synovium were found to express CXCR4 receptors at high levels in the rheumatoid synovium. Its ligand CXCL12 was highly expressed on endothelial cells at the sites of T-cell accumulation. In addition, stromal cell–derived TGF-β is responsible for upregulation of CXCR4 receptors on T cells in the synovium. Evidence also suggests that the stability of lymphocyte infiltrates is reinforced by a positive feedback loop, whereby tissue CXCL12 promotes CD40 ligand expression on T cells, which in turn stimulates further CXCL12 production by CD40 expressing synovial fibroblasts. Furthermore, levels of CXCL12 secreted by synovial fibroblasts have recently been shown to be controlled in part by T cell derived IL-17.

There is therefore clear evidence in support of the hypothesis that aberrant ectopic expression of constitutive chemokines such as CXCL12, CCL19, and CCL21 by synovial stromal cells contributes to the retention of T cells within the RA synovium.

Other cell constituents of the rheumatoid inflammatory infiltrate may be affected by the CXCL12/CXCR4 axis. Blades and colleagues have shown increased expression of CXCL12/CXCR4 by monocyte/macrophage cells in RA compared with osteoarthritis. In addition, using implanted human synovial tissue in SCID mice, they demonstrated that monocytes are recruited into transplanted synovial tissue by CXCL12. Contact-mediated B-cell survival induced by synovial fibroblasts has also been shown to depend on CXCL12 and CD106 (VCAM-1)-dependent mechanisms which are independent of TNF-α. Overexpression of CXCL12 has also been identified as a distinct feature of rheumatoid, as opposed to osteoarthritis synovia using cDNA arrays. Data validating these findings in vivo have come from a collagen-induced arthritis model of RA in DBA/1 (interferon-γ receptor deficient) mice, where administration of the specific CXCR4 antagonist AMD3100 significantly ameliorated disease severity. In another murine collagen-induced arthritis model, the small molecule CXCR4 antagonist 4F-benzoyl-TN14003 ameliorated clinical severity and suppressed DTH (delayed type hypersensitivity) responses. The CXCL12/CXCR4 constitutive chemokine pair therefore seems to play an important role in lymphocyte retention in RA.

Constitutive Chemokines and Lymphoid Neogenesis

Rheumatoid arthritis is one of a number of inflammatory diseases in which the organization of the inflammatory infiltrate shares characteristics of lymphoid tissue. Follicular hyperplasia with germinal center formation can occur in autoimmune thyroid disease, myasthenia gravis, Sjögren's disease, and RA, and may occur during infection with *Helicobacter pylori* and *Borrelia burgdorferi*. The lymphoid infiltrates in the rheumatoid synovium can be divided into at least three distinct histological groupings, varying from diffuse lymphocyte infiltrates through organized lymphoid aggregates to clear germinal center reactions. Moreover, there is conflicting evidence that such distinct histological types correlate with other serum indicators of disease activity. This form of inflammatory lymphoid neogenesis relies upon inappropriate, but highly organized temporal and spatial expression by stromal cells of the constitutive chemokines, particularly CXCL13 and CCL21, which are required for physiological lymphoid organogenesis (Figure 10.6). The elegant choreography of lymphocyte–stromal interactions within lymph nodes is organized by expression of adhesive and chemotactic cues in overlapping and combinatorial fashions. Once they have encountered new antigen, DCs specialized in the presentation of antigen to lymphocytes undergo a process of maturation under the local influence of inflammatory cytokines and bacterial and viral products. As a result inflammatory chemokine receptors are

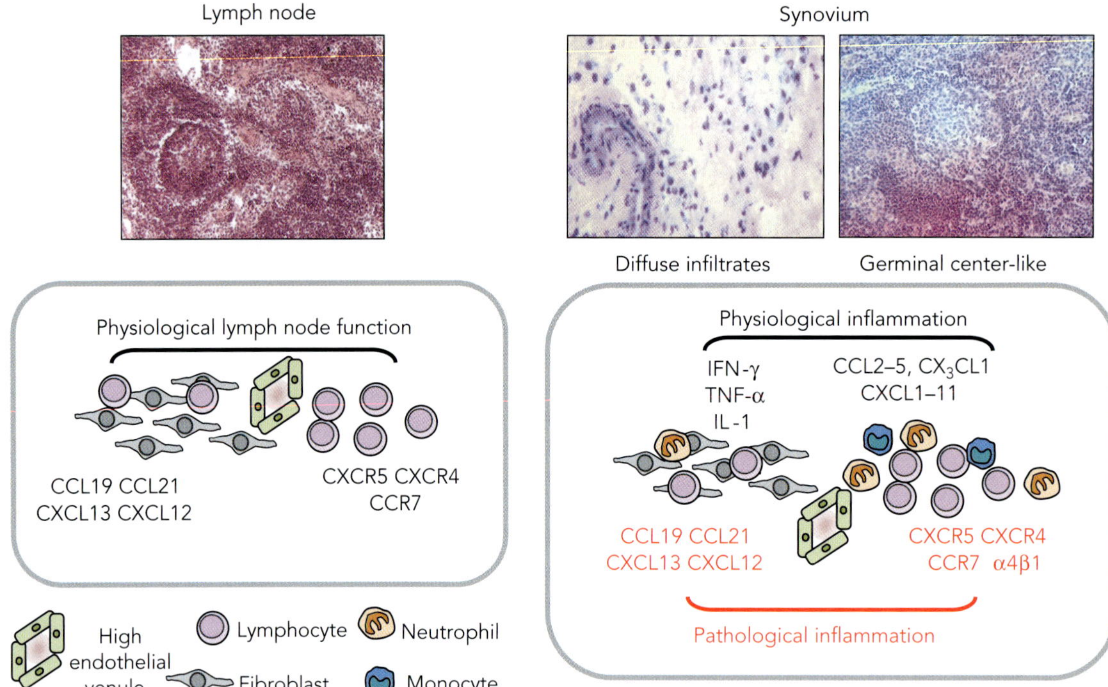

Figure 10.6. Inappropriate expression of constitutive chemokines leads to lymphoid neogenesis. Homeostatic chemokines (CXCL12, CXCL13, CCL19, CCL21) are components of the stromal code that help define niches such as the lymph node and bone marrow, governing leukocyte accumulation, differentiation, and survival. Stromal cells produce/express the appropriate cytokine/chemokine/adhesion receptor that is recognized by cognate receptors on infiltrating leukocytes. During physiological inflammation, inflammatory chemokines (CCL2–5, CX3CL1, and CXCL1–11 and inflammatory mediators such as IFN-γ, TNF-α, and IL-1 are produced by stromal cells and lead to the recruitment of inflammatory cells (lymphocytes, neutrophils, and monocytes). However, in persistent, pathological inflammation occurs in RA, stromal cells begin to aberrantly produce/express components of the physiological stromal code normally associated with lymphoid tissues, leading to lymphoid neogenesis.

downregulated, and upregulation of the constitutive receptors CCR4, CCR7, and CXCR4 occurs, causing DCs to migrate into local draining lymphatics and thereby into peripheral lymph nodes. Trafficking of B and T cells is regulated by CXCL13 (BCA-1, B cell–attracting chemokine 1), its receptor CXCR5, and CCL21 and CCL19 (EBL-1-ligand chemokine, ELC), which are both CCR7 agonists. Within the lymph node CXCR5-bearing B cells are attracted to follicular areas, while T cells and DCs are maintained within parafollicular zones by local expression of CCL21 and CCL19. Some T cells which have been successfully presented with their cognate antigen by DCs then upregulate CXCR5, allowing them to migrate toward and interact with B cells.

The genesis of lymphoid follicular structures in diseases such as diabetes and RA appears to rely upon expression of such constitutive chemokines, in association with the lymphotoxins alpha and beta (LT-α and LT-β) and TNF-α. In this context, it is important to note that transgenic animals overexpressing the TNF-α gene display increased formation of focal lymphoid aggregates and develop a chronic arthritis similar to RA. Clearly one of the many mechanisms of action of anti-TNF therapy may be the dissolution of such aggregates. In transgenic mouse models, expression of CXCL13 in the pancreatic islets was sufficient for the development of T- and B-cell clusters, but as they lacked follicular DCs, was not sufficient for true germinal center formation. CCL21 does appear to be sufficient in some cases for lymph node formation; murine pancreatic islet models have demonstrated formation of lymph node like structures in the presence of CCL21, and lymphoid infiltrates in response to CCL19 expression. Weyand and colleagues used the histological heterogeneity seen in RA to identify those factors critical to the formation of lymphoid microstructures, showing that transcription levels of CXCL13 and CCL21 were increased 10–20 times in tissues with germinal centers compared to tissues with other histological patterns. Multivariate analysis showed that LT-β and CXCL13 were necessary, but not sufficient for lymphoid neogenesis. It has also been shown that CXCR5 is overexpressed in the rheumatoid synovium, consistent with a role in recruitment and positioning of B and T lymphocytes within lymphoid aggregates of the RA synovium. It therefore seems likely that expression of lymphoid constitutive chemokines contributes significantly to the entry, local organization, and exit of lymphocytes in the RA synovium. It also seems that the ectopic expression of chemokines is a general characteristic of a number of chronic rheumatic conditions, since another B-cell attracting chemokine CXCL13 (BCA-1) is inappropriately expressed by stromal cells in the salivary glands of patients with Sjögren's syndrome.

MULTIPLE CELL–CELL INTERACTIONS INVOLVING FIBROBLASTS

A further pathological model in which fibroblasts become shifted into an activated phenotype is that of liver fibrosis. Models of hepatic fibrosis suggest that resident populations of hepatic macrophages are able to modulate the processes of fibrotic progression and subsequent liver recovery. Stellate cells (specialized liver fibroblasts) are known to differentiate during the process of hepatic injury and fibrosis toward a profibrotic, secretory myofibroblast phenotype under the influence of cytokines such as TGF-β1, of which hepatic macrophages are a potentially important source. High expression of tissue inhibitor of metalloproteinase-1 (TIMP-1) by stellate cells/myofibroblasts appears to maintain their survival and inhibit degradation of matrix leading to fibrosis. During recovery from fibrosis in animal models, myofibroblasts undergo apoptosis accompanied by a fall in TIMP-1 expression and degradation of scarring matrix. Expression by macrophages of TNF-related apoptosis-inducing ligand (TRAIL) is one mechanism by which stellate cell apoptosis has been proposed to occur. Using a transgenic mouse model of fibrosis that allowed the conditional depletion of macrophage lineage cells, Duffield et al. were able to demonstrate a fascinating divergence of regulation of these processes. Depletion of macrophages during the induction/progression phase of fibrosis resulted in decreased scarring and fewer myofibroblasts, implying a role for macrophages in the activation of myofibroblasts. However, when macrophages were depleted in the early phases of recovery, sustained accumulation of scarring matrix occurred (Figure 10.7), suggesting that at this stage of the inflammatory response macrophages are required to inactivate myofibroblasts.

These experiments demonstrate that modeling the behavior of stromal cells and leukocytes within microenvironments necessarily requires that we take into account the interactions of both leucocytes and stromal cells. A very elegant example of this approach in vitro is the work of Smith and Lally, who developed a model of cellular recruitment to the rheumatoid synovium. Models of recruitment in which leukocytes adhere and migrate beneath cultured endothelial cells under flow conditions are well established as a means of investigating the mechanisms governing recruitment to tissues from the blood, and will be considered elsewhere. Coculturing fibroblasts from skin and RA synovial membrane with endothelial cells showed that IL-6 released from synovial (but not skin) fibroblasts was able to induce production of chemokines and adhesion molecules, resulting in greater neutrophil recruitment by synovial fibroblasts. Subsequent work interrogating the system using low-density gene arrays demonstrated that the effect of neutrophil attracting chemokines such as CXCL5 released from synovial fibroblasts was dependent on the function of the chemokine transporter molecule DARC (Duffy antigen receptor for chemokines) which was also induced by coculture (Figure 10.8).

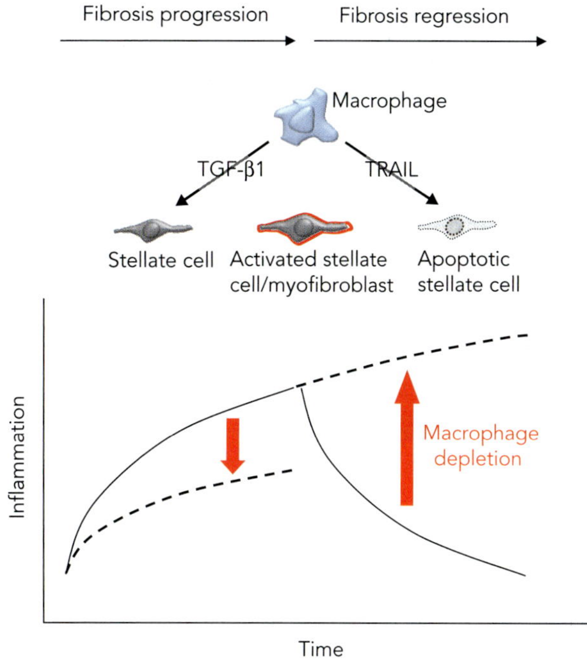

Figure 10.7. Macrophage–stromal cell interactions in the regulation of fibrotic liver disease. Hepatic fibroblasts induce divergent effects on liver stellate cell activation in models of fibrosis. During fibrosis progression, TGF-β1 is a potential macrophage-derived stimulator of stellate cell activation. Depletion of macrophages during this phase results in decreased scarring and fewer myofibroblasts. During fibrosis regression, stellate cell apoptosis may occur under the influence of TRAIL. Depletion of macrophages early during fibrosis regression results in sustained accumulation of scarring matrix. Depletion of macrophages during the healing phase leads to sustained fibrosis.

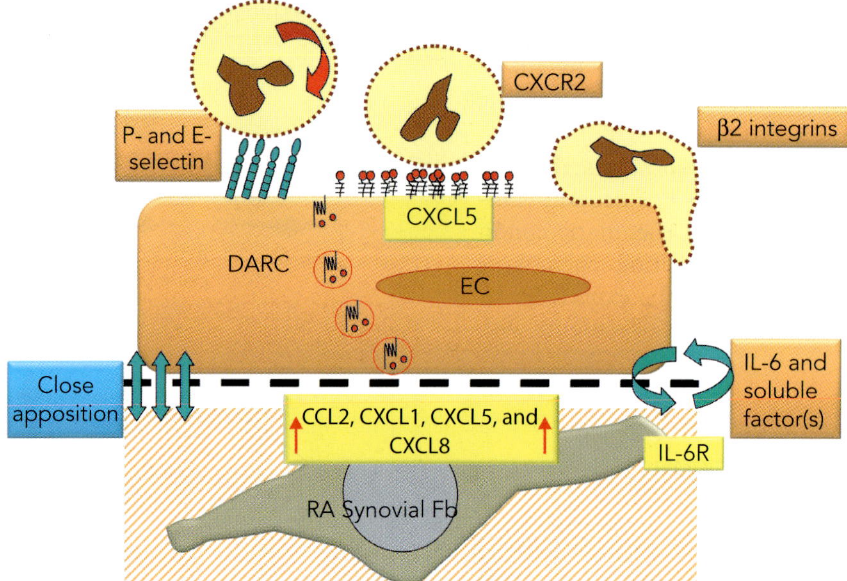

Figure 10.8. Three cell model of recruitment to the inflamed rheumatoid synovium. A multicellular coculture model of RA synovium was used to reconstruct the persistently inflamed RA microenvironment in vitro, by coculturing RA synovial fibroblasts or skin fibroblasts with human endothelial cells (EC). After a period of conditioning, fibroblasts and EC were isolated and screened using microarray analysis. Recapitulating the RA environment upregulated message for several CXC chemokines in ECs and RASFs, an effect that was absent in cocultures of ECs and skin fibroblasts. Flow adhesion assays demonstrated that only CXCL5 was functional on ECs cocultured with RASFs and recruited flowing neutrophils. EC expression of DARC was induced by coculture with RASFs, and antibody-blocking CXCL5 interactions with DARC or small interfering RNA (siRNA) targeting DARC expression abolished neutrophil recruitment. Thus, DARC edits the leukocyte recruitment code in a chemokine-specific manner in a model of human inflammatory disease, demonstrating the impact of inflammatory "cross-talk" between cells of the stromal microenvironment.

SUMMARY

The demonstration of inflammatory "cross-talk" between cells of the stromal microenvironment resulting in recruitment of inflammatory cells illustrates the significant contribution within microenvironments of cells such as fibroblasts. Populations of leukocytes recruited to sites of inflammation should not therefore be considered or studied in isolation, but should be considered in the context of their complementary stromal microenvironment, which provides survival, differentiation, and positioning cues upon which the formation and persistence of leukocyte infiltrates depends.

KEY POINTS

What Is a Stromal Cell?

- Fibroblasts are a heterogeneous population of stromal cells with specialized roles within different organs and tissues.
- Fibroblasts have a role in determining the location of tissue-specific diseases.

Embryological Origins of Fibroblasts

- Stromal cells in tissues may differentiate from a variety of precursor cells.
- Fibroblast precursor cells may have important roles in initiation of disease.

Stromal Cells in Immune Responses

- Stromal cells function as innate immune sentinels elaborating cytokines and chemokines in response to innate *stimuli* such as bacterial products and CD40 ligand.
- Persistent activation of fibroblasts may lead to pathological disease via an inflammatory stromal microenvironment.
- Classical inflammatory signal transduction pathways including NF-κB, MAPK, and PI3K are activated in fibroblasts during inflammation.

Stromal Cell–Leukocyte Interactions and Persistent Inflammation

- Fibroblasts play an active role in recruiting leukocytes to persistently inflamed sites.
- Infiltrating leukocytes are rescued from apoptosis by fibroblasts in persistent inflammation.
- Inappropriate expression of constitutive chemokines by stromal cells results in recruitment and retention of leukocytes within persistently inflamed microenvironments.

- Lymphoid neogenesis in inflammatory disorders results from a stromal defect in chemokine expression.

Multiple Cell–Cell Interactions Involving Fibroblasts

- A shift into a profibrotic fibroblast phenotype regulated by tissue macrophages accounts for disease progression in models of liver fibrosis.
- Inflammatory cross-talk between fibroblasts and endothelial cells drives recruitment of neutrophils in three cell coculture models.

SUGGESTED READINGS

Asahara, T., Murohara, T., Sullivan, A., et al. 1997. Isolation of putative progenitor endothelial cells for angiogenesis. *Science* **275:**964–967.

Bhowmick, N.A., Neilson, E.G., and Moses, H.L. 2004. Stromal fibroblasts in cancer initiation and progression. *Nature* **432:**332–337.

Bjornson, C.R., Rietze, R.L., Reynolds, B.A., Magli, M.C., and Vescovi, A.L. 1999. Turning brain into blood: a hematopoietic fate adopted by adult neural stem cells in vivo. *Science* **283:**534–537.

Blades, M.C., Ingegnoli, F., Wheller, S.K., et al. 2002. Stromal cell-derived factor 1 (CXCL12) induces monocyte migration into human synovium transplanted onto SCID Mice. *Arthritis Rheum* **46:**824–836.

Brouty-Boye, D., Pottin-Clemenceau, C., Doucet, C., Jasmin, C. and Azzarone, B. 2000. Chemokines and CD40 expression in human fibroblasts. *Eur J Immunol* **30:**914–919.

Bucala, R., Spiegel, L.A., Chesney, J., Hogan, M., and Cerami, A. 1994. Circulating fibrocytes define a new leukocyte subpopulation that mediates tissue repair. *Mol Med* **1:** 71–81.

Buckley, C.D., Amft, N., Bradfield, P.F., et al. 2000. Persistent induction of the chemokine receptor CXCR4 by TGF-beta 1 on synovial T cells contributes to their accumulation within the rheumatoid synovium. *J Immunol* **165:** 3423–3429.

Buckley, C.D., Pilling, D., Lord, J.M., Akbar, A.N., Scheel-Toellner, D., and Salmon, M. 2001. Fibroblasts regulate the switch from acute resolving to chronic persistent inflammation. *Trends Immunol* **22:**199–204.

Burger, J.A., Zvaifler, N.J., Tsukada, N., Firestein, G.S. and Kipps, T.J. 2001. Fibroblast-like synoviocytes support B-cell pseudoemperipolesis via a stromal cell-derived factor-1- and CD106 (VCAM-1)-dependent mechanism. *J Clin Invest* **107:**305–315.

Duffield, J.S., Forbes, S.J., Constandinou, C.M., et al. 2005. Selective depletion of macrophages reveals distinct, opposing roles during liver injury and repair. *J Clin Invest* **115:**56–65.

Filer, A., Parsonage, G., Smith, E., et al. 2006. Differential survival of leukocyte subsets mediated by synovial, bone marrow, and skin fibroblasts: site-specific versus activation-dependent survival of T cells and neutrophils. *Arthritis Rheum* **54:**2096–2108.

Haniffa, M.A., Wang, X.N., Holtick, U., et al. 2007. Adult human fibroblasts are potent immunoregulatory cells and functionally equivalent to mesenchymal stem cells. *J Immunol* **179:**1595–1604.

Kalluri, R. and Neilson, E.G. 2003. Epithelial-mesenchymal transition and its implications for fibrosis. *J Clin Invest* **112:**1776–1784.

Kim, K.W., Cho, M.L., Kim, H.R., et al. 2007. Up-regulation of stromal cell-derived factor 1 (CXCL12) production in rheumatoid synovial fibroblasts through interactions with T lymphocytes: role of interleukin-17 and CD40L-CD40 interaction. *Arthritis Rheum* **56:**1076–1086.

Lally, F., Smith, E., Filer, A., et al. 2005. A novel mechanism of neutrophil recruitment in a coculture model of the rheumatoid synovium. *Arthritis Rheum* **52:**3460–3469.

Luther, S.A., Bidgol, A., Hargreaves, D.C., et al. 2002. Differing activities of homeostatic chemokines CCL19, CCL21, and CXCL12 in lymphocyte and dendritic cell recruitment and lymphoid neogenesis. *J Immunol* **169:**424–433.

Manzo, A., Paoletti, S., Carulli, M., et al. 2005. Systematic microanatomical analysis of CXCL13 and CCL21 in situ production and progressive lymphoid organization in rheumatoid synovitis. *Eur J Immunol* **35:**1347–1359.

Marinova-Mutafchieva, L., Williams, R.O., Funa, K., Maini, R.N., and Zvaifler, N.J. 2002. Inflammation is preceded by tumor necrosis factor-dependent infiltration of mesenchymal cells in experimental arthritis. *Arthritis Rheum* **46:**507–513.

Merville, P., Dechanet, J., Desmouliere, A., et al. 1996. Bcl-2+ tonsillar plasma cells are rescued from apoptosis by bone marrow fibroblasts. *J Exp Med* **183:**227–236.

Muller-Ladner, U., Kriegsmann, J., Franklin, B.N., et al. 1996. Synovial fibroblasts of patients with rheumatoid arthritis attach to and invade normal human cartilage when engrafted into SCID mice. *Am J Pathol* **149:**1607–1615.

Parsonage, G., Falciani, F., Burman, A., et al. 2003. Global gene expression profiles in fibroblasts from synovial, skin and lymphoid tissue reveals distinct cytokine and chemokine expression patterns. *Thromb Haemost* **90:**688–697.

Parsonage, G., Filer, A.D., Haworth, O., et al. 2005. A stromal address code defined by fibroblasts. *Trends Immunol* **26:**150–156.

Pierer, M., Rethage, J., Seibl, R., et al. 2004. Chemokine secretion of rheumatoid arthritis synovial fibroblasts stimulated by toll-like receptor 2 ligands. *J Immunol* **172:**1256–1265.

Postlethwaite, A.E., Shigemitsu, H., and Kanangat, S. 2004. Cellular origins of fibroblasts: possible implications for organ fibrosis in systemic sclerosis. *Curr Opin Rheumatol* **16:**733–738.

Rinn, J.L., Bondre, C., Gladstone, H.B., Brown, P.O., and Chang, H.Y. 2006. Anatomic demarcation by positional variation in fibroblast gene expression programs. *PLoS Genet* **2:**e119.

Ritchlin, C. 2000. Fibroblast biology. Effector signals released by the synovial fibroblast in arthritis. *Arthritis Res* **2:**356–360.

Salmon, M., Scheel-Toellner, D., Huissoon, A.P., et al. 1997. Inhibition of T cell apoptosis in the rheumatoid synovium. *J Clin Invest* **99:**439–446.

Sanchez-Pernaute, O., Ospelt, C., Neidhart, M., and Gay, S. 2008. Epigenetic clues to rheumatoid arthritis. *J Autoimmun* **30:**12–20.

Schmeichel, K.L. and Bissell, M.J. 2003. Modeling tissue-specific signaling and organ function in three dimensions. *J Cell Sci* **116:**2377–2388.

Schmidt, M., Sun, G., Stacey, M.A., Mori, L. and Mattoli, S. 2003. Identification of circulating fibrocytes as precursors of bronchial myofibroblasts in asthma. *J Immunol* **171:**380–389.

Smith, E., McGettrick, H.M., Stone, M.A., et al. 2008. Duffy antigen receptor for chemokines and CXCL5 are essential for the recruitment of neutrophils in a multicellular model of rheumatoid arthritis synovium. *Arthritis Rheum* **58:**1968–1973.

Smith, R.S., Smith, T.J., Blieden, T.M. and Phipps, R.P. 1997. Fibroblasts as sentinel cells. Synthesis of chemokines and regulation of inflammation. *Am J Pathol* **151:**317–322.

Sweeney, S.E., and Firestein, G.S. 2004. Rheumatoid arthritis: regulation of synovial inflammation. *Int J Biochem Cell Biol* **36:**372–378.

Szekanecz, Z., Kim, J., and Koch, A.E. 2003. Chemokines and chemokine receptors in rheumatoid arthritis. *Semin Immunol* **15:**15–21.

Takemura, S., Braun, A., Crowson, C., et al. 2001. Lymphoid neogenesis in rheumatoid synovitis. *J Immunol* **167:**1072–1080.

Thalhamer, T., McGrath, M.A., and Harnett, M.M. 2008. MAPKs and their relevance to arthritis and inflammation. *Rheumatology (Oxford)* **47:**409–414.

Thelen, M. 2001. Dancing to the tune of chemokines. *Nat Immunol* **2:**129–134.

Thurlings, R.M., Wijbrandts, C.A., Mebius, R.E., et al. 2008. Synovial lymphoid neogenesis does not define a specific clinical rheumatoid arthritis phenotype. *Arthritis Rheum* **58:**1582–1589.

Tolboom, T.C., van der Helm-van Mil, A.H., Nelissen, R.G., Breedveld, F.C., Toes, R.E., and Huizinga, T.W. 2005. Invasiveness of fibroblast-like synoviocytes is an individual patient characteristic associated with the rate of joint destruction in patients with rheumatoid arthritis. *Arthritis Rheum* **52:**1999–2002.

11 Neutrophil–Endothelial Cell Interactions

János G. Filep and Sean P. Colgan

INTRODUCTION

Neutrophils (polymorphonuclear leukocytes, PMN) have a clearly defined role in inflammation. In response to injury or infection, PMN migration across vascular endothelial cells is a first line of defense against infectious agents, and defects in such PMN–endothelial interactions contributes to fulminate microbial infections, mucosal ulcerations, and delayed tissue healing. The protective aspects of PMN in disease are objectively exemplified by the clinical observation that patients with primary defects in PMN function, including neutropenia and genetic PMN immunopathologies (e.g., leukocyte adhesion deficiencies, chronic granulomatous disease, Chediak–Higashi syndrome, myeloperoxidase deficiency, etc.), exhibit ongoing mucosal infections.

This chapter focuses on our current understanding of how PMN interact with vascular endothelial cells under physiologic and pathophysiologic conditions (Figure 11.1).

MOLECULAR MECHANISMS OF PMN ADHESION AND TRANSMIGRATION

PMN migration across the endothelial surface is a result of an orchestrated series of events, ultimately resulting in PMN accumulation at sites of tissue injury. The recruitment signals, the cell–cell interaction steps, and the regulatory pathways for these events have been an area of extensive exploration in the past two decades. A number of recent reviews have addressed these steps in detail [1–4]. Here, we will summarize some of the major steps and guide the reader to the primary literature for more insight into this dynamic process.

It is now appreciated that adhesion-based interactions involving specific cell adhesion epitopes are the primary means by which PMNs interact with endothelial cells. For example, PMN β2 integrins are required for PMN to migrate across both endothelial and epithelial surfaces. These integrins, like others, are heterodimeric glycoproteins that exist in four forms on the PMN. Each displays a unique α-subunit (CD11a, b, c, or d) and an identical β2-subunit (CD18). These receptors are best demonstrated in the genetic disorder leukocyte adhesion deficiency (LAD), in which patients lack normal expression of the CD18 β-subunit, and as a result, show increased susceptibility to infection due to abnormal leukocyte function [4]. These patients manifest severe mucosal disease, characterized primarily by severe bacterial infections.

At the level of the vasculature, P-selectin (induced in acute injury) and E-selectin (upregulated in inflammatory conditions) mediate the initial capture and subsequent rolling of neutrophils along the wall of postcapillary venules. The selectins bind to specialized fucosylated sialoglycoconjugates, including the tetrasaccharide sialyl Lewis X (sLex), that decorate selected surface glycoproteins. P-selectin glycoprotein ligand-1 (PSGL-1) interacts with all selectins under physiological inflammatory situations. This molecule is expressed as a dimer on most leukocytes, and binding activity of PSGL-1 is conferred by the N-terminal region. Like the integrins, the importance of functional selectins is illustrated by the human genetic disease LAD type II, observed as recurrent infections as a result of deficiency in selectin ligands as a result of a mutation in the GDP-fucose transporter gene [4].

Slow rolling on selectins allows neutrophils to sample chemokines presented at the surface of the endothelium. Chemokines and their receptors provide cell- and tissue-specific activation signals that selectively regulate recruitment. Neutrophils as best characterized to bind to interleukin-8 (CXCL8) through their surface receptors CXCR1/2. Other leukocytes utilize a different chemokine repertoire. T cells, for example, use

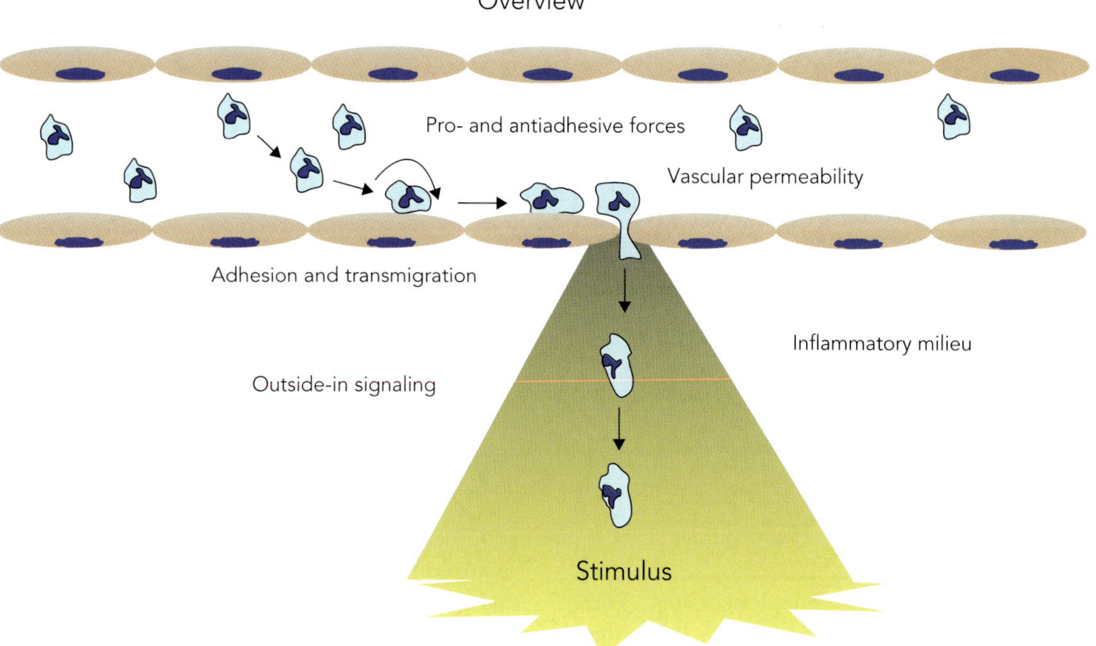

Figure 11.1. Graphical overview of this chapter, highlighting the areas of focus during neutrophil–endothelial interactions.

CCR7 for directed migration toward CCL21 and CCL19, while monocytes respond most avidly to CCL2 through its receptor CCR2, though they also bind CXCL8. The relative importance of chemokine signaling is further substantiated by the genetic disorder LAD type III, a defect wherein chemokine-triggered integrin activation on platelets and leukocytes is impaired due to defective Rap1 activation by the guanine nucleotide exchange factor CalDAG-GEFI [4].

The anatomical location of leukocyte extravasation within the vasculature has been the topic of much debate. Electron microscopy analyses and numerous in vitro studies using human umbilical vein endothelial cells have supported the opinion that leukocytes emigrate between endothelial cells (i.e., the paracellular route). The determinants that mediate paracellular transmigration have been characterized and shown to involve adhesion molecules expressed and distributed at intercellular junctions. These molecules include platelet-endothelial adhesion molecule-1 (PECAM-1), junctional adhesion molecules (JAMs), and CD99 [5]. Conversely, migration directly through individual endothelial cells (i.e., the transcellular route) has been suggested. Electron microscopy of transmigrating PMNs in vivo have indicated that transcellular migration occurs in certain physiological settings, representing the route of emigration for 5%–20% of transmigrating PMNs. These more recent studies have supported the possibility that both the transcellular and paracellular pathways likely coexist. Table 11.1 summarizes the steps and the molecules mediating leukocyte transmigration across endothelial cells.

BALANCE OF PRO- AND ANTIADHESIVE FORCES

Endothelial cells play a critical role in the regulation of leukocyte adhesion and trafficking into inflamed or injured tissues. Alterations in shear patterns and multiple cytokines and growth factors present at the sites of inflammation can potentially influence these events and therefore development and outcome of the inflammatory response. Leukocyte adhesion can be limited by anti-inflammatory counter-regulatory mechanisms that maintain the integrity and homeostasis of the vessel wall. Consistently, impaired protective mechanisms (e.g., decreased endothelial NO production or changes in laminar shear stress) would facilitate leukocyte adhesion even in the presence of low amounts of proinflammatory stimuli. The endothelium integrates the signals generated by proadhesive and antiadhesive forces.

Endothelial cell activation is predominantly mediated through the NF-κB and JNK-AP-1 signaling pathways. Unlike physiological laminar shear stress, low or turbulent blood flow patterns evoke sustained oxidative stress, resulting in activation of NF-κB and redox-sensitive gene expression. Exposure of endothelial cells to proinflammatory cytokines, including IL-1, IL-8,

TABLE 11.1. Leukocyte adhesion cascade

Leukocytes	Endothelial cells	Major role
PSGL-1, E-selectin ligand-1	P-selectin, E-selectin	**Capture and rolling**
Glycosylated CD44		
Sialyl Lewis X		
(neutrophils, monocytes, T cells)		
L-selectin	GlyCam-1	Lymphocyte homing
Selectin signaling	Selectin signaling	Slow rolling
Chemokines → leukocyte activation		
Integrin affinity modulation		
		Migration arrest
LFA-1	ICAM-1	
VLA4	VCAM-1	
α4β2 integrin	MADCAM-1	
outside-in signaling through		
LFA-1 and Mac-1: Scr kinases, PI3K,		Adhesion strengthening and spreading
VAV1, VAV2, VAV3		
		Transendothelial cell migration
Mac-1	ICAM-1	
		Intravascular crawling (neutrophils, monocytes crawl, seeking preferred sites of transmigration)
	Junctional molecules	
LFA-1	JAM-A, JAM-1, ICAM-2	Paracellular route
Mac-1	JAM-C, ICAM-2	
PECAM-1	PECAM-1	
CD99	CD99	
Unknown	ESAM	
Membrane protrusions into endothelial cells	Vesiculo-vacuolar organelles	Transcellular route
	ICAM-1 translocation to caveolae	

The original three steps are shown in bold: (1) rolling that is mediated by selectins; (2) arrest that is mediated by integrins; and (3) transendothelial migration. The table summarizes advances that have been made in defining additional steps.

Abbreviations: ICAM-1, intercellular adhesion molecule-1; ICAM-2, intercellular adhesion molecule-2; ESAM, endothelial cell-selective adhesion molecule; JAM, junctional adhesion molecule; LFA-1, lymphocyte function-associated antigen 1; Mac-1, macrophage receptor 1; PECAM-1, platelet/endothelial cell adhesion molecule 1; PSGL-1, P-selectin glycoprotein ligand 1; PI3K, phosphoinositide 3-kinase; MADCAM1, mucosal vascular addressin adhesion molecule 1; VCAM-1, vascular cell-adhesion molecule 1; VLA4, very late antigen 4.

interferon-γ and tumor necrosis factor (TNF), vasoactive peptides, neuropeptides, minimally oxidized low-density lipoprotein, hyperglycemia and advanced glycosylated end products or smoking enhances production of reactive oxygen species (superoxide in particular), peroxynitrite formation, triggers activation of JNK, leading to transcription of proinflammatory genes (Figure 11.2). Altered hemodynamic forces and proinflammatory signals, such as oxidized lipoproteins are likely to act in concert to induce expression of a proadhesive endothelial cell phenotype as illustrated by permanent NF-κB activation and VCAM-1 expression seen in endothelial cells located in atherosclerosis-prone regions of the aorta of experimental animals.

The antiadhesion, anti-inflammatory mechanisms in the vascular wall involve antiadhesion external signals and intracellular mediators. Physiological laminar shear stress is of particular importance in protecting the endothelium against inflammatory activation. Endothelial cells are constantly exposed to hemodynamic forces generated by the pulsatile blood flow, hydrostatic pressure, cyclic strains, and wall shear

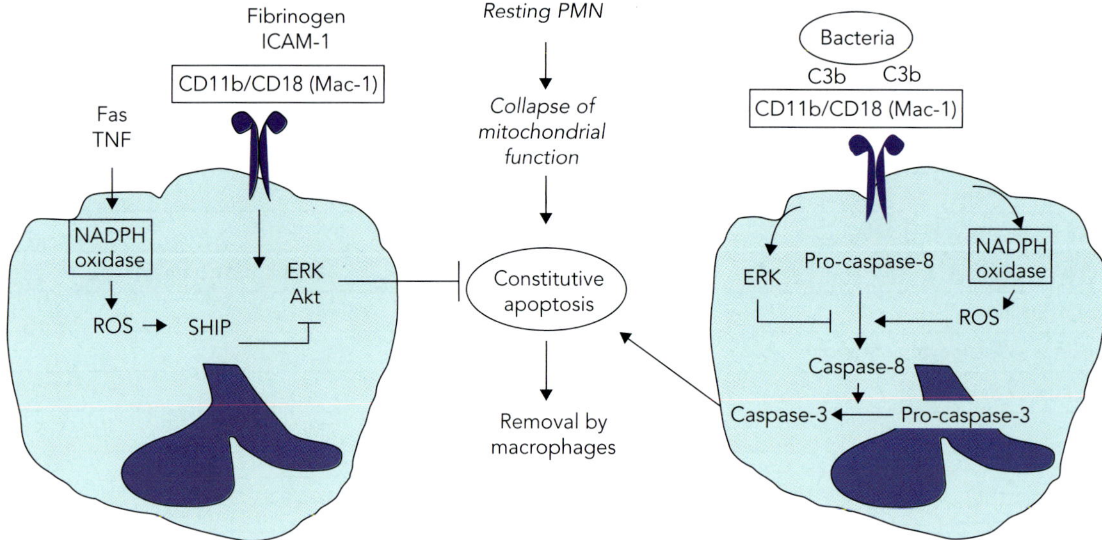

Figure 11.2. Impact of β2-integrin ligation on the fate of neutrophils. Mature neutrophils undergo constitutive apoptosis and removed from the circulation by macrophages. Neutrophil adhesion to β2-integrin Mac-1 (CD11b/CD18) ligands, fibrinogen, and ICAM-1 suppresses the cell death program through activation of the survival proteins Akt and ERK. The additional stimulation of neutrophils with death-inducing agonists, such as Fas or TNF, shifts the balance toward cell death by inducing ROS generation. ROS through lyn activates SHIP, which, in turn, inhibits Akt and induces apoptosis. Mac-1-dependent phagocytosis of complement opsonized bacteria promotes apoptosis. Phagocytosis results in the generation of ROS within the developing phagolysosomes. ROS triggers activation of caspase-8, leading to caspase-3 activation and acceleration of cell death. This overrides survival cues from Mac-1-stimulated ERK activation.

stress [6]. The antiadhesive and anti-inflammatory actions of pulsatile unidirectional flow may also prevail in conditions of activated endothelial cells. Prolonged exposure of endothelial cells to laminar flow occurs in vivo, results in downregulation of ICAM-1, VCAM-1, and E-selectin expression and inhibition of leukocyte adhesion. Molecular mechanisms of antiadhesive actions of laminar shear stress involve protection against oxidative stress and inhibition of the NF-κB and JNK-AP-1 signaling pathways (Figure 11.3). Laminar shear stress evokes expression of Cu/Zn superoxide dismutase (SOD), thereby effectively reducing superoxide. Laminar shear stress is known to be the physiological activator/inducer of NO formation by endothelial NO synthase. Besides its role in the regulation of vascular tone and permeability, endogenous NO inhibits MCP-1 and IL-6 release and cytokine-induced expression of ICAM-1 and VCAM-1 as well as leukocyte rolling and adhesion. These actions of NO are mediated by scavenging superoxide and through inhibition of NF-κB-dependent gene transcription. NO increases transcription, expression, and nuclear translocation of IκBα, resulting in acceleration of p50/p65 nuclear deactivation. Laminar shear stress specifically abrogates cytokine-induced JNK activity (Figure 11.3).

Anti-inflammatory cytokines that exert antiadhesive effects include transforming growth factor-β (TGF-β), interleukin-10 (IL-10), and IL-1 receptor antagonist (IL-1ra). TGF-β is secreted as an inactive complex with a latency-associated peptide (LAP). Cytokine activation of endothelial cells or coculture of endothelial cells with pericytes or smooth muscle cells results in the formation of active TGF-β. TGF-β downregulates cytokine-induced E-selectin, ICAM-1 and VCAM-1 expression, inhibits MCP-1 and IL-8 release, and attenuates the IL-8-dependent adhesion and transmigration of PMNs through the activated endothelial cell monolayer. The TGF-β actions are mediated through three types of cell-surface receptors coupled to their downstream effectors, known as Smad proteins. Smad proteins interact with the cAMP response element-binding protein (CREB)-binding protein (CBP), thereby blocking the association of CBP with the p65 subunit of NF-κB that is required for maximal transcriptional NF-κB activity. IL-10 is a pleiotropic cytokine produced by Th2 lymphocytes, B cells, monocytes, and macrophages. IL-10 is thought to act as a negative feedback to inhibit the production of proinflammatory cytokines, but it can also inhibit leukocyte–endothelial cell interactions in vivo. IL-10 attenuates leukocyte extravasation through decreasing endothelial expression of P-selectin, E-selectin, and ICAM-1. Conversely, IL-10-deficient mice exhibit markedly elevated expression of ICAM-1 and VCAM-1 in the vasculature and increased leukocyte adhesion in the mesenteric circulation in response to bacterial LPS. The protective effects of IL-10 are mediated

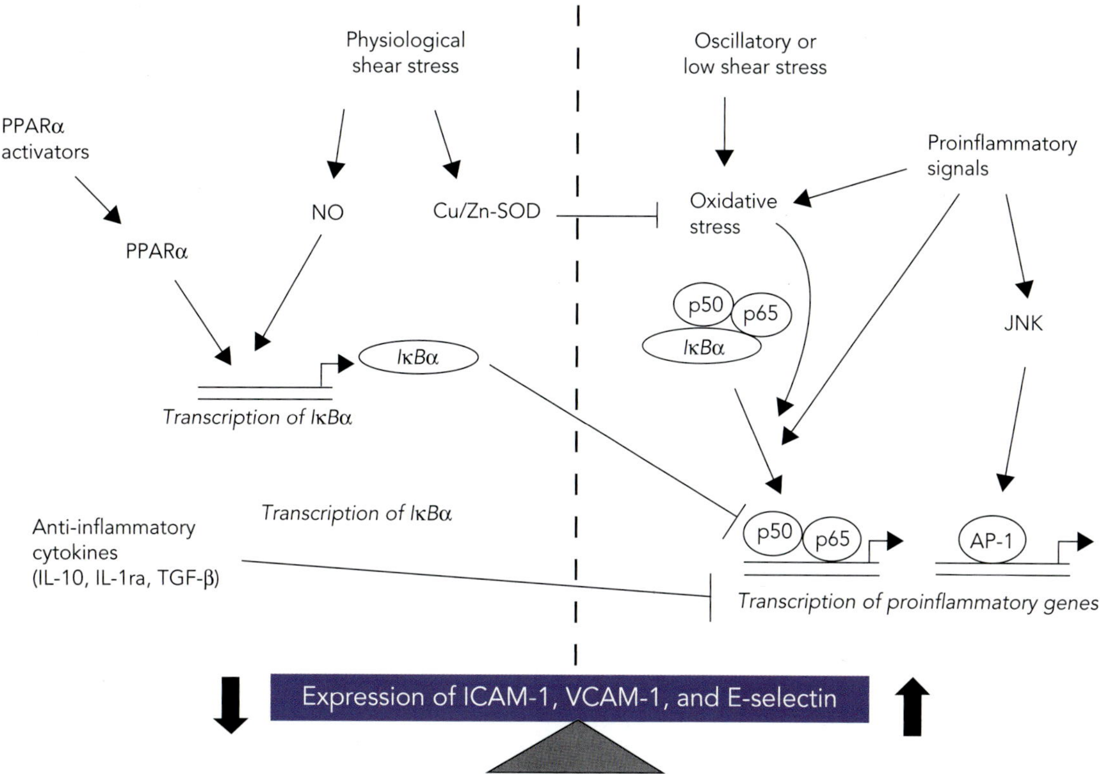

Figure 11.3. Balance of pro- and antiadhesives forces. Oscillatory or low shear stress and proinflammatory signals trigger intracellular oxidative stress and JNK activation that activate the NF-κB- and AP-1-mediated transcription of proinflammatory genes, including ICAM-1, VCAM-1, and E-selectin. Physiological laminar shear stress stimulates the expression of Cu/Zn-SOD and production of NO. Cu/Zn-SOD reduces oxidative stress. NO and activation of PPARα induces IκBα expression, resulting in inhibition of NF-κB activation and deactivation of nuclear p50/p65. Anti-inflammatory cytokines block transcription of NF-κB-driven genes. Laminar shear stress can also block the JNK pathway.

through the prevention of IκB degradation, suppression of NF-κB DNA binding activity [7], and destabilization of mRNA of proinflammatory genes with clustered AU-rich element motifs. IL-1ra is a secreted protein that binds to the IL-1 receptors without signaling. IL-1ra exhibits vascular protective effects in vivo and IL-1ra gene polymorphism is associated with coronary artery disease [8].

Peroxisome proliferators-activated receptors (PPARs) are transcription factors that regulate gene expression by forming a heterodimer with the retinoid receptor RXR that binds to specific DNA sequence elements termed PPAR-responsive elements. Activation of PPARα with its natural ligands n-3 fatty acids, or synthetic ligands, such as the lipid-lowering drugs fibrates, represses cytokine-stimulated expression of VCAM-1 and other proinflammatory mediators in human endothelial cells, resulting in reduced monocyte adhesion. PPARα activators reduce the oxidative stress and induce expression IκBα, and as a result, may inhibit NF-κB activation, in addition to direct protein–protein interactions between PPARα and NF-κB and AP-1 proteins involved in transrepression.

Endothelial cells may also acquire an autoprotective phenotype during inflammation by expressing cytoprotective genes, including the Bcl-2 family member A1, A20, as well as heme oxyganase-1 (HO-1). A1 and A20 inhibit adhesion molecule and IL-8 expression in endothelial cells by suppressing NF-κB activation. Gene transfer of HO-1 protects against hyperoxia-induced PMN infiltration and lung injury [9]. HO-1 deficiency in humans is associated with the presence of severe persistent endothelial damage [10]. Hypoxia or vascular injury induces upregulation and secretion of the glycoprotein Del-1 (developmental endothelial locus-1) that has been implicated in vascular remodeling during angiogenesis. Del-1 has recently been identified as an endogenous inhibitor of inflammatory cell recruitment. Indeed, endothelial Del-1 deficiency increased LFA-1-dependent PMN and monocyte adhesion in vitro and in mice, whereas soluble Del-1 inhibited LFA-1-dependent adhesion of PMN to immobilized ICAM-1 [11].

PMN INFLUENCES ON VASCULAR ENDOTHELIAL PERMEABILITY

Macromolecule transit across blood vessels has evolved to be tightly controlled. Relatively low macromolecular permeability of blood vessels is essential for maintenance of a physiologically optimal equilibrium between intravascular and extravascular compartments. Disturbances of endothelial barrier during disease states can lead to deleterious loss of fluids and plasma protein into the extravascular compartment. Such disturbances in endothelial barrier function are prominent in disorders such as shock and ischemia-reperfusion and contribute significantly to organ dysfunction. The direct relationship between PMN activation and increased endothelial permeability is not clear. While PMN accumulation and increased vascular permeability are often coincidental, PMN activation can occur with limited or no net changes in endothelial permeability. Moreover, while PMN depletion has been demonstrated to decrease organ injury in some models of ischemia and reperfusion, other models have suggested that PMN can exert protective influences in other models. Therefore, further information regarding PMN–endothelial cell interactions and their influence on endothelial permeability may provide a better understanding of the regulation of endothelial permeability.

Approximately 70 million PMN exit the vasculature per minute. These inflammatory cells move into underlying tissue by initially passing between endothelial cells that line the inner surface of blood vessels. This process, referred to as transendothelial migration (TEM), is particularly prevalent in inflamed tissues. Understanding the biochemical details of leukocyte–endothelial cell interactions is currently an area of concentrated investigation. Recent studies of genetically modified animals have suggested that specific molecules may establish "bottlenecks" to the control of the inflammatory response. For example, detailed studies have revealed that the process of leukocyte TEM entails a concerted series of events involving intimate interactions of a series of leukocyte and endothelial glycoproteins that include selectins, β2 integrins, and members of the immunoglobulin supergene family (e.g., ICAM-1). Moreover, histological studies of TEM reveal that PMN initially adhere to the endothelium, move to nearby inter-endothelial junctions via diapedesis, and insert pseudopodia into the interendothelial paracellular space. Successful TEM is accomplished by temporary PMN self-deformation with localized widening of the interendothelial junction. Following TEM, adjacent endothelial cells appear to "reseal," leaving no residual interendothelial gaps. These histological studies are consistent with the observation that leukocyte TEM may result in little or no change in endothelial permeability to macromolecules. In the absence of this tight and dynamic control of endothelial morphology and permeability, interendothelial gap formation during leukocyte TEM could lead to marked increases in endothelial permeability. However, only limited information exists regarding the biochemical events that maintain and dynamically regulate endothelial permeability in the setting of either PMN activation or TEM. A number of studies revealed that activated PMN release soluble factor(s) that support maintenance of endothelial permeability during PMN–endothelial interactions.

The predominant barrier (~90%) to movement of macromolecules across a blood vessel wall is presented by the endothelium [12]. Passage of macromolecules across a cellular monolayer may occur via either a paracellular route (i.e., between cells) or a transcellular route (i.e., through cells). In nonpathologic endothelium, macromolecules such as albumin (molecular weight ~40 kDa) appear to cross the cell monolayer by passing between adjacent endothelial cells (i.e., paracellular) although some degree of transcellular passage may also occur. Endothelial macromolecular permeability is inversely related to macromolecule size. Permeability is also dependent on the tissue of origin. For example, endothelial cells in the cerebral circulation (i.e., blood-brain barrier) demonstrate an exceptionally low permeability. Endothelial permeability may increase markedly upon exposure to a variety of inflammatory compounds (e.g., histamine, thrombin, reactive oxygen species (ROS), leukotrienes, bacterial endotoxins) or adverse conditions (e.g., hypoxia, ischemia). Mechanisms also exist to maintain or balance endothelial permeability during leukocyte TEM. For example, activated PMN release a number of soluble factors, such as ATP, AMP, adenosine, and glutamate, which promote endothelial barrier function.

Endothelial permeability is determined by cytoskeletal mechanisms that regulate lateral membrane intercellular junctions. Tight junctions, also known as zona occludens, comprise one type of intercellular junction and include the proteins zona occludens-1 (ZO-1), ZO-2, cingulin, and occludin. Tight junctions form narrow, cell-to-cell contacts with adjacent cells and comprise the predominant barrier to transit of macromolecules between adjacent endothelial cells. Disruption of cytoskeletal microfilaments (for instance, with cytochalasin B) produces reversible increases in endothelial permeability. JAMs represent another family of proteins important in the transit of PMN across the vascular interface. JAMs immunoglobulin superfamily (IgSF) proteins expressed at cell junctions in epithelial and endothelial cells as well as on the surface of leukocytes, platelets, and erythrocytes. The JAMs (JAM-A, -B, and -C) are variably expressed in endothelial and epithelial cells and mediate homophilic and heterophilic interactions within various tissues. Evidence suggests JAM proteins are important for a variety of cellular processes, ranging from tight junction permeability, leukocyte transmigration, platelet activation, and angiogenesis.

As alluded to earlier, activated PMN release a number of soluble factors that promote structural changes within the endothelium. One of the better understood pathways is nucleotide metabolism at the endothelial surface. Activated PMN can release nanomolar quantities of ATP, and it is now accepted that the major pathway for extracellular hydrolysis of ATP and ADP is the ecto-nucleoside triphosphate diphosphohydrolase (NTPDase), previously identified as ecto-ATPase, ecto-ATPDase or CD39 [13]. CD39 is expressed on the vascular endothelium and its role to date has been to modulate platelet purinoreceptor activity by the sequential hydrolysis of extracellular ATP or ADP to AMP. This thromboregulatory potential of CD39 has been recently demonstrated by the generation of mutant mice with disruption of the CD39 gene, and by a series of experiments where high levels of ATPDase expression are attained by adenoviral vectors in the injured vasculature [13]. Ecto-5'-nucleotidase (CD73) is a membrane bound glycoprotein that functions to hydrolyze extracellular nucleotides into bioactive nucleoside intermediates. Surface-localized CD73 converts adenine nucleotides (e.g., AMP) into adenosine, which in turn, can activate transmembrane adenosine receptors or can be internalized through dipyridamole-sensitive carriers. Activation of these pathways has been shown to enhance endothelial barrier function. Endothelial cells of many origins express constitutive CD73. The primary function attributed to endothelial CD73 has been catabolism of extracellular nucleotides, although CD73 may also mediate lymphocyte binding under some circumstances. Taken together, studies in vitro and in murine models define CD39 and CD73 as gatekeepers for the metabolic fine tuning of endothelial permeability. Such innate protective pathways share the common strategy of increasing extracellular adenosine concentrations and promoting adenosine signaling at the cell surface.

In addition to factors, which preserve barrier function during transmigration (e.g., adenosine), PMN also release factors, which increase endothelial permeability. For example, activation of PMN through $\beta2$ integrins elicits the release of soluble factor(s) that induce endothelial cytoskeletal rearrangement, gap formation, and increased permeability [14]. This PMN-derived permeabilizing factor was subsequently identified as HBP (also called azurocidin and CAP37), member of the serprocidin family of cationic peptides. HBP, but not other PMN granule proteins (e.g., elastase, cathepsin G), was shown to induce Ca^{2+}-dependent cytoskeletal changes in cultured endothelia and to trigger macromolecular leakage in vivo. Interestingly, HBP regulation of barrier may not be selective for PMN, and in fact, endothelial cells themselves are now a reported source of HPB. As such, it is possible that the endothelium may self-regulate permeability through HBP under some conditions, and that mediators found within the inflammatory milieu may also increase endothelial permeability.

OUTSIDE-IN SIGNALING EVENTS IN PMN–ENDOTHELIAL CELL INTERACTIONS

Outside-in signaling via β_2 integrins is required for PMN adhesion and transmigration across the activated endothelium as discussed earlier, as well as for other adhesion-dependent PMN function, including phagocytosis of complement-opsonized pathogens, binding to fibrinogen, immune complexes, and platelets. The biological importance of β_2 integrins is highlighted in patients with a mutation in the β_2 integrin subunit shared by the LFA-1 and Mac-1 integrins. Thus, patients with LAD1 exhibit peripheral blood neutrophilia, increased susceptibility to bacterial infections, and delayed wound healing [15]. PMN adherence to Mac-1 ligands also affects the fate of PMNs.

Mature PMNs are terminally differentiated cells that have the shortest half-life (~7 hours) among leukocytes and die rapidly via apoptosis. This constitutively expressed cell death program renders PMNs unresponsive to proinflammatory stimuli and promotes their removal from inflamed areas by scavenger macrophages with minimal damage to the surrounding tissue, thereby facilitating the resolution of inflammation. The fate of PMNs is profoundly influenced by signals from the inflammatory microenvironment. Inflammatory mediators, such as lipopolysaccharide, IL-8 and acute-phase proteins prolong PMN survival by suppressing apoptosis, whereas proapoptotic stimuli, such as TNF or Fas ligand reduce PMN longevity.

PMN transmigration across the endothelium signals a delay of PMN apoptosis through engagement of Mac-1 with its endothelial cell ligand ICAM-1. Likewise, PMN adherence to another Mac-1 ligand fibrinogen also extends their life span. The delay in apoptosis is mediated through activation of the Ser-Thr kinase Akt, and, to a lesser degree, the MAPK-ERK signaling cascade, leading to inhibition of the mitochondrial pathway of apoptosis.

Although cross-linking Mac-1 alone provides survival signals, engagement of Mac-1 in the presence of proapoptotic TNF or Fas ligand accelerates apoptosis. Phagocytosis of opsonized bacteria by PMNs also evokes programmed cell death by a Mac-1-dependent pathway. This process is referred to as phagocytosis-induced cell death or PICD [16]. Proapoptotic stimuli promote NADPH oxidase–stimulated release of ROS, which leads to activation of SHIP (Src-homology 2 [SH2]-containing inositol 5-phosphatase) that hydrolyzes products of phosphoinositide 3-kinase

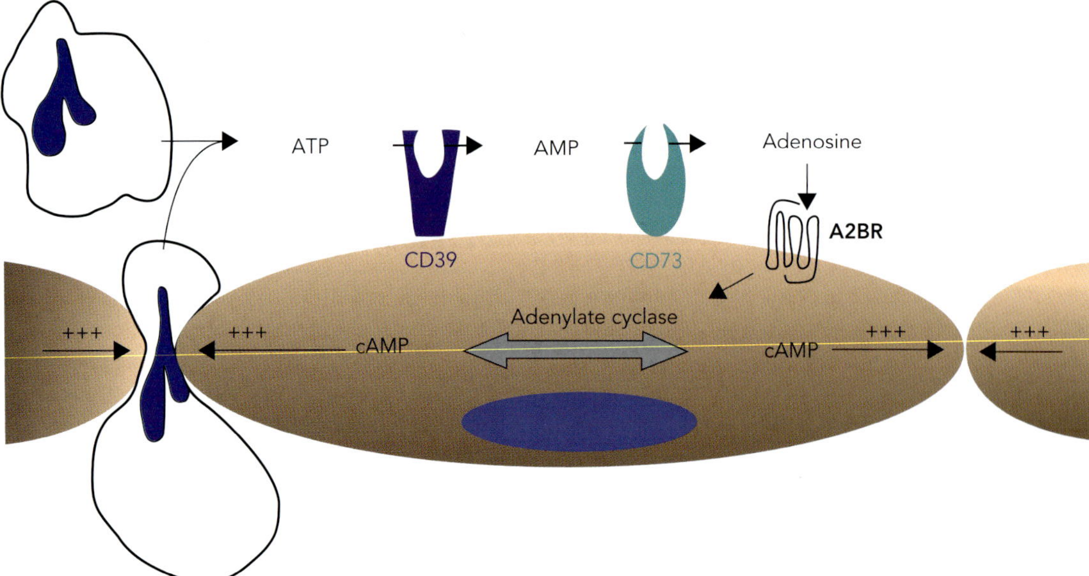

Figure 11.4. Model of coordinated nucleotide metabolism and nucleoside signaling in inflammation. In areas of ongoing inflammation, CD39 and CD73, expressed on the surface of endothelial cells, coordinate the metabolism of ATP to adenosine. Activated PMN provide a readily available extracellular source of ATP that through two enzymatic steps results in the liberation of adenosine. Adenosine generated in this fashion is available for activation of surface endothelial adenosine receptors, particularly the A2BR. Activation of A2BR leads to increases in intracellular cyclic AMP, thereby enhancing endothelial barrier function. As such, this protective mechanism may provide an innate mechanism to preserve vascular integrity and prevent intravascular fluid loss.

through the Src kinase Lyn. This signaling pathway would ultimately lead to decreased activation of the survival factor Akt. This purported mechanism of apoptosis might be particularly relevant in vivo under conditions of high PMN accumulation in tissues. Phagocytosis also results in generation of ROS within the phagolysosomes through NADPH oxidase. NADPH oxidase–derived ROS contribute to microbicidal activity and are also key intracellular trigger of PICD. ROS are required for activation of caspase-8, a signature of receptor-mediated cell death. Caspase-8, in turn, triggers activation of caspase-3, the main effector of apoptosis. Activated caspase-3 and -8 overcome Mac-1 ligation-generated survival signals, eventually favoring cell death. Additional exogenous stimuli, such as TNF or granulocyte/monocyte colony-stimulating factor, present at inflammatory sites, may interfere with phagocytosis-activated intracellular signals to shift the life-death balance of PMNs to either enhanced apoptosis or prolonged survival (Figure 11.4).

Upon leukocyte adhesion, ICAM-1 is capable of initiating outside-in signaling events in endothelial cells. ICAM-1 binding to its ligands β2-integrins or fibrinogen, leads to activation of Src tyrosine kinases, which in turn results in phosphorylation of cortactin, an actin-binding protein that modulates the F-actin cytoskeleton and endothelial cell locomotion. Following engagement, apical ICAM-1 is recruited to caveola- and F-actin–rich regions close to endothelial cell–cell borders, and translocates, with caveolin-1, to the basal plasma membrane. As caveolin-1, ICAM-1 and F-actin surround the transcellular channels around transmigrating lymphocytes, leukocyte-induced ICAM-1 recruitment to caveolae may initiate the formation of a transcellular passage for lymphocytes to cross endothelial cells [17].

Signaling pathways initiated by ICAM-1 binding to PMN LFA-1 or Mac-1 include enhanced ROS production and activation of p38 MAPK, activation of the transcription factors AP-1 and NF-kB, induction of VCAM-1 gene transcription and increased expression of VCAM-1 on the cell surface and increased production of IL-8 and RANTES. Ligation of constitutive or induced ICAM-1 increases intracellular glutathione level, thereby controlling endothelial redox status. Fibrinogen binding to ICAM-1 activates the survival signal ERK that has anti-apoptotic actions. This mechanism may be important in maintaining vascular integrity during ischemia/reperfusion injury, when fibrinogen is deposited on ECs.

INFLUENCE OF THE INFLAMMATORY MILIEU ON PMN–ENDOTHELIAL CELL INTERACTIONS

In intact tissues, endothelial cells lie anatomically positioned adjacent to and physiologically juxtaposed to number of cell types, including leukocytes, fibroblasts, and smooth muscle cells. Such a setting makes paracrine cross-talk pathways an important part of cell–cell communication. Locally generated mediators can bind to endothelial surface receptors, and mediate both physiologic and pathophysiologic functional responses.

A somewhat surprising observation in the past decade has been the finding that inflammatory setting can be quite oxygen deficient (hypoxic). Ongoing inflammatory responses are characterized by dramatic shifts in tissue metabolism. These changes include lactate accumulation with resultant metabolic acidosis and diminished availability of oxygen (hypoxia). Such shifts in tissue metabolism result from profound recruitment of inflammatory cell types, particularly PMN, which when activated can consume copious amounts of oxygen in the formation of oxygen radicals. At the tissue and cellular level, hypoxia induces an array of genes pivotal to survival in low oxygen states. As a global regulator of oxygen homeostasis, the αβ heterodimeric transcription factor hypoxia-inducible factor (HIF) facilitates both oxygen delivery and adaptation to oxygen deprivation. Genes induced by HIF include those necessary for cell, tissue, and whole animal adaptive responses to hypoxia. These proteins include enzymes involved in erythropoiesis, anaerobic metabolism, angiogenesis, and vasodilatation. More recently, it is appreciated that a number of inflammatory genes relevant to PMN–endothelial interactions are also HIF-regulated. For example, the CD18 chain of the β2 integrin complex is transcriptionally regulated by HIF, providing a prime example of how the inflammatory milieu might impact local PMN–endothelial responses.

With regard to paracrine signaling, an area of intense investigation is lipid mediator generation and signaling at inflammatory sites. Important in the regard, both endothelial and nonendothelial cell populations express enzymes (e.g., lipoxygenases, cyclooxygenases) capable of utilizing arachidonic acid substrates to generate bioactive lipid mediators. Such lipid mediators can signal via autocrine or paracrine pathways and depending on the tissue, microenvironment can convey a pro- or anti-inflammatory message. Of particular interest are a group of lipid mediators termed lipoxins and resolvins. Lipoxins are tetraene eicosanoids derived from membrane arachidonic acid through the combined action of 5-lipoxygenase (LO) and 12-LO or 15-LO (i.e., transcellular biosynthesis). A number of in vitro and in vivo studies revealed that lipoxins, and specifically lipoxin A4 (LXA4), function as an innate "stop signals," acting to control local inflammatory processes. At nanomolar concentrations, LXA4 has been demonstrated to inhibit PMN transmigration across confluent epithelia and endothelia. It is likely that the action(s) of LXA4 are predominantly on leukocytes and involve the activation of protein kinase C, since LXA4 inhibition required preincubation and PMN responses were sensitive to the protein kinase C inhibitor staurosporine. Additional mechanistic studies have revealed that LXA4 inhibit PMN β2 integrin (CD11/18) expression, and thus also block Mac-1 outside-in signaling. Lipoxins are rapidly (within minutes) converted to inactive compounds by myeloid cells. For this reason, stable lipoxin analogs have been synthesized and biochemically and functionally studied in detail. In vivo, both native LXA_4 and its metabolically stable have been demonstrated to block PMN trafficking, to facilitate removal of apoptotic neutrophils by macrophages, and to serve as potent anti-inflammatory and proresolution molecules in a number of vascular injury models.

More recent studies have focused on omega-3 fatty acid–derived resolvins [18]. Increasing evidence suggests that resolvin E1 (RvE1: 5*S*,12*R*,18*R*-trihydroxy-eicosapentaenoic acid) contributes to resolution of inflammation via interactions with the resolvin E1 receptor (termed chemR23). RvE1 is generated at sites of inflammation through transcellular biosynthesis and has been shown to potently inhibit PMN TEM and to promote PMN clearance from the epithelial cell surface in vitro. In vivo, RvE1 has been documented to attenuate colonic mucosal inflammation in vivo and to resolve oral inflammation in a rabbit periodontitis model.

PMN–ENDOTHELIAL INTERACTION AS A THERAPEUTIC TARGET

The multistep paradigm of leukocyte recruitment implies that inhibition of any of these steps, in particular rolling and firm adhesion, should interrupt the extravasation process and therefore prevent leukocyte trafficking into inflamed or injured tissues. Indeed, the anti-inflammatory effects of glucocorticoids and many nonsteroidal anti-inflammatory drugs can be attributed, in part, to inhibition of adhesion molecules on PMN or endothelial cells [19,20]. More selective therapeutic approaches include receptor–ligand blockade, allosteric inhibitors, inhibitors of inside-out and outside-in signaling pathways, and targeting expression of immunoglobulin superfamily ligands [19,21]. Thus, inhibition of the rolling and adhesion events using monoclonal antibodies against selectins and integrins resulted in reduced leukocyte accumulation and striking protection against tissue injury in a variety of experimental models, including reperfusion injury, myocardial infarction, stroke, colitis, asthma, and rheumatoid arthritis [21]. Despite promising preclinical results, the outcome of clinical trials that followed in these indications has been disappointingly inconsistent. With the exception of some beneficial effects in asthma, ulcerative colitis, and psoriasis, treatment with humanized antiselectin or anti-CD18 antibodies had no overall benefit compared with placebo in patients with myocardial infarction, stroke, rheumatoid arthritis, traumatic shock, or renal transplant [21]. Considering the redundancy and overlapping functions of adhesion molecules, the negative results point toward alternative pathways that might govern leukocyte recruitment in these disorders. However, more selective targeting of adhesion molecules (e.g., targeting LFA-1 instead of CD18) or leukocyte subtypes resulted in positive clinical results. Thus, efaluzimab, a humanized antibody against LFA-1 (CD11a) decreased severity index by >75% in psoriasis [22] and inhibition of α_4-integrin with natalizumab resulted in significant improvements in the clinical remission rates in patients with Crohn's disease or multiple sclerosis most likely through preventing lymphocyte infiltration [23,24]. Unfortunately, natalizumab therapy longer than 1 year had severe complications associated with lack of lymphocyte surveillance. The therapeutic potential of interfering with cell surface expression of adhesion molecules using antisense oligonucleotides, RNA silencing, or small molecule modifiers of intracellular signaling pathways remains to be investigated in the clinical setting.

Preclinical data indicate that downregulation of CD11b/CD18 expression on PMN and inhibition of PMN transmigration across endothelia and epithelia are among the central events underlying the anti-inflammatory proresolution actions of LXA4, aspirin-triggered 15-epi-LXA4, and RvE1 [18]. Thus, PMN-specific signals, which also include reduced production of proinflammatory cytokines and attenuation of oxidative and peroxidative stress, would lead to dampening of inflammation, whereas monocyte/macrophage-specific signals would promote inflammatory resolution by enhancing removal of apoptotic PMN and production of anti-inflammatory cytokines/chemokines. Accumulating evidence suggests an important role for impaired PMN apoptosis or macrophage phagocytosis in modulating the outcome of inflammation [20,25]. Consistently, induction of PMN apoptosis with cyclin-dependent kinase inhibitors [26] or overriding the potent survival signal generated by the acute-phase protein serum amyloid A with aspirin-triggered 15-epi-LXA4 [27] would enhance resolution of inflammation.

CONCLUSIONS

Our current understanding of PMN interactions with the vascular endothelium has evolved immensely in the past two decades. New technologies, better models, and the identification of human patients with genetic mutations within these pathways have provided unprecedented opportunities to define precise details of how PMN interact with endothelial cells at the molecular level. In particular, the leukocyte adhesion cascade has been augmented by several more steps, and the signaling network linking neutrophil and endothelial cell responses to activation of adhesion molecules is beginning to emerge. More detailed signaling studies on leukocytes and endothelial cells may identify new targets for therapeutic interventions. Indeed, these studies promise to provide a rational basis to develop therapies with higher selectivity for leukocyte subsets for the treatment of inflammatory diseases.

REFERENCES

1. Ley, K., Laudanna, C., Cybulsky, M.I., and Noursargh, S. 2007. Getting to the site of inflammation: the leukocyte adhesion cascade updated. *Nat Rev Immunol* **7**:678–689.
2. Yonekawa, K., and Harlan, J.M. 2005. Targeting leukocyte integrins in human diseases. *J Leukoc Biol.* **77**:129–140.
3. Smith, C.W. 2008. Adhesion molecules and receptors. *J Allergy Clin Immunol* **121**:S375–379.
4. Wagner, D.D., and Frenette, P.S. 2008. The vessel wall and its interactions. *Blood* **111**:5271–5281.
5. Weber, C., Fraemohs, L., and Dejana, E. 2007. The role of junctional adhesion molecules in vascular inflammation. *Nat Rev Immunol* **7**:467–477
6. Gimbrone, M.A., Nagel, T., and Topper, J.N. 1997. Biomechanical activation: an emerging paradigm in endothelial adhesion biology. *J Clin Invest* **99**: 1809–1813.
7. Schottelius, A.J., Mayo, M.W., Sartor, R.B., and Baldwin, A.S., Jr. 1999. Interleukin-10 signaling blocks inhibitor of κB kinase activity and nuclear factor κB DNA binding. *J Biol Chem* **274**:31868–31874.
8. Francis, S.E., Camp, N.J., Dewberry, R.M., et al. 1999. Interleukin-1 receptor antagonist gene polymorphism and coronary artery disease. *Circulation* **99**:861–866.
9. Otterbein, L.E., Kolls, J.K., Mantell, L.L., Cook, J.L., Alam, J., and Choi, A.M. 1999. Exogenous administration of heme oxygenase-1 by gene transfer provides protection against hyperoxia-induced lung injury. *J Clin Invest* **103**:1047–1054.
10. Yachie, A., Niida, Y., Wada, T., et al. 1999. Oxidative stress causes enhanced endothelial cell injury in human heme oxygensae-1 deficiency. *J Clin Invest* **103**:129–135.
11. Choi, E.Y., Chavakis, E., Czabanka, M.A., et al. 2008. Del-1, an endogenous leukocyte-endothelial cell adhesion inhibitor, limits inflammatory cell recruitment. *Science* **322**:1101–1104.
12. Colgan, S.P., Eltzschig, H.K., Eckle, T., and Thompson, L.F. 2006. Physiologic roles for ecto-5'-nucleotidase (CD73). *Purinergic Signal* **2**:351–360.
13. Robson, S.C., Sevigny, J., and Zimmermann, H. 2006. The E-NTPDase family of ectonucleotidases: structure function relationships and pathophysiological significance. *Purinergic Signal* **2**:409–430.
14. Eltzschig, H.K., Macmanus, C.F., and Colgan, S.P. 2008. Neutrophils as sources of extracellular nucleotides: functional consequences at the vascular interface. *Trends Cardiovasc Med* **18**:103–107.
15. Wehrle-Haller, B., and Imhof, B.A. 2003. Integrin-dependent pathologies. *J Pathol* **200**:481–487.

16. Mayadas, T.N., and Cullere, X. 2005. Neutrophil β2 integrins: moderators of life or death decisions. *Trends Immunol* **26**:388–395.
17. Millan, J., Hewlett, L., Glyn, M., Toomre, D., Clark, P., and Ridley, A.J. 2006. Lymphocyte transcellular migration occurs through recruitment of endothelial ICAM-1 to caveola- and F-actin-rich domains. *Nat Cell Biol* **8**:113–123.
18. Serhan, C.N., Chiang, N., and Van Dyke, T.E. 2008. Resolving inflammation: dual anti-inflammatory and pro-resolution lipid mediators. *Nat Rev Immunol* **8**: 349–361.
19. Ulbrich, H., Eriksson, E.E., and Lindbom, L. 2003. Leukocyte and endothelial cell adhesion molecules as targets for therapeutic interventions in inflammatory diseases. *Trends Pharmacol Sci* **24**:640–647.
20. Gilroy, D.W., Lawrence, T., Perretti, M., and Rossi, A.G. 2004. Inflammatory resolution: new opportunities fro drug discovery. *Nat Rev Drug Discov* **2**:965–975.
21. Yonekawa, K., and Harlan, J.M. 2005. Targeting leukocyte integrins in human diseases. *J Leukoc Biol* **77**:129–140.
22. Gordon, K.B., Papp, K.A., Hamilton, T.K., et al. 2003. Efalizumab for patients with moderate to severe plaque psoriasis: a randomized controlled trial. *JAMA* **290**:3073–3080.
23. Keeley, K.A., Rivey, M.P., and Allington, D.R. 2005. Natulizimab for the treatment of multiple sclerosis and Crohn's disease. *Ann Pharmacother* **39**:1833–1843.
24. Rivera-Nieres, J., Gorfu, and Ley, K. 2008. Leukocyte adhesion models of inflammatory bowel disease. *Inflamm Bowel Dis* **14**:1715–1735.
25. El Kebir, D., József, L., Pan, W., and Filep, J.G. 2008. Myeloperoxidase delays neutrophil apoptosis through CD11b/CD18 integrins and prolongs inflammation. *Circ Res* **103**:352–359.
26. Rossi, A.G., Sawatzky, D.A., Walker, A., et al. 2006. Cyclin-dependent kinase inhibitors enhance the resolution of inflammation by promoting inflammatory cell apoptosis. *Nat Med* **12**:1056–1064.
27. ElKebir, D., József, L., Pan, W., Petasis, N.A., Serhan, C.N., and Filep, J.G. 2007. Aspirin-triggered lipoxins override the apoptosis-delaying action of serum amyloid A in human neutrophils: a novel mechanism for resolution of inflammation. *J Immunol* **179**:616–622.

SUGGESTED READINGS

Butcher, E.C. 1991. Leukocyte-endothelial cell recognition: three (or more) steps to specificity and diversity. *Cell* **67**:1033–1036.

Jacobson, K.A., and Gao, Z.G. 2006. Adenosine receptors as therapeutic targets. *Nat Rev Drug Discov* **5**:247–264.

Kong, T., Eltzschig, H.K., Karhausen, J., Colgan, S.P., and Shelley, C.S. 2004. Leukocyte adhesion during hypoxia is mediated by HIF-1-dependent induction of β2 integrin gene expression. *Proc Nat Acad Sci USA* **101**:10440–10445.

Rao, R.M., Yang, L., Garcia-Cardena, G., and Luscinskas, F.W. 2007. Endothelial-dependent mechanisms of leukocyte recruitment to the vascular wall. *Circ Res* **101**:234–247

Serhan, C.N., Brain, S.D., Buckley, C.D., et al. 2007. Resolution of inflammation: state of the art, definitions and terms. *FASEB J* **21**:325–332.

Tedgui, A., and Mallat, Z. 2001. Anti-inflammatory mechanisms in the vascular wall. *Circ Res* **88**:877–887.

12 Lipid Mediators in Acute Inflammation and Resolution: Eicosanoids, PAF, Resolvins, and Protectins

Charles N. Serhan and Jesper Z. Haeggström

INTRODUCTION

Autacoids are locally acting substances that are rapidly biosynthesized in response to specific stimuli, act quickly, and are usually deactivated by metabolism. Eicosanoids are a chemically diverse family of arachidonic acid–derived autacoids that have critical roles in cardiovascular, inflammatory, and reproductive physiology. Pharmacologic interventions in eicosanoid pathways – including the nonsteroidal anti-inflammatory drugs (NSAIDs), COX-2 inhibitors, and leukotriene inhibitors – are useful in the clinical management of inflammation, pain, and fever. Given the many important bioactivities of lipid mediator eicosanoids, resolvins, and protectins, future research may lead to the development of new therapeutics for the treatment of inflammatory conditions, autoimmune diseases, asthma, glomerulonephritis, cancer, sleep disorders, and Alzheimer's disease.

BIOSYNTHESIS OF EICOSANOIDS

Eicosanoids are crucially involved in a number of metabolic pathways that have diverse roles in inflammation and cellular signaling. These pathways center on reactions involving the metabolism of arachidonic acid (Figure 12.1). The following considers the biochemical steps leading to arachidonic acid synthesis, then discusses the cyclooxygenase (COX), lipoxygenase (LOX), epoxygenase, and isoprostane pathways of arachidonic acid metabolism. Arachidonic acid (*cis-,cis-,cis-,cis*-5, 8,11,14-eicosatetraenoic acid), the common precursor to eicosanoids, must be synthesized from the essential fatty acid precursor linoleic acid (*cis,cis*-9,12-octadecadienoic acid), which can be obtained only from dietary sources. In cells, arachidonic acid is esterified to the *sn*2 position of the membrane phospholipids predominantly in phosphatidylcholine phosphatidylethanolamine, and phosphatidyl inositol. Arachidonic acid is released from cellular phospholipids by the enzyme phospholipase A_2 (Figure 12.1), which hydrolyzes the acyl ester bond. This important reaction, which represents the first step in the arachidonic acid cascade, is the overall rate-determining step in the generation of eicosanoids produced by most effector and immune cells.

Membrane-bound and soluble isoforms of phospholipase A_2 are classified as secretory ($sPLA_2$) and cytoplasmic ($cPLA_2$), which are differentiated based on molecular weight, pH sensitivity, regulation and inhibition characteristics, calcium requirements, and substrate specificity. The existence of multiple isoforms allows for tight control and regulation of the enzyme in different tissues to achieve selective biologic responses. Phospholipases relevant in inflammation are stimulated by such cytokines as TNF-α, GM-CSF, and IFN-γ, such growth factors as EGF, and the MAPK–PKC cascade. Glucocorticoids were once thought to act via the direct inhibition of phospholipase A_2 activity. More recent findings indicate that glucocorticoids act by inducing the biosynthesis of lipocortins, a family of phospholipase A_2–regulatory proteins that comprise Annexin 1, which mediates part of the anti-inflammatory actions of glucocorticoids (see later and Chapter 10).

Cyclooxygenases: Biosynthesis of Prostanoids

The unbound form of intracellular arachidonic acid (unesterified: not bound to phospholipids) is rapidly converted in a cell type–specific manner by COX, LOX, or epoxygenase (cytP450) enzymes that dictate the class of local eicosanoids generated. The COX pathway leads to the formation of prostaglandins, prostacyclin, and thromboxanes (Figure 12.2); the LOX pathways lead to leukotrienes and lipoxins; and the epoxygenase pathways lead to epoxyeicosatetraenoic acids (Figure 12.1). Cyclooxygenases (also

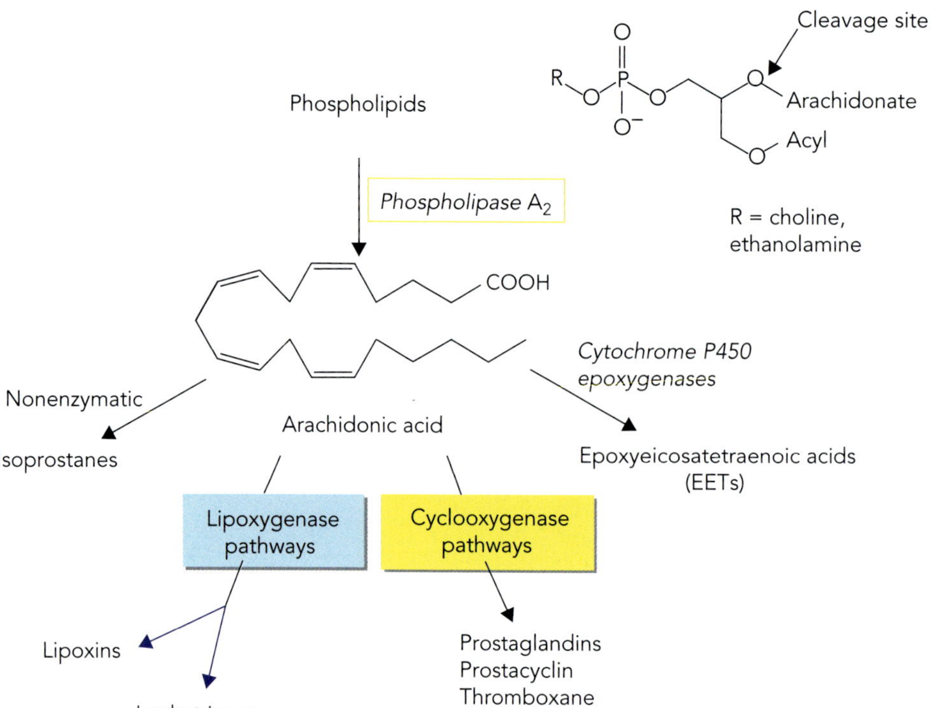

Figure 12.1. Overview of arachidonic acid cascade. Phospholipase A_2 acts on the phospholipids phosphatidylcholine (PC), phosphatidylethanolamine (PE), and phosphatidylinositol (PI) to release arachidonic acid. Phospholipase A_2 cleaves the ester bond marked by the arrow to release arachidonic acid. Unesterified arachidonic acid is substrate for the cyclooxygenase, lipoxygenase, and epoxygenase pathways. The cyclooxygenase pathways produce prostaglandins, prostacyclin, and thromboxane. The lipoxygenase pathways produce leukotrienes and lipoxins. The epoxygenase pathway produces epoxyeicosatetraenoic acids (EETs) and nonenzymatic oxidation of arachidonic acid can produce isoprostanes.

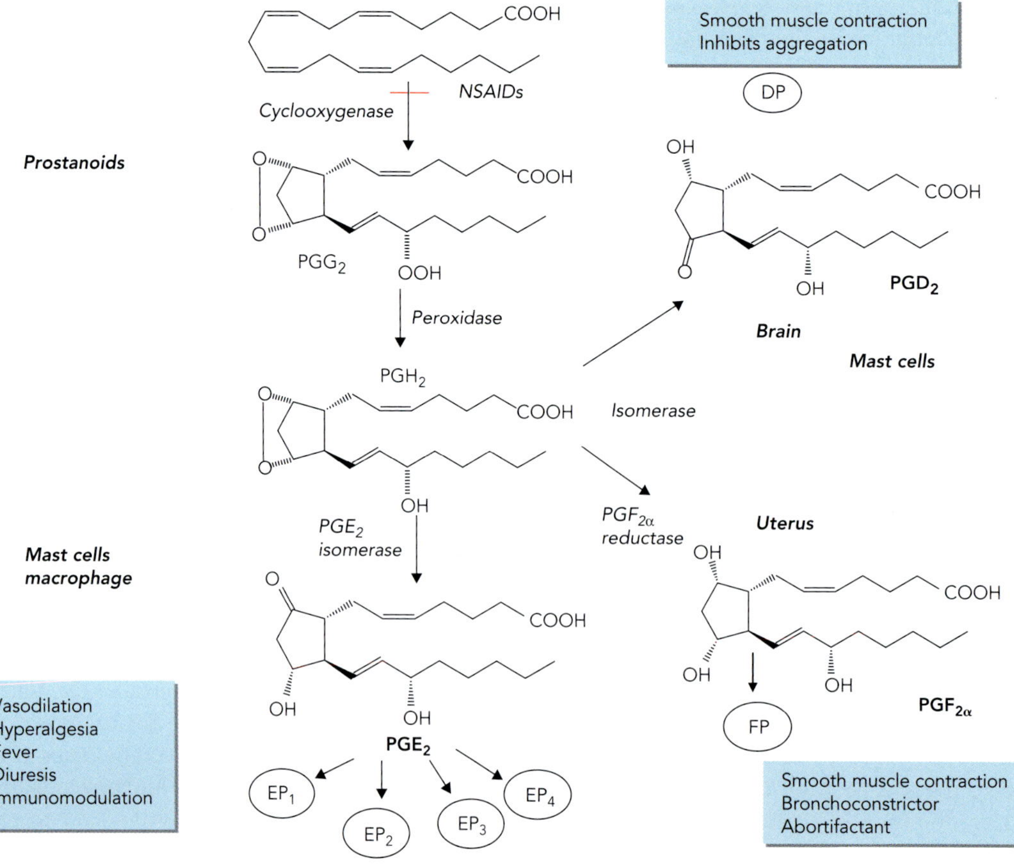

Figure 12.2. *(continued)*.

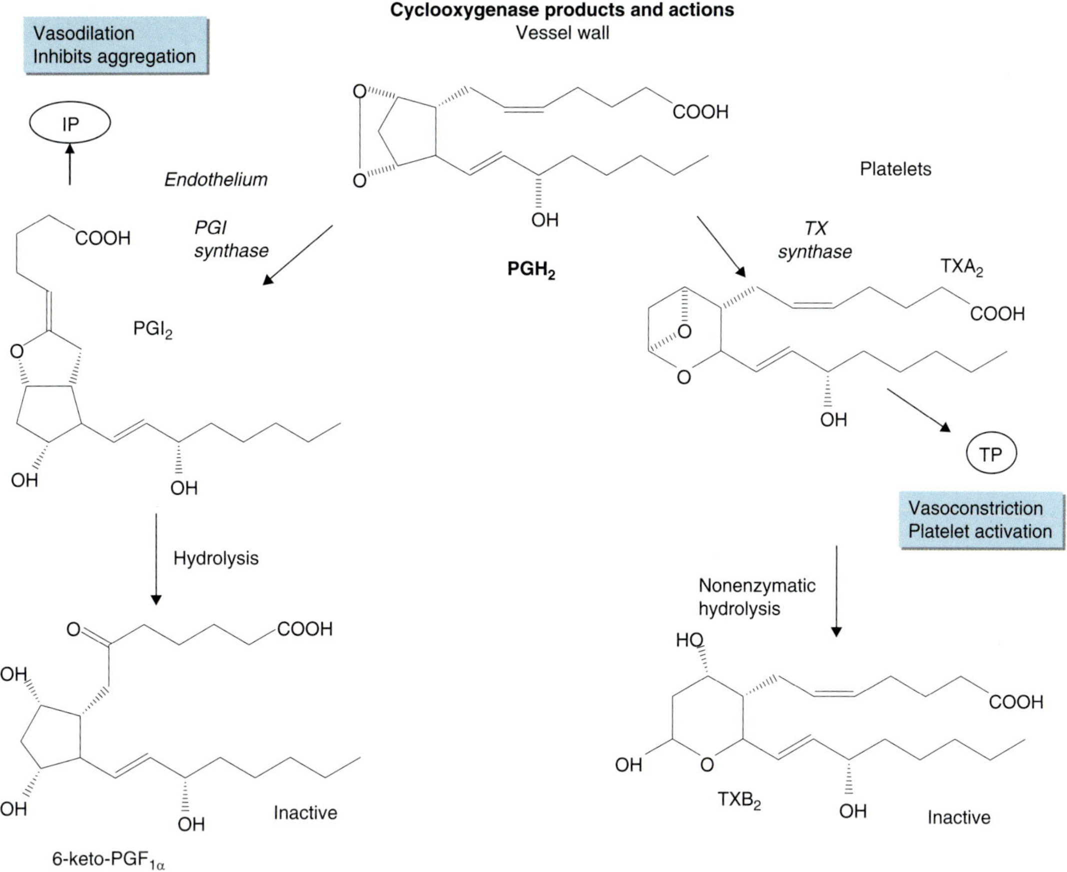

Figure 12.2. Prostaglandin and thromboxane biosynthesis, and major functions. The biosynthetic pathways from arachidonic acid to prostaglandins, prostacyclin, and thromboxane are shown. Note that tissue-specific enzyme expression determines the tissues in which the various PGH_2 products are produced. NSAIDs and COX-2 inhibitors are the most important classes of drugs that modulate prostaglandin and TXA_2 production.

Thromboxane antagonists and PGE_2 synthase inhibitors are each potential new pharmacologic strategies currently in development. COX, cyclooxygenase; PG, prostaglandin; TX, thromboxane.

GPCR specific for eicosanoids: denoted DP, PGD_2 receptor; EP, PGE_2 receptor; FP, $PGF_{2\alpha}$ receptor; IP, PGI_2 receptor; TP, TXA_2 receptor. Each of these receptors specifically and stereoselectively signals with its cognate ligand eicosanoid. NSAID, nonsteroidal antiinflammatory drug.

(A) Prostanoid biosynthesis: cell type specific formation of PGE_2, $PGF_{2\alpha}$, and PGD_2.

(B) Prostacyclin and thromboxane biosynthesis in the vessel wall and platelets.

known as prostaglandin H or endoperoxide synthases) are glycosylated, homodimeric, membrane-bound, heme-containing enzymes that are virtually ubiquitous in animal cells from invertebrates to humans. Two COX isoforms, denoted COX-1 and COX-2, are found in humans. Although COX-1 and COX-2 share 60% sequence homology and near superimposable three-dimensional structures, the genes are located on different chromosomes, and the enzymes differ in cellular, genetic, physiologic, pathologic, and pharmacologic profiles (Table 12.1). Each COX catalyzes two sequential reactions, that is, an oxygen-dependent cyclization of arachidonic acid into prostaglandin G_2 (PGG_2) followed by a peroxidase reaction that reduces PGG_2 to PGH_2.

As a result of differences in cellular localization, regulatory profile, tissue expression, and substrate requirement, COX-1 and COX-2 ultimately produce different sets of eicosanoid products. Constitutively expressed COX-1 is believed to function in physiologic, or "housekeeping," roles such as vascular homeostasis, maintenance of renal and gastrointestinal blood flow, renal function, intestinal mucosal proliferation, platelet function, and antithrombogenesis. A number of "as-needed," or specialized, functions are attributed to the inducible COX-2 enzyme, including roles in inflammation, fever, pain, transduction of pain stimuli in the spinal cord, mitogenesis (particularly in the gastrointestinal epithelium), vascular hemodynamics, deposition of trabecular bone, ovulation, placentation, and uterine contractions of labor. The role of constitutive COX-2 expression in areas of the nervous system as the hippocampus, hypothalamus, and amygdala remains to be elucidated.

TABLE 12.1. COX-1 and COX-2

General properties	COX-1	COX-2
Expression	Constitutive	Inducible Constitutive in parts of nervous and vascular systems
Tissue location	Ubiquitous expression	Inflamed and activated tissues, tumor cells
Subcellular localization	Endoplasmic reticulum	ER and nuclear membrane
Substrate selectivity	Arachidonic acid, eicosapentaenoic acid	Arachidonic acid, γ-linolenate, α-linolenate, linoleate, eicosapentaenoic acid
Role	Gastroprotection and renal maintenance functions	Proinflammatory, mitogenic, and growth factors
Inducers	Generally not induced	LPS, TNF-α, IL-1, IL-2, EGF, IFN-γ mRNA rises 20- to 80-fold upon induction in many tissues regulated within 1–3 hours
Inhibitors	NSAIDs low-dose aspirin	Endogenous glucocorticoids, IL-1β, IL-4, IL-10, IL-13 NSAIDs, COX-2 inhibitors

Prostanoids: The Prostaglandins

Prostaglandins are a relatively large class of structurally similar compounds that each carries potent and stereospecific biological actions that are important in host mechanisms in acute inflammation. The name derives from their initial identification in the genitourinary system of male sheep. Prostaglandins all share a chemical structure, called a prostanoid, consisting of a 20-carbon carboxylic acid containing a cyclopentane ring and a 15-hydroxyl group, and are divided into three major subseries – PG_1, PG_2, and PG_3. The subscript numeral indicates the number of double bonds present in each molecule. The PG_2 series is the most prevalent because these are direct products of arachidonic acid. The PG_1 series derive from the arachidonic acid precursor dihomo-γ-linolenic acid (DHGLA), an eicosatrienoic acid, while the PG_3 series derive from eicosapentaenoic acid (EPA).

Prostaglandin (PG)H_2 is central to the COX pathway (see Figure 12.2) because it is the unstable transient precursor to PGD_2, PGE_2, $PGF_{2\alpha}$, and thromboxane A_2 (TxA_2), as well as prostacyclin (PGI_2). The distribution of these eicosanoids in tissues is determined by the expression pattern of the different enzymes of prostaglandin synthesis (i.e., PG synthases) present in specific cells in the tissues (see Figure 12.2), which gives rise to a cell type–specific biosynthesis profile of eicosanoids.

The prostaglandins are important in many physiologic processes; most are not directly related to inflammation (highlighted in Table 12.2). Note especially the important housekeeping functions of PGE_2, broadly referred to as cytoprotective roles, in which organs such as gastric mucosa, myocardium, and renal parenchyma are shielded from the effects of ischemia by PGE_2-mediated vasodilation and regulation of blood flow. PGE_2 is also involved in inflammatory cell activation, and PGE_2 that is biosynthesized by COX-2 and PGE_2 synthase in endothelial cells at the blood-brain barrier appears to have a role in fever.

Thromboxane and Prostacyclin

Platelets express high levels of the enzyme thromboxane synthase, but do not contain prostacyclin synthase. TxA_2 is the chief eicosanoid product of platelets. TxA_2 has a half-life of only seconds (~10–20 seconds) before it is nonenzymatically hydrolyzed to the inactive form TxB_2. TxA_2, which signals via a 7-transmembrane G-protein–coupled G_q mechanism, is both a strong vasoconstrictor and an agonist of platelet adhesion and aggregation. In contrast, the vascular endothelium lacks thromboxane synthase but expresses prostacyclin synthase. PGI_2 is, therefore, the primary eicosanoid product of the vascular endothelium. PGI_2, which signals via G_s, functions as a vasodilator and inhibitor of platelet aggregation. In other words, PGI_2 is a physiologic antagonist of TxA_2. The vasodilatory actions of PGI_2, as with those of PGE_2, also confer cytoprotective properties (Figure 12.2A and 12.2B).

The local TxA_2 and PGI_2 levels are critical in the regulation of systemic blood pressure and thrombogenesis. Imbalances can lead to hypertension, ischemia, thrombosis, coagulopathy, myocardial infarction, and stroke. In certain populations of the Northern latitudes (including Inuit, Greenlander, Irish, and Danish populations), the incidence of heart disease, stroke, and thromboembolic disorders is less than in Western populations. The diet of these Northern people is richer in whale and fish oils as compared to Western diets and, as a result, it contains relatively smaller amounts of arachidonic acid precursors but relatively

TABLE 12.2. Prostanoid specific enzymes, receptors, and key immune bioactions

Prostaglandin	Synthetic enzyme	Tissues expressing synthetic enzyme	Receptor type and signaling mechanism		Bioactions
PGD_2	PGD_2 isomerase (a) Lipocalin type (b) Hematopoietic	Mast cells Neurons	DP1 DP2	G_s	Bronchoconstriction (asthma), resolution Sleep control functions
PGE_2	PGE_2 isomerase (a) Cytosolic (b) Microsomal 1 (c) Microsomal 2	Many tissues, including macrophages and mast cells	EP1 EP2 EP3 EP4 Other	G_q G_s G_i G_s	Potentiation of responses to painful stimuli Vasodilation Bronchoconstriction Cytoprotective: modulates gastric mucosal acid secretion, mucus, and blood flow Vasodilation Bronchoconstriction Inflammatory cell activation Pyrexia Mucus production
$PGF_{2\alpha}$	$PGF_{2\alpha}$ reductase	Vascular smooth muscle Uterine smooth muscle	FP	G_q	Vascular tone Reproductive physiology (abortifacient) Bronchoconstriction

The prostanoid receptors are G-protein–coupled receptors (see Brink et al. for further details).

larger amounts of EPAs. As a result, the thromboxane–prostacyclin levels tip toward vasodilation, platelet inhibition, and antithrombogenesis. These combined effects and change in balance of TX-PGI represent one possible explanation for the observation that these populations have a lower incidence of cardiovascular diseases and is one rationale for increasing dietary fish consumption in Western diets.

Lipoxygenase Pathways

The LOX pathways (Figure 12.3A and 12.3B) are a major route for converting arachidonic acid to bioactive mediators, namely formation of both leukotrienes and lipoxins. Lipoxygenases are enzymes that catalyze insertion of molecular oxygen into arachidonic acid using nonheme iron generating specific hydroperoxides. Three lipoxygenases, 5-, 12-, and 15-lipoxygenases (5-LOX, etc.), are the major LOX found in humans. The lipoxygenases are named for the position of the inserted molecular O_2 in arachidonic acid. The immediate products of LOX reactions are hydroperoxyeicosatetraenoic acids (HPETEs). HPETEs can be reduced to the corresponding hydroxyeicosatetraenoic acids (HETEs) by enzymes using glutathione peroxidase (GPx). The 5-LOX produces 5-HPETE, the direct precursor to LTA_4, which in turn is the precursor intermediate for the potent bioactive leukotrienes (Figure 12.3). Lipoxygenases are also involved in converting 15-HETE and LTA_4 to lipoxins (see later). 5-LOX requires translocation to the nuclear membrane for activity. The 5-LOX-activating protein (FLAP) helps 5-LOX translocate to the nuclear membrane, form an active enzyme complex, and accept the arachidonic acid substrate.

Leukotriene Biosynthesis

Leukotriene biosynthesis is initiated by the 5-LOX-mediated conversion of arachidonic acid into 5-HPETE and further into leukotriene A_4 (LTA_4). Note that 5-LOX catalyzes both the steps in leukotriene biosynthesis in a concerted manner (Figure 12.3). Depending on the cell type, LTA_4 is next converted to either LTB_4 or LTC_4. The enzyme LTA_4 hydrolase converts LTA_4 to LTB_4 in neutrophils as well as erythrocytes, macrophages, and monocytes. LTA_4 conversion to LTC_4 occurs in mast cells, eosinophils, basophils, and macrophages by the addition of a γ-glutamylcysteinylglycine tripeptide (glutathione), which may be cleaved by peptidases to generate LTD_4 and LTE_4. Together, LTC_4, LTD_4, and LTE_4 represent the cysteinyl leukotrienes (Figure 12.3).

LTB_4 acts via two G-protein–coupled receptors, BLT1 and BLT2. Binding of LTB_4 to BLT1, which is expressed on tissues involved in host-defense and inflammation (leukocytes, thymus, spleen), leads to proinflammatory sequelae, most importantly neutrophil chemotaxis, aggregation, and transmigration across epithelium and endothelium. LTB_4 upregulates neutrophil lysosomal function and free radical production, enhances cytokine production, and potentiates the actions of natural killer cells. The role of the second LTB_4 receptor denoted BLT2 remains unknown.

The cysteinyl leukotrienes (LTC_4, LTD_4, and LTE_4) bind to CysLT1 receptors to cause vasoconstriction,

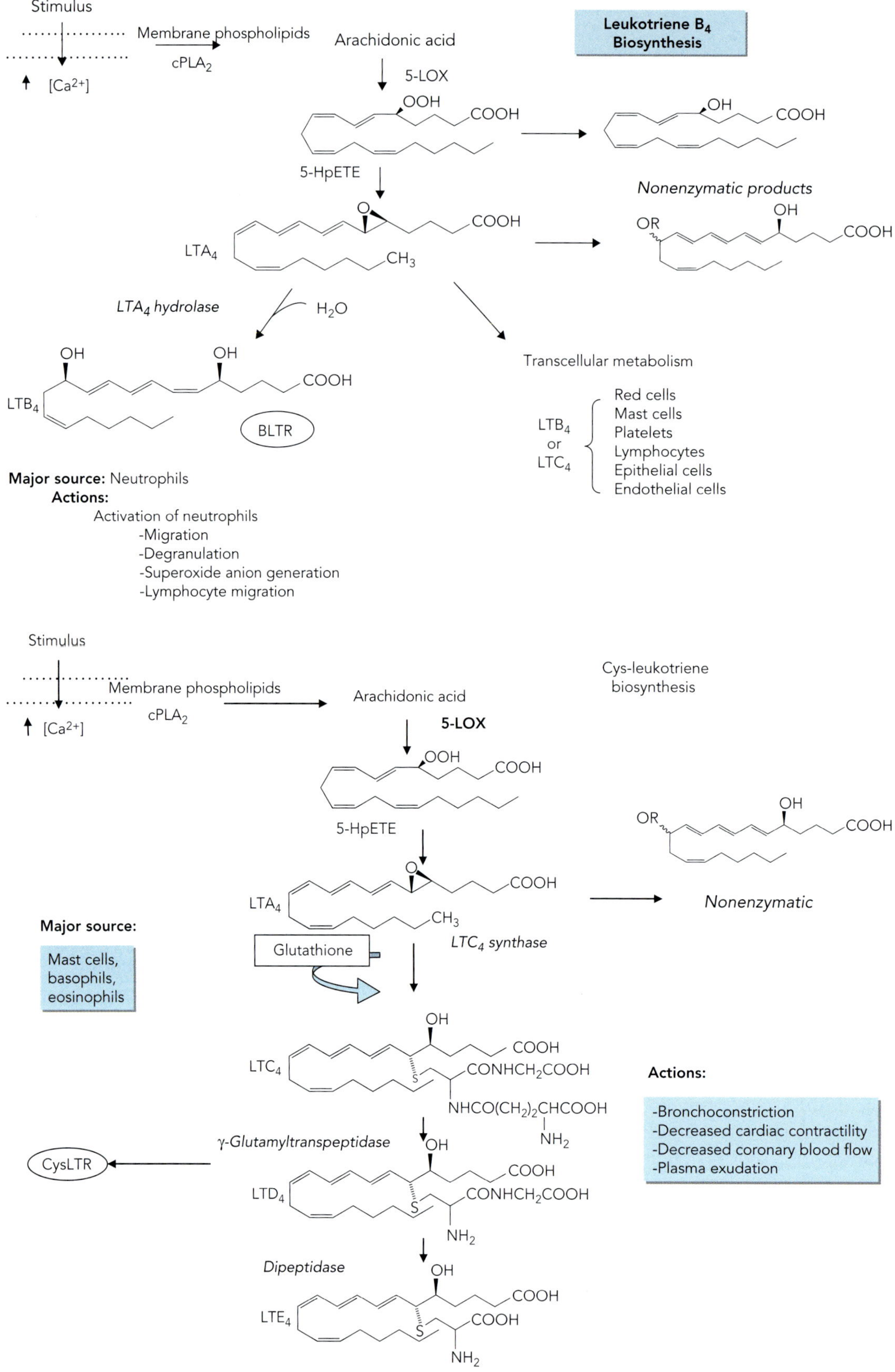

Figure 12.3. Leukotriene biosynthesis and function. The biosynthetic pathways from arachidonic acid to the leukotrienes are shown. (A) Leukotriene B_4 biosynthesis. Zileuton and 5-lipoxygenase-activating protein (FLAP) inhibitors prevent the conversion of arachidonic acid to 5-HPETE and LTA_4; zileuton is used in the chronic management of asthma. BLTR = BLT1 (and BLT2), LTB_4 receptor(s).
(B) Cys-Leukotriene C_4, D_4. Receptors are denoted as CysLTR compromising CysLT1, CysLT2, and GPR17, receptors for cysteinyl-leukotrienes. Zafirlukast and montelukast are antagonists at CysLT1, a receptor for all cysteinyl leukotrienes; these drugs are used in the chronic management of asthma.

bronchospasm, and increased vascular permeability. Cysteinyl leukotrienes are responsible for airway and vascular smooth muscle contraction that occurs in asthma, allergic, and hypersensitivity processes. Leukotrienes are powerful lipid mediators that exert their bioactions already in low nM concentrations, and together, both arms of the leukotriene pathways, that is, LTB_4, and the cysteinyl leukotrienes, are held to play key roles in psoriasis, arthritis, and various inflammatory responses. More recent findings also place them as local mediators in vascular disease as well as atherosclerosis.

Lipoxin Biosynthesis

Lipoxins (LOX interaction products) are derived from arachidonic acid and contain four conjugated double bonds and three hydroxyl groups. The two main lipoxins are LXA_4 and LXB_4 (Figure 12.4). The lipoxins modulate the actions of leukotrienes and cytokines. LX are important in resolution, serving as agonists and counter-regulating mediators.

At sites of inflammation, there is typically an inverse relationship between the amounts of lipoxin and leukotriene present. LXA_4 receptors are present in the lung, spleen, and blood vessels, and on neutrophils. Lipoxins stop neutrophil chemotaxis, adhesion, and transmigration through endothelium (by decreasing P-selectin expression), inhibit eosinophil recruitment, stimulate vasodilation (by inducing synthesis of PGI_2 and PGE_2), inhibit LTC_4- and LTD_4-stimulated vasoconstriction, inhibit LTB_4 inflammatory effects, and inhibit the function of NK cells. Lipoxins are potent (active at similar concentrations as leukotrienes) agonists of resolution mechanisms, which include stimulating the uptake and clearance of apoptotic neutrophils by macrophages. Lipoxin production may, therefore, be important in the resolution of inflammation, and an

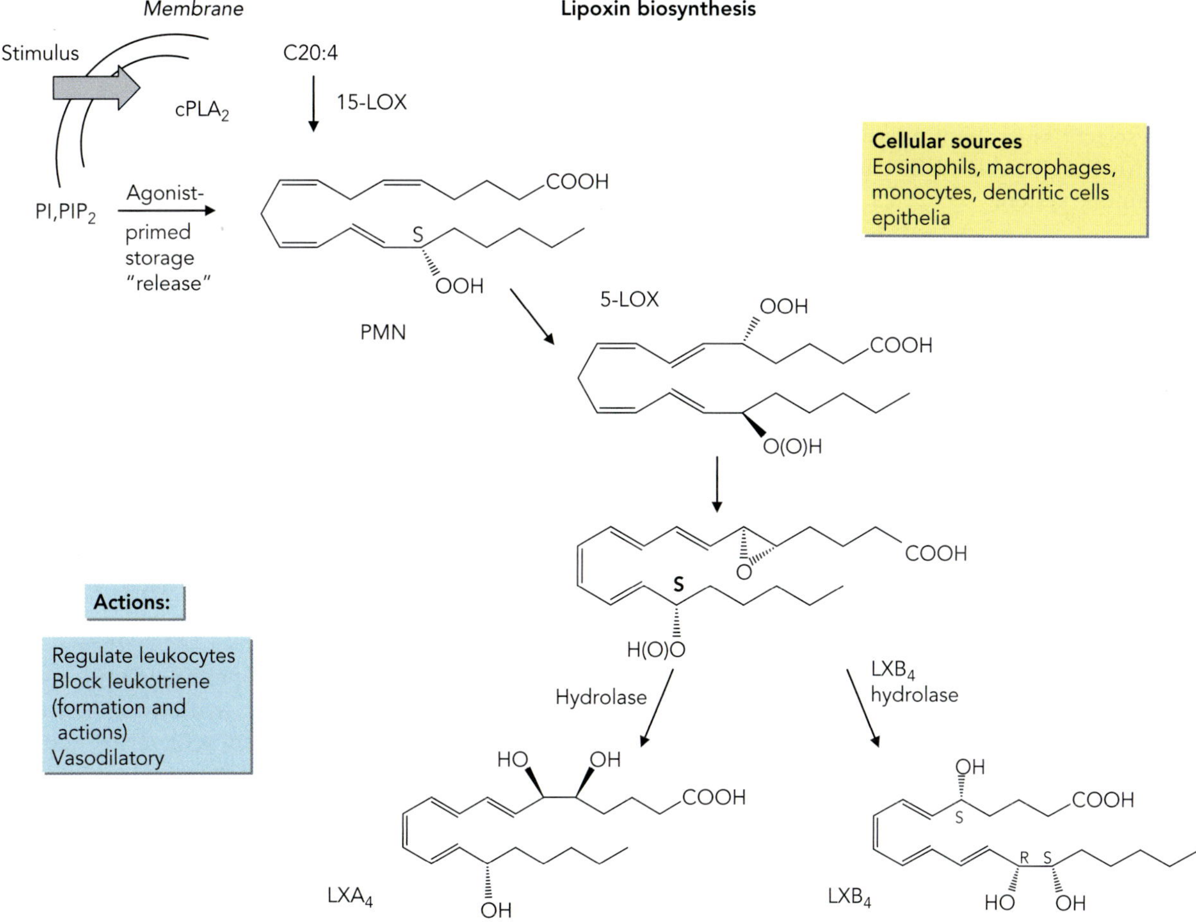

Figure 12.4. Lipoxin biosynthesis. Two main routes lead to biosynthesis of the lipoxins. In each pathway, sequential lipoxygenase reactions are required, followed by hydrolysis. The immediate precursor of the lipoxins is epoxytetraene; hydrolysis of epoxytetraene yields the lipoxins. Left pathway: Arachidonic acid is converted to 15-HETE by sequential activity of 15-LOX and peroxidase. 15-HETE is converted by 5-LOX to the chemical intermediate 5-hydroperoxy, 15-hydroxyeicosatetraenoic acid, and 5-LOX acts on this intermediate to form epoxytetraene. LTA_4 route: Arachidonic acid is converted to 5-HPETE by 5-LOX, and 5-HPETE is converted to LTA_4 by further action of 5-LOX. LTA_4 is converted to epoxytetraene by 15-LOX (see text for further details).

imbalance in lipoxin–leukotriene homeostasis may be a key factor in the pathogenesis of inflammatory disease.

Epoxygenase Products

Microsomal cytochrome P450 epoxygenases are able to oxygenate arachidonic acid, resulting in the formation of epoxyeicosatetraenoic acid (EET) and hydroxyacid derivatives (Figure 12.1). These are major pathways in tissues that do not express COX or LOX, for example, certain cells of the kidney. The epoxygenation of arachidonic acid results in four different EETs, depending on which double bond in arachidonic acid is modified. Dihydroxy derivatives of EETs, formed by hydrolysis, may have roles in the regulation of smooth muscle cells and vascular tone by inhibiting the Na^+/K^+-ATPase, as well as in renal function by regulating ion absorption and secretion. With respect to inflammation, dihydroxy EETs inhibit platelet COX and expression of intercellular adhesion molecules (ICAMs). Downregulation of ICAMs inhibits platelet and inflammatory cell aggregation. Thus, specific EETs (i.e., 11,12-EET at micromolar levels) may play a role in regulating inflammation at certain sites and within select tissues. Future research may reveal more definitive functions for EETs in inflammatory cells and immune system.

Isoprostanes: Markers of Oxidative Stress

Phospholipid-esterified arachidonic acid is susceptible to free-radical–mediated peroxidation; cleavage of these modified lipids from the phospholipid by phospholipase A_2 gives rise to the isoprostanes (Figure 12.1). During oxidative stress, isoprostanes are found in the blood at levels much higher than those of COX products. Two isoprostanes in particular, 8-epi-$PGF_2\alpha$ and 8-epi-PGE_2, are potent vasoconstrictors. Isoprostanes may function in activating NF-κB, phospholipase Cγ, protein kinase C, and calcium flux. Because the formation of isoprostanes depends on cellular oxidation conditions, isoprostane levels may be indicative of oxidative stress and a wide range of pathologies. Urinary isoprostane levels are used as a marker of oxidative stress in ischemic syndromes, reperfusion injury, atherosclerosis, and hepatic diseases. Isoprostanes are not, however, known to have direct roles in host defense.

Metabolic Inactivation of Eicosanoids

Following their local actions, the prostaglandins, leukotrienes, thromboxanes, and lipoxins are inactivated locally by hydroxylation, β-oxidation (resulting in a loss of two carbons), or ω-oxidation (to dicarboxylic acid derivatives). These degradation processes render the molecules more hydrophilic and, therefore, excretable in the urine.

TABLE 12.3. Pathophysiology: human diseases with dysregulated eicosanoids

Asthma	Atherosclerosis
Inflammatory bowel disease	Cardiovascular diseases
Arthritis	Skin disorders
Glomerulonephritis	Reperfusion organ injury
Alzheimer's disease	Periodontal disease
Cancer	

Pathophysiology

The immune response and subsequent inflammation are the body's mechanisms for combating foreign invaders. This overall response is designed to remove the inciting stimulus and resolve tissue damage. In some cases, the response mechanism itself causes local tissue damage, such as when activated neutrophils inadvertently release proteases and reactive oxygen species (ROS) to the local milieu. In other settings, if the inflammatory reactions persist for too long, each counter-regulation, or the immune system misidentifies a part of self as foreign; these abnormal, misdirected immune responses can cause significant and chronic tissue injury. Eicosanoids are implicated in selected inflammatory diseases, including asthma, inflammatory bowel disease, rheumatoid arthritis, atherosclerosis, and glomerulonephritis (see Table 12.3).

MODULATORS AND INHIBITORS

Given the extensive and diverse roles of the many eicosanoids, the complexity of the pathways offers a variety of targets for controlling inflammation and aberrant immune responses. Strategies considered include altering the expression of key enzymes, competitively and noncompetitively inhibiting the activity of specific enzymes (i.e., PGE_2 synthases), activating receptors with exogenous receptor agonists, and preventing receptor activation with exogenous receptor antagonists. As with all other approaches, the therapeutic benefits must be weighed against the potential unwanted side effects.

Phospholipase Inhibitors: Inhibition of phospholipase A_2 prevents the generation of arachidonic acid, the rate-limiting step in eicosanoid biosynthesis. In the absence of proinflammatory mediators derived from arachidonic acid, inflammation is limited. Glucocorticoids (also known as corticosteroids, of which prednisone is a member) are a mainstay of therapy in a multitude of autoimmune and inflammatory diseases. Glucocorticoids induce a family of secreted calcium- and phospholipid-dependent proteins called lipocortins. Lipocortin interferes with the action of phospholipase A_2, and thus limits the release of arachidonic acid. Annexins, such as Annexin 1- and Annexin 1–derived peptides, are also induced by glucocorticoids that, in turn, act at GPCR on leukocytes to block proinflammatory responses and enhance endogenous anti-inflammatory mechanisms including activation of the lipoxin A_4 receptor. Small molecule inhibitors of specific phospholipases are in development; they may offer the potential for reducing some adverse effects associated with glucocorticoid use.

Cyclooxygenase Inhibitors: A number of COX pathway inhibitors are currently in wide use. These are some of the most frequently prescribed over-the-counter drugs. NSAIDs are important because of their combined anti-inflammatory, antipyretic, and analgesic properties. The ultimate goal of most NSAID therapies is to inhibit COX generation of proinflammatory eicosanoids and to limit the extent of inflammation, fever, and pain. The drugs' antipyretic activity is likely related to their decreasing the levels of PGE_2, particularly in the region of the brain surrounding the hypothalamus. Despite the benefits of current NSAIDs, these drugs only suppress the signs of the underlying inflammatory response.

A multitude of NSAIDs have been developed over the past century. All NSAIDs, except aspirin, act as reversible, competitive inhibitors of COX. Generally, these drugs block the hydrophobic channel in COX in which the substrate arachidonic acid binds, thereby preventing access of arachidonic acid to the active site of the enzyme. Traditional NSAIDs inhibit both COX-1 and COX-2 to different degrees. Because of inhibition of COX-1, long-term NSAID therapy has many deleterious effects. The cytoprotective roles of the COX-1 eicosanoid products are eliminated, leading to a spectrum of NSAID-induced gastropathy including dyspepsia, gastrotoxicity, subepithelial damage and hemorrhage, gastric mucosal erosion, frank ulceration, and gastric mucosal necrosis. Regulation of blood flow to the kidney is likewise perturbed, potentially causing renal failure.

Aspirin is the first and the oldest of the NSAIDs and is widely used to treat mild to moderate pain, headache, myalgia, and arthralgia (see Figure 12.5). In contrast to other NSAIDs, aspirin acts in an irreversible manner by acetylating the active site serine residue in both COX-1 and COX-2. Acetylation of COX-1 destroys the enzyme's COX activity and thereby prevents the formation of COX-1–derived prostaglandins, thromboxanes, and prostacyclin.

Daily low-dose aspirin is used as an antithrombotic agent for prophylaxis and postevent management of myocardial infarction and stroke. Recall that aspirin is antithrombotic because of its irreversible inhibition of COX that prevents platelets from biosynthesizing TxA_2. Within an hour of oral aspirin administration, the COX-1 activity in existing platelets is irreversibly destroyed (*vide supra*). The importance of this lies in the fact that platelets circulate in the bloodstream for ~10 days but cannot synthesize new protein. Therefore, irreversibly acetylated COX-1 is not replaced by freshly synthesized proteins. These platelets are irreversibly inhibited for their lifetime. Although aspirin also irreversibly inhibits vascular endothelial cell COX-1, the endothelial cells can synthesize new COX-1 protein, as well as possess constitutively expressed COX-2, and thus can rapidly resume synthesis of PGI_2. A single administration of aspirin decreases the amount of thromboxane that can be generated for several days, shifting the vascular local TxA_2–PGI_2 levels toward PGI_2-mediated vasodilation, platelet inhibition, and antithrombogenesis.

Aspirin-mediated inhibition of COX-2 prevents the generation of prostaglandins. Unlike COX-1 which is totally inactivated, aspirin-modified COX-2 retains a part of its catalytic activity that forms a new product, 15*R*-HETE, from arachidonic acid. By analogy to lipoxin biosynthesis (Figure 12.4), 5-LOX then converts 15*R*-HETE to 15-epi-lipoxins, which are stable epimers (carbon 15 position epimers) of native lipoxins. These epimers are collectively called aspirin-triggered lipoxins (ATLs). 15-Epi-lipoxins mimic the functions of lipoxins as anti-inflammatory and proresolving mediators. 15-Epi-lipoxins or 15*R*-epi-lipoxins may represent an endogenous mechanism of anti-inflammation, and their production is believed to mediate at least part of the anti-inflammatory effects of aspirin. Development of 15-epi-lipoxin analogues could lead to new anti-inflammatory drugs that do not have the unwanted side effects associated with COX-1 inhibition. There are several additional classes of NSAIDs that include ibuprofen, naproxen, ketoprofen, and flurbiprofen. The acetic acid NSAIDs include indole acetic acids – indomethacin, sulindac, and etodolac – and the phenylacetic acids diclofenac and ketorolac (a substituted phenylacetic acid derivative).

Acetaminophen, although sometimes classified along with NSAIDs, is by definition not an NSAID. Although acetaminophen has analgesic and antipyretic effects similar to aspirin, the anti-inflammatory effect of acetaminophen is insignificant because of its weak inhibition of cyclooxygenases. Acetaminophen therapy can, however, be valuable in patients, for example children, who are sensitive to the side effects of aspirin. The key side effect of acetaminophen is hepatotoxicity. Modification of acetaminophen by hepatic cytochrome P450 enzymes produces a highly reactive molecule that is detoxified by conjugation with

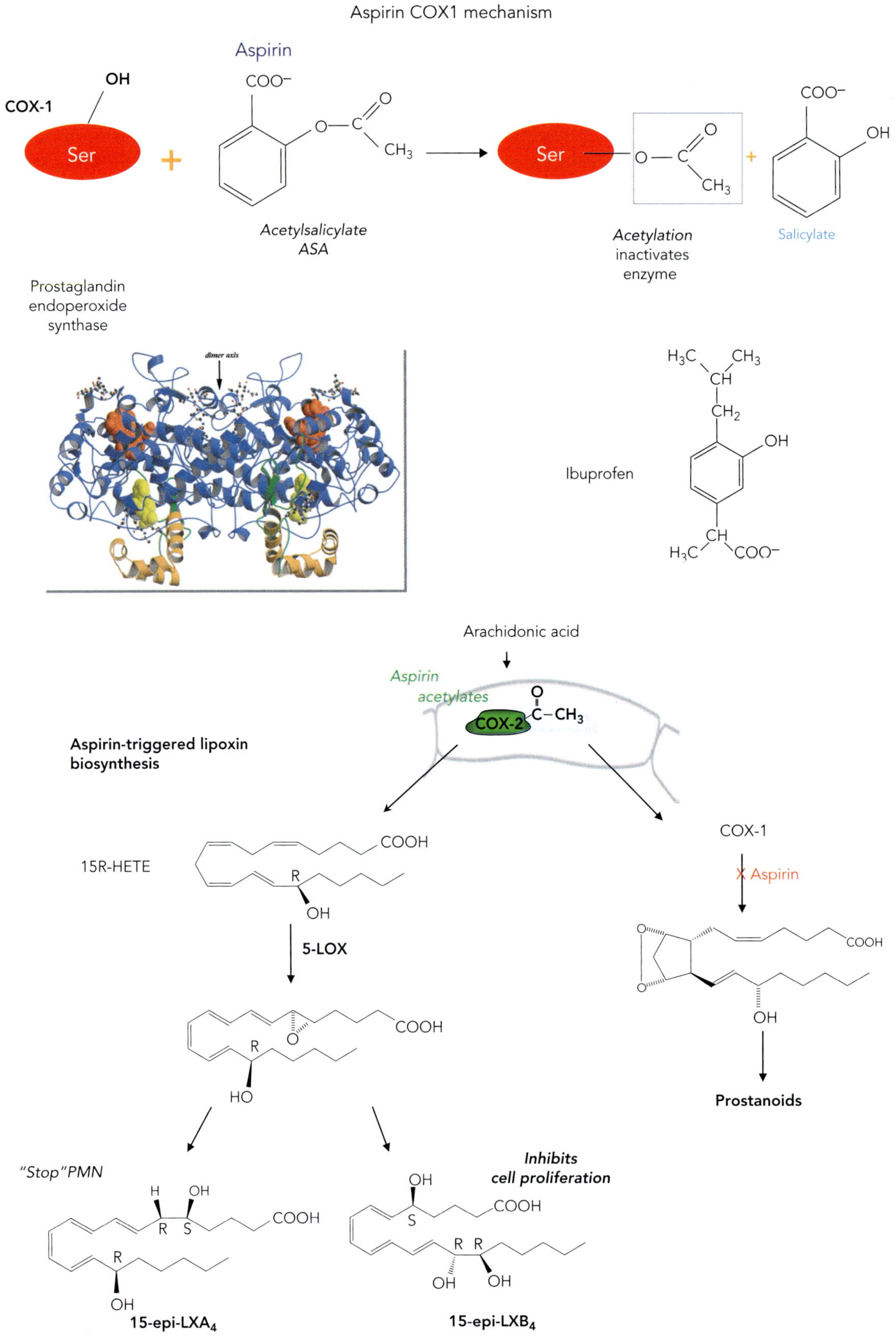
Aspirin COX1 mechanism
Aspirin
COX-1
OH
Ser
COO−
O—C
O
CH3
Acetylsalicylate
ASA
Ser
O—C
O
CH3
Acetylation
inactivates
enzyme
COO−
OH
Salicylate
Prostaglandin
endoperoxide
synthase
dimer axis
Ibuprofen
H3C
CH3
CH
CH2
OH
CH
H3C
COO−
Arachidonic acid
Aspirin
acetylates
COX-2
C–CH3
O
Aspirin-triggered lipoxin
biosynthesis
15R-HETE
COOH
R
OH
5-LOX
COOH
R
HO
O
COX-1
X Aspirin
COOH
OH
Prostanoids
"Stop"PMN
H
OH
R
S
COOH
R
OH
15-epi-LXA4
Inhibits
cell proliferation
OH
S
COOH
R
R
OH
OH
15-epi-LXB4

Figure 12.5. Aspirin mechanism of action. (A) Aspirin mechanism with COX-1: inhibition of prostanoid intermediates. (B) Aspirin and the biosynthesis of the aspirin-triggered Lipoxins (ATL) via COX-2.

glutathione. An overdose of acetaminophen can overwhelm glutathione stores. This can lead to cellular and oxidative damage and, in some settings, to acute hepatic necrosis.

The anti-inflammatory, analgesic, and antipyretic effects of the NSAIDs do vary somewhat among the many drugs in this group. Hence, successful NSAID therapy is still considered more of an art than a science for each patient, and therapy should be directed toward achieving the desired anti-inflammatory, analgesic, and antipyretic effects while minimizing the unwanted side effects from relative inhibition of COX-1 and COX-2.

COX-2 Selective Inhibitors: COX-2 was identified in the early 1990s, and intense research swiftly led to the development of COX-2 selective inhibitors for clinical use. Compared with COX-1, COX-2 has a larger hydrophobic channel through which substrate enters the active site. Subtle structural differences between COX-2 and COX-1 topography allowed development of drugs that act preferentially on COX-2 rather than COX-1. The COX-2 selective inhibitors including celecoxib, rofecoxib, valdecoxib, and meloxicam are sulfonic acid derivatives that display 100-fold selectivity for COX-2 compared to COX-l. The relative inhibition of the two COX isozymes by these drugs in any given tissue is also a function of drug metabolism, pharmacokinetics, and possibly enzyme polymorphisms. The COX-2 selective inhibitors have similar anti-inflammatory, antipyretic, and analgesic properties as the traditional NSAIDs, but they do not share the antiplatelet actions of the COX-1 inhibitors.

Currently, it is not clear if they are less safe than other NSAIDs, and only celecoxib remains an approved drug. The long-term safety profiles of COX-2 inhibitors are in question. There is concern that COX-2 inhibitors – in particular, rofecoxib, a widely marketed COX-2 inhibitor – have deleterious effects on the cardiovascular and renal systems by inducing hypertension, renal failure, and cardiac failure. Prolonged inhibition of vascular COX-2 within endothelial cells reducing PGI_2 formation may account for the increased thrombogenesis uncovered in clinical trials. In addition, inhibition of COX-2 may generate potential problems in resolution, wound healing, and angiogenesis. Also, COX-2 selective inhibitors are much more expensive than equivalent doses of many NSAIDs, especially aspirin.

Lipoxygenase 5-LOX Inhibitors: Selective inhibition of the 5-LOX has the potential to represent a major therapeutic modality in diseases involving leukotriene-mediated pathophysiology, including asthma, inflammatory bowel disease, and possibly atherosclerosis and rheumatoid arthritis. Drugs that impair or alter the ability of LOX to utilize its non-heme iron properly would be expected to inhibit the activity of the enzyme. The only iron chelator drug in clinical use is zileuton, a benzothiophene derivative of N-hydroxyurea that inhibits 5-LOX. In asthma, zileuton induces bronchodilation, improves symptoms, and generates long-lasting improvement in pulmonary function tests. Interfering with the role of FLAP could represent an approach to the selective inhibition of 5-LOX activity and leukotriene function. Recall that 5-LOX is activated after the enzyme localizes to the nuclear membrane and docks with FLAP; FLAP also binds arachidonic acid released by phospholipase A_2 and shuttles it to the 5-LOX active site. FLAP inhibitors that both prevent and reverse LOX binding to FLAP are in clinical development.

CysLT Receptor Antagonists: Leukotriene receptor antagonism represents the receptor-based mechanism for inhibiting leukotriene-mediated bronchoconstriction and smooth muscle actions. Cysteinyl leukotriene receptor (CysLT1) antagonists are effective against antigen-, exercise-, cold-, and aspirin-induced asthma. These agents significantly improve bronchial tone, pulmonary function tests, and asthma symptoms. Montelukast and zafirlukast are the currently available cysteinyl leukotriene receptor antagonists; the main clinical application is in asthma treatment.

Crystal Structures in the Eicosanoid Cascade – Basis for Rational Drug Design

Over the past decade, crystal structures have been determined for several important enzymes in the eicosanoid cascade. In many instances, these structures have been used to perform rational, structure-based drug design, a process in which lead compounds are developed and optimized through repeated cycles of structure determinations of enzyme–inhibitor complexes and chemical refinement.

Starting from the top of the cascade where arachidonic acid is released from membrane phospholipids by the high-molecular weight cPLA2, a "master" enzyme which funnels substrate to both the COX and LOX pathways. The structure of cPLA2 has been solved and revealed an N-terminal calcium-dependent lipid-binding/C2 domain and a catalytic unit with a Ser-Asp dyad. Interestingly, there is a lid that must move to allow substrate access to the active site, explaining the interfacial activation of the enzyme.

As mentioned, crystal structures of the membrane associated COX-1 and COX-2 are available and have been instrumental for development of the second generation COX-2 inhibitors that exploit the small structural differences of a side pocket in the two isoenzymes. In addition, several downstream prostaglandin synthases have been structurally characterized, including PGD, PGI, and PGF synthases, as well as cytosolic and microsomal PGE synthase type 2.

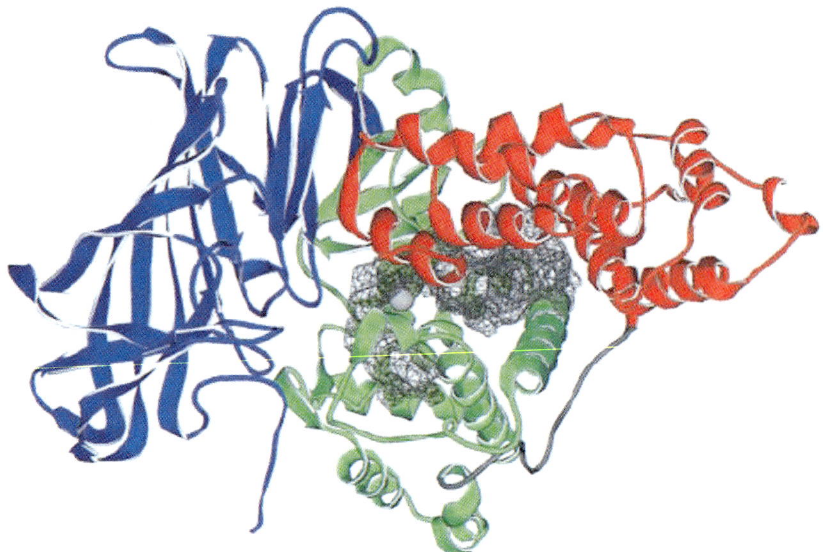

Figure 12.6. Overall structure of LTA_4 hydrolase. The N-terminal, catalytic, and C-terminal domains are colored in blue, green, and red, respectively. The active site is indicated as a gray volume in the central parts of the protein. Figure created by Fredrik Tholander.

To date, only a single mammalian LOX has been structurally characterized, *viz.* rabbit 15-LOX. This enzyme has an N-terminal β-barrel domain resembling the C-terminal domain of certain lipases, and a catalytic domain with an active site pocket starting at the protein surface that puts the fatty acid substrate in an optimal position relative to the catalytic iron. This structure has been used to create model structures of 5-LOX. However, the inherent limitations of these models preclude their use in rational drug design.

Continuing along the 5-LOX/leukotriene pathway, the soluble, 70 kDa, LTA_4 hydrolase has also been crystallized. The protein is folded into three domains and the L-shaped active site goes all the way from the surface deep into the protein (Figure 12.6). The structure reveals the molecular background of the enzyme's ability to convert LTA_4 into LTB_4 and also to cleave peptide bonds. Moreover, the 3-D structure offers a chemical terrain that is almost ideal for structure-based drug design.

The other branch of the leukotriene pathway leading to the cysteinyl-leukotrienes is governed by LTC_4 synthase, an integral membrane protein. This enzyme is a member of a larger protein family denoted MAPEG, which also includes FLAP (five-lipoxygenase activating protein) and mPGES-1 (microsomal PGE synthase type 1), believed to synthesize proinflammatory PGE_2. Recently, high-resolution crystal structures of LTC_4 synthase were presented revealing a trimeric enzyme, with a GSH binding site composed of residues from two neighboring monomers and a hydrophobic, superficial cleft, presumably accommodating the substrate LTA_4 (Figure 12.7). These structures will certainly enable the molecular characterization of the catalytic machineries as well as the design of potent and specific inhibitors of LTC_4 synthase, a hitherto unexploited drug target. Furthermore, crystal structures of FLAP with bound inhibitors have been presented, demonstrating the protein sites at which this class of antileukotrienes interact. Also, another 5-LOX supporting protein, the 16 kDa CLP (coactosin-like protein), was structurally determined by both NMR and crystallography. The next step in the structure biology of the eicosanoid cascade will be the structure determinations of eicosanoid receptors, a huge challenge that requires enormous human and financial efforts. If successful, it will certainly have a profound impact on drug development.

PLATELET-ACTIVATING FACTOR

Biosynthesis and Actions

In the early 1970s, the name *platelet-activating factor* (PAF) was coined to describe a bioactivity or substance(s) that, when released from basophils during IgE-induced anaphylaxis, activated platelets. This bioactive substance was subsequently isolated and its structure elucidated as a group of molecules characterized by the structure 1-*O*-alkyl-2-acyl-*sn*-glycero-3-phosphocholine (Figure 12.8). Individual molecules of PAF differ only within the number of carbons in the alkyl group. Like the eicosanoids, PAFs are derived from membrane phospholipids by phospholipase A_2. Upon cellular activation, phospholipase A_2 also affects the deacylation of 1-*O*-alkyl-2-acyl-glycero-3-phosphocholine yielding lyso-PAF plus unesterifed fatty acid. Lyso-PAF can then be acetylated by

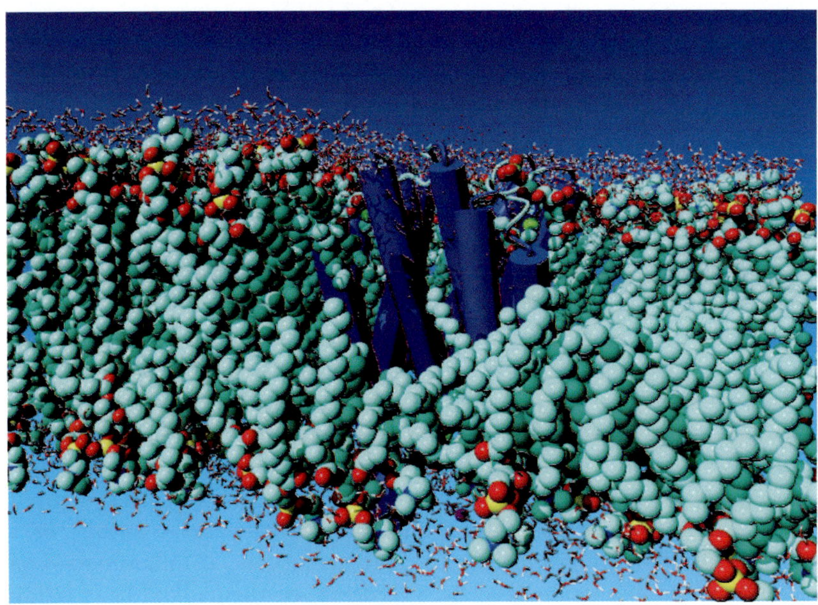

Figure 12.7. Structure of the integral membrane protein LTC_4 synthase. Transmembrane α-helices are shown as dark blue cylinders. The architecture of the lipid bilayer is computer designed. Figure created by Fredrik Tholander.

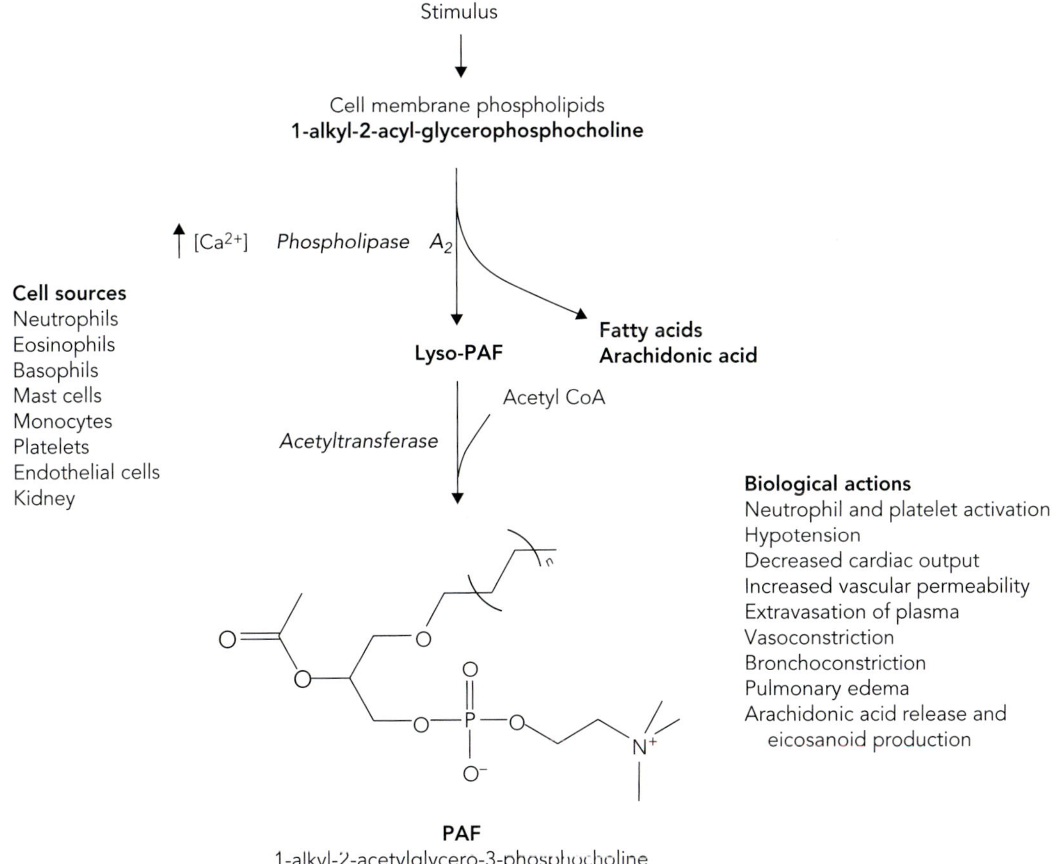

Figure 12.8. Biosynthesis of PAF and its major actions.

acetyl-coenzyme A (acetyl-CoA) transferase to give PAF. As much as 40% of the fatty acid released from the sn-2 position during PAF biosynthesis by human neutrophils is arachidonic acid. Hence, PAF biosynthesis by activated inflammatory cells is often accompanied by the generation of eicosanoids.

A number of different cell types generate PAF, including leukocytes (neutrophils, basophils, and eosinophils), macrophages, and mast cells, as well as platelets. Of interest, PAF biosynthesis can also occur in some noninflammatory and resident cells, such as endothelial and epithelial cells, suggesting that PAF may serve a physiologic function in addition to its role as a mediator of inflammation. A wide range of stimuli can activate PAF biosynthesis. For example, PAF biosynthesis by neutrophils is stimulated by phagocytosis of serum-treated zymosan, C5a fraction of complement and formyl-methionyl-leucyl-phenylalanine (a synthetic peptide that mimics the action of the formylated peptides contained in bacterial cell walls), as well as the divalent cation ionophore A23187. These stimuli increase the levels of intracellular calcium and lead to activation of phospholipase A_2. The stimuli for PAF biosynthesis are cell-type specific. For example, thrombin, angiotensin II, and vasopressin do not trigger the generation of PAF in inflammatory cells, but stimulate PAF formation in vascular endothelial cells.

PAF Actions *in vitro* and *in vivo*: As a mediator, PAF is implicated in the pathogenesis of both acute inflammation and hypersensitivity disorders. PAF is also likely to be an important mediator in several other conditions, including endotoxic shock, vasculitis, and arterial thrombosis. As the name indicates PAF is a potent activator of platelets (activation and aggregation) and of human neutrophils. PAF stimulates neutrophil adhesion, lysosomal enzyme release, the generation of ROS and eicosanoids. PAF causes rapid margination of neutrophils onto endothelial cell walls *in vivo*, and promotes neutrophil migration to the extracellular space. PAF may also be a regulator of lymphocyte function, either directly or via indirect routes that can involve the generation of specific eicosanoids, for example, prostaglandins and leukotrienes. PAF has been shown to inhibit lymphocyte proliferation in response to various mitogens. In turn, lymphocytes can generate PAF upon stimulation, particularly by agonists that increase intracellular calcium.

Local administration of PAF by intradermal injection in laboratory animals or human subjects is associated with neutrophil margination and intravascular thrombosis. There is an associated increase in vascular permeability, edema, and in hyperalgesia. Hence, PAF causes many of the cardinal features of inflammation.

Intravenous injection of PAF has profound effects on bronchial smooth muscle tone and the cardiovascular function of laboratory animals. Intravenous PAF causes a syndrome that is very similar to that of acute anaphylaxis (i.e., severe acute hypersensitivity reaction). PAF induces contraction of bronchial and vascular smooth muscle, including coronary artery smooth muscle. In addition, there is a marked increase in vascular permeability and exudation of plasma into the extravascular space, including the lungs (pulmonary edema). As a result, the animals develop respiratory insufficiency, decreased cardiac output, and hypotension.

Integrated Inflammation Schema: Lipid Mediators in Resolution

As described earlier, eicosanoids are generated locally in numerous complex reactions. It is not necessary to remember every mediator, but rather to understand the general scheme of these synthetic pathways. This section, along with Table 12.4, provides a concise overview of the functions of eicosanoids relevant to host defense, inflammation, and resolution of inflammation.

Acute inflammation is the result of an intricate network of molecular and cellular interactions induced by responses to a variety of stimuli, such as trauma, ischemia, infectious agents, or antibody reactions. Acute superficial inflammation generates local pain, edema, erythema, and heat; inflammation in visceral organs can have similar symptoms and result in severe impairment of organ function.

Leukotrienes and lipoxins, as well as thromboxanes, prostaglandins, and prostacyclins, are critical for generating, maintaining, and mediating inflammatory responses. The inflammatory cascade is initiated when cells in a particular region are exposed to a foreign substance or are damaged. That insult stimulates a local cytokine cascade (including interleukins or TNF) that raises the expression of COX-2 mRNA and enzyme levels. COX-2 then facilitates production of the proinflammatory and vasoactive eicosanoids (see additional illustrations in Chapter 2).

TABLE 12.4. Roles of lipid mediators in the steps of inflammation

Action	Lipid mediators involved
Vasoconstriction	$PGF_2\alpha$, TxA_2, LTC_4, LTD_4, LTE_4
Vasodilation (erythema)	PGI_2, PGE_1, PGE_2, PGD_2, LXA_4, LXB_4, LTB_4
Edema (swelling)	PGE_2, LTB_4, LTC_4, LTD_4, LTE_4
Chemotaxis, leukocyte adhesion	LTB_4, HETE, LXA_4, LXB_4
Increased vascular permeability	LTC_4, LTD_4, LTE_4
Pain and hyperalgesia	PGE_2, PGI_2, LTB_4
Local heat and systemic fever	PGE_2, PGI_2, LXA_4
Eicosanoid class switching	PGE_2, PGD_2
Stimulate resolution	Lipoxins, resolvins, and protectins (see Table 12.6 for specific steps and actions in resolution)

Locally high concentrations of PGE_2, LTB_4, and cysteinyl leukotrienes promote the accumulation and infiltration of inflammatory cells by increasing blood flow and vascular permeability. LTB_4 and 5-HETE are also important in attracting and activating neutrophils. LTB_4, formed by activated neutrophils at the site of inflammation, recruits, and activates additional neutrophils and lymphocytes so that these cells adhere to endothelial surfaces and transmigrate into the interstitial spaces. Increased vascular permeability also results in fluid leakage and cellular infiltration, causing edema.

The aggregation of a multitude of inflammatory cells initiates the cell–cell interactions (as in pus or exudates) that foster transcellular biosynthetic routes that are exploited to generate eicosanoids (Figure 12.9). In transcellular biosynthesis, eicosanoid intermediates are donated from one cell type to another to generate a greater diversity of eicosanoids. This demonstrates the importance of cellular adhesion and cell–cell interaction in inflammatory and immune responses. The body tries to ensure that the inflammatory response cannot proceed unchecked.

Lipoxins help resolve inflammation and promote the return of the tissue to homeostasis. COX-2-derived eicosanoids may also function in wound healing and resolution. Hence, the temporal sequence of events is important in an organized inflammatory response (see Chapter 10). PGE_2 inhibits the functions of B and T lymphocytes and NK cells, while LTB_4 and the cysteinyl leukotrienes regulate T lymphocyte proliferation. PGE_2 and PGI_2 are potent sensitizers in pain and turn on lipoxin production that in turn reduces nociception. These many local factors can coordinate mediating and regulating the transition from acute to chronic inflammation versus the temporal events to resolution and homeostasis.

Lipoxins and ATLs offer the potential to counter-regulate/antagonize the inflammatory actions of leukotrienes and other inflammatory mediators, and as agonists to promote resolution of inflammation. Analogs of these compounds represent new approaches to treatment because they are agonists of endogenous anti-inflammation and proresolving mechanisms rather than direct inhibitors of biochemical pathways or receptor antagonists. Because lipoxins are endogenous regulators, they are expected to have selective actions with few adverse effects. Stable analogues of lipoxins and ATLs are currently being developed, and second-generation LX stable analogues have shown

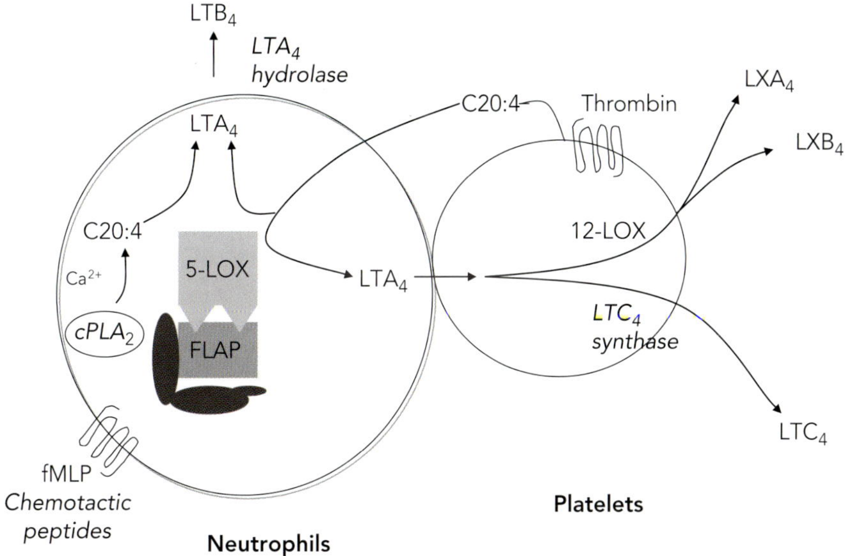

Figure 12.9. Transcellular biosynthesis is used to locally generate lipoxins and cysteinyl leukotrienes. In this example, the leukocyte (neutrophil) obtains arachidonic acid (AA) internally via activation of $cPLA_2$ and also externally from platelets, and uses this AA to biosynthesize LTA_4 and LTB_4. LTA_4 is transferred from the leukocyte to platelets and endothelial cells, which synthesize and secrete LTC_4. Platelets also biosynthesize lipoxins (LXA_4, LXB_4) from LTA_4. Note that the eicosanoids biosynthesized within each cell type are determined by the enzymatic repertoire of that cell type: for example, neutrophils biosynthesize primarily LTA_4 and LTB_4 because they express 5-LOX and LTA_4 hydrolase, whereas platelets biosynthesize LTC_4 and lipoxins because they express LTC_4 synthase and 12-LOX.

efficacy in enhancing the resolution of recurring bouts of acute inflammation in skin inflammation and gastrointestinal inflammation models. This new approach to the treatment of inflammation remains to be established in human trials.

SPECIALIZED LIPID MEDIATORS IN RESOLUTION

Resolvins and Protectins: New Families of Mediators in Resolution

The literature reports that essential omega-3 PUFAs given in high doses (milligrams to grams daily) have beneficial actions in many inflammatory diseases, cancer, and human health in general. The molecular basis of omega-3 fatty acid action in inflammation was not established until recently. To identify potential mechanisms that are actively involved in resolution of inflammation, a systems approach was taken using a new lipid mediator lipidomics and informatics approach, with liquid chromatography–ultraviolet–tandem mass spectrometry-based analyses, to map and profile the appearance and/or loss of mediators in resolving inflammatory exudates. When novel bioactive compounds were encountered, their structures were elucidated and their bioactivity and role(s) confirmed *in vivo*. These studies uncovered two new families of bioactive mediators, termed resolvins and protectins that are biosynthesized from omega-3 essential PUFAs.

Resolvins: The first resolvin was identified in exudates collected from inflamed murine dorsal air pouches in the spontaneous resolution phase and so-named because it proved to be a potent regulator of resolution. Resolvins are derived from EPA and DHA with two chemically unique structural forms, the E-series and D-series of resolvins, respectively. E-series member resolvin E1 reduces inflammation *in vivo* and blocks human neutrophil transendothelial migration. Resolvin E1 can be produced *in vitro*, recapitulating events *in vivo*, by treating human vascular endothelial cells in a hypoxic environment with aspirin. These cells convert EPA to 18*R*-hydroperoxyeicosapentaenoic acid (18*R*-HPEPE) and release 18*R*-hydroxyeicosapentaenoic acid (18*R*-HEPE) that is rapidly transformed by activated human neutrophil 5-LOX. Resolvin E1 is produced in healthy individuals and is increased in plasma of individuals taking aspirin and/or EPA (Figure 12.10).

Resolvin E2 is the second member of the E-series that reduces zymosan-initiated neutrophil infiltration, thereby displaying potent anti-inflammatory actions. These EPA-derived E-series resolvins may contribute to beneficial actions attributed earlier to omega-3 PUFA in human diseases, such as skin inflammation, peritonitis, periodontal disease, and colitis. These results illustrate that 5-LOX in human leukocytes is pivotal for these beneficial effects and is controlled in part by temporal and spatial events *in vivo* to signal production of either leukotrienes or anti-inflammatory mediators, such as lipoxins and resolvins. Of interest, microbial and mammalian cytochrome P450 enzymes convert EPA into 18-HEPE, which can be transformed by human neutrophils to resolvin E1 and resolvin E2. Hence, it is likely that microorganisms at inflamed sites or in the gastrointestinal tract can contribute to production of E-series resolvins in humans.

At least two receptors involved in the actions of resolvin E1 are defined (Figure 12.3B). The GPCR

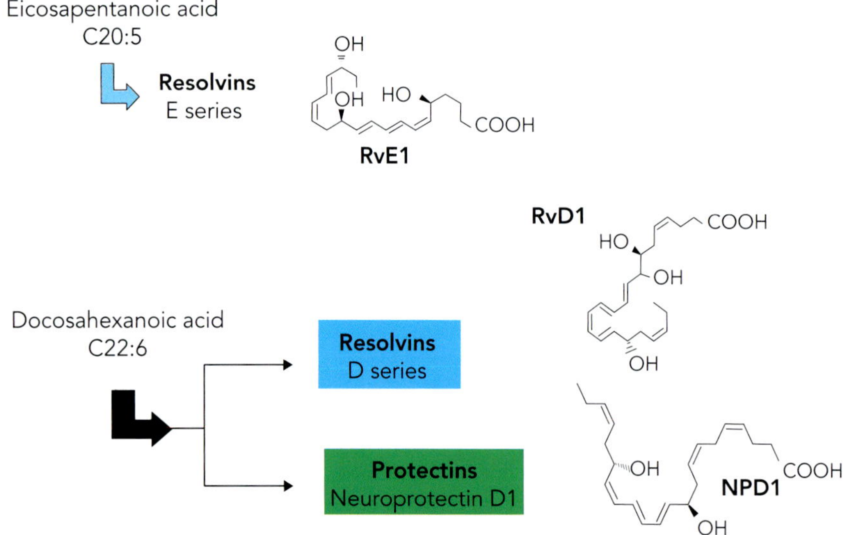

Figure 12.10. Resolvins and protectins: biosynthesis from omega-3 EPA and DHA precursors.

chemokine-like receptor 1 (CMKLR1, a.k.a. ChemR23) attenuates TNF-stimulated nuclear factor-κB activation in response to resolvin E1 binding, indicating a counter-regulatory action of this ligand–receptor pair. Counter-regulation of TNF signaling was used to identify this GPCR because TNF is a key mediator in the early steps of acute inflammation. Recently, a second GPCR that interacts with resolvin E1 was identified, the leukotriene B_4 receptor, denoted BLT1. Resolvin E1 interacts in a stereospecific manner with BLT1 on human neutrophils as a receptor antagonist, attenuating leukotriene-B_4-dependent proinflammatory signals via BLT1. DHA is substrate for two groups of resolvins of interest during resolution in inflammatory exudates produced by different biosynthetic routes, denoted the 17*S* and 17*R* D-series resolvins (Figure 12.10). D-series resolvins display potent anti-inflammatory actions and are particularly interesting because the brain, synapses, and retina are highly enriched in DHA. Endogenous DHA is converted *in vivo* via LOX-initiated mechanisms to the 17*S*-hydroxy-containing series of four resolvins, denoted D1–D4. Each of these potent bioactive resolvins was first isolated in exudates from mice given aspirin and DHA that led to the identification of several new 17*R*-hydroxy-containing products isolated from exudates in the resolution phase. Their ability to stop neutrophil infiltration was used to assess biological function for structural elucidation studies and their biosynthesis was reconstructed to establish their potential origins. Results from these studies showed that human recombinant COX2 converts DHA into a 13-hydro(peroxy)-containing product. In the presence of aspirin, oxygenation at carbon-13 switches to the carbon-17 position with an *R* configuration that is a precursor for potent bioactive aspirin-triggered 17*R* D-series resolvins, denoted aspirin-triggered (AT)-RvD1-D4. These are produced in exudates and in the brain in response to aspirin treatment. Resolvin D1 and aspirin-triggered resolvin D1 have both been shown to be potent regulators of human and mouse neutrophils (Table 12.5). In microglial cells, both 17*S* and 17*R* D-series resolvins block TNF-induced transcripts for proinflammatory cytokine interleukin-1β (IL-1β), which is expressed rapidly in response to neuronal injury. Resolvins control inflammation at multiple levels, reducing peritonitis and skin inflammation, protecting organs from reperfusion injury and neovascularization.

Protectins. DHA is converted in resolving exudates to another new family of mediators named protectins. Mediators of this family are distinguished by the presence of a conjugated triene double bond system and their potent bioactivity. They are biosynthesized via a LOX mechanism that converts DHA to a 17*S*-hydroperoxide-containing intermediate that is converted by human leukocytes into a 16(17)-epoxide intermediate. This epoxide is enzymatically opened in these cells into a 10,17-dihydroxy-containing anti-inflammatory molecule. This bioactive compound, initially coined 10,17*S*-docosatriene, is now known as protectin D1 (10R,17S-dihydroxy-docosahexaenoic acid), owing to its potent protective activity in inflammatory and neural systems. When produced by neural tissues the prefix *neuro* is added to signify its biosynthetic origin, hence the term neuroprotectin D1.

Protectin D1 is stereo-selective and log-orders of magnitude more potent *in vivo* than its precursor DHA. Protectin D1 is also produced by human peripheral blood mononuclear cells in T helper-2-type conditions. Protectin D1 blocks T-cell migration *in vivo*, reduces TNF and interferon-γ secretion, and promotes T-cell apoptosis.

Agonists of Resolution: Specific resolvins, protectins, and lipoxins stereoselectively stimulate resolution and reduce the magnitude of the inflammatory response *in vivo* (Table 12.5). The clearance of apoptotic neutrophils by professional phagocytes, such as macrophages, is a cellular hallmark of tissue resolution, and can be used to quantify resolution using specific indices. Resolvins and protectins initiate resolution, decreasing neutrophils in exudates at earlier times than spontaneous resolution. Disruption of biosynthesis of these proresolution mediators by either COX-2 or LOX inhibitors gives a "resolution deficit" phenotype, characterized by impaired phagocytic removal, delayed resolution, and prolonged inflammation. These findings emphasize a pivotal homeostatic function for lipoxygenase(s) and COX-2 pathways in the timely resolution of acute inflammation. More importantly, proresolution mediators can rescue the deficit in resolution caused by these interventions.

Resolution of Inflammation in Disease Models

Uncontrolled inflammation is now appreciated in the pathogenesis of many diseases that were not previously considered classic inflammatory diseases. These include atherosclerosis, cancer, asthma, and several neurological disorders, such as Alzheimer's disease and Parkinson's disease. Natural proresolution mechanisms involving lipoxins, resolvins and protectins were tested for their ability to promote resolution and control inflammation (Table 12.6). It is now clear that endogenous anti-inflammation alone is not an identical mechanism of action compared to mediators that possess dual anti-inflammatory and proresolution actions. In this regard, lipoxins, resolvins, and protectins have potent multilevel mechanisms of action in disease models and promote resolution in animal models of oral, lung, ocular, kidney, skin, and gastrointestinal inflammation as well as in ischemia-reperfusion injury and angiogenesis.

TABLE 12.5. Key cellular actions of lipoxins, resolvins, and protectins in the innate immune system and resolution*

Mediators	Cell type	Action(s)
Lipoxin A_4 (or aspirin-triggered lipoxin A_4)	Whole blood leukocyte	Downregulates CD11b/CD18 expression, prevents shedding of L-selectin, and reduces peroxynitrite generation on neutrophils, monocytes and lymphocytes
	Neutrophil	Stops chemotaxis, adherence, and transmigration Stops neutrophil-epithelial and neutrophil–endothelial cell interactions Blocks superoxide anion generation Reduces CD11b/CD18 expression and IP_3 formation Inhibits peroxynitrite generation Attenuates AP-1, NF-kB accumulation, inhibits IL-8 gene expression
	Eosinophil	Stops migration and chemotaxis in vivo, and inhibits eotaxin and IL-5 generation
	Monocyte	Stimulates chemotaxis and adhesion to laminin without increase in cytotoxicity Inhibits peroxynitrite generation Reduces IL-8 release by cells from asthma patients
	Macrophage	Stimulates nonphlogistic phagocytosis of apoptotic neutrophils
	T-cell	Inhibits TNF secretion by blocking ERK activation Upregulates CCR5 expression
	Dendritic cell	Blocks IL-12 production
	Epithelia	Inhibits TNF-induced IL-8 expression and release in enterocytes Inhibits *Salmonella typhimurium*–induced IL-8 in enterocytes
	Endothelia	Stimulates protein-kinase-C–dependent prostacyclin formation Blocks reactive oxygen species generation Inhibits VEGF-induced endothelial cell migration
	Fibroblast	Inhibits IL-1β-induced IL-6, IL-8, and MMP3 production Inhibits CTGF-induced proliferation
	Hepatocyte	Reduces PPARa and CINC1 expression levels
	Mesangial cell	Inhibits leukotriene-D_4–induced proliferation Inhibits CTGF-induced chemokine production
	Neural stem cell	Attenuates cell growth
	Astrocyte	Inhibits ERK and JNK activation
Resolvin E1	Neutrophil	Stops transepithelial and transendothelial migration
	Macrophage	Stimulates nonphlogistic phagocytosis of apoptotic neutrophils
	Dendritic cell	Blocks IL-12 production
	T cell	Upregulates CCR5 expression
Resolvin D1	Microglia	Inhibits IL-1β expression
Aspirin-triggered resolvin D1	Neutrophil	Stops transmigration
Protectin D1	Neutrophil	Upregulates CCR5 expression
	Macrophage	Stimulates nonphlogistic phagocytosis of apoptotic neutrophils
	T cell	Inhibits TNF and IFNγ secretion, promotes apoptosis Upregulates CCR5 expression
	Microglia	Inhibits IL-1β expression
	Epithelia	Protects from oxidative stress–induced apoptosis in retinal pigment epithelia

*See Serhan and Chiang (*Br J Pharmacol.* 2007) and Yacoubian and Serhan (*Nat Clin Pract Rheumatol.* 2007) for further details and the review of original literature.

Periodontal diseases such as gingivitis and periodontitis are leukocyte-driven inflammatory diseases characterized by soft tissue and osteoclast-mediated bone loss. As a model of inflammatory diseases, periodontitis has several advantages in that many, if not all, of the tissues involved in inflammatory processes of other organ systems are affected in periodontitis, including the epithelium, connective tissue, and bone. There are many noteworthy similarities in the pathogenesis of periodontitis and arthritis. Results from rabbit periodontitis demonstrate an important role for resolution in disease prevention. Overexpression of 15-lipoxygenase-type 1 in transgenic rabbits increases the levels of endogenous lipoxin A_4 and protects against periodontitis, as well as reduces atherosclerosis. In prevention studies using this model, topical treatment with resolvin E1 prevents >95% of alveolar bone destruction. Histological analysis of resolvin-E1–treated rabbits revealed few, if any, neutrophils and little tissue damage. In addition, osteoclasts responsible for bone resorption are reduced in resolvin-E1–treated rabbits. In established disease, resolvin E1 prevents periodontitis tissue destruction; both soft tissue and bone that were lost during disease were regenerated.

In humans, the differential actions of resolvin E1 were studied using neutrophils from patients with localized aggressive periodontitis (LAP) and healthy individuals. Resolvin E1 reduces neutrophil superoxide generation in response to TNF or the bacterial surrogate peptide *N*-formyl-methionyl-leucyl-phenylalanine. Neutrophils from both healthy subjects and LAP patients produced ~80% less superoxide treated with resolvin E1. In comparison, neutrophils from LAP patients do not exhibit inhibition of superoxide production following treatment with lipoxin A_4, suggesting a molecular basis for excessive inflammation in these patients.

In murine air pouches, nanogram amounts of resolvin E1 (~100 nM) reduce leukocyte infiltration by 50%–70%, levels comparable to those achieved using microgram amounts of dexamethasone (~30 μM) or milligrams of aspirin. Similarly, in spontaneously resolving peritonitis induced by the yeast cell-wall component zymosan, both resolvin E1 and protectin D1 activate and accelerate resolution. Resolvins and protectins reduce neutrophil infiltration and increase nonphlogistic recruitment of monocytes. Notably, resolvin E1, protectin D1, and an ATL analogue each display different kinetics and molecular profiles of action (Tables 12.5 and 12.6).

Summation

Eicosanoids are critical mediators of homeostasis and of many pathophysiologic processes, especially host-defense and inflammation. Arachidonic acid, the primary substrate, is converted into prostaglandins, thromboxanes, prostacyclin, leukotrienes, lipoxins, isoprostanes, and epoxyeicosatetraenoic acids. Prostaglandins have diverse roles in vascular tone regulation, gastrointestinal regulation, uterine physiology, analgesia, and inflammation. Prostacyclin and thromboxane in coordination control vascular tone, platelet activation, and thrombogenesis. Leukotrienes (LTC_4, LTD_4) are the chief activators of bronchoconstriction and airway hyperactivity; LTB_4 is a major activator of mediators of leukocyte chemotaxis and infiltration. Lipoxins antagonize the effects of leukotrienes and reduce the extent of inflammation and active resolution pathways.

Pharmacologic interventions at many critical points in these pathways are useful in limiting inflammatory sequelae. Glucocorticoids downregulate several steps in eicosanoid generation, including the rate-determining step involving phospholipase A_2. However, chronic glucocorticoid use is associated with deleterious side effects, including osteoporosis, muscle wasting, and abnormal carbohydrate metabolism. COX inhibitors block the first step of prostanoid biosynthesis preventing the generation of prostanoid mediators of inflammation. LOX inhibitors, FLAP inhibitors, leukotriene synthesis inhibitors, and leukotriene receptor antagonists prevent leukotriene formation and signaling, thereby limiting inflammation and its deleterious effects. Future drug development efforts will allow selective targeting of eicosanoid pathways involved in many clinical conditions. The discovery of resolvins and protectins opens the possibility of treating inflammatory diseases with agonists of resolution, namely stable resolvins and protectins. Efforts are currently underway for the clinical development of this novel agonist of resolution. Results from clinical trials will, in the near future, determine whether treating inflammation via accelerated resolution is indeed a clinically useful approach to diseases characterized by uncontrolled inflammation.

ACKNOWLEDGMENTS

The authors wish to thank Mary H. Small for assistance and preparation of the manuscript. CNS acknowledges support for studies in his laboratory from the National Institutes of Health, USA GM38765, DK07448, and P50-DE016191 (CNS). JZH acknowledges support from the Swedish Medical Research Council and the Bert Vallee Visiting Professorship from the Vallee Foundation (Harvard Medical School, Boston, Massachusetts).

GLOSSARY

15-HETE

15-hydroxy-5Z,8Z,11Z,14Z-eicosatetraenoic acid

TABLE 12.6. Dual anti-inflammatory and pro-resolution actions of lipoxins, resolvins, and protectins in complex disease models

Mediators	Disease model	Action(s)
Lipoxin A_4/ATL	Rabbit/Periodontitis	Reduce neutrophil infiltration; prevent connective tissue and bone loss
	Mouse/Peritonitis	Stop neutrophil recruitment and lymphatic removal of phagocytes
	Mouse/Dorsal air pouch	Stop neutrophil recruitment
	Mouse/Dermal inflammation	Stop neutrophil recruitment and vascular leakage
	Mouse/Colitis	Attenuate proinflammatory gene expression and reduce severity of colitis. Inhibit weight loss, inflammation, and immune dysfunction
	Mouse/Asthma	Inhibit airway hyperresponsiveness and pulmonary inflammation
	Mouse/Cystic fibrosis	Decrease neutrophilic inflammation, pulmonary bacterial burden, and disease severity
	Mouse/Ischemia-reperfusion	Attenuate hind-limb I/R-induced lung injury Detachment of adherent leukocytes in mesenteric I/R
	Mouse/Cornea	Accelerate cornea re-epithelialization, limit sequelae of thermal injury (i.e., neovascularization, opacity) and promote host defense
	Mouse/Angiogenesis	Reduce angiogenic phenotype:endothelial cell proliferation and migration
	Mouse/Bone-marrow transplant (BMT)	Protect against BMT-induced graft-versus-host diseases
	Murine/Glomerulonephritis	Reduce leukocyte rolling and adherence Decrease neutrophil recruitment
	Rat/Hyperalgesia	Prolong paw withdraw latency, reducing hyperalgesic index Reduce paw edema
	Rat/Pleuritis	Shorten the duration of pleural exudation
Resolvin E1	Rabbit/Periodontitis	Reduces neutrophil infiltration; prevents connective tissue and bone loss; promotes healing of diseased tissues; regeneration of lost soft tissue and bone
	Mouse/Peritonitis	Stops neutrophil recruitment; regulates chemokine/cytokine production Promotes lymphatic removal of phagocytes
	Mouse/Dorsal air pouch	Stops neutrophil recruitment
	Mouse/Retinopathy	Protects against neovascularization
	Mouse/Colitis	Decreases neutrophil recruitment and proinflammatory gene expression; improves survival; reduces weight loss
Resolvin D1	Mouse/Peritonitis	Stops neutrophil recruitment
	Mouse/Dorsal skin air pouch	Stops neutrophil recruitment
	Mouse/Kidney ischemia-reperfusion	Protects from ischemia-reperfusion–induced kidney damage and loss of function; regulates macrophages
	Mouse/Retinopathy	Protects against neovascularization
Protectin D1	Mouse/Peritonitis	Inhibit neutrophil recruitment; regulate chemokine/cytokine production Promote lymphatic removal of phagocytes Regulate T-cell migration
	Mouse/Asthma	Protect from lung damage, airway inflammation, and airway hyperresponsiveness
	Human/Asthma	Protectin D1 is generated in humans and appears to be diminished in asthmatics
	Mouse/Kidney ischemia-reperfusion	Protect from ischemia-reperfusion–induced kidney damage and loss of function; regulate macrophages
	Mouse/Retinopathy	Protect against neovascularization
	Rat/Ischemic stroke	Stop leukocyte infiltration, inhibit NF-κB and cyclooxygenase-2 induction
	Human/Alzheimer's disease	Diminished protectin D1 production in human Alzheimer's disease

AA

Arachidonic acid, *cis-,cis-,cis-,cis*-5,8,11,14-eicosatetraenoic acid

Alzheimer's disease

A degenerative neurological disease that is characterized by progressive deterioration of the brain, dementia, and the presence of senile plaques, neurofibrillary tangles, and neuropil threads. Disease onset can occur at any age, and women seem to be affected more frequently than men.

BLT1

Leukotriene B_4 receptor 1

Cytochrome P450 enzymes

A large and diverse superfamily of hemoproteins. Cytochrome P450 enzymes use a plethora of both exogenous and endogenous compounds as substrates. The most common reaction catalyzed by cytochrome P450 is a monooxygenase reaction, that is, insertion of one atom of oxygen into an organic substrate while the other oxygen atom is reduced to water.

DHA

Docosahexaenoic acid, (4*Z*,7*Z*,10*Z*,13*Z*,16*Z*,19*Z*)-docosa-4,7,10,13,16,19-hexaenoic acid

Eicosanoids

A family of bioactive products that contain 20 (eicos in Greek) carbons. They are biosynthesized from arachidonic acid by the initial activities of either cyclooxygenases (isoforms COX1 or COX2) or lipoxygenases and downstream enzymatic reactions. There are several main classes of eicosanoids: prostaglandins, prostacyclins, thromboxanes, leukotrienes, and lipoxins.

EPA

Eicosapentaenoic acid, (5*Z*,8*Z*,11*Z*,14*Z*,17*Z*)-eicosa-5,8,11,14,17-pentaenoic acid

Exudate

Biological fluids that filter from the circulatory system into lesions or areas of inflammation. Exudate formation is caused by inflammation, and is characterized by a high content of plasma proteins, cells, and cellular debris. Pus is an example of an exudate found in infected wounds that contains bacteria and high concentrations of white blood cells.

G-protein–coupled receptor

GPCR. One of a large family of receptors that bind diverse molecules, including for example lipid mediators, eicosanoids, chemokines, complement components, and neurotransmitters. GPCRs are seven transmembrane–spanning receptors and are coupled to heterotrimeric, GTP-regulated signaling proteins composed of $\alpha\beta$ and $\beta\gamma$ subunits.

Ischemia-reperfusion injury

An injury in which the tissue first suffers from hypoxia as a result of severely decreased or completely arrested blood flow. Restoration of normal blood flow then triggers inflammation, which exacerbates the tissue damage.

Leukotrienes

A family of eicosanoids derived from the metabolism of arachidonic acid by the action of leukocyte 5-lipoxygenase and other enzymes. They have a conjugated triene double-bond structure and various proinflammatory activities, including leukocyte activation (by leukotriene B_4) and bronchoconstriction (by leukotriene C_4 and leukotriene D_4).

Lipoxins

A class of eicosanoids produced via lipoxygenase-mediated conversion of arachidonic acid. They are trihydroxytetraene-containing structures with potent biological activities in the resolution of inflammation.

LOX

Lipoxygenase

LTB_4

Leukotriene B_4, (5*S*,12*R*)-dihydroxy-(6*Z*,8*E*,10*E*,14*Z*)-eicosatetraenoic acid

Murine dorsal air pouch

A well-characterized system for studying inflammatory responses. Air pouches are formed by subcutaneous injection of air in the back of a mouse, into which potential inflammatory stimuli, such as tumor-necrosis factor, can be added. In importance when the stimulus is titrated, inflammatory reactions undergo spontaneous resolution. These structurally contained compartments have been likened to the inflamed synovium (see Chapter 27 from Peretti et al.).

Prostaglandins

Cyclopentane ring–containing lipids derived from the metabolism of arachidonic acid by the action of cyclooxygenases and downstream synthase enzymes. They have a wide range of biological activities and a well-recognized role in inflammation and pain.

Protectins

A family of docosahexaenoic acid (DHA)-derived mediators characterized by the presence of a conjugated

Box 1. The stereochemistries of lipoxins, resolvins, and protectins

- Resolvin D1: 7*S*,8*R*,17*S*-trihydroxy-4*Z*,9*E*,11*E*,13*Z*,15*E*,19*Z*-docosahexaenoic acid
- Resolvin E1: 5*S*,12*R*,18*R*-trihydroxy-6*Z*,8*E*,10*E*,14*Z*,16*E*-eicosapentaenoic acid
- Neuroprotectin D1/Protectin D1: 10*R*,17*S*-dihydroxy-docosa-4*Z*,7*Z*,11*E*,13*E*,15*Z*,19*Z*-hexaenoic acid
- Lipoxin A_4: 5*S*,6*R*,15*S*-trihydroxy-7*E*,9*E*,11*Z*,13*E*-eicosatetraenoic acid
- Aspirin-triggered lipoxin A_4: 5*S*,6*R*,15*R*-trihydroxyl-7,9,13-*trans*-11-*cis*-eicosatetraenoic acid
- Lipoxin B_4: 5*S*,14*R*,15*S*-trihydroxy-6*E*,8*Z*,10*E*,12*E*-eicosatetraenoic acid.
- Aspirin-triggered Resolvin D1: 7*S*,8*R*,17*R*-trihydroxy-4*Z*,9*E*,11*E*,13*Z*,15*E*,19*Z*-docosahexaenoic acid

triene double-bond structure and a 22 carbon backbone structure with six double bonds.

PUFA

Polyunsaturated fatty acid

Rabbit model of periodontitis

A model of periodontitis in rabbits induced by application of *Porphyromonas gingivalis* to ligatures tied to second premolars.

Resolvins

Biosynthesized lipid mediators that are induced in the resolution-phase following acute inflammation. They are synthesized from the essential omega-3 fatty acids eicosapentaenoic acid (EPA) and docosahexaenoic acid (DHA).

Transcellular biosynthesis

The biosynthesis of biologically active mediators that involves two or more cell types, for example, when the necessary enzymes are differentially expressed in two or more cell types.

In this example, a donor cell converts a precursor compound (such as arachidonic acid, EPA and DHA) into an intermediate product. The acceptor cell then converts the intermediate product into the final active product. Transcellular biosynthesis therefore provides a means to produce mediators that neither cell type can generate alone.

SUGGESTED READINGS

Brink, C., Dahlen, S.E., Drazen, J., et al. 2003. International union of pharmacology XXXVII. Nomenclature for leukotriene and lipoxin receptors. *Pharmacol Rev* **55**:195–227. (*International consensus report on eicosanoid receptors and their antagonists.*)

Dahlen, S.E., Haeggstrom, J.Z., Samuelsson, B., et al. 2000. Leukotrienes as targets for treatment of asthma and other diseases: current basic and clinical research. *Am J Respir Crit Care Med* **161**:S1. (*Highlights the potential applications of leukotriene pathway inhibitors.*)

Dwyer, J.H., Allayee, H., Dwyer, K.M., et al. 2004. Arachidonate 5-lipoxygenase promoter genotype, dietary arachidonic acid, and atherosclerosis. *N Engl J Med* **350**:29.

Gilroy, D.W., and Perretti, M. 2005. Aspirin and steroids: new mechanistic findings and avenues for drug discovery. *Curr Opin Pharm* **5**:1–7. (*Review of the anti-inflammatory actions of aspirin-triggered lipoxins and the discovery of annexin and related compounds in the actions of glucocorticoids.*)

Haeggström, J.Z. 2004. Leukotriene A_4 hydrolase/aminopeptidase, the gatekeeper of chemotactic leukotriene B_4 biosynthesis. *J Biol Chem* **279**:50639–50642.

Helgadottir, A., Manolescu, A., Thorleifsson, G., et al. 2004. The gene encoding 5-lipoxygenase activating protein confers risk of myocardial infarction and stroke. *Nat Genet* **36**:233–239.

Peters-Golden, M., and Henderson, W.R., Jr. 2007. Leukotrienes. *N Engl J Med* **357**:1841–1854.

Psaty, B.M., and Furberg, C.D. 2005. COX-2 inhibitors – lessons in drug safety. *N Engl J Med* **352**:1133–1135.

Serhan, C.N. 2007. Resolution phases of inflammation: novel endogenous anti-inflammatory and pro-resolving lipid mediators and pathways. *Annu Rev Immunol* **25**:101–137.

Yacoubian, S., and Serhan, C.N. 2007. New endogenous anti-inflammatory and pro-resolving lipid mediators: implications for rheumatic diseases. *Nat Clin Pract Rheumatol* **3**:570–579.

HISTORICAL OVERVIEW AND REVIEWS OF SPECIAL INTEREST

Bergström, S. 1982. *Les Prix Nobel: Nobel Prizes, Presentations, Biographies and Lectures 129–148*. Stockholm: Almqvist & Wiksell.

Samuelsson, B. 1982. *Les Prix Nobel: Nobel Prizes, Presentations, Biographies and Lectures 153–174*. Stockholm: Almqvist & Wiksell.

Vane, J.R. 1982. *Les Prix Nobel: Nobel Prizes, Presentations, Biographies and Lectures 181–206*. Stockholm: Almqvist & Wiksell.

Vane, J.R., Bakhle, Y.S., and Botting, R.M. 1998. Cyclooxygenases 1 and 2. *Ann Rev Pharmacol Toxicol* **38**:97–120. (*Historical overview of prostaglandin research, including discussion of the pharmacologic manipulation of these pathways.*)

13 Cytokines and Chemokines in Inflammation

Dennis M. Lindell and Nicholas W. Lukacs

SUMMARY

Mediators produced during inflammatory/immune responses dictate the type of response, as well as its magnitude ("quality and quantity"). The profile of cytokines and chemokines produced are responsible for the cell-to-cell communication that facilitates initial recognition of infection or damage. These signals, in turn, communicate with primary lymphoid tissues (the thymus and bone marrow) to mobilize inflammatory cells to the bloodstream. At the tissue site, chemokines and cytokines orchestrate leukocyte adhesion to vascular endothelium, extravasation, and localization of leukocytes at the site of inflammation. Recruitment of leukocyte populations into inflamed tissues is initiated by cytokine-induced expression of adhesion molecules on vascular endothelium. These adhesion molecules play an essential role in capturing and tethering circulating leukocytes from the bloodstream. Chemokines promote the tight adherence of leukocytes to activated endothelium, as well as direct the extravasation of cells into inflamed tissue. Cytokines and chemokines further coordinate the response of inflammatory effector cells as they arrive in inflamed tissues. The sequential activation and cellular recruitment cascades mediated by cytokines and chemokines are essential for successful resolution of infection or other tissue damage. However, overproduction or dysregulation of these inflammatory mediators can also lead to destructive, pathological consequences. This chapter will examine the role played by cytokines and chemokines in the initiation, amplification, and shaping of inflammatory responses.

INTRODUCTION

The actions of cytokines are dictated by the expression of specific receptors. These actions may be on the producing cell itself (autocrine), on neighboring cells (paracrine), or on distant sites carried via circulation (endocrine) (Figure 13.1). Cytokines are very potent at low concentrations and their effects are targeted via directional release, meaning that the cell producing the cytokine often releases it predominantly in one direction, to target the effect. The majority of cytokines can be divided into several major families, including the tumor necrosis factor (TNF) family, hematopoietins, and interferons. Cytokines have a number of characteristics: they are typically pleiotropic and can act upon many different cell types. Second, they are redundant; a number of similar cytokines may produce similar effects. Third, cytokines from a profile may exhibit synergy – that is, the effects of two cytokines are greater than the sum of each individually. Conversely, cytokines from opposing functional categories tend to be antagonistic toward one another – blocking each other's production or activity.

A great deal of insight of the in vivo significance of cytokines has been gained through the use of genetically deficient mice (knockouts), conditional knockouts, transgenics, and neutralizing antibodies. Each of these approaches has its own limitations. In the case of knockout mice, animals tend to develop compensatory pathways, potentially leading to an underestimate of the true in vivo importance of a particular cytokine. Conversely, transgenically overexpressing a protein can lead to artifactual effects due to supraphysiologic concentrations. Neutralizing or depleting antibodies can produce nonspecific effects, or induce antiantibody responses (serum sickness) when given long term making it difficult to assess their effects under chronic disease conditions. Thus, any single approach to the understanding of cytokine function should be regarded with some skepticism. However, these approaches offer a great deal of value for understanding the impact of a particular cytokine in a complex in vivo disease response.

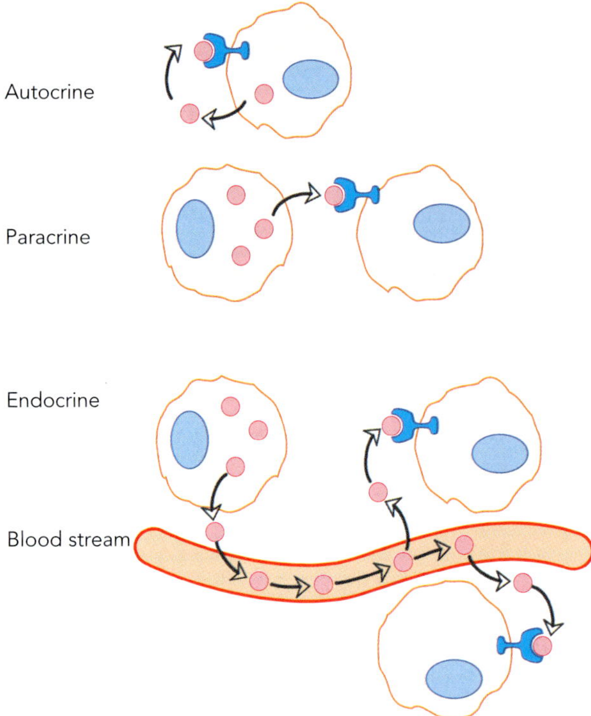

Figure 13.1. Types of cytokine signaling. Autocrine signaling refers to secreted agents that act through specific receptors on the secreting cell. In paracrine signaling, cytokines act on nearby cells. Endocrine signaling refers to action on distant cells following transport by the circulation.

NOMENCLATURE OF CYTOKINES

Many cytokines are given the designation interleukin (IL). Cytokines may be classified by their structural similarity. For instance, the IL-2 family, interferons, and the IL-10 family all share a similar four alpha-helix structure. Other families include the IL-1 family, and the IL-17 family of cytokines. Cytokines may also be classified based on their function. One of these classifications is whether a particular cytokine is Type 1 (T helper 1 associated) or Type 2 (T helper 2 associated). Usually, the classification of cytokines is structural or function based and can be divided depending on a particular disease or class of diseases. Many other functional categories are possible depending on context.

CYTOKINE RECEPTORS/SIGNAL TRANSDUCTION

The cytokine receptors can be classified into several major families, including the TNF receptor superfamily, hematopoietins, interferons, and immunoglobulin (Ig) superfamily. Cytokine receptors are typically composed of several subunits. Many exist as heterodimers or heterotrimers, with one or two subunits primarily responsible for binding and recognition, and the other subunit required for signal transduction through the cell membrane. Often, the signal transduction subunit is common to many different receptors. A prototypical example of this theme is the common gamma chain in the hematopoietin cytokine receptor family.

Hematopoietic cytokine receptors (class I) have similar extracellular domains, and often form multimeric structures with one another. For instance, the IL-2 receptor gamma subunit is shared by IL-4, IL-7, IL-9, and IL-15 receptors, and is thus referred to as the common γ chain (γ_c). A common β chain is shared by the IL-3, IL-5, and GM-CSF receptors. The interferons are also referred to as class II cytokine receptors, and include the receptors for IFN-α and IFN-β, IFN-γ, and the IL-10 receptor. The TNF ligand and receptor family (class III) includes ligands that can be cell associated or soluble. Many members of this family are homotrimers in both the ligand and ligand-bound receptor. Many members of the TNF family are not cytokines, but are membrane-bound signaling molecules important for costimulation of immune responses. In addition, many members of the TNF receptor family have a cytosolic tail that contains a number of death domains, which can activate the caspase cascade, resulting in apoptosis or cell suicide. However, signaling via the TNF receptor through TNF-receptor associated factors (TRAFs) can also promote the activation of NF-κB, resulting in prosurvival and growth promoting activities.

Cytokine receptor signaling results in complex sequential intracellular events [1]. A simplified representation of Class I/II cytokine receptor signaling is depicted in Figure 13.2. Initially, the intracellular domains of the receptor are openly spaced. Janus kinases (JAKs) are associated with the intracellular tail of the receptor. Ligand binding results in recruitment of the JAKs bringing them closer together where they can cross-phosphorylate one another, as well as other tyrosine residues in the cytosolic tail of the receptor. Phosphorylated JAKs in turn can phosphorylate a family of signal transduction and transcription factors (STATs) making them functionally active. The dimerization of STATs results in translocation to the nucleus, where phosphorylated, dimerized STATs bind to the promoters of target genes, activating transcription of the genes. Depending on the cytokine specific STAT pathways are initiated. Several of these pathways have been well-defined and include the IFN (α,β,γ) signaling that activate STAT1, IL-6 signaling initiates STAT3, IL-12 drives the STAT4 pathway, and IL-4 as well as IL-13 that drive the STAT6 pathways. Each of these signaling pathways has been linked to critical aspects of the innate and acquired immune responses via the use of genetically deleted animals as well as rather specific inhibitors. Using in vitro and in vivo experiments the STAT signaling molecules have been demonstrated to influence the direction of the immune responses. The STAT1 pathway initiates signaling that promotes antipathogen responses and without it the immune system cannot properly

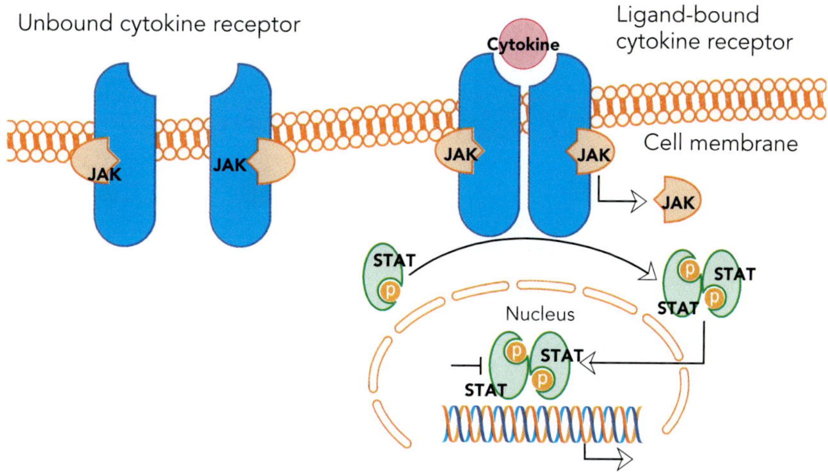

Figure 13.2. A simplified model of cytokine receptor signaling (Type I/II cytokine receptors). Initially, the unbound receptor is present as a dimeric structure, with the subunits spaced. Cytokine receptors are associated with Janus kinases (JAKs). Upon ligation of the receptor by ligand, the subunits are brought into apposition facilitating phosphorylation of Tyr residues on the cytosolic side of the receptor. These in turn are bound via SH2 containing signal transducers and activators of transcription (STATs). STAT phosphorylation leads to dimerization of STATs, which accumulate in the nucleus, activating transcription.

clear pathogens and further increases pathology due to skewed immune responses with increasing Th2 cytokines (IL-4, IL-5, and IL-13). In a similar fashion, the inhibition or deletion of the STAT4 pathway blocks production of IFN and the development of the Th1-mediated cellular immune responses, leading to increased Th2 cytokine production. As indicated earlier, the key Th2 cytokine (IL-4, IL-13) associated signaling protein, STAT6, is essential to initiate the Th2 response and without it there are deficiencies in proper B cell–mediated humoral responses related to isotype switching and extracellular pathogen immune responses. More recently, the development of Th17 subsets of lymphocytes has been shown to be dependent on IL-6-mediated STAT3 responses and links to the severity of autoimmune disease have been established. Thus, these critical signaling molecules have a central role in determining the direction and maintenance of acute and chronic immune responses.

CHEMOKINES (MORE THAN "CHEMOTACTIC CYTOKINES")

Initially, chemotactic cytokines or "chemokines" were viewed as leukocyte chemoattractants that regulate cellular movement from circulation into inflamed tissues (Figure 13.3). However, chemokines also have other functions, including cell activation, hematopoiesis, and angiogenesis. Chemokines have a number of characteristics: (1) Chemokines may be abundantly produced. For instance, mRNA for CCL3/MIP1-α accounts for greater than 20% of the total mRNA produced by LPS-stimulated macrophages. (2) Chemokines tend to be redundant, with many ligand receptor pairs producing seemingly identical effects. (3) Chemokines are (for the most part) promiscuous, meaning that a single ligand can bind multiple receptors, and multiple ligands are bound by a single receptor (Figure 13.4). These characteristics may be misleading, however, as apparent redundancy may be simply due to our lack of understanding of the complex signaling pathways induced by chemokines. Furthermore, specific profiles of chemokines will specifically initiate the migration of different leukocytes to the site of inflammation/immune responses. For example, during acute inflammatory responses early production of CxCR1/2 binding chemokines, such as IL-8 (CxCL8), initiate intense neutrophilic responses since these are the primary cell that expresses the receptors. Thus, the profile of chemokines produced at the site of inflammation dictate the type of leukocytes that migrate due to the profile of receptors that are expressed by those cells. Key findings regarding the diverse roles of chemokines in development, inflammation, as well as resolution and end-stage fibrotic disease has highlighted that chemokines are key players in the regulation of a diverse set of responses. This is likely the reason that there are such a large number of different chemokines that can be produced that allow distinct fidelity and control of a particular inflammatory response.

Nomenclature: Chemokines have been divided into four families based upon their sequence homology

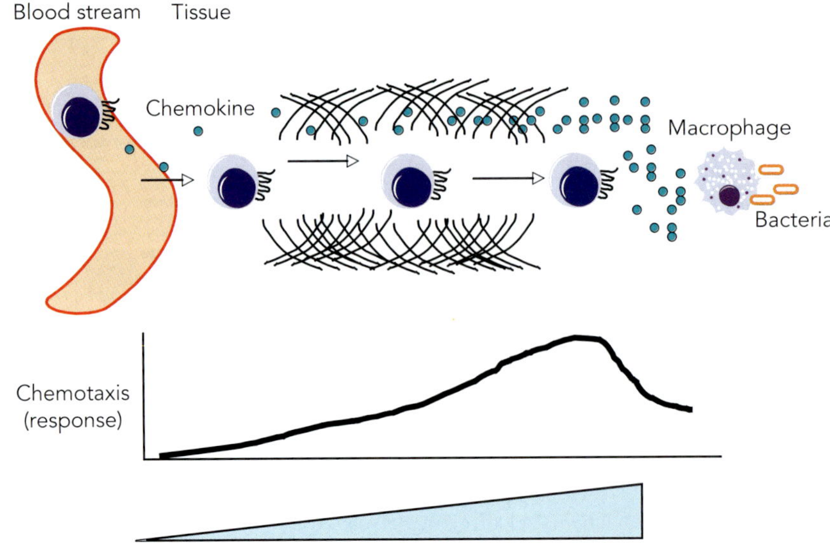

Figure 13.3. Inflammatory cells follow chemokine gradients. Chemokines mediated the homing and extravasation of inflammatory cells through specific receptors via the establishment of chemokine gradients. These gradients are facilitated by the binding of chemokines to extracellular matrix. Inflammatory cells follow the increasing concentration of chemokine through tissues until they reach a threshold concentration, whereupon the cell becomes desensitized.

and the position of the first two conserved cysteine residues, CXC (alpha) and CC (beta), C, and CX3C (Figure 13.4). Both of these latter families have only a single member thus far, XCL1 (lymphotactin) and CX3CL1 (fractalkine), respectively. The nomenclature of chemokines and receptors is, admittedly, confusing. Historically, chemokines were named by the laboratories that discovered them, usually based on the cell type or tissue from which they were isolated (e.g., macrophage inflammatory protein 1 alpha – MIP1-α). While this nomenclature had some advantages, cloning by multiple laboratories resulted in multiple names for a single chemokine. To resolve this issue, nomenclature for chemokine ligands has been replaced by the more systematic CCL, CXCL, XCL, or CX3CL designation (e.g., MIP1-α became CCL3). However, the original names are often still used in the literature, usually in conjunction with the newer, systematic one. The CXC or alpha chemokine family can be further subdivided depending upon whether the chemokines have an amino acid E-L-R motif within its sequence in the N-terminal region. This consensus sequence appears to determine whether the chemokines have the ability to induce chemotaxis of neutrophils and determines whether the CXC chemokine is angiogenic (ELR) or angiostatic (non-ELR). The ELR CXC chemokines are bound by CXCR1 and CXCR2, expressed predominantly by neutrophils. Non-ELR CXC chemokines bind CXCR3–5, and serve as chemoattractants for monocytes, dendritic cells, T, and B lymphocytes. The CC or beta chemokine family is a large group of chemokines with nearly 30 members, and at least 10 receptors. This family serves as recruitment molecules for monocytes and lymphocytes, but also basophils, eosinophils, NK cells, and mast cells. Different inflammatory stimuli result in the induction of different chemokine profiles. Additionally, as will be discussed in more detail, specific cell types typically express a limited array of chemokine receptors. The patterns of chemokines produced and the receptor expression responding cells (in conjunction with cytokines and other inflammatory signals) shapes the trafficking and activation of both resident and inflammatory cells.

CHEMOKINE RECEPTOR SIGNALING

Chemokine receptors are seven transmembrane spanning G-protein coupled receptors. Chemokines mediate their effects predominately via Gi-protein mediated pathways. The activation of chemokine receptors results in a number of cellular events, including Ca^{2+} flux, activation of phosphoinositides, various protein and lipid kinases, and small GTPases (Figure 13.5) [2]. The activated G protein alpha subunit results PLC-mediated cleavage of phosphotidylinositol (4,5)-bisphosphate (PIP_2) into inositol triphosphate (IP_3) and diacylglycerol (DAG), triggering Ca^{2+} flux and protein kinase C activation. Beta/gamma subunits

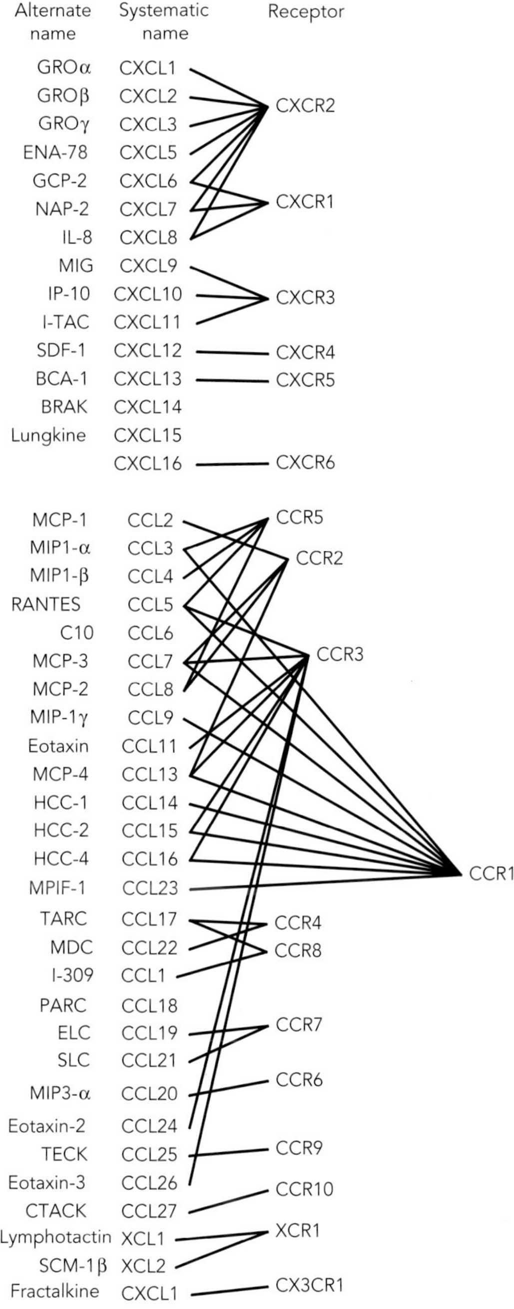

Figure 13.4. Chemokine and chemokine receptor specificity. Receptor–ligand specificity is depicted via connecting lines. Chemokines and chemokine receptor vary significantly in their specificity, with some receptor–ligand pairs being monogamous, whereas others are more promiscuous.

together are responsible for chemotaxis. Signaling through these subunits together is known to activate important pathways including PI3K, MAPK, and FAK. PI3K activation generates phosphatidylinositols (Ptdlns) species, especially Ptdlns (3,4,5)P_3 (PIP3), that are enzymatically modified by a series of kinases and phosphatases to regulate chemotaxis. PI3K isoforms appear to be differentially expressed and activated depending on the chemokine receptor, as well as the responding cell type. For instance, T-cell chemotaxis is dependent on P110γ, whereas B cells may rely upon P110δ. PI3K-dependent signaling can be negatively regulated by PTEN (phosphatase and tensin homology deleted on chromosome 10 protein), as well as by SH2-containing inositol 5-phosphate (SHIP) [3]. Small GTPases Rac, Rho, and Cdc42 play key roles in regulating the morphology of migrating cells via effects on the actin cytoskeleton. During chemotaxis, the cell exhibits polarization, localized to the leading edge during migration; PTEN is found in the trailing edge and is actively excluded from the leading edge. Thus, the polarized regulation of this system facilitates directional leukocyte movement. Thus, many intracellular signaling pathways are stimulated via chemokine receptor signaling. These have both direct chemokine-mediated effects, as well as opportunities for cross talk with other cellular signals.

REGULATION OF HEMATOPOIESIS BY CYTOKINES AND CHEMOKINES

One prominent function of cytokines is the regulation of hematopoiesis. Included in this family are the colony-stimulating factors (e.g., GM-CSF, G-CSF, M-CSF), and a number of interleukins, such as IL-3, summarized in Table 13.1. Many of these cytokines are produced by stromal cells, but can also be produced by hematopoietic cells themselves to provide autocrine signals that modify their own maturation. Hematopoietic cytokines may be lineage restricted, such as erythropoietin (EPO) which acts primarily on red blood cells (RBCs), those that act on multiple lineages (e.g., GM-CSF), or those that act sequentially, such as in the case of SCF and EPO on RBCs. Most hematopoietic cytokines can be made by many cell type and take the form as membrane-bound or cell-secreted factors. They are potent in their actions, functioning at low nM concentrations. They are typically present at low concentrations, but are highly inducible upon stimulation in specific diseases or when needed (e.g., 1,000-fold). They are polyfunctional and can promote survival, proliferation, differentiation, maturation, and cell activation along with other potential functions. Therapeutically, the hematopoietic growth factors have been very useful. For example, GM-CSF and G-CSF are commonly used therapeutically to treat neutropenia, whereas EPO is used to treat anemia.

Chemokines also have effects on the hematopoietic system. Many of these effects are indirect, but some chemokines have been demonstrated to have direct roles, as well. Among the best characterized of the hematopoietic chemokines is CXCL12/SDF-1. CXCL12 is the most abundantly produced chemokine in the

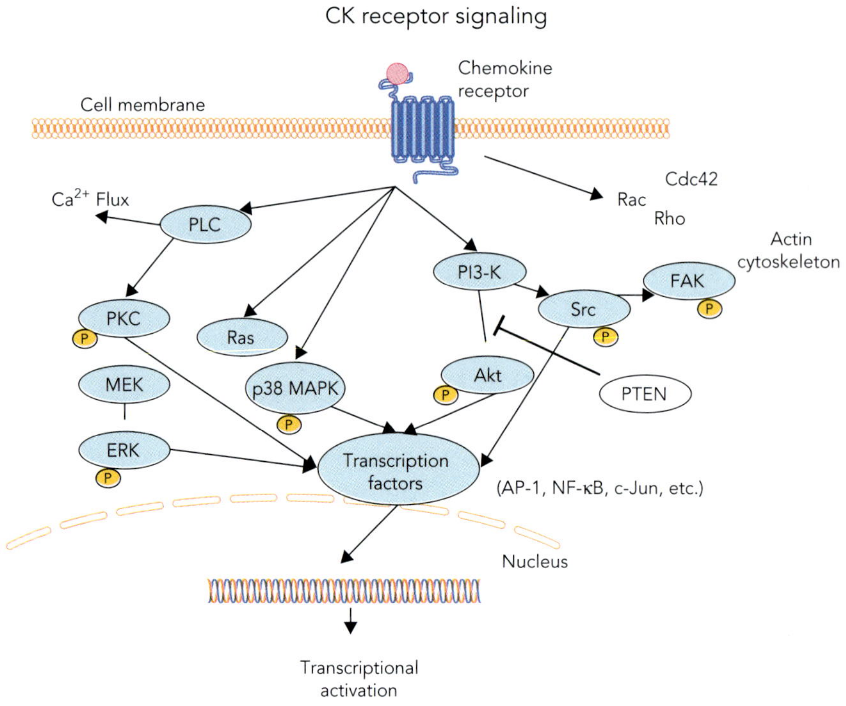

Figure 13.5. Signaling induced by chemokine receptors. Chemokines receptors are seven transmembrane spanning G-coupled receptors. A number of signaling pathways are induced by chemokine ligation, including PI-3 kinases, small GTPases, MAP kinases, and activation of phosphoinositides. These pathways mediate a wide variety of responses, including adhesion, locomotion, chemotaxis, calcium flux, as well as activation of gene transcription.

TABLE 13.1. Prominent/major hematopoietic cytokines

Hematopoietic cytokines
G-CSF (granulocyte-colony stimulating factor)
GM-CSF (granulocyte macrophage-colony stimulating factor)
M-CSF (macrophage-colony stimulating factor)
IL-3
SCF (stem cell factor)
LIF (leukemia inhibitory factor)
Flt3L (FMS-like tyrosine kinase 3 ligand)
IL-11
TPO (thrombopoietin)
IL-2
IL-6
IL-7
OM (oncostatin M)

bone marrow. Although CXCL12 promotes the survival of hematopoietic stem cells (HSCs) and hematopoietic precursor cells (HPCs) in vitro, a number of studies suggest that CXCL12 (through its receptor CXCR4) serves as a bone marrow retention signal in vivo. These functions have relevance to a number of immunologically mediated diseases. CXCR4 also serves as a coreceptor for the HIV-1 virus, which may allow the virus to spread to recent bone marrow emigrants or the bone marrow itself. In cancer, CXCL12 promotes chemotaxis and homing in many tumor cell lines. In addition, CXCL12 promotes the formation of new blood vessels (neoangiogenesis), which tumors require to continue to grow beyond self-limiting size. Other chemokines may promote the mobilization of HSCs or HPCs. One chemokine with this function is CXCL2, which enhances the homing and engrafting capabilities of HSCs. In contrast to CXCL12 and CXCL2, most chemokines are generally believed to have suppressive function on HPCs. CCL2, 3, 19, 20, CXCL4, 5, 8, and 9 all suppress the numbers and proliferation of HPCs when administered in vivo. It is important to note, however, that rarely (if ever) will the host experience a bolus of a single chemokine, in the absence of other signals. Therefore, the effects of chemokines on hematopoiesis during disease are still poorly understood.

INITIATION OF INNATE IMMUNE RESPONSES

Cytokines and chemokines are among the first signals induced by the immune system upon recognition of infection. Initial recognition of microbial infection occurs via pattern recognition receptors (PRRs) that distinguish pathogen-associated molecular patterns

(PAMPs) from normal host associated proteins [4]. Viral infection is accompanied by the induction of IFN-β and IFN-α (type I interferons). This response depends on the recognition of viral nucleic acids and/or viral particles. Toll-like receptors (TLRs) 3, 7, and 8 play important roles in recognition of viral nucleic acids by their ability to recognize either double-stranded RNA (TLR3) or single-stranded RNA (TLR7, 8). In addition, cytosolic sensors of viral replication and double-stranded RNA (such as PKR and RIG-I) also play a role in recognition of cytosolic-replicating viruses. Type I interferons are produced by virus infected cells and facilitate the establishment of an antiviral state in infected (autocrine action) and nearby (paracrine action) cells. Other PRRs recognize other microbial products, such as bacterial lipopolysaccharide (via TLR4), flagellin (via TLR5), CpG DNA (via TLR9), and fungal products (via TLR2 and mannose receptor) [5]. Recognition of microbial products by TLRs and other PRRs initiates the production of proinflammatory cytokines including IL-1, TNF-α, IL-6, as well as a number of chemokines (including CXCL8, CXCL10, CCL2, and CCL3). This is regarded as the acute phase response, and results in fever, increased vascular permeability, edema, and increased expression of adhesion molecules on the vascular epithelium. These early cytokines form an amplification cascade, collectively referred to as the "cytokine storm." These mechanisms serve to clear, or at least limit, infections during the establishment of the more efficient adaptive immune response. This intense inflammatory activity is followed by a compensatory downregulation of inflammation. One anti-inflammatory cytokine, IL-10, plays a critical role in modulating the inflammatory state. Dysregulation of the highly inflammatory state can occur during systemic infection or injury, resulting in sepsis. The highly inflamed state (systemic inflammatory response syndrome, or SIRS) can be lethal. Potentially life-threatening, as well, is the downregulation of severe inflammation. During this time the immune system is reflexive to further challenges and this is referred to as the compensatory antiinflammatory response syndrome (CARS). At this time, the host exhibits increased susceptibility to infections. For example, patients who survive an initial septic episode often shortly succumb to pneumonia. Cytokines and chemokines play an essential role in the establishment of the inflammatory state of the host, which is essential to combat infection, but may be detrimental if not properly controlled. Thus, many types of insults, bacterial infection, head trauma, burns, and so on can result in a similar life-threatening condition due to the overproduction and/or dysregulation of cytokine responses, leading to multiorgan shutdown.

CHEMOKINES AND THE REGULATION OF LEUKOCYTE TRAFFICKING

The acute phase response establishes conditions that favor the rapid migration of leukocytes into affected tissues. The vascular endothelium upregulates a series of adhesion molecules ("activated endothelium"), which in turn, serve to trap circulating leukocytes. The initial adhesion is mediated by E and P selectins that facilitate rolling of the leukocytes on the activated endothelium. In response to chemokine signals, leukocytes firmly adhere to the activated endothelium via the upregulation of beta integrins. Subsequently, the firm adhesion allows the leukocytes to spread along the endothelium and begin extravasating into the inflamed tissues. Leukocytes are guided to the site of inflammation via a chemokine gradient (Figure 13.3). However, the nature of these chemokine gradients have been difficult to assess in an in vivo setting. Multiple studies have suggested that chemokines, in particular, have specialized binding attributes that allow them to be deposited into the tissue so that they are not simply diluted or "washed" away. Extracellular matrix proteins are important for the maintenance of chemokine gradients. Most members of the chemokine family have the ability to bind to glycosyaminoglycans (GAGs), a component of extracellular matrix. This function serves to immobilize chemokines, and provides a scaffolding for migrating leukocytes. In addition, the oligomerization of chemokines on GAGs may be important for migration as leukocytes or even hypertrophic structural cells allowing them to haplotax (crawl) along the solid phase chemokine gradients. At high concentrations of chemokines (signaling arrival at the target site), leukocytes no longer migrate. Instead, other cell activating functions, such as degranulation of mast cells and eosinophils, take place. Thus, chemokines play essential roles in the firm adherence, extravasation, and chemoattractant-direct trafficking of leukocytes to inflammatory sites, as well as the initiation of the effector function once at the site of the inciting agent. While most research has been performed targeting a single chemokine or receptor at a time, it is clear that an individual inflammatory response initiates a series of chemokines to be produced. Since leukocytes have a profile of receptors, there are coordinated mechanisms that allow cells to respond, traffic, and be activated by multiple chemokines during any one response. Thus, a chemokine that initiates the firm adhesion event at the vascular endothelial cell surface is different from the chemokine that eventually localizes the migrating cell at the site of inflammation. This complex system along with the redundancy of chemokine function enables more fidelity and control within the system as well as increased protection if a signal chemokine/receptor system is perturbed.

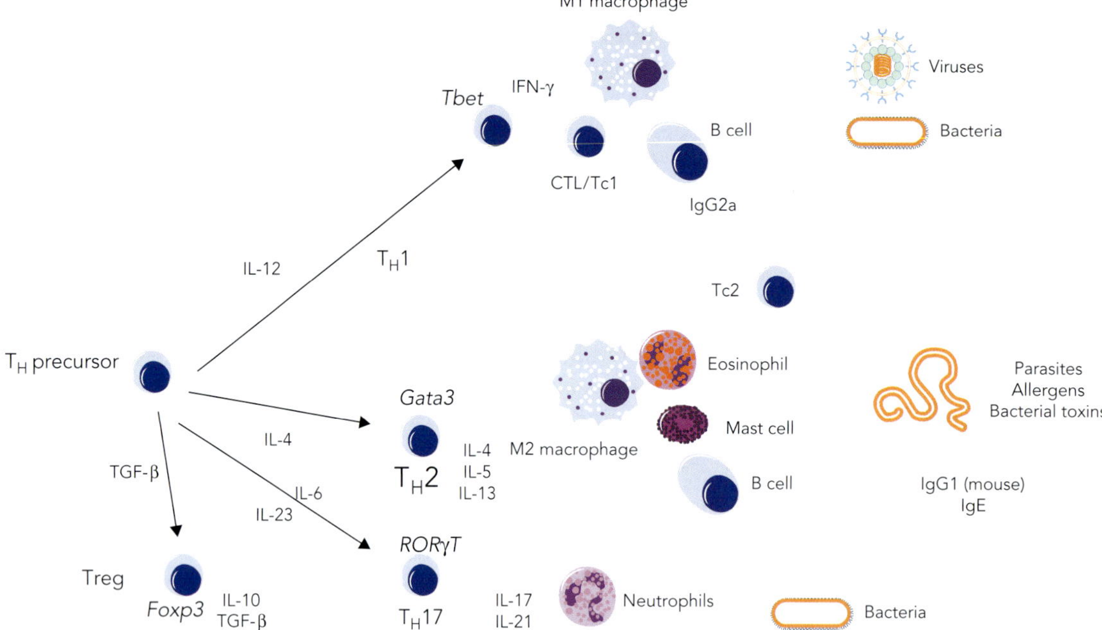

Figure 13.6. Adaptive immune responses are controlled by the actions of cytokines. The adaptive immune response is tailored to the type of pathogen via the actions of cytokines. Cytokine signals from innate immune cells and non-immune cells promote one of several pathways of CD4+ T cells (T_H) cell differentiation. IL-12 promotes the generation of IFNγ-producing T_H1 cells, important for the clearance of viruses and intracellular bacteria. IL-4 promotes the generation of T_H2 responses, which mediated antibody-dependent (or "humoral") immunity. T_H2 responses are generated in response to bacterial toxins, parasites, but also drive allergic responses. IL-6, TGF-β, and IL-23 favor the generation of T_H17 responses, important for the clearance of extracellular bacteria. T_H17 responses are also involved in autoimmunity. The immunosuppressive regulatory T-cell lineage (Treg) is promoted by TGF-β and IL-10 and serves to negatively regulate many types of immune responses.

CYTOKINES SHAPE THE ADAPTIVE IMMUNE RESPONSES

A critical function of cytokines is in shaping the development of adaptive immunity. Different types of microbial infection (i.e., viruses, bacteria, fungi, and parasites) are effectively controlled by different effector cell types. The cytokine profiles generated initially by resident cells in the infected tissue, and via antigen-presenting cells (APC) in the lymph nodes, dictate the type of immune response generated by T and B lymphocytes. Ideally, the immune response serves to activate the proper cell populations for antimicrobial effector functions. The role of cytokines in controlling adaptive immunity is exemplified in the differentiation and polarization CD4+ or helper T-cell populations (Figure 13.6) [6]:

- T helper type 1 (Th1) responses are generated in response to most viruses and intracellular bacteria. Early production of IL-12 from macrophages and dendritic cells in response to PRR engagement promotes the development of Th1 responses. IL-12 promotes induction of the transcription factor regulating Th1 development, T-bet. Th1 cells in turn produce large amounts of IFN-γ, which promotes the classical pathway of macrophage activation, inducing phagocytosis and intracellular killing via reactive oxygen intermediates.
- T helper type 1 (Th2) responses are generated primarily in response to extracellular pathogens, such as parasites, some fungi, as well as allergens. The development of Th2 responses is promoted by the cytokine IL-4, and Th2 lineage development is controlled via the transcription factor Gata-3. Prominent effector functions of Th2 responses include antibody production, basophil, mast cell, and eosinophil activation. Since most of these pathogens are too large to be engulfed by phagocytes, effector cells dump antimicrobial mediators on these organisms, or literally expel them from the host. Effector cytokines produced by Th2 cells include IL-4 IL-5, IL-13, as well as others.
- T helper 17 (Th17) cells, thus named for their production of IL-17, are involved in the elimination of extracellular bacteria (from mucosal surfaces especially), but are also thought to be involved in the development of chronic diseases, including autoimmune disease [7]. Th17 cells develop in the simultaneous presence of TGF-β and IL-6, and differentiation is mediated via the transcription factor

RORγt. The major target cells of Th17 cells are thought to be neutrophils.

- Regulatory T cells (Tregs) are critical for controlling excessive inflammation, as well as maintaining tolerance to self-antigens [8,9]. Their development is dependent on the cytokine TGF-β, and controlled via the transcription factor Foxp3. In the absence of functional Tregs, lethal autoimmunity results. In addition to TGF-β, IL-10 is also a critical factor in Treg responses.

In addition to promoting the development of a particular lineage directly, cytokines also serve to antagonize the development of other lineages. For instance, in addition to promoting Th2 development through its actions on STAT 3, IL-4 also negatively regulates the development of Th1 cells in part via downregulation of the IL-12 receptor. By both promoting one response, and antagonizing others, cytokines shape the character of adaptive immunity.

The cytokine environment determined via inflamed tissues, as well as inflammatory cells, regulates B-cell activation, proliferation, and antibody production. IL-4 is secreted in a directional fashion by Th2 cell to target antigen-specific B cells. B-cell activation in the presence of IL-4, as well as other signals derived from direct T-cell contact, leads to B-cell proliferation. In addition, IL-4 promotes isotype switching in B cells to IgG1 and IgE, important for mast cell and eosinophil-mediated defense against parasites. IL-5 and TGF-β can both promote the switch to IgA secretion, critical for defense of the mucosa. IFN-γ from Th1 cells promotes isotype switching to IgG2a and IgG3, important for opsonic phagocytosis. Thus, each of these antibody subtypes is specialized for a different effector function. The antibody isotypes promoted via the cytokine environment serve to promote the antibody effector functions appropriate for the particular pathogen encountered.

The cytokine environment that predominates during an immune response affects many other cell types, including macrophages. Macrophages can serve as both initiators of adaptive immunity, as well as effector cells. In the presence of microbial products (such as LPS), IFN-γ is critical for the classical activation of macrophages (also referred to as M1 macrophages). This pathway serves to support Th1 responses by promoting IL-12, IL-18, and IL-15 secretion, as well as inducing the upregulation of phagocytosis and the respiratory burst. Activated M1 macrophages undergo efficient phagolysosomal fusion, exposing intracellular and/or recently ingested extracellular microbes to a variety of microbicidal lysosomal enzymes. In contrast, alternatively activated macrophages (or M2 macrophages) are activated under conditions of Th2 cytokines IL-4 and IL-13, as well as other conditions such as TGF-β. M2 macrophages are associated with Th2 responses to eukaryotic pathogens and allergens, resolution of inflammation, and tissue repair. M2 macrophages exhibit suppression of nitric oxide, and the enhancement of arginase, IL-10 production, and numerous gene products associated with tissue remodeling and repair.

DIRECTION OF INNATE AND ADAPTIVE IMMUNITY BY CHEMOKINES

Although the precise factors that govern the expression of chemokine receptors are incompletely understood, it is clear that certain chemokines and receptors direct the trafficking of specific cell types [10,11]. During the innate immune response, inflammatory infiltrate is typically composed of neutrophils, followed by macrophages. Among the first inflammatory signals induced, the ELR CXC chemokines, and their receptors, CXCR1 and CXCR2, play primary roles in the recruitment of neutrophils. The recruitment of macrophages tends to be more context-specific, but CCR1, CCR2, and CCR5 (and their ligands) play predominant roles in the recruitment of macrophages in most settings. Most leukocyte subtypes express CCR1, which binds CCL3, CCL4, and CCL5. CCR2 binds all of the CCL2 family members (CCL2, 7, 8, 12, and 13). Under both Type 1 and Type 2 conditions, CCR2 and CCR6 play important roles in the trafficking of monocytes and dendritic cells to inflamed tissues.

G-protein–coupled receptors have historically served as good targets for pharmaceuticals. Consequently, a significant amount of investigation has sought to identify the chemokine receptor (or receptors) required for Th1 versus Th2 inflammatory diseases. This effort has met with limited success. In general, differential chemokine receptor expression tends to be present under Th1 and Th2 inflammatory conditions. Under Th1 conditions, CXCR3 and, to a lesser extent, CCR5 predominate. Conversely, under Th2 conditions, CCR3, CCR4, and CCR8 are preferentially expressed. Thus, under Th1 promoting conditions, CXCR3 ligands including IP-10/CXCL9, MIG/CXCL10, and I-TAC/CXCL11 preferentially attract Th1 T cells. CCR3, CCR4, and CCR8 ligands, including eotaxins/CCL11, 24, 26, MDC/CCL22, TARC/CCL17, and TCA-3/CCL1 preferentially attract Th2 cells, eosinophils, mast cells, and basophils.

CXCR3 is expressed by a number of cell types, but preferentially on Th1 and Tc1 cells. CXCR3 ligands CXCL9–11 were all originally identified as IFN-γ inducible chemokines, so it comes as no surprise that CXCR3 ligands predominate under Th1 conditions. However, CXCR3 is expressed by dendritic cells, both myeloid dendritic cells and plasmacytoid dendritic cells (Type I IFN producing innate immune cells), so while CXCR3 may be preferentially expressed during Th1 conditions, its expression is not limited to Th1 cells. Perhaps the best-characterized receptor is CCR5 due to its use as a coreceptor for HIV infectivity. This receptor binds CCL3, CCL4, and CCL5 and when ligated with these chemokines can inhibit HIV infectivity in vitro. CCR5 antagonists have been tested as anti-HIV therapeutics with minimal success.

Although a number of studies demonstrated that Th2 cells preferentially express CCR3, CCR4, and CCR8, these chemokine receptors may not be specific for Th2-mediated diseases. Although CCR3 is highly expressed on Th2 cell and eosinophils (a hallmark of allergic disease), the biology of CCR4 appears to be more complex. CCR4 has been linked to both Tregs and modulation of M1/M2 phenotype in macrophages.

Clearly, the regulation of chemokines dictates the nature of cellular recruitment. On the basis of numerous signaling pathways induced by chemokine receptor ligation, it is likely that chemokines also play a role in the differentiation of responding cells. However, the precise relationship between chemokine production and cellular differentiation is still unclear. Chemokines can aid in directing the immune response either directly by activating APCs and T cells or indirectly by recruiting the proper cell populations.

DEVELOPMENT AND ORGANIZATION OF LYMPHOID TISSUES

Chemokines and cytokines mediate the development and organization of lymphoid tissues [12]. Lymphoid tissues such as thymus, spleen, and lymph nodes are distinct organized tissues. One cytokine in particular, lymphotoxin, is integrally important for the development of lymph nodes, and LT-deficient mice do not develop lymph nodes. B-cell–rich follicles require the chemokine receptor/ligand pair CXCL13/CXCR5. Conversely, T-cell zones develop in response to stromal and dendritic cell production of the chemokines CCL19 and CCL21, which both bind CCR7. The position of B cells and T cells within secondary lymphoid tissues depends on their differential responsiveness to CCR7 and CXCR5 to first enter the lymph node and subsequently localize to the follicle, respectively. This is also a dynamic process. For instance, upon antigen exposure, B cells decrease responsiveness to CXCR5 and upregulate responsiveness to CCR7, which promotes localization to the B-cell/T-cell margin. In general, cells in the lymph nodes migrate through these opposing chemokine gradients by modulation of receptors, facilitating distribution and redistribution of leukocytes within the node. While this process is not completely understood, the manipulation of these chemokine receptor systems are central to proper positioning of immune cells for activation during immune events.

When dendritic cells encounter pathogen-associated antigen in peripheral tissues, they upregulate CCR7 on their surface. This facilitates the trafficking and entry into draining lymph nodes. At the same time naive and central memory T cells express CCR7 allowing entry of the lymph node from peripheral circulation. The expression of CCR7 on dendritic cells and T cells allows the localization in the proper zone (T-cell zone) of the lymph node and therefore antigen presentation is optimal. When T cells are primed in the lymph nodes, downregulation of CCR7, and the upregulation of other chemokine receptors, such as CXCR3, CCR4, or CCR5 mediates their sensitivity to chemokines produced in the inflamed tissues. Subsequent egression from the lymph node and recirculation to the inflamed tissue allows extravasation to the site of immune response and allows the activated T cell to perform their effector function in the local tissue. Thus, the coordinated expression of chemokine receptors mediates the T cells to be properly activated and subsequently perform their effector function in the most appropriate manner to rid the host of the inciting pathogens.

CYTOKINES IN REPAIR AND REGENERATION OF TISSUE

Whether an acute or chronic inflammatory response occurs in tissue to pathogens or any other inciting agent the damage induced by the responses requires that tissue repair occur eventually allowing normal tissue function. A number of key cytokines are required for the repair process to proceed. Those cytokines studied in this process include TGF-β, TNF-α, platelet-derived growth factor (PDGF), basic fibroblasts growth factor (βFGF), monocyte chemoattractant protein-1 (MCP-1), macrophage inflammatory protein-1 α (MIP-1α) and IL-1, IL-13 and IL-8. Within hours of injury, reepithelialization is initiated and the release of numerous growth factors, such as EGF, TGF-α, and FGF stimulate epithelial cell migration and proliferation allowing the wound to be closed.

Perhaps the most defined repair cytokine has been TGF-β. The most prominent effect of TGF is its ability to induce fibroblast production of extracellular matrix (ECM), especially collagen. This process is most important during tissue repair and allows the activation local fibroblasts to produce ECM, which provides a foundation for epithelial cell proliferation to restructure the tissue. Once the wound is closed the re-epithelialization can occur. In an optimal situation TGF-β is produced for a defined period of time allowing both matrix deposition for basement membrane organization and epithelial cell growth and a stimulus for other important processes including angiogenesis as well as immune cell regulation [13]. Together, these important mechanisms that are induced by TGF-β signal the end of the immune response and the beginning of tissue reorganization and repair. However, the overproduction of repair cytokines, such as TGF-β, can have grave detrimental effects on the host system leading to end-stage disease [14]. The chronic overexpression of TGF-β has been linked to almost all fibrotic diseases in the kidney, lung, liver, as well as in cardiac infarct repair leading to increased lesion size. Thus, many therapeutic approaches have attempted to target this cytokine but it has been fraught with complications since it is central to the several important biologic processes.

CONCLUSION

The coordination of inflammatory and immune responses is a complex, ongoing balance between the clearance of pathogens and simultaneous damage control for the host. The production of cytokines and chemokines during these responses participate in all aspects of the regulation, including initiation and maintenance of inflammation while inciting agents are present, as well as resolution and repair of damaged tissue after the response has subsided. A better understanding of the differential regulation of immune responses by particular cytokines and chemokines facilitates the development of therapies to target immune/inflammatory mediators during chronic disease. Thus, our continued research and application in appropriate models of disease will enhance our ability to identify clinically relevant targets.

REFERENCES

1. Shuai, K., and Liu, B. 2003. Regulation of JAK-STAT signaling in the immune system. *Nat Rev Immunol* **3**(11):900–911.
2. Chang, F., Steelman, L.S., Lee, J.T., et al. 2003. Signal transduction mediated by the Ras/Raf/MEK/ERK pathway from cytokine receptors to transcription factors: potential targeting for therapeutic intervention. *Leukemia* **17**(7):1263–1293.
3. Tamura, M., Gu, J., Matsumoto, K., Aota, S., Parsons, R., and Yamada, K.M. 1998. Inhibition of cell migration, spreading, and focal adhesions by tumor suppressor PTEN. *Science* **280**(5369):1614–1617.
4. Janeway, C.A., Jr., and Medzhitov, R. 2002. Innate immune recognition. *Annu Rev Immunol* **20**:197–216.
5. Trinchieri, G., and Sher, A. 2007. Cooperation of Toll-like receptor signals in innate immune defense. *Nat Rev Immunol* **7**(3):179–190.
6. Mosmann, T.R., and Coffman, R.L. 1989. TH1 and TH2 cells: different patterns of lymphokine secretion lead to different functional properties. *Annu Rev Immunol* **7**:145–173.
7. Weaver, C.T., Harrington, L.E., Mangan, P.R., Gavrieli, M., and Murphy, K.M. 2006. Th17: an effector CD4 T cell lineage with regulatory T cell ties. *Immunity* **24**(6):677–688.
8. Schubert, L.A., Jeffery, E., Zhang, Y., Ramsdell, F., and Ziegler, S.F. 2001. Scurfin (FOXP3) acts as a repressor of transcription and regulates T cell activation. *J Biol Chem* **276**(40):37672–37679.
9. Hori, S., Nomura, T., and Sakaguchi, S. 2003. Control of regulatory T cell development by the transcription factor Foxp3. *Science* **299**(5609):1057–1061.
10. Sallusto, F., and Lanzavecchia, A. 2000. Understanding dendritic cell and T-lymphocyte traffic through the analysis of chemokine receptor expression. *Immunol Rev* **177**:134–140.
11. Ward, S.G., and Marelli-Berg, F.M. 2009. Mechanisms of chemokine and antigen-dependent T-lymphocyte navigation. *Biochem J* **418**(1):13–27.
12. Cyster, J.G. 2003. Lymphoid organ development and cell migration. *Immunol Rev* **195**:5–14.
13. Li, M.O., Wan, Y.Y., Sanjabi, S., Robertson, A.K., and Flavell, R.A. 2006. Transforming growth factor-beta regulation of immune responses. *Annu Rev Immunol* **24**:99–146.
14. Border, W.A., and Ruoslahti, E. 1992. Transforming growth factor-beta in disease: the dark side of tissue repair. *J Clin Invest* **90**(1):1–7.

14 Adenosine Receptors: Therapeutic Aspects for Inflammatory and Immune Diseases

György Haskó and Bruce Cronstein

Metabolic stress, hypoxia, and cell damage cause adenosine to accumulate in the extracellular space, and increases in extracellular adenosine levels are observed in hypoxia, ischemia, inflammation, and trauma [1,2]. Extracellular adenosine levels accumulate following the release of adenosine from cells or as a consequence of extracellular degradation of released ATP and ADP. Intracellular adenosine, which can originate from increased intracellular metabolism of ATP during cellular stress or *S*-adenosyl homocysteine, is released through equilibrative nucleoside transporters ENT1 and ENT2. Extracellular ATP and ADP are catabolized by a cascade of ectoenzymes which consists of CD39 (ENTPD1 [ectonucleoside triphosphate diphosphohydrolase-1]), an enzyme that hydrolyzes ATP and ADP to AMP, and CD73 (ecto-5'nucleotidase), which in turn, rapidly dephosphorylates AMP to adenosine [3]. Owing to the widespread expression of equilibrative nucleoside transporters, adenosine derived from extracellular ATP is rapidly recycled from the extracellular space by uptake into cells. Adenosine in the cytosol is then either phosphorylated by adenosine kinase to AMP or metabolized by adenosine deaminase to inosine [4,5]. As a net result of these various metabolic processes, adenosine levels in the extracellular space are maintained in a range of 10–200 nM in normal, healthy tissues. In contrast, under pathophysiological conditions adenosine is generated at a rate that is higher than the rate of degradation leading to markedly increased extracellular adenosine levels that can range between 10 and 100 μM.

Adenosine produces its biological effects by binding to and activating one or more of four membrane spanning adenosine receptors, designated A_1, A_{2A}, A_{2B}, and A_3. All four adenosine receptors contain seven transmembrane domains and couple to intracellular GTP binding proteins (G proteins). Adenosine elicits activation of A_1, A_{2A}, and A_3 receptors with EC_{50}s that range from 0.01 to 1 μM, and A_{2B} receptor activation occurs at adenosine levels that exceed 10 μM (EC_{50}: 24 μM) [6]. Since physiological adenosine concentrations are less than 1 μM, physiological levels of adenosine can activate only A_1, A_{2A}, and A_3 receptors, and A_{2B} receptor activation requires pathophysiological conditions [5]. In addition to adenosine concentrations at the cell surface, receptor density and the functionality of the intracellular signaling pathways coupled to adenosine receptors are also key factors in dictating the nature and magnitude of the effect of adenosine on the cell. For example, A_{2A} receptor activation inhibits production of the T helper (Th)1-inducing cytokine interleukin-12 (IL-12) more potently by human monocytes that are pretreated with the proinflammatory cytokine IL-1 or tumor necrosis factor-α (TNF-α), mediators that also increase A_{2A} receptor expression in these cells [7]. In addition, the effect of adenosine can also be affected by the polarized localization of adenosine receptors: A_3 adenosine receptors accrue at the leading edge of migrating neutrophils and are instrumental in directing the movement of cells in response to chemotactic mediators [8]. Finally, it is important to keep in mind that results regarding adenosine receptor function in one species cannot be readily extrapolated to another one, because sequence differences in cloned adenosine receptors have been shown to be associated with differential pharmacological responses to selective agonists and antagonists. In this regard, A_3 receptors were not even discovered until they were cloned [9], because the A_3 receptor, especially in rodents, is insensitive to xanthines such as caffeine and theophylline, antagonists, which had been pivotal in identifying adenosine receptor-mediated effects.

ADENOSINE RECEPTOR SIGNALING

Adenosine receptors in general dictate cell function through coupling to G proteins, but some G-protein–independent actions have also been reported [5]. Adenosine receptors were initially classified as A_1

(cAMP-decreasing) or A_2 receptors (cAMP-increasing) [10]. Subsequently, the cAMP-increasing A_2 receptors were divided into two groups: high-affinity A_{2A} receptors and low-affinity A_{2B} receptors [11]. More recent studies have revealed that in addition to A_1 receptors, A_3 receptors also decrease intracellular cAMP concentrations [12]. In addition to signaling via the adenylyl cyclase-cAMP system, adenosine receptors can signal through a variety of other pathways. For example, A_1 receptor activation can be linked to various kinase pathways, including PI3-kinase, protein kinase C (PKC), and mitogen-activated protein (MAP) kinases [13]. In addition, A_1 receptor engagement can directly activate K^+ channels and inhibit Q-, P-, and N-type Ca^{2+} channels. A_{2A} receptors, as other Gs protein–coupled receptors, signal via chiefly the adenylate cyclase–cAMP–protein kinase A (PKA) canonical pathway, but they can also activate exchange factor directly activated by cAMP (Epac) [5]. Signaling downstream from PKA occurs through phosphorylation of the transcription factor CREB on serine residue 133 leading to direct CREB-mediated transcriptional activation [14]. Activated CREB can also modulate gene expression indirectly by competing with NF-κB or other transcription factors for an important cofactor, CBP [15]. In other cell types adenosine A_{2A} receptors stimulate collagen production via MAP kinases [16] and inhibit neutrophil superoxide production through activation of protein phosphatases [17]. Furthermore, recent results implicated c/EBPβ in the stimulatory effect of A_{2A} receptor agonists on IL-10 production by *Escherichia coli*–challenged macrophages [18]. A_{2B} receptor stimulation can induce both adenylyl cyclase activation via Gs and phospholipase C activation via Gq [19]. Interaction between these two pathways is important for upregulation of IL-4 production by mast cells upon A_{2B} receptor activation [20]. Specifically, Gq-mediated activation of phospholipase Cβ causes calcium mobilization and an increase in NFATc1-dependent IL-4 transcription, which response is further facilitated by Gs-mediated NFATc1 protein accumulation. Traditionally, A_3 receptor activation is linked to Gi-mediated inhibition of adenylyl cyclase and Gq-mediated stimulation of PLC [21] and A_3 receptors can activate PLD, RhoA, WNT, MAP kinase, and PI3 kinase pathways in governing cell function. For example, A_3 receptor–mediated augmentation of histamine released in sensitized murine mast cells was blocked by inactivating Gi proteins with pertussis toxin and by using pharmacological PI3 kinase inhibitors [22].

EFFECTS OF ADENOSINE ON THE CELLS OF INFLAMMATION

Neutrophils

Neutrophils are the most abundant white blood cell in the peripheral blood and are generally the earliest responders to trauma, infection, and bacterial invasion or other inflammatory stimuli. Adenosine is a potent regulator of neutrophil function and it has long been known that adenosine, acting at its receptors, regulates stimulated production of reactive oxygen species by these cells [23–26]. Adenosine, acting at A_{2A} receptors, also regulates phagocytosis [27,28]. In general neutrophils do not secrete large quantities of cytokines but because there are so many of these cells the cumulative contribution to proinflammatory cytokine levels at a given site are large and adenosine, acting at A_{2A} receptors regulate the production of a variety of cytokines including TNF-α, MIP-1α/CCL3, MIP-1β/CCL4, MIP-2α/CXCL2, and MIP-3α/CCL20 [29]. Neutrophils are recruited to an inflamed site by the postcapillary venular endothelium, which alters the expression of adhesive molecules on their surface in order to "capture" neutrophils from the circulation. Adenosine, acting at A_{2A} receptors, inhibits the adhesion of neutrophils to endothelium by diminishing the expression and "stickiness" of the neutrophil's adhesive molecules [30–38]. Interestingly, adenosine A_1 receptors increase neutrophil adhesion to different adhesive molecules on the endothelium and on other surfaces [30]. Once in the tissue neutrophils follow a gradient of chemoattractants to their source. There are many chemoattractants including activated complement components (C5a), bacterial products (formylated peptides), and chemokines and recent studies have indicated that adenosine receptors promote directed migration of neutrophils via A_1 and A_3 receptors [8,39,40]. More strikingly, neutrophils cluster their A_3 receptors at the leading edge of the cell and release ATP that is converted at the cell surface to adenosine, which then acts in an autocrine fashion to promote migration [8]. At inflamed sites, neutrophils undergo apoptosis and this process can be further stimulated by such cytokines as TNF-α. Adenosine, acting at A_{2A} receptors, prevents neutrophils from undergoing apoptosis [41–43]. Thus, nearly every function carried out by neutrophils is regulated by adenosine and its receptors.

Macrophages

The effect of adenosine on cytokine production by macrophages has been extensively investigated, for macrophage-derived cytokines are pivotal initiators and orchestrators of immune responses. The earliest studies concentrated on the ability of adenosine receptor ligands to suppress TNF-α production by activated macrophages. These early results have recently been complemented with data obtained by utilizing adenosine receptor KO mice, allowing a better elucidation of the receptor subtypes involved. On the basis of these results, it appears that the A_{2A} receptor is the dominant adenosine receptor subtype mediating the inhibition of TNF-αI secretion by adenosine [44–46]. In addition, since adenosine, NECA, and IB-MECA can each inhibit, albeit to a lesser extent, TNF-α production even in A_{2A} receptor KO mice [44,46], other adenosine receptors can also contribute to the inhibition of TNF-α production. By employing a combination approach consisting of A_{2A} receptor KO mice and the A_{2B} receptor antagonist MRS 1754, a recent study demonstrated a role for A_{2B} receptors as the other inhibitory receptor [44]. Nevertheless, A_{2B} receptors become functional only when their effect is not overshadowed by A_{2A} receptors, for both MRS 1754 and knockout of A_{2B} receptors in the presence of A_{2A} receptors fails to affect the inhibition of TNF-α production [44,45].

The regulation of IL-10 production also appears to be dependent primarily on A_{2A} receptors, with A_{2B} receptors having a minor role. In this context, adenosine failed to upregulate *E. coli*–induced IL-10 release by macrophages lacking A_{2A} receptors but not WT macrophages [18]. In RAW264.7 macrophages, which express low copy numbers of A_{2A} receptors, adenosine upregulates IL-10 production through an A_{2B} receptor-mediated posttranscriptional mechanism [47]. While A_{2B} receptor activation has a marginal effect on TNF-α and IL-10 production, it is central to the stimulatory effect of adenosine on IL-6 production, as NECA fails to induce the release of IL-6 in A_{2B} receptor KO but not WT macrophages [45].

Finally, it is noteworthy that the question of which adenosine receptors regulate cytokine production by human monocytes/macrophages is even less clear, with data supporting a role for all four receptors [48]. However, because these human results are derived from pharmacological approaches, they must be interpreted with caution, as many of the ligands used are not particularly selective. However, the recent demonstration of elevated concentrations of adenosine contained in human newborn plasma when compared to adult plasma coupled with the observation that degrading adenosine with adenosine deaminase or interrupting adenosine signaling by adenosine receptor blockade augments TNF-α production by neonatal but not adult blood, indicate the clinical importance of the suppressive effect of adenosine on TNF-α production in human immunity [49].

Dendritic Cells

The only available human studies demonstrate a relatively consistent role for adenosine receptors in regulating dendritic cell function. Gi-coupled (A_1, A_3, or both depending on the experimental system) adenosine receptors are present on both immature myeloid [50] and plasmocytoid dendritic cells [51], and their activation leads to the liberation of intracellular calcium from its stores and reorganization of the actin cytoskeleton. This actin reorganization is instrumental in causing immature dendritic cells to migrate along different concentration gradients of adenosine, indicating that adenosine receptor activation triggers chemotaxis in immature dendritic cells [50,51]. This chemotactic effect can be seen only in immature dendritic cells, as the Gi-coupled adenosine receptors undergo downregulation during dendritic cell maturation [50,51].

Although immature dendritic cells express A_{2A} receptors, they appear to be nonfunctional, as their activation is unable to trigger intracellular signaling events such as increase of intracellular cAMP [50]. Dendritic cell maturation, however, is associated with the appearance of A_{2A} receptor-mediated signaling responses, due to both increased expression and coupling of A_{2A} receptors [50,51]. A_{2A} receptor activation on mature dendritic cells results in a shift in the pattern of cytokine release from a proinflammatory to an anti-inflammatory profile, characterized by decreased IL-12, IL-6, and IFN-α production and increased IL-10 production [50–52]. Owing to this adenosine-induced shift in dendritic cell cytokine production away from the Th1-inducing IL-12 toward the Th2-inducing IL-10, dendritic cells in the presence of adenosine have a decreased ability to elicit T helper (Th)1 versus Th2 polarization of naive CD4+ cells [52].

Overall, the available data demonstrate a dual role for adenosine in regulating dendritic cell function: adenosine promotes the migration of immature dendritic cells to inflammatory sites through A_1 or A_3 receptors, where adenosine triggers, via A_{2A} receptors, an anti-inflammatory dendritic cell phenotype shifting T-cell responses toward a Th2 direction.

Mast Cells

As a result of the demonstration that inhaled adenosine elicits bronchoconstriction in patients afflicted with asthma but not healthy volunteers, adenosine, and adenosine receptors have emerged as potential therapeutic targets in asthma [53]. Although the bronchospastic mechanisms of adenosine are incompletely understood, there is an increasing body of evidence incriminating mast cell–derived mediators such as cytokines and histamine as having a central role in provoking airway constriction following adenosine inhalation. Although A_{2A}, A_{2B}, and A_3 receptors are expressed on the cell membrane of mast cells [54], it is unclear which receptor(s) account for the augmented mast cell activation in response to adenosine.

Adenosine augments release of the pivotal asthma-inducing cytokine IL-13 by murine mast cells from WT but not A_{2B} receptor KO mice indicating that A_{2B} receptors are responsible for the proinflammatory effects of adenosine [55]. Furthermore, studies with adenosine receptor KO mice demonstrated that A_3 but not A_{2B}

receptor activation can trigger the release of histamine from naive lung mast cells [22]. Thus, A_{2B} and A_3 receptors subserve different proinflammatory roles in murine mast cells in the absence of specific antigenic stimulation. The role of A_{2B} receptors in mediating antigen-stimulated histamine release by murine mast cells is subject to controversy. Although mast cells isolated from WT animals exhibit decreased histamine release in response to antigenic stimulation when compared to mast cells from A_{2B} receptor KO mice [56], exogenous adenosine retains its stimulatory effect in A_{2B} receptor KO mast cells [20]. Therefore, it is plausible that the decreased release of histamine from mast cells obtained from A_{2B} receptor WT versus KO mice indicates an impaired responsiveness of A_{2B} receptor WT versus KO cells to antigen that may originate from altered development of mast cells in A_{2B} receptor KO mice. In contrast to A_{2B} receptors, A_3 receptor activation directly augments histamine release in antigen-stimulated mouse mast cells, as the stimulatory effect of added adenosine that can be observed in WT mast cells is not seen in A_3 receptor KO mast cells [57].

Unlike results noted with rodent cells, in human and canine mast cells, adenosine-induced degranulation or cytokine release is mediated predominantly by A_{2B} receptor activation [58,59]. Stimulation of A_{2B} receptors on human mastocytoma HMC-1 cells elicits production of the T helper 2 cytokines IL-4 and IL-13, in addition to a several other proinflammatory cytokines including IL-1β and IL-8. Although adenosine receptor activation is incapable of triggering histamine release from human mast cells [56], the demonstration of augmented IL-4 and IL-13 production in response to adenosine receptor stimulation indicates that human mast cell adenosine receptors are important factors in the pathophysiology of asthma.

Lymphocytes

Several recent studies utilizing adenosine receptor KO mice have examined the effect of adenosine receptor activation on various lymphocyte functions, and dominant view based on these studies is that A_{2A} receptors are the primary adenosine receptors in shaping lymphocyte responses. Studies employing A_{2A} KO mice have shown that A_{2A} receptor activation limits IL-2 secretion by naive CD4+ T-cells [60] leading to their reduced proliferation [61] following T-cell receptor engagement. In addition, A_{2A} receptor activation decreases the production of both IFN-γ and IL-4 by both naive CD4+ T cells [60,62] and polarized Th1 and Th2 cells [63], casting doubt on the hypothesis that A_{2A} receptor stimulation in lymphocytes might switch Th cell responses toward a Th2 profile. A_{2A} receptor stimulation also upregulates the expression of negative costimulatory molecules including cytotoxic T-lymphocyte antigen-4 (CTLA-4) and programmed cell death-1 (PD-1), and downregulates expression of the positive costimulatory molecule CD-40L [61].

As with CD4+ cells, adenosine limits IL-2 production by both polarized Tc1 and Tc2 CD8+ cytotoxic cells, which effect is mediated through A_{2A} receptors [64]. However, the production of neither Tc1 (IFN-γ) nor Tc2 (IL-4 and IL-5) cytokines was affected by A_{2A} receptor stimulation. Moreover, A_{2A} receptor stimulation failed to inhibit Tc1 or Tc2 cell cytotoxicity [64], which indicates that another subtype, possibly the A_3 receptor may dictate the anticytotoxic effect of adenosine observed in earlier studies [65,66]. Recent analysis of A_1, A_{2A}, and A_3 receptor KO mice, however, indicate a key role for A_{2A} receptors in decreasing the cytolytic activity of IL-2-activated natural killer (NK) cells [67].

Adenosine has emerged as a major mediator of the immune suppressive effects of regulatory T (Treg) cells, which have a key role in harnessing the immune system, thereby preventing excessive immune-mediated tissue injury. Studies show that Tregs, as defined by expression of CD4+/CD25+/Foxp3+ and a capacity to limit the proliferation of CD4+/CD25– cells, express high levels of both CD39 [68] and CD73 [68,69]. On the basis of these results CD39 and CD73 were proposed to serve as novel-specific markers of Treg cells. Recent studies confirmed the intimate relationship between Foxp3+ expression and CD39 expression by demonstrating that Foxp3+ triggers the expression of CD39 [70].

Evidence indicates that CD73 and CD39 have important roles in mediating the immune suppressive properties of Tregs by hydrolyzing exogenous ATP/ADP to generate adenosine [68]. Specifically, Tregs obtained from CD39 KO mice lost their ability to suppress proliferation of CD4+/CD25– cells, which effect was reinstituted by introducing soluble exogenous NTPDases [68]. In addition, a role for adenosine was further highlighted by studies demonstrating that A_{2A} receptor KO target cells (CD4+/CD25–) proliferated more than WT cells when cultured with WT Tregs. The intimate relationship between A_{2A} receptors and Tregs is also indicated by the observation that A_{2A} receptor activation augments Foxp3+ expression in T cells [71].

A subset of T cells known as invariant NKT cells (iNKT) are important for the innate immune response that induces very rapid host responses to infection [72,73]. A_{2A} receptors are expressed on iNKT cells and markedly suppress the production of proinflammatory cytokines such as INF-γ [74]. Since NKT cell activation has been implicated in a wide range of disease processes including atherosclerosis, type 1 diabetes, arthritis, and various allergic diseases [75], the presence of A_{2A} receptors on NKT cells indicates novel therapeutic applications of A_{2A} receptor ligands.

Endothelial Cells

Endothelial cells express predominantly A_{2A} and A_{2B} adenosine receptors although A_{2B} receptors appear to play a larger role in regulation of the microvasculature [76]. At inflamed sites the vascular endothelium expresses or upregulates the expression of adhesive molecules, including E-selectin, VCAM-1, and ICAM-1, on their surface, which mediates recruitment of leukocytes. Adenosine A_{2A} receptors inhibit the expression of E-selectin and VCAM-1 but do not affect ICAM-1 upregulation [77], effects apparently mediated by inhibition of NF-κB activation in the endothelium [78]. More recently, similar effects have been ascribed to the A_{2B} receptor as well [79]. In addition, the vascular endothelium secretes a variety of chemokines and cytokines at inflamed sites, including IL-8 and IL-6, and adenosine, acting at A_{2A} and A_{2B} receptors, diminishes their expression [77]. Thus, adenosine diminishes the endothelial contribution to inflammation and leukocyte recruitment to inflamed sites.

The vascular endothelium is also an important source of adenosine [80–82], particularly during inflammation, by extracellularly converting adenine nucleotides to adenosine via CD39/CD73 expressed on their surface [83–86]. Inflammatory cytokines such as IFN-α, promote the expression of CD73 in a form of feedback regulation of inflammation as well [87]. The adenosine released by the endothelium not only protects the endothelium and the underlying tissue from attack by stimulated leukocytes [85,88] but inhibits the vascular leakage (edema) that is so prominent at inflamed sites via activation of adenosine A_{2B} receptors [84,89–91].

As discussed later, adenosine and its receptors play an important role in granulation tissue formation in wounded tissues. The central event in granulation tissue formation is angiogenesis and adenosine, acting at both A_{2A} and A_{2B} receptors, stimulates angiogenesis. Application of an adenosine A_{2A} receptor agonist to wounds increases blood vessel formation [92–95] and enhances wound healing. This effect is mediated, in part, by promotion of angiogenic factors by endothelial cells and macrophages and downregulation of antiangiogenic factors [96–98]. In other tissues, such as the retina, adenosine A_{2B} receptors mediate this function [99–101].

OVERALL EFFECTS OF ADENOSINE AND ITS RECEPTORS ON INFLAMMATION

Asthma and Chronic Obstructive Pulmonary Disease

Inhaled adenosine can trigger bronchospasm in patients with asthma and chronic obstructive pulmonary disease (COPD), but not healthy patients, and adenosine receptor antagonists can block this bronchospastic response [54]. These observations coupled with the findings that both adenosine concentrations and adenosine receptor abundance on immune cells in the lung are increased in individuals with asthma and COPD link endogenous adenosine to pathophysiological mechanisms in lung diseases. Because the bronchospastic response to adenosine can be prevented by agents that stabilize the mast cell membrane and prevent the release of injurious mediators, and because histamine receptor blockers can also prevent adenosine-mediated bronchoconstriction, mast cells have been proposed as major targets of the bronchoconstrictive effects of adenosine [54]. Rodent models of asthma have confirmed this, and have also incriminated A_{2B} [55] and A_3 [22,57] receptors as contributing to the stimulatory effect of adenosine on mast cell activation. In contrast, studies employing human mast cells have demonstrated that A_{2B} receptors mediate the proinflammatory effects of adenosine in humans. A_{2B} stimulation is proinflammatory not only in mast cells, but also in human bronchial smooth muscle cells [102], human bronchial epithelial cells [103], and human lung fibroblasts [104], which produce increased amounts of IL-6 [102,104] and IL-19 [103] in response to adenosine. A_{2B} receptors promote the differentiation of human lung fibroblasts into myofibroblasts, a major cell type that produces extracellular matrix [22]. This suggests that adenosine may participate in the fibrosis and remodeling of the lung during asthma and COPD.

Preclinical data showing the effectiveness of A_{2B} receptor antagonism in preventing disease progression in rodent animal models [105,106] suggest that selective A_{2B} antagonists may be a feasible option for the treatment of human patients suffering from asthma and COPD. This notion is underlined by evidence that enprofylline, an antiasthmatic agent, is a relatively selective, albeit not potent A_{2B} receptor antagonist [107]. In this regard it is noteworthy that CVT-6883 [105,106], a selective A_{2B} receptor antagonist that has demonstrated efficacy in preventing disease in rodent models of asthma and COPD appeared to be safe and well tolerated in a recent human Phase 1 study.

A_{2A} agonists have been found to have widespread anti-inflammatory effects in pulmonary inflammation [108]. Bone marrow transfer experiments demonstrated that A_{2A} receptor activation only decreased lung inflammation when the A_{2A} receptor was present on bone marrow–derived cells. In addition, utilizing A_{2A} receptor KO mice, Nadeem et al. showed that the adenosine–A_{2A} receptor–cAMP axis is a potent endogenous anti-inflammatory signaling mechanism that prevents airway hyperreactivity and inflammatory cell sequestration following ragweed sensitization [109]. Thus, A_{2A} agonists appear to be effective at harnessing inflammatory lung tissue injury. However, in clinical trials, the utility of the A_{2A} receptor agonist GW328267X was limited by cardiovascular side effects [110].

Reperfusion Injury

A_{2A} receptors on bone marrow–derived cells can protect organs from reperfusion injury following an ischemic episode. The organs that have been shown to be protected in this way include liver [111], kidney [112], heart [113], skin [114], spinal cord [115,116], and lung [117]. In liver [74] and kidney [118], the proinflammatory activities of iNKT cells have been found to be inhibited by A_{2A} receptor activation, and to be critical targets of the organ protective effects of A_{2A} adenosine receptor activation.

Inflammatory Arthritis

Methotrexate is the mainstay of therapies for inflammatory arthritis. At low doses methotrexate enters cells and becomes polyglutamated to a long-lasting metabolite [119]. AICAR transformylase, an enzyme which is part of the de novo purine synthetic pathway, is inhibited by methotrexate polyglutamates [120] causing, the intracellular accumulation of AICAR [121,122]. AICAR competitively inhibits AMP deaminase causing, ultimately enhanced release of adenosine from cells [121]. Although it is difficult to measure adenosine levels in blood and other fluids because of the short half-life (2–8 seconds) of adenosine [123], in patients suffering from rheumatoid arthritis treated with methotrexate there is strong evidence that methotrexate therapy increases adenosine release [124,125]. Experiments employing rodents show that the anti-inflammatory effects of methotrexate are mediated by adenosine and that the anti-inflammatory effect of methotrexate is absent if animals are administered adenosine receptor antagonists or if their adenosine A_{2A} or A_3 receptors have been knocked out [122,126–132]. Similarly, methotrexate loses its efficacy in patients with rheumatoid arthritis who consume significant quantities of caffeine, an adenosine receptor antagonist [133]. Thus, by increasing adenosine release at inflamed sites methotrexate attenuates inflammation in patients with rheumatoid arthritis. These observations that methotrexate exerts some of its anti-inflammatory effects through A_3 receptors together with the demonstration that the selective A_3 receptor agonist IB-MECA suppresses the course of arthritis in collagen-induced arthritis in mice [134], led to clinical trials with IB-MECA (CFA101) in patients with rheumatoid arthritis. IB-MECA was safe and well tolerated in this Phase II trial, and patients that were given this drug attained a moderate attenuation of symptoms [135].

Sepsis

Mouse models employing both knockout and pharmacological approaches have delineated the role of the different adenosine receptors in governing the physiological response of the host to sepsis [136]. Blockade of A_1 receptors augmented mortality in intraperitoneal sepsis induced by cecal ligation and puncture in mice, which was associated with increased inflammation-induced hepatic and renal injury [137]. Similar to these results, both knockout and pharmacological inactivation of A_3 receptors enhanced mortality, in mice challenged with cecal ligation and puncture [138]. In contrast to the deleterious effects of blockade of A_1 and A_3 receptors, A_{2A} receptor blockade by either gene deletion or administration of ZM241385 diminished cecal ligation and puncture-induced lethality. This protective effect occurred by a mechanism that involved decreased bacterial growth, which was due to preserved immune system function [139]. Surprisingly, A_{2A} receptor activation, when combined with antibiotics, has been shown to decrease lethality from sepsis induced by injecting with live *E. coli*, possibly by suppressing an exaggerated inflammatory response that can be caused by the rapid drug-induced killing of large number of bacteria [140]. Finally, it appears that A_{2B} receptor inactivation in cecal ligation and puncture results in an augmented inflammatory response (Csoka, Hasko, Nemeth, and Pacher, unpublished observation).

Inflammatory Bowel Disease

Ulcerative colitis and Crohn's disease are chronic, relapsing diseases characterized by dysfunctional mucosal T lymphocytes, imbalances in cytokine production and inflammation leading to damage of the intestinal mucosa. Although the etiology of inflammatory bowel disease (IBD) is elusive, the consensus is that that disease is the result of a dysregulated immune response to luminal antigens in a genetically susceptible host [141,142]. A_{2A} receptor activation has been shown to be protective in several animal models of IBD, and these protective effects can be attributed to two main mechanisms: attenuation of inflammatory cell sequestration and function in the mucosa, and enhanced activity of Tregs. Contribution of the former mechanism is based on results that A_{2A} receptor stimulation with ATL146e diminishes both leukocyte infiltration and the release of inflammatory cytokines by disease-promoting effector T cells [143]. The latter mechanism was predicated on data that co-transfer of CD25+ CD4+ T regulatory cells obtained from A_{2A} receptor WT mice could prevent disease induction SCID mice that were transferred adoptively with disease-inducing CD45RB^{high} CD4+ T cells, while co-administration of CD25+ CD4+ Tregs obtained from A_{2A} receptor KO mice together with disease-inducing CD45RB^{high} CD4+ T cells was inefficient in preventing the development of disease [60].

Stimulation of A_{2B} receptors on intestinal epithelial cells increases IL-6 production by these cells leading to increased neutrophil activation [144]. In addition, A_{2B} receptors expression increases in epithelial cells isolated from murine or human colitis [145]. On the basis of these observations, recent studies have assessed the effect of A_{2B} receptor blockade in IBD. Treating mice with a selective A_{2B} antagonist, ATL-801, strongly decreased IL-6 production and neutrophil infiltration, reduced mucosal damage, and ameliorated the course of disease in a model of IBD [146]. These are encouraging results, because blockade of A_{2B} receptor might be a potentially advantageous treatment option that selectively targets the gut of patients suffering from IBD. This idea is based on results that A_{2B} receptor expression is highest in the gut as compared to other organs and that concentrations of endogenous adenosine that seem to propagate the disease process through A_{2B} receptors are elevated exceedingly in the gut [147].

Finally, it is noteworthy that in addition to endogenous adenosine, its metabolite inosine can also ameliorate the course of experimental IBD [148]. Since inosine can potently deactivate macrophages, in part, through A_{2A} receptors [149], one possibility is that inosine exerts its beneficial effect by preventing production of macrophage-derived inflammatory cytokines in the gut [150,151].

Wound Healing

Inflammation in the skin is part of a continuum in which damaged tissue is repaired and replaced, leading to wound healing. Montesinos et al. first showed that applying adenosine A_{2A} receptor agonists topically increases the rate of wound healing in normal and diabetic mice and rats [92]. Adenosine receptor agonists increased both matrix and vessel formation and their effects were superior to that induced by recombinant platelet-derived growth factor [92–94,152]. Activation of A_{2A} and A_{2B} receptors enhances angiogenesis both directly and indirectly: adenosine receptors directly promote proliferation of microvascular endothelial cells in part through the autocrine production of VEGF, a central stimulus for endothelial proliferation and angiogenesis [7,99–101,153–159] and suppress the production of thrombospondin 1, an antiangiogenic matrix protein [97]. Furthermore, adenosine receptor stimulation upregulates VEGF production by macrophages [96,152,160]. In addition, A_{2A} and A_{2B} receptor stimulation enhances matrix production by fibroblasts, a central step in tissue repair [104,161–163]. In this regard, it is worth mentioning that A_{2A} receptor KO mice fail to generate granulation tissue confirming that endogenous adenosine has a central role in wound healing [93]. Sonedenoson (MRE0094, King Pharmaceuticals), an A_{2A} agonist that regulates the inflammatory response and enhances tissue regeneration, is currently undergoing trials for the treatment of diabetic foot ulcers.

CONCLUSIONS

A recent increase in our understanding of the role of the various adenosine receptors in regulating tissue injury have led to the identification of novel pharmacological targets to restore tissue function in various diseases. Targeting adenosine receptors with agonists, antagonists, and interfering with the function of enzymes and transporters that are responsible for the accumulation of extracellular adenosine represent a wide range of approaches to alter adenosine signaling at inflammatory sites.

REFERENCES

1. Linden, J. 2001. Molecular approach to adenosine receptors: receptor-mediated mechanisms of tissue protection. *Annu Rev Pharmacol Toxicol* **41**:775–787.
2. Fredholm, B.B., AP, I.J., Jacobson, K.A., Klotz, K.N., and Linden, J. 2001. International union of pharmacology. XXV. Nomenclature and classification of adenosine receptors. *Pharmacol Rev* **53**:527–552.
3. Yegutkin, G.G. 2008. Nucleotide- and nucleoside-converting ectoenzymes: Important modulators of purinergic signalling cascade. *Biochem Biophys Acta* **1783**:673–694.
4. Hasko, G., and Cronstein, B.N. 2004. Adenosine: an endogenous regulator of innate immunity. *Trends Immunol* **25**:33–39.
5. Fredholm, B.B. 2007. Adenosine, an endogenous distress signal, modulates tissue damage and repair. *Cell Death Differ* **14**:1315–1323.
6. Fredholm, B.B., Irenius, E., Kull, B., and Schulte, G. 2001. Comparison of the potency of adenosine as an agonist at human adenosine receptors expressed in Chinese hamster ovary cells. *Biochem Pharmacol* **61**:443–448.
7. Khoa, N.D., Montesinos, M.C., Reiss, A.B., Delano, D., Awadallah, N., and Cronstein, B.N. 2001. Inflammatory cytokines regulate function and expression of adenosine A(2A) receptors in human monocytic THP-1 cells. *J Immunol* **167**:4026–4032.
8. Chen, Y., Corriden, R., Inoue, Y., et al. 2006. ATP release guides neutrophil chemotaxis via P2Y2 and A3 receptors. *Science* **314**:1792–1795.
9. Zhou, Q.Y., Li, C., Olah, M.E., Johnson, R.A., Stiles, G.L., and Civelli, O. 1992. Molecular cloning and characterization of an adenosine receptor: the A3 adenosine receptor. *Proc Natl Acad Sci USA* **89**:7432–7436.
10. van Calker, D., Muller, M., and Hamprecht, B. 1979. Adenosine regulates via two different types of receptors, the accumulation of cyclic AMP in cultured brain cells. *J Neurochem* **33**:999–1005.
11. Bruns, R.F., Lu, G.H., and Pugsley, T.A. 1986. Characterization of the A2 adenosine receptor labeled by [3H]NECA in rat striatal membranes. *Mol Pharmacol* **29**:331–346.

12. Jin, X., Shepherd, R.K., Duling, B.R., and Linden, J. 1997. Inosine binds to A3 adenosine receptors and stimulates mast cell degranulation. *J Clin Invest* **100**:2849–2857.
13. Jacobson, K.A., and Gao, Z.G. 2006. Adenosine receptors as therapeutic targets. *Nat Rev Drug Discov* **5**:247–264.
14. Nemeth, Z.H., Leibovich, S.J., Deitch, E.A., et al. 2003. Adenosine stimulates CREB activation in macrophages via a p38 MAPK-mediated mechanism. *Biochem Biophys Res Commun* **312**:883–888.
15. Fredholm, B.B., Chern, Y., Franco, R., and Sitkovsky, M. 2007. Aspects of the general biology of adenosine A2A signaling. *Prog Neurobiol* **83**:263–276.
16. Che, J., Chan, E.S., and Cronstein, B.N. 2007. Adenosine A2A receptor occupancy stimulates collagen expression by hepatic stellate cells via pathways involving protein kinase A, Src, and extracellular signal-regulated kinases 1/2 signaling cascade or p38 mitogen-activated protein kinase signaling pathway. *Mol Pharmacol* **72**:1626–1636.
17. Revan, S., Montesinos, M.C., Naime, D., Landau, S., and Cronstein, B.N. 1996. Adenosine A2 receptor occupancy regulates stimulated neutrophil function via activation of a serine/threonine protein phosphatase. *J Biol Chem* **271**:17114–17118.
18. Csoka, B., Nemeth, Z.H., Virag, L., et al. 2007. A2A adenosine receptors and C/EBPbeta are crucially required for IL-10 production by macrophages exposed to Escherichia coli. *Blood* **110**:2685–2695.
19. Feoktistov, I., and Biaggioni, I. 1997. Adenosine A2B receptors. *Pharmacol Rev* **49**:381–402.
20. Ryzhov, S., Goldstein, A.E., Biaggioni, I., and Feoktistov, I. 2006. Cross-talk between G(s)- and G(q)-coupled pathways in regulation of interleukin-4 by A(2B) adenosine receptors in human mast cells. *Mol Pharmacol* **70**:727–735.
21. Gessi, S., Merighi, S., Varani, K., Leung, E., Mac Lennan, S., and Borea, P.A. 2008. The A3 adenosine receptor: an enigmatic player in cell biology. *Pharmacol Ther* **117**:123–140.
22. Zhong, H., Shlykov, S.G., Molina, J.G., et al. 2003. Activation of murine lung mast cells by the adenosine A3 receptor. *J Immunol* **171**:338–345.
23. Cronstein, B.N., Rosenstein, E.D., Kramer, S.B., Weissmann, G., and Hirschhorn, R. 1985. Adenosine; a physiologic modulator of superoxide anion generation by human neutrophils. Adenosine acts via an A2 receptor on human neutrophils. *J Immunol* **135**:1366–1371.
24. Cronstein, B.N., Kramer, S.B., Weissmann, G., and Hirschhorn, R. 1983. Adenosine: a physiological modulator of superoxide anion generation by human neutrophils. *J Exp Med* **158**:1160–1177.
25. Visser, S.S., Theron, A.J., Ramafi, G., Ker, J.A., and Anderson, R. 2000. Apparent involvement of the A(2A) subtype adenosine receptor in the anti-inflammatory interactions of CGS 21680, cyclopentyladenosine, and IB-MECA with human neutrophils. *Biochem Pharmacol* **60**:993–999.
26. Varani, K., Gessi, S., Dionisotti, S., Ongini, E., and Borea, P.A. 1998. [3H]-SCH 58261 labeling of functional A2A adenosine receptors in human neutrophil membranes. *Br J Pharmacol* **123**:1723–1731.
27. Nishida, Y., Honda, Z., and Miyamoto, T. 1987. Suppression of human polymorphonuclear leukocyte phagocytosis by adenosine analogs. *Inflammation* **11**:365–369.
28. Salmon, J.E., and Cronstein, B.N. 1990. Fc gamma receptor-mediated functions in neutrophils are modulated by adenosine receptor occupancy. A1 receptors are stimulatory and A2 receptors are inhibitory. *J Immunol* **145**:2235–2240.
29. McColl, S.R., St-Onge, M., Dussault, A.A., et al. 2006. Immunomodulatory impact of the A2A adenosine receptor on the profile of chemokines produced by neutrophils. *FASEB J* **20**:187–189.
30. Cronstein, B.N., Levin, R.I., Philips, M., Hirschhorn, R., Abramson, S.B., and Weissmann, G. 1992. Neutrophil adherence to endothelium is enhanced via adenosine A1 receptors and inhibited via adenosine A2 receptors. *J Immunol* **148**:2201–2206.
31. Lesch, M.E., Ferin, M.A., Wright, C.D., and Schrier, D.J. 1991. The effects of (R)-N-(1-methyl-2-phenylethyl) adenosine (L-PIA), a standard A1-selective adenosine agonist on rat acute models of inflammation and neutrophil function. *Agents Actions* **34**:25–27.
32. Firestein, G.S., Bullough, D.A., Erion, M.D., et al. 1995. Inhibition of neutrophil adhesion by adenosine and an adenosine kinase inhibitor. The role of selectins. *J Immunol* **154**:326–334.
33. Derian, C.K., Santulli, R.J., Rao, P.E., Solomon, H.F., and Barrett, J.A. 1995. Inhibition of chemotactic peptide-induced neutrophil adhesion to vascular endothelium by cAMP modulators. *J Immunol* **154**:308–317.
34. Bullough, D.A., Magill, M.J., Firestein, G.S., and Mullane, K.M. 1995. Adenosine activates A2 receptors to inhibit neutrophil adhesion and injury to isolated cardiac myocytes. *J Immunol* **155**:2579–2586.
35. Jordan, J.E., Zhao, Z.Q., Sato, H., Taft, S., and Vinten-Johansen, J. 1997. Adenosine A2 receptor activation attenuates reperfusion injury by inhibiting neutrophil accumulation, superoxide generation and coronary endothelial adherence. *J Pharmacol Exp Ther* **280**:301–309.
36. Zhao, Z.Q., Sato, H., Williams, M.W., Fernandez, A.Z., and Vinten-Johansen, J. 1996. Adenosine A2-receptor activation inhibits neutrophil-mediated injury to coronary endothelium. *Am J Physiol* **271**:H1456–H1464.
37. Okusa, M.D., Linden, J., Huang, L., Rieger, J.M., Macdonald, T.L., and Huynh, L.P. 2000. A(2A) adenosine receptor-mediated inhibition of renal injury and neutrophil adhesion. *Am J Physiol Renal Physiol* **279**: F809–F818.
38. Sullivan, G.W., Lee, D.D., Ross, W.G., et al. 2004. Activation of A2A adenosine receptors inhibits expression of alpha 4/beta 1 integrin (very late antigen-4) on stimulated human neutrophils. *J Leukoc Biol* **75**: 127–134.
39. Rose, F.R., Hirschhorn, R., Weissmann, G., and Cronstein, B.N. 1988. Adenosine promotes neutrophil chemotaxis. *J Exp Med* **167**:1186–1194.
40. Cronstein, B.N., Daguma, L., Nichols, D., Hutchison, A.J., and Williams, M. 1990. The adenosine/neutrophil paradox resolved: human neutrophils possess both A1 and A2 receptors that promote chemotaxis and inhibit O2 generation, respectively. *J Clin Invest* **85**: 1150–1157.
41. Mayne, M., Fotheringham, J., Yan, H.J., et al. 2001. Adenosine A2A receptor activation reduces proinflammatory events and decreases cell death following intracerebral hemorrhage. *Ann Neurol* **49**:727–735.
42. Yasui, K., Agematsu, K., Shinozaki, K., et al. 2000. Theophylline induces neutrophil apoptosis through adenosine A2A receptor antagonism. *J Leukoc Biol* **67**:529–535.
43. Walker, B.A., Rocchini, C., Boone, R.H., Ip, S., and Jacobson, M.A. 1997. Adenosine A2a receptor activation delays apoptosis in human neutrophils. *J Immunol* **158**:2926–2931.

44. Kreckler, L.M., Wan, T.C., Ge, Z.D., and Auchampach, J.A. 2006. Adenosine inhibits tumor necrosis factor-alpha release from mouse peritoneal macrophages via A2A and A2B but not the A3 adenosine receptor. *J Pharmacol Exp Ther* **317**:172–180.
45. Ryzhov, S., Zaynagetdinov, R., Goldstein, A.E., et al. 2008. Effect of A2B adenosine receptor gene ablation on adenosine-dependent regulation of proinflammatory cytokines. *J Pharmacol Exp Ther* **324**:694–700.
46. Hasko, G., Kuhel, D.G., Chen, J.F., et al. 2000. Adenosine inhibits IL-12 and TNF-[alpha] production via adenosine A2a receptor-dependent and independent mechanisms. *FASEB J* **14**:2065–2074.
47. Nemeth, Z.H., Lutz, C.S., Csoka, B., et al. 2005. Adenosine augments IL-10 production by macrophages through an A2B receptor-mediated posttranscriptional mechanism. *J Immunol* **175**:8260–8270.
48. Hasko, G., Pacher, P., Deitch, E.A., and Vizi, E.S. 2007. Shaping of monocyte and macrophage function by adenosine receptors. *Pharmacol Ther* **113**:264–275.
49. Levy, O., Coughlin, M., Cronstein, B.N., Roy, R.M., Desai, A., and Wessels, M.R. 2006. The adenosine system selectively inhibits TLR-mediated TNF-alpha production in the human newborn. *J Immunol* **177**:1956–1966.
50. Panther, E., Idzko, M., Herouy, Y., et al. 2001. Expression and function of adenosine receptors in human dendritic cells. *FASEB J* **15**:1963–1970.
51. Schnurr, M., Toy, T., Shin, A., et al. 2004. Role of adenosine receptors in regulating chemotaxis and cytokine production of plasmacytoid dendritic cells. *Blood* **103**:1391–1397.
52. Panther, E., Corinti, S., Idzko, M., et al. 2003. Adenosine affects expression of membrane molecules, cytokine and chemokine release, and the T-cell stimulatory capacity of human dendritic cells. *Blood* **101**:3985–3990.
53. Cushley, M.J., Tattersfield, A.E., and Holgate, S.T. 1983. Inhaled adenosine and guanosine on airway resistance in normal and asthmatic subjects. *Br J Clin Pharmacol* **15**:161–165.
54. Polosa, R., and Holgate, S.T. 2006. Adenosine receptors as promising therapeutic targets for drug development in chronic airway inflammation. *Curr Drug Targets* **7**:699–706.
55. Ryzhov, S., Zaynagetdinov, R., Goldstein, A.E., et al. 2008. Effect of A2B adenosine receptor gene ablation on proinflammatory adenosine signaling in mast cells. *J Immunol* **180**:7212–7220.
56. Hua, X., Kovarova, M., Chason, K.D., Nguyen, M., Koller, B.H., and Tilley, S.L. 2007. Enhanced mast cell activation in mice deficient in the A2b adenosine receptor. *J Exp Med* **204**:117–128.
57. Salvatore, C.A., Tilley, S.L., Latour, A.M., Fletcher, D.S., Koller, B.H., and Jacobson, M.A. 2000. Disruption of the A(3) adenosine receptor gene in mice and its effect on stimulated inflammatory cells. *J Biol Chem* **275**:4429–4434.
58. Feoktistov, I., and Biaggioni, I. 1995. Adenosine A2b receptors evoke interleukin-8 secretion in human mast cells. An enprofylline-sensitive mechanism with implications for asthma. *J Clin Invest* **96**:1979–1986.
59. Auchampach, J.A., Jin, X., Wan, T.C., Caughey, G.H., and Linden, J. 1997. Canine mast cell adenosine receptors: cloning and expression of the A3 receptor and evidence that degranulation is mediated by the A2B receptor. *Mol Pharmacol* **52**:846–860.
60. Naganuma, M., Wiznerowicz, E.B., Lappas, C.M., Linden, J., Worthington, M.T., and Ernst, P.B. 2006. Cutting edge: critical role for A2A adenosine receptors in the T cell-mediated regulation of colitis. *J Immunol* **177**:2765–2769.
61. Sevigny, C.P., Li, L., Awad, A.S., et al. 2007. Activation of adenosine 2A receptors attenuates allograft rejection and alloantigen recognition. *J Immunol* **178**:4240–4249.
62. Lappas, C.M., Rieger, J.M., and Linden, J. 2005. A2A adenosine receptor induction inhibits IFN-gamma production in murine CD4+ T cells. *J Immunol* **174**:1073–1080.
63. Csoka, B., Himer, L., Nemeth, Z.H., et al. 2008. Adenosine A2A receptor activation inhibits T helper 1and T helper 2 cell development and effector function. *FASEB J* **22**:3491–3499.
64. Erdmann, A.A., Gao, Z.G., Jung, U., et al. 2005. Activation of Th1 and Tc1 cell adenosine A2A receptors directly inhibits IL-2 secretion in vitro and IL-2-driven expansion in vivo. *Blood* **105**:4707–4714.
65. Koshiba, M., Kojima, H., Huang, S., Apasov, S., and Sitkovsky, M.V. 1997. Memory of extracellular adenosine A2A purinergic receptor-mediated signaling in murine T cells. *J Biol Chem* **272**:25881–25889.
66. Hoskin, D.W., Butler, J.J., Drapeau, D., Haeryfar, S.M., and Blay, J. 2002. Adenosine acts through an A3 receptor to prevent the induction of murine anti-CD3-activated killer T cells. *Int J Cancer* **99**:386–395.
67. Raskovalova, T., Huang, X., Sitkovsky, M., Zacharia, L.C., Jackson, E.K., and Gorelik, E. 2005. Gs protein-coupled adenosine receptor signaling and lytic function of activated NK cells. *J Immunol* **175**:4383–4391.
68. Deaglio, S., Dwyer, K.M., Gao, W., et al. 2007. Adenosine generation catalyzed by CD39 and CD73 expressed on regulatory T cells mediates immune suppression. *J Exp Med* **204**:1257–1265.
69. Kobie, J.J., Shah, P.R., Yang, L., Rebhahn, J.A., Fowell, D.J., and Mosmann, T.R. 2006. T regulatory and primed uncommitted CD4 T cells express CD73, which suppresses effector CD4 T cells by converting 5'-adenosine monophosphate to adenosine. *J Immunol* **177**:6780–6786.
70. Borsellino, G., Kleinewietfeld, M., Di Mitri, D., et al. 2007. Expression of ectonucleotidase CD39 by Foxp3+ Treg cells: hydrolysis of extracellular ATP and immune suppression. *Blood* **110**:1225–1232.
71. Zarek, P.E., Huang, C.T., Lutz, E.R., et al. 2008. A2A receptor signaling promotes peripheral tolerance by inducing T-cell anergy and the generation of adaptive regulatory T cells. *Blood* **111**:251–259.
72. Zajonc, D.M., Maricic, I., Wu, D., et al. 2005. Structural basis for CD1d presentation of a sulfatide derived from myelin and its implications for autoimmunity. *J Exp Med* **202**:1517–1526.
73. Kaneko, S., Okumura, K., Numaguchi, Y., et al. 2000. Melatonin scavenges hydroxyl radical and protects isolated rat hearts from ischemic reperfusion injury. *Life Sci* **67**:101–112.
74. Lappas, C.M., Day, Y.J., Marshall, M.A., Engelhard, V.H., and Linden, J. 2006. Adenosine A2A receptor activation reduces hepatic ischemia reperfusion injury by inhibiting CD1d-dependent NKT cell activation. *J Exp Med* **203**:2639–2648.
75. Yamamura, T., Sakuishi, K., Illes, Z., and Miyake, S. 2007. Understanding the behavior of invariant NKT cells in autoimmune diseases. *J Neuroimmunol* **191**:8–15.
76. Sands, W.A., and Palmer, T.M. 2005. Adenosine receptors and the control of endothelial cell function in inflammatory disease. *Immunol Lett* **101**:1–11.

77. Bouma, M.G., van den Wildenberg, F.A., and Buurman, W.A. 1996. Adenosine inhibits cytokine release and expression of adhesion molecules by activated human endothelial cells. *Am J Physiol* **270**:C522–C529.
78. Sands, W.A., Martin, A.F., Strong, E.W., and Palmer, T.M. 2004. Specific inhibition of nuclear factor-kappaB-dependent inflammatory responses by cell type-specific mechanisms upon A2A adenosine receptor gene transfer. *Mol Pharmacol* **66**:1147–1159.
79. Yang, D., Zhang, Y., Nguyen, H.G., et al. 2006. The A2B adenosine receptor protects against inflammation and excessive vascular adhesion. *J Clin Invest* **116**:1913–1923.
80. Deussen, A., Moser, G., and Schrader, J. 1986. Contribution of coronary endothelial cells to cardiac adenosine production. *Pflugers Arch* **406**:608–614.
81. Deussen, A., Bading, B., Kelm, M., and Schrader, J. 1993. Formation and salvage of adenosine by macrovascular endothelial cells. *Am J Physiol* **264**:H692–H700.
82. Zernecke, A., Bidzhekov, K., Ozuyaman, B., et al. 2006. CD73/ecto-5'-nucleotidase protects against vascular inflammation and neointima formation. *Circulation* **113**:2120–2127.
83. Narravula, S., Lennon, P.F., Mueller, B.U., and Colgan, S.P. 2000. Regulation of endothelial CD73 by adenosine: paracrine pathway for enhanced endothelial barrier function. *J Immunol* **165**:5262–5268.
84. Thompson, L.F., Eltzschig, H.K., Ibla, J.C., et al. 2004. Crucial role for ecto-5'-nucleotidase (CD73) in vascular leakage during hypoxia. *J Exp Med* **200**:1395–1405.
85. Eltzschig, H.K., Thompson, L.F., Karhausen, J., et al. 2004. Endogenous adenosine produced during hypoxia attenuates neutrophil accumulation: coordination by extracellular nucleotide metabolism. *Blood* **104**:3986–3992.
86. Niemela, J., Henttinen, T., Yegutkin, G.G., et al. 2004. IFN-alpha induced adenosine production on the endothelium: a mechanism mediated by CD73 (ecto-5'-nucleotidase) up-regulation. *J Immunol* **172**:1646–1653.
87. Kalsi, K., Lawson, C., Dominguez, M., Taylor, P., Yacoub, M.H., and Smolenski, R.T. 2002. Regulation of ecto-5'-nucleotidase by TNF-alpha in human endothelial cells. *Mol Cell Biochem* **232**:113–119.
88. Cronstein, B.N., Levin, R.I., Belanoff, J., Weissmann, G., and Hirschhorn, R. 1986. Adenosine: an endogenous inhibitor of neutrophil-mediated injury to endothelial cells. *J Clin Invest* **78**:760–770.
89. Rosengren, S., Bong, G.W., and Firestein, G.S. 1995. Anti-inflammatory effects of an adenosine kinase inhibitor. Decreased neutrophil accumulation and vascular leakage. *J Immunol* **154**:5444–5451.
90. Eckle, T., Faigle, M., Grenz, A., Laucher, S., Thompson, L.F., and Eltzschig, H.K. 2008. A2B adenosine receptor dampens hypoxia-induced vascular leak. *Blood* **111**:2024–2035.
91. Eckle, T., Krahn, T., Grenz, A., et al. 2007. Cardioprotection by ecto-5'-nucleotidase (CD73) and A2B adenosine receptors. *Circulation* **115**:1581–1590.
92. Montesinos, M.C., Gadangi, P., Longaker, M., et al. 1997. Wound healing is accelerated by agonists of adenosine A2 (G alpha s-linked) receptors. *J Exp Med* **186**:1615–1620.
93. Victor-Vega, C., Desai, A., Montesinos, M.C., and Cronstein, B.N. 2002. Adenosine A2A receptor agonists promote more rapid wound healing than recombinant human platelet-derived growth factor (Becaplermin gel). *Inflammation* **26**:19–24.
94. Montesinos, M.C., Desai, A., Chen, J.F., et al. 2002. Adenosine promotes wound healing and mediates angiogenesis in response to tissue injury via occupancy of A(2A) receptors. *Am J Pathol* **160**:2009–2018.
95. Montesinos, M.C., Shaw, J.P., Yee, H., Shamamian, P., and Cronstein, B.N. 2004. Adenosine A(2A) receptor activation promotes wound neovascularization by stimulating angiogenesis and vasculogenesis. *Am J Pathol* **164**:1887–1892.
96. Leibovich, S.J., Chen, J.F., Pinhal-Enfield, G., et al. 2002. Synergistic up-regulation of vascular endothelial growth factor expression in murine macrophages by adenosine A(2A) receptor agonists and endotoxin. *Am J Pathol* **160**:2231–2244.
97. Desai, A., Victor-Vega, C., Gadangi, S., Montesinos, M.C., Chu, C.C., and Cronstein, B.N. 2005. Adenosine A2A receptor stimulation increases angiogenesis by down-regulating production of the antiangiogenic matrix protein thrombospondin 1. *Mol Pharmacol* **67**: 1406–1413.
98. Nguyen, D.K., Montesinos, M.C., Williams, A.J., Kelly, M., and Cronstein, B.N. 2003. Th1 cytokines regulate adenosine receptors and their downstream signaling elements in human microvascular endothelial cells. *J Immunol* **171**:3991–3998.
99. Grant, M.B., Davis, M.I., Caballero, S., Feoktistov, I., Biaggioni, I., and Belardinelli, L. 2001. Proliferation, migration, and ERK activation in human retinal endothelial cells through A(2B) adenosine receptor stimulation. *Invest Ophthalmol Vis Sci* **42**:2068–2073.
100. Grant, M.B., Tarnuzzer, R.W., Caballero, S., et al. 1999. Adenosine receptor activation induces vascular endothelial growth factor in human retinal endothelial cells. *Circ Res* **85**:699–706.
101. Feoktistov, I., Goldstein, A.E., Ryzhov, S., et al. 2002. Differential expression of adenosine receptors in human endothelial cells: role of A2B receptors in angiogenic factor regulation. *Circ Res* **90**:531–538.
102. Zhong, H., Belardinelli, L., Maa, T., Feoktistov, I., Biaggioni, I., and Zeng, D. 2004. A(2B) adenosine receptors increase cytokine release by bronchial smooth muscle cells. *Am J Respir Cell Mol Biol* **30**:118–125.
103. Zhong, H., Wu, Y., Belardinelli, L., and Zeng, D. 2006. A2B adenosine receptors induce IL-19 from bronchial epithelial cells, resulting in TNF-alpha increase. *Am J Respir Cell Mol Biol* **35**:587–592.
104. Zhong, H., Belardinelli, L., Maa, T., and Zeng, D. 2005. Synergy between A2B adenosine receptors and hypoxia in activating human lung fibroblasts. *Am J Respir Cell Mol Biol* **32**:2–8.
105. Sun, C.X., Zhong, H., Mohsenin, A., et al. 2006. Role of A2B adenosine receptor signaling in adenosine-dependent pulmonary inflammation and injury. *J Clin Invest* **116**:2173–2182.
106. Mustafa, S.J., Nadeem, A., Fan, M., Zhong, H., Belardinelli, L., and Zeng, D. 2007. Effect of a specific and selective A(2B) adenosine receptor antagonist on adenosine agonist AMP and allergen-induced airway responsiveness and cellular influx in a mouse model of asthma. *J Pharmacol Exp Ther* **320**:1246–1251.
107. Holgate, S.T. 2005. The Quintiles Prize Lecture 2004. The identification of the adenosine A2B receptor as a novel therapeutic target in asthma. *Br J Pharmacol* **145**:1009–1015.
108. Reutershan, J., Cagnina, R.E., Chang, D., Linden, J., and Ley, K. 2007. Therapeutic anti-inflammatory effects of myeloid cell adenosine receptor A2a stimulation in lipopolysaccharide-induced lung injury. *J Immunol* **179**:1254–1263.

109. Nadeem, A., Fan, M., Ansari, H.R., Ledent, C., and Jamal Mustafa, S. 2007. Enhanced airway reactivity and inflammation in A2A adenosine receptor-deficient allergic mice. *Am J Physiol Lung Cell Mol Physiol* **292**:L1335–1344.
110. Luijk, B., van den Berge, M., Kerstjens, H.A., et al. 2008. Effect of an inhaled adenosine A2A agonist on the allergen-induced late asthmatic response. *Allergy* **63**:75–80.
111. Day, Y.J., Li, Y., Rieger, J.M., Ramos, S.I., Okusa, M.D., and Linden, J. 2005. A2A adenosine receptors on bone marrow-derived cells protect liver from ischemia-reperfusion injury. *J Immunol* **174**:5040–5046.
112. Day, Y.J., Huang, L., McDuffie, M.J., et al. 2003. Renal protection from ischemia mediated by A2A adenosine receptors on bone marrow-derived cells. *J Clin Invest* **112**:883–891.
113. Yang, Z., Day, Y.J., Toufektsian, M.C., et al. 2006. Myocardial infarct-sparing effect of adenosine A2A receptor activation is due to its action on CD4+ T lymphocytes. *Circulation* **114**:2056–2064.
114. Peirce, S.M., Skalak, T.C., Rieger, J.M., Macdonald, T.L., and Linden, J. 2001. Selective A(2A) adenosine receptor activation reduces skin pressure ulcer formation and inflammation. *Am J Physiol Heart Circ Physiol* **281**:H67–H74.
115. Li, Y., Oskouian, R.J., Day, Y.J., et al. 2006. Mouse spinal cord compression injury is reduced by either activation of the adenosine A2A receptor on bone marrow-derived cells or deletion of the A2A receptor on non-bone marrow-derived cells. *Neuroscience* **141**:2029–2039.
116. Reece, T.B., Kron, I.L., Okonkwo, D.O., et al. 2006. Functional and cytoarchitectural spinal cord protection by ATL-146e after ischemia/reperfusion is mediated by adenosine receptor agonism. *J Vasc Surg* **44**:392–397.
117. Reece, T.B., Ellman, P.I., Maxey, T.S., et al. 2005. Adenosine A2A receptor activation reduces inflammation and preserves pulmonary function in an in vivo model of lung transplantation. *J Thorac Cardiovasc Surg* **129**:1137–1143.
118. Li, L., Huang, L., Sung, S.S., et al. 2007. NKT cell activation mediates neutrophil IFN-gamma production and renal ischemia-reperfusion injury. *J Immunol* **178**:5899–5911.
119. Chabner, B.A., Allegra, C.J., Curt, G.A., et al. 1985. Polyglutamation of methotrexate. Is methotrexate a prodrug? *J Clin Invest* **76**:907–912.
120. Allegra, C.J., Drake, J.C., Jolivet, J., and Chabner, B.A. 1985. Inhibition of phosphoribosylaminoimidazolecarboxamide transformylase by methotrexate and dihydrofolic acid polyglutamates. *Proc Natl Acad Sci USA* **82**:4881–4885.
121. Baggott, J.E., Vaughn, W.H., and Hudson, B.B. 1986. Inhibition of 5-aminoimidazole-4-carboxamide ribotide transformylase, adenosine deaminase and 5'-adenylate deaminase by polyglutamates of methotrexate and oxidized folates and by 5-aminoimidazole-4-carboxamide riboside and ribotide. *Biochem J* **236**:193–200.
122. Cronstein, B.N., Naime, D., and Ostad, E. 1993. The antiinflammatory mechanism of methotrexate. Increased adenosine release at inflamed sites diminishes leukocyte accumulation in an in vivo model of inflammation. *J Clin Invest* **92**:2675–2682.
123. Moser, G.H., Schrader, J., and Deussen, A. 1989. Turnover of adenosine in plasma of human and dog blood. *Am J Physiol* **256**:C799–806.
124. Riksen, N.P., Barrera, P., van den Broek, P.H., van Riel, P.L., Smits, P., and Rongen, G.A. 2006. Methotrexate modulates the kinetics of adenosine in humans in vivo. *Ann Rheum Dis* **65**:465–470.
125. Baggott, J.E., Morgan, S.L., Sams, W.M., and Linden, J. 1999. Urinary adenosine and aminoimidazolecarboxamide excretion in methotrexate-treated patients with psoriasis. *Arch Dermatol* **135**:813–817.
126. Cronstein, B.N., Eberle, M.A., Gruber, H.E., and Levin, R.I. 1991. Methotrexate inhibits neutrophil function by stimulating adenosine release from connective tissue cells. *Proc Natl Acad Sci USA* **88**:2441–2445.
127. Morabito, L., Montesinos, M.C., Schreibman, D.M., et al. 1998. Methotrexate and sulfasalazine promote adenosine release by a mechanism that requires ecto-5'-nucleotidase-mediated conversion of adenine nucleotides. *J Clin Invest* **101**:295–300.
128. Montesinos, M.C., Yap, J.S., Desai, A., Posadas, I., McCrary, C.T., and Cronstein, B.N. 2000. Reversal of the antiinflammatory effects of methotrexate by the nonselective adenosine receptor antagonists theophylline and caffeine: evidence that the antiinflammatory effects of methotrexate are mediated via multiple adenosine receptors in rat adjuvant arthritis. *Arthritis Rheum* **43**:656–663.
129. Montesinos, M.C., Desai, A., Delano, D., et al. 2003. Adenosine A2A or A3 receptors are required for inhibition of inflammation by methotrexate and its analog MX-68. *Arthritis Rheum* **48**:240–247.
130. Delano, D.L., Montesinos, M.C., Desai, A., et al. 2005. Genetically based resistance to the antiinflammatory effects of methotrexate in the air-pouch model of acute inflammation. *Arthritis Rheum* **52**:2567–2575.
131. Montesinos, M.C., Desai, A., and Cronstein, B.N. 2006. Suppression of inflammation by low-dose methotrexate is mediated by adenosine A2A receptor but not A3 receptor activation in thioglycollate-induced peritonitis. *Arthritis Res Ther* **8**:R53.
132. Montesinos, M.C., Takedachi, M., Thompson, L.F., Wilder, T.F., Fernandez, P., and Cronstein, B.N. 2007. The antiinflammatory mechanism of methotrexate depends on extracellular conversion of adenine nucleotides to adenosine by ecto-5'-nucleotidase: findings in a study of ecto-5'-nucleotidase gene-deficient mice. *Arthritis Rheum* **56**:1440–1445.
133. Nesher, G., Mates, M., and Zevin, S. 2003. Effect of caffeine consumption on efficacy of methotrexate in rheumatoid arthritis. *Arthritis Rheum* **48**:571–572.
134. Szabo, C., Scott, G.S., Virag, L., et al. 1998. Suppression of macrophage inflammatory protein (MIP)-1alpha production and collagen-induced arthritis by adenosine receptor agonists. *Br J Pharmacol* **125**:379–387.
135. Silverman, M.H., Strand, V., Markovits, D., et al. 2008. Clinical evidence for utilization of the A3 adenosine receptor as a target to treat rheumatoid arthritis: data from a phase II clinical trial. *J Rheumatol* **35**:41–48.
136. Law, W.R. 2006. Adenosine receptors in the response to sepsis: what do receptor-specific knockouts tell us? *Am J Physiol Regul Integr Comp Physiol* **291**:R957–R958.
137. Gallos, G., Ruyle, T.D., Emala, C.W., and Lee, H.T. 2005. A1 adenosine receptor knockout mice exhibit increased mortality, renal dysfunction, and hepatic injury in murine septic peritonitis. *Am J Physiol Renal Physiol* **289**:F369–F376.
138. Lee, H.T., Kim, M., Joo, J.D., Gallos, G., Chen, J.F., and Emala, C.W. 2006. A3 adenosine receptor activation decreases mortality and renal and hepatic injury

in murine septic peritonitis. *Am J Physiol Regul Integr Comp Physiol* **291**:R959–R969.

139. Nemeth, Z.H., Csoka, B., Wilmanski, J., et al. 2006. Adenosine A2A receptor inactivation increases survival in polymicrobial sepsis. *J Immunol* **176**:5616–5626.
140. Sullivan, G.W., Fang, G., Linden, J., and Scheld, W.M. 2004. A2A adenosine receptor activation improves survival in mouse models of endotoxemia and sepsis. *J Infect Dis* **189**:1897–1904.
141. Bamias, G., and Cominelli, F. 2007. Immunopathogenesis of inflammatory bowel disease: current concepts. *Curr Opin Gastroenterol* **23**:365–369.
142. Izcue, A., Coombes, J.L., and Powrie, F. 2006. Regulatory T cells suppress systemic and mucosal immune activation to control intestinal inflammation. *Immunol Rev* **212**:256–271.
143. Odashima, M., Bamias, G., Rivera-Nieves, J., et al. 2005. Activation of A2A adenosine receptor attenuates intestinal inflammation in animal models of inflammatory bowel disease. *Gastroenterology* **129**:26–33.
144. Sitaraman, S.V., Merlin, D., Wang, L., et al. 2001. Neutrophil-epithelial crosstalk at the intestinal lumenal surface mediated by reciprocal secretion of adenosine and IL-6. *J Clin Invest* **107**:861–869.
145. Kolachala, V., Asamoah, V., Wang, L., et al. 2005. TNF-alpha upregulates adenosine 2b (A2b) receptor expression and signaling in intestinal epithelial cells: a basis for A2bR overexpression in colitis. *Cell Mol Life Sci* **62**:2647–2657.
146. Kolachala, V.L., Ruble, B.K., Vijay-Kumar, M., et al. 2008. Blockade of adenosine A(2B) receptors ameliorates murine colitis. *Br J Pharmacol* **155**:127–137.
147. Kolachala, V.L., Bajaj, R., Chalasani, M., and Sitaraman, S.V. 2008. Purinergic receptors in gastrointestinal inflammation. *Am J Physiol Gastrointest Liver Physiol* **294**:G401–G410.
148. Mabley, J.G., Pacher, P., Liaudet, L., et al. 2003. Inosine reduces inflammation and improves survival in a murine model of colitis. *Am J Physiol Gastrointest Liver Physiol* **284**:G138–G144.
149. Gomez, G., and Sitkovsky, M.V. 2003. Differential requirement for A2a and A3 adenosine receptors for the protective effect of inosine in vivo. *Blood* **102**:4472–4478.
150. Hasko, G., Sitkovsky, M.V., and Szabo, C. 2004. Immunomodulatory and neuroprotective effects of inosine. *Trends Pharmacol Sci* **25**:152–157.
151. Hasko, G., Kuhel, D.G., Nemeth, Z.H., et al. 2000. Inosine inhibits inflammatory cytokine production by a posttranscriptional mechanism and protects against endotoxin-induced shock. *J Immunol* **164**:1013–1019.
152. Macedo, L., Pinhal-Enfield, G., Alshits, V., Elson, G., Cronstein, B.N., and Leibovich, S.J. 2007. Wound healing is impaired in MyD88-deficient mice: a role for MyD88 in the regulation of wound healing by adenosine A2A receptors. *Am J Pathol* **171**:1774–1788.
153. Meininger, C.J., Schelling, M.E., and Granger, H.J. 1988. Adenosine and hypoxia stimulate proliferation and migration of endothelial cells. *Am J Physiol* **255**:H554–H562.
154. Ethier, M.F., Chander, V., and Dobson, J.G., Jr. 1993. Adenosine stimulates proliferation of human endothelial cells in culture. *Am J Physiol* **265**:H131–H138.
155. Sexl, V., Mancusi, G., Baumgartner-Parzer, S., Schutz, W., and Freissmuth, M. 1995. Stimulation of human umbilical vein endothelial cell proliferation by A2-adenosine and beta 2-adrenoceptors. *Br J Pharmacol* **114**:1577–1586.
156. Sexl, V., Mancusi, G., Holler, C., Gloria-Maercker, E., Schutz, W., and Freissmuth, M. 1997. Stimulation of the mitogen-activated protein kinase via the A2A-adenosine receptor in primary human endothelial cells. *J Biol Chem* **272**:5792–5799.
157. Takagi, H., King, G.L., Ferrara, N., and Aiello, L.P. 1996. Hypoxia regulates vascular endothelial growth factor receptor KDR/Flk gene expression through adenosine A2 receptors in retinal capillary endothelial cells. *Invest Ophthalmol Vis Sci* **37**:1311–1321.
158. Takagi, H., King, G.L., Robinson, G.S., Ferrara, N., and Aiello, L.P. 1996. Adenosine mediates hypoxic induction of vascular endothelial growth factor in retinal pericytes and endothelial cells. *Invest Ophthalmol Vis Sci* **37**:2165–2176.
159. Merighi, S., Benini, A., Mirandola, P., et al. 2007. Caffeine inhibits adenosine-induced accumulation of hypoxia-inducible factor-1alpha, vascular endothelial growth factor, and interleukin-8 expression in hypoxic human colon cancer cells. *Mol Pharmacol* **72**:395–406.
160. Pinhal-Enfield, G., Ramanathan, M., Hasko, G., et al. 2003. An angiogenic switch in macrophages involving synergy between Toll-like receptors 2, 4, 7, and 9 and adenosine A(2A) receptors. *Am J Pathol* **163**:711–721.
161. Chan, E.S., Fernandez, P., Merchant, A.A., et al. 2006. Adenosine A2A receptors in diffuse dermal fibrosis: pathogenic role in human dermal fibroblasts and in a murine model of scleroderma. *Arthritis Rheum* **54**:2632–2642.
162. Chunn, J.L., Mohsenin, A., Young, H.W., et al. 2006. Partially adenosine deaminase-deficient mice develop pulmonary fibrosis in association with adenosine elevations. *Am J Physiol Lung Cell Mol Physiol* **290**:L579–L587.
163. Chen, Y., Epperson, S., Makhsudova, L., et al. 2004. Functional effects of enhancing or silencing adenosine A2b receptors in cardiac fibroblasts. *Am J Physiol Heart Circ Physiol* **287**:H2478–H2486.

SUGGESTED READINGS

Khoa, N.D., Montesinos, C.M., Williams, A.J., Kelly, M., and Cronstein, B.N. 2003. Th1 cytokines regulate adenosine receptors and their downstream signalling elements in human microvascular endothelial cells. *J Immunol* **171**:3991–3998.

Lappas, C.M., Day, Y.J., Marshall, M.A., Engelhard, V.H., and Linden, J. 2006. Adenosine A2A receptor activation reduces hepatic ischemia reperfusion injury by inhibiting CD1d-dependent NKT cell activation. *J Exp Med* **203**:2639–2648.

Nemeth, Z.H., Lutz, C.S., Csoka, B., et al. 2005. Adenosine augments IL-10 production by macrophages through an A2B receptor-mediated posttranscriptional mechanism. *J Immunol* **175**:8260–8270.

15 Leukocyte Generation of Reactive Oxygen Species

William M. Nauseef

INTRODUCTION

The rapid response of the innate immune system of humans depends in large part on the behavior of a subset of leukocytes known as polymorphonuclear leukocytes or neutrophils. Constituting the majority of white blood cells in the circulation, neutrophils provide continuous surveillance for microbial and noninfectious threats to the integrity of the host. Once recruited to an area of infection, neutrophils ingest microbes and thereby confine them in the phagosome, a membrane-bound compartment where an array of antibacterial factors can be delivered to kill and degrade the trapped microorganism. In addition to proteins that exhibit direct antimicrobial action, cytoplasmic granules in neutrophils contain enzymes that utilize microbial constituents as substrate, thereby compromising the structural integrity of the organism and indirectly contributing to microbial killing. Under normal circumstances, these complex and synergistic events occur in the presence of reactive oxygen species (ROS) generated in situ by activation of a multicomponent enzyme, the phagocyte NADPH-dependent oxidase. Like the prefabricated granule constituents, ROS damage targets directly and act in concert with granule proteins to generate second derivative active agents. The concomitant release of granule proteins and activation of the NADPH oxidase thus provide phagocytic cells such as neutrophils with the capacity to respond rapidly to potentially life-threatening challenges and thus provide the initial wave of host defense.

Although monocytes, macrophages, and eosinophils share many of the fundamental features of the phagocytosis-coupled response exhibited by neutrophils, the discussion will use human neutrophils as the paradigm for presenting the biochemical principles of ROS generation by stimulated leukocytes.

CONTEXT OF THE PHAGOCYTE NADPH OXIDASE

Appreciation of the role of ROS in normal host defense against infection serves as another example in modern medicine of how basic research and clinical medicine reciprocally rely on each other for critical insights as well as for identifying the proper biological context. As investigators were probing the origins and significance of the "burst" in oxygen consumption that phagocytes exhibited when challenged with microbes, clinicians were characterizing the phenotype of a new clinical syndrome associated with a compromise in normal immune function.

Whereas resting neutrophils rely on glycolysis for energy and consume little to no O_2, stimulated neutrophils exhibit a rapid and robust burst of respiration, often referred to as the "respiratory burst," and thereby generate a large flux of ROS. Nearly all of the oxygen consumed by stimulated neutrophils can be recovered as H_2O_2 produced, demonstrating that the phagocyte oxidase represents the sole mechanism for oxygen utilization in phagocytes. The respiratory burst oxidase was not constitutively active but required stimulation to initiate activity, resisted inhibition by cyanide or azide, thus excluding a mitochondrial source, and required flavin adenine dinucleotide (FAD) as well as NADPH for activity. Over the same time span that investigators were characterizing the respiratory burst oxidase, clinicians described a new inherited disorder, chronic granulomatous disease (CGD), that was manifested as frequent and severe pyogenic infections. Patients with CGD were first recognized in the 1950s when four boys with recurrent infections, lymphadenopathy, and granulomatous reactions in liver and lymph nodes were reported. The concurrent pursuits to elucidate the enzymology and significance of the respiratory burst of phagocytes and to understand the basis of the increased susceptibility to infection seen in patients with CGD merged with recognition

(1) that the biochemical phenotype of CGD reflected the absence of oxidase activity from neutrophils, monocytes, macrophages, and eosinophils and (2) that the failure of CGD phagocytes to generate ROS resulted in depressed antimicrobial activity and the resultant increased occurrence of pyogenic infection.

Collectively, ROS refers to molecules generated by the incomplete reduction of oxygen, which include, in the context of the phagocyte oxidase, superoxide anion (O_2^-), hydrogen peroxide (H_2O_2), and hydroxyl radical (HO).

O_2^- is relatively unstable and because of its negative charge does not diffuse efficiently through membranes. However, because of its negative charge, O_2^- avidly oxidizes charged redox centers such as iron–sulfur clusters in bacteria. In contrast, H_2O_2 is relatively stable and readily diffuses through membranes to oxidize susceptible targets, especially sulfur-containing amino acids such as cysteine and methionine residues. HO· ranks as the most toxic of these three ROS and reacts indiscriminately with lipids, proteins, DNA, and virtually all potential biologic substrates. As discussed in the following, the interactions among these ROS and between the ROS collectively and granule proteins create a hostile environment within the phagosome for the ingested microbe.

MEMBRANE COMPONENTS OF THE PHAGOCYTE OXIDASE

As noted earlier, the phagocyte NADPH-dependent oxidase is inactive in resting neutrophils but converted into an active form by exposure of the cells to a suitable agonist. The activation state of the phagocyte oxidase is regulated by the spatial disposition of its components. The essential elements of this multicomponent enzyme in resting cells are segregated either in membranes or as soluble proteins within the cytoplasm. Exposure to effective agonists triggers assembly of the oxidase at the plasma membrane or the phagosomal membrane. Once assembled, the enzyme becomes active, converting molecular oxygen to O_2^-.

The membrane component of the phagocyte NADPH oxidase is flavocytochrome b_{558}, a heterodimeric integral membrane protein composed of a glycosylated 91-kDa protein, gp91*phox* (where *phox* indicates *ph*agocyte *ox*idase), and a nonglycosylated 22-kDa protein, p22*phox*. Encoded by the *CYBB* gene on the X-chromosome and once thought to be uniquely expressed in phagocytes, gp91*phox* is the electron transferase component of the phagocyte oxidase and has recently been recognized to be the patriarch of a family of NADPH oxidase proteins, referred to now as the NOX protein family (*vide infra*). Neutrophils, monocytes, eosinophils, and macrophages express relatively large amounts of gp91*phox*, which is referred to as NOX2 in the nomenclature of the NOX protein family. Ongoing studies of the tissue distribution of NOX protein family members have revealed detectable amounts of both gp91*phox* message and protein in a wide variety of cells, although in many cases the physiologic significance of the findings awaits further elucidation. The integral membrane protein p22*phox* is encoded by the *CYBA* gene on chromosome 16q24 and associates with gp91*phox* in 1:1 stoichiometry to create functional flavocytochrome b_{558}. During the biosynthesis of the individual subunits of flavocytochrome b_{558} in the endoplasmic reticulum (ER), the heme-containing 65-kDa precursor of gp91*phox* associates with p22*phox* to form the nascent heterodimer (Figure 15.1). In the absence of either subunit, as occurs in some types of CGD, the lone subunit undergoes degradation, mediated in part in the proteasome. Furthermore, heme acquisition in the ER is a prerequisite for maturation of the heterodimer in the secretory protein pathway and proper targeting to membranes at the neutrophil surface or in intracellular vesicles. In the absence of heme synthesis, heterodimers do not associate in the ER and flavocytochrome b_{558} biosynthesis is aborted, demonstrating the critical contribution of heme incorporation to the integrity and stability of the heterodimer.

In the absence of a solved crystal structure, the topography of flavocytochrome b_{558} and some of its structural features have been derived from sequence analysis, mapping by monoclonal antibodies, and the application of a host of analytical techniques (Figure 15.2). Data suggest that gp91*phox* possesses six transmembrane helices and an extended C-terminus of ~300 amino acids in the cytoplasm. Several features of gp91*phox* that are shared by all members of the NOX protein family are suited to its role as an electron transferase, shuttling electrons across the membrane from cytoplasmic NADPH to molecular oxygen. The C-terminal cytoplasmic tail of gp91*phox* possesses binding sites for both NADPH and FAD and the transmembrane α-helices contain two nonequivalent heme groups, each ligated to histidine residues in parallel helices and stacked atop each other perpendicular to the plane of the membrane. This arrangement is reminiscent of that suggested for the iron reductase of *Saccharomyces cerevisiae*, where the individual ligating histidines appear 12–13 residues apart in the linear sequence. Direct data on neutrophils indicate that the midpoint potentials for the inner and outer heme groups are –225 and –265 mV, respectively, thus providing a favorable pathway for sequential electron movement from NADPH to FAD to the hemes and then to molecular oxygen, with the ultimate production of superoxide anion.

Whereas gp91*phox* serves as the catalytic subunit of flavocytochrome b_{558}, supporting electron movement from NADPH to oxygen, p22*phox* has not been implicated

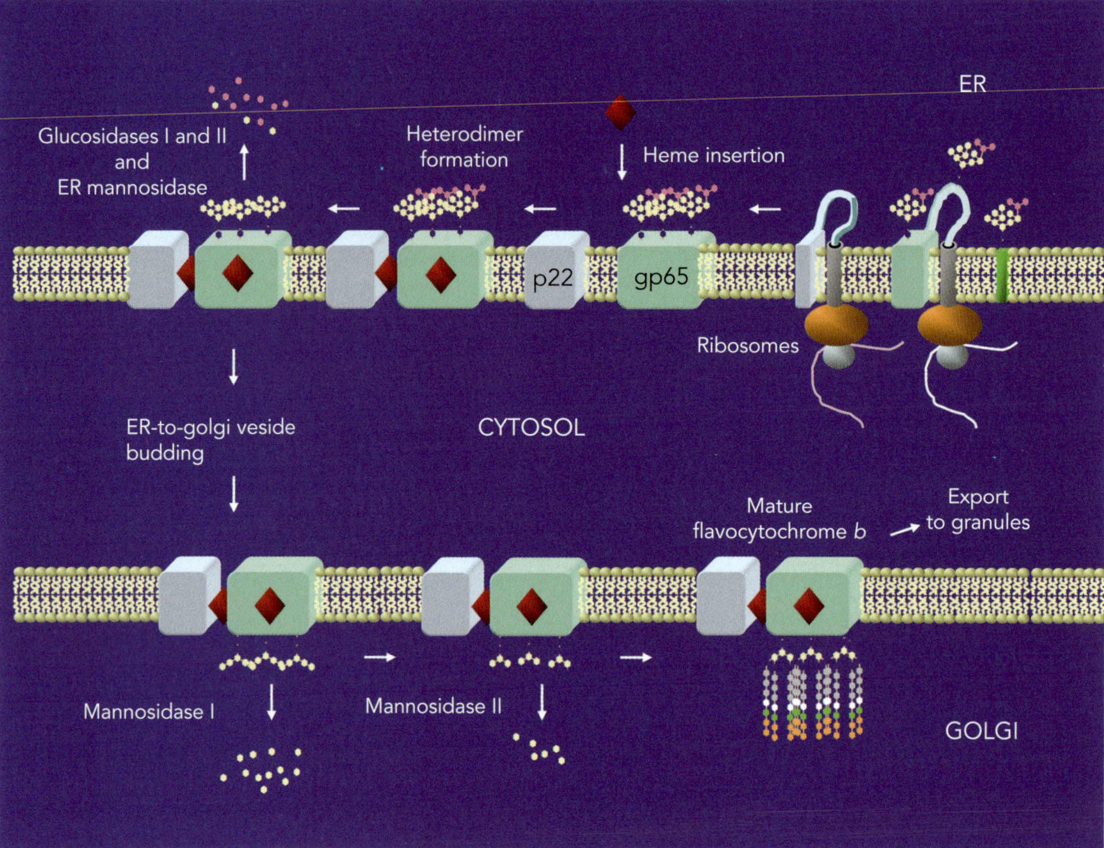

Figure 15.1. Biosynthesis of flavocytochrome b_{558}. Both subunits of flavocytochrome b_{558} are synthesized in the ER, with the primary translation product for gp91*phox* undergoing cotranslational glycosylation to generate a 65-kDa precursor, gp65. The nascent subunits associate and acquire the heme groups in the ER, with any unassociated subunits undergoing degradation, in part mediated by the proteasome. The heme-containing gp65-p22*phox* heterodimer has posttranslational modification of carbohydrates by ER glucosidases and mannosidases before exiting the ER and entering the Golgi, where significant modification of the oligosaccharides occurs. The net result of extensive glycosylation is the appearance of gp91*phox*. The mature heterodimer exists the Golgi and transferred into the membranes of the secretory vesicles, secondary granules, and cell surface.

in directly contributing to enzymatic activity of the heterodimer. Rather, p22*phox* is essential for the stability of gp91*phox* during its biosynthesis and thus to the structural and functional integrity of flavocytochrome b_{558}; patients with absent or abnormal forms of p22*phox* have the biochemical and clinical phenotype of CGD and suffer the same frequent and life-threatening infectious complications as do patients with CGD secondary to a lack of normal gp91*phox*. In addition to the association with gp91*phox* in the membrane, p22*phox* supports interactions with cytosolic oxidase components that are essential to assembly of a functional oxidase (*vide infra*). It is noteworthy that the stabilizing effect of p22*phox* on gp91*phox* is not a feature shared by all NOX protein family members; neither NOX5 nor the two Duox proteins is associated with p22*phox*. As discussed later, these same NOX isoforms also function independently of cytosolic cofactors, thus highlighting the specific contribution p22*phox* makes to the infrastructure-supporting oxidase assembly.

CYTOPLASMIC COMPONENTS OF THE PHAGOCYTE OXIDASE

In addition to the membrane-associated flavocytochrome b_{558}, activity of the phagocyte NADPH oxidase requires three cytosolic proteins, p47*phox*, p67*phox*, and either Rac1 or Rac2 from the Rho family of low-molecular-weight GTPases. The absence of any of these proteins from phagocytes results in the biochemical and clinical phenotype of CGD. In addition to these three proteins, other cytosolic proteins have been implicated in modulating activity of the phagocyte oxidase. The essential proteins exist in the cytoplasm as multicomponent complexes; p47*phox*, p67*phox*, and p40*phox* are associated in one cytosolic complex, and Rac is coupled with RhoGDI, the GDP dissociation inhibitor for Rho, in the other.

Serving as an adaptor protein that organizes assembly of the multiple oxidase components at the membrane, p47*phox* contains several domains that have

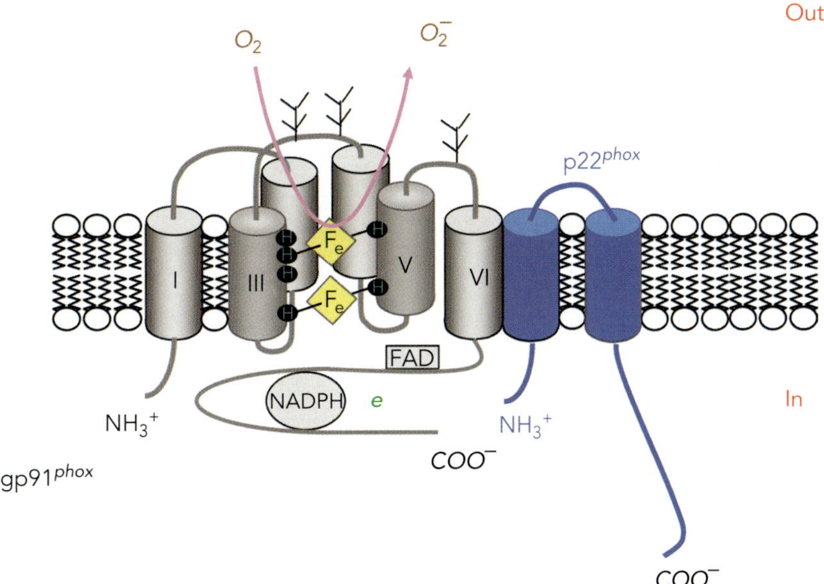

Figure 15.2 Electron transfer via flavocytochrome b_{558}. Substantial complementary data support a model for gp91^{phox} with six transmembrane α-helices, with both a short N-terminus and extended C-terminus extending into the cytoplasm. The cytoplasmic C-terminus possesses binding sites for NADPH and FAD. Stacked in the membrane, coupled in bis-histidine linkages between helices III and V, are two nonequivalent heme groups. When the oxidase is assembled and activated, electrons shuttle sequentially from NADPH to FAD, across the two hemes in the membrane, to reduce molecular oxygen to superoxide anion.

been implicated in mediating intermolecular interactions (Figure 15.3). Near the N-terminus of p47^{phox} is a PX domain, originally identified in Phox proteins and demonstrated to associate with specific phosphoinositides in or at the membrane. In addition, p47^{phox} possesses tandem Src homology 3 (SH3) domains and a proline-rich region (PRR), motifs known to support protein–protein interactions. However, the conformation of p47^{phox} in the cytoplasm of unstimulated phagocytes renders these domains cryptic and inaccessible for interactions with their potential binding partners. In addition to the protein domains that support intermolecular associations, p47^{phox} has a region rich in arginine and in serines that are susceptible to phosphorylation. Referred to as an autoinhibitory region, this cationic region of p47^{phox} includes 8–10 sites for agonist-dependent phosphorylation. During phagocyte stimulation, several kinases are activated and mediate phosphorylation of p47^{phox} in the autoinhibitory region, thereby releasing the conformational constraints on its potentially interactive domains. As a consequence of these posttranslational modifications in p47^{phox}, its SH3 domains become accessible to associate with the PRR in p22^{phox} and its PX motif can interact with targets in the stimulated plasma or phagosomal membrane. The translocation of p47^{phox} to the membrane is required to promote the movement of p67^{phox} and assembly of a functioning oxidase; in neutrophils of patients with CGD secondary to the absence of p47^{phox}, p67^{phox} remains in the cytoplasm after exposure to agonists. Thus, p47^{phox} contributes to oxidase activation in its role as an organizing element, providing a platform that is essential for the orderly translocation of p67^{phox} to the membrane and the stable association there with flavocytochrome b_{558}.

Whereas p47^{phox} provides critical organizational input into phagocyte oxidase assembly and action, p67^{phox} exhibits intrinsic catalytic activity, regulating the reduction of FAD by NADPH and interacting with Rac. In fact, in an experimental cell-free system using flavocytochrome b_{558}–enriched neutrophil membranes and high concentrations of recombinant p67^{phox}, superoxide anion can be generated in a Rac-dependent fashion in the absence of p47^{phox}. The catalytic activity of p67^{phox} depends as well on interactions with Rac, with an effector domain within the latter demonstrated to associate directly with p67^{phox}. In cytoplasm of resting phagocytes, isoprenylated Rac is in its GDP-bound inactive state, associated with RhoGDI. However, stimulation results in phosphorylation of RhoGDI, generation of the GTP bound form of Rac, and translocation of Rac-GTP to the membrane (Figure 15.4). The latter translocation is independent of concurrent translocation of the p47^{phox}–p67^{phox}–p40^{phox} complex, although

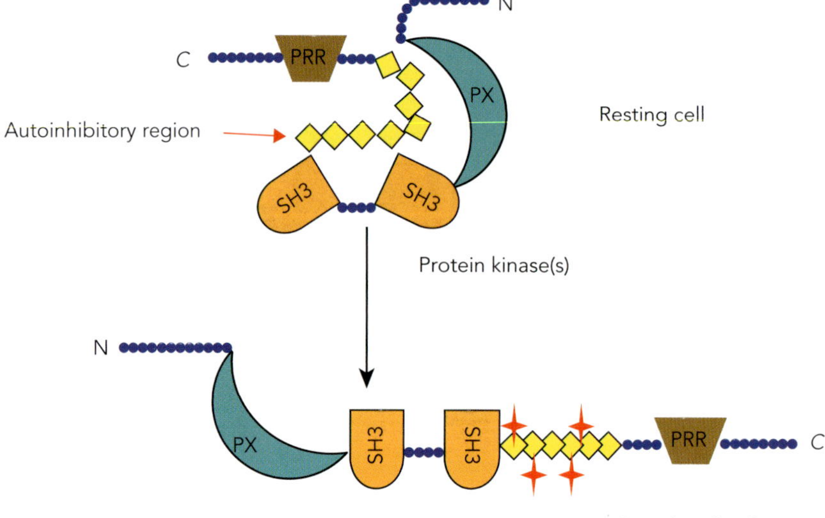

Figure 15.3. Agonist-triggered conformational changes in p47*phox*. In the cytoplasm of resting phagocytes, p47*phox* exists in a conformational state wherein domains that possess the capacity to support protein–protein and protein–lipid interactions are cryptic and inaccessible. There is a PX domain, which binds to specific phosphoinositides; tandem Src homology 3 (SH3) domains, which interact with proline-rich regions (PRR); and a C-terminal PRR. Very importantly, there is a cationic region between the second SH3 domain and the PRR known as the autoinhibitory region that is rich in arginine and serine residues. Stimulation of phagocytes activates several protein kinases that phosphorylate 8–10 serines in the autoinhibitory region. Consequently, there is a conformational change in p47*phox*, rendering its interactive domains available for binding. SH3 domains interact with a PRR in p22*phox*, PX domain binds phosphoinositides on the inner leaflet of the phagocyte membrane, and other interactions between cytosolic elements and flavocytochrome b_{558} stabilize the assembled complex.

evidence supports interactions of the membrane-bound Rac-GTP with translocated p67*phox* as well as with flavocytochrome b_{558}. The efficient redistribution of the two cytosolic complexes upon phagocyte stimulation and the stable association of oxidase complexes to support a functional oxidase require multiple protein–protein and protein–lipid interactions, the precise details of which are still being elucidated.

In the most reductionist view, phagocyte oxidase activity can be fully reconstituted in vitro with lipidated flavocytochrome b_{558}, p47*phox*, p67*phox*, and Rac-GTP. However, additional proteins and lipids likely contribute to modulation or regulation of phagocyte oxidase activity in vivo. The lipid environment of flavocytochrome b_{558} influences its conformation and activity. Furthermore, the phospholipid composition of the membranes at which oxidase assembly occurs contributes to overall activity of the enzyme complex. Like p47*phox*, p40*phox* possesses a PX domain but may recognize phosphoinositide species that differ or overlap with those to which p47*phox* binds. The PX domain of p40*phox* binds phosphatidylinositol 3-phosphate [$PI_{(3)}P$], a species generated on the cytoplasm-oriented leaflet of nascent phagosome membranes. Experimental evidence indicates that robust oxidase activation during phagocytosis requires $PI_{(3)}P$-dependent binding of p40*phox* at the phagosome, thereby providing another level of spatial control of the phagocyte oxidase. Several reports of flavocytochrome b_{558} purification have noted the co-isolation of Rap1A, a low-molecular-weight G protein, although a functional link with the phagocyte oxidase remains to be further defined. Of note, neutrophils from mice lacking Rap1A exhibit slightly reduced oxidase activity. There is recent evidence as well that the abundant cytosolic calcium-binding proteins MRP8 and MRP14 associate with flavocytochrome b_{558} in neutrophils and modulate oxidase activity in intact cells.

OXIDANT GENERATION WITHIN THE PHAGOSOME

The clinical consequences of CGD dramatically highlight the important contribution of oxidant-dependent antimicrobial activity to human defense against infectious agents. Discussions of neutrophil antimicrobial action often juxtapose such activity with microbial killing that occurs in the absence of O_2, the so-called O_2-independent system. The granule proteins delivered

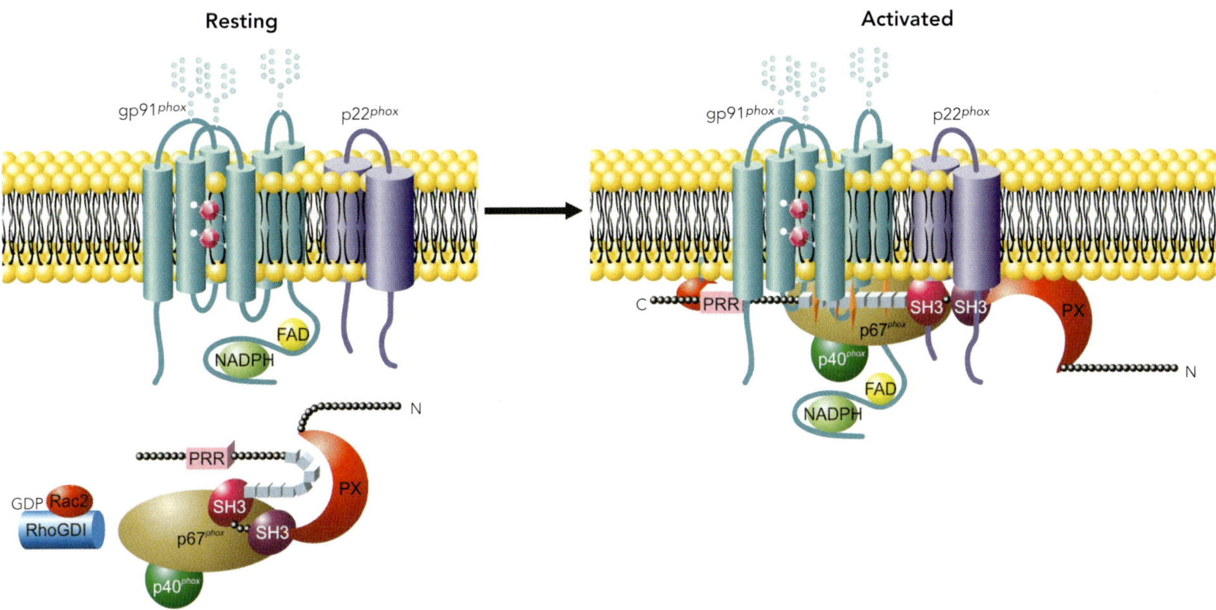

Figure 15.4. Assembly of the phagocyte NADPH oxidase. In resting phagocytes, the NADPH oxidase is unassembled and inactive. Upon stimulation, phosphorylation of p47^{phox} and RhoGDI relaxes conformational constraints in the complexes. The phosphorylation of p47^{phox} permits the p47^{phox}–p67^{phox} complex to associate with the membrane and with flavocytochrome b_{558} (Figure 15.3). Conversion of Rac-GDP to Rac-GTP permits the isoprenylated protein to translocate to the membrane, independently of the p47^{phox}–p67^{phox} complex, where interactions with gp91^{phox}, p67^{phox}, or both occur.

to attack the ingested microbe by granule–phagosome fusion include a variety of proteases with a broad array of substrate specificities that react with exposed or secreted microbial products. Some granule proteins are not enzymes but exert direct antimicrobial activity. For example, cationic antimicrobial peptides present in azurophilic granules of neutrophils, such as α-defensins or bactericidal-permeability increasing protein (BPI), lack protease activity but disrupt the integrity of the microbial surface upon binding and thereby compromise the viability of the organism. The intrinsic activity of both of these antimicrobial agents is independent of the presence of O_2, and their initial engagement with the target is at the microbial surface, dictated, in part, by electrostatic interactions between the cationic antimicrobial peptide and the anionic outer structures on the organism. Consequently, any posttranslational events that change charge or other intrinsic physical properties of the microbial surface will influence subsequent associations with such granule antimicrobial proteins in the phagosome. Such considerations are important to include in a discussion of phagocytic antimicrobial action, as the conventional classification of events as "oxygen-dependent" versus "oxygen-independent" underestimates the complex synergy that occurs in the physiologic setting of an infectious challenge.

As detailed earlier, the assembled oxidase serves as an electron transferase, sequentially moving electrons from cytoplasmic NADPH, across the membrane (either plasma or phagosomal), to molecular oxygen, and thereby generating O_2^- as the proximal product of oxidase activity. Although the magnitude of oxidase activity parallels the intensity of the stimulation, the phagocytosis-triggered oxidase generally generates O_2^- within the phagosome, with very little detected extracellularly unless the number of particles in the challenge is high or closure of the phagosome is incomplete. The turnover number of the phagocyte oxidase is ~150–160 electrons/heme/s, and a soluble agonist such as phorbol myristate acetate can stimulate extracellular generation of 7–10 nmol O_2^-/min/10^6 neutrophils.

As the proximal ROS generated by the phagocyte oxidase, O_2^- has several potential fates. Within the phagosome, O_2^- can undergo protonation to yield the perhydroxy radical OH_2, although the pK_a for dissociation is 4.8. Furthermore, O_2^- can serve either to donate or to accept electrons and thus can react with itself to generate O_2 and H_2O_2. The latter dismutation reaction can occur spontaneously, optimally when the concentrations of the two reactive species are identical (i.e., pH 4.8). In neutrophils, the pH in the phagosome reaches 6–6.5, which would seem to limit spontaneous dismutation of O_2^-. However, the influx of protons during oxidase activation, an obligate response to mediate charge compensation for the shuttling of electrons into the phagosome, likely accounts for conversion of O_2^- to H_2O_2. Catalytic promotion of H_2O_2 generation by superoxide dismutase (SOD) may occur when ingested microbes release their SOD into the phagosome, but no experimental data suggest that host-derived SOD contributes in this chemistry. Nearly all of the O_2 consumed by stimulated neutrophils can be recovered as H_2O_2.

Reactive nitrogen species such as nitric oxide (NO) likewise participate in inflammation and in host defense against microbes. Produced from arginine by nitric oxide synthase (NOS), NO has been linked to a wide variety of immune activities, including cell signaling (both within and between cells), tumor killing, and antimicrobial action. Within the context of intracellular events in phagocytes, NO has been most extensively characterized in murine systems. In contrast to the general recognition that NO contributes to murine host defense and participates in both acute and chronic inflammatory states, there is considerable controversy as to the capacity of human phagocytes to produce enzymatically NO in vivo. In the case of monocyte-macrophage production of NO, the story is relatively complicated. In vitro, production of NO by isolated monocytes or monocyte-derived macrophages has not been demonstrated after exposure to a wide array of potential agonists and cytokines. However, macrophages recovered from patients with mycobacterial infection exhibit the capacity to kill mycobacteria in a fashion that is ablated by inhibition of NOS, suggesting that mycobactericidal activity of the macrophages is NO-mediated. It is possible that the failure of the in vitro studies to mirror what is implied in vivo reflects shortcomings in our understanding of human macrophage differentiation in vitro rather than the absence of biologically relevant NOS activity. Elucidation of the role of NO in human macrophage-mediated antimicrobial activity will be anticipated to allow insights into cell models of human macrophage differentiation.

In the case of human neutrophils, evidence supporting the presence of significant NOS activity is likewise controversial. Regardless of the source of NO, there is general agreement that NO participates in influencing the behavior of neutrophils and thus contributes to the biological phenotype during inflammation. NO has been implicated in modulating neutrophil recruitment to sites of inflammation and NO from endothelial cells can interact with neutrophil-derived superoxide anion to generate peroxynitrite, a reactive species with a broad range of biological targets. In this light, one can ascribe to NO a role in influencing neutrophil-dependent antimicrobial activity, even if the inputs are indirect.

Since phagosomes contain large amounts and a wide array of potential substrates derived from both ingested microbes as well as released granule contents, it is impossible to provide a comprehensive list of all the reactive species generated. The complexity of the biochemistry in the phagosome is magnified by recognizing that the microbial contribution will vary as a function of the specific identity as well as the growth phase of the ingested organism, and that the individual phagosome itself is very heterogeneous. Oxidant production is localized to the site on the membrane where the oxidase has assembled, thus providing a point source for oxidant dispersal, the extent of which will reflect the intrinsic diffusion capacity of the specific ROS as well as the probability of its collision with suitable substrates in its environment. Granules likewise fuse at discrete sites on the phagosome and release contents locally. Some granule proteins bind avidly to targets, thus delivering antimicrobial action in close proximity to the potential targets, whereas others may act in solution. Collectively, these factors magnify the inherent challenge of precisely defining the identity of products in a given phagosome or assigning essential antimicrobial activity to any specific constituent.

The direct toxic action of H_2O_2 on microbes is relatively modest at concentrations generated in vivo but is potentiated as much as 100-fold, depending on the organism, in the presence of the azurophilic granule protein myeloperoxidase (MPO). H_2O_2 reacts with the ferric iron in MPO to generate Compound I, a complex in which the heme group of MPO forms a double bond with O_2 and serves as the catalytically active form of MPO (Figure 15.5). Compound I catalyzes a two-electron oxidation of halide ions to produce hypohalous acids. O_2^- can also react directly with MPO and this interaction has been implicated in maintaining optimal activity of the enzyme in the phagosomal environment. In the case of the reaction of Compound I with chloride ion, the predominant halide in the phagosome, Cl^- is oxidized to Cl^+ to yield hypochlorous acid (HOCl), or bleach. Depending on the agonist used for activation and the analytical techniques employed to assess reaction products, ~11%–40% of the O_2 consumed by stimulated human neutrophils can be recovered as a chlorinating species such as HOCl, and stimulated neutrophils can produce ~50 nmoles HOCl/30 minutes/10^6 cells. HOCl is a very potent oxidant and although it can dissociate in a pH-dependent fashion to hypochlorite or generate Cl_2, much of the HOCl generated in the phagosome attacks any potentially oxidizable group to which it has access. Candidate substrates include the sulfur-containing amino acids cysteine and methionine, amino side-groups on all amino acids, carbohydrates, lipids, phospholipids, and nucleic acids. In some cases, the resulting change in the microbe will directly compromise the viability of the organism, whereas in other situations the induced chemical modifications will alter susceptibility to other reactive species in the phagosome. Vulnerable targets are not restricted to the ingested microbe but include as

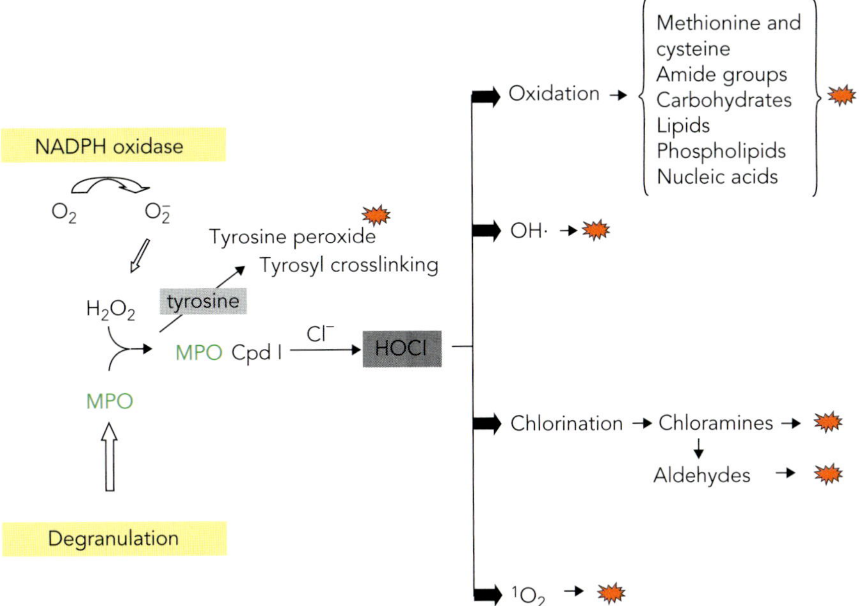

Figure 15.5. Potential MPO-based reactions in the phagosome. Concurrent activation of the NADPH oxidase and release of azurophilic granules into the phagosome delivers reactive oxygen species and granule products, respectively, in close proximity of the ingested microbe. Among the many interactions that occur among oxidants and granule proteins to produce an environment inhospitable to microbes are those dependent directly or indirectly on MPO. MPO and H_2O_2 combine to form Compound I that in turns supports the two-electron oxidation of Cl^- to Cl^+ and thereby generate HOCl, or bleach. MPO and H_2O_2 also directly oxidize tyrosine to generate tyrosyl radicals, a substrate for production of tyrosine peroxide as well as tyrosyl cross-linking of free tyrosines or tyrosyl residues in proteins. HOCl is a very potent oxidant and directly attacks amino acids, especially methionine and cysteine, as well as amide sidechains on peptides and proteins, and other biological substrates. HOCl also supports generation of other reactive species, including hydroxyl radical ($OH^\bullet$), singlet oxygen ($^1O_2$), and chloramines. The latter products include those that are relatively long lived and exert significant antimicrobial activity. Some chloramines decay and release aldehydes, some of which are antimicrobial in their own right. Many of these various derivatives of MPO-dependent HOCl generation mediate antimicrobial activity (indicated by the red starburst).

well host proteins delivered to the phagosome during degranulation.

The HOCl generated by the MPO-H_2O_2-Cl system promotes several other chemical reactions in addition to mediating the direct oxidation of susceptible targets. The downstream products from reactive HOCl create a host of long-lived agents by attacking nitrogen and generating chloramines. Some chloramines retain oxidizing capacity and thus extend the antimicrobial influence of MPO-derived products in the phagosome, whereas others decompose to release aldehydes, some of which possess intrinsic biological activity. The MPO-H_2O_2 system in the absence of chloride can also directly oxidize the aromatic amino acid tyrosine to produce tyrosyl radicals, which in turn promote dityrosine linkages between proteins. Furthermore, tyrosyl radicals and O_2^- can interact to generate tyrosine peroxide, as has been directly demonstrated for stimulated neutrophils.

The MPO-H_2O_2-Cl system supports modification of ROS initially produced by the phagocyte NADPH oxidase. For example, stimulated neutrophils generate low levels of hydroxyl radical ($OH^\bullet$) in an MPO-dependent fashion, via a reaction between HOCl and O_2^-. $OH^\bullet$ reacts readily and rapidly and, in the presence of CO_2, can generate $HCO_3^\bullet$, a radical which would thus be present under physiologic conditions and has demonstrated antimicrobial capacity. In addition, MPO-generated HOCl can interact with H_2O_2 to produce singlet oxygen ($^1O_2$). In both cases, very sensitive analytical systems are required to detect $OH^\bullet$ and $^1O_2$, raising questions about their contribution to the overall toxic environment in the phagosome and physiologic relevance to antimicrobial action. Nonetheless, the species are highly reactive and can be detected in biological settings.

In addition to MPO, other granule proteins exhibit synergistic activity with each other and with oxidants.

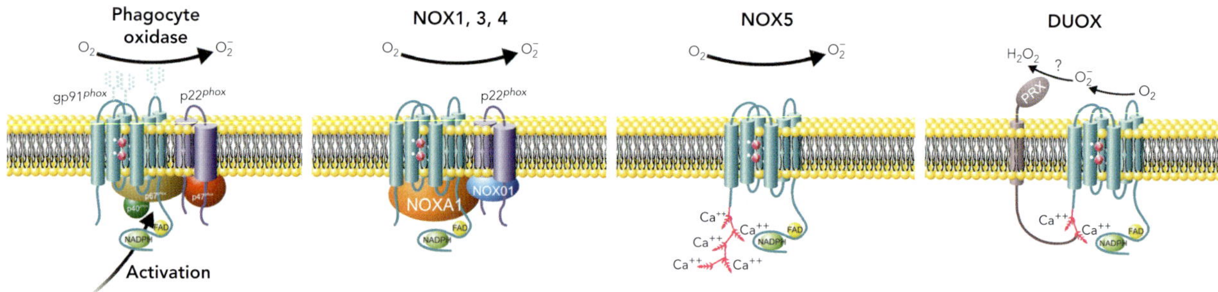

Figure 15.6. Members of the NOX family of NADPH oxidases. The NOX family of NADPH oxidases includes seven proteins (NOX1–5, DUOX1, and DUOX2) that are identified by sequence homology with gp91*phox* (aka NOX2) and share dependence on NADPH and FAD for activity. Distributed broadly in a nearly all tissues and cell types in both plant and animal kingdoms, the nonphagocyte oxidases serve diverse functions. The organization of the nonphagocyte oxidases differs in several ways from that of the phagocyte oxidase. Only NOX1, 3, and 4 pair with p22*phox* in the membrane. NOXO1 and NOXA1, homologues of p47*phox* and p67*phox* respectively, interact with NOX1, 3, and 4, whereas NOX5 and the DUOX proteins function independently of cytosolic protein cofactors. In contrast to the agonist-dependent assembly and activation seen in the phagocyte oxidase, the nonphagocyte systems function constitutively, with modulation by exposure to various agonists. Both NOX5 and the DUOX proteins possess EF-hands in their cytoplasmic domains and are regulated by changes in intracellular calcium.

The antimicrobial activities of elastase or cathepsin G are potentiated by the MPO-H_2O_2-chloride system by a mechanism that is dependent on the cationic nature of the granule proteins; neither requires enzymatic activity to kill target bacteria. The same synergy probably occurs as well with the other granule antimicrobial proteins, including the defensins and BPI. However, it is important to recognize as well that many of the granule proteins exert potent and effective antimicrobial action even in the absence of a functioning oxidase. Neutrophils from patients with CGD, cells that lack the capacity to generate any ROS, can kill *Escherichia coli*, *Pseudomonas aeruginosa*, and many catalase-negative streptococci normally, thus demonstrating *bona fide* effective antimicrobial action mediated exclusively by the granule proteins.

Given the quantity and expansive functional capacity of granule proteins released into the phagosome, the varied and highly reactive oxidants generated by the phagocyte NADPH oxidase in situ, and the dynamic ways in which these constituents of the phagosome synergize, the phagocyte provides comprehensive and at times redundant biochemical agents to kill and degrade ingested microbes. In addition to supporting antimicrobial activity, neutrophil-derived oxidants contribute to proinflammatory events outside the phagocyte when produced in excess, at inopportune times, or in inappropriate tissue sites. Granule proteins such as MPO and proteinase 3 have been linked to systemic vasculitides and MPO-mediated oxidation of lipoproteins, glycerophospholipids, proteins, and amino acids has been associated with a wide variety of inflammatory diseases. In plasma at physiologic pH, released MPO and H_2O_2 can support the production of HOBr and HOSCN in addition to HOCl, each of which can target the full range of biological molecules. The relationships of MPO and HOCl in modifying lipoproteins and thereby contributing to the development and progression of atherosclerosis are currently being widely studied. MPO can catalyze nitration of targets, most notably tyrosine, either free or as tyrosyl residues in proteins. Although nitration does not occur within the phagosome, MPO-dependent nitration of extracellular targets likely contributes to nitric oxide–mediated tissue injury in inflammation. Thus, much of the potent biochemistry that serves to target microbes captured in the phagosome can induce significant tissue damage as part of the inflammatory response.

NADPH OXIDASE BEYOND PHAGOCYTES

As mentioned earlier, it has been recently appreciated that the phagocyte NADPH oxidase is a member of the NOX protein family, a group defined by homology with gp91*phox*, aka NOX2 (Figure 15.6). Several excellent reviews of the NOX protein family describe the incredibly wide biological contexts in which they operate, extending through both the plant and animal kingdoms. Although gp91*phox* serves as the organizing principle for the family, several distinctive features of the nonphagocyte members merit comment.

First, the critical dependence of gp91*phox* on cofactors is not shared by all NOX proteins. Only NOX 1, 3, and 4 resemble gp91*phox* in requiring association with p22*phox* for stability, activity, and proper targeting into membranes. Homologues of p47*phox* and p67*phox*, designated as NOXO1 and NOXA1 respectively, serve as cofactors for NOX1 and 3, whereas NOX4, NOX5, and the DUOXes do not depend on cytosolic protein cofactors. Instead, NOX5 and the DUOXes, all EF-hand containing proteins, are regulated by cytosolic calcium and may represent very ancient members of the protein family. The extent to which these differences with respect to supporting cofactors provides mechanistic insights into how p22*phox* precisely influences gp91*phox* activity or into the intricacies of cytosolic factor modulation of oxidase activity awaits further study. Second, ROS production in the phagocyte system is relatively robust, suitable for damaging microbial targets. In their native setting, the nonphagocyte NOX proteins generally generate lower levels of ROS. With respect to NOX1, 3, and 4, this lower activity seems better suited to mediating signal transduction than rendering irreversible modifications in targets. Third, whereas the phagocyte system is unassembled and inactive in unstimulated cells, some of the NOX proteins are active constitutively but subject to modulation by various factors. For NOX proteins that interact with NOXO1, the constitutive activity may reflect, in part, the absence of an autoinhibitory region in NOXO1. As a consequence, NOXO1 is likely always in an open conformation, supporting continuous interaction with complementary domains at the membrane. Last, studies of the DUOXes detect only H_2O_2 as the proximal product of its activation; no O_2^- is recovered. The biochemical basis for this activity remains undefined, but its eventual elucidation will likely reveal novel biochemistry, either at the level of FAD-supported electron transfer or in the context of novel superoxide dismutase activity in DUOX itself.

Definition and characterization of similarities and contrasts among the NOX proteins with respect to the phagocyte NADPH oxidase promises to expand understanding of the role of ROS in inflammation in its myriad manifestations.

SUGGESTED READINGS

Bedard, K., and Krause, K.-H. 2007. The NOX family of ROS-generating NADPH oxidases: physiology and pathophysiology. *Physiol Rev* **87**:245–313.

Hampton, M.B., Kettle, A.J., and Winterbourn, C.C. 1998. Inside the neutrophil phagosome: oxidants, myeloperoxidase, and bacterial killing. *Blood* **92**:3007–3017.

Klebanoff, S.J. 2005. Myeloperoxidase: friend or foe. *J Leuk Biol* **77**:598–625.

Nauseef, W. M. 2007. How human neutrophils kill and degrade microbes: an integrated view. *Immunol Rev* **219**:88–102.

Nauseef, W.M. 2008. Biological roles for the NOX family NADPH oxidases. *J Biol Chem* **283**:16961–16965.

Pattison, D.I., and Davies, M.J. 2006. Reactions of myeloperoxidase-derived oxidants with biological substrates: gaining chemical insight into human inflammatory diseases. *Curr Med Chem* **13**:3271–3290.

16 Cell Adhesion Molecules

Lucy V. Norling, Giovanna Leoni, Dianne Cooper, and Mauro Perretti

BIOLOGICAL FUNCTIONS OF ADHESION MOLECULES

Cell adhesion molecules are endowed with many biological functions, often crucial to life. For instance, in gut or lung epithelia they assure cell-to-cell contacts, thus making and maintaining barrier functions, while allowing passage of nutrients and other solutes. Besides maintaining body form and structure, cell adhesion molecules play important roles in tissue development as well as helping to determine its final organization. A common feature of adhesion molecules is their ability to sustain strong bonds between adjacent cells on either a permanent or intermittent basis.

Apart from these *static functions* of cell adhesion molecules, there are many *mobile functions*, whereby they are able to induce intermittent cell-to-cell bonds hence assuring particular cell movements either in the blood vessel (as during inflammation; topic to be investigated in this chapter) or within tissues and organs.

It is therefore evident that, contrary to previous simplistic visions and predictions, cells *are never free floating* but rather they move on surfaces such as the inflamed joint or gut subepithelial strata, penetrate into injured organs or circuit the body (think of T cells traveling the blood lymphatic circulation, and so forth) by activating processes **all** relating to cell adhesion molecule-mediated phenomena.

ADHESION MOLECULES IN INFLAMMATION

Multistep Leukocyte Adhesion Cascade

Leukocyte recruitment in both homeostatic and inflammatory situations follows a generic paradigm characterized by a sequential cascade of interactions between leukocytes and endothelial cells (ECs). Insight into the cellular and molecular processes involved in each step of the cascade have been provided by a range of experimental approaches performed both in vitro and in vivo. These include antibody inhibition studies, static adhesion assays, parallel-plate flow chamber models, as well as the use of intravital microscopy to visualize live interactions of leukocytes with the vessel wall.

Collectively, these studies have helped to elucidate that (i) initial leukocyte–endothelial interactions (capture and rolling) are instigated primarily by a family of molecules called selectins interacting with oligosaccharide residues expressed by counter-ligands, and (ii) firm adhesion and transmigration are mediated by leukocyte integrins interacting with the endothelial immunoglobulin superfamily of adhesion molecules (outlined in Figure 16.1 and described later in detail).

Leukocytes are a heterogeneous population of immune cells with specialized functions. The type of leukocyte recruited to an inflammatory site depends on the stimulus and hence the differential expression of adhesion molecules on the endothelium. In acute inflammatory episodes, the predominant leukocyte influx is neutrophils, with the remaining leukocytes consisting of monocytes or macrophages and, in allergic responses, eosinophils. Comparatively, in chronic inflammatory responses lymphocytes predominate, as well as monocytes and macrophages whereas neutrophils are present to a lesser extent.

Adhesion Molecules Promoting Cell Rolling: The Selectins

Selectins are the main initiators of leukocyte tethering and rolling, although this process can – in some cases – be facilitated by several integrins (especially when selectin deficiency occurs). The selectins are a three-member family (L-, P-, and E-selectin) that share a highly conserved extracellular N-terminal Ca^{2+}-dependent C-type lectin domain. All selectins contain a transmembrane domain and a cytoplasmic

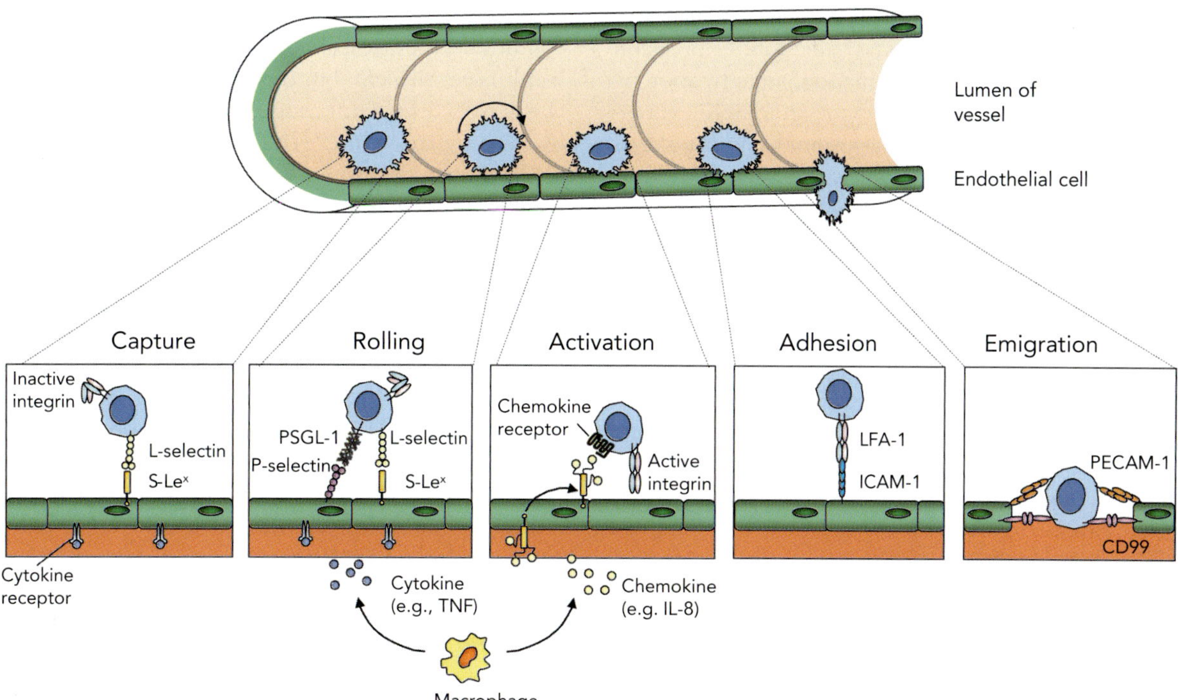

Figure 16.1. Leukocyte recruitment cascade. Following an inflammatory stimulus, tissue-resident macrophages release inflammatory mediators such as the cytokine TNF-α, which induces the rapid expression of preformed P-selectin (and transcription-dependent E-selectin expression) on the endothelium. The interaction between selectins and their glycoprotein ligands initiates leukocyte rolling. Activation by chemokines – and other leukocyte activators (e.g., leukotriene B_4 or platelet-activating factor) – presented on endothelial cells causes leukocyte integrin activation, thus resulting in transition from cell rolling to cell adhesion, in view of the strength of integrin-mediated binding with endothelial immunoglobulin superfamily members. Leukocytes can then transmigrate through the endothelial monolayer and chemotactically move toward the inflammatory stimulus. Examples of adhesion molecules involved in each step are depicted.

tail, the major structural difference between the family members being the number of consensus repeat units. All selectins bind, through their lectin domain, to sialylated forms of oligosaccharides displayed on various mucin-like glycoproteins, for example, sialyl lewis X (sLeX). One of the most extensively characterized selectin ligands is P-selectin glycoprotein ligand-1 (PSGL-1), a highly sialyated mucin found on all leukocytes and a ligand for all three selectins.

Selectin ligands need several carbohydrate modifications before leukocytes will bind; the enzymes involved in these posttranslational modifications include the glycosyltransferases, fucosyltransferases, and the sulfotransferases. Insights into the contributions of these enzymes on ligand structure and hence leukocyte rolling has been demonstrated using enzymatic treatments to cleave certain chains and more specifically in mutant mice with targeted deficiencies.

A rare congenital disease in humans, namely leukocyte adhesion deficiency II (LAD-II), underscores the importance of posttranslational modifications on selectin-ligand biosynthesis. These individuals have a defective Golgi GDP-fucose transporter, resulting in a hypofucosylation of glycoproteins. Patients exhibit persistent leukocytosis, defects in selectin-mediated leukocyte rolling and consequently suffer from recurring bacterial infections because blood-borne leukocytes fail to reach the site of infection/inflammation.

The intrinsic affinity of selectins for their ligands is rather low, and they display extremely rapid association and dissociation rates which assures a transient interaction of the leukocyte with the endothelium, insufficient to permit cell arrest. However, once initial bonds have formed on the endothelium, shear stress from the blood flow *pushes* the leukocyte forward and new bonds occur before the initial bonds break, manifesting in the rolling motion of the leukocyte. Selectins require shear stress to support this interaction, and without flow the cells detach. This counter-intuitive phenomenon is related to the catch bond characteristic of selectins, whereby the bond becomes stronger with applied force.

P-Selectin (CD62P) is preformed in cytoplasmic secretory α-granules of platelets and Weibel-Palade bodies of ECs. Stimulation with vasoactive mediators such as thrombin or histamine results in rapid mobilization of endothelial P-selectin to the external plasma membrane allowing leukocytes to interact. Thus, it is not surprising that P-selectin contributes to recruitment early in inflammation and infection. Expression at the cell surface is transient due to internalization by endocytosis for reuse. In addition to this rapid expression, P-selectin can be transcriptionally upregulated in human ECs by cytokines such as IL-4 whereas in murine ECs, TNF-α and LPS can induce its expression. Interactions between P-selectin on activated platelets and its main counterligand PSGL-1 on leukocytes results in leukocyte–platelet interactions that are pertinent for wound healing and contribute to leukocyte recruitment during inflammation.

E-selectin (CD62E) expression is mostly restricted to activated ECs, with the exception of skin microvessels and bone marrow where it is constitutively expressed. E-selectin requires de novo mRNA and protein synthesis and can be induced on cultured EC by cytokines such as IL-1β or TNF-α with maximal expression at 4 hour, reducing toward baseline levels by 24 hours. In comparison to P-selectin, the rate of E-selectin internalization is much slower and it is directed to lysosomes for degradation instead of being recycled.

L-selectin (CD62L) is exclusively and constitutively expressed on leukocytes, particularly neutrophils, monocytes, and naive lymphocytes (but not memory lymphocytes). It serves as a key receptor for lymphocyte homing to secondary lymphoid organs, and at peripheral sites of injury and inflammation. Its location on the tips of microvilli facilitates its interaction with EC, allows high receptor concentrations and also evades the electrostatic repulsion of the negative cell surface. Upon leukocyte activation, L-selectin is rapidly shed; although this mechanism is not fully elucidated, it is known that shedding can be blocked using inhibitors of zinc-based metalloproteases, and that a disintegrin and metalloprotease-17 (ADAM-17) is one of the enzymes responsible for this cleavage.

Essential physiological roles of the selectins have been reinforced by studies using gene-targeted mice and bone marrow–transplanted chimeric mice. In all cases, mice homozygous deficient in selectins are viable and fertile, and display only mild phenotypes suggesting that these molecules are functionally redundant. In sharp contrast, E/P-selectin double null mice exhibit severe defects in leukocyte rolling, with a complete abrogation of trauma-induced rolling in venules 10–120 minutes after exteriorization of the cremaster muscle. These double knockout mice, as well as the triple knockout E/P/L/-selectin triple null mice display an increased susceptibility to spontaneous mucocutaneous infections, which can be lethal.

Creation of mice lacking multiple selectins has helped show overlapping and unique features of the selectins. In addition, generation of quadruple adhesion molecule (E/P/L-selectin and ICAM-1) deficient mice resulted in viable, seemingly healthy mice in the absence of any challenge. Under inflammatory conditions, they exhibited a significantly impaired neutrophil recruitment, but mononuclear cell recruitment was barely affected, because rolling is also facilitated by α4-integrins on these cells. Treating these mice with a blocking monoclonal antibody for the α4-integrin concomitantly with the inflammatory stimulus TNF-α, completely eliminated rolling, and significantly reduced cell adhesion, although a small *residual* level of cell adhesion persisted suggesting again the involvement of additional recruitment mechanisms.

Two α4-integrins, α4β1 (very late antigen-4; VLA-4) and α4β7 (lymphocyte Peyer's patch HEV adhesion molecule-1; LPAM-1), mediate both lymphocyte rolling and firm adhesion, using the same ligand for both processes, notably VCAM-1 and mucosal addressin cell adhesion molecule-1 (MAdCAM-1, a vascular addressin expressed on HEV), respectively.

Adhesion Molecules in Cell Adhesion

Following the initial rolling process mediated by selectins and their ligands, activation of leukocytes coincides with a decrease in their rolling velocity and the transition to firm adhesion. Adherent leukocytes can become elongated and flatten onto the endothelium, decreasing their protrusion into the vessel lumen, and thus their tendency to detach.

Cell adhesion is established by the interaction of integrins with binding partners that belong to the immunoglobulin superfamily, referred to as cellular adhesion molecules (CAMs), including intercellular (I)-CAM-1 (CD54), vascular (V)-CAM-1 (CD106), and platelet-endothelial (PE)-CAM-1 (CD31). These immunoglobulins are expressed basally on the resting endothelium, and can be upregulated during inflammation, as is the case for ICAM-1 and VCAM-1, whilst PECAM-1 is redistributed to the site of leukocyte transmigration and is involved in homophilic PECAM-1 interactions with the leukocyte.

VCAM-1 binds to VLA-4/α4β1 expressed on lymphocytes and monocytes and at very low levels on neutrophils. Furthermore, VCAM-1 can contribute to leukocyte transendothelial migration (TEM) via the paracellular pathway (discussed later in more detail) by causing gap formation between adjacent endothelial cells.

Integrins are a family of type I transmembrane cell surface receptors, each consisting of a noncovalently associated α and β subunit. At present, 18 α-subunits and 8 β-subunits are known, yielding 24 distinct integrins. The overall shape and dimensions of integrins have been determined by electron microscopy, illustrating at least three distinct conformational states. The N-terminal regions of both subunits together form a globular structure that resembles a "head" (containing the ligand-binding site), connected to the two "legs." Essential for ligand binding is the metal-ion-dependent adhesion site, where divalent metals normally reside.

Unstimulated leukocytes are nonadherent, yet become adherent in response to an activating stimulus for example, chemokines, cytokines, or antigens. These stimuli activate pathways that act on the cytoplasmic portions of integrins altering their affinity for ligands; key to this process is a conformational change in the integrin structure. In addition, stimuli can cause clustering on the cell surface, altering integrin avidity. GPCR-triggered integrin activation occurs through an "inside-out" signaling process, whereas downstream signaling from ligand binding is referred to as "outside-in" signaling; either process can contribute to the stabilization of adhesion.

Integrins normally exist in a low-affinity conformation in which the leg region is bent, subsequently bringing the ligand-binding head-piece into close proximity to the cell membrane. Activation signals induce a switchblade-like extension of the leg that extends the structure, allowing an intermediate affinity conformation, and an opening of the head-piece resulting in a high-affinity state, and hence increasing the propensity for ligand binding. It is predicted that certain integrins only extend about 5 nm above the plasma membrane in a low-affinity conformation, and can protrude 20–25 nm in an extended higher-affinity form, therefore making them more likely to be topologically accessible to ligands.

Leukocyte integrins are expressed on the cell body rather than microvilli, where L-selectin is expressed. Integrins most relevant to leukocyte arrest include the β1 and β2 subfamilies of integrin receptors. All leukocytes express the β2 integrin (CD18), which pairs with distinct α partners (CD11) on different leukocyte subsets to form different glycoprotein complexes, namely, αLβ2 (LFA-1; CD11a), αMβ2 (Mac-1; CD11b), αXβ2 (p150, 95; CD11c), and αDβ2 (CD11d), binding partners of which are summarized in Table 16.1.

In vivo studies using either CD18-specific monoclonal antibodies or studies in CD18 null mice exhibit a decrease in the percentage of adherent leukocytes in inflamed postcapillary venules. The importance of β2 integrin–mediated adhesion is best exemplified in humans which lack a functional gene for the β2 integrin, referred to as LAD-I genetic deficiency. These individuals have severe infections, and develop spontaneous skin lesions; moreover, the life expectancy of the severely deficient patients is limited.

Adhesion Molecules in Cell Emigration

Following a period of stationary adhesion, a leukocyte may leave the postcapillary venule and pull itself into the subendothelial space. This event called leukocyte extravasation, TEM, or diapadesis is dependent on an array of cellular processes including adhesion molecule expression and activation, cytoskeletal reorganization, and alterations in membrane fluidity. Molecules such as junctional adhesion molecules (JAMs), PECAM-1 (CD31), ICAM-1 (CD54), ICAM-2 (CD102), VCAM-1 (CD106), CD99, and endothelial cell–selective adhesion molecule (ESAM) can be found at endothelial junctions and regulate leukocyte migration into underlying tissues, indeed inflamed endothelial cells reorganize these junctional proteins such that transmigration is favored (Table 16.1).

Two routes of leukocyte diapadesis have been noted so far: a *paracellular* route that dominates most extravasation processes, and a *transcellular* route (Figure 16.2). Recent evidence suggests that the route taken may be cell type specific with neutrophils preferring a paracellular route and lymphocytes a transcellular exit.

Endothelial cells are not merely passive participants in leukocyte transmigration, but are actively involved in this process. Leukocyte adhesion promotes remodeling of the apical endothelial plasma membrane into projections that surround the adherent leukocyte. These transmigratory cups or docking structures are enriched in ICAM-1 and VCAM-1 and may initiate migration by either a transcellular or paracellular route (Figure 16.2).

Transmigration via the paracellular route is a lot more frequent and therefore has been more extensively studied. During this process, leukocytes must cross endothelial junctions that contribute to the barrier function of the blood vessel, and which consist of transmembrane adhesion proteins connected to their intracellular partners. These are composed of two types of junctional structures, tight junctions, which include the claudins, occludins, and junctional adhesion molecules (JAMs) and adherens junctions, the main protein being vascular endothelial (VE)-cadherin. It is thought that ligation of molecules such as ICAM-1 may lead to activation of Rho which then results in opening of endothelial junctions.

Whereas capture, rolling and adhesion all involve heterophilic interactions of adhesion molecules, homophilic interactions between PECAM-1 and CD99 play a major role in transmigration via a paracellular route. Endothelial cell lateral junctions are enriched with these two proteins, which are also expressed on most leukocyte subtypes. Strong in vitro evidence for a role of PECAM-1 in transmigration of leukocytes was originally obtained using anti-PECAM-1 blocking antibodies. In vivo studies using intravital microscopy in PECAM-1 null mice highlighted an additional role for PECAM-1 in the migration of leukocytes through the perivascular basement membrane. PMNs were

TABLE 16.1. Summary of adhesion molecules involved in leukocyte–endothelial cell interaction

Adhesion molecule	Distribution	Ligands and counter-receptors	Function
L-selectin (CD62L)	All leukocytes except effector and memory effector T cells	PNAd, MAdCAM-1, PSGL-1, E-selectin, P-selectin	Rolling
E-Selectin (CD62E)	Endothelial cells	PSGL-1, ESL-1, CD44*, CD43* (*CLA decorated)	Rolling
P-selectin (CD62P)	Endothelial cells, platelets	PSGL-1, PNAd	Rolling
Selectin ligands			
sLe^x	Myeloid cells, some memory T cells, HEVs	All selectins	Rolling
PSGL-1	All leukocytes	All selectins (essential for P-selectin)	Rolling
PNAd	HEV, some sites of chronic inflammation	L-selectin, P-selectin	Rolling
CLA	Skin-homing T cells, DCs, granulocytes	E-selectin	Rolling
Integrins			
$\alpha_L\beta_{2\ (LFA\text{-}1;\ CD11a/CD18)}$	All leukocytes	ICAM-1, ICAM-2, JAM-A	Adhesion, transmigration
$\alpha_M\beta_{2\ (MAC\text{-}1;\ CD11b/CD18)}$	Granulocytes, monocytes, some activated T cells	ICAM-1, fibrinogen, C3b, JAM-C	Adhesion, transmigration
$\alpha_x\beta_{2\ (p150,95;\ CD11c/CD18)}$	DCs	Fibrinogen, C3b	adhesion
$\alpha_D\beta_{2\ (CD11d/CD18)}$	Monocytes, macrophages, eosinophils	ICAM-1, VCAM-1	adhesion
$\alpha_4\beta_{1\ (VLA\text{-}4)}$	Most leukocytes	VCAM-1, fibrinogen, JAM-B	Rolling, Adhesion
$\alpha_4\beta_{7\ (LPAM\text{-}1)}$	Lymphocytes, NKCs, mast cells, monocytes	MAdCAM-1, fibronectin, VCAM-1	Rolling, Adhesion
Immunoglobulin superfamily			
ICAM-1 (CD54)	Most types of cells	LFA-1 Mac-1, fibrinogen	Adhesion, transmigration
ICAM-1 (CD102)	Endothelial cells, platelets	LFA-1 Mac-1	Adhesion, transmigration
VCAM-1 (CD106)	Endothelial cells	VLA-4, $\alpha_4\beta_7\alpha_D\beta_2$	Rolling, adhesion
MAdCAM-1	HEVs in PP and MLN	$\alpha_4\beta_{7,\ L\text{-}selectin}$	Rolling
PECAM-1	Endothelial cells, platelets, leukocytes	PECAM-1	Transmigration
JAM-A	Endothelial cells, epithelial cells, platelets, most leukocytes	JAM-A	Transmigration
JAM-B	Endothelial cells, HEVs	JAM-B, JAM-C	Transmigration
JAM-C	Endothelial cells, HEVs platelets, monocytes, DCs, some T cells	JAM-C, JAM-B	Transmigration
Miscellaneous			
CD99	Endothelial cells, leukocytes	CD99	Transmigration

trapped between the vascular endothelium and the underlying basement membrane, although the number of emigrated leukocytes was unaltered suggesting compensation by other adhesion molecules. Blocking antibodies demonstrated that CD99 is involved in a step in transmigration that is distal to PECAM-1, and blocking both of these adhesion molecules results in an additive inhibition of monocyte transmigration.

The JAM family of adhesion molecules, members of the immunoglobulin superfamily, are targeted to cell–cell contacts and are also implicated in transmigration (Table 16.1 and Figure 16.3). Since JAMs can be expressed on ECs, epithelial cells, platelets, and leukocytes, they function in cell–cell interactions not only between the same types of cells but also among distinct types of cells through homophilic and heterophilic interactions. Furthermore, JAMs exhibit heterophilic interactions via their extracellular domains with the leukocyte integrins LFA-1, VLA-4, and Mac-1 (Figure 16.3). In the context of adhesion and

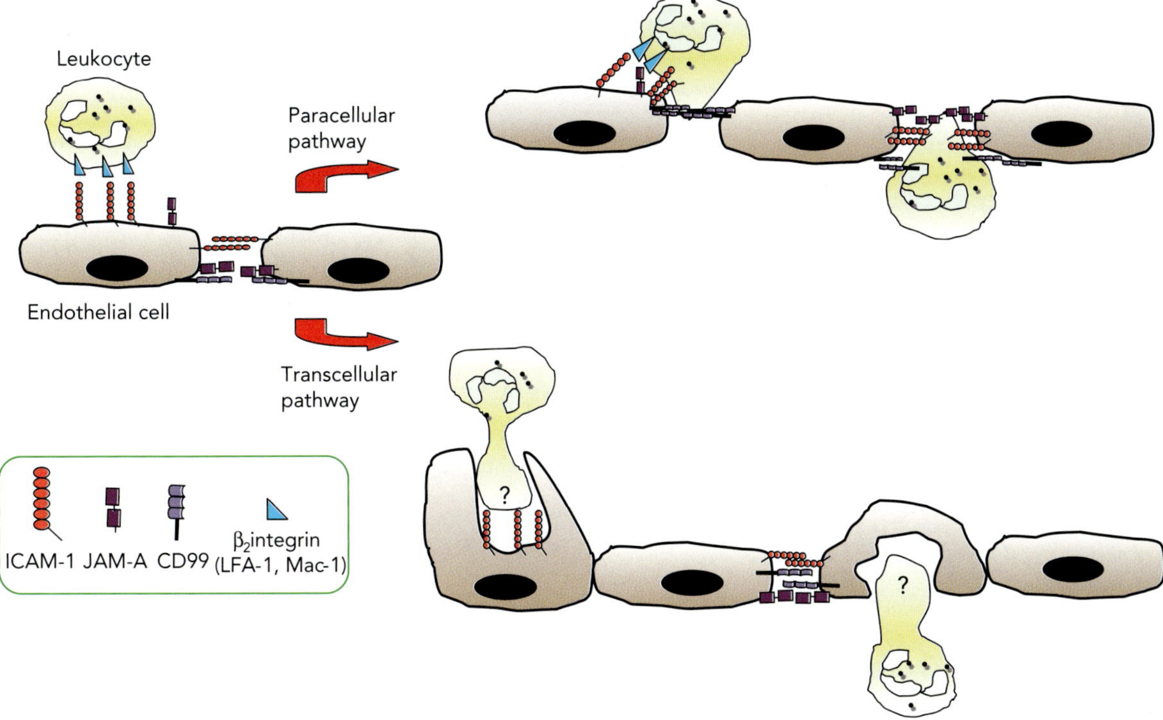

Figure 16.2. Pathways of leukocyte transmigration across the endothelium. To cross the vascular endothelium, leukocytes can take two different pathways. Adherent cells can disengage from the integrin-mediated adhesion and migrate by the paracellular route, passing through interendothelial junctions, squeezing through between adjacent endothelial cells; evidence for leukocytes transmigrating through the transcellular route also exist, whereby the cell will migrate across an individual endothelial cell body.

transmigration, JAM-A and JAM-C have been shown to play a functional role. Blocking JAM-A diminishes monocyte recruitment induced by MCP-3 into subcutaneous air pouches, while JAM-A deficient mice demonstrate reduced neutrophil infiltration in models of peritonitis and ischemia. JAM-C transfected ECs support increased transmigration of lymphocytes in vitro, and in parallel, transmigration could be reduced by 50% using a monoclonal antibody specific for JAM-C or by a soluble form of JAM-C. A prominent role for JAM-C in neutrophil transmigration has also been reported in vitro and in a mouse model of acute inflammatory peritonitis.

ESAM-1 appears to play a particular role in neutrophil transmigration, with diminished or delayed transmigration of these leukocytes in contrast to normal lymphocyte infiltration in ESAM null mice.

Migration via the transcellular route is thought to occur in areas of the endothelial cell that are rich in ICAM-1. Ligation of ICAM-1 by integrins on the surface of the leukocyte results in signaling events leading to the formation of channels in the endothelial cell, through which leukocytes can migrate. Many of the cell adhesion molecules involved in paracellular migration are also thought to play a role in transcellular migration, these being ICAM-1, ICAM-2, PECAM-1, CD99, JAMs, ESAM, VE-cadherin and LFA-1.

Although it appears that leukocyte subtypes may have a preferred route of migration, this may be influenced by the stimulus for migration and may also depend on the vascular bed involved. For example, neutrophil migration in the skin appears to occur largely by a paracellular route in response to high concentrations of FMLP, whereas in the cremaster, chemokine superfusion results in largely transcellular migration.

ASSESSMENT OF ADHESION MOLECULE FUNCTION UNDER FLOW

In Vitro (Flow Chamber)

The parallel-plate flow chamber is an in vitro system widely used to study leukocyte recruitment under defined laminar flow conditions, which more closely resembles the in vivo environment. The flow chamber assay was first performed in late 1980s to examine neutrophil–endothelial interactions, and has since allowed significant advances in the contributions attributed to different adhesion molecules in the process of leukocyte recruitment, demonstrating that

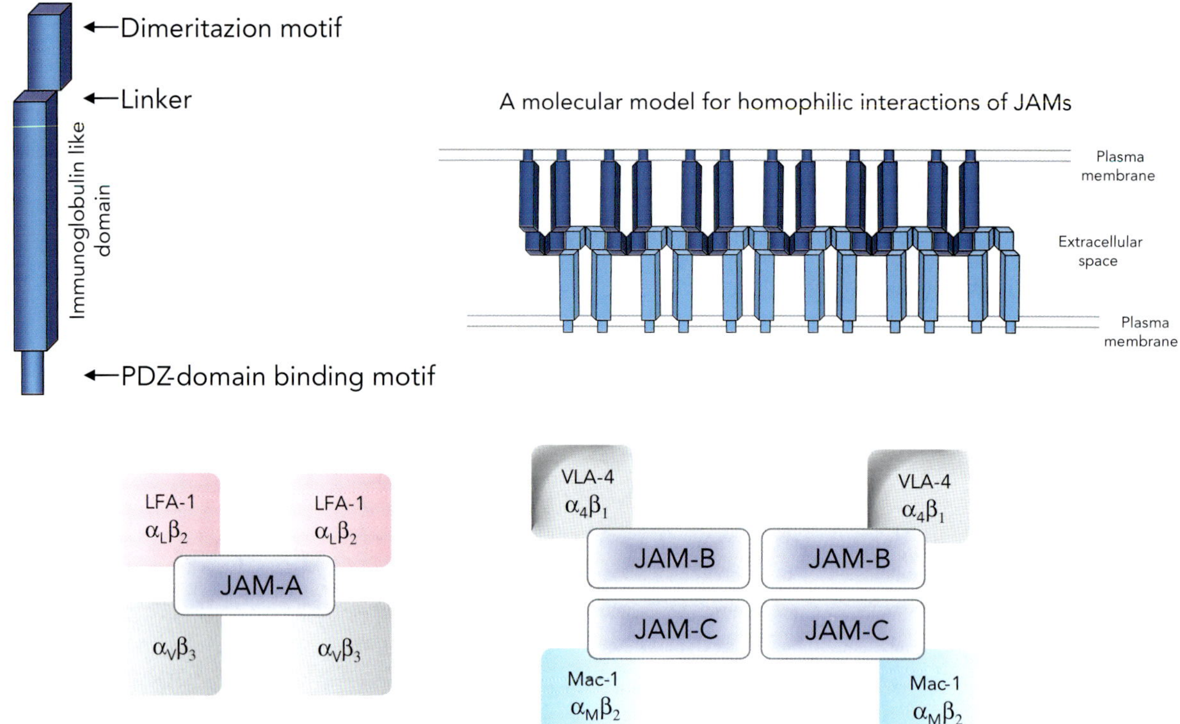

Figure 16.3. Structure of JAM-family members and binding partners. (A) *Upper panel*: JAMs are characterized by two immunoglobulin-like domains in the extracellular portion, a single transmembrane segment and a short cytoplasmatic tail with a PDZ-domain-binding motif (Phe-Leu-Val). A short linker sequence Val-Leu-Val connects the two immunoglobulin domains to impose a bent conformation. A dimerization motif Arg- (Val-Leu-Ile)-Glu in the membrane-distal domain is essential for homodimer formation. *Bottom panel:* Model of JAM homodimerization. (B) JAMs can form homo- and heterodimers, then forming specific *cis*- or *trans*- interactions with other CAMs.

multiple receptor–ligand pairs function in a sequential and orchestrated manner.

In brief, this assay involves perfusing leukocytes (or platelets) in suspension between two flat surfaces. The bottom surface can be coated with a variety of adhesive substrates such as a monolayer of endothelial cells or purified/recombinant adhesion molecules. Confluent endothelial monolayers can be activated with various inflammatory stimuli to delineate the roles of different adhesion molecules, for example, TNF-α (4h) causes an upregulation of E-selectin, ICAM-1, and VCAM-1, whereas histamine (15 minutes stimulation) results in a transient release of P-selectin. Blocking monoclonal antibodies can then be used to confirm the specificity of interactions. Alternatively, stable cell lines can be transfected with specific adhesion molecules to assess their particular function.

Typically, studies using the flow chamber are used to assess the number of interactions with their substrate in terms of quantification of initial capture of cells from the stream of flow, and the proportion of rolling and adherent cells. Once the first few cells start rolling, these cells themselves are also capable of instigating rolling (termed secondary tethering), a process more likely to occur in this assay due to leukocyte gravity sedimentation in the flow chamber. If an endothelial monolayer is used, leukocyte transmigration can also be quantified. Video 16.1 (http://cambridge.org/9780521887298) shows a short clip of neutrophils interacting with a monolayer of human umbilical vein endothelial cells: rolling, adhesion, and transmigration are evident.

A range of defined shear rates and stresses can be produced to facilitate the study of interactions under physiologically relevant conditions. These can be calculated based on a number of known parameters including the chamber geometry, according to the following equation:

$$\tau_w = \frac{6\mu Q}{a^2 b}$$

where τ_w is the wall shear stress (dynes/cm^2); μ is the coefficient of viscosity (Poise); Q is the volumetric flow rate (mL/s); a is the channel height (i.e., gasket thickness, cm); b is the channel width (i.e., gasket width, cm). Certainly, this has proved useful in determining the shear threshold and bond lifetimes of adhesion molecules involved in cell rolling. For instance, E-selectin permits the slowest rolling velocities, followed by P-selectin, then L-selectin; this is consistent with the fact that E-selectin demonstrates longer bond lifetimes.

The flow chamber has therefore proved a versatile tool for determining various aspects of the leukocyte recruitment cascade under conditions of shear stress, and the latter point is important and worth remembering.

In Vivo (Intravital Microscopy)

Intravital microscopy (IVM), literally, *microscopy inside life*, is a sophisticated research tool used to visualize and directly observe the microcirculation of animals in vivo. Early pioneers, including Robert Hooke and Antonie van Leeuwenhoek, described microscopic observations of living creatures in the 17th and 18th centuries, and Virchow (1821–1902), Conheim (1839–1884), and Metchnikoff (1845–1916) used microscopy of living blood vessels to advance our understanding of inflammation. Since many immunological disease mechanisms are reflected by primary interactions at the microcirculatory level, IVM can be used to study the underlying trigger and effector mechanisms, their contributions to tissue damage and implications to the immune response of leukocyte trafficking.

Early descriptions of leukocyte rolling and transmigration were provided in the first half of the 19th century and the entire leukocyte recruitment cascade (rolling, firm adhesion, and transmigration) was described well before the 20th century. In the 20th century, photography, film, and video cameras enabled more information to be collected and even more sophisticated and convenient methods are available today. Gus Born was the first (in 1972) to report quantification of rolling velocity of white blood cells on postcapillary venule endothelium.

The basic set-up for intravital fluorescence microscopy requires a high-energy light source (usually a mercury or xenon lamp), a microscope suited for epi- or transillumination, and a camera. Using digital cameras, high-resolution images can be acquired and recorded easily and cheaply for later, off-line analysis. Conventional microscopes illuminate subjects using light from a bulb (halogen, mercury, and xenon) that is either transmitted through or reflected off the subject. All points and focal planes of the subject are illuminated simultaneously and light is collected indiscriminately to form an image. Intravital microscopy allows detailed quantitative analysis of a wide range of microvascular parameters related to leukocyte–endothelium interactions. However, a problem with this method is that the most commonly used variants of intravital microscopy involve surgical preparation and/or exteriorization of the tissue to be examined, for example, mesentery or cremaster. This could induce mild stimulation of the tissue and causes rapid and pronounced upregulation of rolling apparently via partial degranulation of perivascular mast cells and endothelial expression of P-selectin, although this effect has been exploited for studies investigating selectin-dependent recruitment. A further limitation has been the inability to differentiate granulocyte responses from those of mononuclear cells by intravital microscopy. However, this problem has been overcome recently with the use of leukocyte subtype-specific antibodies.

The use of fluorescent dyes such as rhodamine 6G and calcein AM and ever more sophisticated microscopes has broadened the range of vascular beds studied by IVM from thin transparent tissues such as the cremaster to microcirculations of organs such as the brain, lung, and liver.

Table 16.2 depicts the parameters that are quantifiable by IVM pertaining to the leukocyte recruitment cascade. Oedema and arteriolar vasodilation responses are also commonly measured by IVM.

Video 16.2 (http://cambridge.org/9780521887298) shows an inflamed vessel after an ischemic event, with large numbers of leukocytes interacting with a neatly seen postcapillary venule endothelium: cell rolling (including platelets), adhesion, and migration are very distinguishable.

CONCLUSIONS

The fundamental response to tissue damage during an inflammatory response is the recruitment of blood leukocytes to the underlying tissue. Thus pharmacological manipulation of adhesion molecules involved in the cell recruitment cascade represents a promising strategy for therapeutic intervention. Ultimately, the aim is to stop leukocytes in their tracks and break the path of the multistep cascade, insuring rapid inflammatory resolution and restoration to homeostasis. Indeed, compelling data in a multitude of animal models have demonstrated that interfering with key adhesion molecules results in significantly attenuated inflammatory responses. However, despite these findings in mice, the transition to clinical trials in humans, targeting specific adhesion molecules has had a mixed success rate. Generally, endeavors to alleviate chronic relapsing inflammatory diseases (such as multiple sclerosis, psoriasis, and irritable bowel syndrome) have been far more successful than attempts to treat severe acute inflammatory responses.

The major approach of targeting adhesion molecules has been via direct blockade of receptor–ligand interactions through the development of small molecule

TABLE 16.2. Measurement parameters used in IVM

Parameter	Description
Rolling velocity	Distance traveled/time taken (units are μm/s)
Cell flux	Number of cells passing a fixed point in the vessel per minute
Cell adhesion	Cells that remain stationary for 30 seconds or longer
Cell emigration	Number of cells that had emigrated up to 50 μm from the wall of the 100 μm vessel
Wall shear rate (WSR)	Calculated by the Newtonian definition; WSR = 8000 × (V_{mean}/diameter) (units are s^{-1})

inhibitors or using monoclonal antibodies (either chimaeric or humanized). However, caution must be taken as blocking studies carry a significant risk associated with an impaired immune response. Nevertheless, recent successful trials have been documented using a humanized monoclonal antibody directed against the α_4 integrin chain that blocks binding of $\alpha_4\beta_1$ to VCAM-1 on brain-infiltrating lymphocytes and binding of $\alpha_4\beta_7$ to MadCAM-1 on gut-infiltrating lymphocytes for treatment of multiple sclerosis and Crohn's disease. In addition, a FDA-approved humanized monoclonal antibody (eflizumab) specific for the integrin subunit CD11a is currently being developed for the treatment of plaque-type psoriasis.

Future antiadhesion drug development is aimed at determining which molecular targets offer the best specificity and the least overall risk of reducing host defense mechanisms.

SUGGESTED READINGS

Atherton, A. and Born, G.V. 1973. Quantitative investigations of the adhesiveness of circulating polymorphonuclear leucocytes to blood vessel walls. *J Physiol* 222:447–474.

Chiang, E.Y., Hidalgo, A., Chang, J., and Frenette, P.S. 2007. Imaging receptor microdomains on leukocyte subsets in live mice. *Nat Methods* **4**:219–222.

Dejana, E. 2004. Endothelial cell-cell junctions: happy together. *Nat Rev Mol Cell Biol* **5**:261–270.

Ley, K., Laudanna, C., Cybulsky, M.I., and Nourshargh, S. 2007. Getting to the site of inflammation: the leukocyte adhesion cascade updated. *Nat Rev Immunol* **7**:678–689.

Luster, A.D., Alon, R., and von Andrian, U.H. 2005. Immune cell migration in inflammation: present and future therapeutic targets. *Nat Immunol* **6**:1182–1190.

Petri, B., Phillipson, M., and Kubes, P. 2008. The physiology of leukocyte recruitment: an in vivo perspective. *J Immunol* **180**:6439–6446.

Wagner, D.D., and Frenette, P.S. 2008. The vessel wall and its interactions. *Blood* **111**:5271–5281.

LIST OF SYNONYMS AND ABBREVIATIONS

Cell adhesion molecules = CAM
$\alpha_4\beta_1$ integrin = VLA-4= CD49d/CD29
$\alpha_5\beta_1$ integrin = VLA-5= CD49e/CD29
$\alpha_6\beta_1$ integrin = VLA-6= CD49f/CD29
$\alpha_L\beta_2$ integrin = LFA-1= CD11a/CD18
$\alpha_M\beta_2$ integrin = MAC-1= CR3= CD11b/CD18
$\alpha_X\beta_2$ integrin = p150, 95= CR14 receptor
$\alpha_4\beta_7$ integrin = LPAM-1= CD49d/CD-
LFA-2 = CD2
LFA-3 = CD58
VCAM = CD106
ICAM-1 = CD54
ICAM-2 = CD102
PE-CAM-1 = CD31
CD36 = "leukocyte differentiation antigen"
MAdCAM-1
E-selectin = CD62E=ELAM-1
P-selectin = CD62P=GMP=140=PADGEM
L-selectin = CD62L=LAM-1=Mel-14

17 Mediators and Mechanisms of Inflammatory Pain

Tony L. Yaksh

After tissue injury or exposure of the organ to a foreign body or infectious entity, a complex cascade of cellular and humoral events is initiated. These events, observed in virtually every organ system, including skin, muscle, meninges, dentition, bone, and visceral tissues, fall broadly under the rubric of "inflammation." In general, this cascade serves to protect and maintain the functional integrity of the systems in the face of insult. The process activates the immune system and directs chemotactic agents, neutrophils, and mononuclear cells to migrate from the vascular bed to the injury site. The cardinal clinical signs of this process, *rubor/calor* (increased local blood flow) and *tumor* (swelling secondary to local plasma extravasation) are manifested to varying degrees in all of these tissues. The fourth cardinal sign, *dolor* (pain), typically accompanies such cascades. The biological processes whereby these inflammatory cascades serve to initiate and sustain a pain state are the focus of this commentary.

INFLAMMATORY PAIN PHENOTYPE

The acute application of a thermal or mechanical stimulus of such intensity as to *potentially* produce tissue injury will typically evoke a somatic escape response (e.g., a withdrawal of the stimulated limb) and an autonomic response (e.g., hypertension and tachycardia), a syndrome classically referred to by Sherrington as a nociceptive reflex [1]. These acute responses typically display four characteristics: (i) the magnitude of these responses varies directly with stimulus intensity; (ii) the latency varies inversely with stimulus intensity; (iii) the focus of the response is referred to as the specific site of stimulation (e.g., it is homotopic); and (iv) in the absence of tissue injury, removal of the acute stimulus results in a rapid attenuation of the sensation and the attendant behaviors.

In the face of local tissue injury and inflammation, a distinct pattern of aversive sensations is reported in the human and parallel effects are noted in nonverbal organisms as defined by the attendant behaviors. This behavioral phenotype typically characterized as a dull, throbbing, aching sensation in the injured area of skin, soft tissue, or joint. Inflammation of the linings of the abdominal or thoracic cavity (e.g., pleura or pericardial sac) or visceral organs such as kidney, pancreas, or gastrointestinal (GI) tract will lead to ongoing pain sensations that are referred to specific somatic dermatomes, for example, where coronary ischemia is referred to the left arm and shoulder or an appendicitis to the lower right abdomen in humans [2]. Such referred pain is commonly reported to be burning or cramping in character. As discussed further later, in nonverbal models, the ongoing pain sensation is intuited from guarding of the limb, reduced weight bearing, impaired ambulation and ongoing vocalization, both sonic and ultrasonic [3,4]. Abnormal grooming or licking of the affected body part is also noted. In the face of visceral inflammation, abnormal postures such as stretching of abdomen and/or increased abdominal tone may be present [5]. In the face of injury and inflammation of somatic or visceral tissues, there is typically evidence of enhanced autonomic activity (e.g., increased blood pressure and heart rate, pupillary dilation) [6].

Facilitated Pain States

Application of evocative stimuli to the injury site such as a mechanical distortion or thermal stimulation of the tissue or flexion of the joint in a manner that is not normally considered to be noxious will lead to reports of pain sensations. This lowered magnitude of stimulus intensity required to elicit an aversive response to a stimuli applied to the injury site is referred to as primary hyperalgesia [7]. In the face of local injury and inflammation, it is also appreciated that stimuli applied to injured tissue will be responded to in an exaggerated fashion. Thus, low-intensity mechanical or thermal stimuli applied to the skin, modest flexion of the inflamed joint, or moderate distention of the gastrointestinal (GI) tract will lead to an augmentation of the ongoing behavioral signs [8]. In addition, low-intensity stimulation applied to regions *adjacent* to the injury site may also produce a pain condition. This sensitization of adjacent noninjured tissue to low-intensity stimuli is then referred to as 2° hyperalgesia or allodynia [9]. In human experimental models, these response components have been readily identified with a focal injury applied to the forearm or hand [10]. In summary, this property of sensory "sensitization" with local injury is generalized to virtually all organ systems. Thus, inflammation of the skin by sunburn leads to extreme sensitivity to warm water, pleural inflammation may lead to pain secondary to chest expansion (inhalation), irritation of the cornea lead to pain secondary to eyelid closure. In the case of inflammation of the bowel, light touch applied to the abdomen reveals reported "tenderness" and an increased abdominal muscle tone [11,12].

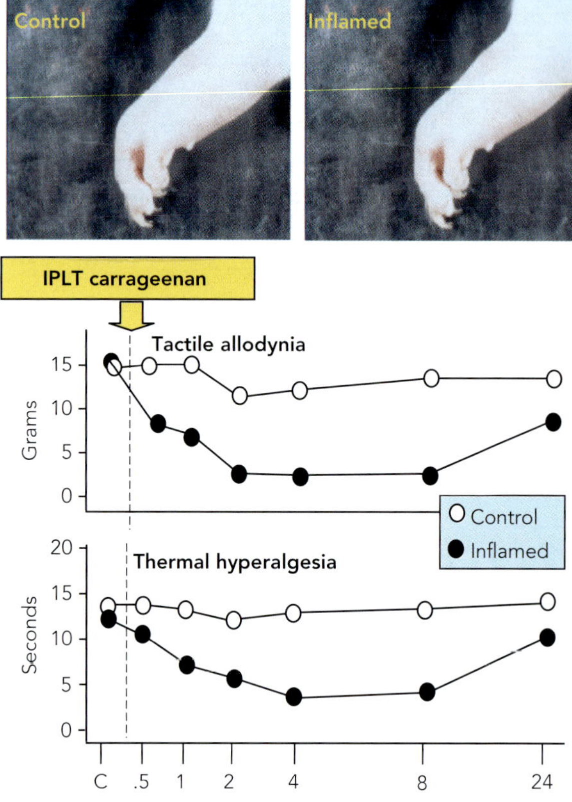

Figure 17.1. Unilateral intraplantar injection of carrageenan will initiate a potent inflammatory response (top). Measuring the thermal escape latency on the paw contralateral to the carrageenan shows ho change in the tactile thresholds (middle) or thermal escape latencies (bottom). In contrast, over the 24 hr interval, there is a robust fall in tactile and thermal thresholds.

Preclinical behavioral models and inflammation/injury initiated pain states.

Defining the role of inflammatory mechanisms in "pain" or nociception depends on the effects of manipulating these various components of the pain behaviors otherwise generated by the inflammatory state. As reviewed earlier, such pain states may be composed of spontaneous behavior and changes in thresholds that occur secondary to the injury (e.g., hyperalgesia). Animal models involving local tissue injury or inflammation at specific target sites, for example, the paw or the GI tract, has been systematically used to assess pain behavior over time after injury. One such material carrageenan (linear sulfated polysaccharides extracted from seaweed) has been classically used as a model of acute inflammation [13]. As shown in Figure 17.1, the intraplantar injection of carrageenan into one hind paw will lead to (i) a time-dependent inflammatory response in the injected paw; (ii) a corresponding decrease in the magnitude of the mechanical (or thermal) stimulus required to evoke a withdrawal of the injured hind paw; and (iii) a reduced response latency. The latency/mechanical threshold of the contralateral paw in this model is little changed. Here the thermal escape latency is measured using a system that permits targeting the radiant thermal stimulus independently to either hind paw (a Hargreaves device) [14], while the mechanical threshold is determined using von Frey filaments [15]. This model demonstrates the prominent and long lasting thermal and mechanical hyperalgesia of this paw. Similar behavioral effects can be observed after the delivery of a variety of proinflammatory treatments such as (i) complete Freund's adjuvant (evoking a disseminated autoimmune response to a joint or connective tissue antigen) [16,17]; (ii) local injuries, for example, focal thermal injury [18]/incision [19] of the paw; (iii) intraarticular injection of irritants (carrageenan-kaolin [20]); (iv) delivery of products that deliver or initiate antibodies against joint components (such as collagen [21]); (v) inflammation of the GI tract produced by the instillation of products such as trinitrobenzene sulfonic acid (TNBS) (a hapten that induces colitis [22]) or iodoacetamide [23]; (vi) bladder inflammation (cystitis) as produced by instillation of irritants such as cyclophosphamide [24]; or (vii) pancreatitis produced by dibutyltin dichloride (DBTC) [25]. All of these treatments in various representative organ systems will commonly produce evident behavioral signs of ongoing discomfort and hypersensitivity. Inflammation or injury of visceral organs will typically

TABLE 17.1. Classification of primary afferents

Fiber class	Size (μ)	Myelin	Velocity (m/sec)	DRG neuron	Transmitter	Characteristic epitopes	Typical stimuli	Psychophysical correlate of acute stimulus
Aß	12–20	Yes	>40–50	Lg	Glutamate	Neurofilament 200	Low threshold Mechanical/tactile	Light touch proprioception
Aδ	1–4	Yes	10 < 40	Lg	Glutamate	IB-4	Subpopulations: low threshold (thermal or mechanical) or high threshold (thermal or mechanical)	Warmth Light touch Sharp-stinging Thermal crush
C	0.5–1.5	No	<2	Small	Glutamate peptides (sP/CGRP)	TRPV1-r	Subpopulations: polymodal	Dull, throbbing burning

be accompanied by signs such as abdominal stretching and increases in the tone of the abdominal musculature (as measured by increased abdominal EMGs), tachycardia, and hypertension. In addition, GI inflammation is often accompanied by marked increases in the aversiveness displayed by the animals to the application of mechanical stimuli to the body surface (corresponding to referred pain sites and to distention of the GI tract with inflatable balloons [26]). In short, preclinical inflammatory models will yield in animals a constellation of behavioral and physiological effects that parallel the verbal pain descriptors of ongoing discomfort and evocative hyperpathia observed in conditions of acute and chronic inflammation in the human patient (see [12,27]).

PERIPHERAL COMPONENTS CONTRIBUTING TO INFLAMMATORY PAIN STATES

Primary Afferent

With few exceptions, the sensory correlates of the body are communicated to the brain via the activation of specific populations of primary afferents projecting to the medullary (head and neck), spinal (body regions below the neck), and dorsal horn (see [28,29]). In visceral pain, information may also be trafficked through cranial nerves such as the vagus (Xth nerve) [30,31]. The sensory afferents, the cell bodies of which are found in the trigeminal (head and neck) and dorsal root ganglion (below the neck), are subdivided into classes reflective of their size, state of myelination, and the modality of stimulation that most effectively results in activity in the associated afferent axon (Table 17.1). On the basis of psychophysical responses evoked by stimuli that can activate these respective populations of axons, specific categories of sensations have been affiliated with the information carried by these afferents. Conversely, in the absence of injury, the range of stimulus intensities that most effectively activate a population of afferents is typically reflective of the presumed role played by these axons in initiating a sensory state. Thus, light touch is typically associated with the large mechanically sensitive afferents (Aß) that display sensitivity to low threshold mechanical stimuli. Aδ axons form subpopulations which may respond to low- or high-intensity mechanical or thermal stimuli, while C fibers respond predominately to high-intensity stimuli of thermal, mechanical and, importantly, chemical modalities. On the basis of differential sensitivity to blockade and the marked differences in conduction velocity, it is considered that the high-threshold Aδs may carry the initial stinging sensation generated by an acute high-intensity stimulus (e.g., first pain), while the slower C-fiber mediates the delayed dull, throbbing, aching sensation (2nd pain) [32].

Initiation of Small Afferent Activity

The primary afferent terminals of peptidergic C fibers are morphologically characterized as "free nerve endings." These peripheral terminals are highly ramified in most tissues and frequently appear as a triad of terminal, capillary, and mast cells [33]. Under normal conditions, in the absence of injury or inflammation, sensory afferents, particularly those with high-stimulus thresholds, show little or no spontaneous activity. Mechanical and/or thermal stimuli applied to the peripheral terminal of the afferent is converted by transducer proteins on the terminal into an intensity-dependent depolarization leading to an

TABLE 17.2. Afferent transducer proteins

Channel	Activating parameters	Chemical sensitivity
TRPV1	>43°C	Lipids, capsaicin H^+
TRPV2	>52°C	
TRPV3	>34–38°C	
TRPV4	>27–35°C	
TRPM8	<25–28°C	Menthol
TRPA1	<17°C	Mustard oil
ASIC	H+	Acid
P2X	ATP (mechanical?)	

Abbreviations: TRP, transient receptor potential; ATP, adenosine triphosphate.

intensity-linked increase in the discharge frequency of these afferents. The mechanical and thermal stimuli sensitivity of a given afferent fiber is defined by the constituent transducer proteins that are present on the afferent terminal (see Table 17.2). These transducer proteins are typically effectively activated over a narrow range of intensities of the respective stimulus modality. Increasing stimulus intensities typically lead to the activation of higher threshold transducers and an increasing cation permeability, all leading to a progressive depolarization of the terminal [34]. The local depolarization then leads to the opening of local voltage-sensitive sodium channels and this leads in turn to an increasing frequency of afferent spike generation. An important property of these gated channels is that they may also be activated by specific chemical entities such as the TRPV1, which is activated by capsaicin and the TRP M8 channel by menthol. These activities parallel the psychophysical elements that such stimuli normally elicit, for example, sensation of heat and cold, respectively. Table 17.2 summarizes other such chemical correspondences.

Conduction of Sensory Afferent Activity

Voltage-sensitive sodium channels mediate the conducted potential. Cloning shows the presence of multiple populations of sodium channels. These multiple sodium channels have been identified based on structure (NaV 1.1-NaV 1.9), whether they are tetrodotoxin-sensitive or insensitive (TTX) and their activation kinetics. Some of the subtypes are geographically restricted in their distribution. Thus, NaV1.8 and 1.9 are largely limited to small primary afferents [35]. These NaV channels have several consensus sites that can be phosphorylated, serving to reduce their activation threshold and prolong their opening. As will be discussed later, these phosphorylation sites are important for regulating the excitability of the terminals by inflammatory products.

Primary Afferent Terminals and Inflammation

The preceding comments on the primary afferent and transduction mechanisms emphasized the transduction of acute high-intensity stimuli and an absence of spontaneous activity. As suggested earlier, increasing stimulus intensities (thermal, mechanical) will result in a monotonic increase in the discharge frequency of these axons. Local injury or inflammation in the environment of these peripheral terminals will, however, characteristically lead to (i) the appearance of high levels of ongoing (spontaneous) afferent activity; and (ii) a left shift in the stimulus discharge curve of these afferents such that there is a significant augmentation in discharge frequency at stimulus intensities that are of lower magnitude. Figure 17.2 summarizes such comments for thermal stimuli applied to the skin and for rotation of the joint. Here it is portrayed that the magnitude of spontaneous afferent traffic may mirror that which is otherwise observed after a moderate stimulus or an extreme joint rotation. It is appreciated that this exaggerated response reflects the inflammatory events that occur secondary to local tissue injury and inflammation.

The Local Inflammatory Milieu

Local tissue injury such as an incision or a burn results to varying degrees in a variety of distinguishable events that reflect activation of an innate immune response. These events, discussed in great detail in other chapters for various organ systems, may be heuristically subdivided as described in Figure 17.3. (i) Acutely, the stimulus serves to activate small high threshold sensory afferents. Such activation yields orthodromic and antidromic activity on the primary afferent. The antidromic activity (e.g., afferent traffic along collaterals toward the distal terminal away from the spinal cord) yields the local terminal release of sP and CGRP. These peptides result in a prominent local dilation of arterioles and plasma extravasation [36,37]. (ii) The damage to local tissue integrity yields cellular debris and an increase in extracellular [K^+] and [H^+]. Thus, in injured tissues, μM concentrations of H^+ may be found [38]. (iii) Disruption of local capillaries leads to local accumulation of serum products such as complement and products of hemolysis such as serotonin [39,40]. This injury leads to activation of the clotting cascade and kininogens, resulting in the formation of bradykinin [41]. (iv) These products (such as the sP released from the local C fibers) serve to activate mast cells. The mast cells secrete histamine, tryptase, and chymase, which, along with macrophages, secrete a large number of

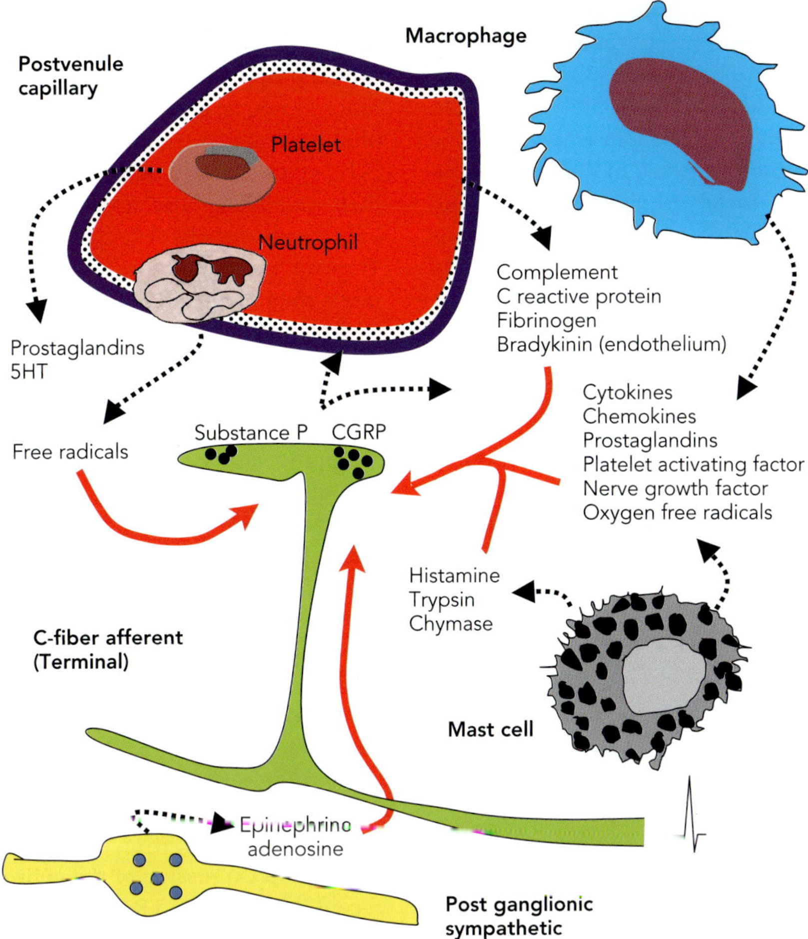

Figure 17.2. Summary of the injury milieu of the primary afferent, including the local vasculature, macrophages, and mast cells and sympathetic terminals and products that are released secondary to that injury which can influence afferent activity. See text for additional details.

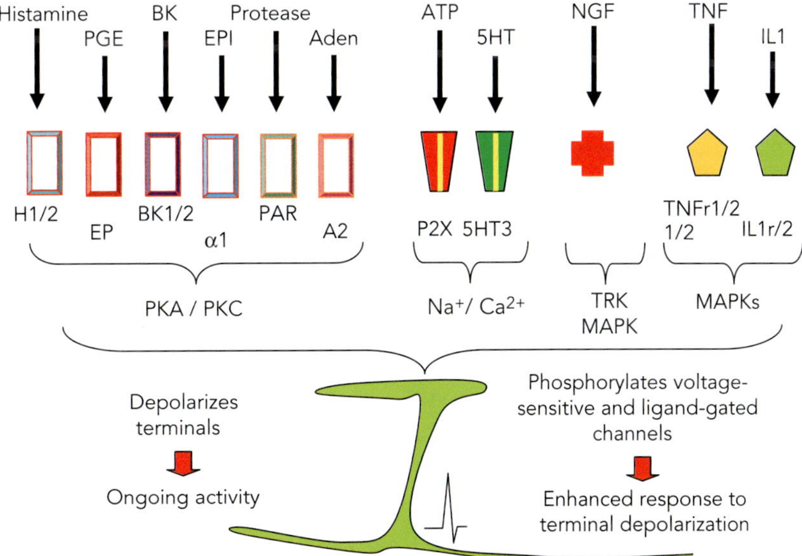

Figure 17.3. Summary of receptor proteins present on the terminals of small primary afferents. Occupancy of the receptor protein by the respective ligand leads to activation of the indicated intracellular messages and/or channel permeability. See text for additional commentary.

cytokines (IL-1, TNF), chemokines, lipid mediators (prostaglandins, platelet activating factor), and growth factors (NGF) [42,43]. (v) These products also activate integrins, which promote local neutrophil adhesion to the vascular wall and their subsequent diapedesis. The neutrophils migrate in the interstitial fluid along gradients established by a variety of chemotactic stimuli generated by the injured tissue and foreign materials [44]. Neutrophils generate reactive oxygen species and hydrolytic enzymes [45]. These injury events and the antidromic activity on the C-fibers leading to sP and CGRP release are mechanistically responsible for the "triple response of Lewis" noted after tissue injury: a red flush around the site of the stimulus (local arterial dilation), a local edema (increased capillary permeability) and, as will be discussed further later, a local reduction in the magnitude of the stimulus required to elicit a pain response, that is, a hyperalgesia. The chemical complexity of the local injury environment has prompted it to be designated as an "inflammatory soup" [46]. These local cascades associated with tissue injury represent the initiating events associated with the ongoing pain state and the enhanced responsiveness of the afferent axon to a given stimulus.

Our understanding of the cellular and molecular bases of transduction of painful stimuli has burgeoned in the past year, mainly as a result of studies on isolated sensory neurons in culture. The ion channels' underlying neuronal responses to noxious heat, to protons, and to ATP have recently been characterized. The typical increase in nociceptor sensitivity produced by tissue damage has been found to be mediated by at least two distinct mechanisms. In the first, bradykinin and other mediators augments the current activated by heat through a mechanism that involves activation of protein kinase C. In a second sensitization mechanism, prostaglandin E2 for example alters the voltage threshold of several ion channels, including a novel tetrodotoxin-insensitive Na^+ channel, in such a way that initiation of action potentials is facilitated [47,48].

Terminal Receptors for Inflammatory Products

Based on immunohistochemistry as well as physiological studies defining the effect of many of these mediators released by local tissue injury and inflammation, there are eponymous receptors that can be identified on the afferent terminal or in the dorsal root ganglion cells (Figure 17.4). Several examples can be cited.

pH. H^+ ions can directly depolarize afferents [49] through the TRPV1 cation channel [50] and several acid-sensing (ASIC) Na-channels [50]. Thus, TRPV1 channels display a sustained nonselective cationic current at a pH < 6 and are present on a limited population of peptidergic C fibers [51].

Prostaglandins: Agents are synthesized by lipoxygenase (leukotrienes) or cyclooxygenase (prostanoids) upon the release of cell membrane–derived arachidonic acid secondary to the activation of phospholipase A_2. A number of prostanoids, including PGE_2, can directly activate C fibers. Others, such as PGI_2 and TXA_2, and several leukotrienes, can markedly facilitate the excitability of C fibers [52,53]. These effects are also mediated by specific eponymous membrane receptors.

Kinins: Bradykinin released from endothelium can activate two receptors, (constitutively expressed B2 receptors and inducible B1 receptors) to activate free nerve endings of C polymodal nociceptors and generate pain behavior [54]. Bk1/2 receptors act through Gaq and Gai to activate phospholipase A2 (PLA2) and protein kinase C (PKC) indirectly through phospholipase Cb (PLCb) [55]. When given intradermally, Bk produces pain, inflammation and hyperalgesia [56].

Growth factors: Nerve growth factor (NGF) released from fibroblasts, keratinocytes, Schwann cells, and a variety of immune cells can act upon TRK receptors identified in small DRG cells [57]. This action can lead to depolarization of the terminal. SQ NGF can produce a potent sensitization, while anti-NGF strategies can diminish the injury-induced facilitated state. Importantly, over the longer term, NGF serves to regulate expression of the neuropeptides, as well as receptors including TRPV-1 and bradykinin receptors, and voltage-gated sodium channels (see [58]). Glial derived nerve growth factor (GDNF) has been shown to sensitize small afferent terminals in vitro and produce thermal hyperalgesia [59].

Proteinases: Serine proteases are released from mast cells (tryptase), epithelial cells, and/or from the local vasculature. These proteases activate several families of G protein–coupled protein receptors (PAR 1–4) by cleaving a local N-terminal tethered ligand, which binds to and activates receptors present on small afferent terminals [60]. Intradermal injections will activate free nerve endings and produce an enhanced behavioral sensitivity in the animal to subsequent stimulation, for example hyperalgesia [61].

Cytokines: Cytokines such as tumor necrosis factor, interleukin-1 (IL-1), IL-6, and the chemokine IL-8 are released from a variety of inflammatory cells and have been shown to exert powerful sensitizing effects on C-fibers [62,63]. Cytokines appear to induce sensitization via receptor-associated kinases and phosphorylation of ion channels. In chronic inflammation, transcriptional upregulation of receptors has been observed [64]. Intradermal injections of these agents generally produce both mechanical and thermal hyperalgesia, while binding proteins directed against TNF and IL-1β reduce hyperalgesia in inflammatory models [65].

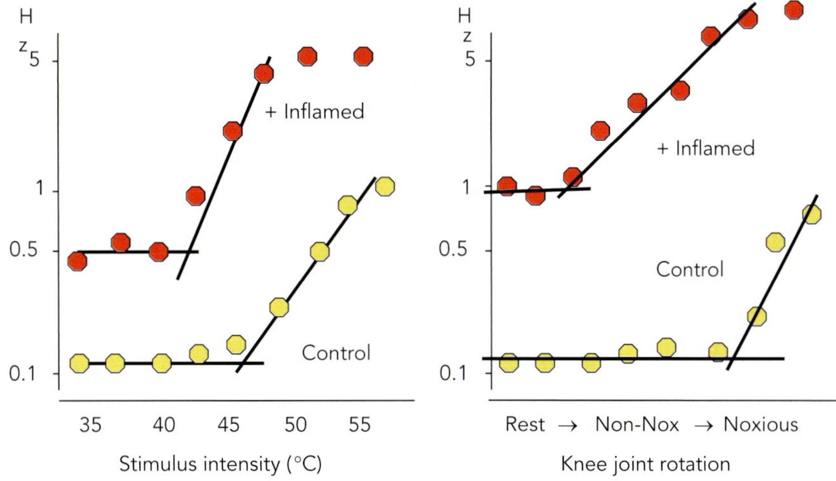

Figure 17.4. Firing of small afferents in the skin (left) occurs with increasing frequency with increasing temperatures. Following carrageenan injection into skin, afferent shows increasing spontaneous activity, a left shift, and an increase in slope of stimulus-response curve, indicating a facilitated response to the thermal stimulus (Right). Firing of articular afferent under normal state and in presence of nonnoxious and then noxious rotation of knee. After injection of irritant into joint, mild rotation results in significant discharge.

Intraterminal Coupling

These peripheral terminal receptors are coupled to a variety of terminal protein kinases (protein kinases A and C as well as mitogen-activated protein kinases, MAPK). These kinases phosphorylate consensus sites found in a variety of transducer proteins (e.g., the TRPV1) and ion channels (e.g., voltage-sensitive sodium channels-NaV) [66]. NGF activation of trkA activation leads to tyrosine phosphorylation of intracellular targets, including ion channels, leading to their enhanced activation by any given degree of terminal activation. These intraterminal cascades represent an important link in the events that lead to terminal sensitization in the face of local injury and inflammation in various organ systems (e.g., skin and joint), such that modest stimuli lead to a pronounced responsiveness [67].

Role of Terminal Sensitization in Pain Behavior

The prominent role of these inflammatory products in the generation of post injury/inflammation pain cannot be overstated. The majority of C-fibers are referred to as "silent nociceptors," as they have little or no spontaneous activity and thresholds so high as to be activated only by extreme physical stimuli. Preclinical studies have emphasized that signs of ongoing pain behavior (paw lifting, paw guarding) is dependent on ongoing activity in these C-fibers [68]. In the presence of the injury products, these terminals can be activated by relatively mild stimuli. Such sensitizing motifs have been described for all tissues including skin [29], bone [69], joint [70], GI mucosa [71], bladder [72], meninges [73], and tooth pulp (see [74]). Accordingly, it can be considered to be an important unifying principle for understanding the sensory components associated with injury. Consistent with these stimulatory and facilitatory effects on afferent function, the local delivery of these products into the skin or viscera have typically been shown to evoke spontaneous pain behavior and a behaviorally defined hyperalgesia. Conversely, local inhibition of the synthesis of many of these factors or their receptors may attenuate the proinflammatory pain state. It is important to note that while any single agent may serve to initiate sensitization of the terminal, it does not necessarily follow, given the complexity of the inflammatory soup, that blocking any single product will significantly attenuate the observed injury/inflammation-induced hyperalgesia.

Time Dependence of Inflammation-Evoked Effects

Injury and inflammation can lead to a relatively acute sensitization (e.g., minutes to hours). However, it should be additionally noted that afferent-mediated downstream effects initiated by inflammation may evolve over time. Thus, inflammation may result in an enhanced expression of a number of terminal channels and receptors that may lead to enhanced sensitivity. Several examples are channels such as voltage-sensitive sodium [75] and calcium channels [76] and afferent neurotransmitters such as sP and CGRP.

The mechanisms by which peripheral inflammation drives changes in terminal transducer protein expression may result from an increase in afferent traffic and DRG depolarization, an effect of local mediators to which the DRG may be exposed [77], and/or transport from the periphery to the DRG of a variety of factors from the injury site. Such transported factors may include NGF (nerve growth factor) or cytokines such as TNF released from Schwann cells and inflammatory cells. All of these products may initiate increased DRG transcription and transport of a variety of proteins related to the inflammatory response [78].

CENTRAL COMPONENTS CONTRIBUTING TO INFLAMMATORY PAIN STATES

In the preceding section, the generation of afferent traffic initiated by tissue injury and inflammation was emphasized. This afferent traffic activates dorsal horn neurons. Given the change in afferent transduction in the face of local inflammation, it is reasonable to consider that such peripheral changes may account for the majority of the inflammatory pain phenotype. While important, it is evident that in the face of such peripheral injury, the response of the second-order neuron itself undergoes a significant alteration in responsiveness. Broadly speaking this effect upon second-order excitability reflects events mediated by afferent input as well as (more speculatively) circulating factors that directly effect neuraxial elements (such as microglia).

Afferent Evoked Excitation

Acute input: In the absence of a peripheral stimulus or inflammation, there is minimal if any small afferent traffic. With an adequate stimulus, small afferent input into the spinal dorsal horn is initiated and this input will activate specific populations of second-order neurons, some lying superficially in lamina I (marginal cells) and some lying more deeply in lamina V (wide dynamic range neurons) (Figure 17.5). Many of these neurons that project from the spinal cord to the brain travel in the contralateral-ventrolateral spinal cord (see [28]). The magnitude of the response of the dorsal horn neuron, either wide dynamic range or nociceptive-specific, is related to the frequency (and identity) of the afferent input. As reviewed earlier, the frequency of the afferent input is typically proportional to the magnitude of the acutely applied stimulus.

Persistent input: The organization of this neuraxial response to an acute peripheral stimulus is thus typically modeled in terms of a monotonic relationship between activity in the peripheral afferent generated by the local stimulus and the activity of neurons that project out of the spinal cord to the brain. However, in the face of persistent small afferent input, many dorsal horn neurons will display an enhanced excitability. Intracellular recording has indicated that the facilitated state is represented by a progressive and long-sustained partial depolarization of the cell, rendering

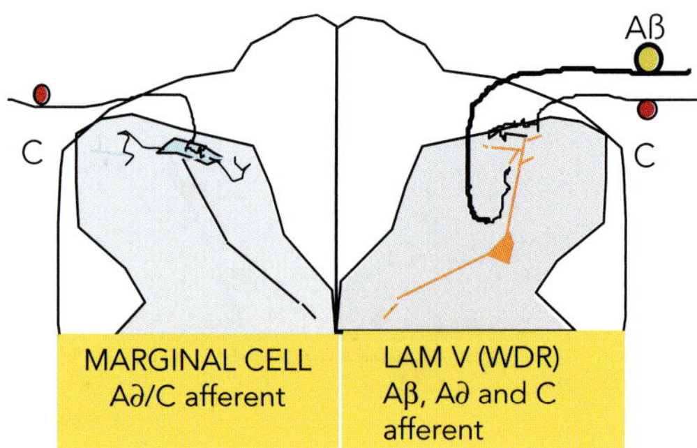

Figure 17.5. Organization of small afferent input into the dorsal horn. Left: small high threshold afferents terminate in the superficial dorsal horn specifically in Rexed Lamina I where they activate marginal lamina I cells. Right: For lamina 5 neurons, there is excitatory input from high threshold afferents that project superficially and from large Aβ axons that terminate deep to the dorsal horn.

the membrane increasingly susceptible to afferent input [79,80]. The pharmacology and organization of this spinal facilitation will be considered further in the following.

Systemic Routes of CNS Activation viz Pain States

In the preceding section, the common thesis was that peripheral injury/inflammation led to the activation of specific populations of sensory afferents innervating the injury site and their input drove a spinal sensitization mediated by a variety of spinal cascades. Such connectivity clearly accounts for the somatotopy associated with local injury, for example, injury to the left hind paw results in a sensitization of the left hind paw mediated in part by the initiation of a facilitated state in the dorsal horn segments receiving input from the left hind paw. An important additional component in the inflammation-induced changes in sensory processing arises from circulating factors. After local trauma or infection, cytokines (such as IL-1β/TNF) are generated and act locally to initiate local inflammatory cascades that can, among their many effects designed to destroy pathogens, serve to sensitize local nerve terminals. In addition, these agents may reach the systemic circulation and generate a system-wide effect, which may be amplified by the subsequent generation of a variety of proinflammatory products such as liver-derived acute phase proteins, pancreatic enzymes, GI products, and granulocytosis [81]. This septic cascade may have potent deleterious effects upon cardiac, pulmonary, and visceral organ function. In addition, these circulating products may have potent effects upon central nervous system (CNS) function, such as temperature regulation (fever), metabolism (cachexia), and pain. The functional linkage between circulating inflammatory products and CNS function is not clear. However, ample evidence has pointed to linkages that may involve nonneuronal perivascular cells such as astrocytes and microglia [82,83]. As will be noted later, these non-neuronal cells play an important constitutive role in spinal nociceptive processing after peripheral tissue injury. Such systemic effects could account for bilateral (even whole body) effects. Thus, IV LPS or IL-1β will produce bilateral increases in acute and chronic signs of CNS effects such as bilateral increases in cFOS expression and bilateral increases in the inducible spinal COX-2. Such systemic effects of circulating products may account for reports of bilateral changes after unilateral peripheral inflammation [84].

Functional Consequences of Spinal Sensitization

Sensory transmission: There are three functional consequences of this facilitated spinal state. (i) There is a nonlinear increase in the spinal input–output function, for example, a higher frequency and more persistent discharge for any given afferent input. (ii) Low threshold tactile stimulation also becomes increasingly effective in driving these neurons. (Note that the so-called wide dynamic range Lamina V neurons noted above receive both small and large afferent input.) (iii) In addition to the augmented response of the neuron, the conditioning has the added effect of increasing the receptive field size of the neurons. This occurs because neurons in any given spinal segment receive depolarizing input not only from the spinal root projecting to that segment, but collateral afferent input from adjacent spinal segments which may provide subeffective degrees of depolarization. In the face of sensitization, this subeffective input becomes sufficient to drive depolarization of that neuron. Hence, the effective receptive field of that neuron is increased.

Given the likelihood that the discharge frequency of these dorsal horn neurons contributes to the encoding of a high threshold stimulus as aversive, and that many of these neurons project through the ventrolateral quadrant of the spinal cord (i.e., spinobulbar or spinothalamic projections), this augmented response to a given stimulus and the enlarged receptive field is believed to contribute to the hyperalgesic states observed after nerve injury [28]. Dorsal horn systems that lead to this facilitated state are complex but reflect upon cascades that are in part initiated by the persistent small afferent input and perhaps to a lesser degree by circulating factors.

Inflammation: The sensitization of the spinal systems leads to an enhanced pain response. What is less appreciated is that sensory afferents conduct both orthromically (from the outside to the spinal cord) and antidromically (from the spinal cord to the outside). As will be reviewed later, the peripheral terminal of the primary afferent is able to release a variety of products such as peptide transmitters (substance P/CGRP). Many of these products are believed to control the local inflammatory response. It is appreciated that central sensitization can enhance the excitability of the central terminal leading to primary afferent depolarization (PAD) and significant amounts of antidromic traffic [85,86]. Although there is not enough space to discuss the ramifications of this phenomenon, there is ample data to suggest that such antidromic linkages may additionally contribute to the local inflammatory response and sensitization [87,88].

COMPONENTS OF THE DORSAL HORN SYSTEMS CONTRIBUTING TO SPINAL SENSITIZATION

As noted earlier, persistent small afferent input as generated by small afferent input initiated by tissue injury and inflammation will lead to a sensitized behavioral state. This sensitization results from a series of cascades, which reflect the organization of dorsal horn systems that regulate the activity in neurons (such as the Lam I – marginal neurons and Lam 5 – wide dynamic range neurons) that project to higher centers. In the following narrative, representative examples of the contribution played by these several systems will be reviewed [89].

Local Neuronal Excitability

Afferent evoked membrane depolarization: As indicated in Figure 17.6, small afferents contain and release into the spinal dorsal horn excitatory amino acids such as glutamate and peptides such as substance P. Acute activation yields second-order neuronal depolarization mediated largely by glutamate acting through an AMPA receptor Na^+ ion-selective channel, which produces a potent but transient depolarization of the membrane [90,91]. As reviewed earlier, repetitive small afferent input will yield a facilitated state in the dorsal horn as exemplified by the phenomena of "wind up." The pharmacology of this central facilitation suggests that the spinal phenomenon reflects more than just the repetitive activation of a simple excitatory system. Thus, while the acute excitation is unaffected by N-methyl-D-aspartate (NMDA) receptor antagonists, these agents will prevent spinal wind-up [92]. In parallel with these physiological observations, behavioral work demonstrated that the spinal delivery of such drugs will have no effect upon acute pain behavior, but reduce the facilitated states induced after tissue injury [93]. This differential effect upon persistent versus acute activity results from the fact that the NMDA ionophore under resting membrane potential is blocked by a Mg^{2+} ion in the channel. This block is removed in the face of persistent membrane depolarization, such as that which occurs with repetitive afferent input. Activation of the NMDA ionophore leads to a marked increase in intracellular Ca^{2+} that initiates a variety of downstream facilitatory events. A variant of the AMPA channel occurs in a small proportion of dorsal horn channels and creates an AMPA-calcium permeable ionophore that can also contribute to downstream facilitation [94]. In addition to the classic transmitters, growth factors such as brain-derived nerve growth factor (BDNF) is synthesized by small DRGs and released from spinal terminals, packaged in dense-cored vesicles, and transported within axons into terminals in the dorsal horn of the spinal cord. BDNF has a potent sensitizing effect on spinal neurons [95] (see Figure 17.6).

Activation of kinases: Persistent afferent input leads to a marked increase in intracellular Ca^{2+} through several mechanisms: (i) voltage-gated ion channels (e.g., N and T type channels), (ii) ion-gated channels (e.g., NMDA/calcium-permeable AMPA channels), and (iii) G-protein–coupled receptors that act to mobilize intracellular calcium (e.g., neurokinin-1 receptors). The increased calcium leads in turn to the activation of a wide variety of phosphorylating enzymes, including protein kinases A and C, calcium calmodulin-dependent protein kinases, as well as mitogen-activated kinases (MAPKs) including p38 MAP kinase and ERK (see [66]), which increases the excitability of a variety of spinal linkages. Several examples of such local intracellular cascades will be noted.

(i) The functionality of many channels is regulated by phosphorylation. One such channel is the NMDA receptor. Such phosphorylation of spinal NMDA receptors leads to increases, and dephosphorylation leads to decreases in NMDA currents [96]. These events initiated by repetitive afferent activation serves to reduce the threshold for activation, leading to an enhanced response of the NMDA ionophore to further depolarization. Block of spinal PKA and PKC activity can reduce behaviorally defined hyperalgesia otherwise observed after tissue injury [97].

(ii) P38 MAPK is activated (phosphorylated) by an increase in intracellular Ca. This activation leads to several events. The first is to phosphorylate phospholipase A2 (PLA2), which initiates the release of arachidonic acid and provides the substrate for cyclooxygenase (COX) to synthesize prostaglandins. The second is that this MAPK activates a variety of transcription factors (such as NF-κB), which activates the synthesis of a variety of proteins, including COX-2. The spinal delivery of P38 MAPK inhibitors will thus reduce acutely initiated hyperalgesia and reduce the upregulation of COX-2 otherwise produced by injury [98,99].

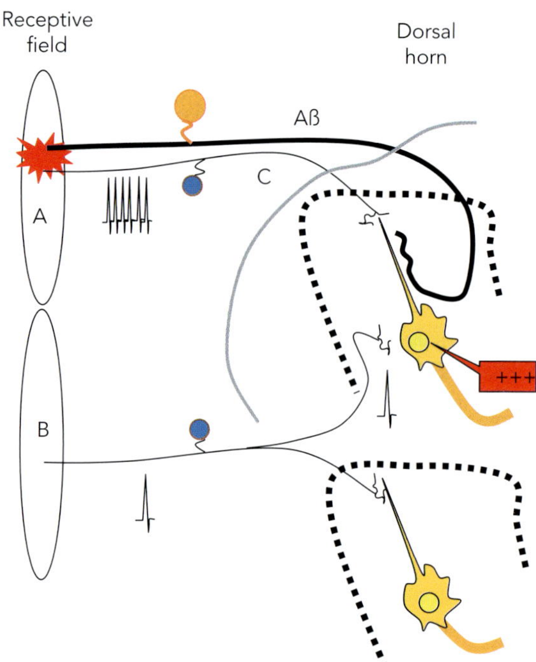

Figure 17.6. Injury to receptive field (RF) A leads to a persistent afferent barrage which strongly excites and sensitizes the second order neuron. This sensitization has three consequences to the response properties of Neuron A. (i) Neuron A gives a greater response for any given stimulus. (ii) This cell (a WDR neuron) receives large low threshold (Aß) input and that input now evokes a greater response. (iii) Collateral input from receptive field B now is sufficient to activate effects to activate the sensitized neuron A. This means that the effective receptive field of neuron A is now enlarged to include RF A and RF B, for example, stimuli now applied to RF-A and B will activate Neuron A. This connectivity likely accounts for the enlarged receptive field observed after peripheral injury, for example, 2° hyperalgesia.

Lipid cascades: A variety of phospholipases (c, i, and s-PLA2) and cyclooxygnease (COX-1 and 2) are constitutively expressed in the dorsal horn in both neuronal and nonneuronal cells) [100]. Not surprisingly, downstream lipid products such as prostaglandins are synthesized and released into the spinal dorsal horn and this release can be markedly enhanced by small afferent input generated by tissue injury and by the direct activation of second-order neurons [101]. These agents diffuse extracellularly to act presynaptically to enhance the opening of voltage-sensitive calcium channels, thus augmenting transmitter release [102]. In addition, prostaglandins act postsynaptically to *block* glycinergic inhibition on second-order dorsal horn neurons [103]. Such a reduction in the activation of inhibitory glycine or GABA interneuron regulation can lead to a potent facilitation of dorsal horn excitability (see later). The spinal delivery of PGE will increase, while PLA2 or COX-2 inhibitors will reduce spinal PGE2 release and reduce injury-induced hyperalgesia.

Nitric oxide synthase (NOS): NOS forms diffusible nitric oxide (NO) from arginine. There are three principal NOS isoforms: endothelial, neuronal, and inducible. The neuronal and inducible forms have been found to play a facilitatory role in the spinal cord through the formation of NO, which acts presynaptically through cGMP to enhance transmitter release. Spinal NOS inhibitors can reduce posttissue injury hyperalgesia [104,105].

Modulation of Tonic Inhibitory Circuits

Second-order dorsal horn neurons (Lamina V WDR neurons) receive robust excitatory input from large (Aß) afferents. In spite of this afferent input onto dorsal horn neurons that are believed to play a role in nociceptive processing, such large afferent input does not lead normally to a pain state (Figure 17.7). As reviewed earlier, after local tissue injury such tactile stimulation may lead to a sensitization to light touch (tactile allodynia). An important component of this sensitization is believed to be related to a loss of local GABAergic or glycinergic inhibition. Thus, dorsal horn GABAergic and glycinergic interneurons receive excitatory drive from low-threshold afferents and in turn form synapses pre- and postsynaptic to the large primary afferent. Accordingly, the postsynaptic excitation by large afferent input is modulated by the concurrent activation of these inhibitory interneuons. The blocking of spinal GABA A and glycine receptors by spinal receptor blockers will lead to a marked exaggeration of the excitation evoked by Aß input and will induce a behavioral phenotype in which low-threshold tactile stimulation will induce a prominent pain state [106,107]. Three circumstances may alter this GABAergic/glycinergic modulation. (i) Loss of GABA and glycine neurons or a downregulation in the respective receptors. While early work did suggest such changes may occur after nerve injury, such changes have not been demonstrated to be robust and are unlikely to occur after inflammation [108]. (ii) As reviewed earlier, small afferent input will lead to a release of a number of lipidic acids including PGE_2. PGE_2 will reduce the glycine-mediated opening of the glycine ionophore, leading to a reduction in this local modulation [103]. (iii). The GABA A and glycine receptors are Cl^- ionophores. At the postsynaptic membrane increasing Cl^- conductance typically leads to a mild hyperpolarization of the membrane. It is now appreciated that in the face of peripheral tissue (or nerve injury) there is an alteration in the relative Cl^- exporter/importer activity leading to an increase in the Cl^- equilibrium potential and an enhanced depolarization with increased Cl^- permeability [109]. These events would lead to the enhanced response to low-threshold afferent input observed after peripheral tissue injury and inflammation.

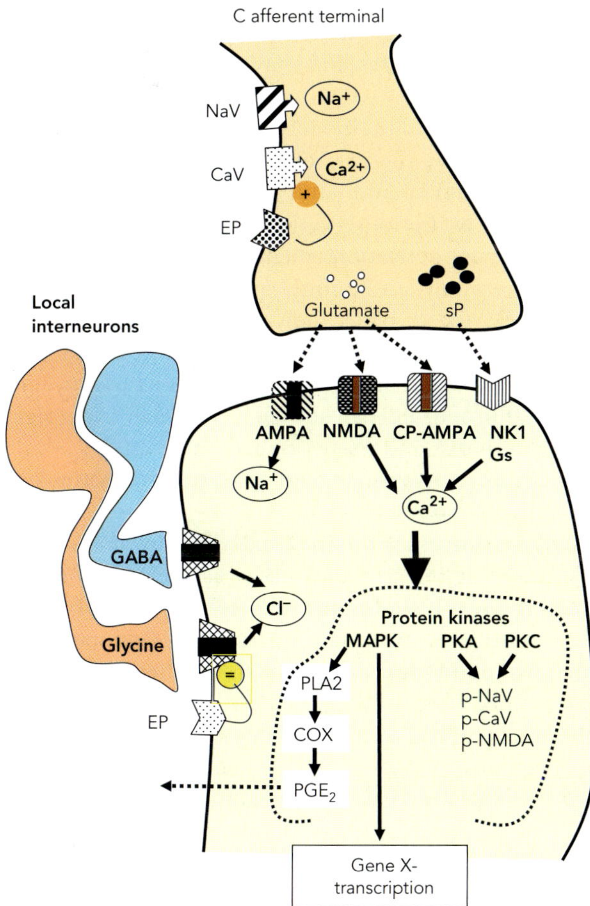

Figure 17.7. First-order synapse for a small afferent axon in the spinal dorsal horn. As indicated, there is the release of small afferent transmitters which depolarize the second order neurons through glutamate (AMPA/NMDA0 and sP (Neurokinin1: NK1) receptors). These serve to depolarize the membrane by increasing cation conductance. This in turn activates downstream cascades that serve to sensitize the membrane to additional depolarization through a variety of mechanisms. See text for additional commentary.

Bulbospinal Systems

Serotonergic pathways (originating in the midline raphe) and noradrenergic pathways (originating in several brain nuclei including the locus corueleus) project into the spinal dorsal horn (see Figure 17.8). Serotonergic systems act through a variety of dorsal horn receptors that may be either inhibitory (5HT1a/b) or directly excitatory (5HT2/3). As indicated schematically in Figure 17.8, small afferent input will activate superficial Lamina I dorsal horn neurons that project into the medulla to activate raphe spinal projections that excite deep dorsal horn (lamina 5) neurons by 5HT3 receptors. Blocking of this circuit has been shown to prominently reduce dorsal horn facilitatory conditioning by small afferent input [110].

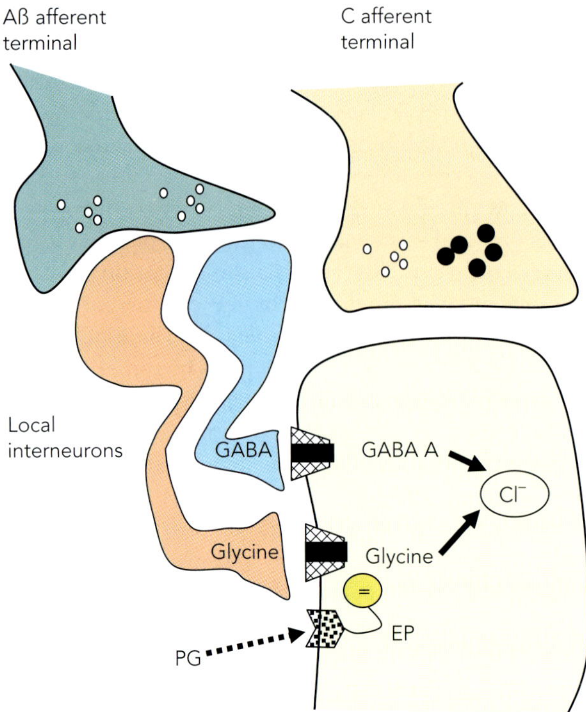

Figure 17.8. Small afferent input activates dorsal horn nociceptive neurons. Large afferent input also activates nociceptors, but as indicated these large low-threshold primary afferents additionally activate inhibitor interneurons (GABA and glycine) which serve to attenuate the excitatory drive produced by large afferent input. Though the interneurons are shown on the postsynaptic membrane they may also be presynaptic on the large myelinated afferent. The outcome is to diminish the excitation otherwise produced by the large afferents. Loss of that inhibition would lead to a prominent increase in large afferent-evoked excitation.

Non-Neuronal Cells

In the CNS, there are a variety of nonneuronal cells. Among these are astrocytes and microglia. Microglia are resident macrophages that appear in the brain from the circulation during development. While classically these non-neuronal cells have been shown to play important roles in a variety of trophic functions such as blood-brain barrier integrity, extracellular water balance, glucose transport, and phagocytic functions in the face of injury, current thinking has begun to emphasize their contributions to the excitability of local neuronal circuits. These astrocytes, microglia and neurons form a complex network in which each can acutely influence the excitability of synaptic transmission [111]. Thus, non-neuronal cells can influence synaptic transmission by the release of a wide variety of proexcitatory products (such as ATP, superoxide/free radicals, nitric oxide, and cytokines). Glial cells can also regulate extracellular parenchymal glutamate by their glutamate transporters. This can serve to increase extracellular glutamate activating neuronal glutamate receptors [112] (see Figure 17.9). These non-neuronal networks can be activated by transmitters from primary afferents, and intrinsic neurons releasing glutamate, sP and ATP can overflow from the synaptic cleft to these adjacent non-neuronal cells and lead through eponymous receptors to their activation. Astrocytes can form local networks which may communicate over a distance by the spread of excitation through local nonsynaptic contacts ("gap" junctions) [113]. Further, astrocytes may communicate with other astrocytes and microglia by releasing a number of products including glutamate/cytokines and "S100" protein [114]. Neurons may activate microglia by the specific release of a membrane chemokine (fractalkine), which acts on specific receptors found on microglia. In addition to afferent input after tissue injury and inflammation, *circulating* cytokines (such as IL-1β/TNF can activate perivascular astrocytes/microglia. As noted, microglia are in fact brain resident macrophages. Many circulating products referred to as "acute phase reactants" (TNF, prostaglandins (PGE_2), sympathetic amines and leukotriene-B_4) can initiate sensitized pain states and may account for mechanical hypersensitivity which arises, for example, in allergenic reactions in rat models [64,115].

Importantly, the role of these non-neuronal cells in inflammation and injury-induced pain states is supported by the observations that spinal inhibitors of microglial activation, such as minocycline (a second generation tetracycline) and pentoxyfilline, have been reported to block indices of acute and chronic glial activation and diminish hyperalgesic states [82]. Similar metabolic inhibitors that block astrocyte activation (flurocitrate) can similarly diminish hyperalgesia after nerve and tissue injury. These agents, while not clinically useful, suggest important directions in drug therapy development.

These processes outlined earlier represent a complex cascade referred to broadly as "neuroinflammation" (Figure 17.9). Importantly, while indices of glial activation (e.g., increased expression of epitopes such as GFAP and OX42) are considered hallmarks of neuroinflammation as observed after local injury, current thinking emphasizes that astrocytes and microglia are indeed *constitutively* active and contributing to acute changes in spinal network excitability [116].

In conclusion, tissue injury and inflammation lead to a surprisingly complex behavioral phenotype characterized by ongoing pain and the appearance of states wherein moderate noxious or frankly innocuous stimuli will lead to an augmented pain state at the injury site (primary) and adjacent to the injury site (secondary). The mechanisms underlying these behavioral states are equally complex, reflecting the release of active factors at the injury site, which initiates afferent traffic and produces a sensitization of the afferent terminal leading to an enhanced response to a given stimulus. In addition, the ongoing activity initiated

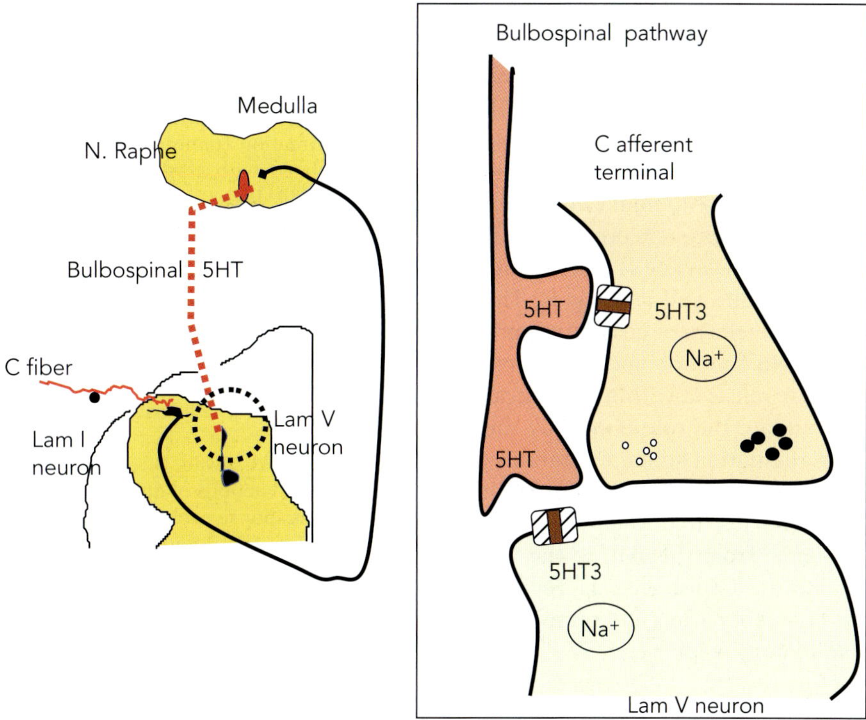

Figure 17.9. Small afferent input activates ascending projections for Lamina V neurons that activate raphe spinal neurons that send serotonin projections to the spinal cord. At the spinal level 5HT released from the bulbospinal projection activates the second-order lamina V neuron through a $5HT_3$ receptor.

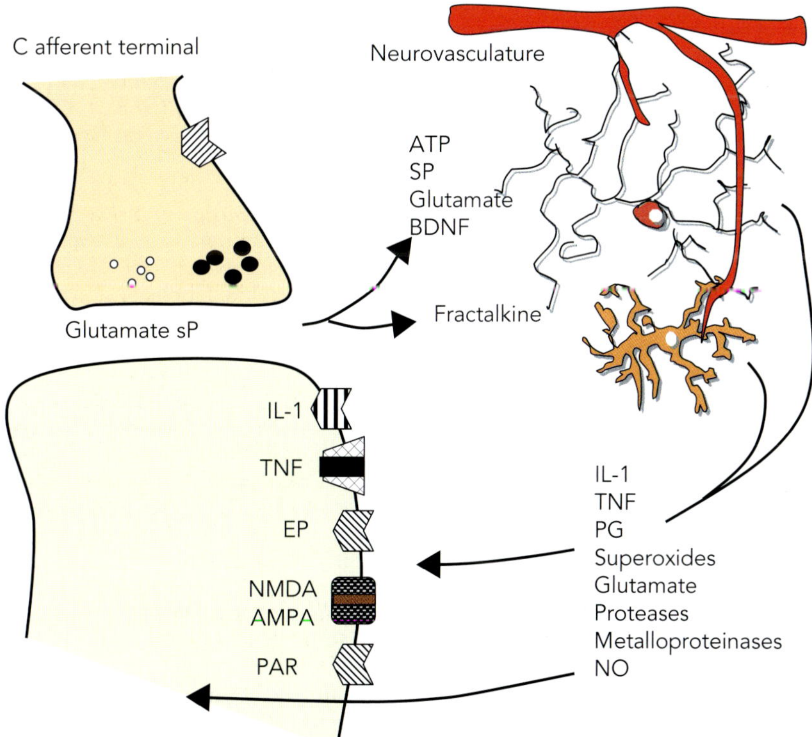

Figure 17.10. Dorsal horn non-neuronal cells (microglia/astrocytes) are activated by overflow from dorsal horn synapses, from second-order neuron, from each other, and through products in the cerebral vasculature. Such activation leads to the secretion of a large variety of inflammatory products that can influence the excitability of neuronal transmission.

by injury leads to a powerful spinal facilitation of spinal sensory networks that is mediated by increases in local circuit excitability, activation of non-neuronal cells and extraspinal feedback loops (Figure 17.10). Interestingly, the role of spinal non-neuronal cells provides an important link between circulating factors generated by the body's response to injury and inflammation and the central nervous system processing. It is of particular interest in that a variety of Toll receptors are present on both microglia and astrocytes. These recptors likely mediate the efects of circulating products (such as LPS) on CNS function and it is increasingly likely that products released within the neuraxis may also serve to modulate the role played by these non neuronal cells in regulating spinal excitability in the face of persistent afferent input and tissue injury. Importantly, these scenarios find their expression in *all* innervated systems, from the skin to the inner lumen of visceral organs. Conversely, as reviewed, antidromic activity on sensory afferents can modulate the peripheral inflammatory response.

REFERENCES

1. Sherrington, C. 1947. *The Integrative Action of the Nervous System*. New Haven, CT: Yale University Press. 228 pp.
2. Cervero, F. 1994. Sensory innervation of the viscera: peripheral basis of visceral pain. *Physiol Rev* **74**:95–138.
3. Neugebauer, V., Han, J.S., Adwanikar, H., Fu, Y., and Ji, G. 2007. Techniques for assessing knee joint pain in arthritis. *Mol Pain* **3**:8.
4. Vermeirsch, H., Biermans, R., Salmon, P.L., and Meert, T.F. 2007. Evaluation of pain behavior and bone destruction in two arthritic models in guinea pig and rat. *Pharmacol Biochem Behav* **87**:349–359.
5. Christianson, J.A., and Gebhart, G.F. 2007. Assessment of colon sensitivity by luminal distension in mice. *Nat Protocol* **2**: p. 2624–31.
6. Delgado-Aros, S., and Camilleri, M. 2005. Visceral hypersensitivity. *J Clin Gastroenterol* **39**:S194–S203; discussion S210.
7. Johanek, L., Shim, B., and Meyer, R.A. 2006. Chapter 4 Primary hyperalgesia and nociceptor sensitization. *Handb Clin Neurol* **81**:35–47.
8. Schaible, H.G., Ebersberger, A., and Von Banchet, G.S. 2002. Mechanisms of pain in arthritis. *Ann N Y Acad Sci* **966**:343–354.
9. Dougherty, P.M. 2003. Central sensitization and cutaneous hyperalgesia. *Semin Pain Med* **1**:121–131.
10. Treede, R.D., Meyer, R.A., Raja, S.N., and Campbell, J.N. 1992. Peripheral and central mechanisms of cutaneous hyperalgesia. *Prog Neurobiol* **38**:397–421.
11. Mulak, A. 2003. Testing of visceral sensitivity. *J Physiol Pharmacol* **54**(Suppl 4):55–72.
12. Sandkuhler, J. 2009. Models and mechanisms of hyperalgesia and allodynia. *Physiol Rev* **89**:707–758.
13. Morris, C.J. 2003. Carrageenan-induced paw edema in the rat and mouse. *Methods Mol Biol* **225**:115–121.
14. Dirig, D.M., Salami, A., Rathbun, M.L., Ozaki, G.T., and Yaksh, T.L. 1997. Characterization of variables defining hindpaw withdrawal latency evoked by radiant thermal stimuli. *J Neurosci Methods* **76**:183–191.
15. Chaplan, S.R., Bach, F.W., Pogrel, J.W., Chung, J.M., and Yaksh, T.L. 1994. Quantitative assessment of tactile allodynia in the rat paw. *J Neurosci Methods* **53**:55–63.
16. Oliveira, P.G., Brenol, C.V., Edelweiss, M.I., Meurer, L., Brenol, J.C., and Xavier, R.M. 2007. Subcutaneous inflammation (panniculitis) in tibio-tarsal joint of rats inoculated with complete Freund's adjuvant. *Clin Exp Med* **7**:184–187.
17. Waksman, B.H. 2002. Immune regulation in adjuvant disease and other arthritis models: relevance to pathogenesis of chronic arthritis. *Scand J Immunol* **56**:12–34.
18. Nozaki-Taguchi, N., and Yaksh, T.L. 1998. A novel model of primary and secondary hyperalgesia after mild thermal injury in the rat. *Neurosci Lett* **254**:25–28.
19. Brennan, T.J. 1999. Postoperative models of nociception. *Ilar J* **40**:129–136.
20. Nagasaka, H., Awad, H., and Yaksh, T.L. 1996. Peripheral and spinal actions of opioids in the blockade of the autonomic response evoked by compression of the inflamed knee joint. *Anesthesiology* **85**:808–816.
21. Staines, N.A., and Wooley, P.H. 1994. Collagen arthritis – what can it teach us? *Br J Rheumatol* **33**:798–807.
22. Morris, G.P., Beck, P.L., Herridge, M.S., Depew, W.T., Szewczuk, M.R., and Wallace, J.L. 1986. Hapten-induced model of chronic inflammation and ulceration in the rat colon. *Gastroenterology* **96**:795–803.
23. Lamb, K., Kang, Y.M., Gebhart, G.F., and Bielefeldt, K. 2003. Gastric inflammation triggers hypersensitivity to acid in awake rats. *Gastroenterology* **125**:1410–1418.
24. Chuang, Y.C., Yoshimura, N., Huang, C.C., Wu, M., Chiang, P.H., and Chancellor, M.B. 2008. Intravesical botulinum toxin A administration inhibits COX-2 and EP4 expression and suppresses bladder hyperactivity in cyclophosphamide-induced cystitis in rats. *Eur Urol* **56**(1):159–167.
25. Glawe, C., Emmrich, J., Sparmann, G., and Vollmar, B. 2005. In vivo characterization of developing chronic pancreatitis in rats. *Lab Invest* **85**:193–204.
26. Mayer, E.A., and Gebhart, G.F. 1994. Basic and clinical aspects of visceral hyperalgesia. *Gastroenterology* **107**:271–293.
27. Yaksh, T.L. 1998. Preclinical models of nociception. In *Cousins & Bridenbaugh's Neural Blockade in Clinical Anesthesia and Pain Medicine*, Cousins, M.J., and Bridenbaugh, P.O. (eds.), 3rd ed., pp. 685–718. Philadelphia: Lippincott-Raven Publisher.
28. Willis, W.D., Jr. 2007. The somatosensory system, with emphasis on structures important for pain. *Brain Res Rev* **55**:297–313.
29. Yaksh, T.L. 2009. Physiologogic and pharmacologic substrates of nociception after tissue and nerve injury. In *Cousins & Bridenbaugh's Neural Blockade in Clinical Anesthesia and Pain Medicine*, Cousins, M.J., Carr, D.B., Horlocker, T.T., and Bridenbaugh, P.O. (eds), 4th ed., pp. 693–751. Philadelphia: Lippincott Williams & Wilkins.
30. Berthoud, H.R., and Neuhuber, W.L. 2000. Functional and chemical anatomy of the afferent vagal system. *Auton Neurosci* **85**:1–17.
31. Maier, S.F., Goehler, L.E., Fleshner, M., and Watkins, L.R. 1998. The role of the vagus nerve in cytokine-to-brain communication. *Ann N Y Acad Sci* **840**:289–300.
32. Price, D.D., Barrell, J.J., and Rainville, P. 2002. Integrating experiential-phenomenological methods and neuroscience to study neural mechanisms of pain and consciousness. *Conscious Cogn* **11**:593–608.

33. Luger, T.A. 2002. Neuromediators – a crucial component of the skin immune system. *J Dermatol Sci* **30**:87–93.
34. Julius, D., and Basbaum, A.I. 2001. Molecular mechanisms of nociception. *Nature* **413**:203–210.
35. England, S. 2008. Voltage-gated sodium channels: the search for subtype-selective analgesics. *Expert Opin Investig Drugs* **17**:1849–1864.
36. Schaible, H.G., Del Rosso, A., and Matucci-Cerinic, M. 2005. Neurogenic aspects of inflammation. *Rheum Dis Clin North Am* **31**:77–101, ix.
37. Steinhoff, M., Stander, S., Seeliger, S., Ansel, J.C., Schmelz, M., and Luger, T. 2003. Modern aspects of cutaneous neurogenic inflammation. *Arch Dermatol* **139**:1479–1488.
38. Reeh, P.W., and Steen, K.H. 1996. Tissue acidosis in nociception and pain. *Prog Brain Res* **113**:143–151.
39. Rueff, A., and Dray, A. 1992. 5-Hydroxytryptamine-induced sensitization and activation of peripheral fibres in the neonatal rat are mediated via different 5-hydroxytryptamine-receptors. *Neuroscience* **50**:899–905.
40. Schmelz, M., Schmidt, R., Weidner, C., Hilliges, M., Torebjork, H.E., and Handwerker, H.O. 2003. Chemical response pattern of different classes of C-nociceptors to pruritogens and algogens. *J Neurophysiol* **89**:2441–2448.
41. Mayer, S., Izydorczyk, I., Reeh, P.W., and Grubb, B.D. 2007. Bradykinin-induced nociceptor sensitisation to heat depends on COX-1 and COX-2 in isolated rat skin. *Pain* **130**:14–24.
42. Russell, F.A., and McDougall, J.J. 2009. Proteinase activated receptor (PAR) involvement in mediating arthritis pain and inflammation. *Inflamm Res* **58**:119–126.
43. Weidner, C., Klede, M., Rukwied, R., et al. 2000. Acute effects of substance P and calcitonin gene-related peptide in human skin – a microdialysis study. *J Invest Dermatol* **115**:1015–1020.
44. Perretti, M. 1998. Lipocortin 1 and chemokine modulation of granulocyte and monocyte accumulation in experimental inflammation. *Gen Pharmacol* **31**: 545–552.
45. Burgos, R.A., Hidalgo, M.A., Figueroa, C.D., Conejeros, I., and Hancke, J.L. 2009. New potential targets to modulate neutrophil function in inflammation. *Mini Rev Med Chem* **9**:153–168.
46. Michaelis, M., Vogel, C., Blenk, K.H., and Janig, W. 1997. Algesics excite axotomised afferent nerve fibres within the first hours following nerve transection in rats. *Pain* **72**:347–354.
47. Cheng, J.K., and Ji, R.R. 2008. Intracellular signaling in primary sensory neurons and persistent pain. *Neurochem Res* **33**:1970–1978.
48. Wang, H., Ehnert, C., Brenner, G.J., and Woolf, C.J. 2006. Bradykinin and peripheral sensitization. *Biol Chem* **387**:11–14.
49. Bevan, S., and Yeats, J. 1991. Protons activate a cation conductance in a sub-population of rat dorsal root ganglion neurones. *J Physiol* **433**:145–161.
50. Krishtal, O. 2003. The ASICs: signaling molecules? Modulators? *Trends Neurosci* **26**:477–483.
51. Kollarik, M., Ru, F., and Undem, B.J. 2007. Acid-sensitive vagal sensory pathways and cough. *Pulm Pharmacol Ther* **20**:402–411.
52. Cesare, P., and McNaughton, P. 1997. Peripheral pain mechanisms. *Curr Opin Neurobiol* **7**:493–499.
53. Hucho, T., and Levine, J.D. 2007. Signaling pathways in sensitization: toward a nociceptor cell biology. *Neuron* **55**:365–376.
54. Banik, R.K., Kozaki, Y., Sato, J., Gera, L., and Mizumura, K. 2001. B2 receptor-mediated enhanced bradykinin sensitivity of rat cutaneous C-fiber nociceptors during persistent inflammation. *J Neurophysiol* **86**:2727–2735.
55. Couture, R., Harrisson, M., Vianna, R.M., and Cloutier, F. 2001. Kinin receptors in pain and inflammation. *Eur J Pharmacol* **429**:161–176.
56. Manning, D.C., Raja, S.N., Meyer, R.A., and Campbell, J.N. 1991. Pain and hyperalgesia after intradermal injection of bradykinin in humans. *Clin Pharmacol Ther* **50**:721–729.
57. Mendell, L.M., Albers, K.M., and Davis, B.M. 1999. Neurotrophins, nociceptors, and pain. *Microsc Res Tech* **45**:252–261.
58. Pezet, S., and McMahon, S.B. 2006. Neurotrophins: mediators and modulators of pain. *Annu Rev Neurosci* **29**:507–538.
59. Malin, S.A., Molliver, D.C., Koerber, H.R., et al. 2006. Glial cell line-derived neurotrophic factor family members sensitize nociceptors in vitro and produce thermal hyperalgesia in vivo. *J Neurosci* **26**:8588–8599.
60. Ossovskaya, V.S., and Bunnett, N.W. 2004. Protease-activated receptors: contribution to physiology and disease. *Physiol Rev* **84**:579–621.
61. Vergnolle, N. 2004. Modulation of visceral pain and inflammation by protease-activated receptors. *Br J Pharmacol* **141**:1264–1274.
62. Murphy, P.G., Borthwick, L.A., Altares, M., Gauldie, J., Kaplan, D., and Richardson, P.M. 2000. Reciprocal actions of interleukin-6 and brain-derived neurotrophic factor on rat and mouse primary sensory neurons. *Eur J Neurosci* **12**:1891–1899.
63. Schafers, M., Sommer, C., Geis, C., Hagenacker, T., Vandenabeele, P., and Sorkin, L.S. 2008. Selective stimulation of either tumor necrosis factor receptor differentially induces pain behavior in vivo and ectopic activity in sensory neurons in vitro. *Neuroscience* **157**:414–423.
64. Cunha, F.Q., and Ferreira, S.H. 2003. Peripheral hyperalgesic cytokines. *Adv Exp Med Biol* **521**:22–39.
65. Coutaux, A., Adam, F., Willer, J.C., and Le Bars, D. 2005. Hyperalgesia and allodynia: peripheral mechanisms. *Joint Bone Spine* **72**:359–371.
66. Ji, R.R., Gereau, R.W.T., Malcangio, M., and Strichartz, G.R. 2009. MAP kinase and pain. *Brain Res Rev* **60**:135–148.
67. Nicol, G.D., and Vasko, M.R. 2007. Unraveling the story of NGF-mediated sensitization of nociceptive sensory neurons: ON or OFF the Trks? *Mol Interv* **7**:26–41.
68. Djouhri, L., Koutsikou, S., Fang, X., McMullan, S., and Lawson, S.N. 2006. Spontaneous pain, both neuropathic and inflammatory, is related to frequency of spontaneous firing in intact C-fiber nociceptors. *J Neurosci* **26**:1281–1292.
69. Mantyh, P.W. 2004. A mechanism-based understanding of bone cancer pain. *Novartis Found Symp* **261**:194–214; discussion 214–219, 256–261.
70. Boettger, M.K., Hensellek, S., Richter, F., et al. 2008. Antinociceptive effects of tumor necrosis factor alpha neutralization in a rat model of antigen-induced arthritis: evidence of a neuronal target. *Arthritis Rheum* **58**:2368–2378.
71. Holzer, P. 2007. Role of visceral afferent neurons in mucosal inflammation and defense. *Curr Opin Pharmacol* **7**:563–569.
72. Nazif, O., Teichman, J.M., and Gebhart, G.F. 2007. Neural upregulation in interstitial cystitis. *Urology* **69**:24–33.

73. Zimmermann, K., Reeh, P.W., and Averbeck, B. 2002. ATP can enhance the proton-induced CGRP release through P2Y receptors and secondary PGE(2) release in isolated rat dura mater. *Pain* **97**:259–265.
74. Hermanstyne, T.O., Markowitz, K., Fan, L., and Gold, M.S. 2008. Mechanotransducers in rat pulpal afferents. *J Dent Res* **87**:834–838.
75. Waxman, S.G., Cummins, T.R., Dib-Hajj, S.D., and Black, J.A. 2000. Voltage-gated sodium channels and the molecular pathogenesis of pain: a review. *J Rehabil Res Dev* **37**:517–528.
76. Lu, S.G., and Gold, M.S. 2008. Inflammation-induced increase in evoked calcium transients in subpopulations of rat dorsal root ganglion neurons. *Neuroscience* **153**:279–288.
77. Maingret, F., Coste, B., Padilla, F., Clerc, N., Crest, M., Korogod, S.M., and Delmas, P. 2008. Inflammatory mediators increase Nav1.9 current and excitability in nociceptors through a coincident detection mechanism. *J Gen Physiol* **131**:211–225.
78. Delcroix, J.D., Valletta, J.S., Wu, C., Hunt, S.J., Kowal, A.S., and Mobley, W.C. 2003. NGF signaling in sensory neurons: evidence that early endosomes carry NGF retrograde signals. *Neuron* **39**:69–84.
79. D'Mello, R., and Dickenson, A.H. 2008. Spinal cord mechanisms of pain. *Br J Anaesth* **101**:8–16.
80. Herrero, J.F., Laird, J.M., and Lopez-Garcia, J.A. 2000. Wind-up of spinal cord neurones and pain sensation: much ado about something? *Prog Neurobiol* **61**:169–203.
81. Gabay, C., and Kushner, I. 1999. Acute-phase proteins and other systemic responses to inflammation. *N Engl J Med* **340**:448–454.
82. Milligan, E.D., and Watkins, L.R. 2009. Pathological and protective roles of glia in chronic pain. *Nat Rev Neurosci* **10**:23–36.
83. Ren, K., and Dubner, R. 2008. Neuron-glia crosstalk gets serious: role in pain hypersensitivity. *Curr Opin Anaesthesiol* **21**:570–579.
84. Wieseler-Frank, J., Maier, S.F., and Watkins, L.R. 2005. Central proinflammatory cytokines and pain enhancement. *Neurosignals* **14**:166–174.
85. McDonald, D.M., Bowden, J.J., Baluk, P., and Bunnett, N.W. 1996. Neurogenic inflammation. A model for studying efferent actions of sensory nerves. *Adv Exp Med Biol* **410**:453–462.
86. Sorkin, L.S., Moore, J., Boyle, D.L., Yang, L., and Firestein, G.S. 2003. Regulation of peripheral inflammation by spinal adenosine: role of somatic afferent fibers. *Exp Neurol* **184**:162–168.
87. Heppelmann, B. 1997. Anatomy and histology of joint innervation. *J Peripher Nerv Syst* **2**:5–16.
88. Konttinen, Y.T., Tiainen, V.M., Gomez-Barrena, E., Hukkanen, M., and Salo, J. 2006. Innervation of the joint and role of neuropeptides. *Ann N Y Acad Sci* **1069**:149–154.
89. Todd, A.J. 2002. Anatomy of primary afferents and projection neurones in the rat spinal dorsal horn with particular emphasis on substance P and the neurokinin 1 receptor. *Exp Physiol* **87**:245–249.
90. Todd, A.J., and Spike, R.C. 1993. The localization of classical transmitters and neuropeptides within neurons in laminae I-III of the mammalian spinal dorsal horn. *Prog Neurobiol* **41**:609–645.
91. Willis, W.D. 2001. Role of neurotransmitters in sensitization of pain responses. *Ann N Y Acad Sci* **933**:142–156.
92. Bleakman, D., Alt, A., and Nisenbaum, E.S. 2006. Glutamate receptors and pain. *Semin Cell Dev Biol* **17**:592–604.
93. Yaksh, T.L. 1999. Spinal systems and pain processing: development of novel analgesic drugs with mechanistically defined models. *Trends Pharmacol Sci* **20**:329–337.
94. Luo, C., Seeburg, P.H., Sprengel, R., and Kuner, R. 2008. Activity-dependent potentiation of calcium signals in spinal sensory networks in inflammatory pain states. *Pain* **140**:358–367.
95. Merighi, A., Salio, C., Ghirri, A., et al. 2008. BDNF as a pain modulator. *Prog Neurobiol* **85**:297–317.
96. Black, I.B. 1999. Trophic regulation of synaptic plasticity. *J Neurobiol* **41**:108–118.
97. Jones, T.L., and Sorkin, L.S. 2005. Activated PKA and PKC, but not CaMKIIalpha, are required for AMPA/Kainate-mediated pain behavior in the thermal stimulus model. *Pain* **117**:259–270.
98. Svensson, C.I., Marsala, M., Westerlund, A., et al. 2003. Activation of p38 mitogen-activated protein kinase in spinal microglia is a critical link in inflammation-induced spinal pain processing. *J Neurochem* **86**:1534–1544.
99. Sorkin, L., Svensson, C.I., Jones-Cordero, T.L., Hefferan, M.P., and Campana, W.M. 2009. Spinal p38 mitogen-activated protein kinase mediates allodynia induced by first-degree burn in the rat. *J Neurosci Res* **87**:948–955.
100. Svensson, C.I., and Yaksh, T.L. 2002. The spinal phospholipase-cyclooxygenase-prostanoid cascade in nociceptive processing. *Annu Rev Pharmacol Toxicol* **42**:553–583.
101. Svensson, C.I., Hua, X.Y., Protter, A.A., Powell, H.C., and Yaksh, T.L. 2003. Spinal p38 MAP kinase is necessary for NMDA-induced spinal PGE(2) release and thermal hyperalgesia. *Neuroreport* **14**:1153–1157.
102. Meves, H. 2006. The action of prostaglandins on ion channels. *Curr Neuropharmacol* **4**:41–57.
103. Harvey, R.J., Depner, U.B., Wassle, H., et al. 2004. GlyR alpha3: an essential target for spinal PGE2-mediated inflammatory pain sensitization. *Science* **304**:884–887.
104. Malmberg, A.B., and Yaksh, T.L. 1993. Spinal nitric oxide synthesis inhibition blocks NMDA-induced thermal hyperalgesia and produces antinociception in the formalin test in rats. *Pain* **54**:291–300.
105. Tang, Q., Svensson, C.I., Fitzsimmons, B., Webb, M., Yaksh, T.L., and Hua, X.Y. 2007. Inhibition of spinal constitutive NOS-2 by 1400W attenuates tissue injury and inflammation-induced hyperalgesia and spinal p38 activation. *Eur J Neurosci* **25**:2964–2972.
106. Cronin, J.N., Bradbury, E.J., and Lidierth, M. 2004. Laminar distribution of GABAA- and glycine-receptor mediated tonic inhibition in the dorsal horn of the rat lumbar spinal cord: effects of picrotoxin and strychnine on expression of Fos-like immunoreactivity. *Pain* **112**:156–163.
107. Loomis, C.W., Khandwala, H., Osmond, G., and Hefferan, M.P. 2001. Coadministration of intrathecal strychnine and bicuculline effects synergistic allodynia in the rat: an isobolographic analysis. *J Pharmacol Exp Ther* **296**:756–761.
108. Polgar, E., Hughes, D.I., Riddell, J.S., Maxwell, D.J., Puskar, Z., and Todd, A.J. 2003. Selective loss of spinal GABAergic or glycinergic neurons is not necessary for development of thermal hyperalgesia in the chronic constriction injury model of neuropathic pain. *Pain* **104**:229–239.

109. Price, T.J., Cervero, F., and de Koninck, Y. 2005. Role of cation-chloride-cotransporters (CCC) in pain and hyperalgesia. *Curr Top Med Chem* **5**:547–555.
110. Suzuki, R., Rygh, L.J., and Dickenson, A.H. 2004. Bad news from the brain: descending 5-HT pathways that control spinal pain processing. *Trends Pharmacol Sci* **25**:613–617.
111. Cao, H., and Zhang, Y.Q. 2008. Spinal glial activation contributes to pathological pain states. *Neurosci Biobehav Rev* **32**:972–983.
112. Scholz, J., and Woolf, C.J. 2007. The neuropathic pain triad: neurons, immune cells and glia. *Nat Neurosci* **10**:1361–1368.
113. Scemes, E., Suadicani, S.O., Dahl, G., and Spray, D.C. 2007. Connexin and pannexin mediated cell-cell communication. *Neuron Glia Biol* **3**:199–208.
114. Abbadie, C., Bhangoo, S., De Koninck, Y., Malcangio, M., Melik-Parsadaniantz, S., and White, F.A. 2009. Chemokines and pain mechanisms. *Brain Res Rev* **60**:125–134.
115. Clark, A.K., Gentry, C., Bradbury, E.J., McMahon, S.B., and Malcangio, M. 2007. Role of spinal microglia in rat models of peripheral nerve injury and inflammation. *Eur J Pain* **11**:223–230.
116. Hald, A. 2009. Spinal astrogliosis in pain models: cause and effects. *Cell Mol Neurobiol* **29**:609–619.

SUGGESTED READINGS

Abbadie, C., Bhangoo, S., De Koninck, Y., Malcangio, M., Melik-Parsadaniantz, S., and White, F.A. 2009. Chemokines and pain mechanisms. *Brain Res Rev* **60**:125–134.

Coutaux, A., Adam, F., Willer, J.C., and Le Bars, D. 2005. Hyperalgesia and allodynia: peripheral mechanisms. *Joint Bone Spine* **72**:359–371.

Hucho, T., and Levine, J.D. 2007. Signaling pathways in sensitization: toward a nociceptor cell biology. *Neuron* **55**:365–376.

Milligan, E.D., and Watkins, L.R. 2009. Pathological and protective roles of glia in chronic pain. *Nat Rev Neurosci* **10**:23–36.

Yaksh, T.L. 2006. Central pharmacology of nociceptive transmission. In *Wall and Melzack's Textbook of Pain*, McMahon, S.B., and Koltzenburg, M. (eds), 5th ed., pp. 371–414. Philadelphia: Elsevier Churchill Livingstone.

18 Nonsteroidal Anti-Inflammatory Drugs

Samir S. Ayoub and Roderick Flower

INTRODUCTION

Nonsteroidal anti-inflammatory drugs (NSAIDs), sometimes called the *"aspirin-like drugs,"* are among the most widely used of all drugs. Aspirin itself was introduced by Bayer in 1898 as a replacement for salicylic acid, which had been available in synthetic form since the 1870s and as the active constituent of plant (e.g., willow bark) preparations for many centuries before that. Since the beginning of the 20th century, the number of NSAIDs has grown substantially: phenylbutazone was introduced in the 1940s, the fenamates in the 1950s, indomethacin in the 1960s, the proprionates in the 1970s, and the oxicams in the 1980s. The 1990s saw a radical new development – the introduction of "coxibs" and the new millennium has ushered in an era of reappraisal and reassessment of our understanding of the pharmacology of these drugs and their therapeutic and unwanted effects. There are now more than 50 different NSAIDs on the global market and some of the more prominent examples are listed in Figure 18.1. (Chemical structures of NSAIDs shown, grouped according to chemical structure.)

NSAIDs had been in clinical use for a long time prior to an understanding of their pharmacological mechanism of action. The real breakthrough in NSAID pharmacology was made in the early 1970s when John Vane identified that NSAIDs work by inhibition of prostaglandin (PG)-forming cyclo-oxygenase (COX) activity resulting in the reduction of prostanoid synthesis. The second breakthrough came about in the early 1990s with the discovery of COX-2 that eventually lead to the development of the COX-2-selective NSAIDs that lacked some of the major side effects of classical NSAIDs.

Clinically, the NSAIDs provide symptomatic relief from pain and swelling in chronic joint disease such as occurs in osteo- and rheumatoid arthritis as well as in more acute inflammatory conditions such as sports injuries, fractures, sprains, and other soft tissue injuries; can also alleviate postoperative, dental, and menstrual pain, as well as headaches and migraine. Many NSAIDs are available in a variety of different formulations such as tablets, injections, and gels and several NSAIDs are available from pharmacies "over-the-counter" without prescription. The latter are taken in large quantities for many types of minor aches and pains – and occasionally abused. Aspirin itself is still consumed in prodigious amounts around the world and new uses are continually being found for this drug. Virtually all NSAIDs, particularly the "classical" NSAIDs, can have significant unwanted effects, especially in the elderly. Newer agents have less severe adverse actions.

PRINCIPAL PHARMACOLOGICAL ACTIONS

NSAIDs are sometimes known as the aspirin-like drugs because their pharmacology is broadly similar to that of aspirin the "archetypal" NSAID. Their three main therapeutic effects are an *anti-inflammatory* effect – reducing the symptoms of inflammation; an *analgesic* effect – reduction of certain types of (especially inflammatory) pain and an *antipyretic* effect – lowering the body temperature when this is raised in disease (i.e., fever). It is important to note that whilst NSAIDs relieve many of the symptoms of inflammation, they have little or no action on the actual progress of underlying chronic diseases itself. As a class, they are generally without effect on other aspects of inflammation such as leukocyte migration, T-cell function, lysosomal enzyme release, toxic oxygen radical production, and so on that contribute to tissue damage in chronic inflammatory conditions such as rheumatoid arthritis, vasculitis, and nephritis.

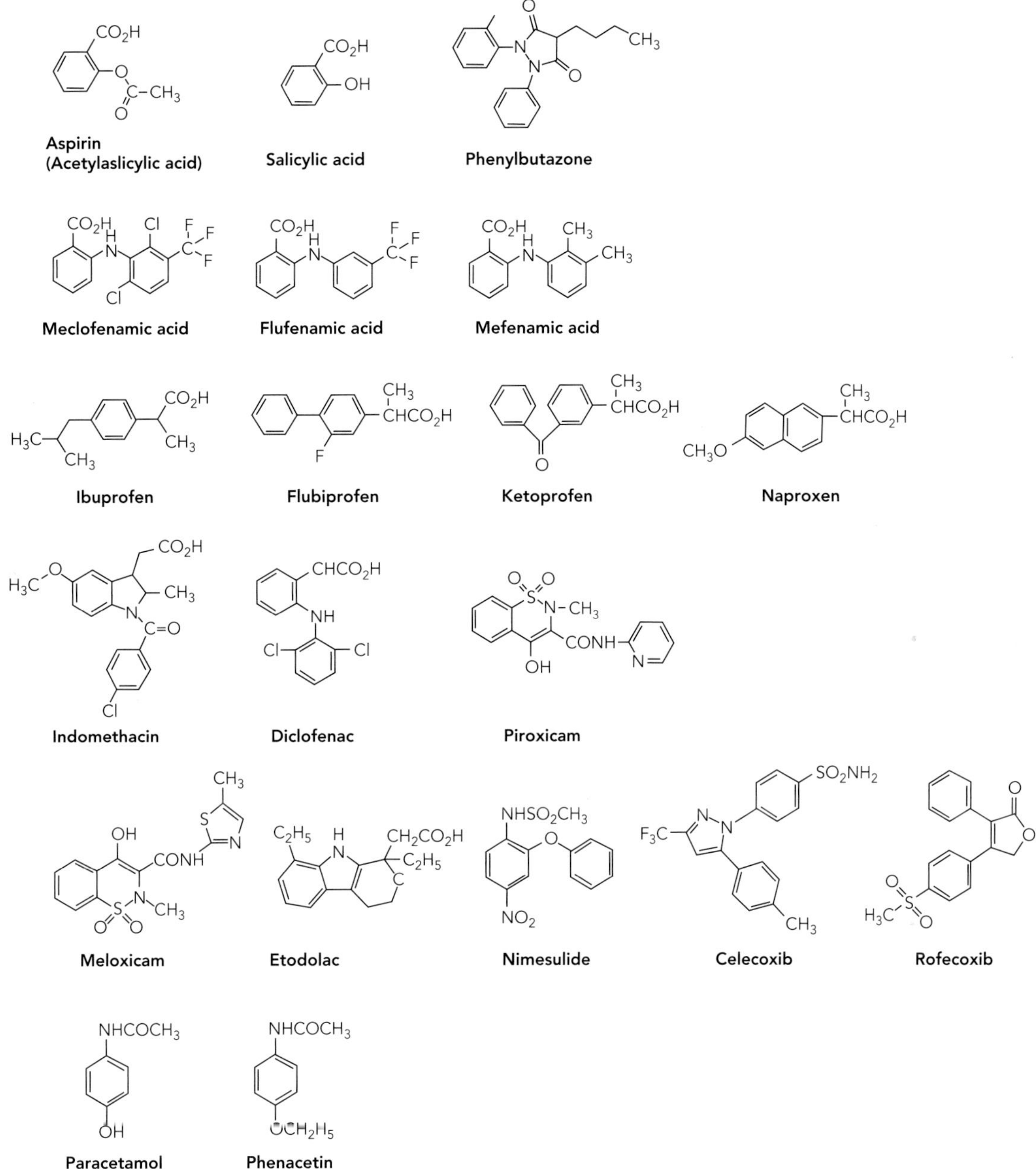

Figure 18.1. The chemical structures of some NSAIDs and "coxibs" representative of the main classes of these drugs. Note that whilst NSAIDs are often carboxylic acids, the coxibs are not. This has some relevance to their isoform selectivity (see Figure 18.2).

We Will Review Each of Their Principal Actions in Turn

Anti-Inflammatory Effects

Many mediators act to initiate and coordinate acute inflammatory and allergic reactions. Whilst some are produced in response to specific stimuli (e.g., histamine in allergic inflammation) there is considerable redundancy and each facet of the response – vasodilatation, increased vascular permeability, cell accumulation, and so forth – can generally be produced by several separate mechanisms.

NSAIDs reduce mainly those components of the inflammatory and immune response in which prostaglandins, predominantly derived from COX-2 (see later), play a significant part. These include *vasodilatation*, in which the blood flow to inflamed tissues is increased by a direct vasodilator action of prostanoids (giving rise to the "redness" of inflammation); and *oedema*, this is brought about through an indirect potentiation of the effect of other agents (e.g., histamine) which directly increase the permeability of postcapillary venules leading to increased fluid movement through the microvasculature and giving rise to the "swelling" of inflammation.

In experimental animals, NSAIDs have been shown to prevent resolution of the inflammatory reaction if administered toward the end of the acute inflammatory process. These findings provide supportive evidence for an anti-inflammatory role for some prostaglandins, which have been shown to increase at the latter phase of acute inflammation.

Analgesic Effects

The NSAIDs are suitable for the treatment of many types of mild-moderate pain including inflammatory and postoperative pain. They are generally without action in neuropathic pain. NSAIDs provide a different quality of analgesia to that of the opiates in the sense that they do not affect other sensory modalities. Their analgesic "ceiling" is also different. Two sites of action have been identified: peripherally, NSAIDs decrease production of the prostaglandins that sensitize nociceptors to the direct action of inflammatory mediators such as bradykinin. In doing so they relieve the hyperalgesia in conditions that are associated with increased local prostaglandin synthesis such as arthritis, bursitis, pain of muscular and vascular origin, toothache, dysmenorrhoea, the pain of postpartum states, and the pain of cancer metastases in bone. They may be administered in combination with opioids and in some cases can substantially reduce the requirement for opioids required to produce an analgesic effect. Their ability to relieve some types of headache (less frequently migraine) may be related to the abrogation of the vasodilator effect of prostaglandins on the cerebral vasculature.

In addition to these peripheral effects, there is a less well-characterized central action possibly in the spinal cord. Inflammation increases COX-2 expression in the cord and local prostaglandin release facilitates transmission by afferent nociceptive fibers, by enhancing the release of excitatory mediators such as substance P and glutamate, to relay neurons in the dorsal horn. Prostaglandins have an inhibitory effect on descending noradrenergic and glycinergic inhibitory pathways that terminate in the spinal cord, thus causing dis-inhibition of spinal nociceptive pathways enhancing facilitation further. Prostaglandins may also activate microglia in the cord that destroy inhibitory interneurones leading to a remodeling and permanent activation of pain pathways.

Other sites of action have been mooted: NSAIDs and COX-2-selective inhibitors injected directly to the preoptic area of the hypothalamus have been shown to induce analgesia in response to systemically administered noxious agents.

Antipyretic Effect

Body temperature is regulated by the preoptic area of the hypothalamus, which functions in a way similar to a thermostat by controlling the balance between heat loss and heat production, thereby regulating the overall temperature of the organism around a "set point" (e.g., normal body temperature). Fever occurs when there is a disturbance of this homeostatic system, which leads to the set-point of body temperature being raised. During an inflammatory reaction, bacterial endotoxins, viruses, or other foreign substances acting through TOLL receptors, cause the release from macrophages and other cells of cytokines such as TNF and IL-1. These substances can directly stimulate the generation, in the hypothalamus, of E-type prostaglandins which elevate the temperature set-point. COX-2 is induced by IL-1 in vascular endothelium in the hypothalamus. NSAIDs exert their antipyretic action largely through inhibition of prostaglandin production in (or close to) the hypothalamus. There is some evidence that prostaglandins are not the only mediators of fever; hence, NSAIDs may have an additional antipyretic effect through mechanisms as yet unknown. Once there has been a return to the normal hypothalamic set-point, the temperature-regulating mechanisms (dilatation of superficial blood vessels, sweating, etc.) then operate to reduce temperature to normal. Normal body temperature in humans is not affected by NSAIDs.

UNWANTED EFFECTS

The NSAIDs also share, to a greater or lesser degree, the same types of mechanism-based side effects although there may be other additional unwanted effects peculiar to individual members of the group. The main effects include *gastric irritation*, which may range from simple discomfort to ulcer formation; an effect on *renal blood flow*, particularly in the compromised kidney; and a tendency to prolong bleeding through inhibition of *platelet function* and *skin reactions*. More controversially, it is argued that all NSAIDs – but especially COX-2 selective drugs – increase the likelihood of thrombotic events such as myocardial infarction, by inhibiting PGI_2 (Prostacyclin) synthesis.

Overall, the burden of unwanted NSAID side effects is high. When used in joint diseases (necessitating fairly large doses and long-continued use), there is a high incidence of side effects – not only in the gastrointestinal (GI) tract but also in liver, kidney, spleen, blood, and bone marrow. COX-2-selective drugs generally have less GI toxicity (see later).

We Will Again Look at Each Area in Turn

Gastrointestinal Disturbances

Adverse GI events are the commonest unwanted effects of the NSAIDs and are believed to result mainly from inhibition of the gastric COX-1 responsible for the synthesis of the prostaglandins that normally exert an inhibitory tone on acid secretion and protect the mucosa. Common GI side effects include gastric discomfort, dyspepsia, diarrhea (but sometimes constipation), nausea and vomiting and, in some cases, gastric bleeding and frank ulceration. It has been estimated by some that 34%–46% of users of NSAIDs will sustain some GI damage that, whilst it may be asymptomatic, carries a risk of serious hemorrhage and/or perforation. Severe GI effects alone (perforations, ulcers, or bleeding) are said to result in the hospitalization of over 100,000 people per year in the United States. Some 15% of these patients may die from this iatrogenic disease. These figures probably reflect the fact that NSAIDs are used extensively in the elderly, and often for extended periods of time. The mechanism is dependent on inhibition of COX in the gastric mucosa and damage is seen whether the drugs are given orally or systemically. However, in some cases (aspirin being a good example) *local* damage to the gastric mucosa caused directly by the drug itself may compound the damage. Oral administration of prostaglandin analogues such as misoprostol can diminish the gastric damage produced by these agents and NSAIDs are also frequently given together with H_2 blockers or proton pump inhibitors for the same reason.

On the basis of extensive experimental evidence, it had been predicted that COX-2-selective agents would provide good anti-inflammatory and analgesic actions with less gastric damage and some older drugs (e.g., meloxicam) that were believed to be better tolerated in the clinic, turned out to have some COX-2 selectivity. Two large prospective studies compared celecoxib and rofecoxib with standard comparator NSAIDs in patients with arthritis and showed some benefit, although the results were not as clear-cut as had been hoped.

Skin Reactions

Rashes are common idiosyncratic unwanted effects of NSAIDs, particularly with mefenamic acid (10%–15% frequency) and sulindac (5%–10% frequency). The underlying mechanism is uncertain. Symptoms range from mild erythematous, urticarial, and photosensitivity reactions to more serious and (rarely) potentially fatal diseases including Stevens–Johnson syndrome.

Haemostatic System

All NSAIDs (except COX-2 inhibitors) prevent platelet aggregation because COX-1 is important in generating proaggregatory TxA_2 (thromboxane A_2) and therefore may prolong bleeding. Again, aspirin is the main problem in this regard although this action is also the basis of aspirin's cardiovascular utility (see later).

Renal Effects

Therapeutic doses of NSAIDs in healthy individuals pose little threat to kidney function, but in susceptible patients they cause acute renal insufficiency, which is reversible on stopping the drug. This occurs through the inhibition of the biosynthesis of those prostanoids (PGE_2 and PGI_2) involved in the maintenance of renal blood flow, specifically in the PGE_2-mediated compensatory vasodilatation that occurs in response to the action of noradrenaline or angiotensin II. Neonates and the elderly are especially at risk as are patients with heart, liver, or kidney disease, or a reduced circulating blood volume.

It seems that adverse cardiovascular events especially during prolonged use or in high cardiovascular risk patients may be a feature of all drugs that inhibit COX activity. Hypertension, caused by inhibition of renal COX-2, seems to be a strong possibility. The notion that COX-2 inhibitors selectively increase cardiovascular risk through a selective action on vascular COX-2, thereby diminishing PGI_2 production (the "Fitzgerald hypothesis"), seems less likely.

Chronic NSAID consumption, especially NSAID "abuse," can cause *analgesic nephropathy* characterized by chronic nephritis and renal papillary necrosis. Historically, phenacetin, now withdrawn, was the main culprit; paracetamol, one of its major metabolites, is much less toxic. Regular use of prescribed doses of NSAIDs is less hazardous for the kidney in this respect than very heavy and prolonged use of over-the-counter analgesics.

Other Unwanted Effects

Other, much less common, unwanted effects of NSAIDs include CNS effects, bone marrow disturbances, and liver disorders, the latter being more likely if renal impairment is already present. Paracetamol overdose may cause liver failure if not treated promptly. Approximately 5% of patients exposed to NSAIDs may experience *aspirin sensitive asthma*. The exact mechanism is unknown but inhibition of COX is implicated. Aspirin is the worst offender but there is cross-reaction with all other class members except, possibly, COX-2 inhibitors.

MECHANISM OF PHARMACOLOGICAL ACTION

As a group, the NSAIDs are structurally diverse, although most are carboxylic acids (Figure 18.1). The question that went unanswered for almost seven decades was how the apparently disparate therapeutic and side effects were mechanistically linked.

There were several early suggestions, but writing about his breakthrough discovery in 1971, Vane tells us that the idea that the aspirin-like drugs blocked the conversion of substrate arachidonic acid to prostaglandins came to him while reviewing experiments in which aspirin blocked the release of "rabbit aorta contracting substance" (RCS) from guinea-pig and dog lung. Believing that RCS (later shown to be a mixture of TxA_2 and prostaglandin endoperoxides) was an intermediate in prostaglandin synthesis, he wrote *"a logical corollary was that aspirin might well be blocking the synthesis of prostaglandins."*

A quartet of papers appeared that year from Vane and members of his group, showing that aspirin itself, indomethacin and (less effectively) sodium salicylate blocked prostaglandin synthesis in a crude cell-free COX preparation and in isolated perfused spleen of dogs. In humans, therapeutic doses of aspirin taken by volunteers reduced prostaglandin generation by aggregating platelets ex vivo or in seminal plasma samples collected during the course of the treatments. The overall message was clear – at least some NSAIDs were able to prevent the generation of prostaglandins by direct action on the COX enzyme, and did so in humans in clinical doses. But how was this linked to their therapeutic actions?

The late 1960s and early 1970s had seen an explosion of interest in prostaglandin biology. It was observed that prostaglandins were generated during platelet aggregation, and that they produced fever, hyperalgesia, and inflammation. Prostaglandins had also been detected in the gastric mucosa and been shown to inhibit ulcer formation in rodent models of gastric damage and cytoprotection. In other words, the ability of NSAIDs to block COX provided the much sought after link between the therapeutic and side effects of these drugs.

Over the next couple of years, it was shown that the entire gamut of NSAIDs inhibited COX at concentrations well within their therapeutic plasma range and that the overall order of potency corresponded with their therapeutic activity. Other types of anti-inflammatories, such as the glucocorticoids and the so-called disease modifying drugs such as gold and penicillamine, were inactive in these cell-free assays, providing further evidence for the specificity of the NSAID effect. Using active/inactive enantiomeric pairs of NSAIDs, such as naproxen, an exquisite correlation was observed between the anti-inflammatory and anti-COX activity, and many more studies confirmed the notion that this is a fundamental mechanism of this class of drugs.

Other actions besides inhibition of COX may contribute to the anti-inflammatory effects of some NSAIDs. Reactive oxygen radicals produced by neutrophils and macrophages are implicated in tissue damage in some conditions, and some NSAIDs (e.g., sulindac) have oxygen-radical-scavenging effects as well as COX-inhibitory activity so may decrease tissue damage. Aspirin also inhibits expression of the transcription factor NF-κB, which has a key role in the transcription of the genes for inflammatory mediators. A direct toxic action of NSAIDs on the gut may be at least in part mediated through an "uncoupling" effect on epithelial mitochondria.

By 1974, Vane's concept was firmly established as the most powerful explanatory idea in the field. One effect of this was that the pharmaceutical industry now possessed, probably for the first time, a simple and robust in vitro screen for putative anti-inflammatory compounds. Probably as a result of this, the number of chemical abstracts dealing with potential inhibitors of the COX enzyme rose markedly, with more than 2,500 per year recorded within a decade of these ideas taking hold.

However, several anomalies were noted by early workers in the field. For example, paracetamol (acetaminophen) – another hugely popular drug that was also introduced in the 1890s possessed antipyretic and analgesic activity like all other aspirin-like drugs but, unlike most, had little anti-inflammatory activity and was virtually devoid of gastric or platelet side effects. Apparently in accord with its therapeutic profile, paracetamol was found to have a different pattern of inhibitory activity, being more effective against crude brain COX preparations than those prepared from peripheral tissues such as the spleen. At the time (1972), this observation was put forward as a putative explanation for the selectivity of its therapeutic action and the idea that there were several forms of the enzyme was formulated. Wide variations in the inhibitory potency of indomethacin against COX enzymes prepared from a range of tissues was subsequently reported and the "isoenzyme" idea was further elaborated in several early reviews.

DISCOVERY OF OTHER COX ISOFORMS

Despite this early, presumptive, evidence for alternative forms of COX, little hard evidence emerged for several years probably because technical problems delayed progress. The structural features of COX, which has dual hydroperoxidase and COX activity, were not well understood and with the notable exception of the seminal vesicles of some species, the enzyme is usually expressed in tissues in low abundance. As a dimeric membrane-bound protein, COX posed many challenges to purification and was not sequenced until 1988. Other factors which caused confusion included the wide variations in assay conditions used by different laboratories as well as differences in the kinetics of the COX inhibitors. Many of these issues are still relevant today.

The 1980s, however, saw some exciting advances in our understanding of COX pharmacology. Morrison et al. [1] observed that the inflamed kidney unexpectedly developed an enormous capacity to generate prostaglandins that was due to de novo synthesis of fresh enzyme and in 1982, the group published evidence suggesting the presence of two distinct forms of COX in brain tissue with differing sensitivities to indomethacin. Other studies in GI tissue also supported the selectivity of action of NSAIDs in different tissues.

By the end of the 1980s it was clear that exposure of cells of various types to growth factors, cytokines, endotoxin, or some hormones induced the de novo synthesis of enzyme which, when partially sequenced, strongly resembled the bovine seminal vesicle COX.

While investigating the expression of early-response genes in fibroblasts transformed with Rous sarcoma virus, Simmons and his colleagues [2] identified a novel mRNA transcript that coded for a protein that was not identical but which had a high sequence similarity to the seminal vesicle COX enzyme and suggested that a COX isozyme had been discovered. Herschman and colleagues [3] also observed a novel cDNA species encoding a protein with a predicted structure similar to COX-1, while studying phorbol-ester–induced genes in Swiss 3T3 cells.

Masferrer et al. [4] identified an inflammation-inducible form of COX as the species that both Simmons and Herschman had cloned and the two COX enzymes were renamed *COX-1*, referring to the original enzyme isolated from seminal vesicles and subsequently found to be distributed almost ubiquitously, and *COX-2*, denoting the "inducible" form of the enzyme (although it was expressed basally in the brain and elsewhere). COX-1/COX-2 mRNA was differentially expressed in human tissues and promoter analysis confirmed a fundamental difference between the two isozymes, with the COX-2 promoter containing elements characteristic of genes that are switched on during cellular stress and downregulated by glucocorticoids, whereas COX-1 appeared to be a "housekeeping" gene.

WHICH ISOFORM IS MOST RELEVANT TO INFLAMMATION?

Histological and other studies confirmed this apparent division of labor between the two enzymes, showing, for example, that COX-1 seemed to be the predominant isoform in healthy GI tissue from rat, dog, and monkeys whereas COX-2 was seen mainly in inflamed tissues. This suggested that COX-2 was the predominant inflammatory species, and by implication, the best target for NSAIDs. So was COX-2 inhibition the true therapeutic modality of NSAIDs? If so, COX-1 inhibition might account for the side effects such as gastric irritation and depression of platelet aggregation. This was the notion put forward by more than one group and which came to be known as the "COX-2-bad:COX-1-good" hypothesis. If true, then the obvious corollary was that a selective COX-2 inhibitor should be an ideal drug, possessing anti-inflammatory actions but lacking gastric and other side effects.

At the time, the sequence of the catalytic domains of the two isozymes appeared to be so similar that the prospect of finding a specific inhibitor initially seemed remote and most of the nonsteroidal drugs available at that time were found to inhibit both enzymes to a greater or lesser degree. However, some selectivity of action was seen with experimental drugs such as 6-MNA, BF389, and DuP697.

An experimental compound (DuP697), produced by the Dupont Company in 1990, was observed to be an effective anti-inflammatory agent with reduced GI toxicity. Significantly, it showed only feeble activity in vitro using the COX-1 enzyme, but was more effective against rat brain COX preparations. The authors originally attributed the GI safety of this compound to its chemical structure (a nonacidic thiophene), which was presumed to produce a different pharmacokinetic profile to other COX inhibitors. However, other experimental compounds with similar properties were subsequently discovered and in the light of the discovery of COX-2, it was not long before it was realized that these drugs might have exhibit unusual behavior because they acted predominantly on the COX-2 isozyme. The two companies who were the first to exploit these ideas and to market the new "coxibs" were Searle Monsanto (now Pfizer) and Merck. The former focused on sulphonamide-substituted 1,S-diaryl pyrazole compounds that lead eventually to the discovery of celecoxib, whereas Merck scientists settled on a series of methylsulphonylphenyl compounds which, in turn, culminated in the manufacture and introduction of rofecoxib (since withdrawn).

In support of the COX-1/COX-2 concept, the pharmacology of some other selective COX-2 inhibitors were described in the literature. SC558 (a celecoxib prototype), for example, showed good efficacy in rodent models of inflammation, fever, and pain, whereas the structurally related SC560, a selective COX-1 inhibitor, was ineffective. Celecoxib itself (then known as SC58635) reversed carrageenan-induced hyperalgesia and local prostaglandin production in rats, and a related compound was active intrathecally.

The meloxicam (Boehringer Ingleheim) and etodolac (Wyeth/Shire) were already under development during the time that progress was being made with COX-2 biology, and nimesulide (Helsinn) had been on the market in Europe since 1985. In each case, evidence already indicated that these agents were different from other NSAIDs in terms of their GI tolerance and these drugs turned out to be effective COX-2 inhibitors. In an authoritative survey, Warner et al. (1999) found, for example, that meloxicam and etodolac showed almost the same order of selectivity for COX-1/COX-2 as some of the newer agents.

BIOCHEMISTRY OF COX ENZYMES AND MECHANISMS OF NSAID INHIBITION

Both COX isoforms are bifunctional enzymes, having two distinct catalytic activities: the first, *dioxygenase* step incorporates two molecules of oxygen into the arachidonic (or other fatty acid substrate) chain at C-11 and C-15 giving rise to the highly unstable endoperoxide intermediate *PGG*$_2$ with a hydroperoxy group at C-15. A second *peroxidase* function of the enzyme converts this to *PGH*$_2$ with a hydroxy group at C-15, which can then transformed in a cell-specific manner by separate *isomerase*, *reductase*, or *synthase* enzymes into other prostanoids. Both COX-1 and COX-2 are heme-containing enzymes which probably exist as homodimers attached to intracellular membranes.

Most NSAIDs inhibit only the initial dioxygenation reaction. They are generally "competitive-reversible" inhibitors but there are differences in their kinetics. Generally, these drugs inhibit COX-1 rapidly but the inhibition of COX-2 is time-dependent and often irreversible, as explained earlier. To block the enzymes, NSAIDs must enter the hydrophobic channel forming hydrogen bonds with an *Arg* residue at position 120, thus preventing substrate fatty acids from entering into the catalytic domain. The solution of the crystal structures of COX-1 in 1994 and COX-2 in 1996 made a substantial impact on the drug discovery effort. Despite their high homology, detailed examination of the structure of the catalytic sites revealed the substrate binding "channel" in the two enzymes to be quite different. A single amino-acid change, from the comparatively bulky *Ile* in COX-1 to *Val* at position 523 in COX-2 (equivalent to position 509 in COX-1), and the conformational changes that this produced resulted in enhanced access to a "side pocket" that allowed the binding of COX-2-selective inhibitors by providing a docking site for the phenylsulphonamide residue of drugs such as SC558. In an elegant demonstration of how crucial this single change was, Gierse et al. (1996) showed that mutation of this residue in COX-2 back to *Ile* largely prevented selective COX-2 inhibitors from working (see Figure 18.2).

These structural data also helped explain differences in the inhibitory kinetics of COX-1 and COX-2. There are, as stated earlier, several distinct mechanisms by which COX-1 inhibitors can inhibit the enzyme, but many are of the competitive-reversible type. By contrast, the COX-2 inhibitors are irreversible, time-dependent inhibitors, partly as a result of the binding of the sulphonamide (or related) moiety into the enzyme "side pocket." An analysis of the kinetic behavior of several COX inhibitors subsequently discerned four separate modes of enzyme inhibition, ranging from the simple competitive inhibition of drugs such as ibuprofen, through the "weak binding, time-dependent" mechanism of naproxen and the oxicams and the "tight binding, time-dependent" inhibition of indomethacin, to the covalent modification produced by aspirin. Alone among the NSAIDs, aspirin irreversibly acetylated a serine residue (*Ser* 530) within the COX active site thus blocking substrate access and inhibiting catalytic activity.

RELATIVE POTENCIES OF NSAIDS ON COX ENZYMES

Most "traditional" NSAIDs are inhibitors of both isoenzymes, though they vary in the degree to which they inhibit each isoform. It is believed that the anti-inflammatory action (and probably most analgesic actions) of the NSAIDs is generally related to inhibition of COX-2 whilst their unwanted effects – particularly those affecting the GI tract – are largely a result of their inhibition of COX-1. A broad scheme for classifying the relative selectivity for COX-1/2 of the currently available NSAIDs is given in Table 18.1.

ASPIRIN IN NONINFLAMMATORY CONDITIONS

Aspirin – previously thought of as an old anti-inflammatory workhorse – is now approaching the status of wonder drug that is not only of benefit in inflammation, but in an increasing number of other conditions. Most significant is its role in the therapy of *cardiovascular disorders*. The long-lasting antiplatelet action of low-dose aspirin may be use prophylactically to great clinical benefit in patients at high risk from (for example) myocardial infarction. Other diseases in which the drug is being (or has been) trialed include *colonic and rectal cancer, Alzheimer's disease*, and *radiation-induced diarrhea*.

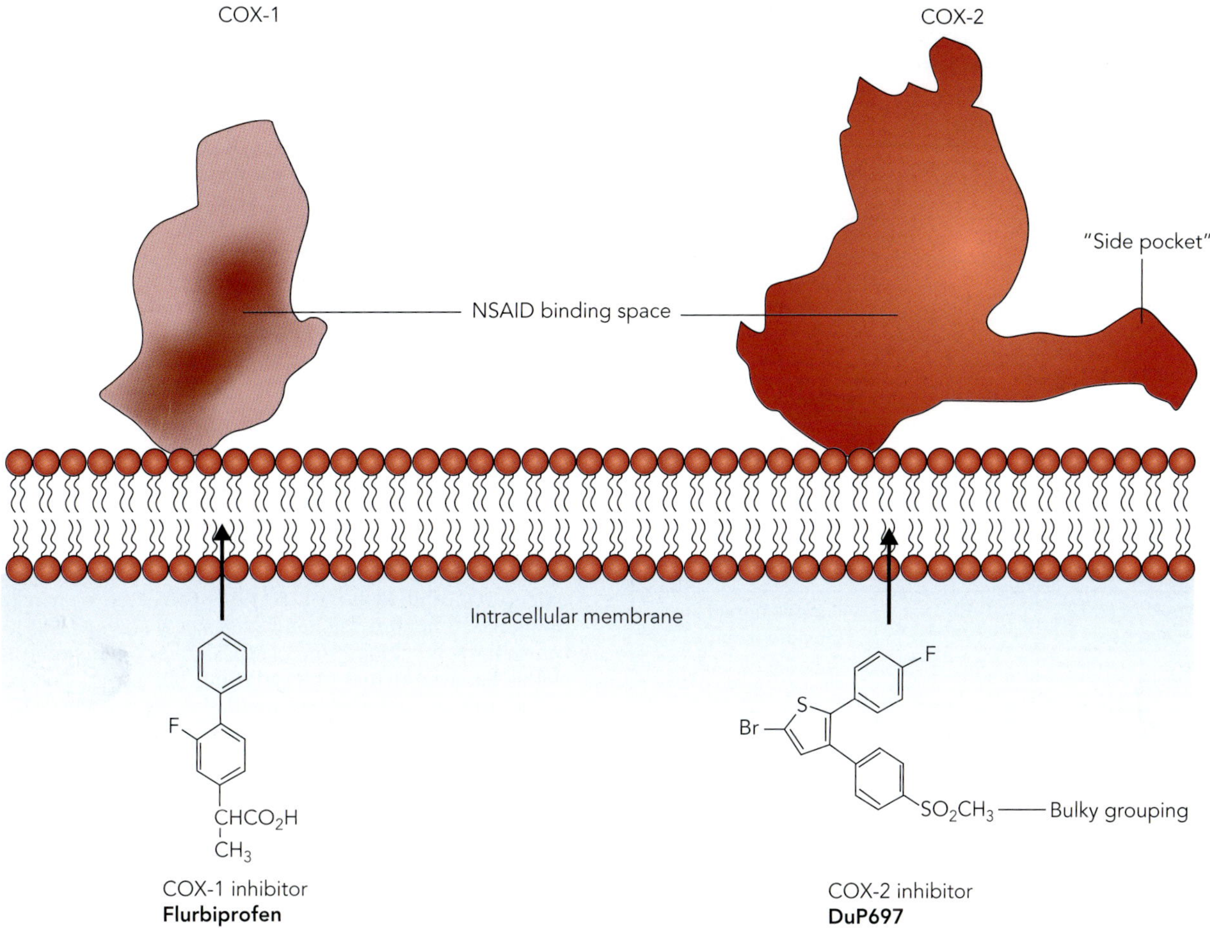

Figure 18.2. Comparison of drug binding sites in COX-1 and COX-2. This schematic diagram (modified from Flower 2003) shows how the presence of a "side pocket" allows access to the hydrophobic channel in the COX-2 enzyme, of inhibitory compounds (exemplified here by the drug DuP697), that cannot gain access to the channel in the COX-1 isoform.

PARACETAMOL

Whilst it is an excellent analgesic and antipyretic, the anti-inflammatory activity of paracetamol is very low and seems to be restricted to a few special cases (e.g., inflammation following dental extraction). As already mentioned paracetamol potently reduced prostaglandin synthesis in the central nervous systems (CNS), but not in the periphery. In experimental animals, analgesia and antipyresis with paracetamol were accompanied by the reduction in prostaglandin synthesis in the CNS. As paracetamol is a weak inhibitor of COX-1 and COX-2 activities, its pharmacological actions cannot be explained by inhibition of these enzymes. COX-3 identified as a splice variant of COX-1 in 2002 in canine tissues was shown to be inhibited by paracetamol. It was thus hoped that the discovery of COX-3 might provide a neat explanation for its pharmacological actions. However, the existence of COX-3 in rodent and human tissues is disputed because retention of intron-1 results in an out-of-reading frame sequence, despite some evidence in the literature on its possible expression in these species. A COX-1 variant, which may be related, is involved in thermoregulation in normothermic animals. Whilst being a weak anti-inflammatory drug, paracetamol produced the same effect as NSAIDs at the latter phase of acute inflammation, in that it prevented resolution of the inflammatory reaction (our unpublished observations).

SUMMARY

The NSAID field has undergone major transformation since the mechanism of action of these drugs was discovered, but what of the future? At the time of writing, the unique role of COX-2 selective inhibitors in therapy remains to be fully clarified and the propensity of the entire range of NSAIDs to cause hypertension requires elucidation. It is ironic that the precise mechanism of paracetamol action continues to elude us, as indeed does the mechanism of action of salicylate itself which is the major metabolite of aspirin and which has anti-inflammatory actions in its own right. Clearly there is

TABLE 18.1. Relative inhibitory potency of some common NSAIDs and Coxibs on the two COX isoforms

Description	Selectivity ratio	Examples
Highly COX-1 selective	100–1,000	Ketorolac
Very COX-1 selective	10–100	Flurbiprofen
Weakly COX-1 selective	1–10	Indomethacin, aspirin, naproxen, ibuprofen
Nonselective; full inhibition of both isoforms	1	Fenoprofen
Nonselective; incomplete inhibition of both isoforms	1	Salicylate
Weakly COX-2 selective	1–10	Diflunisal, piroxicam, meclofenamate. Sulindac, diclofenac, celecoxib
Very COX-2 selective	10–100	Valdecoxib*, etoricoxib
Highly COX-2 selective	100–1,000	Rofecoxib*

Source: Warner and Mitchell, 2004.
*These drugs have been withdrawn from clinical use and are shown here for illustration purposes only.

much work still to be done and quite possibly further COX or COX-like enzymes to discover.

REFERENCES

1. Nishikawa, K., Morrison, A., and Needleman, P. 1977. Exaggerated prostaglandin biosynthesis, and its influence on renal resistance in the isolated hydronephrotic rabbit kidney. *J Clin Invest* **59**(6):1143–1150.
2. Xie, W.L., Chipman J.G., Robertson, D.L., Erikson, R.L., and Simmons, D.L. 1991. Expression of a mitogen-responsive gene encoding prostaglandin synthase is regulated by mRNA splicing. *Proc Natl Acad Sci USA* **88**(7):2692–2696.
3. Kujubu, D.A., Fletcher, B.S., Varnum, B.C., Lim, R.W., and Herschman, H.R. 1991. TIS10, a phorbol ester tumor promoter-inducible mRNA from Swiss 3T3 cells, encodes a novel prostaglandin synthase/cyclooxygenase homologue. *J Biol Chem* **266**(20):12866–12872.
4. Masferrer, J.L., Seibert, K., Zweifel, B., and Needleman, P. 1992. Endogenous glucocorticoids regulate an inducible cyclooxygenase enzyme. *Proc Natl Acad Sci USA* **89**(9):3917–3921.

SUGGESTED READINGS

Abe, M., Oka, T., Hori, T., and Takahashi, S. 2001. Prostanoids in the preoptic hypothalamus mediate systemic lipopolysaccharide-induced hyperalgesia in rats. *Brain Res* **916**(1–2):41–49.

Ahmadi, S., Lippross, S., Neuhuber, W.L., and Zeilhofer, H.U. 2002. PGE(2) selectively blocks inhibitory glycinergic neurotransmission onto rat superficial dorsal horn neurons. *Nat Neurosci* **5**(1):34–40.

Ayoub, S.S., Botting, R.M., Goorha, S., Colville-Nash, P.R., Willoughby, D.A., and Ballou, L.R. 2004. Acetaminophen-induced hypothermia in mice is mediated by a prostaglandin endoperoxide synthase 1 gene-derived protein. *Proc Natl Acad Sci USA* **101**(30):11165–11169.

Ayoub, S.S., Colville-Nash, P.R., Willoughby, D.A., and Botting, R.M. 2006. The involvement of a cyclooxygenase 1 gene-derived protein in the antinociceptive action of paracetamol in mice. *Eur J Pharmacol* **538**(1–3):57–65.

Bombardier, C., Laine, L., Reicin, A., et al. 2000. Comparison of upper gastrointestinal toxicity of rofecoxib and naproxen in patients with rheumatoid arthritis. VIGOR Study Group. *N Engl J Med* **343**(21):1520–1528, 1522 p following 1528.

Chandrasekharan, N.V., Dai, H., Roos, K.L., et al. 2002. COX-3, a cyclooxygenase-1 variant inhibited by acetaminophen and other analgesic/antipyretic drugs: cloning, structure, and expression. *Proc Natl Acad Sci USA* **99**(21):13926–13931.

Collier, J.G., and Flower, R.J. 1971. Effect of aspirin on human seminal prostaglandins. *Lancet* **2**(7729):852–853.

Cronstein, B.N., Montesinos, M.C., and Weissmann, G. 1999. Salicylates and sulfasalazine, but not glucocorticoids, inhibit leukocyte accumulation by an adenosine-dependent mechanism that is independent of inhibition of prostaglandin synthesis and p105 of NFkappaB. *Proc Natl Acad Sci USA* **96**(11):6377–6381.

DeWitt, D.L., Meade, E.A., and Smith, W.L. 1993. PGH synthase isoenzyme selectivity: the potential for safer nonsteroidal antiinflammatory drugs. *Am J Med* **95**(2A):40S–44S.

Dinchuk, J.E., Liu, R.Q., and Trzaskos, J.M. 2003. COX-3: in the wrong frame in mind. *Immunol Lett* **86**(1):121.

Feldberg, W., Gupta, K.P., Milton, A.S., and Wendlandt, S. 1972. Effect of bacterial pyrogen and antipyretics on prostaglandin activity in cerebrospinal fluid of unanaesthetized cats. *Br J Pharmacol* **46**(3):550P–551P.

Ferreira, S.H., and Lorenzetti, B.B. 1996. Intrathecal administration of prostaglandin E2 causes sensitization of the primary afferent neuron via the spinal release of glutamate. *Inflamm Res* **45**(10):499–502.

Ferreira, S.H., Moncada, S., and Vane, J.R. 1971. Indomethacin and aspirin abolish prostaglandin release from the spleen. *Nat New Biol* **231**(25):237–239.

FitzGerald, G.A., and Patrono, C. 2001. The coxibs, selective inhibitors of cyclooxygenase-2. *N Engl J Med* **345**(6):433–442.

Flower, R.J. 1974. Drugs which inhibit prostaglandin biosynthesis. *Pharmacol Rev* **26**(1):33–67.

Flower, R.J. 2003. The development of COX2 inhibitors. *Nat Rev Drug Discov* **2**(3):179–191.

Flower, R.J., and Vane, J.R. 1972. Inhibition of prostaglandin synthetase in brain explains the anti-pyretic activity of paracetamol (4-acetamidophenol). *Nature* **240**(5381):410–411.

Flower, R., Gryglewski, R., Herbaczynska-Cedro, K., and Vane, J.R. 1972. Effects of anti-inflammatory drugs on prostaglandin biosynthesis. *Nat New Biol* **238**(82):104–106.

Gans, K.R., Galbraith, W., Roman, R.J., et al. 1990. Anti-inflammatory and safety profile of DuP 697, a novel orally effective prostaglandin synthesis inhibitor. *J Pharmacol Exp Ther* **254**(1):180–187.

Gierse, J.K., Koboldt, C.M., Walker, M.C., Seibert, K., and Isakson, P.C. 1999. Kinetic basis for selective inhibition of cyclo-oxygenases. *Biochem J* **339**(Pt 3):607–614.

Gierse, J.K., McDonald, J.J., Hauser, S.D., Rangwala, S.H., Koboldt, C.M., and Seibert, K. 1996. A single amino acid difference between cyclooxygenase-1 (COX-1) and -2 (COX-2) reverses the selectivity of COX-2 specific inhibitors. *J Biol Chem* **271**(26):15810–15814.

Gilroy, D.W., Colville-Nash, P.R., Willis, D., Chivers, J., Paul-Clark, M.J., and Willoughby, D.A. 1999. Inducible cyclooxygenase may have anti-inflammatory properties. *Nat Med* **5**(6):698–701.

Hingtgen, C.M., Waite, K.J., and Vasko, M.R. 1995. Prostaglandins facilitate peptide release from rat sensory neurons by activating the adenosine 3',5'-cyclic monophosphate transduction cascade. *J Neurosci* **15**(7 Pt 2):5411–5419.

Kurumbail, R.G., Stevens, A.M., Gierse, J.K., et al. 1996. Structural basis for selective inhibition of cyclooxygenase-2 by anti-inflammatory agents. *Nature* **384**(6610):644–648.

Lysz, T.W., and Needleman, P. 1982. Evidence for two distinct forms of fatty acid cyclooxygenase in brain. *J Neurochem* **38**(4):1111–1117.

Malmberg, A.B., and Yaksh, T.L. 1994. Capsaicin-evoked prostaglandin E2 release in spinal cord slices: relative effect of cyclooxygenase inhibitors. *Eur J Pharmacol* **271**(2–3):293–299.

Masferrer, J.L., Zweifel, B.S., Manning, P.T., et al. 1994. Selective inhibition of inducible cyclooxygenase 2 in vivo is anti-inflammatory and nonulcerogenic. *Proc Natl Acad Sci USA* **91**(8):3228–3232.

Masferrer, J.L., Zweifel, B.S., Seibert, K., and Needleman, P. 1990. Selective regulation of cellular cyclooxygenase by dexamethasone and endotoxin in mice. *J Clin Invest* **86**(4):1375–1379.

Meade, E.A., Smith, W.L., and DeWitt, D.L. 1993. Differential inhibition of prostaglandin endoperoxide synthase (cyclooxygenase) isozymes by aspirin and other non-steroidal anti-inflammatory drugs. *J Biol Chem* **268**(9):6610–6614.

Merlie, J.P., Fagan, D., Mudd, J., and Needleman, P. 1988. Isolation and characterization of the complementary DNA for sheep seminal vesicle prostaglandin endoperoxide synthase (cyclooxygenase). *J Biol Chem* **263**(8):3550–3553.

Mitchell, J.A., Akarasereenont, P., Thiemermann, C., Flower, R.J., and Vane, J.R. 1993. Selectivity of nonsteroidal antiinflammatory drugs as inhibitors of constitutive and inducible cyclooxygenase. *Proc Natl Acad Sci USA* **90**(24):11693–11697.

Morrison, A.R., Moritz, H., and Needleman, P. 1978. Mechanism of enhanced renal prostaglandin biosynthesis in ureter obstruction. Role of de novo protein synthesis. *J Biol Chem* **253**(22):8210–8212.

Muth-Selbach, U.S., Tegeder, I., Brune, K., and Geisslinger, G. 1999. Acetaminophen inhibits spinal prostaglandin E2 release after peripheral noxious stimulation. *Anesthesiology* **91**(1):231–239.

Nishihara, I., Minami, T., Uda, R., Ito, S., Hyodo, M., and Hayaishi, O. 1995. Effect of NMDA receptor antagonists on prostaglandin E2-induced hyperalgesia in conscious mice. *Brain Res* **677**(1):138–144.

Ouellet, M., and Percival, M.D. 2001. Mechanism of acetaminophen inhibition of cyclooxygenase isoforms. *Arch Biochem Biophys* **387**(2):273–280.

Qin, N., Zhang, S.P., Reitz, T.L., Mei, J.M., and Flores, C.M. 2005. Cloning, expression, and functional characterization of human cyclooxygenase-1 splicing variants: evidence for intron 1 retention. *J Pharmacol Exp Ther* **315**(3):1298–1305.

Roth, G.J., and Siok, C.J. 1978. Acetylation of the NH2-terminal serine of prostaglandin synthetase by aspirin. *J Biol Chem* **253**(11):3782–3784.

Roth, G.J., Stanford, N., and Majerus, P.W. 1975. Acetylation of prostaglandin synthase by aspirin. *Proc Natl Acad Sci USA* **72**(8):3073–3076.

Schwab, J.M., Beiter, T., Linder, J.U., et al. 2003. COX-3 – a virtual pain target in humans? *FASEB J* **17**(15):2174–2175.

Silverstein, F.E., Faich, G., Goldstein, J.L., et al. 2000. Gastrointestinal toxicity with celecoxib vs. nonsteroidal anti-inflammatory drugs for osteoarthritis and rheumatoid arthritis: the CLASS study: a randomized controlled trial. Celecoxib Long-term Arthritis Safety Study. *JAMA* **284**(10):1247–1255.

Smith, J.B., and Willis, A.L. 1971. Aspirin selectively inhibits prostaglandin production in human platelets. *Nat New Biol* **231**(25):235–237.

Taiwo, Y.O., and Levine, J.D. 1990. Effects of cyclooxygenase products of arachidonic acid metabolism on cutaneous nociceptive threshold in the rat. *Brain Res* **537**(1–2):372–374.

Vane, J.R. 1971. Inhibition of prostaglandin synthesis as a mechanism of action for aspirin-like drugs. *Nat New Biol* **231**(25):232–235.

Warner, T.D., Giuliano, F., Vojnovic, I., Bukasa, A., Mitchell, J.A., and Vane, J.R. 1999. Nonsteroid drug selectivities for cyclo-oxygenase-1 rather than cyclo-oxygenase-2 are associated with human gastrointestinal toxicity: a full in vitro analysis. *Proc Natl Acad Sci USA* **96**(13):7563–7568.

Warner, T.D. and Mitchell, J.A. 2004. Cyclooxygenases: new forms, new inhibitors, and lessons from the clinic. *FASEB J* **18**:790–804.

19 Cytokines and Chemokines in Inflammation and Cancer

Thorsten Hagemann and Toby Lawrence

Cytokines and chemokines are peptide mediators that regulate a broad range of processes involved in the pathogenesis of inflammatory diseases and cancer. It is well established that an imbalance cytokine or chemokine activities can favor chronic inflammation leading to organ failure. Chemokines and cytokines are also implicated in malignant disease with links to tumor progression, angiogenesis, and invasion. Biological therapies targeting cytokines and chemokines have already improved outcomes of inflammatory disease and clinical trials are ongoing in cancer patients. Targeting tumor necrosis factor (TNF)-α represents a major success story for this approach. Anti-TNF-α was the first antibody against an inflammatory cytokine demonstrated to be efficacious in human disease and it showed over the years to be effective in a range of inflammatory diseases such as rheumatoid arthritis (RA), inflammatory bowel disease (IBD), and more recently, cancer.

CANCER AND INFLAMMATION

The cells and mediators of inflammation also form a major part of the tumor microenvironment. In some cancers, inflammatory conditions precede development of malignancy; in others, oncogenic changes drive a tumor-promoting inflammatory milieu. Whatever its origin, this "smoldering" inflammation aids proliferation and survival of malignant cells, angiogenesis, and metastasis; subverts adaptive immunity, and alters response to hormones and chemotherapeutic agents [1,2]. The cytokine (Figure 19.1) and chemokine network (Figure 19.2) is of great importance in the processes of cancer-related inflammation regulating both host and malignant cells in the tumor microenvironment [3]. The link between inflammation and cancer was first suggested by Rudolph Virchow in 1863, when he demonstrated leukocytes in neoplastic tissue. Virchow's original hypothesis has been revisited by many research groups [1]. The epidemiological data available are very impressive and show a clear association between chronic inflammatory conditions and subsequent malignant transformation in the inflamed tissue (Table 19.1).

CYTOKINES IN INFLAMMATION AND CANCER

It is known that neoplastic cells often overexpress proinflammatory mediators including proteases, eicosanoids, cytokines, and chemokines. Several cytokines, such as TNF-α, IL-6, IL-17, IL-12, IL-23, IL-10, and TGF-β, have been linked with both experimental and human cancers and can either promote or inhibit tumor development. Here we will describe some of the cytokines that may be potential targets for cancer therapy.

TNF-α is the prototypical proinflammatory cytokine, the critical role of TNF-α in chronic inflammatory diseases is well known [4], although originally shown to be toxic to tumor cells in high doses (hence tumor *necrosis* factor) the tumor-promoting function of TNF-α have been clearly demonstrated in mice [5]. TNF-α can be produced by malignant cells or inflammatory cells in the tumor microenvironment and TNF-α signaling can promote cell survival, invasion, and angiogenesis [6]. In mesothelioma phagocytosis of asbestos fibers by macrophages leads to TNF-α release that promotes cell survival and thereby reduces asbestos-induced cytotoxicity [7]. TNF-α has also been proposed to contribute to tumor initiation by stimulating the production of genotoxic reactive nitrogen (RNS) and oxygen (ROS) species [8]. Genetic polymorphisms that enhance TNF-α production are associated with increased risk of multiple myeloma (MM), bladder cancer, hepatocellular carcinoma (HCC), gastric cancer, and breast cancer, as well as poor prognosis in various hematological malignancies [9]. Other actions of TNF-α that might promote tumorigenesis, as opposed

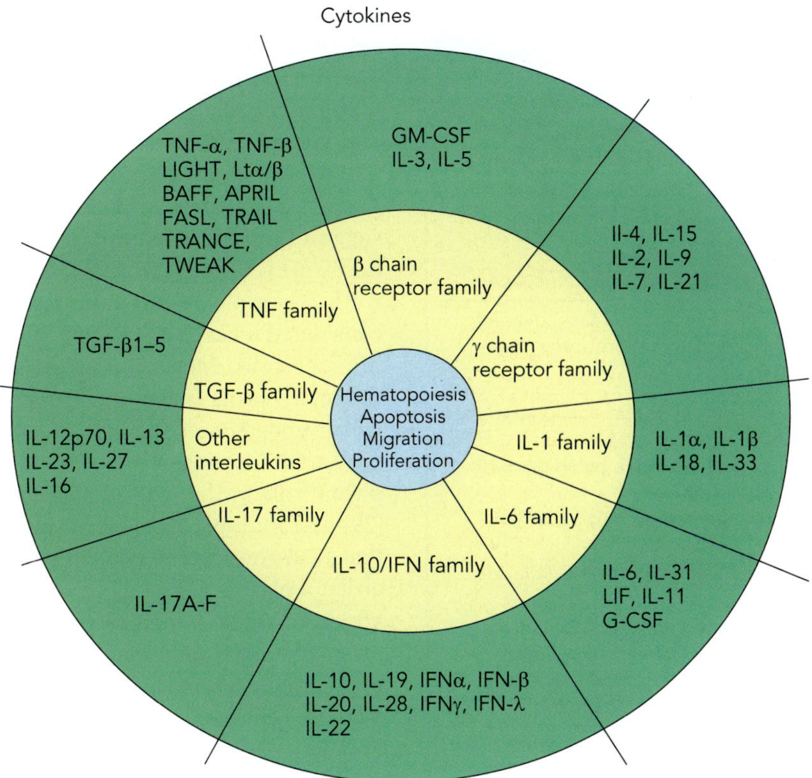

Figure 19.1. The cytokine network.

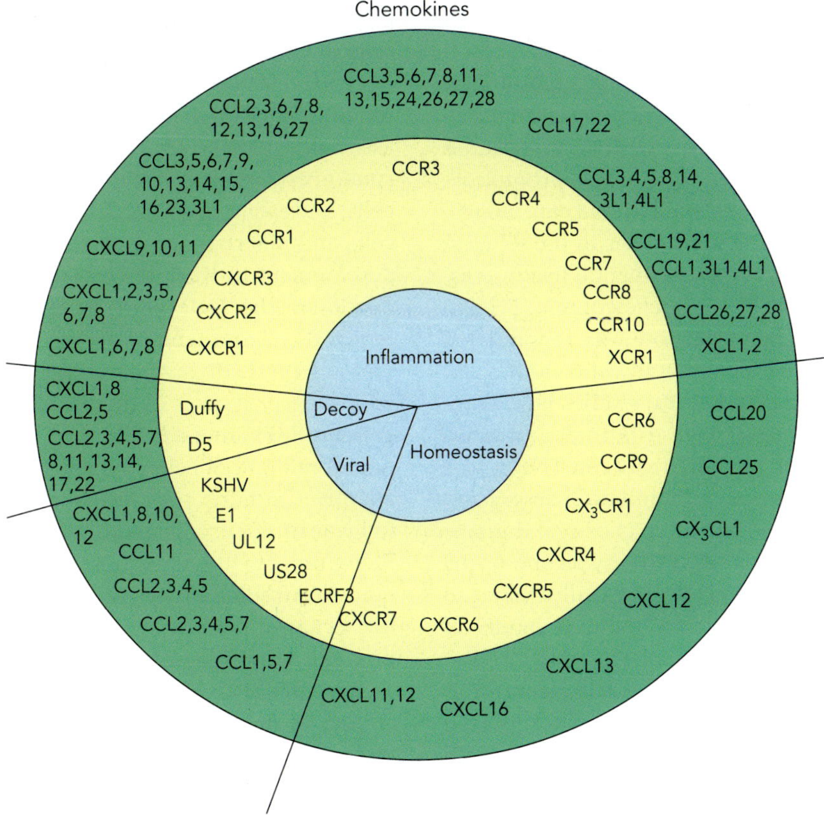

Figure 19.2 The chemokine network.

TABLE 19.1. The association between inflammation and cancer in different organ systems

System	Inflammation
Respiratory	Asbestos and mesothelioma silica, Cigarette smoke and bronchial cancer Chronic asthma and bronchial cancer
Genitourinary	Human papilloma virus and penile carcinoma Schistosomiasis and bladder cancer
Reproductive	Pelvic inflammatory disease, ovarian epithelial inflammation, and ovarian cancer
Gastrointestinal	Epstein–Barr virus and nasopharyngeal cancer Barrett's metaplasia and esophageal cancer *H. pylori* gastritis and gastric cancer Chronic pancreatitis and pancreatic cancer Chronic cholecystitis and gallbladder cancer Hepatitis and hepatocellular cancer Inflammatory bowel disease, and colorectal cancer Human papilloma virus and ano-genital cancer
Hematopoietic	Epstein–Barr virus and Burkitt's lymphoma HTLV1 and T-cell leukemia/lymphoma

to tumor initiation, include promotion of angiogenesis and impairment of immune surveillance through T-cell suppression and inhibiting the cytotoxic activity of activated macrophages [10].

TNF-α binds to two receptors, the ubiquitously expressed TNFR1, and TNFR2, which is restricted to expression on hematopoietic cells. Previous studies show that TNFR1 is important in tumor promotion. TNFR1 knockout mice are resistant to skin carcinogenesis [11], experimental lung metastases in the RENCA renal cancer model [12], and experimental liver metastases were attenuated in TNFR1-deficient mice [13]. In chimeric mice whose bone marrow was repopulated with TNFR1–/– cells, development of colitis and colon cancer was attenuated [14]. Several studies have demonstrated that stromal cell TNF-α is tumor promoting. In a genetic model, bone marrow cell TNF-α was implicated in promotion of inflammation-associated liver tumors [15]. In a model where chemical damage led to liver cancer, Kupffer cell TNF-α was one of the mitogens driving proliferation of genetically damaged hepatocytes [16]. In a chemically induced model of colitis and colorectal cancer, mononuclear cell TNF-α was implicated in inflammation and subsequent tumor development [14]. Several studies have also reported therapeutic activity of anti-TNF-α antibodies, or a TNF receptor fusion molecule, in genetic models of liver and colorectal cancer [15,17]; a carcinogen-induced model of colorectal cancer [14]; and pancreatic cancer xenografts [18], although the exact mechanisms of action are not understood.

In many different experimental and human cancers, malignant cells produce TNF-α during tumor growth and spread [5,6,18–24]. Preclinical experiments with TNF-α antagonists and early phase clinical trials of TNF-α antagonists in patients with advanced cancer suggest that this inflammatory cytokine may be a useful target [25–28].

IL-6 is a pleiotropic inflammatory cytokine that is considered a key growth factor for both malignant and immune cells. Most IL-6 target genes are involved in cell cycle progression and suppression of apoptosis, which underscores the importance of IL-6 in cell survival and tumorigenesis. IL-6 is suggested to have a pivotal role in the pathogenesis of Castleman's disease [29] and multiple myeloma (MM) [30]. Clear evidence that IL-6 governs the growth of MM, a malignant disorder of plasma cells, has come from studies using IL-6 knockout mice, which were found to be resistant to plasmacytoma induction [30]. In MM, IL-6 is produced by stromal cells in the bone marrow and its synthesis by these cells can be further enhanced by their interaction with malignant plasma cells. New IL-6 antagonists are being evaluated for treatment of MM such as small molecules and monoclonal antibodies. IL-6 is also a key mediator of inflammatory disease, IBD and colitis is associated with high concentrations of IL-6 [31], in experimental models of colitis-associated colon cancer (CAC) and HCC IL-6 production by myeloid cells is also critical for carcinogenesis [32–34].

IL-10 is an immunosuppressive and anti-inflammatory cytokine also linked with inflammation-associated cancer [35]. IL-10-deficient mice develop spontaneous colitis, due to hyperactivation of immune cells, and eventually CAC. Expression of IL-10 by tumor cells and macrophages is also thought to promote the development of Burkitt's lymphoma through the production of the TNF family member BAFF, which promotes B cell and lymphoma survival [36]. An elevated amount of IL-10 in the plasma has been correlated with poor prognosis in diffuse large B cell lymphoma patients [37]. In addition to direct growth modulation of malignant cells, the ability of IL-10 to suppress adaptive immune responses has also been suggested to favor tumor escape from immune surveillance [38]. Therefore, IL-10 has complex effects on tumor development. In some experimental systems, IL-10 is found to exert antitumor activity, but in other cases it can be promote tumorigenesis.

Recently, a new T cell subset named "Th17" characterized by the production of IL-17 was identified as an important player in inflammatory responses [39]. IL-17 induces the recruitment of immune cells and leads to the induction of many proinflammatory factors, including TNF-α, IL-6, and IL-1β [40]. Studies have begun to address the role of IL-17 in chronic inflammation and cancer [41]. IL-17 overexpressing human cervical cancer cells and nonsmall cell lung carcinoma cells show substantially greater ability to form tumors in immunocompromised mice compared with control cells not overexpressing IL-17 [42,43]. In primary NSCLC samples, IL-17 expression has frequently been detected in tumor-infiltrating inflammatory cells and was associated with increased tumor vascularity [44]. However, there is also evidence that IL-17 might be involved in tumor surveillance in immunocompetent mice [45].

This presents a caveat for cytokines as targets in cancer therapy, although inflammation is clearly linked with tumorigenesis, both the innate and adaptive immune systems have the capacity to recognize and eliminate malignant cells [46,47]. Many proinflammatory cytokines may function in tumor immune surveillance and it is vital to determine whether the function of potential therapeutic targets, such as IL-17, contributes to antitumor immunity.

CHEMOKINES IN INFLAMMATION AND CANCER

Chemokines are chemotactic cytokines that regulate cell trafficking in development and disease. In inflammation chemokines regulate leukocyte recruitment, diapedesis, activation, and retention within the inflamed tissue. Other chemokines, such as CCL2, CCL5, CCL21, and CXCL13, control lymphoid organogenesis. Chemokines also regulate the retention and release of stem cells from the bone marrow. In all these activities, chemokines function together with cytokines that regulate their expression and that of their reciprocal receptors. Chemokine expression and their functional importance in inflammation have provoked several therapeutic targeting studies, although with limited success so far [46].

Chemokine receptors and their ligands influence cell motility, invasion, and survival in experimental models of cancer [3]. Malignant cells and tumors express a large number of chemokines and chemokine receptors; accordingly, tumor cells are able to recruit leukocytes and endothelial or mesenchymal cell progenitors from the circulation into the tumor microenvironment. Many chemokines are produced downstream of oncogenic mutations; Ras mutations induce CXCL1 (Gro-1) [48], c-myc activation induces CC chemokines [49], and the RET/PTC gene fusion induces several chemokines including CXCL12 [50].

These chemokines control the extent and phenotype of the leukocyte infiltrate in cancer [51–53]. Ablation of leukocyte function, their infiltration into tumors [54], or switch of their phenotype [55] can inhibit tumor growth and metastases.

Until recently, the evidence the role of chemokines in carcinogenesis was largely inferred from correlative studies and experiments with antagonistic antibodies. Genetic evidence for a tumor-promoting role of chemokines eventually came from the discovery of the D6 decoy receptor that functions to block chemokine function [56]. D6 null mice are inefficient in removing chemokines from inflamed sites and therefore show exaggerated responses to otherwise innocuous inflammatory agents [56]. These mice were also recently shown to have increased susceptibility to skin tumors [57].

There are at least 40 different human chemokines and 20 chemokine receptors [58]. They are divided into CC, CXC, XC, and CX3C subfamilies according to the nature of the conserved cysteine motif [59]. Chemokines bind to seven transmembrane-spanning (7TM) receptors typically coupled to G proteins for signaling [60]. There are 11 CC chemokine receptors, seven CXC chemokine receptors, and single receptors for XC and CX3C chemokines [59,61]. These can be broadly subdivided into constitutive and inflammatory [59,61]. There are two nonsignaling receptors, Duffy and D6, thought to attenuate the actions of inflammatory chemokines [62], and viral chemokine and chemokine receptor homologues [63] (Table 19.2).

Positive preclinical data in animal cancer models show that chemokine/receptor antagonists reduce the leukocyte infiltrate into tumors, leading to reduced angiogenesis and cancer growth, and that targeting chemokine receptors on malignant cells inhibits survival, metastasis, and peritoneal dissemination (Table 19.3).

Therapeutic antibodies are likely to provide the most successful way to inhibit the chemokine/chemokine receptor axis. Anticytokine antibodies have had clinical success in inflammatory diseases [70–73]

TABLE 19.2. Drugs targeting chemokines or their receptors currently in clinical development or recently licensed

Target	Disease	Stage
CCR5	CCR5-trophic HIV	Licensed
CXCR4	Stem cell mobilization in cancer	Licensed
CXCR4	Metastatic cancer	Phase I/II
CCR9	Crohn's disease	Phase II
CCR4*	Adult T cell leukemia, severe asthma	Phase I
CXCR2	COPD	Phase II
CCL2*	Advanced cancer and COPD	Phase I

*Therapeutic antibody.
Abbreviation: COPD, chronic obstructive pulmonary disease.

TABLE 19.3. Targeting chemokines and their receptors in mouse cancer models

Target	Cancer model	Action	Reference
CCR5	Breast cancer	Inhibited leukocyte infiltrate and tumor growth	64
CCL2*	Prostate cancer	Inhibited tumor growth and metastasis	65
CC chemokines	Skin cancer	Inhibited leukocyte infiltrate and tumor development	57
CXCR4	Lymphoma	Inhibited tumor growth	66
CXCR4	Glioma	Inhibited tumor growth	67
CXCL8	Ovarian cancer	Inhibition of tumor growth	68
CXCR7	Breast and lung cancer	Inhibition of angiogenesis and tumor growth	69

*Therapeutic antibody.

TABLE 19.4. Antibody therapeutics in the clinic

Target	Disease	Stage
TNF-α	Rheumatoid arthritis, psoriatic arthritis, ankylosing spondylitis, Crohn's disease	Marketed
B Lys	Systemic lupus erythematosus, rheumatoid arthritis	Phase II, III trial
IL-12	Crohn's disease, psoriasis	Phase II trial
TRAIL-R1	Variety of tumors	Phase II trial
TRAIL-R2	Advanced solid tumors	Phase II trial
IL-13	Severe asthma	Phase I trial
TGF-β	Idiopathic pulmonary fibrosis	Phase I trial
GDF-8	Muscle wasting disease	Phase I trial
Eotaxin*	Allergic conjunctivitis	Phase I trial

* Eoxtaxin is a chemokine (CCL11).

and there are promising preclinical [15,18,74,75] and early-phase clinical trial results in cancer [26–28]. Antibodies against chemokine receptors are likely to have significant advantage because of the redundancy of chemokine ligand binding to some receptors.

RISKS OF TARGETED THERAPEUTICS

The functions of cytokines and chemokines in normal physiology have an important role in the healthy host. Foremost, TNF, IL-1, and IL-6 have critical roles in host defense in both the innate and adaptive immune systems. Therefore, one should expect an increase in the rates of infections, which is sometimes the case [76,77]. For example, TNF plays a pivotal role in granuloma formation and, therefore, in the defense against intracellular pathogens. Indeed, reactivation of tuberculosis has been observed with TNF inhibitors. In contrast, IL-6 may have an inhibiting role on granuloma formation [78]; therefore, targeting IL-6 therapeutically might not induce reactivation of, or promote infections with, some intracellular pathogens. The side effects of these targeted therapeutics on host defense may be particularly important in cancer patients who may already have seriously compromised immune systems.

Other side effects have been reported with IL-6 inhibition, such as increases in plasma lipid levels, but it remains to be seen whether cardiovascular risk increases. IL-6 is also a growth factor for hepatocytes, and tocilizumab (IL-6R neutralizing antibody) leads to increases in hepatic enzymes, which are usually of transient nature and not associated with hepatitis [77]. However, long-term follow up will have to show whether this might be associated with liver damage. Given that TNF inhibition also leads to normalization of IL-6 [79] but not liver enzyme elevations, it will also be of interest to learn whether a monoclonal anti-IL-6, rather than an antibody to the IL-6R, will induce a similar adverse event profile. Although antibodies offer exquisite specificity and high avidity, often these biological agents are chimeric or humanized monoclonal antibodies, therefore sensitization might occur, which can lead to both allergic reactions and reduction in efficacy. However, a number of candidate therapeutic antibodies (Table 19.4) have been identified and successfully tested in the clinic.

FUTURE DIRECTIONS

There are many additional molecules, both cytokines and chemokines, that may be future targets for effective therapy in inflammation and cancer we have merely highlighted a few. Targeting of pro-inflammatory cytokines such as TNF-α or IL-6 is highly efficacious in inflammatory disease such as RA and, at least for anti-TNF agents as well as other chronic inflammatory disorders. There are also very positive early signs for anti-TNF therapy in cancer. It is clear the full potential of these agents in cancer therapy may only be realized in combination with tolerable regimes of existing chemotherapy. The challenge now is to identify a rationale for which combinations will be most effective in which patients. Further mechanistic studies on the role of cytokines and chemokines in carcinogenesis will undoubtedly reveal new potential targets with increased efficacy and reduced side effects.

Cytokines are structurally diverse molecules with diverse and cell type–specific activities. Cytokine functions are mediated by binding specific receptors and their activities include regulating cell activation, hematopoiesis, apoptosis, migration, and proliferation. By virtue of their broad spectrum of activity in the hematopoietic system, they are implicated in virtually all aspects of innate and adaptive immunity. White blood cells are the primary source of cytokines, although they may also be produced by other cell types in the context of disease. This figure illustrates the scope of the cytokine network grouped into families of related molecules (Figure 19.1).

Chemokines are a large superfamily chemotactic cytokines that function to regulate leukocyte trafficking and activation. Chemokines are divided into subfamilies based on conserved amino acid (CC, CXC, CX_3C) sequence motifs. Most family members have at least four conserved cysteine residues that form two intramolecular disulfide bonds. The subfamilies are defined by the position of the first two cysteine residues. The inflammatory chemokines regulate leukocyte recruitment during inflammation, allergic responses, infection, and autoimmune diseases. There are also constitutive chemokines that regulate cell migration in homeostasis, hematopoiesis, and angiogenesis. Both inflammatory and homeostatic chemokines have been implicated in tumor growth and metastasis (adapted from F Balkwill, *Nat Rev Cancer* 2004) (Figure 19.2).

REFERENCES

1. Balkwill, F. 2005. Immunology for the next generation. *Nat Rev Immunol* **5**:509–512.
2. Bonecchi, R., Borroni, E.M., Anselmo, A., et al. 2008. Regulation of D6 chemokine scavenging activity by ligand- and Rab11-dependent surface up-regulation. *Blood* **112**:493–503.
3. Balkwill, F., and Coussens, L.M. 2004. Cancer: an inflammatory link. *Nature* **431**:405–406.
4. Feldmann, M., and Maini, S.R. 2008. Role of cytokines in rheumatoid arthritis: an education in pathophysiology and therapeutics. *Immunol Rev* **223**:7–19.
5. Moore, R.J., Owens, D.M., Stamp, G., et al. 1999. Mice deficient in tumor necrosis factor-alpha are resistant to skin carcinogenesis. *Nat Med* **5**:828–831.
6. Kulbe, H., Thompson, R., Wilson, J.L., et al. 2007. The inflammatory cytokine tumor necrosis factor-alpha generates an autocrine tumor-promoting network in epithelial ovarian cancer cells. *Cancer Res* **67**:585–592.
7. Yang, H., Bocchetta, M., Kroczynska, B., et al. 2006. TNF-alpha inhibits asbestos-induced cytotoxicity via a NF-kappaB-dependent pathway, a possible mechanism for asbestos-induced oncogenesis. *Proc Natl Acad Sci USA* **103**:10397–10402.
8. Hussain, S.P., Hofseth, L.J., and Harris, C.C. 2003. Radical causes of cancer. *Nat Rev Cancer* **3**:276–285.
9. Mocellin, S., Rossi, C.R., Pilati, P., and Nitti, D. 2005. Tumor necrosis factor, cancer and anticancer therapy. *Cytokine Growth Factor Rev* **16**:35–53.
10. Elgert, K.D., Alleva, D.G., and Mullins, D.W. 1998. Tumor-induced immune dysfunction: the macrophage connection. *J Leukoc Biol* **64**:275–290.
11. Arnott, C.H., Scott, K.A., Moore, R.J., Robinson, S.C., Thompson, R.G., and Balkwill, F.R. 2004. Expression of both TNF-alpha receptor subtypes is essential for optimal skin tumour development. *Oncogene* **23**:1902–1910.
12. Tomita, Y., Y.X., Ishida, Y., et al. 2004. Spontaneous regression of lung metastasis in the absence of tumour necrosis factor p55. *Int J Cancer* **112**:927–933.
13. Kitakata, H., Nemoto-Sasaki, Y., Takahashi, Y., Kondo, T., Mai, M., and Mukaida, N. 2002. Essential roles of tumor necrosis factor receptor p55 in liver metastasis of intrasplenic administration of colon 26 cells. *Cancer Res* **62**:6682–6687.
14. Popivanova, B.K., Kitamura, K., Wu, Y., et al. 2008. Blocking TNF-alpha in mice reduces colorectal carcinogenesis associated with chronic colitis. *J Clin Invest* **118**:560–570.
15. Pikarsky, E., Porat, R.M., Stein, I., et al. 2004. NF-kappaB functions as a tumour promoter in inflammation-associated cancer. *Nature* **431**:461–466.
16. Maeda, S., Kamata, H., Luo, J.L., Leffert, H., and Karin, M. 2005. IKKbeta couples hepatocyte death to cytokine-driven compensatory proliferation that promotes chemical hepatocarcinogenesis. *Cell* **121**:977–990.
17. Rao, V.P., Poutahidis, T., Ge, Z., et al. 2006. Proinflammatory CD4+CD45RBhi lymphocytes promote mammary and intestinal carcinogenesis in *ApcMin/+* mice. *Cancer Res* **66**:57–61.
18. Egberts, J.-H., Cloosters, V., Noack, A., et al. 2008. Anti-tumor necrosis factor therapy inhibits pancreatic tumor growth and metastasis. *Cancer Res* **68**:1443–1450.
19. Naylor, M.S., Stamp, G.W.H., Foulkes, W.D., Eccles, D., and Balkwill, F.R. 1993. Tumor necrosis factor and its receptors in human ovarian cancer. *J Clin Invest* **91**:2194–2206.
20. Harrison, M.L., Obermueller, E., Maisey, N.R., et al. 2007. Tumour necrosis factor (TNF-a) as a new target for renal cell carcinoma: two sequential phase II trials of infliximab at standard and high dose. *J Clin Oncol* **25**(29):4542–4549.
21. Galban, S., Fan, J., Martindale, J.L., et al. 2003. von Hippel-Lindau protein-mediated repression of tumor necrosis factor alpha translation revealed through use of cDNA arrays. *Mol Cell Biol* **23**:2316–2328.
22. Stathopoulos, G.T., Kollintza, A., Moschos, C., et al. 2007. Tumor necrosis factor-a promotes malignant pleural effusion. *Cancer Res* **67**:9825–9834.
23. Petersen, S.L., Wang, L., Yalcin-Chin, A., et al. 2007. Autocrine TNFa signaling renders human cancer cells

susceptible to Smac-mimetic-induced apoptosis. *Cancer Cell* **12**:445–456.

24. Zins, K., Abraham, D., Sioud, M., and Aharinejad, S. 2007. Colon cancer cell-derived tumor necrosis factor-a mediates the tumor growth-promoting response in macrophages by up-regulating the colony-stimulating factor-1 pathway. *Cancer Res* **67**:1038–1045.
25. Madhusudan, S., Foster, M., Muthuramalingam, S.R., et al. 2004. A phase II study of Etanercept (Enbrel), a tumour necrosis factor-a inhibitor in patients with metastatic breast cancer. *Clinical Cancer Res* **10**:6528–6534.
26. Madhusudan, S., Muthuramalingam, S.R., Braybrooke, J.P., et al. 2005. A phase II study of Ethanercept (ENBREL) a tumour necrosis factor-a inhibitor in recurrent ovarian cancer. *J Clin Oncol* **10**:6528–6534.
27. Harrison, M.L., Obermueller, E., Maisey, N.R., et al. 2007. Tumor necrosis factor alpha as a new target for renal cell carcinoma: two sequential phase II trials of infliximab at standard and high dose. *J Clin Oncol* **25**:4542–4549.
28. Brown, E.R., Charles, K.A., Hoare, S.A., et al. 2008. A clinical study assessing the tolerability and biological effects of infliximab, a TNF-alpha inhibitor, in patients with advanced cancer. *Ann Oncol* **19**(7):1340–1346.
29. Screpanti, I., Musiani, P., Bellavia, D., et al. 1996. Inactivation of the IL-6 gene prevents development of multicentric Castleman's disease in C/EBP beta-deficient mice. *J Exp Med* **184**:1561–1566.
30. Bommert, K., Bargou, R.C., and Stuhmer, T. 2006. Signalling and survival pathways in multiple myeloma. *Eur J Cancer* **42**:1574–1580.
31. Mudter, J., Amoussina, L., Schenk, M., et al. 2008. The transcription factor IFN regulatory factor-4 controls experimental colitis in mice via T cell-derived IL-6. *J Clin Invest* **118**:2415–2426.
32. Greten, F.R., Eckmann, L., Greten, T.E., et al. 2004. IKKbeta links inflammation and tumorigenesis in a mouse model of colitis-associated cancer. *Cell* **118**:285–296.
33. Weigmann, B., Lehr, H.A., Yancopoulos, G., et al. 2008. The transcription factor NFATc2 controls IL-6-dependent T cell activation in experimental colitis. *J Exp Med* **205**:2099–2110.
34. Naugler, W.E., Sakurai, T., Kim, S., et al. 2007. Gender disparity in liver cancer due to sex differences in MyD88-dependent IL-6 production. *Science* **317**:121–124.
35. Mantovani, A., Allavena, P., Sica, A., and Balkwill, F. 2008. Cancer-related inflammation. *Nature* **454**:436–444.
36. Ogden, C.A., Pound, J.D., Batth, B.K., et al. 2005. Enhanced apoptotic cell clearance capacity and B cell survival factor production by IL-10-activated macrophages: implications for Burkitt's lymphoma. *J Immunol* **174**:3015–3023.
37. Lech-Maranda, E., Bienvenu, J., Michallet, A-S., et al. 2006. Elevated IL-10 plasma levels correlate with poor prognosis in diffuse large B-cell lymphoma. *Eur Cytokine Netw* **17**:60–66.
38. Mocellin, S., Marincola, F.M., and Young, H.A. 2005. Interleukin-10 and the immune response against cancer: a counterpoint. *J Leukoc Biol* **78**:1043–1051.
39. Mangan, P.R., Harrington, L.E., O'Quinn, D.B., et al. 2006. Transforming growth factor-beta induces development of the T(H)17 lineage. *Nature* **441**:231–234.
40. Park, H., Li, Z., O'Yang, X., et al. 2005. A distinct lineage of CD4 T cells regulates tissue inflammation by producing interleukin 17. *Nat Immunol* **6**:1133–1141.
41. Moseley, T.A., Haudenschild, D.R., Rose, L., and Reddi, A.H. 2003. Interleukin-17 family and IL-17 receptors. *Cytokine Growth Factor Rev* **14**:155–174.
42. Numasaki, M., Watanabe, M., Suzuki, T., et al. 2005. IL-17 enhances the net angiogenic activity and in vivo growth of human non-small cell lung cancer in SCID mice through promoting CXCR-2-dependent angiogenesis. *J Immunol* **175**:6177–6189.
43. Tartour, E., Fossiez, F., Joyeux, I., et al. 1999. Interleukin 17, a T-cell-derived cytokine, promotes tumorigenicity of human cervical tumors in nude mice. *Cancer Res* **59**:3698–3704.
44. Numasaki, M., Fukushi, J., Ono, M., et al. 2003. Interleukin-17 promotes angiogenesis and tumor growth. *Blood* **101**:2620–2627.
45. Benchetrit, F., Ciree, A., Vives, V., et al. 2002. Interleukin-17 inhibits tumor cell growth by means of a T-cell-dependent mechanism. *Blood* **99**:2114–2121.
46. Dranoff, G. 2004. Cytokines in cancer pathogenesis and cancer therapy. *Nat Rev Cancer* **4**:11–22.
47. Lawrence, T., Hageman, T., and Balkwill, F. 2007. Cancer. Sex, cytokines, and cancer. *Science* **317**:51–52.
48. Yang, G., Rosen, D.G., Zhang, Z., et al. 2006. The chemokine growth-regulated oncogene 1 (Gro-1) links RAS signaling to the senescence of stromal fibroblasts and ovarian tumorigenesis. *Proc Natl Acad Sci USA* **103**:16472–16477.
49. Soucek, L., Lawlor, E.R., Soto, D., Shchors, K., Swigart, L.B., and Evan, G.I.. 2007. Mast cells are required for angiogenesis and macroscopic expansion of Myc-induced pancreatic islet tumors. *Nat Med* **13**:1211–1218.
50. Borrello, M.G., Alberti, L., Fischer, A., et al. 2005. Induction of a proinflammatory program in normal human thyrocytes by the RET/PTC1 oncogene. *Proc Natl Acad Sci USA* **102**:14825–14830.
51. Mantovani, A., Sozzani, S., Locati, M., Allavena, P., and Sica, A. 2002. Macrophage polarization: tumor-associated macrophages as a paradigm for polarized M2 mononuclear phagocytes. *Trends Immunol* **23**:549–555.
52. Bingle, L., Brown, N.J., and Lewis, C.E. 2002. The role of tumour-associated macrophages in tumour progression: implications for new anticancer therapies. *J Pathol* **196**:254–265.
53. Condeelis, J., and Pollard, J.W. 2006. Macrophages: obligate partners for tumor cell migration, invasion, and metastasis. *Cell* **124**:263–266.
54. Kassiotis, G., and Kollias, G. 2001. Uncoupling the proinflammatory from the immunosuppressive properties of tumor necrosis factor (TNF) at the p55 TNF receptor level: implications for pathogenesis and therapy of autoimmune demyelination. *J Exp Med* **193**:427–434.
55. Hagemann, T., Lawrence, T., McNeish, I., et al. 2008. "Re-educating" tumor-associated macrophages by targeting NF-kappaB. *J Exp Med* **205**:1261–1268.
56. Jamieson, T., Cook, D.N., Nibbs, R.J.B., et al. 2005. The chemokine receptor D6 limits the inflammatory response in vivo. *Nat Immunol* **6**:403–411.
57. Nibbs, R.J., Gilchrist, D.S., King, V., et al. 2007. The atypical chemokine receptor D6 suppresses the development of chemically induced skin tumours. *J Clin Invest* 117:1752–1755.
58. Rot, A., and von Andrian, U.H. 2004. Chemokines in innate and adaptive host defense: basic chemokinese grammar for immune cells. *Annu Rev Immunol* **22**:891–928.
59. Mantovani, A. 1999. The chemokine system: redundancy for robust outputs. *Immunol Today* 20:254–257.
60. Murphy, C.A., Hoek, R.M., Wiekowski, M.T., Lira, S.A., and Sedgwick, J.D. 2002. Interactions between hemopoietically derived TNF and central nervous system-resident

glial chemokines underlie initiation of autoimmune inflammation of the brain. *J Immunol* **169**:7054–7062.
61. Allen, S.J., Crown, S.E., and Handel, T.M. 2006. Chemokine: receptor structure, interactions, and antagonism. *Annu Rev Immunol* **25**:787–820.
62. Graham, G.J., and McKimmie, C.S. 2006. Chemokine scavenging by D6: a movable feast? *Trends Immunol* **27**:381–386.
63. Alcami, A. 2007. New insights into the subversion of the chemokine system by poxviruses. *Eur J Immunol* **37**:880–883.
64. Robinson, S.C., Scott, K.A., Wilson, J.L., Thompson, R.G., Proudfoot, A.E.I., and Balkwill, F.R. 2003. A chemokine receptor antagonist inhibits experimental breast tumor growth. *Cancer Res* **63**:8360–8365.
65. Loberg, R.D., Ying, C., Craig, M., et al. 2007. Targeting CCL2 with systemic delivery of neutralizing antibodies induces prostate cancer tumor regression in vivo. *Cancer Res* **67**:9417–9424.
66. Bertolini, F., Ying, C., Craig, M., et al. 2002. CXCR4 neutralization, a novel therapeutic approach for non-Hodgkin's lymphoma. *Cancer Res* **62**:3106–3112.
67. Rubin, J.B., Kung, A.L., Klein, R.S., et al. 2003. A small-molecule antagonist of CXCR4 inhibits intracranial growth of primary brain tumors. *Proc Natl Acad Sci USA* **100**:13513–13518.
68. Merritt, W.M., Lin, Y.G., Spannuth, W.A., et al. 2008. Effect of interleukin-8 gene silencing with liposome-encapsulated small interfering RNA on ovarian cancer cell growth. *J Natl Cancer Inst* **100**:359–372.
69. Miao, Z., Luker, K.E., Summers, B.C., et al. 2007. CXCR7 (RDC1) promotes breast and lung tumor growth in vivo and is expressed on tumor-associated vasculature. *Proc Natl Acad Sci USA* **104**:15735–15740.
70. Strand, V., Kimberly, R., and Isaacs, J.D. 2007. Biologic therapies in rheumatology: lessons learned, future directions. *Nat Rev* **6**:75–91.
71. Feldmann, M. 2002. Development of anti-TNF therapy for rheumatoid arthritis. *Nat Rev Immunol* **2**:364–371.
72. Sands, B.E., Anderson, F.H., Bernstein, C.N., et al. 2004. Infliximab maintenance therapy for fistulizing Crohn's disease. *N Engl J Med* **350**:876–885.
73. Scheinecker, C., Redlich, K., and Smolen, J.S. 2008. Cytokines as therapeutic targets: advances and limitations. *Immunity* **28**:440–444.
74. Rao, V.P., Poutahidis, T., Ge, Z., et al. 2006. Innate immune inflammatory response against enteric bacteria *helicobacter hepaticus* induces mammary adenocarcinoma in mice. *Cancer Res* **66**:7395–7400.
75. Scott, K.A., Moore, R.J., Arnott, C.H., et al. 2003. An anti-tumor necrosis factor-alpha antibody inhibits the development of experimental skin tumors. *Mol Cancer Ther* **2**:445–451.
76. Smolen, J.S., Aletaha, D., Koeller, M., Weisman, M.H., and Emery, P. 2007. New therapies for treatment of rheumatoid arthritis. *Lancet* **370**:1861–1874.
77. Smolen, J.S., Beaulieu, A., Rubbert-Roth, A., et al. 2008. Effect of interleukin-6 receptor inhibition with tocilizumab in patients with rheumatoid arthritis (OPTION study): a double-blind, placebo-controlled, randomised trial. *Lancet* **371**:987–997.
78. Nagabhushanam, V., Solache, A., Ting, L-M., Escaron, C.J., Zhang, J.Y., and Ernst, J.D. 2003. Innate inhibition of adaptive immunity: mycobacterium tuberculosis-induced IL-6 inhibits macrophage responses to IFN-gamma. *J Immunol* **171**:4750–4757.
79. Charles, P., Elliott, M.J., Davis, D., et al. 1999. Regulation of cytokines, cytokine inhibitors, and acute-phase proteins following anti-TNF-alpha therapy in rheumatoid arthritis. *J Immunol* **163**:1521–1528.

SUGGESTED READINGS

Egberts, J.-H., Cloosters, V., Noack, A., et al. 2008. Anti-tumor necrosis factor therapy inhibits pancreatic tumor growth and metastasis. *Cancer Res* **68**:1443–1450.

Mantovani, A. 1999. The chemokine system: redundancy for robust outputs. *Immunol Today* **20**:254–257.

Nibbs, R.J., Gilchrist, D., King, V., et al. 2007. The atypical chemokine receptor D6 suppresses the development of chemically induced skin tumours. *J Clin Invest* **117**:1752–1755.

Rot, A., and von Andrian, U.H. 2004. Chemokines in innate and adaptive host defense: basic chemokinese grammar for immune cells. *Annu Rev Immunol* **22**:891–928.

Tomita, Y., Yang, X., Ishida, Y., et al. 2004. Spontaneous regression of lung metastasis in the absence of tumour necrosis factor p55. *Int J Cancer* **112**:927–933.

20 Lung

Bruce D. Levy

Inflammation in the lung is common in both physiologic responses as well as many respiratory illnesses. In particular, chronic inflammation is associated with a variety of prevalent disorders, including asthma, chronic obstructive pulmonary disease, bronchiectasis, and interstitial lung diseases [1–3]. For purposes of host defense, an overexuberant inflammatory response can also lead to respiratory disorders. For example, inhalation of pathogens, toxins, or specific allergens initiates an acute inflammatory response that characterizes acute exacerbations of bronchiectasis, COPD, and asthma [1,4]. Perhaps the most extensively investigated example of acute inflammation and its spontaneous resolution is pneumonia. In this chapter, a common clinical presentation of pneumonia is provided with examples of its radiographic appearance and histology during both the initiation and resolution phases of the illness.

Bea Coffin is a 56-year-old woman who presents with a new cough and dyspnea. She has felt ill for about 3 days. Her cough is productive of blood-tinged green phlegm. She has also had fevers, chills, and sweats that are getting worse. The symptoms began with the sudden onset of right sided chest pain that makes it difficult to take a deep breath. She has tried acetaminophen and an expectorant, but these interventions have not been successful in controlling her symptoms. She is a lifelong nonsmoker and has no significant past medical history.

On physical examination, she appears ill. Her vital signs are notable for a temperature of 103.5°F, respiratory rate of 26 with shallow respirations, blood pressure of 122/60 mmHg and pulse of 115 beats per minute. She is warm to the touch and appears plethoric. Skin is moist without rash. Pupils are anicteric. There is no sinus tenderness and tympanic membranes are within normal limits. No nasal discharge. Her oropharynx is only notable for erythema, but no exudates or tonsilar swelling. Her jugular venous pressure is 5 cm H_2O and there is no significant lymphadenopathy. Her chest examination reveals tachypnea with diminished air movement, but symmetric expansion of the chest wall with inspiration. There is no dullness to percussion. She has inspiratory rales in the right mid-posterior lung zones with egophony. A soft pleural rub is also present in this region of the chest. Cardiovascular exam is only notable for tachycardia. There is no hepatosplenomegaly and her abdomen is soft and nontender. Extremities have symmetric pulses and there is no clubbing, cyanosis or edema.

Her laboratory examination uncovered an increased white blood cell count with a leukemoid reaction, normal hematocrit and mild thrombocytosis. Renal and liver function was within normal limits. Her chest x-ray and computed tomography of the chest demonstrated a right lower lobe infiltrate (Figure 20.1). She was given a diagnosis of community acquired pneumonia and antibiotics were begun. One day later, both sputum and blood cultures grew *Streptococcus pneumoniae*.

More than any other infection, pneumonia carries the highest morbidity and mortality. Each year in the United States, it is estimated that 3–4 million people will contract pneumonia leading to as many as 1 million hospitalizations and approximately 50,000 deaths (reviewed in reference [5]). Unfortunately, there has been little progress in decreasing the pneumonia-associated mortality over the past several decades [6]. Most commonly, pneumonia is community-acquired and self-resolving over a period of weeks to months [5]. The outcome of an acute pneumonia depends on the organism and host response. Below the vocal cords, the respiratory tract is sterile, so the presence of any bacteria is abnormal. When small numbers of bacteria gain access to the lower respiratory tract, the innate immune host defense mechanisms can effectively contain the pathogens. Several mechanisms are operative in the lower respiratory tract to clear bacteria, including the mucociliary escalator in the large airways, phagocytosis by alveolar macrophages and direct killing by antimicrobial proteins. Deficiencies in mucosal integrity and immunity increase host susceptibility to inhaled or aspirated pathogens [7]. Moreover, larger

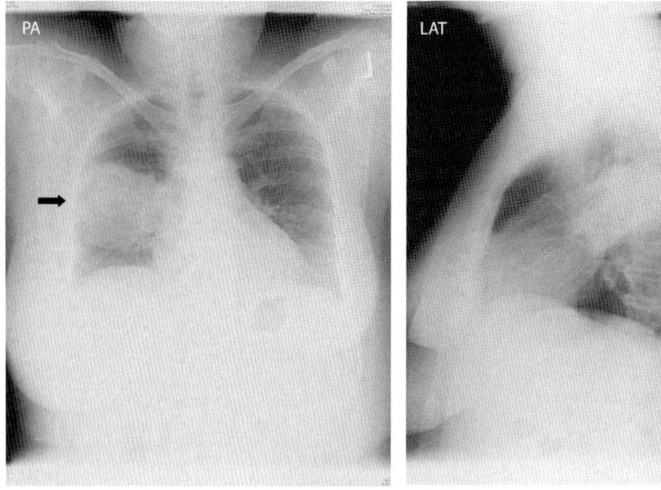

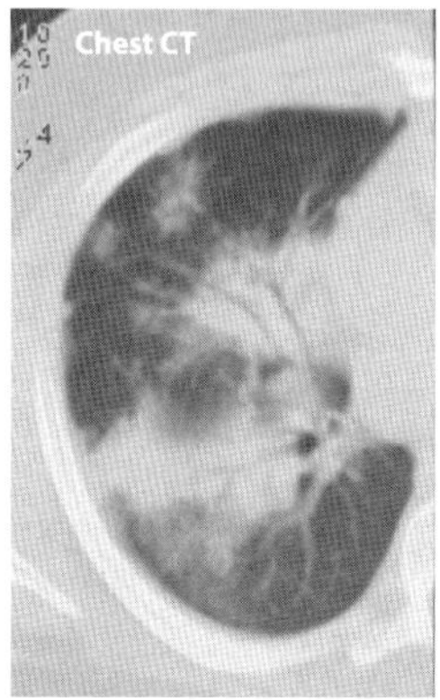

Figure 20.1. Chest imaging of acute bacterial pneumonia. PA (left) and Lateral (LAT, right) chest X-rays as well as computed tomography of the chest (Chest CT) were performed on presentation to the Emergency Department and revealed consolidation in the right lower lobe (arrows), consistent with a diagnosis of pneumonia. (Images are courtesy of Christopher H. Fanta.)

numbers of bacteria or particularly virulent pathogens can overwhelm the usual host responses and trigger a substantial inflammatory response.

The microbiology of pneumonia is well studied and closely monitored. The incidence of specific pathogens leading to community acquired pneumonia varies in clinical series, but most commonly demonstrates a rank order frequency of *Streptococcus pneumonia* (incidence 20%–70%), *Haemophilus influenzae* (3%–18%), *Mycoplasma pneumoniae* (2%–29%), *Chlamydia pneumoniae* (2%–8%), *Legionella* species (1%–8%), and viruses (1%–10%) [5]. Nosocomial pathogens can also include *Staphylococcus aureus* and enteric Gram-negative rods, especially in the setting of aspiration of gastric contents leading to pneumonia. Immunocompromised hosts can have select sensitivity to pathogens based on their immune defect, such as the increased susceptibility for *Pneumocystis carinii* with impaired T cell numbers or function (e.g., human immunodeficiency virus). Emergence of new pathogens and antibiotic-resistant organisms is not uncommon.

The host responds to pathogen invasion with acute inflammation that enhances microbial killing. This pathogen-mediated inflammation is often exuberant and the release of antimicrobial effectors, such as proteases and reactive oxygen species, leads to lung tissue bystander injury in the name of host defense [8]. Gross pathology of pneumonia findings demonstrate a "hepatization" of the lungs that transitions from an early red to a later gray hepatization phase [9]. During the initiation of the pneumonia, microscopic findings are most notable for numerous polymorphonuclear leukocytes (i.e., neutrophils) (Figure 20.2). In addition, edema fluid, erythrocytes, intraalveolar fibrin, and even hyaline membranes can be present. In pneumococcal pneumonia, gram stain often reveals intracellular Gram-positive diplococci in both neutrophils and macrophages. Of interest, murine models of acute pneumococcal pneumonia can serve as useful experimental models of human pneumonia with very similar findings [10] (Figure 20.3). Although some patients may experience pneumonia-induced acute respiratory distress syndrome [8], most patients with intact immune function will resolve their infection – a process that can be accelerated with appropriate antibiotics [5].

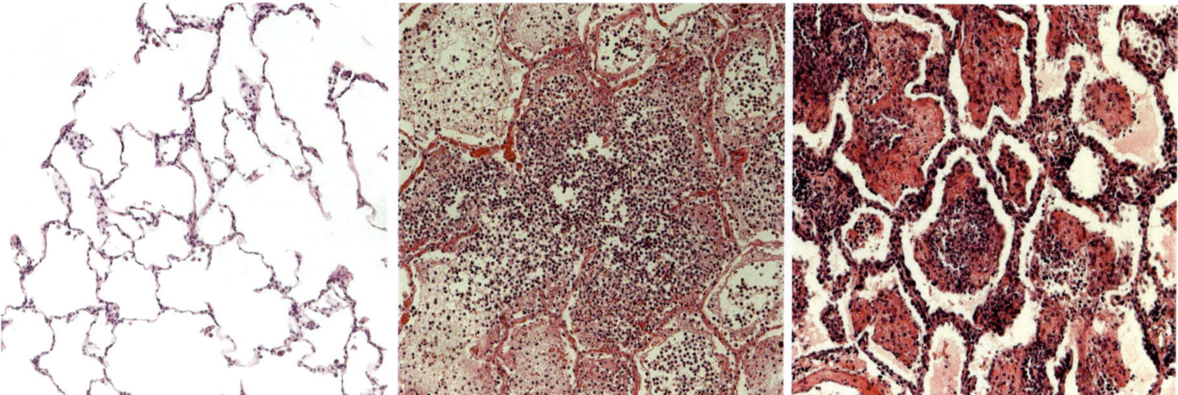

Figure 20.2. Histopathology of acute pneumonia. In acute bacterial pneumonia, healthy alveolar air spaces for gas exchange (left) become filled with edema and leukocytes that are predominantly neutrophils (middle), as well as blood, fibrin, and hyaline membranes (right) that together consolidate the lung and impair oxygenation. (Images are courtesy of Les Kobzik.)

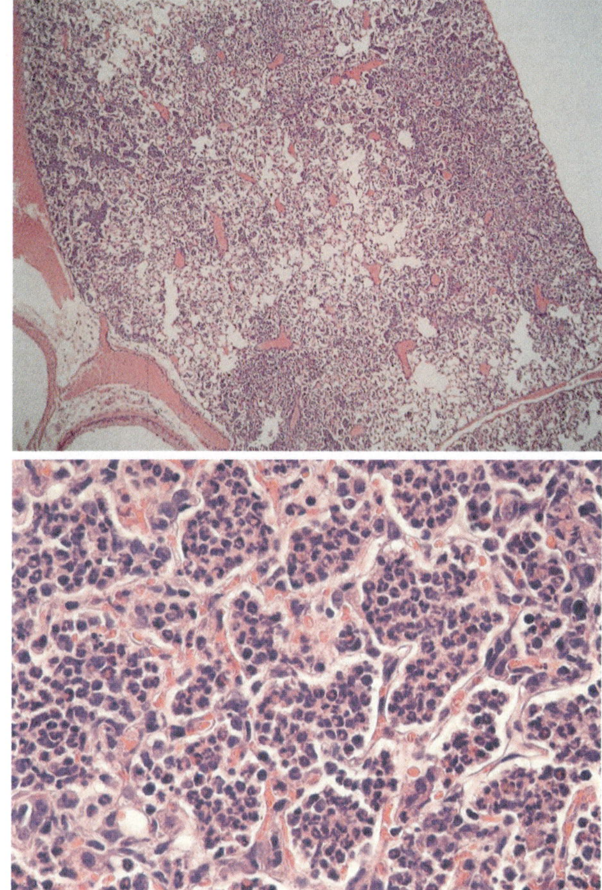

Figure 20.3. Histopathology of murine pneumococcal pneumonia. Two days after the introduction of *Streptococcus pneumoniae* into the airways of Balb/c mice there is an exuberant, neutrophil-rich infiltrate that is evident at both low (top) and high (bottom) resolution. (Images are courtesy of Joseph P. Mizgerd.)

After receiving antibiotics for 3 days, Bea began to feel better and her fevers and pleuritic chest pain disappeared. After 2 weeks, Bea's cough improved. She continued to experience an occasional cough productive of yellow phlegm for a total of 4 weeks. Follow-up chest x-ray demonstrated complete resolution of the right lower lobe infiltrate (Figure 20.4). 6 weeks after her initial symptoms, Bea's energy returned, the cough completely resolved and she finally felt healthy again.

Of interest, community acquired pneumonia spontaneously resolves in most instances, suggesting the existence of endogenous, host-protective signaling pathways [5]. The resolution of acute pneumonia generally unfolds over several days to weeks with lung histopathology that reflects specific cellular events that dampen acute inflammation and promote microbial and leukocyte clearance. Within days, the leukocytic infiltrate changes from neutrophil-rich to predominantly lymphocytic and monocytic. The inflammatory macrophages can clear apoptotic neutrophils, microbial and protein debris by phagocytosis. Of interest, the anti-inflammatory events during the resolution phase of inflammation differ from immunosuppression in part by the promotion of host defense [11]. Recently, endogenous mediators, such as lipoxins, resolvins, and protectins, have been identified that can inhibit neutrophil activation and tissue entry, promote monocyte/macrophage tissue accumulation, and stimulate these phagocytes to clear apoptotic neutrophils and microbial products [12]. Early events in acute inflammation serve to engage biosynthetic circuits for later formation of these natural resolution mediators [13]. Disruption of these counter-regulatory signaling pathways can lead to prolonged lung inflammation and delayed resolution [14]. Lipoxins are the first class of mediators that were identified with both anti-inflammatory and proresolving effects [12]. In addition to controlling pathogen-induced inflammatory responses [15], lipoxins can also enhance mucosal host defense in the gastrointestinal tract by inducing the expression of potent antimicrobial proteins, such as bactericidal permeability inducing protein [16].

For resolution of pneumonia, the chest x-ray will usually indicate clearance of the pulmonary infiltrate within 3–4 weeks in those under 50 and within 8–12

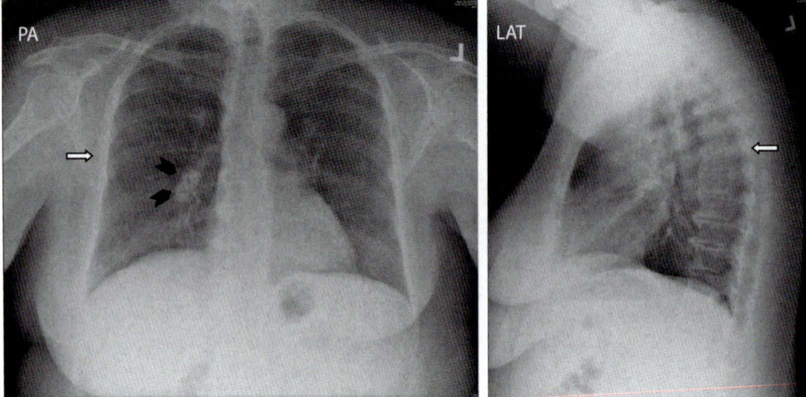

Figure 20.4. Chest radiographic evidence for resolution of pneumonia. After 4 weeks of treatment for acute bacterial pneumonia, the PA (left) and LAT (right) chest x-ray reveal resolution of the prior parenchymal infiltrate (block arrows) with only minimal peribronchial cuffing/inflammation (arrowheads). (Images are courtesy of Christopher H. Fanta.)

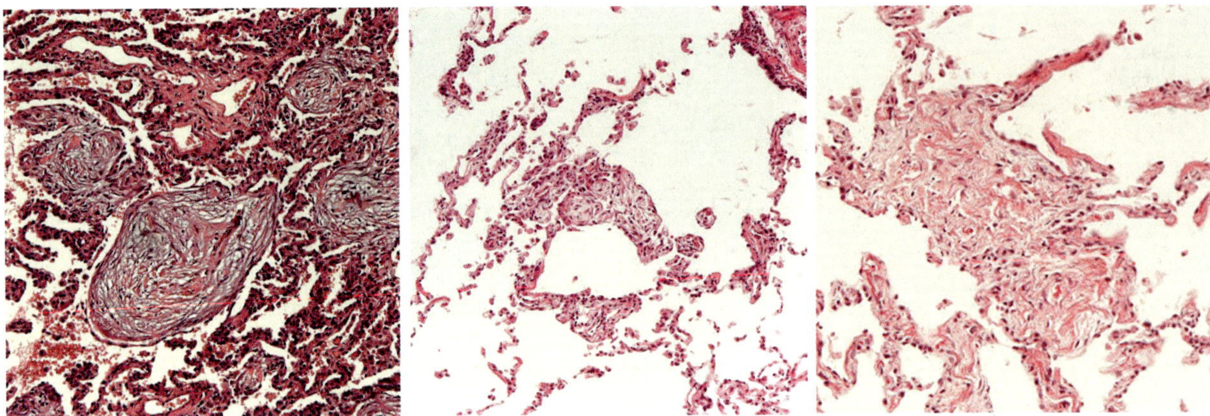

Figure 20.5. Histopathology of the resolution phase of pneumonia. The alveolar exudates become organized with deposition of connective tissue (left) that will be resorbed and remodeled during resolution (middle and right) to promote a return to normal alveolar architecture (not shown). (Images are courtesy of Les Kobzik.)

weeks in those older than 50 [17]. During the resolution phase, microscopic findings are notable for organization of the intraalveolar exudates with transient fibrosis [9] (Figure 20.5). With time, the architecture of the lung and its functional capacity return to normal. Patients are expected to return completely to their premorbid condition.

In most cases, acute community acquired bacterial pneumonia provides a nice example of a clinical pathological correlation for an acute, spontaneously resolving inflammatory response. Antibiotics can shorten the duration of symptoms, but complete resolution of the process takes weeks to months. Host defense initiates acute inflammation in the lung as well as its resolution. These are both active processes that are governed by distinct pro- and anti-inflammatory signals. The cellular events that are evident in pneumonia histopathology nicely illustrate the distinct nature of the initiation and resolution phases of acute inflammation. These observations are not unique to pneumonia, but are also evident in other forms of acute, self-limited inflammation in the lung.

KEY POINTS

- Both chronic and acute inflammation in the lung has been linked to common human diseases.
- Pneumonia is an example of an acute inflammatory response (to microbial invasion) that usually completely resolves.
- Resolution of inflammation is an active process with specific pro-resolving signaling pathways.
- The initiation and resolution phases of acute inflammation in the lung have distinct cellular and molecular effectors.

REFERENCES

1. Cowburn, A.S., Condliffe, A.M., Farahi, N., Summers, C., and Chilvers, E.R. 2008. Advances in neutrophil biology: clinical implications. *Chest* **134**:606–612.
2. Galli, S.J., Tsai, M., and Piliponsky, A.M. 2008. The development of allergic inflammation. *Nature* **454**:445–454.

3. Sutherland, E.R., and Cherniack, R.M. 2004. Management of chronic obstructive pulmonary disease. *N Engl J Med* **350**:2689–2697.
4. Kim, E.Y., Battaile, J.T., Patel, A.C., et al. 2008. Persistent activation of an innate immune response translates respiratory viral infection into chronic lung disease. *Nat Med* **14**:633–640.
5. Mizgerd, J.P. 2008. Acute lower respiratory tract infection. *N Engl J Med* **358**:716–727.
6. Armstrong, G.L., Conn, L.A., and Pinner, R.W. 1999. Trends in infectious disease mortality in the United States during the 20th century. *JAMA* **281**:61–66.
7. Waite, S., Jeudy, J., and White, C.S. 2006. Acute lung infections in normal and immunocompromised hosts. *Radiol Clin North Am* **44**:295–315, ix.
8. Ware, L.B., and Matthay, M.A. 2000. The acute respiratory distress syndrome. *N Engl J Med* **342**:1334–1349.
9. Farver, C.F. 2008. Bacterial diseases. In *Pulmonary Pathology*, Zander, D.S. and Farver, C.F. (eds.), pp. 167–203. Philadelphia: Churchill Livingstone Elsevier.
10. Mizgerd, J.P., and Skerrett, S.J. 2008. Animal models of human pneumonia. *Am J Physiol Lung Cell Mol Physiol* **294**:L387–L398.
11. Serhan, C.N., Brain, S.D., Buckley, C.D., et al. 2007. Resolution of inflammation: state of the art, definitions and terms. *FASEB J* **21**:325–332.
12. Serhan, C.N., Chiang, N., and Van Dyke, T.E. 2008. Resolving inflammation: dual anti-inflammatory and pro-resolution lipid mediators. *Nat Rev Immunol* **8**:349–361.
13. Levy, B.D., Clish, C.B., Schmidt, B., Gronert, K., and Serhan, C.N. 2001. Lipid mediator class switching during acute inflammation: signals in resolution. *Nat Immunol* **2**:612–619.
14. Fukunaga, K., Kohli, P., Bonnans, C., Fredenburgh, L.E., and Levy, B.D. 2005. Cyclooxygenase 2 plays a pivotal role in the resolution of acute lung injury. *J Immunol* **174**:5033–5039.
15. Machado, F.S., Johndrow, J.E., Esper, L., et al. 2006. Anti-inflammatory actions of lipoxin A4 and aspirin-triggered lipoxin are SOCS-2 dependent. *Nat Med* **12**:330–334.
16. Canny, G., Levy, O., Furuta, G.T., et al. 2002. Lipid mediator-induced expression of bactericidal/ permeability-increasing protein (BPI) in human mucosal epithelia. *Proc Natl Acad Sci USA* **99**:3902–3907.
17. Bruns, A.H., Oosterheert, J.J., Prokop, M., Lammers, J.W., Hak, E., and Hoepelman, A.I. 2007. Patterns of resolution of chest radiograph abnormalities in adults hospitalized with severe community-acquired pneumonia. *Clin Infect Dis* **45**:983–991.

SUGGESTED READINGS

Levy, B.D., Clish, C.B., Schmidt, B., Gronert, K., and Serhan, C.N. 2001. Lipid mediator class switching during acute inflammation: signals in resolution. *Nat Immunol* **2**:612–619.

Mizgerd, J.P. 2008. Acute lower respiratory tract infection. *N Engl J Med* **358**:716–727.

Schwab, J.M., Chiang, N., Arita, M., and Serhan, C.N. 2007. Resolvin E1 and protectin D1 activate inflammation-resolution programmes. *Nature* **447**:869–874.

Serhan, C.N., Brain, S.D., Buckley, C.D., et al. 2007. Resolution of inflammation: state of the art, definitions and terms. *FASEB J* **21**:325–332.

21 Neural Inflammation, Alzheimer's Disease, and Stroke

Andrew P. Lieberman and Constance D'Amato

Inflammation in the central nervous system (CNS) occurs after brain injury. The proximal cause may be an anoxic event such as a stroke, chronic neurodegeneration, or an infectious process. In each instance, glial cells including astrocytes and bone marrow–derived microglia, the resident macrophages of the CNS, are activated and participate in the inflammatory response. Additional components of neural inflammation, such as polymorphonuclear leukocytes, lymphocytes, and macrophages, arrive from the periphery. The following cases illustrate common causes of CNS inflammation.

CASE 1: STROKE

Clinical history: A 71-year-old man with hypertension, noninsulin-dependent diabetes mellitus, and peripheral vascular disease had a stroke resulting in right arm and leg weakness about 3 months before he died. He had another stroke that caused left sided paralysis about 2 1/2 weeks before death (Figure 21.1).

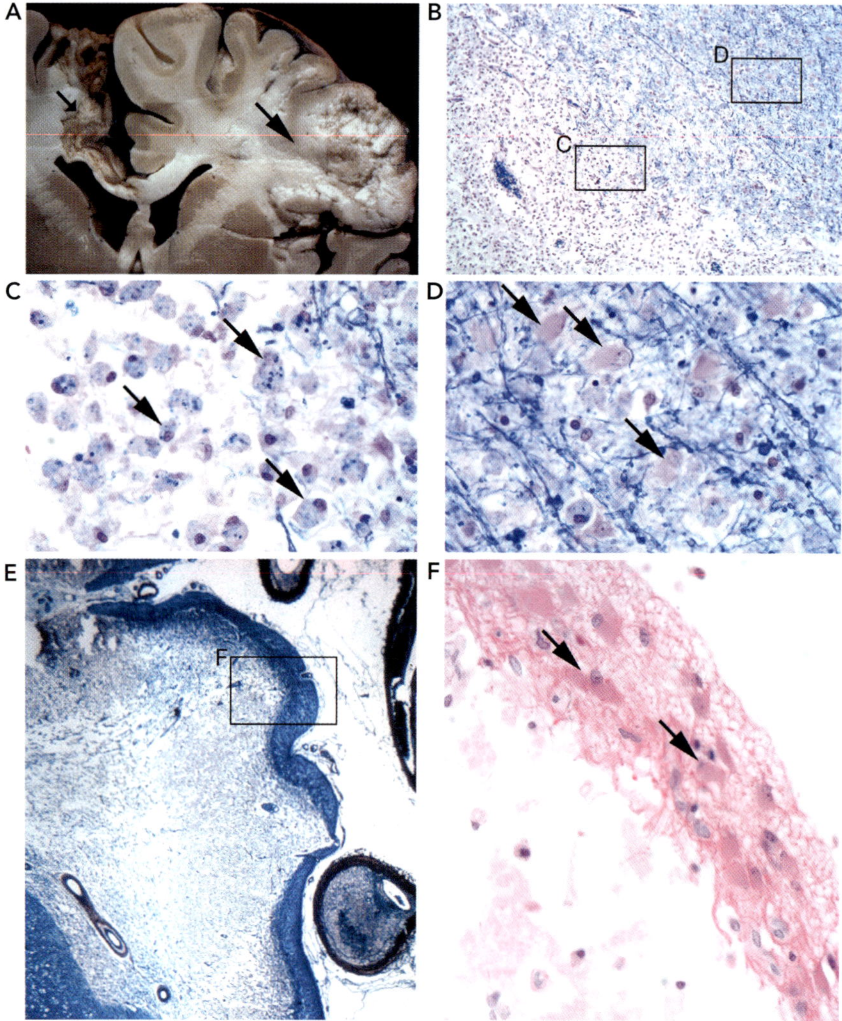

Figure 21.1. Subacute and chronic infarctions. (A) (Top row, left) A sharply demarcated, chronic infarction is present in the distribution of the left anterior cerebral artery (small arrow), and a large, poorly delineated subacute infarction is seen in the distribution of the right middle cerebral artery (large arrow). (B–D) Subacute infarction. (B) (Top row, right) A dense infiltrate of foamy macrophages is present at the lesion's center (bottom left) (Luxol fast blue–cresyl violet–eosin stain). (C) (Middle row, left) Macrophages (representative cells at arrows) are adjacent to disrupted myelinated axons (blue lines), and their cytoplasm contains myelin debris after phagocytosis (Luxol fast blue–cresyl violet–eosin stain). (D) (Middle row, right) Reactive astrocytes, with smooth, eosinophilic cytoplasm, and eccentric nuclei (representative cells at arrows), proliferate at the lesion's edge starting 10–14 days after injury (Luxol fast blue–cresyl violet–eosin stain). (E–F) Chronic infarction. (E) (Bottom row, left) A rim of reactive astrocytes surrounds the cavitary lesion of an old infarction. The lumen of the anterior cerebral artery is largely occluded by atherosclerotic plaque. (PTAH stain). (F) (Bottom row, right) Reactive astrocytes (representative cells at arrows) overlie acellular debris and macrophages within the cavitary lesion of the old infarction (Luxol fast blue–cresyl violet–eosin stain).

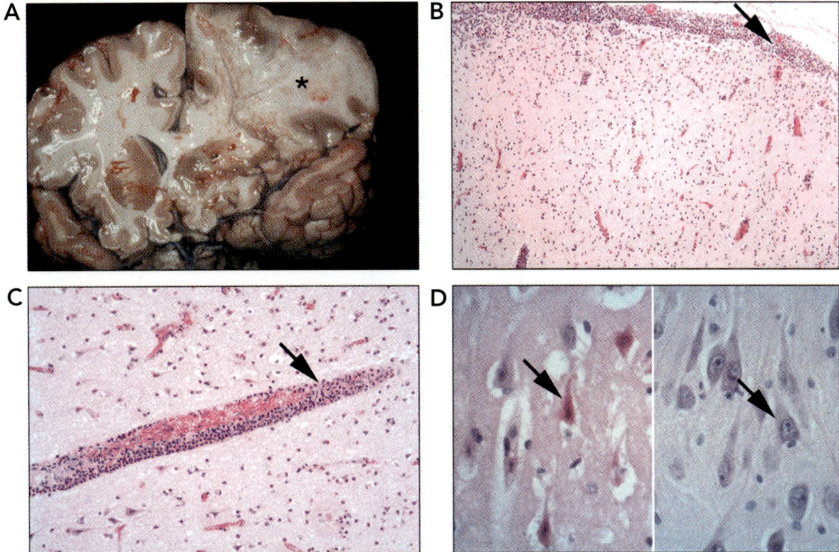

Figure 21.2. Acute infarction. (A) (Top row, left) An acute infarction (asterisk) in the distribution of the right middle cerebral artery causes edema (widened gyri, effaced sulci, and compressed right lateral ventricle) and blurring of the grey-white matter junctions. (B) (Top row right) and (C) (Bottom row, left) In acutely infarcted regions, polymorphonuclear leukocytes coat the pial surface (arrow), rim blood vessels (arrow), and spill out into the adjacent brain tissue (H&E stain). (D) (Bottom row, right) Acute anoxia/ischemia (on left) causes the neuronal cytoplasm to turn pink and the chromatin to condense (representative cell at arrow). These initial morphologic changes occur within hours of ischemia and precede infiltration by polymorphonuclear leukocytes. In contrast, healthy neurons (on right, representative cell at arrow) are characterized by cytoplasm with a slight purple color and large nuclei with open chromatin and prominent nucleoli (H&E stain).

CASE 2: ALZHEIMER'S DISEASE

Clinical history: A 76-year-old man died after a 7-year course of progressive cognitive decline initially characterized by memory loss. His intellectual function slowly deteriorated, and by the time he was near death, he was no longer able to care for himself or eat without assistance (Figure 21.3).

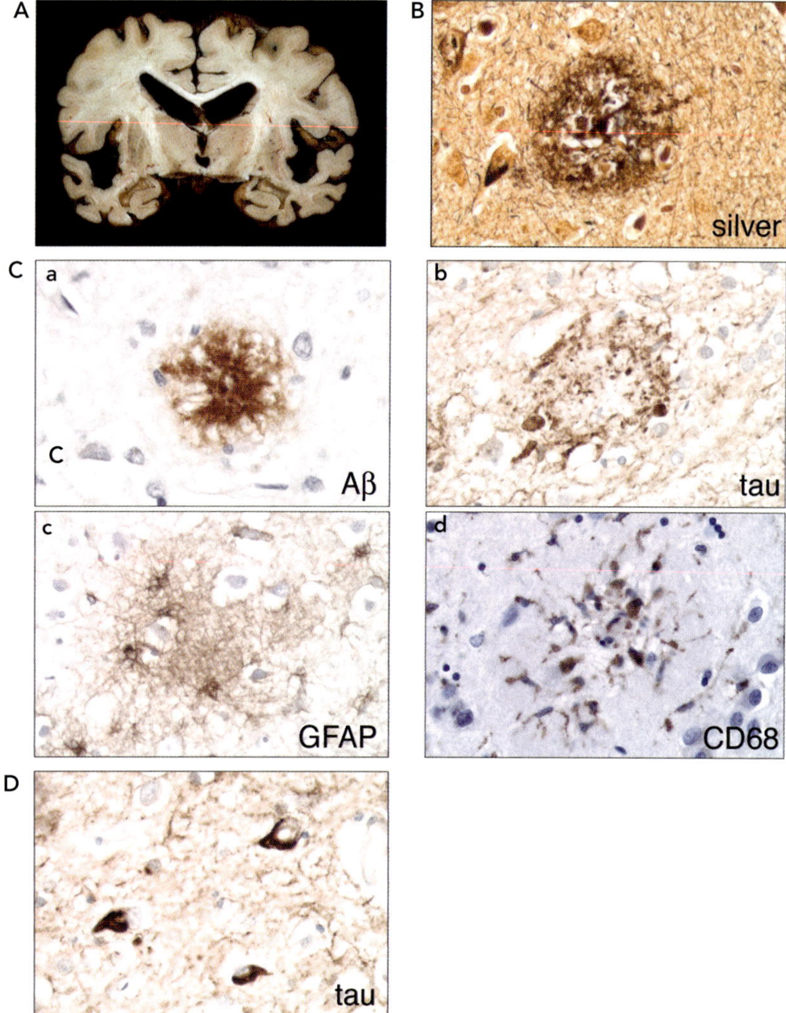

Figure 21.3. Alzheimer's disease. (A) (Top row, left) Neuron loss occurs preferentially in the cerebral cortex, hippocampus, and amygdala of Alzheimer's disease patients. These result in atrophy of the affected sites: narrowed gyri and widened sulci in the cortex, and shrunken limbic system structures. The ventricles enlarge secondarily due to loss of brain tissue. (B) (Top row, right) Characteristic protein aggregates define Alzheimer's disease pathology. A neuritic senile plaque is present in the neuropil (at center) and a neurofibrillary tangle is seen in the soma and proximal axon of an adjacent neuron (bottom left, Bielschowky silver stain). (C) (Second and third rows) Components of a senile plaque include a central core of the amyloid protein Aβ (panel a, second row, left), surrounded by a halo of swollen or dystrophic neurites (panel b, second row, right), reactive astrocytes (panel c, third row, left), and activated microglia (panel d, third row, right). (D) (Bottom row) Neurofibrillary tangles are intraneuronal accumulations of the microtubule binding protein tau. Hyperphosphorylated tau is arrayed in paired helical filaments that coalesce to form characteristically flame-shaped aggregates.

CASE 3: MULTIPLE SCLEROSIS

Clinical history: A 34-year-old woman had an 8-year history of multiple sclerosis. She initially presented with blurred vision in one eye, and subsequently developed double vision, an unsteady gait, tremors, and swallowing difficulties. She died from aspiration pneumonia (Figure 21.4).

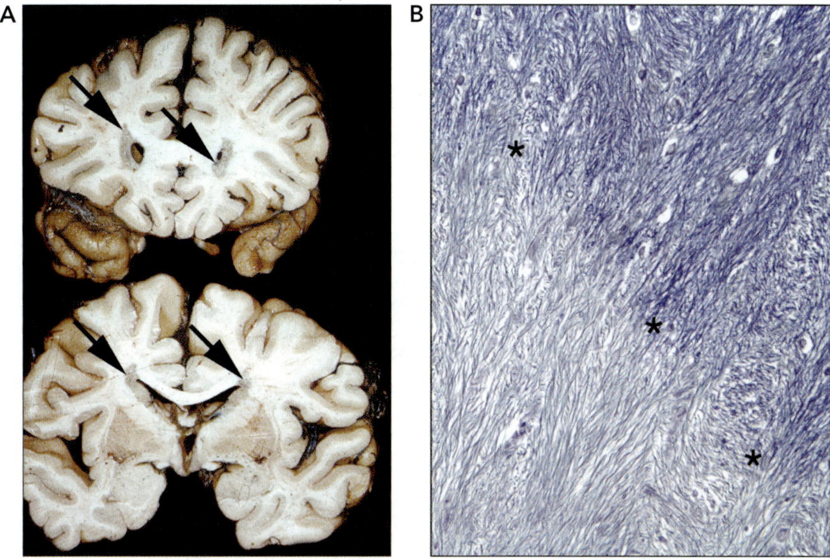

Figure 21.4. Chronic multiple sclerosis. (A) (Left) Chronic demyelinating plaques at the edge of the lateral ventricles (at arrows) appear as sharply demarcated grey lesions. (B) (Right) Chronic inactive plaques are characterized by an abrupt transition in which the myelin shealth (stained blue by Luxol stain) is lost, yet axons (stained grey by silver stain) are relatively preserved (transition indicated by asterisks). Active plaques have macrophages and lymphocytes at their edges, whereas inactive plaques, such as this one, have reactive astrocytes at their periphery and over time may partially remyelinate.

CASE 4: ACUTE MENINGITIS

Clinical history: A 4-month-old girl was born full term, but was small and failed to gain weight. She developed a fever, and after 5 days of increasing lethargy and weakness, she was brought to the hospital where she was found to have a bulging anterior fontanel. In the hospital, she developed seizures and died a few hours later (Figure 21.5).

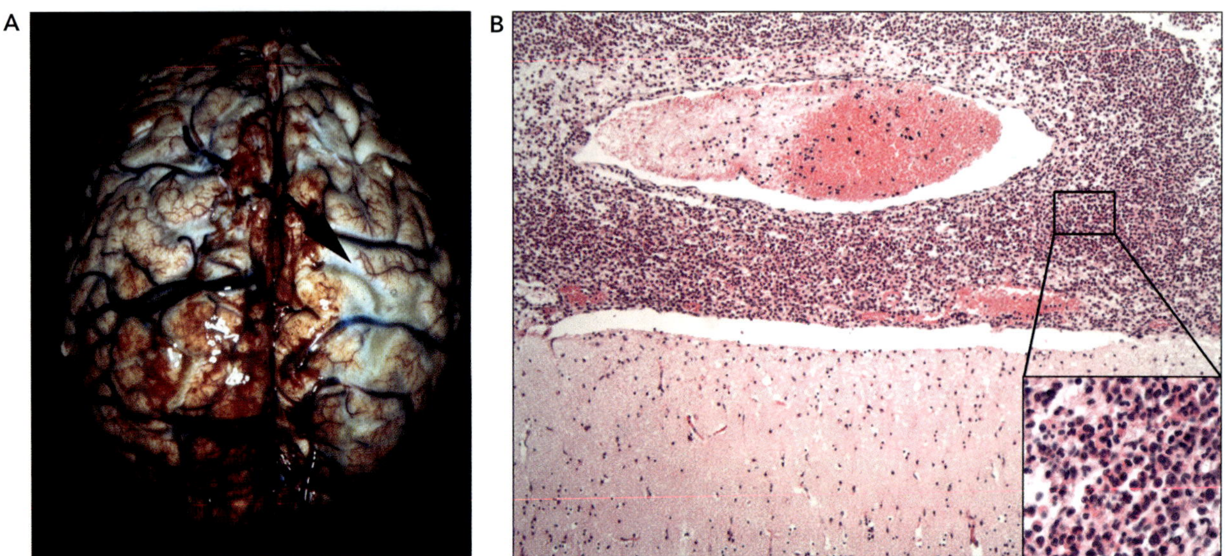

Figure 21.5. Acute purulent meningitis. (A) (Left) Engorged blood vessels course through the subarachnoid space, which is filled with purulent exudate (arrow). (B) (Right) Polymorphonuclear leukocytes and fibrin fill the space between the pia mater, at the cortical surface, and the arachnoid. *Inset:* Acute inflammatory infiltrate in the subarachnoid space (H&E stain).

CASE 5: BRAIN ABSCESS

Clinical history: A 52-year-old man developed bronchopneumonia after a bout with the flu. He was admitted to the hospital with a high fever and confusion. Brain imaging studies revealed an abscess in the right temporal lobe. Despite aggressive treatment, the patient died 2 days after hospital admission (Figure 21.6).

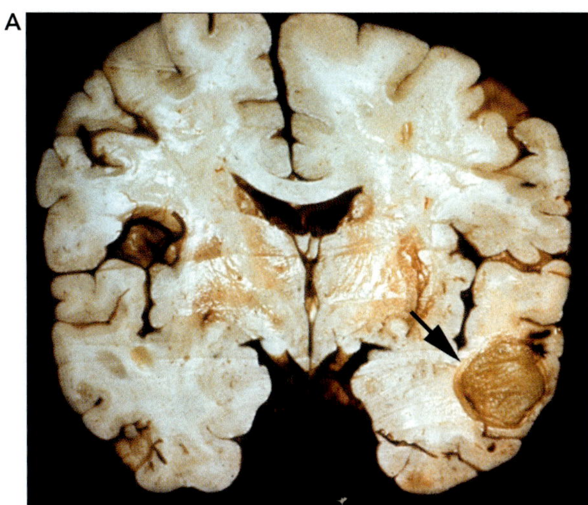

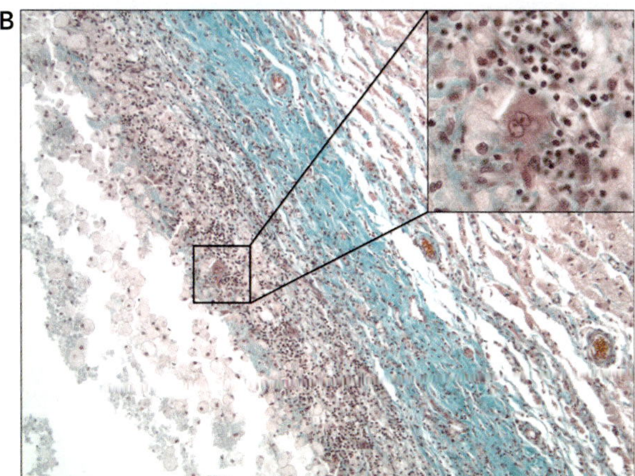

Figure 21.6. Chronic abscess. (A) (Top) A well-encapsulated, mature abscess is present in the temporal lobe (arrow). (B) (Bottom) The lumen of the abscess (at lower left) contains foamy macrophages admixed with acellular debris, and its wall contains macrophages, lymphocytes, and foreign body giant cells (*inset:* high magnification). A collagenous capsule (stained green on trichrome stain) surrounds the chronic inflammatory infiltrate, and reactive astrocytes are present in the adjacent brain tissue.

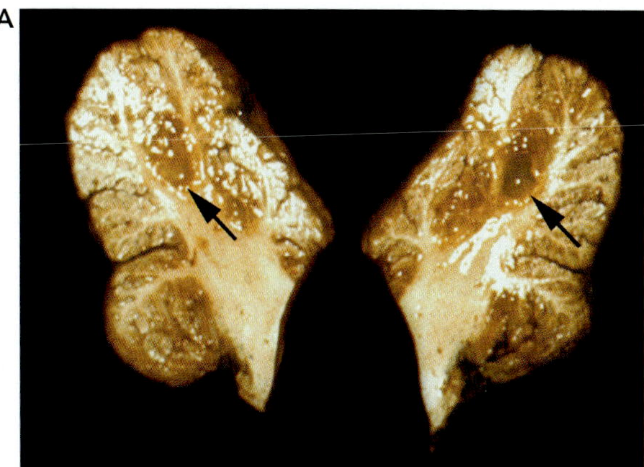

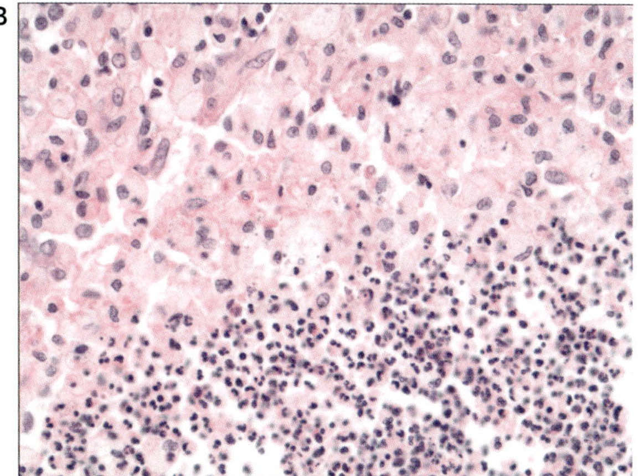

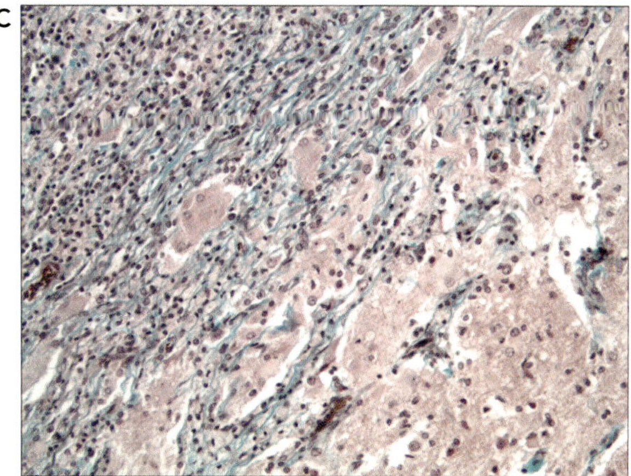

Figure 21.7. Recent abscess. (A) (Top) Poorly demarcated, centrally necrotic brain lesions (arrows). (B) (Middle) Polymorphonuclear leukocytes predominate at the center of the abscess (lower right), and are surrounded by a mixture of foamy macrophages and proliferative fibroblasts (H&E stain). (C) (Bottom) The collagenous capsule (green on trichrome stain), deposited by fibroblasts, is still just being formed.

CASE 6: VIRAL ENCEPHALITIS

Clinical history: A 9-year-old girl became irritable, hyperactive and hard to manage about 5 weeks before death. Her symptoms progressed, and she developed incessant flailing of limbs and thrashing movements, high fever and rapid respiration, and subsequently died (Figure 21.8).

A

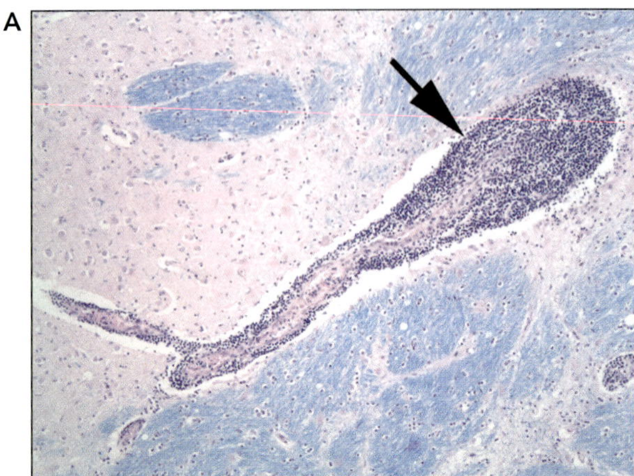

B

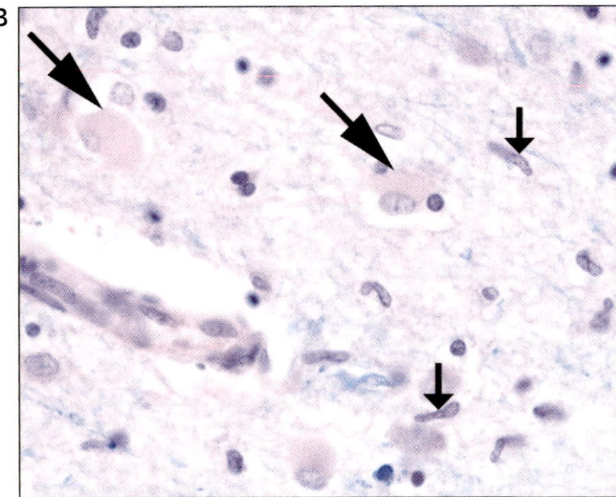

Figure 21.8. Viral encephalitis. (A) (Top) A cuff of reactive lymphocytes (arrow) surrounds a blood vessel within the basal ganglia (Luxol fast blue–cresyl violet–eosin stain). (B) (Bottom) Reactive astrocytes (with eccentric nuclei and pink cytoplasm; large arrows) and microglia (with rod shaped nuclei; small arrows) are present in the grey matter. Neither microglial nodules nor viral inclusions were identified in this case, though both are occasionally seen in viral encephalitis (Luxol fast blue–cresyl violet–eosin stain).

KEY POINTS

1. Brain injury, irrespective of mechanism, triggers the proliferation and hypertrophy of astrocytes in a process known as gliosis.
2. These reactive astrocytes and microglial cells can release inflammatory mediators and participate in neural inflammation.
3. Polymorphonuclear leukocytes, lymphocytes, and macrophages also enter the CNS from the periphery to participate in immune reactions.
4. Only in certain instances, such as abscess formation, do perivascular CNS fibroblasts proliferate and deposit collagen to form fibrosis.

SUGGESTED READINGS

Gray, F., Poirier, J., and De Girolami, U. *Escourolle and Poirier's Manual of Basic Neuropathology*, 4th Ed. *This is a beautifully illustrated introductory textbook to neuropathology.*

22 Rheumatoid Arthritis/SLE

Karim Raza and Caroline Gordon

RHEUMATOID ARTHRITIS

Rheumatoid Arthritis: Clinical Case

A 60-year-old man presented with a 4-week history of a gradually worsening ankle pain and swelling. He gave no history of morning stiffness, or symptoms in any other joints. There was no preceding history of infection (including of the gastrointestinal or genitourinary tract) and no history of previous inflammatory disease (including the skin and eye). There was a family history of rheumatoid arthritis with his daughter having been diagnosed with this condition at the age of 30. He smoked 10 cigarettes per day and drank 6 units of alcohol per week. He had been treated with diclofenac by his primary care physician but had derived little benefit from this.

On examination, he had tenderness with clinically apparent synovial swelling at the left ankle (Figure 22.1A). The remainder of the physical examination was normal.

The differential diagnosis of an inflammatory mono arthritis includes septic arthritis and crystal arthritis and to exclude these diagnoses the patient underwent an ultrasound guided joint aspiration (Figure 22.2). No organisms were identified on synovial fluid microscopy or culture and no crystals were seen on polarized light microscopy. Further investigations revealed the following: ESR 16 mm/h, CRP 17 mg/L (normal < 5 mg/L), rheumatoid factor 72 IU/mL (positive > 20 IU/mL), anti-CCP antibody 81 U/mL (positive >10 U/mL), chest radiograph normal.

The patient was diagnosed as having an unclassified inflammatory monoarthritis and was treated with an injection of 40 mg depomedrone to the left ankle joint. However, at review 6 weeks later the patient's condition was significantly worse. He had developed widespread synovitis affecting multiple joint areas in a symmetric fashion (Figure 22.1B) and now fulfilled classification criteria for rheumatoid arthritis (RA) [1].

He was commenced on oral prednisolone (10 mg daily) and methotrexate (initially at 7.5 mg per week and increased to 20 mg per week over 3 months). Three months later, his synovitis remained very active and sulfasalazine was started (2 g per day in divided doses). Two months later, he was commenced on hydroxychloroquine (400 mg per day). Despite this therapeutic combination, together with further intraarticular and intramuscular injections of depomedrone, his synovitis persisted and seven months after the initial onset of symptoms he had a DAS28 score [2] of 7.8 and was commenced on infliximab (3 mg/kg) (Figure 22.3). His disease improved with both his DAS28 score and his CRP falling. However his RA remained active with a DAS28 being categorized as "high" according to widely accepted threshold levels ("high" DAS28 > 5.1) [2]. In view of this, 11 months after having commenced infliximab, this drug was replaced with etanercept (25 mg twice weekly) (Figure 22.3). His CRP fell further and his disease activity continued to improve though never entered the range categorized as representing remission (DAS28 < 2.6) (Figure 22.3). At follow-up assessment, he had developed radiological evidence of destruction to bone as evidenced by the development of erosions at the metacarpophalyngeal joints.

Rheumatoid Arthritis: Discussion

A new onset of inflammatory arthritis is a remarkably common event. However, in at least half of patients the disease resolves spontaneously over a few months [3,4]. In the rest, the processes driving the natural resolution of inflammation are disrupted, leading to a switch to chronic persistent disease characterized by the accumulation of large numbers of leukocytes and stromal cells in the synovium. Rheumatoid arthritis (RA) is the most prevalent of the persistent inflammatory arthritides, affecting 0.81% of adults in the UK [5]. The disease typically manifests as a symmetrical

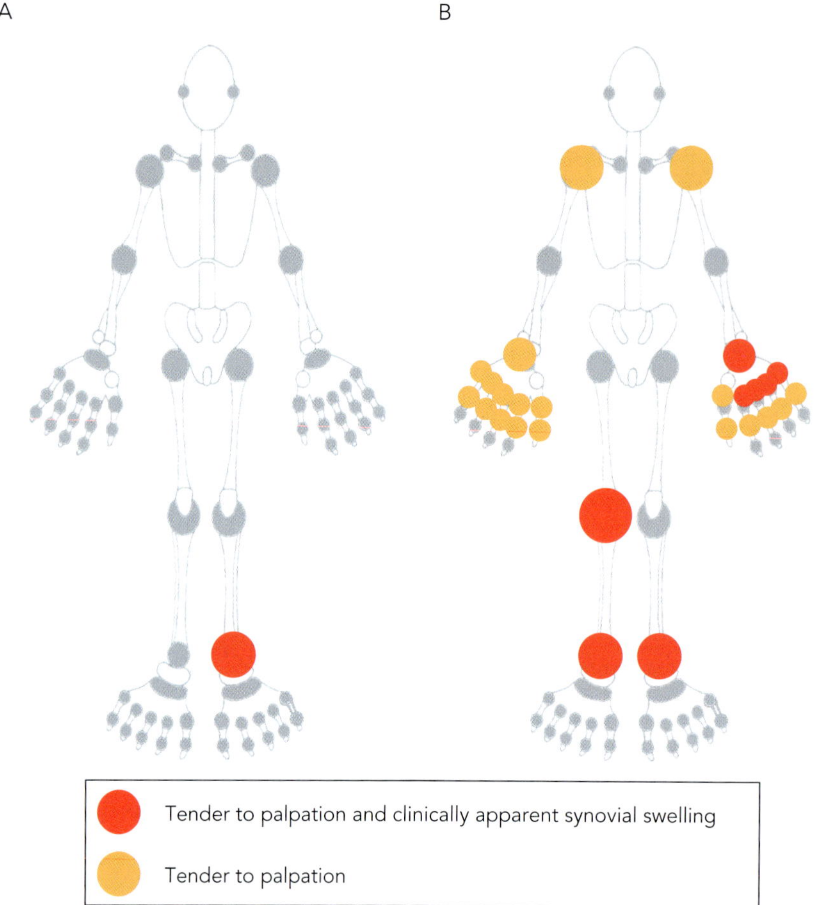

Figure 22.1. Joints that were tender to palpation, and joints that were both tender to palpation and had clinically apparent synovial swelling, at presentation to secondary care (A) and 6 weeks later (B).

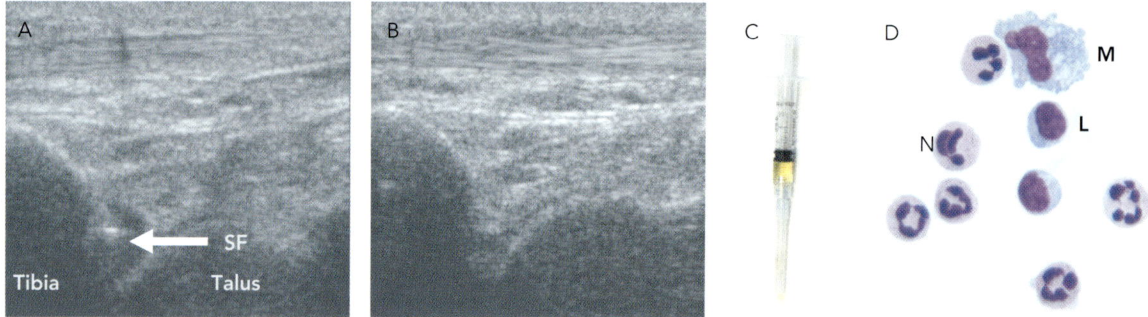

Figure 22.2. (A) Ultrasound scan of the inflamed ankle joint at clinical presentation showing a joint effusion that was aspirated under ultrasound guidance. (B) Image of ankle joint following aspiration of synovial fluid. (C) Image of synovial fluid aspirated from ankle. (D) Diff-quik stained cytospin preparation of synovial fluid cells showing a mixed population of neutrophils (N), lymphocytes (L), and macrophages (M).

peripheral inflammatory polyarthritis (Figure 22.4) that leads to destruction of articular cartilage and bone (as illustrated by the development of erosions in Figure 22.5) and may be associated with extraarticular features including inflammatory disease within the skin, lungs, and vascular systems. Rheumatoid arthritis causes significant disability [6,7] and enhanced mortality, predominantly related to accelerated ischemic heart disease [8,9].

Whilst many patients present with features that will allow their immediate classification as having RA, some patients (including the case reported

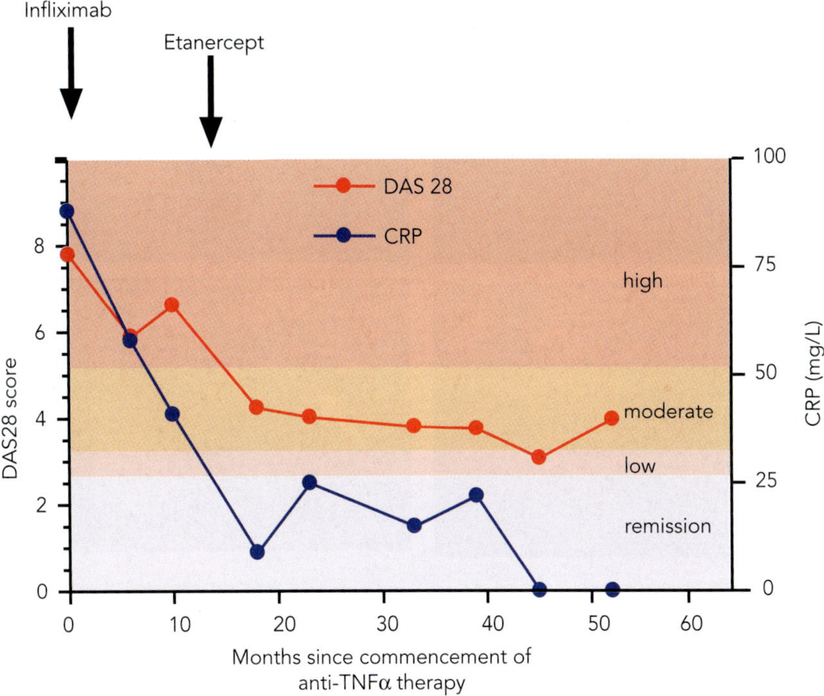

Figure 22.3. Change in DAS28 score and CRP level over time following the introduction of anti-TNFα therapy. DAS28 thresholds for high (DAS28 > 5.1), moderate (3.2 < DAS28 ≤ 5.1), low (2.6 ≤ DAS28 ≤ 3.2) disease activity and for remission (DAS 28 < 2.6) are shown.

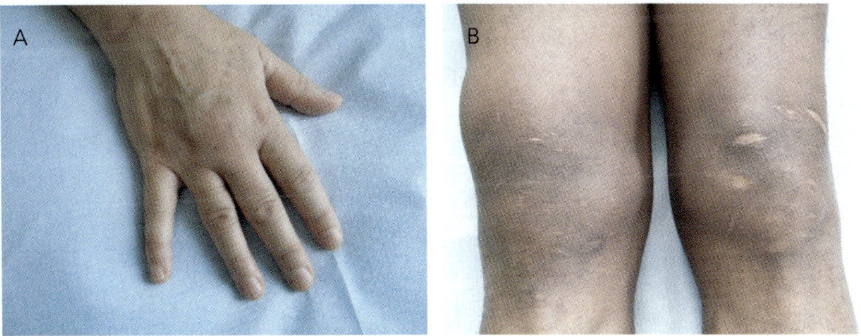

Figure 22.4. (A) Synovial swelling of the metacarpophalyngeal (MCP) and proximal interphalyngeal (PIP) joints of the right hand in patients with rheumatoid arthritis. (B) Bilateral synovial swelling of the knee joints in a patients with rheumatoid arthritis, with a very large synovial effusion in the right knee.

earlier) have a disease that gradually evolves into RA. Recent work has highlighted the role of demographic, clinical, and serological features in the prediction of the development of RA [10,11]. In particular, seropositivity for both rheumatoid factor and the anti-CCP antibody (as in the case reported earlier) has a very high specificity for the development of RA [12,13]. The ability to accurately predict the development of RA in patients with early unclassified arthritis is important as early aggressive therapy in patients with RA has been shown to significantly improve articular outcomes [14,15].

After the onset of clinically apparent joint disease, the normally hypocellular synovial membrane becomes hyperplastic. This inflamed synovium contains a superficial lining layer of synovial fibroblasts and macrophages overlying a layer that contains an intense cellular infiltrate including macrophages, T cells (both CD4+ve and CD8+ve), B cells, plasma cells, natural killer (NK) cells, NKT cells, mast cells, and fibroblasts. This cellular infiltrate drives the process of bone and cartilage destruction, which is mediated predominantly by fibroblasts and osteoclasts. Cytokines derived from these infiltrating cells integrate many of

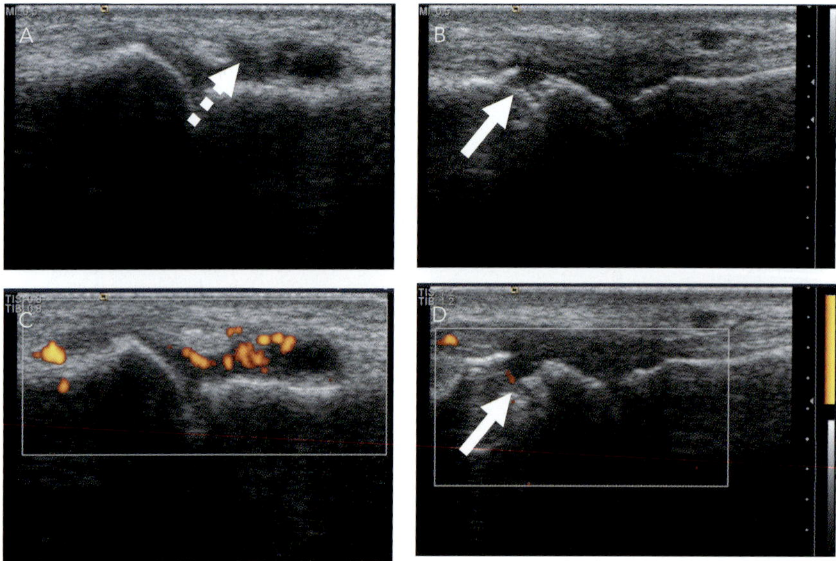

Figure 22.5. Longitudinal (A and C) and oblique longitudinal (B and D) ultrasound views of the second metacarpophalyngeal joint showing synovitis, which appears as a hypoenhocic area (dashed arrow), power Doppler positivity (C and D), and (in oblique longitudinal view) an erosion (arrow).

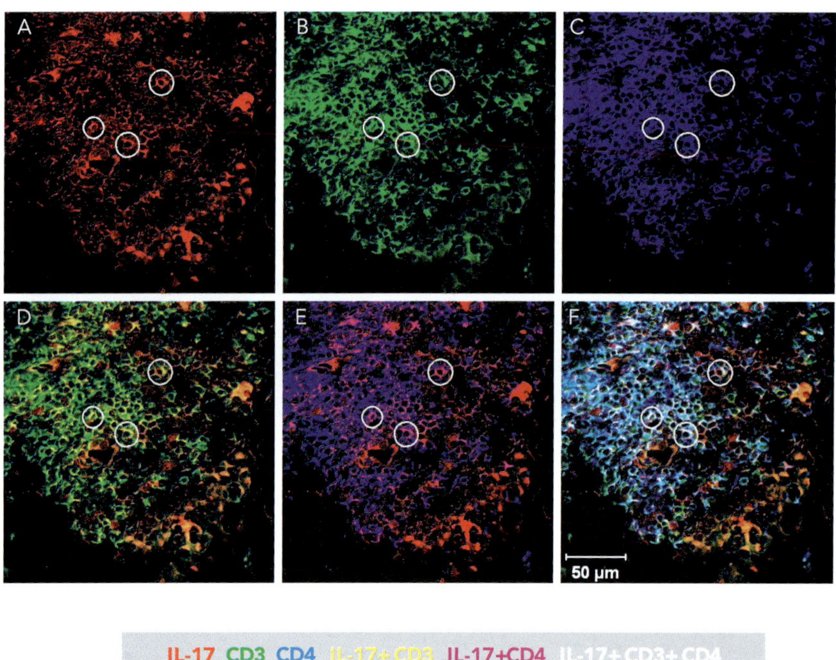

Figure 22.6. IL-17 is expressed by CD4 positive T cells within the rheumatoid synovium. Sections of synovial tissue were stained with (A) anti-17 mAb (red), (B) anti-CD3 mAb (green), and (C) anti-CD4 mAb (blue). Overlay images are shown in D, E, and F.

the inflammatory and destructive events that characterize the joint disease of RA (reviewed in [16]) and, like some of the cellular components of the inflammatory infiltrate, have been identified as important therapeutic targets.

Synovial macrophages are activated via a number of routes including the binding of immune complexes to FcγR, the ligation of Toll-like receptors [17] and direct T cell contact [18]. Such activated macrophages are an important source of proinflammatory cytokines including TNFα, IL-1, IL-6, IL-15, and IL-23. Activated T cells play a direct role in macrophage activation and also produce IL-17 (Figure 22.6) which can itself activate macrophages as well as fibroblasts and

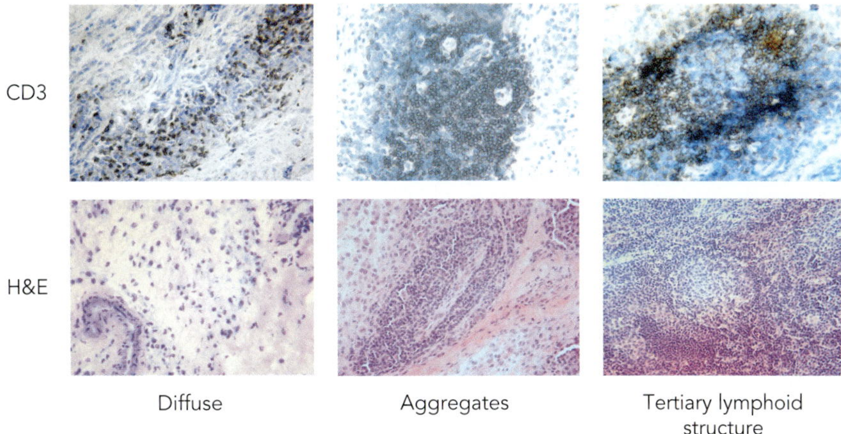

Figure 22.7. Staining for CD3 by immunohistochemistry and hematoxylin and eosin staining of rheumatoid synovium showing the different patterns of inflammatory infiltrate.

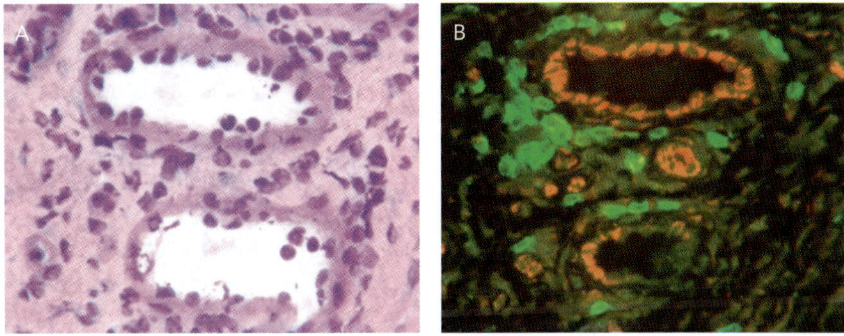

Figure 22.8. SDF-1 is expressed on endothelial cells and in a perivascular distribution within the rheumatoid synovium. Sections of synovial tissue were stained with (A) Diff-Quik. (B) Anti-SDF-1 mAb (red) and anti-CD3 mAb (green).

osteoclasts. B cells and plasma cells produce proinflammatory cytokines (e.g., IL-6) and autoantibodies (e.g., Rheumatoid factor, anti-citrullinated protein antibodies, and anti-type II collagen antibodies) and may play a role in local T-cell activation via the presentation of peptide on MHC class II molecules. The patterns of lymphocytic infiltrate vary between patients. Cells can be diffusely distributed, loosely aggregated, or form organized germinal center like structures (which are found in approximately 20% of RA patients) (Figure 22.7). Synovial fibroblast play a central role in orchestrating this inflammatory process (reviewed in [19]). Through soluble and contact dependent mechanism they maintain the survival of T cell and B cells [20,21]. Furthermore, they facilitate the retention of leukocytes within the inflamed synovium. In the case of T cells, for example, the interaction between the chemokine receptor CXCR4 on T cells (whose expression is upregulated by fibroblast-derived TGFβ) and the chemokine SDF-1 on endothelial cells and synovial fibroblasts (Figure 22.8) restricts the emigration of T cells from the synovium [22,23]. In addition, fibroblasts are able to directly invade cartilage and bone and contribute to the joint destruction that characterizes RA [24]. The hyperplastic synovium is sustained by a vascular network (Figure 22.5C and 22.5D) that develops as a consequence of angiogenesis though the synovial environment remains profoundly hypoxic [25].

Increased understanding of pathological processes operating within the rheumatoid joint has led to the identification of new therapeutic targets and has already led to real clinical benefit. The anti-TNFα agents infliximab, etanercept, and adalimumab are associated with significant improvements in disease activity with approximately 50% of patients achieving an ACR50 response with any of the drugs. As in the case reported, a poor response to one agent does not necessarily mean that the entire class will be ineffective [26]. Other components of the inflammatory milieu that have been effectively targeted include IL-6 [27], B cells [28], and T cells [29].

SYSTEMIC LUPUS ERYTHEMATOSUS

Systemic Lupus Erythematosus: Clinical Case

A 20-year-old Afro-Caribbean woman weighing 60 kg presented with a 3-month history of symmetrical joint pain and early morning stiffness accompanied by tenderness but little swelling affecting her wrists, proximal interphalyngeal joints and knees in a migratory pattern. There was no preceding infection, no associated systemic features and no relevant past medical or family history. The inflammatory arthritis improved with ibuprofen but recurred each time she tried to stop the medication. At presentation, her blood pressure was 129/75 and urinalysis was negative. However, her initial full blood count showed leucopenia with predominantly neutropenia (total white count 2.5×10^9/L with neutrophils 1.1×10^9/L) and platelets were close to the lower limit of normal at 152×10^9/L. She had an ESR of 105 mm/h and a CRP of 11 mg/L (upper limit of normal 5 mg/L). Further investigation showed that she was rheumatoid factor negative, ANA 1 in 1,600 and that she had anti-ds DNA antibodies that were Crithidia positive with a titer on ELISA of 133 ku/L. She was also found to have anti-Ro, anti-La, anti-RNP, and anti-Sm antibodies but no anticardiolipin antibodies and lupus anticoagulant was negative. Her complement levels were low with C3 0.50, C4 0.05 mg/L.

She missed her next appointment but returned 6 months later feeling more unwell with increasing pain and swelling in the joints, disabling fatigue, diffuse alopecia, lymphadenopathy with multiple cervical lymph nodes measuring 1–2 cm, weight loss of 8 kg and urinalysis was strongly positive for protein on dipstick testing. Blood pressure was 121/66 and the protein:creatinine ratio was raised at 54 mg/mmol (normal < 2). She had evidence of pancytopenia with Hb 9.4 g/dl, white count reduced to 1.7×10^9/L, neutrophils down to 1.0×10^9/L, lymphocytes low at 0.6×10^9/L, and platelets reduced at 100×10^9/L. Her anti-dsDNA antibodies had risen to 4,523 ku/L and complement levels had fallen further to C3 0.32 and C4 0.02 g/L. She was started on prednisolone, 40 mg daily, and had three pulses of intravenous (IV) methylprednisolone while she had a breast lump investigated. This was a benign breast adenoma. Meanwhile, her proteinuria deteriorated with a protein:creatinine ratio of 84 mg/mol. She developed red and white cells in the urine with no evidence of infection but she had red cell casts and her blood pressure rose to 148/73.

She was referred for a renal biopsy.

Renal biopsy showed mild mesangial hypertrophy (Figure 22.9A and 22.9B) with slightly thickened basement membranes (Figure 22.10A and 22.10B). In addition, there were some glomeruli with sharply outlined segmental lesions ranging from thrombosis and tuft disruption through groups of cells in Bowman's space to poorly cellular sclerosed areas (Figure 22.9C and 22.9D). The tubules, interstitial tissues, and blood vessels were mostly normal though there were a few areas with lymphocytic infiltration. Immunoperoxidase showed heavy deposition of IgG (Figure 22.11A), C9 (Figure 22.11B), and IgA (Figure 22.11C) in mesangium and as regular granules on the outside of glomerular capillary loops with IgM in the mesangium (Figure 22.11D). There was no chronic damage. The features were consistent with active lupus nephritis with a mixture of proliferative and membranous glomerulonephritis consistent with ISN/RPS class V and III (see Table 22.1 and the subsequent text for details of this classification system).

She was treated with IV pulse methylprednisolone and IV cyclophosphamide 3–4 weekly for 6 months, together with a slowly reducing course of oral prednisolone from 40 to 20 mg daily. She felt much better and her symptoms all resolved. Her blood pressure settled to 119/71 with an angiotensin converting enzyme inhibitor (lisinopril 5 mg daily). Her pancytopenia improved and the anti-dsDNA antibodies resolved with improvement of C3 to 0.70 g/L; and C4 to 0.05 g/L. Protein:creatinine ratio slowly improved to 17 mg/mmol and the urinary cells and casts resolved. However the protein:creatinine ratio later rose to 111 mg/mmol after stopping cyclophosphamide pulses despite being switched to oral ciclosporin, possibly because she increased the dose more slowly than recommended and yet reduced prednisolone to 10 mg daily. After 4 months on ciclosporin (but only 1 month on the target dose of 2 mg/kg/day) she rapidly developed a disease flare with a diffuse maculo-papular lupus rash on her limbs after sun exposure, digital infarcts, fatigue, and recurrence of lymphadenopathy over 3 weeks. The protein:creatinine ratio rose to 209, the anti-dsDNA antibodies to 194 ku/L, the C3 fell to 0.48 g/L, and the C4 to 0.03 g/L. She improved with an increase in prednisolone to 30 mg daily. Ciclosporin was increased to 3 mg/kg/day, the maximum dose that we would use for lupus and the steroids were gradually reduced to 10 mg daily.

Unfortunately, 1 year after starting ciclosporin she presented again with widespread rash, extreme fatigue, anorexia with weight loss, and leg edema to the knee complicated by a chest infection. She was diagnosed as having nephrotic syndrome as her protein:creatinine ratio had increased to 613 mg/mol and she had become hypertensive with a BP of 160/95, so her lisinopril was increased. Her serum creatinine had risen steadily from 75 µmol/L to 108 micromol/L over 3 months while her creatinine clearance fell from 97 to 66 mL/min. Her anti-dsDNA antibodies had risen to 284 ku/L and her C3 and C4 fell to 0.46 and

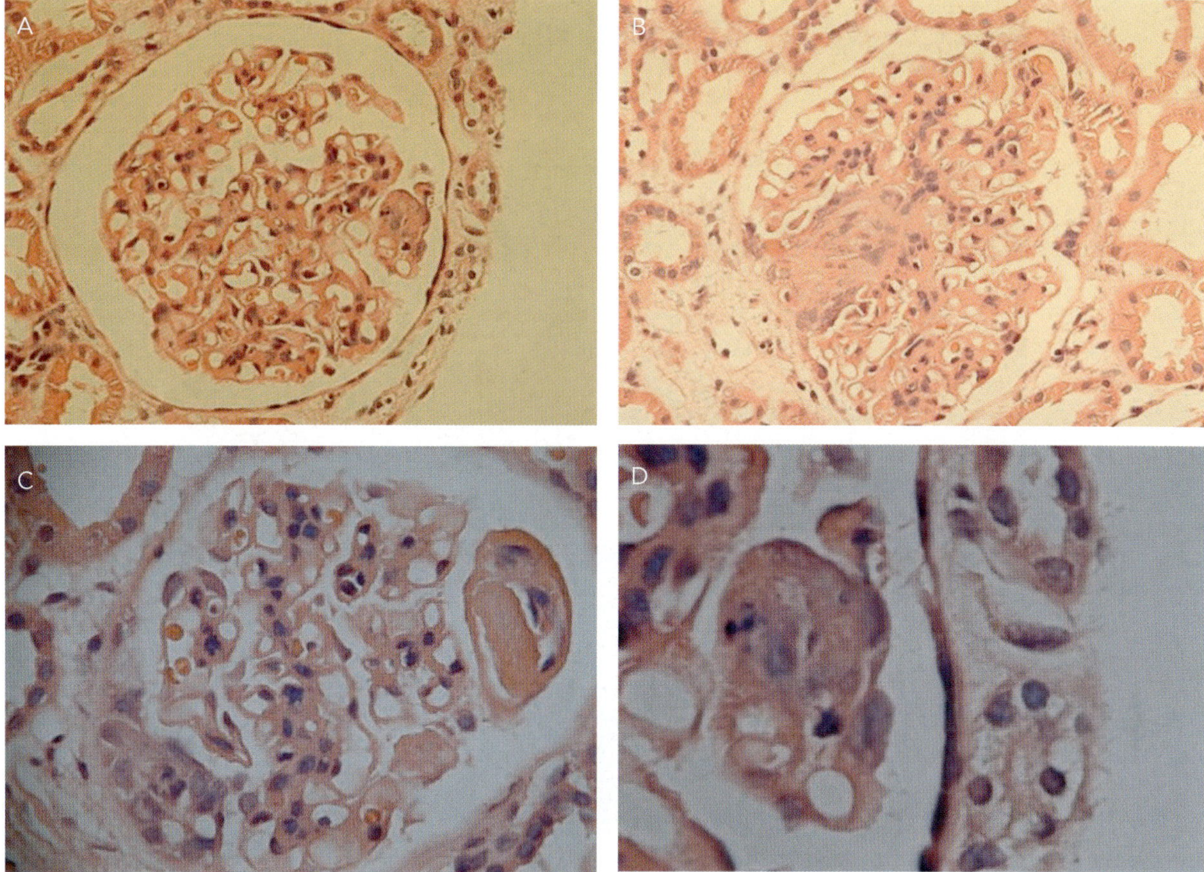

Figure 22.9. (A) First renal biopsy (1 of 21 glomeruli, H&E, low power). (B) First renal biopsy (1 of 21 gomeruli, H&E, low power). (C) First renal biopsy (1 of 21 gomeruli, H&E, high power). (D) First renal biopsy (1 of 21 gomeruli, H&E, high power).

0.03 g/L, respectively, but the C3 became normal at 0.87 g/L, with low C4 0.09 g/L and high C3d and CRP 124 with the infection. The CRP settled to <5 g/L after antibiotics.

The ciclopsorin was stopped and she was treated with mycophenolate mofetil 2 g/day and a reducing course of prednisolone starting at 40 mg a day with benefit for about 9 months but she deteriorated again on prednisolone 15 mg daily. She had a recurrence of facial rash, alopecia, and arthritis. Her proteinuria had dropped to a protein:creatinine ratio of 138 mg/mmol but then rose back up to 394 mg/mmol though renal function remained in the normal range. Her anti-dsDNA antibodies that had disappeared became positive again at 119 ku/L, and her C3 and C4 that had normalized fell again to 0.58 and 0.05 g/L, respectively. She underwent repeat a further renal biopsy.

The biopsy consisted of 18 glomeruli, 5 of which were globally sclerosed and the remainder had thickened basement membranes (Figure 22.12A–22.12E identified with silver stain). There was focal and segmental endocapillary proliferation, disruption of capillary loops and karyorrhectic debris in some glomeruli, and segmental sclerosis in others (Figure 22.13A and 22.13B). There was mild tubular atrophy and interstitial fibrosis with some acute tubular injury. Immunostaining showed positive membranous and mesangial staining with IgG, IgM, IgA, and C9 (Figure 22.14A–22.14D). These findings were consistent with the ISN/RPS Class V with III (A/C). In view of the persisting disease activity she received full dose mycophenolate mofetil (3 g/day) but as she still did not go in to renal remission she was subsequently given two doses of rituximab therapy with a course of increased oral prednisolone. Following this, she was asymptomatic with normal anti-dsDNA antibodies, C3, C4, renal function, and proteinuria fell to a protein:creatinine ratio of less than 100 mg/mmol for the first time since initial remission after the course of cyclophosphamide.

Systemic Lupus Erythematosus: Discussion

SLE affects about 1 in 2,000 adult women and is more common in patients of Afro-Caribbean origin than in Caucasians [30]. The disease most often presents between the ages of 25 and 45 but may develop at any age from 1 to over 80 years. Renal disease occurs in 35%–65% of patients, depending on the ethnic

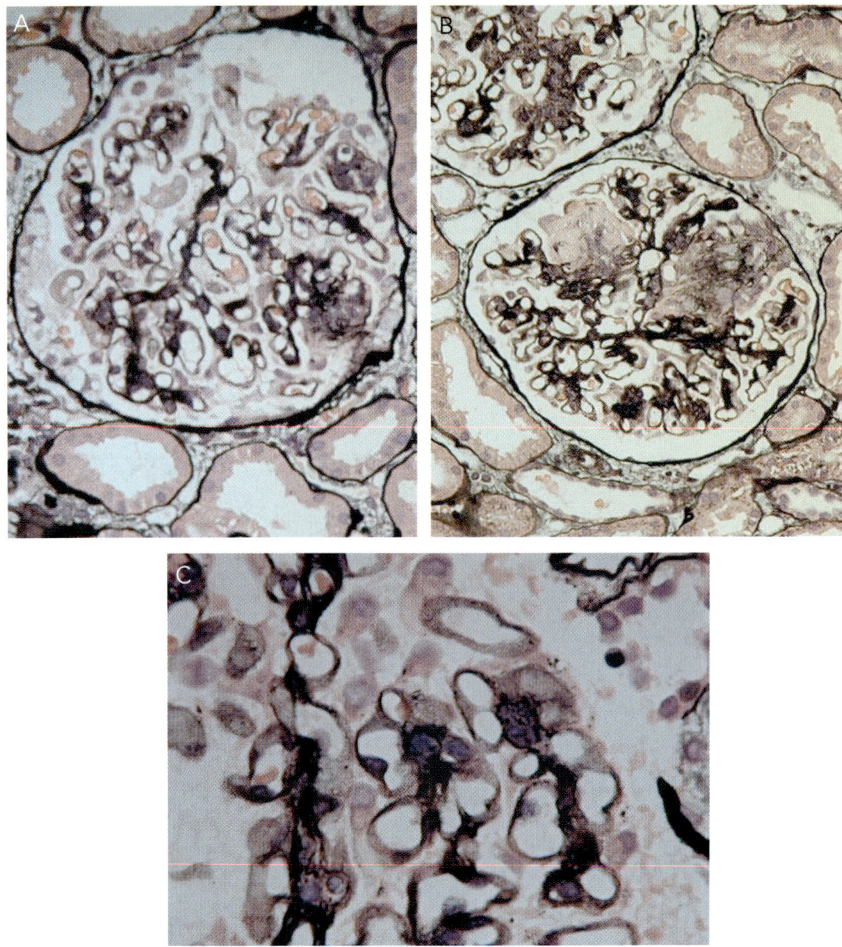

Figure 22.10. (A) CG: first renal biopsy (1 of 21 gomeruli, silver stain, low power). (B) CG: first renal biopsy (1 of 21 gomeruli, silver stain, low power). (C) CG: first renal biopsy (1 of 21 gomeruli, silver stain, high power).

composition of the cohort studied. Preceding skin and joint involvement is seen in many but not all patients. There have also been a number of studies showing that it tends to present at a younger age and to more often involve the kidneys in patients of Afro-Caribbean and Afro-Americans origin [31]. The disease is associated with the development of autoantibodies, the formation of immune complexes, and the activation of complement and other inflammatory pathways. Many autoantibodies are produced in lupus patients but only a few are tested for in the clinic [32].

The classic screening test is the antinuclear antibody test which is sensitive but not specific for lupus. For monitoring of increasing disease activity, particularly in patients with renal involvement, the most useful tests are rising anti-dsDNA antibodies and falling complement C3 and C4 levels [33] as illustrated by this patient. Sometimes, the anti-dsDNA antibody level will fall after rising as a flare develops [34], possibly due to tissue deposition and/or excretion in the urine. Anti-dsDNA antibodies are only positive in about 60% of patients but are positive in most patients with lupus nephritis. It is particularly important to screen patients for renal involvement if they have anti-dsDNA antibodies as there are no symptoms initially (see the subsequent text). CRP is usually normal or only slightly elevated in lupus patients with active disease even when the ESR is high, in contrast to RA patients in which both tests will be significantly elevated. However, in lupus patient with infection the CRP rises as usual, as seen in this patient when she had a chest infection.

Almost any part of the body can be affected but involvement of the skin (mucocutaneous features), the joints (synovitis) and cytopenias are common, particularly at onset (see Table 22.1). The diagnosis is made most easily when there are typical rashes, but it may be delayed in their absence. Misdiagnosis of the arthritis may occur as the distribution is often the same as that of RA as in this case. Nodules may occur and lupus patients are rheumatoid factor positive in about 50% of cases but they are rarely anti-CCP antibody positive [12,13]. In contrast to RA, the joints in lupus are

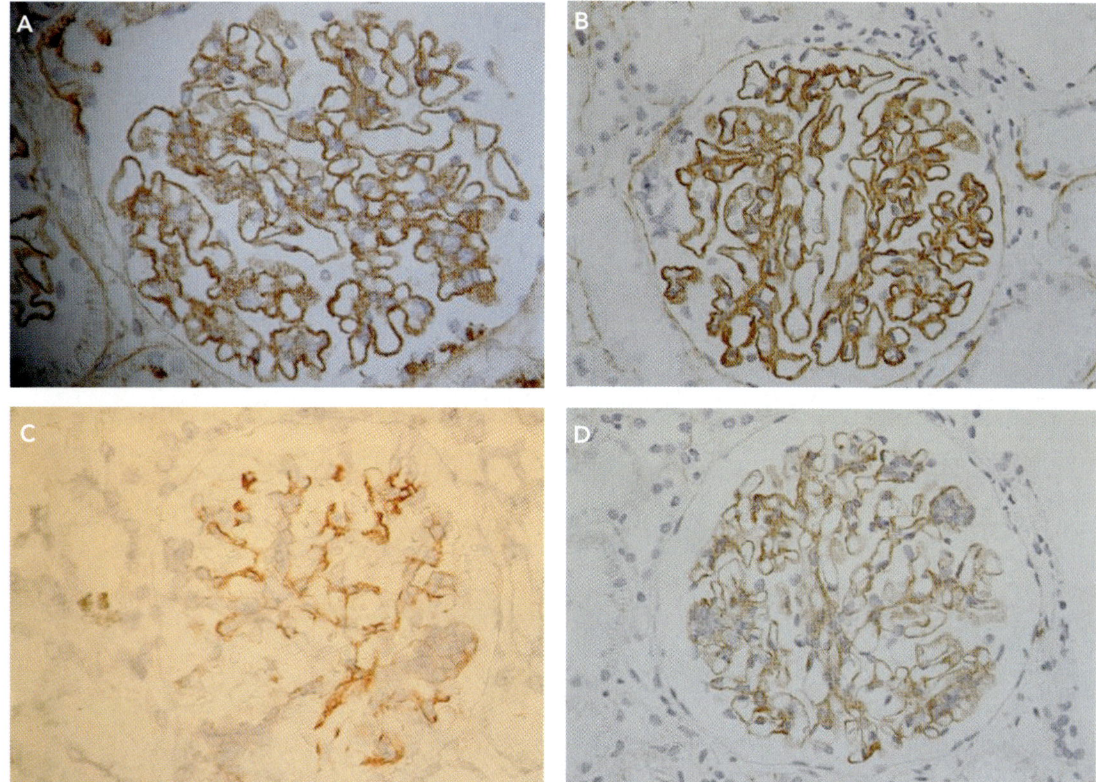

Figure 22.11. (A) CG: first renal biopsy (1 of 21 gomeruli, peroxidase stain to demonstrate IgG deposition, low power). (B) CG: first renal biopsy (1 of 21 gomeruli, peroxidase stain to demonstrate C9 deposition, low power). (C) CG: first renal biopsy (1 of 21 gomeruli, peroxidase stain to demonstrate IgM deposition, low power). (D) CG: first renal biopsy (1 of 21 gomeruli, peroxidase stain to demonstrate IgA deposition, low power).

TABLE 22.1. International Society of Nephrology/Renal Pathology Society Classification of lupus nephritis (2003)

Class I	**Minimal mesangial lupus nephritis** Normal at light microscopy Mesangial deposits on immunofluorescence
Class II	**Mesangial proliferative lupus nephritis** Mesangial hypercellularity or expansion with mesangial immune deposits Some subepithelial or subendothelial deposits on immunofluorescence by electron microscopy
Class III	**Focal lupus nephritis** Involves <50% glomeruli. Active or inactive lesions typically with subendothelial deposits **A**: Active lesions; focal proliferative lupus nephritis **A/C**: Active and chronic lesions; focal proliferative and sclerosing lupus nephritis **C**: Chronic inactive lesions with glomerular scars; focal sclerosing lupus nephritis
Class IV	**Diffuse lupus nephritis** Involves >50% glomeruli. Active or inactive diffuse, segmental or global endo-or extracapillary glomerulomephritis. Typically with subendothelial deposits. Divided into diffuse segmental (S) when >50% of involved glomeruli have segmental lesions and diffuse global when >50% of involved glomeruli have global lesions. **S(A)**: Active lesions; diffuse segmental proliferative lupus nephritis **G(A)**: Active lesions; diffuse global proliferative lupus nephritis **S(A/C)**: Active and chronic lesions; diffuse segmental proliferative and sclerosing lupus nephritis **G(A/C)**: Active and chronic lesions; diffuse global proliferative and sclerosing lupus nephritis **S(C)**: Chronic inactive lesions with scars; diffuse segmental sclerosing lupus nephritis **G(C)**: Chronic inactive lesions with scars; diffuse global sclerosing lupus nephritis
Class V	**Membranous lupus nephritis** Global or segmental subepithelial immune deposits by light microscopy and immunofluorescence or electron microscopy, with or without mesangial changes. Class V lupus nephritis may occur in combination with class III or class IV disease in which case both are diagnosed. Class V disease may show advanced sclerosis
Class VI	**Advanced sclerosis lupus nephritis** >90% of glomeruli globally sclerosed without residual activity

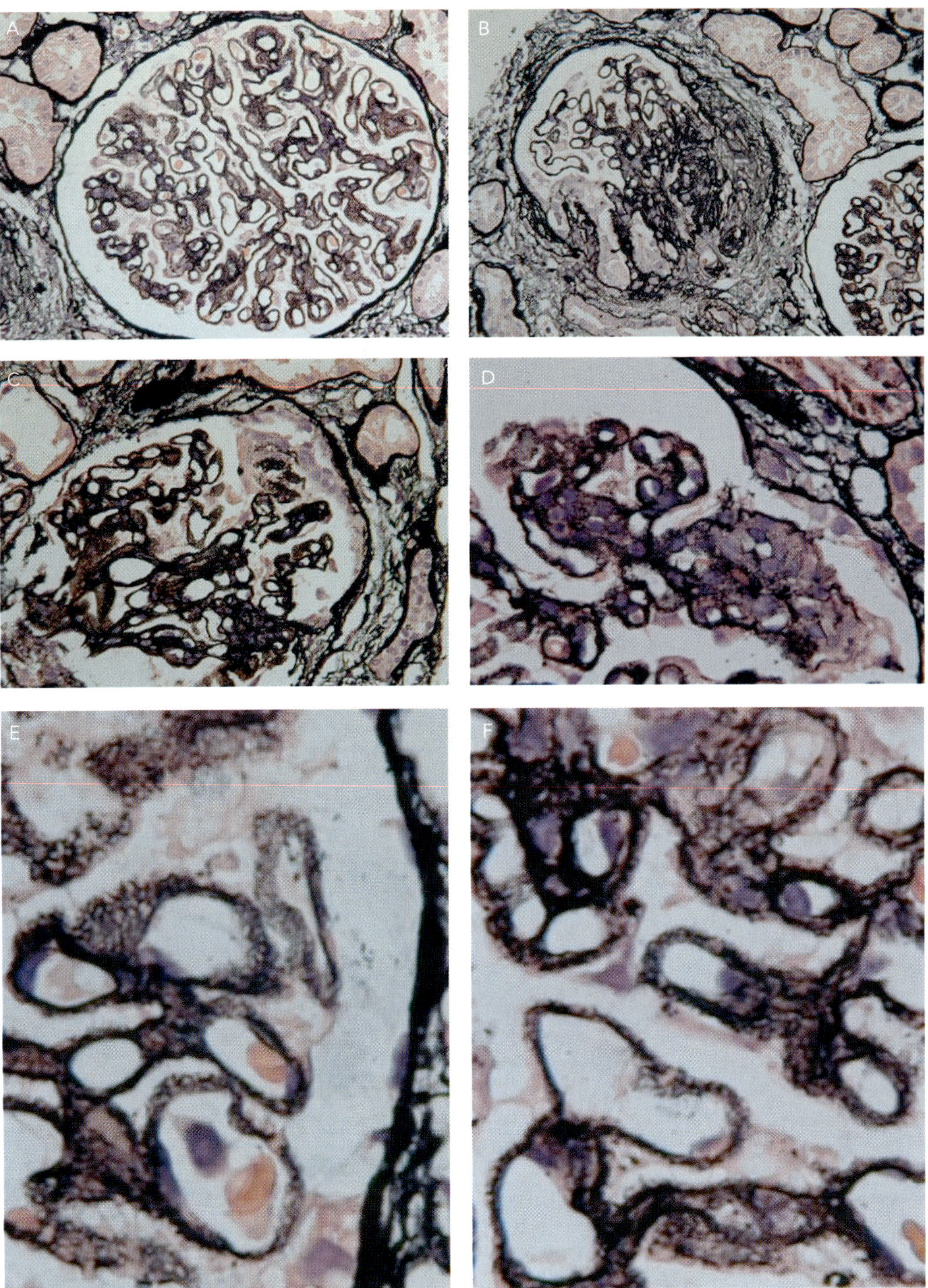

Figure 22.12. (A) CG: second renal biopsy (1 of 17 glomeruli, silver stain, low power). (B) CG: second renal biopsy (1 of 17 glomeruli, silver stain, low power). (C) CG: second renal biopsy (1 of 17 glomeruli, silver stain, low power). (D) CG: second renal biopsy (1 of 17 glomeruli, silver stain, high power). (E) CG: second renal biopsy (1 of 17 glomeruli, silver stain, high power). (F) CG: second renal biopsy (1 of 17 glomeruli, silver stain, high power).

usually painful, stiff, and tender but not necessarily swollen. Arthritis, although it may occur in the same distribution as that of RA can be persistent, is more often migratory with one or more joints affected for only a few days at a time, but a variety of joints are involved in turn if not treated.

Various rashes are associated with lupus, particularly discoid rash, subacute cutaneous lupus, cutaneous vasculitis, and malar erythema (the butterfly rash). Other common mucocutaneous features include mucosal ulceration and alopecia. Fatigue is a frequent complaint but other constitutional features including

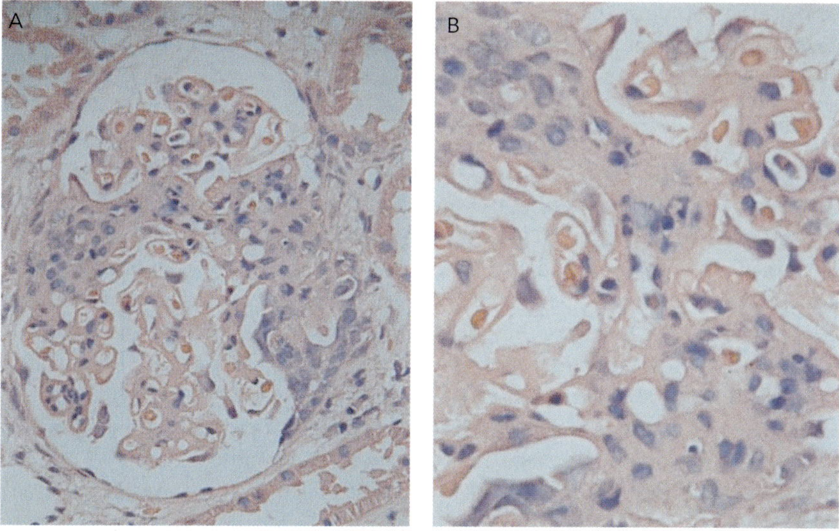

Figure 22.13. (A) CG: second renal biopsy (1 of 17 glomeruli, H&E, low power). (B) CG: second renal biopsy (1 of 17 glomeruli, H&E, high power).

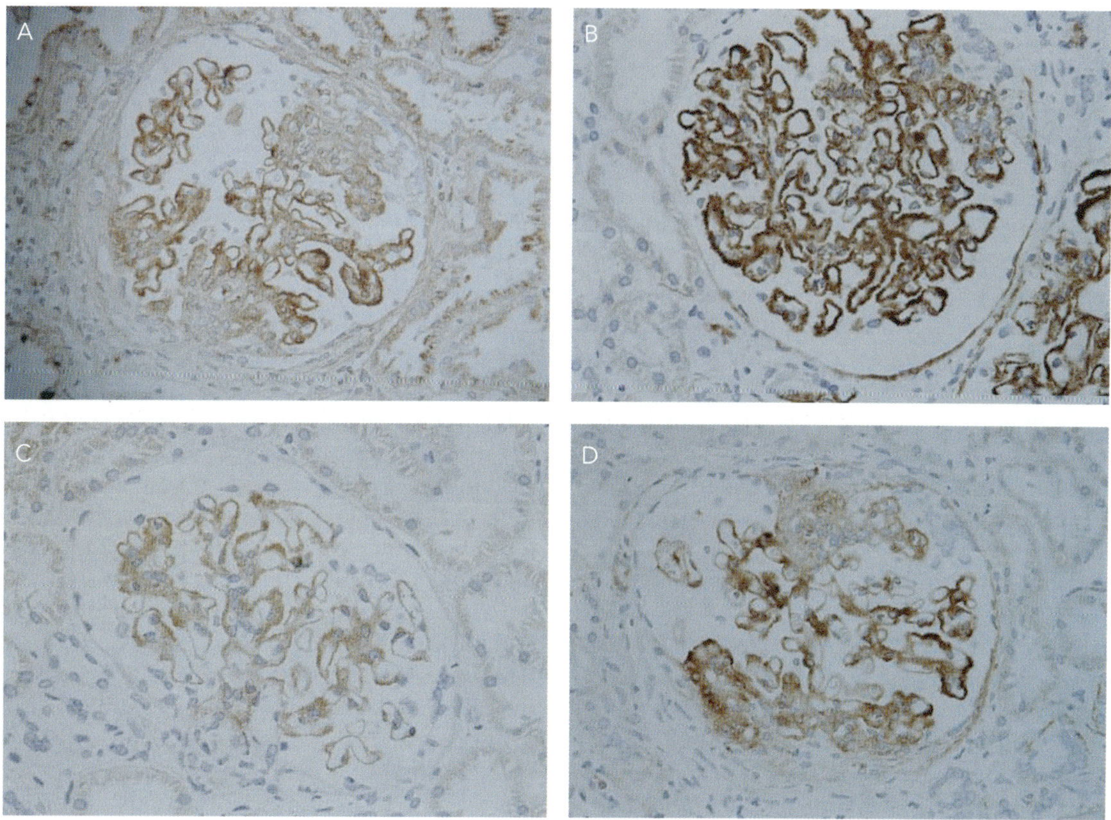

Figure 22.14. (A) CG: second renal biopsy (1 of 21 gomeruli, peroxidase stain to demonstrate IgG deposition, low power). (B) CG: second renal biopsy (1 of 21 gomeruli, peroxidase stain to demonstrate C9 deposition, low power). (C) CG: second renal biopsy (1 of 21 gomeruli, peroxidase stain to demonstrate IgA deposition, low power). (D) CG: second renal biopsy (1 of 21 gomeruli, peroxidase stain to demonstrate IgM deposition, low power).

fever, lymphadenopathy, and weight loss may occur, particularly early in the disease course and in younger patients. Patients with these nonspecific features need to be investigated carefully for infection, malignancy, and other comorbid disease before attributing the features to lupus. This also applies to the cytopenias, as despite being directly antibody mediated, it is not often possible to confirm the presence of antibodies to the white blood cell subsets or to platelets in clinical practice, though Coombs' test is available as part of the assessment of hemolytic anemia. Of the cytopenias, lymphopenia is the most common but hemolytic anemia and thrombocytopenia can be the most serious and difficult to manage. An anemia of chronic disease is more common than hemolytic anemia and leucopenia may be drug induced, particularly in patients on cytotoxic agents. The presence of lymphopenia or leucopenia in a patient with inflammatory arthritis and low CRP is strongly suggestive of lupus and further testing for lupus should be undertaken.

Other important manifestations in lupus, but not illustrated by this patient's history, include pleuropericarditis which is sometimes associated with effusions, abdominal serositis which is less common but often misdiagnosed, neuropsychiatric manifestations which are varied [35] and include seizures, stroke, aseptic meningitis, psychosis, and mononeuritis multiplex. Patients with antiphospholipid syndrome may present with venous or arterial thrombosis, and recurrent miscarriages including second and third trimester fetal loss.

It is important to look for early renal involvement in lupus patients because lupus nephritis is a strong predictor of atherosclerotic disease and death as well as of end stage renal disease [36,37]. There are three main types of renal disease in general terms: proliferative glomerulonephritis, membranous nephropathy, and thrombotic microangiopathy related to antiphospholipid syndrome [38]. It is particularly important to distinguish the latter from the inflammatory forms of lupus nephritis, as thrombotic disease does not require immunosuppression. Clinical presentation does not accurately predict renal biopsy findings so the patient, as in our case, should be referred for biopsy if they have persistent proteinuria >0.5 g in 24 hour or protein:creatinine ratio >50 mg/mmol, red cells, or red casts in the urine. In patients with new onset proteinuria and/or cells in the urine, it is important to exclude infection and other potential causes of the renal parameters before undertaking biopsy. Renal function should be monitored but the decision of when to biopsy is not dependent on finding impaired renal function. Indeed, it is preferable to biopsy and to institute definitive treatment with an appropriate immunosuppressive regime before there is chronic damage resulting in irreversible renal impairment. It has been shown that increased time to definitive cytotoxic treatment (which usually follows biopsy) is associated with an increased risk of poor renal outcome and that early reduction in proteinuria is associated with a better prognosis [39–41]. Hypertension is common in patients with lupus nephritis and needs close monitoring, as this has been a predictor of increased mortality in many cohorts. Treatment usually involves angiotensin converting enzyme inhibitors or angiotensin receptor blockers. These drugs have the advantage that they reduce proteinuria and help to protect the kidneys from renal damage due to proteinuria and hypertension.

Renal biopsy provides more information about the type of renal involvement than urinalysis, urinary sediment, and serum creatinine, and establishes that the patient does have lupus nephritis and not an alternative renal diagnosis such as IgA nephropathy, which occasionally is found in SLE patients [42]. Renal biopsy does have some limitations as only a very small sample of tissue is obtained and lupus nephritis does not affect the kidney uniformly. The extent and pattern of immune deposits and inflammation in the glomeruli that are detected by immunohistochemistry on light microscopy determine the classification of lupus glomerulonephritis as shown in Figures 22.3A–22.3D and 22.6A–22.6D using immunoperoxidase. The characteristic finding in lupus nephritis, which is an immune complex–mediated glomerulonephritis, is positive staining for IgG, IgM, and IgA together with staining for C1q, C3, C4, and C9, but not all of these are present in all patients. In focal proliferative lupus nephritis as in the case described, there may be segmental areas of fibrinoid necrosis and karyorrhexis. In diffuse proliferative lupus nephritis, the glomeruli are globally involved with endocapillary proliferation, extensive fibrinoid necrosis and karyorrhexis. There may be irregular changes among different segments with varying numbers of wire loop lesions and hyaline thrombi reflecting subendothelial and intraluminal deposits of immune complexes with variable degrees of mesangial proliferation. The International Society of Nephrology/Renal Pathology Society (ISN/RPS) have published a classification system (see Table 22.1) that is now widely used and incorporates definitions of biopsy activity, chronicity, tubulo-interstitial disease and includes vascular pathology [43]. As with the previous WHO system, it is based on glomerular pathology but differs in that Class I disease now denotes minimal mesangial lupus nephritis. The ISN/RPS classification is summarized in Table 22.1. Repeat biopsy is recommended to confirm the presence of a clinically suspected renal flare that requires additional immunosuppression when there is a deterioration in proteinuria and/or recurrence of red cells or red cell cases, and to identify the development of chronic damage and progression that may limit the capacity for the renal signs to resolve with immunosuppression, particularly if the patient has renal impairment.

Traditionally patients with proliferative glomerulonephritis have been treated with monthly intravenous cyclophosphamide based on the NIH trials [44], but more recently there has been increasing interest and evidence supporting the use of shorter and lower dose cyclophosphamide regimes for induction of remission followed by azathioprine for maintenance therapy (the Eurolupus regime) [45]. Over the past 10 years, an alternative approach to treatment has been developed consisting of oral mycophenolate mofetil for induction and maintenance therapy in lupus nephritis [46–48]. Trial data suggest that this approach is as efficacious and may be safer than cyclophosphamide but there is no long-term outcome data yet to confirm that this regime can prevent end-stage renal disease. The problem with cyclophosphamide is that it causes infertility, particularly in women aged over 30 years and it has been associated with bladder carcinomas, so there is a need to find a safer therapy for patients with lupus nephritis [49,50]. At present, physicians often use the biopsy to help them decide which cytotoxic treatment to use with corticosteroids in lupus nephritis, with many preferring to continue with IV cyclophosphamide for those with class IV disease but they will use mycophenolate mofetil for those with class III disease, and for class V disease that has not responded to corticosteroids alone. However, it should be recognized that there is considerable variation in practice and in local policy. Neither drug is licensed for lupus nephritis but cyclophosphamide has been the standard of care for class IV lupus nephritis in most centers for at least the past 20 years. For patients that have failed these "conventional" treatments there have been case reports and case series suggesting that B-cell depletion therapy with the anti-CD20 monoclonal antibody, Rituximab may provide an effective way of switching off disease activity for 6–12 months with only two doses of therapy [51,52]. The results of clinical trials assessing the effects of mycophenolate mofetil and rituximab in lupus nephritis are eagerly awaited but it will be a long time before there is evidence that these new drugs (or other biologicals currently being developed for this indication) can prevent end-stage renal disease and/or improve mortality from the disease or its complications, such as the predisposition to premature atherosclerosis.

REFERENCES

1. Arnett, F.C., Edworthy, S.M., Bloch, D.A., et al. 1988. The American Rheumatism Association 1987 revised criteria for the classification of rheumatoid arthritis. *Arthritis Rheum* **31:**315–324.
2. Fransen, J., and van Riel, P.L. 2005. The Disease Activity Score and the EULAR response criteria. *Clin Exp Rheumatol* **23:**S93–S99.
3. Tunn, E.J., and Bacon, P.A. 1993. Differentiating persistent from self-limiting symmetrical synovitis in an early arthritis clinic [see comments]. *Br J Rheumatol* **32:**97–103.
4. Harrison, B.J., Symmons, D.P., Brennan, P., Barrett, E.M., and Silman, A.J. 1996. Natural remission in inflammatory polyarthritis: issues of definition and prediction. *Br J Rheumatol* **35:**1096–1100.
5. Symmons, D., Turner, G., Webb, R., et al. 2002. The prevalence of rheumatoid arthritis in the United Kingdom: new estimates for a new century. *Rheumatology (Oxford)* **41:**793–800.
6. Wolfe, F., and Cathey, M.A. 1991. The assessment and prediction of functional disability in rheumatoid arthritis. *J Rheumatol* **18:**1298–1306.
7. Wolfe, F., and Hawley, D.J. 1998. The long-term outcomes of rheumatoid arthritis: Work disability: a prospective 18-year study of 823 patients. *J Rheumatol* **25**:2108–2117.
8. del Rincon, I.D., Williams, K., Stern, M.P., Freeman, G.L., and Escalante, A. 2001. High incidence of cardiovascular events in a rheumatoid arthritis cohort not explained by traditional cardiac risk factors. *Arthritis Rheum* **44:**2737–2745.
9. Bacon, P.A., and Townend, J.N. 2001. Nails in the coffin: increasing evidence for the role of rheumatic disease in the cardiovascular mortality of rheumatoid arthritis. *Arthritis Rheum* **44:**2707–2710.
10. van der Helm-van Mil, A.H., le Cessie, S., van Dongen, H., Breedveld, F.C., Toes, R.E., and Huizinga, T.W. 2007. A prediction rule for disease outcome in patients with recent-onset undifferentiated arthritis: how to guide individual treatment decisions. *Arthritis Rheum* **56**:433–440.
11. van der Helm-van Mil, A.H., Detert, J., le, C.S., et al. 2008. Validation of a prediction rule for disease outcome in patients with recent-onset undifferentiated arthritis: moving toward individualized treatment decision-making. *Arthritis Rheum* **58:**2241–2247.
12. Raza, K., Breese, M., Nightingale, P., et al. 2005. Predictive value of antibodies to cyclic citrullinated peptide in patients with very early inflammatory arthritis. *J Rheumatol* **32:**231–238.
13. Nell, V.P., Machold, K.P., Stamm, T.A., et al. 2005. Autoantibody profiling as early diagnostic and prognostic tool for rheumatoid arthritis. *Ann Rheum Dis* **64:**1731–1736.
14. Quinn, M.A., Conaghan, P.G., O'Connor, P.J., et al. 2005. Very early treatment with infliximab in addition to methotrexate in early, poor-prognosis rheumatoid arthritis reduces magnetic resonance imaging evidence of synovitis and damage, with sustained benefit after infliximab withdrawal: results from a twelve-month randomized, double-blind, placebo-controlled trial. *Arthritis Rheum* **52:**27–35.
15. Goekoop-Ruiterman, Y.P., Vries-Bouwstra, J.K., Allaart, C.F.. et al. 2005. Clinical and radiographic outcomes of four different treatment strategies in patients with early rheumatoid arthritis (the BeSt study): a randomized, controlled trial. *Arthritis Rheum* **52:**3381–3390.
16. McInnes, I.B., and Schett, G. 2007. Cytokines in the pathogenesis of rheumatoid arthritis. *Nat Rev Immunol* **7**:429–442.
17. Brentano, F., Kyburz, D., Schorr, O., Gay, R., and Gay, S. 2005. The role of Toll-like receptor signalling in the pathogenesis of arthritis. *Cell Immunol* **233**:90–96.
18. Brennan, F.M., Hayes, A.L., Ciesielski, C.J., Green, P., Foxwell, B.M., and Feldmann, M. 2002. Evidence that rheumatoid arthritis synovial T cells are similar to cytokine-activated T cells: involvement of phosphatidylinositol 3-kinase and nuclear factor kappaB pathways in tumor necrosis factor alpha production in rheumatoid arthritis. *Arthritis Rheum* **46**:31–41.

19. Parsonage, G., Filer, A.D., Haworth, O., et al. 2005. A stromal address code defined by fibroblasts. *Trends Immunol* **26**:150–156.
20. Pilling, D., Akbar, A.N., Girdlestone, J., et al. 1999. Interferon-beta mediates stromal cell rescue of T cells from apoptosis. *Eur J Immunol* **29**:1041–1050.
21. Reparon-Schuijt, C.C., van Esch, W.J., van Kooten, C., et al. 2000. Regulation of synovial B cell survival in rheumatoid arthritis by vascular cell adhesion molecule 1 (CD106) expressed on fibroblast-like synoviocytes. *Arthritis Rheum* **43**:1115–1121.
22. Bradfield, P.F., Amft, N., Vernon-Wilson, E., et al. 2003. Rheumatoid fibroblast-like synoviocytes overexpress the chemokine stromal cell-derived factor 1 (CXCL12), which supports distinct patterns and rates of CD4+ and CD8+ T cell migration within synovial tissue. *Arthritis Rheum* **48**:2472–2482.
23. Buckley, C.D., Amft, N., Bradfield, P.F., et al. 2000. Persistent induction of the chemokine receptor CXCR4 by TGF-beta 1 on synovial T cells contributes to their accumulation within the rheumatoid synovium. *J Immunol* **165**:3423–3429.
24. Muller-Ladner, U., Kriegsmann, J., Franklin, B.N., et al. 1996. Synovial fibroblasts of patients with rheumatoid arthritis attach to and invade normal human cartilage when engrafted into SCID mice. *Am J Pathol* **149**:1607–1615.
25. Taylor, P.C., and Sivakumar, B. 2005. Hypoxia and angiogenesis in rheumatoid arthritis. *Curr Opin Rheumatol* **17**:293–298.
26. Hyrich, K.L., Lunt, M., Dixon, W.G., Watson, K.D., and Symmons, D.P. 2008. Effects of switching between anti-TNF therapies on HAQ response in patients who do not respond to their first anti-TNF drug. *Rheumatology (Oxford)* **47**:1000–1005.
27. Smolen, J.S., Beaulieu, A., Rubbert-Roth, A., et al. 2008. Effect of interleukin-6 receptor inhibition with tocilizumab in patients with rheumatoid arthritis (OPTION study): a double-blind, placebo-controlled, randomised trial. *Lancet* **371**:987–997.
28. Edwards, J.C., Szczepanski, L., Szechinski, J., et al. 2004. Efficacy of B-cell-targeted therapy with rituximab in patients with rheumatoid arthritis. *N Engl J Med* **350**:2572–2581.
29. Genovese, M.C., Becker, J.C., Schiff, M., et al. 2005. Abatacept for rheumatoid arthritis refractory to tumor necrosis factor alpha inhibition. *N Engl J Med* **353**: 1114–1123.
30. Johnson, A.E., Gordon, C., Palmer, R.G., and Bacon, P.A. 1995. The prevalence and incidence of systemic lupus erythematosus in Birmingham, England. Relationship to ethnicity and country of birth. *Arthritis Rheum* **38**:551–558.
31. Samanta, A., Roy, S., Feehally, J., and Symmons, D.P. 1992. The prevalence of diagnosed systemic lupus erythematosus in whites and Indian Asian immigrants in Leicester city, UK. *Br J Rheumatol* **31**:679–682.
32. Gordon, C., and Salmon, M. 2001. Update on systemic lupus erythematosus: autoantibodies and apoptosis. *Clin Med* **1**:10–14.
33. Ho, A., Barr, S.G., Magder, L.S., and Petri, M. 2001. A decrease in complement is associated with increased renal and hematologic activity in patients with systemic lupus erythematosus. *Arthritis Rheum* **44**:2350–2357.
34. Ho, A., Magder, L.S., Barr, S.G., and Petri, M. 2001. Decreases in anti-double-stranded DNA levels are associated with concurrent flares in patients with systemic lupus erythematosus. *Arthritis Rheum* **44**:2342–2349.
35. The American College of Rheumatology nomenclature and case definitions for neuropsychiatric lupus syndromes. *Arthritis Rheum* 1999; **42**:599–608.
36. Frostegard, J. 2008. Systemic lupus erythematosus and cardiovascular disease. *Lupus* **17**:364–367.
37. Bono, L., Cameron, J.S., and Hicks, J.A. 1999. The very long-term prognosis and complications of lupus nephritis and its treatment. *QJM* **92**:211–218.
38. Moroni, G., Pasquali, S., Quaglini, S., et al. 1999. Clinical and prognostic value of serial renal biopsies in lupus nephritis. *Am J Kidney Dis* **34**:530–539.
39. Houssiau, F.A., Vasconcelos, C., D'Cruz, D., et al. 2004. Early response to immunosuppressive therapy predicts good renal outcome in lupus nephritis: lessons from long-term follow-up of patients in the Euro-Lupus Nephritis Trial. *Arthritis Rheum* **50**:3934–3940.
40. Fraenkel, L., MacKenzie, T., Joseph, L., Kashgarian, M., Hayslett, J.P., and Esdaile, J.M. 1994. Response to treatment as a predictor of long-term outcome in patients with lupus nephritis. *J Rheumatol* **21**:2052–2057.
41. Esdaile, J.M., Joseph, L., MacKenzie, T., Kashgarian, M., and Hayslett, J.P.. 1994. The benefit of early treatment with immunosuppressive agents in lupus nephritis. *J Rheumatol* **21**:2046–2051.
42. Moroni, G., Pasquali, S., Quaglini, S., et al. 1999. Clinical and prognostic value of serial renal biopsies in lupus nephritis. *Am J Kidney Dis* **34**:530–539.
43. Weening, J.J., D'Agati, V.D., Schwartz, M.M., et al. 2004. The classification of glomerulonephritis in systemic lupus erythematosus revisited. *Kidney Int* **65**:521–530.
44. Boumpas, D.T. 2004. Sequential therapies with intravenous cyclophosphamide and oral mycophenolate mofetil or azathioprine are efficacious and safe in proliferative lupus nephritis. *Clin Exp Rheumatol* **22**:276–277.
45. Houssiau, F.A., Vasconcelos, C., D'Cruz, D., et al. 2002. Immunosuppressive therapy in lupus nephritis: the Euro-Lupus Nephritis Trial, a randomized trial of low-dose versus high-dose intravenous cyclophosphamide. *Arthritis Rheum* **46**:2121–2131.
46. Walsh, M., James, M., Jayne, D., Tonelli, M., Manns, B.J., and Hemmelgarn, B.R. 2007. Mycophenolate mofetil for induction therapy of lupus nephritis: a systematic review and meta-analysis. *Clin J Am Soc Nephrol* **2**:968–975.
47. Zhu, B., Chen, N., Lin, Y., et al. 2007. Mycophenolate mofetil in induction and maintenance therapy of severe lupus nephritis: a meta-analysis of randomized controlled trials. *Nephrol Dial Transplant* **22**:1933–1942.
48. Moore, R.A., and Derry, S. 2006. Systematic review and meta-analysis of randomised trials and cohort studies of mycophenolate mofetil in lupus nephritis. *Arthritis Res Ther* **8**:R182.
49. Ginzler, E.M., Dooley, M.A., Aranow, C., et al. 2005. Mycophenolate mofetil or intravenous cyclophosphamide for lupus nephritis. *N Engl J Med* **353**:2219–2228.
50. Houssiau, F.A., and Ginzler, E.M. 2008. Current treatment of lupus nephritis. *Lupus* **17**:426–430.
51. Gunnarsson, I., Sundelin, B., Jonsdottir, T., Jacobson, S.H., Henriksson, E.W., Van Vollenhoven, R.F. 2007. Histopathologic and clinical outcome of rituximab treatment in patients with cyclophosphamide-resistant proliferative lupus nephritis. *Arthritis Rheum* **56**: 1263–1272.
52. Vigna-Perez, M., Hernandez-Castro, B., Paredes-Saharopulos, O., et al. 2006. Clinical and immunological

effects of Rituximab in patients with lupus nephritis refractory to conventional therapy: a pilot study. *Arthritis Res Ther* **8**:R83.

SUGGESTED READINGS

D'Cruz, D.P., Khamashta, M.A., and Hughes, G.R. 2007. Systemic lupus erythematosus. *Lancet* **369**:587–596.

Klareskog, L., Catrina, A.I., and Paget, S. 2009. Rheumatoid arthritis. *Lancet* **373**(9664):659–672.

McInnes, I.B., and Schett, G. 2007. Cytokines in the pathogenesis of rheumatoid arthritis. *Nat Rev Immunol* **7**:429–442.

Smolen, J.S., Aletaha, D., Koeller, M., Weisman, M.H., and Emery, P. 2007. New therapies for treatment of rheumatoid arthritis. *Lancet* **370**(9602):1861–1874.

23 Gastrointestinal Inflammation and Ulceration: Mediators of Induction and Resolution

Linda Vong, Paul L. Beck, and John L. Wallace

The gastrointestinal (GI) tract is composed of a series of organs whose collective aim is to extract and absorb nutrients from food, and expel undigested matter as waste. As the lumen of the GI tract is essentially outside of the body, it is not surprising to find that the physical interface between "self" and the outside world is quite complex. The gut must be able to withstand fluctuations associated with changes in temperature, pH, and osmolarity, as well as overcome and respond to microbiota residence and the associated release of immunostimulatory molecules and detergent-like substances. As such, the mucosa has evolved to act as a first line of defense, its repertoire including (1) the spatial secretion of protective substances (acid, bicarbonate, and mucus); (2) the presence of tight junctions to limit the passive diffusion of harmful substances; (3) the mucosal microcirculation which enables rapid dilution/removal of noxious stimuli; and (4) the mucosal immune system where resident and blood-borne immunocytes and granulocytes are mobilized to initiate an appropriate inflammatory response. This so-called mucosal defense is supported by a catalogue of endogenous mediators; together these act to dampen the inflammatory response, which results from the aggressive accumulation of tissue-damaging metabolites.

The aim of this chapter is to review current knowledge regarding mediators involved in the pathogenesis of GI disease, with particular reference to nonsteroidal anti-inflammatory drug (NSAID)-induced gastropathy and colitis. We will address mediators that play a key role in the proresolution phase of inflammation, and discuss some of the new therapeutic approaches aimed at promoting mucosal healing and tissue homeostasis.

HOMEOSTATIC MECHANISMS OF MUCOSAL PROTECTION

"Mucosal defense" defines the innate ability of the GI system to protect itself from the onslaught of harmful stimuli that it is exposed to from day-to-day. Such stimuli must overcome a number of physical and chemical barriers, whose hierarchical organization normally presents an effective mechanism against deviations in mucosal homeostasis.

GI integrity is maintained by a number of factors. In the stomach, gastric acid functions to kill ingested bacteria and is very effective in minimizing bacterial colonization (with the notable exception of *Helicobacter pylori*) (Figure 23.1). Additionally, mucus and bicarbonate secreted by epithelial cells contribute to the maintenance of a physical barrier. These act both to diminish bacterial infection and to reduce luminal acid from coming into direct contact with the epithelium. The mucus layer is a hydrophobic surface rich in glycoproteins, sialic acid, and sulfated oligosaccharides, and under normal circumstances acts to reduce physical abrasion. Luminal acid that diffuses through the mucus can be effectively neutralized by bicarbonate secreted by the epithelium. Fluid secretion across the GI epithelium also contributes to mucosal defense by reducing bacterial adherence, and diluting or flushing away luminal toxins. Microbes that are able to bypass these initial defensive mechanisms next encounter the tight junctions between adjacent epithelial cells that further limit bacterial translocation.

The microcirculation along the length of the GI tract plays a pivotal role in the maintenance of mucosal integrity. Lying beneath the epithelium is a dense network of capillaries, which supply nutrients and oxygen, as well as diluting and removing toxins that have traversed the mucosa from the lumen. In the stomach, topical irritation results in a profound and rapid increase in mucosal blood flow (the hyperemic response). This response is aimed at diluting and neutralizing noxious substances. If the hyperemic response is impaired, significant mucosal injury can result. The hyperemic response to irritants is controlled by sensory afferent nerves. Upon detection of back-diffusing acid, sensory

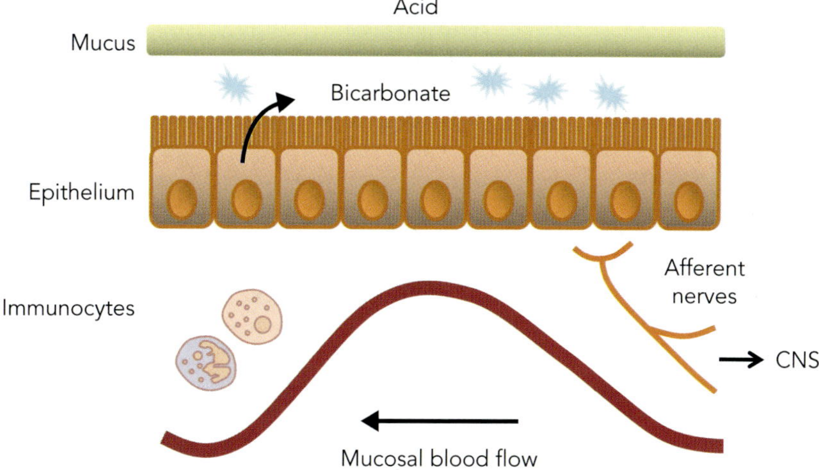

Figure 23.1. Mechanisms of mucosal protection. The function of the mucosal defence system is to limit mucosal damage incurred by acid and/or luminal antigens and microbes. The hierarchical organization of a number of physical and chemical barriers limits significant deviations in gut mucosal homeostasis. In the stomach, acid secretion serves to kill ingested bacteria, while mucus and bicarbonate secretion limits acid back-diffusion. The epithelium is adapted to resist against acid damage, and rapidly repairs itself through the process of restitution. Entry of acid into the lamina propria is sensed by afferent neurons and leads to a rapid increase in mucosal blood flow. In addition to supplying oxygen and nutrients, the increase in blood flow acts to flush and dilute any toxins that have traversed the mucosa. Resident immunocytes (e.g., mast cells and macrophages) are important for the sensing of foreign pathogens, and help coordinate an appropriate inflammatory response.

afferent nerves release calcitonin gene–related peptide (CGRP), a potent, nitric oxide–dependent vasodilator. Dilation of the submucosal arterioles, which are in close proximity to the sensory afferent nerves, leads to a rapid and substantial increase in mucosal blood flow. If the sensory afferent nerves are ablated, which can be achieved experimentally using capsaicin, the hyperemic response does not occur when the mucosa is exposed to an irritant, and extensive mucosal damage develops. Sensory afferent nerves also release substance P, which in the gastric mucosa can activate mast cells to release histamine, thereby further promoting the mucosal hyperemic response. In addition, mast cells and macrophages resident within the lamina propria act as "sentinels" – detection of a foreign substance stimulates the mobilization of inflammatory mediators and cytokines that alter mucosal blood flow, and increase the recruitment of granulocytes from the microcirculation.

Damage to the GI epithelium is not uncommon; however, mechanisms exist to ensure the rapid replacement of damaged cells, limiting the exposure of underlying tissue to luminal acid. The basement membrane is very sensitive to acid-induced damage, and exposure of this mucosal layer initiates the rapid migration of healthy cells from the gastric pits. This process of "restitution" is facilitated by the presence on the epithelial surface of a "mucoid cap," consisting of cellular debris, mucus, and plasma proteins. The mucoid cap is formed over the denuded region when epithelial cells are damaged, and it acts as a physical barrier to luminal content as well as maintaining a near-neutral microenvironment (thus protecting the exposed basement membrane from damage induced by luminal acid). This microenvironment is conducive to restitution, requiring both bicarbonate and an uninterrupted supply of blood to the region. Restitution can be accomplished in vivo within 15–60 minutes.

When the above-mentioned components of mucosal defense are overwhelmed, true ulceration can occur; that is, the damage extends through the entire mucosa and the muscularis mucosa. In this instance, the process of repair requires initial removal of necrotic tissue (as occurs during the acute inflammatory response) and subsequent revascularization of the damaged area. Revascularization (angiogenesis) is a prerequisite for the reestablishment of gastric glands. The healing process begins at the margins of the ulcers and gradually moves from all sides to the center of the previously ulcerated tissue.

INFLAMMATION AND ULCERATION OF THE GI TRACT

Gastric Ulceration

NSAIDs

The gastric toxicity of nonsteroidal anti-inflammatory drugs (NSAIDs) is well recognized. Given the efficacy

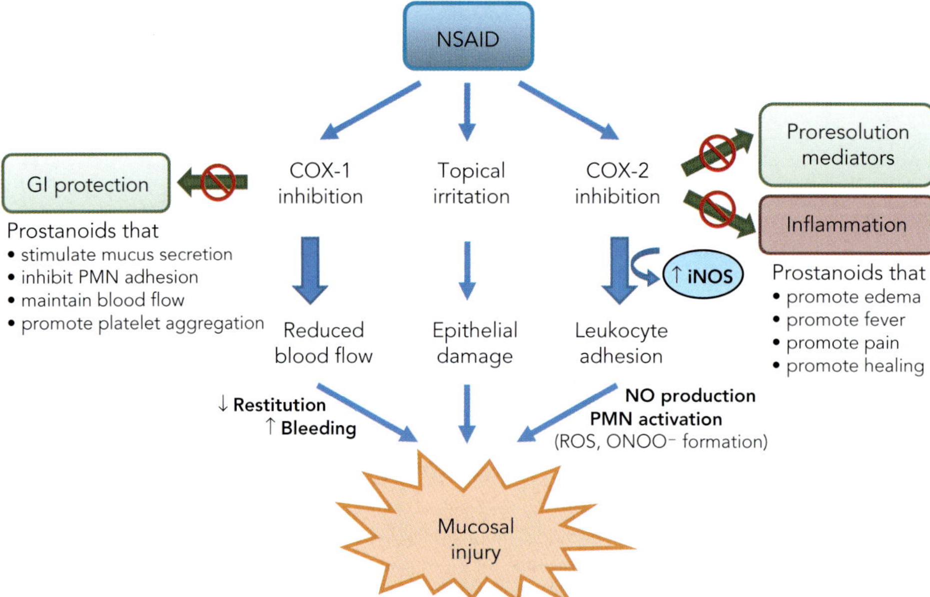

Figure 23.2. Pathogenesis of NSAID-induced gastric mucosal damage. COX-1 and COX-2 inhibition by NSAIDs leads to an imbalance in pro- versus anti-inflammatory prostanoids and eicosanoid metabolites. Whereas COX-1-derived prostanoids are involved in GI protection, COX-2-derived prostanoids are involved in the promotion of pain and fever. However, emerging evidence suggests that COX-2-derived prostanoids and lipoxins make an important contribution to GI mucosal defense. Inhibition of COX-1 reduces mucosal blood flow, leading to an impairment of restitution and ulcer repair. Inhibition of COX-2 induces leukocyte adhesion to the vascular endothelium. Alongside the release of reactive oxygen species (ROS) from degranulated neutrophils, the induction of iNOS and release of NO forms tissue-damaging peroxynitrite ($ONOO^-$). The topical irritant properties of NSAIDs also contribute to their ability to elicit gastric damage.

of NSAIDs in reducing pain and inflammation, it is almost impossible to eliminate their use; nevertheless, NSAID-induced gastropathy (gastric erosions, bleeding, ulcers, and perforations) still presents a significant burden to the healthcare system. Approximately 50% of people who take NSAIDs (including aspirin) develop gastric erosions, while an estimated 2%–4% of these people develop clinically significant gastric ulcers, which can lead to bleeding and/or perforation.

The effectiveness of NSAIDs in reducing pain and swelling lies in their ability to inhibit cyclooxygenase (COX) enzymes. These are rate-limiting enzymes for the synthesis of prostaglandins, which are small lipid molecules that regulate numerous processes in the body. In mammals, at least two isoforms of COX exist – COX-1 and COX-2. It is widely written that COX-1 performs housekeeping functions while COX-2 is an inducible enzyme involved in pathological processes, such as inflammation. However, this is an oversimplification. In fact, both COX-1 and COX-2 can be induced by certain cytokines and both perform several important physiological roles. In the healthy stomach, most of the prostaglandins that are produced are derived from COX-1. However, COX-2 can be very rapidly induced in the stomach in response to quite subtle challenges, such as administration of aspirin, a short period of ischemia, or topical application of an irritant. In these and other circumstances, COX-2-derived prostaglandins make an important contribution to gastric mucosal defense. Moreover, COX-2-derived prostaglandins are extremely important in ulcer repair in the stomach.

Prostaglandins modulate many aspects of gastric mucosal defense, including mucus and bicarbonate secretion, mucosal blood flow, epithelial cell turnover, and mucosal immunocyte function (Figure 23.2). By suppressing mucosal prostaglandin synthesis (as well as through direct topical irritant effects on the epithelium), NSAIDs can impair mucosal defense and render the stomach more susceptible to injury. Indeed, the ability of NSAIDs to cause gastric damage correlates well with their ability to suppress gastric prostaglandin synthesis. While conventional NSAIDs inhibit both COX-1 and COX-2, selective inhibition of the COX-2 isoform was considered more therapeutically desirable – insomuch as blocking COX-2 activity would reduce the production of prostaglandins at sites of inflammation while sparing COX-1-mediated prostaglandin synthesis in the stomach. Indeed, selective COX-2 inhibitors have been shown to produce severe GI complications less frequently than conventional NSAIDs. However, this has been misinterpreted as evidence that only

COX-1-derived prostaglandins contribute to gastric mucosal defense. It is now clear that both isoforms are important for mucosal defense, and both COX-1 and COX-2 must be inhibited for gastric damage to be elicited in an otherwise healthy stomach.

NSAID administration reduces mucosal blood flow, likely via the inhibition of COX-1-derived prostaglandins with vasodilatory actions (such as prostaglandin E_2). Moreover, one of the earliest events in the pathogenesis of NSAID-induced gastric mucosal damage is the adhesion of neutrophils to the vascular endothelium. This likely contributes to NSAID-induced gastropathy by (1) physically obstructing capillary flow, and (2) through the release, subsequent to adhesion, of their granule contents (including proteolytic enzymes) and the generation of reactive oxygen and nitrogen species. Functional inhibition of this process, either by rendering animals neutropenic or by administering antibodies targeting neutrophil or endothelial cell adhesion molecules, has been shown to reduce the severity of NSAID-induced experimental gastropathy. COX-2 inhibition accounts for the NSAID-induced leukocyte adherence to the vascular endothelium. As mentioned earlier, COX-2-derived prostaglandins contribute to ulcer repair, a process requiring the formation of granular tissue at the ulcer base, angiogenesis, and re-establishment of glandular architecture. This enzyme is strongly expressed at both the ulcer margin (the site where epithelial cell proliferation occurs) and in endothelial cells of the ulcer bed (the site of new vessel growth). Administration of selective COX-2 inhibitors significantly delays ulcer healing in animals with gastric ulcers.

Helicobacter pylori

Helicobacter pylori colonize more than 50% of the world's population and represent the main risk factor for peptic ulceration, gastric adenocarcinoma, and gastric lymphoma. The importance of its discovery is underscored by the fact that peptic ulcer disease is now approached as an infectious disease, whereby elimination of the causative agent (i.e., *H. pylori*) cures the condition. Infection is acquired by oral ingestion of the bacterium, usually during early childhood, and causes continuous gastric inflammation. Left untreated, *H. pylori* infection can persist for decades. Patients with the more common form of *H. pylori* gastritis, antral-predominant gastritis, are predisposed to duodenal ulcers, whilst patients with corpus-predominant gastritis and multifocal atrophy are more likely to have gastric ulcers, gastric atrophy, intestinal metaplasia, and gastric carcinoma. Antral gastritis results in hypergastrinemia and increased gastric acidity, whereas corpus gastritis can result in acid hyposecretion. It is estimated that among all individuals infected with *H. pylori*, 15%–20% develop the more severe forms of gastric disease including peptic ulcers and cancer. There is clear evidence that severe complications associated with peptic ulcer disease are often linked with NSAID consumption. However, while both *H. pylori* and NSAIDs increase the risk of GI ulcer development (via independent mechanisms), their actions are additive, rather than synergistic.

Despite the gastric mucosa being well protected against bacterial infection, *H. pylori* have evolved to be highly adapted to this microenvironment. Mucosal protective mechanisms are consistently overcome. *H. pylori* have flagella, rendering the bacterium highly motile, and thus easily gaining entry to the mucosal layer. It is able to neutralize the acidic milieu by way of urease activity, which converts urea to carbon dioxide and ammonia. Infection also reduces bicarbonate secretion from the inflamed mucosa, and mucosal surface hydrophobicity. The bacterium has a number of surface components to enable tight association with epithelial cells, such as the blood group antigen-binding adhesin (BabA). Expression of BabA enables binding to Lewis b, a blood group antigen expressed on gastric epithelial cells. Upon cell association, *H. pylori* can secrete VacA – a vacuolating cytotoxin that induces channel formation and likely provides the bacterium with nutrients. Furthermore, most strains of *H. pylori* possess the *cag* pathogenicity island – a 37 kb genomic fragment that encodes for, among others, CagA. This protein is translocated to the host cell upon infection, leading to activation of a number of cell signaling pathways.

The damage to the gastric mucosa results from the host response to *H. pylori* infection rather than from the bacterium itself. During the innate immune response to infection, the bacterium elicits a rapid recruitment of neutrophils, followed later by T and B lymphocytes, plasma cells, and macrophages. Activated neutrophils contribute to epithelial-cell damage through release of proteolytic enzymes and reactive oxygen species. Pro-inflammatory cytokines such as interleukin (IL)-1β, IL-2, IL-6, IL-8, and tumor necrosis factor (TNF)-α are also upregulated. *H. pylori* stimulates both T_H1 and T_H2 responses (the extent of this varies among people due to unknown factors), although T_H1 responses seem to predominate, and may be more important in eliciting severe gastritis.

Small Intestine and Colon Damage

NSAIDs

While damage to the stomach is most frequently associated with NSAID use, these drugs can provoke injury and inflammation throughout the GI tract. Indeed, the small intestine is the major site of NSAID-induced bleeding. Damage to the more distal regions of the GI tract may develop either in the healthy gut or at preexisting sites of bowel disease. In the small intestine, side effects of NSAID usage include the manifestation of enteropathy, chronic iron-deficiency anemia, hypoalbuminemia, acute bleeding and perforation, stricture formation, and villous atrophy. The damage is likely mediated through mechanisms different from those of the damage induced by NSAIDs in the stomach; thus, suppression of prostaglandin synthesis is unlikely to be the major contributor to mucosal injury. Instead, NSAIDs directly damage enterocytes, possibly via uncoupling of oxidative phosphorylation. Repeated exposure of the mucosa to NSAIDs, via enterohepatic cycling, is a crucial aspect of the pathogenesis of NSAID-enteropathy, since bile duct ligation prevents such damage in animal models and NSAIDs that undergo little or no enterohepatic cycling do not cause significant intestinal injury.

NSAID-induced intestinal lesions are exacerbated by enteric bacteria, the numbers of which increase substantially following NSAID administration to rats. The contribution of enteric bacteria to NSAID-induced intestinal injury is evident from the protective effects of broad-spectrum antibiotics. Furthermore, markedly less intestinal damage is induced by NSAIDs in animals raised in a germ-free environment. Endotoxins released by translocated bacteria trigger a local inflammatory response, resulting in the recruitment of neutrophils and the liberation of proinflammatory cytokines (such as IL-1 and TNF-α), causing further mucosal damage. NSAIDs have also been reported to upregulate inducible nitric oxide synthase (iNOS). Release of high concentrations of nitric oxide (NO) can increase epithelial cell permeability and damage surrounding cells by reactive nitrosation or generation of the reactive nitrogen species such as peroxynitrite.

NSAIDs have the capacity to induce colorectal lesions in patients with no history of bowel disease, as well as increasing the severity of preexisting disorders (such as inflammatory bowel disease; IBD). Side effects include the manifestation of colonic lesions and ulcers, rectitis, diaphragm-like strictures, eosinophilic colitis, colonic diverticular disease, and IBD. In contrast to the small intestine, enterohepatic recirculation of the NSAID does not seem to play a role in the development of such lesions; instead, inhibition of prostaglandin synthesis is likely involved in NSAID-mediated exacerbation of colitis.

Inflammatory Bowel Disease

Crohn's disease and ulcerative colitis represent the two main types of IBD. The highest incidence rates for IBD have been reported in North America and Europe, with 10–200 cases per 100,000 individuals being affected. While race and ethnicity are contributing factors, it is clear that disease incidence is highest in developed, urbanized countries. The etiology of IBD is unknown, although a number of factors contribute to disease pathogenesis – including genetic, environmental, and microbial, as well as a dysregulated immune system. Despite an overlap in clinical and pathological characteristics, there is sufficient evidence to indicate that distinct mechanisms regulate the pathogeneses of these two conditions. For example, differing cytokine patterns are elicited in Crohn's disease versus ulcerative colitis, suggesting alternate T-cell contributions. Crohn's disease is characterized by a strong T_H1 response, involving increased secretion of IL-12, interferon (IFN)-γ, and/or TNF-α. While less clear, it is likely that ulcerative colitis is driven by a T_H2 dominant phenotype, involving increased secretion of IL-5, IL-6, and/or IL-13. However, the recent discovery of a new T-cell subset, the T_H17 cells, has led to a reevaluation of the dichotomous characterization of Crohn's disease and ulcerative colitis as T_H1 versus T_H2.

Crohn's disease can affect any part of the GI tract, but is more common to the terminal ileum, ceacum, peri-anal area, and colon. Ulcerated regions (or "skip" lesions) are interspersed between normal or edematous mucosa, resulting in a cobblestone-like appearance. Inflammation is transmural, affecting all layers of the bowel wall, and often there is a dense infiltration of lymphocytes and macrophages into the tissue. Granuloma formation, fissuring ulceration, and submucosal fibrosis are common in these tissues. Clinical features of Crohn's disease include diarrhea, pain, narrowing of the gut lumen (leading to stricture formation), abscess formation, and fistulization to the skin and internal organs. In ulcerative colitis, inflammation is confined to the colon, extending proximally from the rectum in a continuous fashion. Only the mucosal layer is affected, which is infiltrated by lymphocytes and granulocytes. Erosion of the goblet cells and crypt abscesses are common. Clinical features of ulcerative colitis include severe diarrhea, blood loss, and progressive loss of peristaltic function. In more severe cases, this can lead to toxic megacolon and perforation.

The gut harbors the largest and most diverse population of microbiota, and it is the inappropriate response of the gut mucosal system to such indigenous gut flora and other luminal antigens that is thought to be one of the main causes of IBD. Oral tolerance (which is normally established early in life) acts to dampen the mucosal immune response to intestinal antigens which, when administered by any other route, would cause an inflammatory response. A defect in oral tolerance may exacerbate the responses of an already compromised immune system. Indeed, in experimental models of colitis, inflammation is substantially reduced when mice are kept in a germ-free environment, suggesting that the normal mucosal microflora is required to initiate or maintain the inflammatory response. In IBD, the intestinal immune system is disturbed at all levels. A leaky epithelial barrier allows luminal antigens access to the submucosa and thus exposure to granulocytes and regulatory lymphocytes. Degranulation of neutrophils and mast cells liberates tissue-damaging proteases, proinflammatory cytokines and vasoactive peptides. The epithelial layer itself expresses an alternate profile of pattern recognition receptors (Toll-like receptors) which are important for the detection of bacterial components. The mechanisms involved in antigen recognition are also compromised in IBD – dendritic cells incorrectly recognize commensal bacteria as pathogens, triggering activation of naive T cells into effector (T_H1/T_H2) or natural killer cells. Furthermore, atypical antigen-presenting cells become potent T cell activators; in the surrounding milieu of proinflammatory cytokines, epithelial cells acquire an activated phenotype – increasing the expression of MHC molecules. Activated T cells also show a delayed ability to undergo apoptosis, and persist in the mucosa. Such inappropriate T-cell activation and accumulation contributes to inflammation of the mucosa.

Studies in both human IBD and experimental colitis models have identified an upregulation and accumulation of numerous proinflammatory mediators; including iNOS-derived nitric oxide, leukotriene B_4, thromboxane A_2, platelet activating factor (PAF), as well as cytokines such as TNF-α, IL-1β, IL–6, IL-8, and IL-16.

MEDIATORS OF MUCOSAL INJURY AND INFLAMMATION

Inflammation of the GI tract can be considered a mechanism of preservation, a way in which the host can protect itself from invading pathogens and noxious stimuli. The inflammatory response acts to remove and inactivate the damaging substance, and is aided by an array of cell-derived proteases and reactive oxygen products, as well as soluble mediators. Together, these are coordinated by the milieu of cytokines and chemokines released by granulocytes, lymphocytes, and epithelial cells resident within and/or mobilized to the lamina propria. Inflammation is normally self-limiting, and is aided by a number of proresolution factors to ensure its timely cessation. However, in some circumstances, such as when the factor that initiates the inflammatory response persists, inflammation can be unrelenting – leading to excessive tissue injury.

The susceptibility of the GI mucosa to injury is dependent on the balance between defensive factors (those that maintain mucosal protection) and aggressive factors present within the lumen. For the most part, eicosanoid derivatives in the gut are proinflammatory, with the notable exception of a few prostaglandins and lipoxins. Here, we discuss some of the major inflammatory mediators that weaken the mucosal defense system and contribute to mucosal injury.

Thromboxane

Thromboxane is the major arachidonate metabolite derived from the sequential actions of COX and thromboxane synthase (Figure 23.3). Cellular sources of thromboxane include platelets, neutrophils, endothelial cells, epithelial cells, and mononuclear cells. While platelets account for 95% of thromboxane that is detectable in the serum, infiltrating neutrophils are the principal source of this mediator during an acute inflammatory response. Thromboxane has potent proinflammatory actions – both direct (involving diapedesis and activation of neutrophils, and a reduction in suppressor T-cell activity) and indirect (vasoconstriction and platelet activation). Thromboxane stimulates leukotriene B_4 release, augmenting the adherence of leukocytes to the vascular endothelium. Moreover, in patients with IBD, platelets circulate in an activated state, and aggregates of these cells have been identified in capillaries of inflamed rectal biopsies. Altered thrombogenesis is also observed in Crohn's disease patients. Given the importance of mucosal blood flow in protecting against GI injury, it is not surprising that thromboxane is thought to be a key player in the mediation of GI ulceration.

Thromboxane production is increased both in human IBD and experimental models of colitis. In experimental colitis, thromboxane production increases during the course of chronic inflammation, with levels persisting long after other arachidonate metabolites (such as prostaglandin E_2 and leukotriene B_4) have declined.

Leukotrienes

Leukotrienes (LT) are products of the lipoxygenase pathway, requiring the oxygenation of arachidonic acid by 5-lipoxygenase (Figure 23.3). These metabolites can be divided into two main subclasses: leukotriene B_4 and the cysteinyl-leukotrienes (LTC_4, LTD_4, and LTE_4), with the latter group consisting of both fatty acid and amino acid moieties. Leukotriene synthesis occurs mainly in immunocytes, epithelial cells, and endothelial cells. In the mucosa, neutrophils are the predominant source of LTB_4, while mast cells appear to be responsible for cysteinyl-leukotriene production.

LTB_4 is an immunomodulatory mediator and a highly potent chemotaxin for neutrophils, monocytes, and effector T-cells. LTB_4 induces neutrophil adherence to the endothelium, degranulation, and the generation of reactive oxygen species. Ultimately,

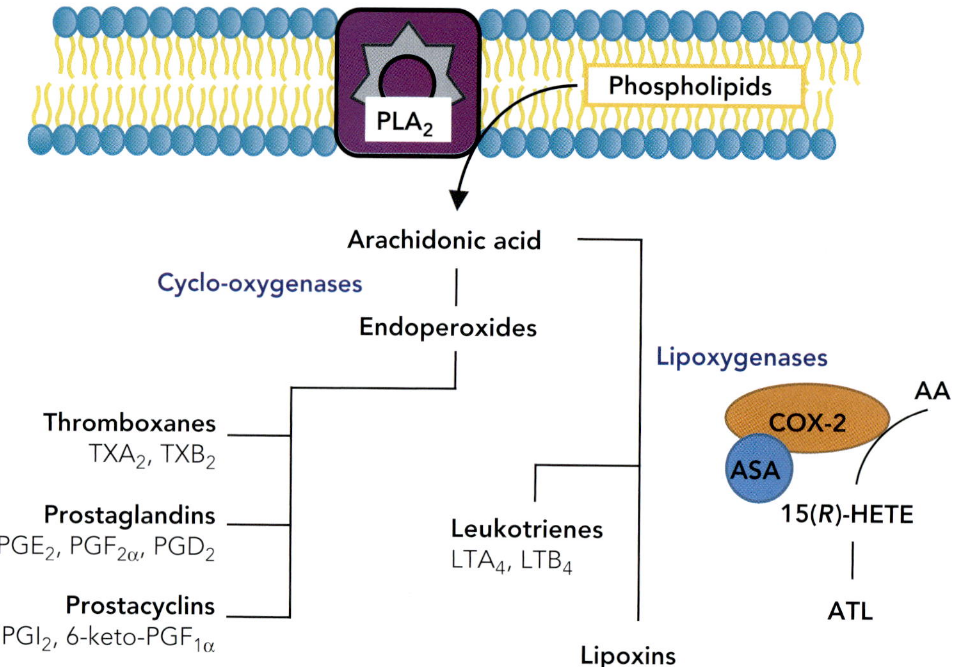

Figure 23.3. The arachidonic acid cascade. Arachidonic acid (AA) is liberated from phospholipids by phospholipase A_2 (PLA_2). Collectively, the metabolites of arachidonic acid (eicosanoids) are formed by the actions of cyclo-oxygenases (COX) or lipooxygenases (LO). COX-1 and COX-2 metabolize arachidonic acid to endoperoxides (PGH); thromboxane synthase and prostaglandin synthases (e.g., PGE synthase, PGD synthase, PGI synthase, etc.) are required for the production of thromboxanes and the various prostaglandins. The lipoxygenase pathway produces leukotrienes and lipoxins, via the enzymes 5-LO, 12-LO, and 15-LO. Acetylation of COX-2 by aspirin (ASA) directs the metabolism of arachidonic acid to 15(*R*)-hydroxyeicosatetranoic acid (15*R*-HETE), which can undergo further metabolism to form aspirin-triggered lipoxin epimers (ATL).

these processes result from the binding of LTB_4 to the G-protein-coupled receptor BLT-1. NSAID-induced gastric ulceration is associated with leukocyte adherence to the vascular endothelium; as such, LTB_4 has been implicated as a potential pathogenic mediator. Elevated levels of LTB_4 have been reported in patients taking NSAIDs. Moreover, receptor antagonists for LTB_4 and inhibitors of 5-lipoxygenase have been shown to attenuate NSAID-induced leukocyte adherence, reducing the severity of NSAID-induced mucosal damage. LTB_4 may also play a role in gastric ulceration associated with *H. pylori* infection. In experimental models of colitis, intracolonic administration of LTB_4 exacerbates tissue injury, whereas 5-lipoxygenase inhibitors have been shown to accelerate healing. Increased production of LTB_4 is observed in the colon mucosa of IBD patients. Furthermore, standard treatments for IBD, such as 5-aminosalicylic acid and glucocorticoids, have been shown to reduce LTB_4 levels in the colon mucosa and lumen. Despite such evidence supporting a role for LTB_4 in IBD, when tested in clinical trials, inhibitors of leukotriene synthesis have failed to maintain clinical remission in ulcerative colitis patients.

In contrast to the actions of LTB_4, cysteinyl-leukotrienes increase microvascular permeability and selectively attract eosinophils. These arachidonate metabolites also increase P-selectin expression, promoting leukocyte capture and rolling. Intra-arterial infusion of cysteinyl-leukotrienes increases the susceptibility of the rat stomach to injury induced by topical irritants, most likely through a reduction in mucosal blood flow. Overproduction of cysteinyl-leukotrienes has been reported in IBD patients, and is thought to mediate mucosal injury and intestinal damage associated with mucosal mast cell activation – a major source of these metabolites.

Platelet-Activating Factor

Platelet-activating factor (PAF) is a lipid mediator generated by the action of phospholipases on membrane phospholipids. It is produced and released by a variety of cells, including leukocytes, macrophages, platelets, vascular endothelial cells, and epithelial cells. PAF induces a range of effects similar to those observed in shock, including increased vascular permeability, hypotension, neutrophil aggregation, and degranulation. It has been shown to stimulate the release of proinflammatory eicosanoid mediators, such as thromboxane and cysteinyl-leukotrienes, and can also serve as a novel signal for leukocyte–endothelial

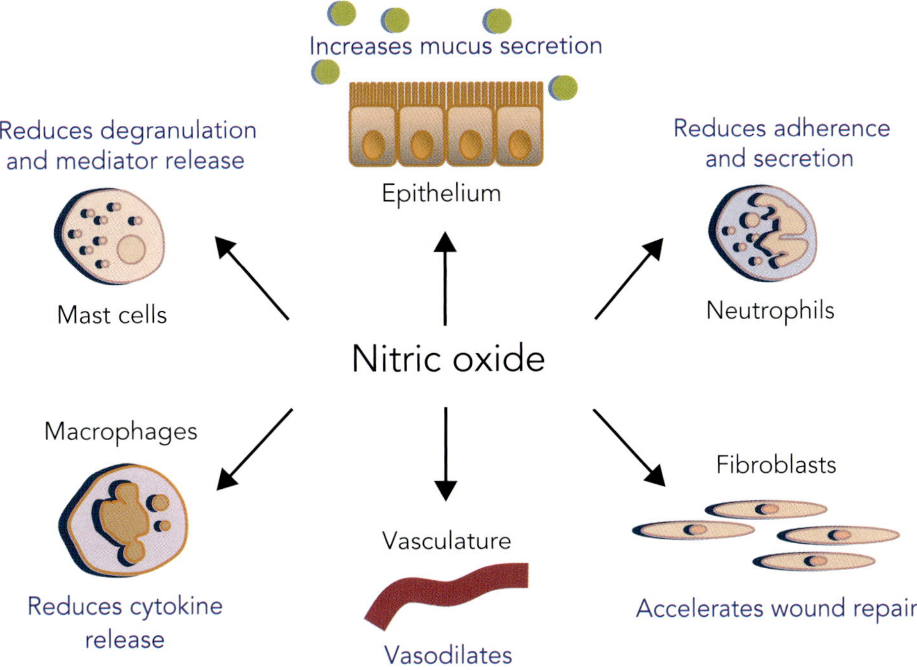

Figure 23.4. The role of nitric oxide in mucosal defence. NO downregulates inflammation and protects against mucosal damage in a number of ways. In the epithelium, NO stimulates mucus and bicarbonate secretion, and regulates fluid secretion. NO also regulates the activation and function of immunocytes; it reduces the adhesion of neutrophils to endothelial cells, reduces mast cell degranulation, and the release of cytokines from macrophages. Release of NO by vascular endothelial cells triggers vasodilation in a guanylate cyclase–dependent manner. NO also accelerates wound repair by stimulating angiogenesis, and enhancing collagen deposition by fibroblasts.

cell interactions. Indeed, PAF has been shown to be responsible for much of the tissue damage observed in the GI tract during experimental septic shock, hemorrhagic shock, and ischemia-reperfusion; in each case PAF receptor antagonists exerted a protective effect.

Intravenous infusion of PAF in the rat induces extensive damage to the gastric mucosa. While a reduction in mucosal blood flow (a consequence of PAF's hypotensive effects) contributes in part to its ulcerogenic effects, its potent ability to damage the gastric mucosa is likely mediated by stimulatory effects on leukocyte–endothelial cell adhesion and degranulation. PAF production is increased in patients with IBD, with the predominate source of this mediator being the intestinal epithelium. Glucocorticoids commonly used to treat IBD have been shown to inhibit PAF production in the colon. PAF levels are increased in experimental models of colitis. Its actions were defined by the ability of PAF receptor antagonists to inhibit colonic mucosal injury, the incidence of neutrophil adhesion and thus accumulation within the mucosa, and diarrhea. PAF production has also been shown to be elevated during helminth infections, and mediates some of the GI immune responses to *H. pylori* infection.

Nitric Oxide

NO is a multifaceted mediator, with roles pertaining to both GI protection and damage (Figure 23.4). The constitutive forms of NO synthase (NOS), neuronal NOS (nNOS), and endothelial NOS (eNOS) are important for homeostatic regulation of the GI tract – including modulation of mucus and fluid secretion, mucosal blood flow, and the mucosal immune system. In contrast, the inducible NOS isoform (iNOS) has been linked to mucosal injury and dysfunction. The paradoxical actions of NO can be explained in part by the ability of different concentrations of NO to elicit different effects in the same tissue. Production in small amounts generally exerts beneficial effects in the GI tract, whereas production in high amounts can be detrimental. iNOS-derived NO generation occurs at several orders of magnitude greater than that of other NOS isoforms, and in inflamed tissues, the predominant source of NO is from infiltrating leukocytes and epithelial cells. Exposure of epithelial cells to NO donors impairs epithelial barrier function, by increasing cell permeability. The effects of IFN-γ on intestinal hyperpermeability have also been shown to be mediated by NO. Functionally, rats treated with NOS inhibitors or those deficient in iNOS show reduced levels of bacterial translocation.

NO is a lipophilic free radical, with the ability to react with other free radicals, oxygen, and transition metals. Interaction of NO with neutrophil-derived superoxide (O_2^-) forms the potent oxidant peroxynitrite ($ONOO^-$). This prototypical nitrating species is highly toxic to both mammalian cells and bacteria. Administration of peroxynitrite into the colon produces widespread injury and inflammation similar to that seen in several experimental models of IBD, and in the rat, induction of colitis and subsequent local peroxynitrite formation induces epithelial cell apoptosis – a phenomenon reversible by selective iNOS inhibition. Indeed, a decrease in cell viability is seen in epithelial cells overexpressing iNOS, suggesting a cause-and-effect relationship between NO and cytotoxicity. Nitrotyrosine (a marker of nitrosative stress) is a stable nitration product formed by peroxynitrite and myeloperoxidase-nitrite catalyzed reactions. In human ulcerative colitis, serum nitrate and nitrotyrosine levels are increased, and at least in this condition support a pathogenic role for NO. Elevated iNOS activity has been shown in colon adenomas, colon cancer and metastases, and it has been suggested that this contributes to the increased incidence of cancer associated with chronic ulcerative colitis.

Increased iNOS activity and NO production have been reported in patients with IBD, as well as in models of experimental colitis. Under these conditions, overproduction of NO seems to be associated with reduced colonic motility, and may contribute to secretory diarrhea. Studies from early on indicated that inhibition of NO synthesis could reduce the severity of gut injury and inflammation. Despite this, there is still conflicting evidence for the pro- versus anti-inflammatory roles of NO. In animal studies, administration of NO donors at very high doses did not compromise mucosal or microvascular barrier integrity. Moreover, agents that release small amounts of NO over a prolonged period of time have been shown to greatly reduce inflammation and accelerate healing in experimental colitis. The administration of NO donors also protects the stomach from injury. Overall, the balance of evidence suggests that NO has a *net* protective effect in the GI tract.

Hydrogen Sulfide

Hydrogen sulfide (H_2S) has long been known for its toxic effects. It is a colorless gas with the distinct smell of rotten eggs, and is commonly associated with industrial pollution. In recent years, H_2S has become increasingly recognized as a regulator of various physiological processes. H_2S is highly lipid-soluble and therefore passes freely through the lipid bilayer. Regulatory functions have been ascribed to H_2S in the neuronal, cardiovascular, GI, and hepatic systems. Furthermore, H_2S likely modulates inflammatory responses. Experimental data have shown H_2S to be a potent vasodilator and regulator of a number of leukocyte functions, as well as a modulator of leukocyte–endothelial cell interactions, neutrophil apoptosis, transcription factor activation, and edema formation (Figure 23.5).

Given the susceptibility of the human body to toxic effects of H_2S (in the brain, toxic levels are only two-fold that of the basal H_2S concentrations), it is not surprising that cellular levels of this mediator are tightly regulated. H_2S is rapidly metabolized by oxidation in the mitochondria or by methylation in the cell cytosol. Moreover, H_2S can be scavenged by methemoglobin or by metallo- and disulfide-containing molecules such as oxidized glutathione. H_2S also binds to hemoglobin, which acts as a common "sink" for it and other gaseous transmitters (NO and carbon monoxide). Indeed, while endogenous levels of H_2S are reported to be as high as 160 µM in the brain, with serum levels varying between 30 and 100 µM, administration of H_2S donors rarely alters these basal levels significantly, indicative of efficient clearance and inactivation mechanisms.

H_2S appears to be produced in most tissues. It is synthesized largely by two pyridoxal-5'-phosphate-dependent enzymes: cystathionine β-synthase (CBS) and cystathionine γ-lyase (CSE). Substrates for these enzymes include L-cysteine, cystathionine, homocysteine, and cystine. Genetic deficiency of CBS is the major cause of homocystinuria, a disease characterized by dysfunction of the endothelium and an adverse cardiovascular prognosis. While some tissues require both CSE and CBS for H_2S synthesis, others require only one of these enzymes. In the heart and vasculature CSE appears to be the predominant source of H_2S, whereas CBS is more important in the central nervous system. Some actions of H_2S are mediated by the activation of ATP-sensitive K^+ (K_{ATP}) channels.

In the GI tract, H_2S plays an important role in the regulation of mucosal defense. As outlined earlier, the susceptibility of the gastric mucosa to injury is intimately associated with mucosal blood flow and the adherence of leukocytes to the vascular endothelium. In the rat, H_2S administration increases gastric mucosal blood flow. Inhibition of endogenous H_2S synthesis, through administration of a CSE inhibitor results in an increase in leukocyte adhesion to the vascular endothelium in the mesenteric microcirculation, as well as increased leukocyte extravasation and edema formation. Administration of H_2S donors has the opposite effects (i.e., anti-inflammatory). Expression of TNF-α, a potent activator of lymphocytes and granulocytes, has been shown to be attenuated by H_2S donors, possibly via inhibition of NF-κB activation. H_2S can also inhibit the functional response of leukocytes (and thus tissue damage) by scavenging peroxynitrite.

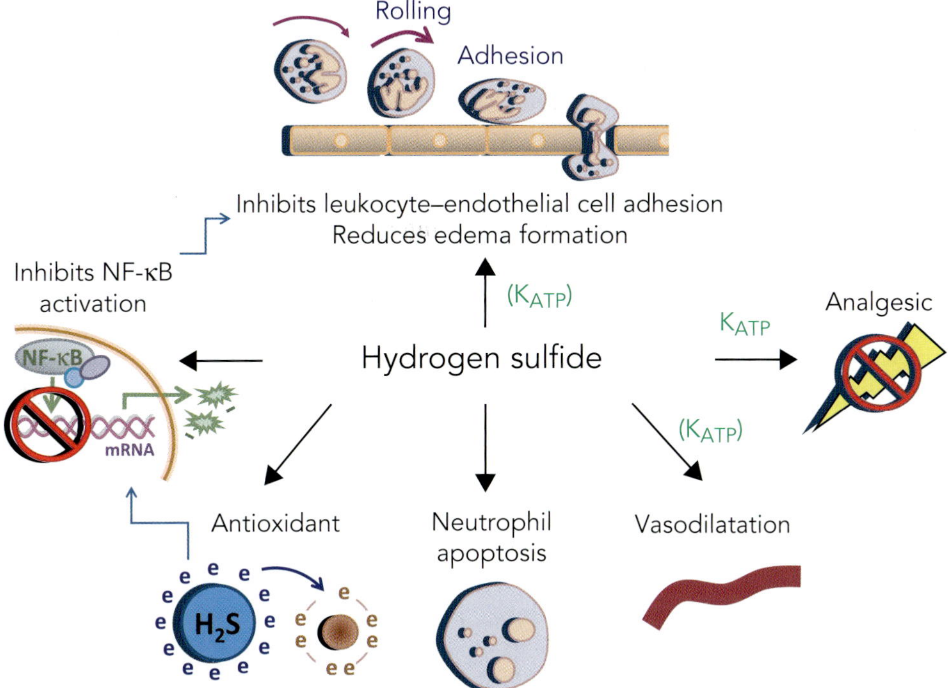

Figure 23.5. The role of hydrogen sulfide in mucosal defence. The gaseous transmitter H_2S regulates a number of components of mucosal defense. H_2S inhibits leukocyte rolling and adhesion, by modulating selectin- and integrin-dependent interactions. Like nitric oxide, it is also a potent vasodilator, and pertinent to inflammatory bowel disease, H_2S has analgesic properties and promotes repair of mucosal injury. Some of these effects may be mediated via the activation of ATP-sensitive K^+ channels (K_{ATP}). H_2S neutralizes peroxynitrite by acting as an antioxidant, and can also induce neutrophil apoptosis and inhibit the activation of the nuclear transcription factor, NF-κB.

Both CBS and CSE are expressed in the gastric mucosa, but CSE accounts for the majority of H_2S generation in the stomach. Inhibition of CSE exacerbates the gastric injury caused by NSAIDs. On the other hand, administration of H_2S donors reduces the severity of damage induced by NSAIDs, as well as effects of the NSAIDs on mucosal blood flow and adhesion molecule expression.

Cytokines

Cytokines play a central role in the regulation of the mucosal immune system, and thus the local immune response. Owing to their involvement in the pathogenesis of IBD, our knowledge of cytokine responses is skewed mainly toward the small and large intestine. A number of cytokines, both pro- and anti-inflammatory, are intimately linked to the gut. As such, the cytokines discussed within this chapter are by no means an exhaustive list.

In both Crohn's disease and ulcerative colitis, there is a characteristic increase in levels of proinflammatory cytokines such as TNF-α, IL-1β, IL-6, IL-8, IL-12, IL17, and IL-23. The synthesis of these cytokines by lymphocytes, monocytes, and epithelial cells of the gut is regulated, at least in part, by transcription factors that bind to gene promoter regions. Among these, NF-κB activity has been shown to be enhanced in the mucosa of patients with active IBD. NF-κB activity can be upregulated by IL-1β and TNF-α, as well as components of bacterial cell walls such as lipopolysaccharide. The presence of bacterial products is detected through Toll-like receptors, and in patients with IBD, an increase in Toll-like receptor-4 expression has been demonstrated. TNF-α is able to recruit and activate circulating inflammatory cells, induce edema, and contribute to granuloma formation. An increase in the number of TNF-α producing cells is evident in patients with IBD; this is restricted to subepithelial macrophages in patients with ulcerative colitis, while in Crohn's disease patients, TNF-α-producing cells can be detected throughout the mucosa. Increased TNF-α levels are also detectable in the plasma and stool of patients with Crohn's disease, as are serum concentrations of the soluble TNF-α receptors. Indeed, the importance of TNF-α in the pathogenesis of Crohn's disease is underscored by the efficacy of anti-TNF-α antibodies in the induction of clinical remission.

TNF-α also contributes to gastric mucosal injury associated with NSAID usage or *H. pylori* infection. Polymorphisms of the TNF-α gene are associated with increased risk of gastric ulcers and gastric cancer, and have been shown to increase patient susceptibility to *H. pylori* infection and peptic ulcer disease.

Alongside TNF-α, the proinflammatory cytokine IL-1β is also released early in an inflammatory reaction. Many of the cells that produce IL-1 (including monocytes, macrophages, neutrophils, endothelial cells, and fibroblasts) also produce an endogenous IL-1 antagonist (IL-1a). Recombinant forms of this antagonist have been shown to inhibit many of the biological activities of IL-1 both in vitro and in vivo. Elevated levels of IL-1 have been detected in both plasma and tissue of patients with IBD, as well as in experimental models of colitis. Furthermore, the ratio of IL-1:IL-1a is increased in Crohn's disease and ulcerative colitis. Paradoxically, IL-1 appears to have protective effects on the GI tract in some circumstances. It has been shown to reduce the severity of gastroduodenal damage in various inflammatory models, perhaps through inhibition of leukocyte adherence. Furthermore, IL-1 has also been shown to inhibit gastric acid secretion and the release of ulcer-promoting mediators, such as PAF and histamine, from mast cells.

More recently, a role of IL-23 in the pathogenesis of IBD has been suggested. Now identified as part of the IL-12 family of cytokines, the heterodimeric cytokine IL-23 is composed of the subunits p19 and p40. Historically, it was the p40 subunit, present also in IL-12, which was targeted and neutralized to show the importance of IL-12 in the induction and maintenance of the T_H1 immune response. Indeed, IL-12 drives the differentiation and proliferation of naive T cells along the T_H1 pathway, leading to the production of IFN-γ. In turn, IFN-γ alters epithelial permeability and promotes phagocytic cell activation and transmigration through the epithelial barrier. While the importance of IL-12 in the adaptive immune response should not be underestimated, the emerging role of IL-23 in immune-mediated inflammatory diseases suggests this to be a key cytokine whose targeted blockade may be therapeutically promising. Interestingly, many of the actions that were initially attributed to IL-12 have been found to be related to IL-23. In human studies, although antibodies directed against IL-12 were shown to be therapeutic in IBD patients, it was later realized that their effectiveness was attributable to actions on IL-23 (since IL-23 and IL-12 share the so-called 'IL12p40' subunit). In experimental models, deficiency of IL-12p40 significantly protects animals from developing colitis, possibly implicating roles for both IL-12 and IL-23. These cytokines differ in that they target alternate cell populations during the T_H1 response: IL-12 induces naive CD4+ T cells and stimulates clonal expansion of early committed T_H1 cells, while IL-23 seems to be a late-phase factor, inducing the proliferation of memory CD4+ T cells and enhancing cytotoxic T-lymphocyte function. IL-23 also stimulates a unique class of T cells, namely T_H17. Such cells are characterized by the production of the effector cytokine IL-17. This proinflammatory molecule stimulates various cell types to produce other cytokines, such as IL-6 and IL-8, and synergizes with TNF-α and IL-1β to further induce chemokine expression.

An uncommon variant of the gene encoding the IL-23 receptor has been shown to confer protection against Crohn's disease, while several noncoding variants of the gene were independently associated with disease susceptibility. In animal models of IBD, antibodies targeting p19 (a subunit thought to be specific to IL-23) have been shown to suppress intestinal inflammation, with a concomitant reduction in levels of the cytokines TNF-α, IL-6, and IFN-γ. Furthermore, the actions of IL-23 appear to be "mucosal-selective," with an exacerbation of inflammation being demonstrated upon administration of recombinant forms of this cytokine.

Considering the biphasic nature of IL-12 (low doses of IL-12 and IFN-γ are proinflammatory, while high doses of IL-12 are anti-inflammatory), a paradigm has emerged whereby the IL-12/IFN-γ pathway is seen as largely protective whereas the IL-23/IL-17 pathway is proinflammatory and leads to the manifestations of disease.

RESOLUTION OF MUCOSAL INFLAMMATION AND INJURY

In the context of an inflammatory response, the earliest change that heralds the move from a proinflammatory microenvironment to one conducive for resolution is the replacement of invading polymorphonuclear leukocytes and eosinophils by monocytes and phagocytosing macrophages (Figure 23.6). These cells undergo programmed cell death, and are cleared by phagocytosis, with the macrophages themselves being later cleared via the lymphatic system. Mediators that complement this process, for example those that inhibit neutrophil chemotaxis or are able to modulate the surrounding milieu of cytokines/chemokines, are considered anti-inflammatory and thus support this resolution phase. With respect to chronic inflammation, such as that seen during IBD, there is considerable evidence to show that a number of mediators enforce this process.

Prostaglandin D_2

While prostaglandins were initially thought to contribute to the pathogenesis of IBD, a large body of evidence now shows that these arachidonate metabolites mainly exert anti-inflammatory and protective roles. Indeed NSAIDs, which inhibit COX activity and thus prostaglandin synthesis, have been reported to both

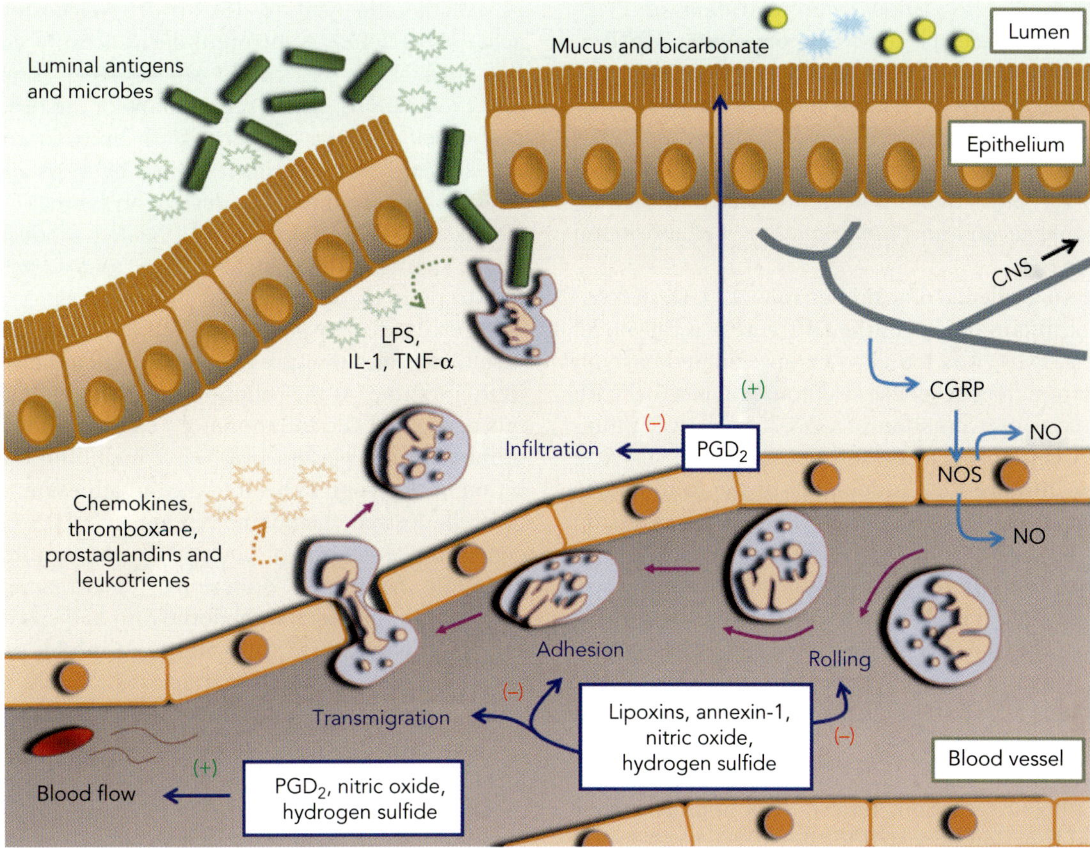

Figure 23.6. Mediators of resolution. Luminal antigens and microbes that have bypassed the gut epithelium are detected by resident immunocytes, triggering the release of cytokines and chemokines (such as IL-1 and TNF-α) to activate the inflammatory cascade. Neutrophils travel toward the chemotactic gradient, undergoing cell rolling, adhesion, and transmigration. Together with the epithelium, these cells release a number of proinflammatory eicosanoid metabolites, including thromboxane, prostaglandins, and leukotrienes, to further recruit and activate cells. The supply of blood to the area is increased by stimulation of sensory afferent neurons underlying the gut epithelium. Release of calcitonin gene-related peptide (CGRP) stimulates the subsequent release of eNOS-derived NO. Mucosal blood flow is also increased by the actions of prostaglandins, NO, and H_2S. Resolution of inflammation is characterized by the replacement of invading neutrophils and eosinophils with monocytes and macrophages. Mediators that complement this process, for example those that inhibit neutrophil chemotaxis or are able to modulate the surrounding milieu of cytokines/chemokines, are considered anti-inflammatory and thus support this resolution phase. Lipoxins, annexin-1, NO, and H_2S modulate the interaction of neutrophils with endothelial cells, limiting their infiltration into the tissue. PGD_2 also enhances the resolution of inflammation by limiting granulocyte infiltration.

exacerbate IBD, and cause a reactivation of quiescent disease. The discovery of the second COX isoform (COX-2) was met with much excitement. According to the "COX-2 hypothesis," most of the prostaglandins involved in inflammatory processes were derived from this isoform.

In the late 1990s, a number of studies set out to delineate the role of COX-2-derived prostaglandins in the resolution of inflammation. Using a carrageenan-induced paw edema model (characterized by hind paw swelling within 4–6 hours, and subsidence within 24–48 hours), it was demonstrated that while swelling in COX-2-deficient mice could be inhibited by NSAIDs, in those that were given no treatment, the swelling failed to spontaneously resolve. Moreover, significant leukocyte infiltration was observed. This indicated that COX-1-derived prostaglandins contributed to the inflammatory response, and importantly, COX-2-derived prostaglandins played a significant role in the resolution process. Around the same time, the profile of COX-2 and its eicosanoid derivatives in inflammation was clarified in a rat model of pleurisy. It was reported that leukocyte infiltration into the pleural cavity was maximal within 24 hours, and the period thereafter (24–36 hours) was marked by a reduction in leukocyte infiltration – indicative of resolution. Accompanying this was the observation that COX-2 expression was at its greatest when leukocyte clearance was taking place, with a concomitant elevation in PGD_2 levels. Administration of a selective COX-2 inhibitor abolished the PGD_2 production and impaired the reduction of leukocyte numbers inside the pleural cavity, whereas exogenous administration of PGD_2 reversed these effects. It was reported that a hydration product of PGD_2, termed 15-deoxy-$\Delta^{12-14}PGJ_2$, was responsible for driving the resolution of the inflammatory response in the pleural cavity. 15-deoxy-$\Delta^{12-14}PGJ_2$ is an endogenous ligand for PPARγ, one of a family of nuclear receptors that has been implicated in the regulation of various anti-inflammatory mediators and angiogenesis.

There is evidence of a similar role of PGD_2 in resolution of inflammation in the GI tract. In a rat model of colitis, PGD_2 was found to be upregulated early in the inflammatory response (1–3 hours) in parallel with an increase in expression of COX-2. Selective inhibition of COX-2 abrogated the production of PGD_2 and resulted in more than a doubling of the granulocyte infiltration. Thus, COX-2-derived PGD_2 acted as a "stop signal" for granulocyte infiltration.

Interestingly, in the same colitis model, PGD_2 levels remained sustained long after the inflammation had subsided and the ulceration had healed, in contrast to the return to basal levels of the synthesis of other prostanoids (e.g., PGE_2). Once again, the elevated production of PGD_2 occurred via COX-2. This was shown to correlate with changes in mucosal barrier function and epithelial secretion. Indeed, this impairment of mucosal function persisted for weeks after resolution of intestinal inflammation, resulting in an elevated level of bacterial translocation. An increase in epithelial cell proliferation was also observed with prolonged upregulation of PGD_2, and this appeared to contribute to neoplastic changes occurring in the colon. This may have a clinical correlate, as it is well established that individuals with ulcerative colitis have an increased susceptibility to the development of colon cancer.

Lipoxins

Lipoxins (lipoxygenase interaction products) are a specific series of eicosanoids derived from aracidonic acid (described in detail elsewhere in this volume). Such lipid mediators (including lipoxin-A_4, LXA_4; and lipoxin-B_4, LXB_4) are generated via the lipoxygenase pathway, requiring the sequential actions of two different lipoxygenases (LO) (Figure 23.3). In human tissues these are produced predominately via transcellular mechanisms – such as leukocyte-platelet (involving 5-LO and 12-LO, respectively) and leukocyte–endothelial/epithelial cell (involving 5-LO and 15-LO, respectively) interactions. Aspirin-triggered lipoxins (ATL) represent another mechanism of transcellular lipoxin synthesis, and may be a means by which aspirin mediates some of its beneficial actions. Acetylation of COX-2 by aspirin induces a conformational change in the enzyme. This inhibits the conversion of arachidonic acid to PGH_2, instead directing the metabolism of arachidonic acid to 15-*R*-hydroxyeicosatetranoic acid (15*R*-HETE). Neutrophil-derived 5-LO converts this product to a series of lipoxin isomers, 15-epi-LXA_4 and 15-epi-LXB_4, which retain many of the functional properties of native lipoxins. Both lipoxins and ATL are subject to rapid inactivation by prostaglandin dehydrogenase; thus, the design of synthetic LXA_4, LXB_4, and ATL analogues has yielded products that retain significant biological activity, and yet are resistant to rapid metabolism.

Lipoxins display potent anti-inflammatory and pro-resolution roles that are described in detail elsewhere in this volume. At the cellular level, they regulate various aspects of the inflammatory cascade as well as a number of the players involved – including the inhibition of neutrophil chemotaxis and adhesion, epithelial cell–derived chemokine release, and COX product generation. In terms of neutrophil function, lipoxins attenuate neutrophil adhesion and transmigration across both the vascular endothelium and epithelium. This is in part due to their ability to inhibit selectin- and integrin-dependent interactions. Stable lipoxin analogues mimic these effects. Furthermore, lipoxins also inhibit neutrophil activation–induced release of peroxynitrite, IL-1, IL-8, and superoxide anion. A so-called class switch from pro- to anti-inflammatory eicosanoids has been demonstrated, whereby initial neutrophil influx was shown to be accompanied by the generation of leukotrienes and prostaglandins, whereas subsequent lipoxin biosynthesis was accompanied by spontaneous resolution. Indeed, lipoxin and ATL biosynthesis has been demonstrated both in experimental models of inflammatory disease and in human disease states, and are considered to be powerful stop signals for inflammation. Such lipid mediators signal via the activation of lipoxin receptors – known as ALX or FPRL-1, and are members of the G-protein class of receptors. That these receptors are expressed on a number of epithelial, lymphoid, and myeloid cells is reflected by the ability of lipoxins to modulate cell activation and associated downstream responses. As mentioned earlier, an important aspect of inflammatory resolution is the influx of monocytes and macrophages to remove apoptotic neutrophils. Whereas lipoxins attenuate neutrophil function, these bioactive mediators can potently activate monocytes to stimulate chemotaxis and adhesiveness without eliciting degranulation or oxyradical production. Furthermore, lipoxins promote actin reorganization and the formation of cytoplasmic extensions/pseudopodia in monocytes and macrophages. Such remodeling of the cell architecture is important for phagocytosis to take place. Altogether, lipoxins and their analogues counteract the effects of proinflammatory mediators, whilst promoting the migration of monocytes and macrophages (essential for the resolution of inflammation).

In patients with IBD, the degree of neutrophil transepithelial migration correlates well with patient symptoms and the degree of intestinal epithelial barrier dysfunction. Under these conditions, it is likely that lipoxins modulate intestinal inflammation by attenuating the interaction of leukocytes with gut epithelial cells. Studies in in vitro systems have demonstrated the ability of lipoxins to dose-dependently inhibit neutrophil adhesion and transmigration across intestinal epithelial cell monolayers. Furthermore, these effects are physiologically significant, as transmigration is influenced in the basolateral-to-apical direction – which is the route taken by neutrophils that have exited the circulation. In animal models of IBD, administration of lipoxin analogues resolves inflammation. While it is more difficult to survey the role of lipoxins in human disease, lipoxin levels have been reported to be reduced in patients with ulcerative colitis, and this correlated with a reduced expression of 15-LO. With respect to the epithelial cells, lipoxins potently inhibit inflammatory cytokine release both in vitro and in vivo. Such modulation of the surrounding milieu closely influences the behavior of surrounding cell types. Chemical mediators such as TNF-α, leukotrienes, and PAF have been implicated in the pathogenesis of GI injury. Lipoxins, at nanomolar concentrations, inhibit intestinal epithelial cell release of TNF-α, thus hampering activation of these cells and subsequent neutrophil transmigration. Inhibition of epithelial cell–derived IL-8, macrophage chemoattractant peptide-1, and RANTES has also been reported.

ATL can play an important role in the protection of the stomach against chemical injury. As outlined earlier, maintenance of mucosal blood flow is pivotal for mucosal defense and repair, whilst neutrophil adhesion within the gastric microcirculation contributes to gastric damage. Experimental evidence suggests that COX-2-dependent production of ATL protects against the mucosal-damaging properties of aspirin, by markedly reducing the number of neutrophils that adhere to the vascular endothelium. Furthermore, these effects could be counteracted by the administration of LXA_4 or enhanced following addition of an LXA_4 receptor antagonist. Of interest, elevated levels of ATL in the inflamed stomach confers gastroprotection, with less mucosal damage being observed in response to aspirin when compared to that of the noninflamed stomach. Although still unclear, some of the actions of ATL may also be mediated via NO. NO modulates gastric mucosal resistance to injury in part by augmenting mucosal blood flow. Inhibitors of NO synthesis attenuate the initial hyperemic response to aspirin. In the rat, aspirin-induced gastric damage can be reduced by prior administration of LXA_4 – an effect that could be reversed by pretreatment with an NO synthase inhibitor. In the clinical setting, selective COX-2 inhibitors produce less GI ulcer complications compared to conventional NSAIDs. However, when examining the effect of patients taking a selective COX-2 inhibitor (celecoxib) alongside low-dose aspirin, there is no reported difference in ulcer complication rates when compared to patients taking aspirin and a conventional NSAID (diclofenac and ibuprofen). A possible explanation for this may be the blockade of ATL production that occurs as a consequence of COX-2 inhibition.

Annexin-1

Annexin-1 (previously known as lipocortin 1) is a calcium- and phoshopolipid-binding protein with potent anti-inflammatory actions. It belongs to a larger family of annexin proteins, characterized by the presence of two principal domains (1) the N-terminal, which differs in both length and sequence and is believed to confer the particular biological function of each annexin, and (2) the C-terminal containing core, which consists of a variable number of 70 amino acid "annexin-repeats." Annexin-1 was originally described as a mediator of glucocorticoid action, following intensive research efforts to identify an intermediate proteinaceous factor that was both sensitive to glucocorticoid induction, and could inhibit eicosanoid release from isolated lung preparations. Such actions of annexin-1 were attributable to its ability to inhibit membrane phospholipase A_2 activity and thus arachidonic acid release.

Annexin-1 exhibits diverse biological actions – from anti-inflammatory, antihyperalgesic, and antipyretic activities, to regulation of cell proliferation, differentiation, and apoptosis. It is particularly abundant in cells of the neuroendocrine system and those involved in the host immune system, including macrophages, monocytes, and neutrophils. Annexin-1 activates members of the formyl peptide receptor family (FPR) in an autocrine/paracrine manner.

Annexin-1 potently inhibits leukocyte adhesion and transmigration. Immunoneutralization of annexin-1 results in a blockade of inflammogen-induced leukocyte recruitment, and reverses glucocorticoid-mediated inhibition of leukocyte migration. Its effects on leukocyte migration appear homogeneous, despite the use of a variety of inflammogens. Furthermore, glucocorticoid administration to a hamster cheek pouch preparation was shown to alter the fate of adherent leukocytes in the postcapillary venules. There, the leukocyte content of annexin-1 was increased in response to dexamethasone, resulting in both the detachment of neutrophils and of the subset that continued to pass the endothelial cell barrier, this event proceeded at a 3–4-fold reduction in rate. Immunoneutralization of annexin-1 was able to reverse these effects. Indeed, annexin-1-null mice display a higher degree of leukocyte recruitment compared to wild-type littermate mice, and this is attributable to enhanced cell emigration.

Relatively little is known about annexin-1 with regard to GI mucosal defense, however, given the glucocorticoid-sensitive actions of this protein, it is not surprising that several studies have identified a gastroprotective role. Glucocorticoid administration to rats prior to indomethacin challenge has been shown to greatly reduce leukocyte adherence in mesenteric postcapillary venules and the extent of gastric damage produced by NSAIDs. This effect is mediated, at least in part, by annexin-1, as immunoneutralization of this protein reversed the protective effects afforded by glucocorticoids. Furthermore, this could be replicated with the administration of annexin-1 receptor antagonists.

Annexin-1 has also been shown to contribute to healing of lesions in the gastric mucosa. Experimental induction of gastric ulcers results in a marked upregulation of annexin-1 expression at the ulcer margin. Administration of an annexin-1 mimetic acclerated ulcer healing, and this could be blocked by co-administration of an antagonist to the annexin-1 receptor (FPRL-1). Several studies have also demonstrated the active secretion annexin-1 from inflamed colonic tissue. During relapses of Crohn's disease, an increase in release of PGE_2 has been correlated with a decrease in the expression of annexin-1 in intestinal inflammatory tissues. Despite the release of small quantities of PGE_2 being beneficial, release of larger amounts correspond to activation of the inflammatory cascade and increased severity of inflammation. The inverse correlation between PGE_2 and annexin-1 is in line with the anti-inflammatory properties of annexin-1. Altogether, such evidence demonstrates an important role for annexin-1 in the modulation of GI inflammation and repair.

Nitric Oxide

As noted earlier, NO can exert protective and damaging effects in the GI tract, depending on the concentrations of this mediator produced. Here, we define the roles of NO that contribute significantly to gut mucosal defense.

As in the case of epithelial restitution, maintenance of adequate mucosal blood flow is very important during ulcer healing. The ulcer margin receives a high rate of blood flow, and any reduction has been shown to retard ulcer healing. NO is a particularly important vasodilatory during ulcer repair. Moreover, NO can enhance ulcer repair through its ability to stimulate angiogenesis and enhance collagen deposition by fibroblasts. Not surprisingly, therefore, inhibitors of NO production retard gastric ulcer healing, while NO donors have been shown to accelerate this process.

NO also modulates the activity of a number of immunocytes within the GI lamina propria, which can contribute to the resolution of inflammation. Mast cells act as alarm cells, alerting the immune system to the presence of foreign pathogens or noxious substances. They secrete a number of factors that modulate the surrounding milieu; these can be proinflammatory, such as leukotrienes, PAF, endothelin, and TNF-α, or anti-inflammatory, such as NO. NOS inhibitors have been shown to elicit mast cell degranulation, suggesting that NO acts as a tonic inhibitor of mast cell activation. Indeed, NO released from the mast cell itself can feedback to inhibit histamine release from those cells. NO also regulates the production of other proinflammatory mediators. Of these, PAF and histamine both have the capacity to regulate epithelial barrier permeability. NO is also essential for the antimicrobial and cytotoxic activities of macrophages, and can inhibit release of cytokines, such as IL-12 and IL-1, by macrophages.

Hydrogen Sulfide

H_2S appears to be an important endogenous anti-inflammatory substance, although, like NO, at high concentrations it may exert proinflammatory effects. Suppression of H_2S synthesis results in an increase in leukocyte adherence and emigration, and an enhancement of edema formation. Administration of H_2S donors has the opposite (anti-inflammatory) effects. Like NO, H_2S donors have been shown to accelerate the healing of experimental gastric ulcers. Recent studies suggest that H_2S can inhibit the activation of the nuclear transcription factor, NF-κB. Thus, H_2S donors have been shown to reduce expression of TNF-α, IFN-γ, and various other proinflammatory cytokines and chemokines in experimental colitis and endotoxemia. Such actions are consistent with a role of H_2S as a mediator of the resolution of inflammation and injury.

Cytokines and Growth Factors

Within the gut mucosa, the effects of proinflammatory cytokines are normally counter-balanced by regulatory/anti-inflammatory mediators, such as the cytokines IL-10, IL-4, IL-5, IL-13 and growth factors such as transforming growth factor beta (TGF-β). In the context of IBD, the therapeutic utility of recombinant anti-inflammatory cytokines or related gene therapies are currently being explored.

IL-10 is a T_H2 cell–derived cytokine, which functions to dampen the inflammatory response by blocking synthesis of several cytokines (e.g., IL-1β, TNF-α, IL-4,

IL-5, and IL-6) and inhibiting antigen presentation. It has also been shown to regulate the differentiation and proliferation of immune T and B cells, natural killer cells, and mast cells, and increase the expression of anti-inflammatory proteins (such as IL-1a). IL-10-deficient mice spontaneously develop intestinal inflammation, and in patients with ulcerative colitis, a deficiency of IL-10 has been noted. However, clinical trials of recombinant IL-10 therapies have so far yielded disappointing results.

TGF-β is a pleiotropic growth factor that regulates the growth, differentiation, and function of immune and nonimmune cells. In the GI tract, it has been shown to enhance gastric ulcer healing and epithelial restitution, as well as to enhance epithelial barrier function in the presence of cytokines that otherwise increase intestinal permeability (such as IFN-γ and IL-4). Despite the regulatory roles of TGF-β in mucosal inflammation (it serves in part to balance the effects of IFN-γ), elevated levels of this cytokine are present in patients with IBD.

Intestinal adaptation, particularly after injury or surgical removal of damaged intestinal tissue, is important to ensure sufficient absorption of nutrients, water, and essential vitamins. Growth factors such as glucagon-like peptide-2, intestinal trefoil factor (ITF), and kerotinocyte growth factor (KGF) have been shown to stimulate intestinal mucosal growth and enhance repair following damage or surgical resection. Such repair processes during the resolution of inflammation facilitate tissue remodeling and help restore normal intestinal architecture.

The trefoil factors are a family of mucin-associated peptides involved in the regulation of epithelial motility and mucosal protection. ITF, which is abundantly expressed in the goblet cells of the intestinal mucosa, has been shown to enhance intestinal epithelial repair and restitution. Oral administration of recombinant ITF protects against ethanol- and NSAID-induced gastric injury. Moreover, mice deficient in ITF display impaired epithelial migration and develop more severe colitis in response to oral dextran sodium sulfate.

KGF is a potent and highly specific mitogen for epithelial cells. Its significance during wound repair is highlighted by its overexpression in the mesenchyme both below and at the wound edge. In animal models of IBD, KGF administration significantly ameliorates mucosal damage. Animals deficient in KGF are more susceptible to colitis as well as chemotherapy- and radiation-induced intestinal injury. In human IBD tissues, KGF is strikingly upregulated, and likely acts in a paracrine manner to stimulate the proliferation of intestinal epithelial cells. Such repair mechanisms act as a defensive maneuver to protect against ongoing inflammatory assault. KGF is presently in clinical use for the prevention of radiation and/or chemotherapy induced intestinal injury.

THERAPEUTIC APPLICATIONS

In recent years, a number of approaches have been taken to develop anti-inflammatory drugs that exploit the anti-inflammatory (pro-resolution) actions of some of the above-mentioned mediators. NSAID-induced gastric mucosal damage has been shown to be a neutrophil-dependent process, and a reduction in gastric mucosal blood flow has been shown to increase the susceptibility to mucosal damage, as well as impairing the ability of the mucosa to undergo repair. Given the inhibitory effects of both NO and H_2S on leukocyte–endothelial cell adherence, as well as their potent vasodilatory effects, it was postulated that modification of NSAIDs such that they release small amounts of either NO or H_2S would greatly reduce the GI toxicity of this class of compounds. Thus, COX-inhibiting NO donating drugs (CINODs or NO-NSAIDs) and H_2S-releasing NSAIDs have been developed and extensively evaluated. The ability of lipoxins to potently drive the resolution of inflammation has been exploited in the development of stable lipoxin analogs. These approaches are briefly outlined later.

NO-NSAIDs

Like conventional NSAIDs, the NO-NSAIDs retain the ability to inhibit both COX-1 and COX-2, and thus prostaglandin synthesis. However, their modification to include an NO-releasing moiety has been shown to produce a superior pharmacological profile. In rodents and humans, these drugs produce much less gastric and intestinal damage, and display increased anti-inflammatory and analgesic potencies relative to the parent NSAIDs. Moreover, whereas NSAIDs elevate systemic blood pressure, the NO-NSAIDs have the opposite effect. This can be attributed to the slow release of NO into the systemic circulation from these drugs. The improved GI profile of NO-NSAIDs may be attributed, at least in part, to their ability to inhibit leukocyte–endothelial cell adhesion and to maintain mucosal blood flow.

NSAIDs, including selective COX-2 inhibitors, have well characterized inhibitory effects on gastric ulcer healing. NO-NSAIDs, on the other hand, do not delay ulcer healing, and in some studies have been found to accelerate the healing of preexisting ulcers.

An NO-releasing derivative of naproxen, called "naproxcinod," has been evaluated extensively in clinical trials. Phase II trials involving patients with hip or knee osteoarthritis demonstrated that naproxcinod reduced the incidence of gastroduodenal ulcers by 40% (compared to naproxen), with fewer stomach and duodenal ulcers being detected. Given the concern over adverse cardiovascular events of NSAIDs, naproxcinod has been examined for its ability to affect blood pressure. In contrast to the selective COX-2 inhibitors and conventional NSAIDs, which significantly elevate systemic blood pressure, naproxcinod exerts a significant antihypertensive effect.

H_2S-Releasing NSAIDs

Like NO, H_2S makes a significant contribution to gastric mucosal defense, as outlined earlier. As was the case for the NO-NSAIDs, the H_2S-releasing NSAIDs retain the same ability to inhibit prostanoid synthesis, while showing an improved GI tolerability, and an increase in anti-inflammatory potency. Also like NO-NSAIDs, H_2S-releasing NSAIDs inhibit leukocyte adherence and do not reduce gastric mucosal blood flow.

H_2S-releasing NSAIDs have been shown to suppress mucosal expression of proinflammatory cytokines, such as TNF-α, most likely through inhibition of NF-κB activation. This may contribute significantly to the improved anti-inflammatory effects of these drugs as compared to the parent drugs, as well as contributing to their improved GI tolerability. H_2S-releasing NSAIDs have not yet been evaluated in humans.

Lipoxin Analogs

Lipoxins are quite unstable, and therefore not very suitable as therapeutic agents. However, stable lipoxin analogs have been developed and are being evaluated in a variety of animal models and in clinical trials. In models of colitis, lipoxin analogs were shown to exert beneficial effects, which included more rapid resolution of the mucosal inflammation and enhanced repair of lesions.

SUMMARY

A great deal has been learned in the past decade regarding the processes involved in the resolution of inflammatory reactions and healing of tissue injury. In the GI tract, it is clear that a number of mediators are crucial both to mucosal defense against injury induced by luminal agents, and the repair of mucosal injury. There appears to be considerable redundancy in terms of the effects of some of the key mediators in this process (e.g., prostaglandins, NO, H_2S), and there may be some cooperative interactions among these mediators to promote gastrointestinal integrity. In recent years, the knowledge about the roles of these mediators in mucosal defense and repair has been exploited in the development of novel anti-inflammatory drugs, which show great promise in preclinical studies.

SUGGESTED READINGS

Fiorucci, S., Distrutti, E., Cirino, G., and Wallace, J.L. 2006. The emerging roles of hydrogen sulfide in the gastrointestinal tract and liver. *Gastroenterology* **131**:259–271.

Gilroy, D.W., and Perretti, M. 2005. Aspirin and steroids: new mechanistic findings and avenues for drug discovery. *Curr Opin Pharmacol* **5**:405–411.

Serhan, C.N. 2007. Resolution phase of inflammation: novel endogenous anti-inflammatory and proresolving lipid mediators and pathways. *Annu Rev Immunol* **25**:101–137.

Wallace, J.L., and Del Soldato, P. 2003. The therapeutic potential of NO-NSAIDs. *Fundam Clin Pharmacol* **17**:11–20.

Wallace, J.L., and Devchand, P.R. 2005. Emerging roles for cyclooxygenase-2 in gastrointestinal mucosal defence. *Br J Pharmacol* **145**:275–282.

Xavier, R.J., and Podolsky, D.K. 2007. Unravelling the pathogenesis of inflammatory bowel disease. *Nature* **448**:427–434.

24 Inflammatory Skin Diseases

Gayathri K. Perera and Frank O. Nestle

PSORIASIS

Psoriasis is a common, chronic inflammatory skin disease with an overall prevalence of 2%–3% of the world population with a higher prevalence in the United States and Canada (4.6% and 4.7%, respectively) compared to a lower prevalence of 0.4–0.7% in Africans, African Americans, and Asians [1].

Psoriasis can appear at any age but there is a tendency toward a peak at ages 20–30 years and then again at ages 50–60 years. It has a chronic relapsing and remitting nature in response to a variety of stressors. It is rarely life threatening but is associated with a high degree of morbidity. Psoriasis can be limited to skin but can also be associated with debilitating arthritis in 5%–30% of sufferers. Its burden on health care economy is on par with cardiovascular disease, diabetes, and depression.

Its aetiology is multifactorial, combining genetics, immunity, and environmental triggers. There is a 2–3-fold increase in risk of developing psoriasis in monozygotic versus dizygotic twins, and results of WGAS (whole genome wide association scans) show that psoriasis is a complex genetic disease [2]. At least 19 gene loci have been implicated but replication of a single locus has been provided for only a few of these. One of these regions lies on a 210-kb stretch of DNA on the short arm of chromosome 6 (termed PSORS1) which includes genes coding for HLA-Cw6, a potential immunological candidate gene, and corneodesmosin, a potential epidermal structure related candidate gene [3].

In a genetically primed individual, there occurs a complex interplay between the innate and adaptive arms of the immune system in response to as yet an unidentified antigen. The immune cells thought to contribute toward disease aetiology include T cells, dendritic cells, and monocytes/macrophages. Of the T-cell subsets, CD8 (cytotoxic) T cells are mainly positioned in the epidermis close to keratinocytes and Langerhans cells, while CD4 T helper cells are mainly located in the upper papillary dermis. The role of epidermal T cells including clonality of T-cell receptors is an important concept when regarding

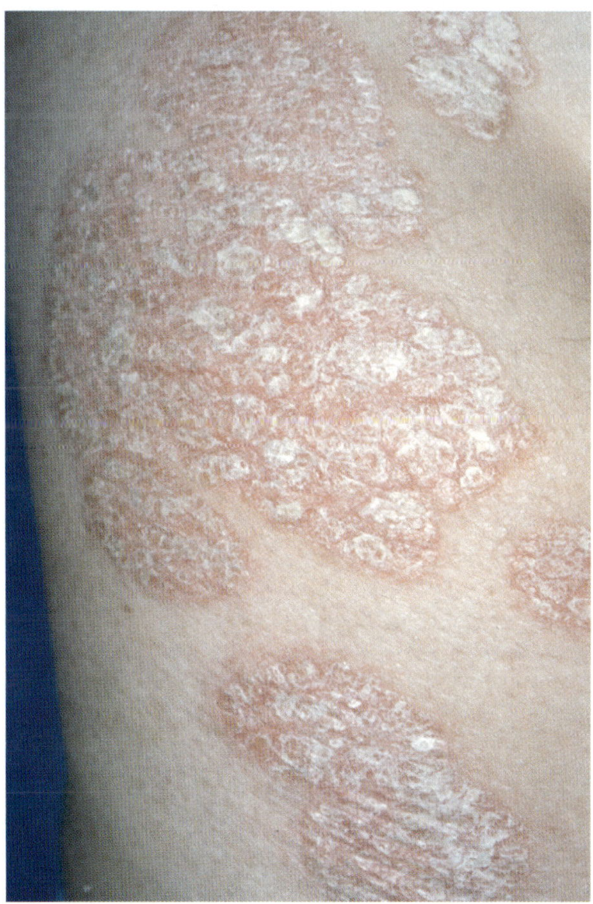

Figure 24.1. A typical chronic psoriatic plaque. A thickened, sharply demarcated plaque with overlying silvery scale. (Courtesy of St. John's Institute of Dermatology, UK.)

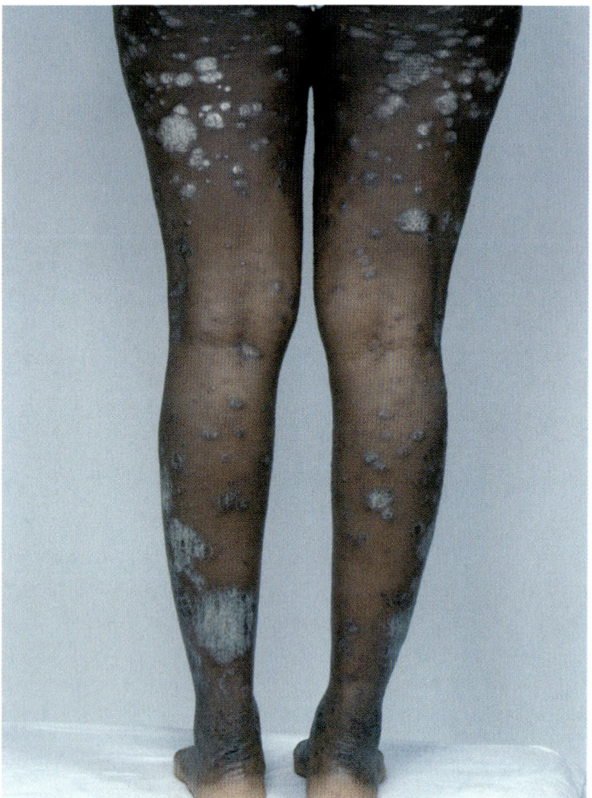

Figure 24.2. Symmetrical distribution of psoriatic plaques. (Courtesy of St. John's Institute of Dermatology, UK.)

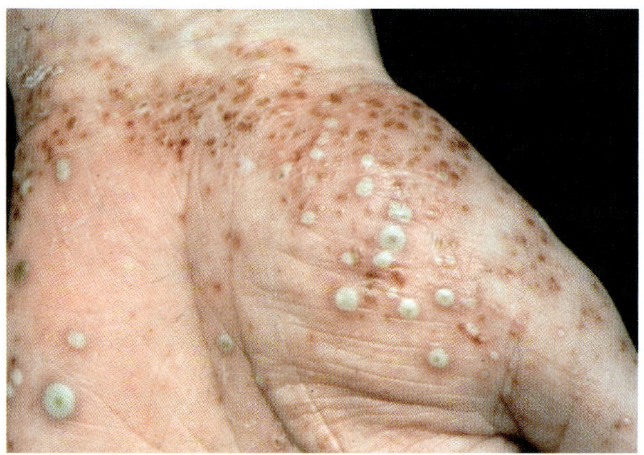

Figure 24.3. Pustular psoriasis of the palms. Multiple acute sterile pustules are interlaced with dark brown macules, representing healing pustules. (Courtesy of St. John's Institute of Dermatology, UK.)

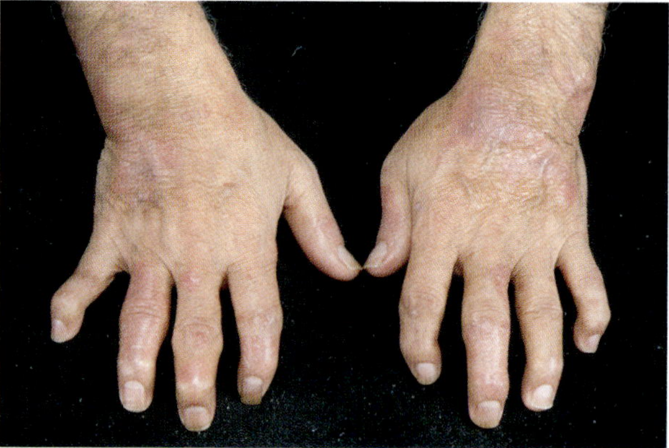

Figure 24.4. Psoriatic arthritis (PsA) is typically an asymmetric arthritis involving the distal (DIP) and proximal interphalangeal (PIP) joints. When both DIP and PIP joints are involved a "sausage" finger is formed (arrow). In its most severe form PsA results in *arthritis mutilans*. (Courtesy of St. John's Institute of Dermatology, UK.)

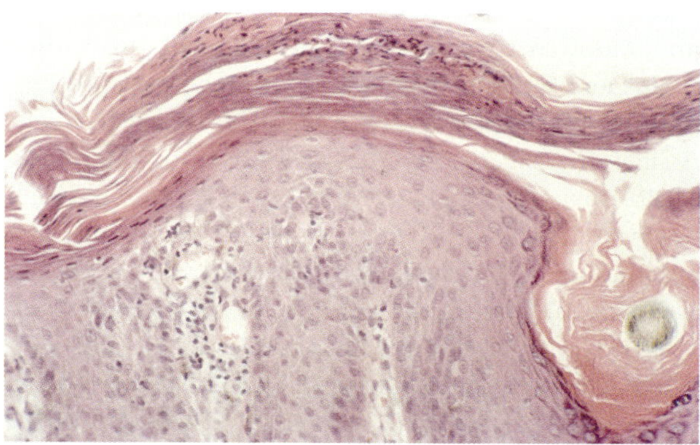

Figure 24.5. A well established psoriatic plaque showing hyperkeratosis and parakeratosis. There is also a microabscess of Munro (accumulation of neutrophil remnants in the stratum corneum surrounded by parakeratosis) accompanied by spongiform pustules of Kogoj (collection of neutrophils within the stratum spinosum). Hall marks of psoriasis. (Courtesy of St. John's Institute of Dermatology, UK.)

the pathogenesis of psoriasis [4]. Lesional psoriatic T cells are of an activated memory phenotype expressing the CD25 receptor [5] secreting predominantly interferon-γ (IFN-γ) and classified as T helper 1 (Th1) cells. The recent uncovering of T helper 17 (Th17) cells, which produce IL-17 and IL22 (a cytokine involved in epithelial hyperplasia) [6], provides another important element with a potential impact on psoriasis immunopathogenesis. The functional role of Th17 cells in psoriasis needs to be defined.

Antigen-presenting dendritic cells (DCs) are an important armament of the innate immune system, which have the capacity to direct the adaptive immune system by activating and/or priming T cells and other effector cells such as B and NK cells. Dermal DCs are increased in psoriasis lesions and increase the production Th-1 cytokines in autologous T cells [7]. Major subsets of DCs include myeloid DC such as dermal DC or plasmacytoid DC (pDC). PDCs produce IFN-α and have been shown to be increased in psoriasis lesions. By blocking PDCs or IFN-α in xenotransplantation models with anti-TNF-α therapies, the development of psoriasis is arrested [8,9]. Recent evidence has shown that self-DNA is able to activate pDCs to produce IFN-α after forming complexes with the antimicrobial peptide LL-37 [10]. LL-37 has been shown to be elevated in psoriasis.

Clinically psoriasis represents a spectrum of disease entities, ranging from small thin plaques, to sterile pustules, to thick, silvery scaly plaques. One individual may show different forms of the disease at any one time (Figures 24.1–24.6).

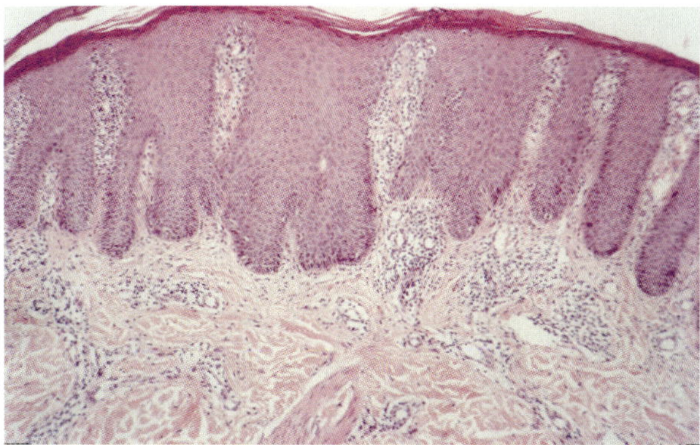

Figure 24.6. Stable plaque psoriasis showing acanthosis (thickening of the epidermis) and elongation of the rete ridges (dermal papillae). (Courtesy of St. John's Institute of Dermatology, UK.)

ATOPIC DERMATITIS (FIGURES 24.7–24.12)

Atopic dermatitis (AD), more commonly known as eczema, is a chronic, invariably pruritic, relapsing, and remitting inflammatory skin disease arising from disruption of the epidermal barrier. It is often associated with other atopic conditions such as asthma and allergic rhinitis. The prevalence of AD has risen rapidly in industrialized countries over the past 30–40 years [11].

The aetiology of AD is complex and the high monozygotic twin rate of 77% compared to dizygotic twins (15%) suggests that genetic susceptibility genes are involved along with environmental triggers [12]. The involvement of the immune system is far from straightforward and there has been a division of AD into IgE-mediated and non-IgE–mediated disease, implying two different diseases. An approach to unifying these concepts has led to the theory that the natural evolution of AD has three phases starting with the non-IgE–mediated, nonsensitization phase seen in infancy. This is rapidly followed by the IgE-mediated phase where sensitization under genetic influences then predominates (classical AD). The third phase arises as a result of itch leading to scratching, the latter leading to the release of possible autoantigens from skin cells which then induce the production of IgE autoantibodies [13].

The clinical manifestations vary with age. There appear to be three peaks of incidence, with one in infancy (more than 50% of all cases begin after 2 months of age), one in early childhood (more than 90% before the age of 5 years), and one in adulthood. Although some of the skin manifestations can vary from papules and vesicles to thickened scaly plaques, depending on the phase of the disease, whether acute or chronic, the symptom common to all is the incessant itching, which is exacerbated at night, and, in infancy, if untreated, can lead to a failure to thrive. The distribution of eczematous lesions in infancy is centered on the cheeks and scalp, whereas the classical flexural lesions are seen commonly in childhood AD. Adult AD is often associated with chronicity and results in lichenified plaques of the flexures, nape of neck, and limbs [14]. Diagnostic criteria are based on clinical signs and symptoms along with history and the exclusion of other disease entities (see Table 24.1) [15].

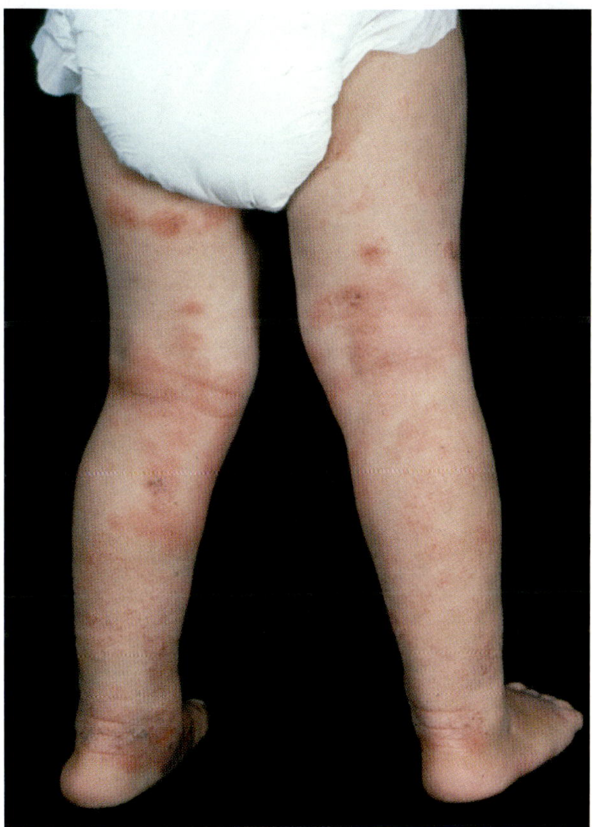

Figure 24.7. Atopic dermatitis in a toddler showing a generalized distribution over the lower limbs. Note the excoriations around areas within reach. (Courtesy of St. John's Institute of Dermatology, UK.)

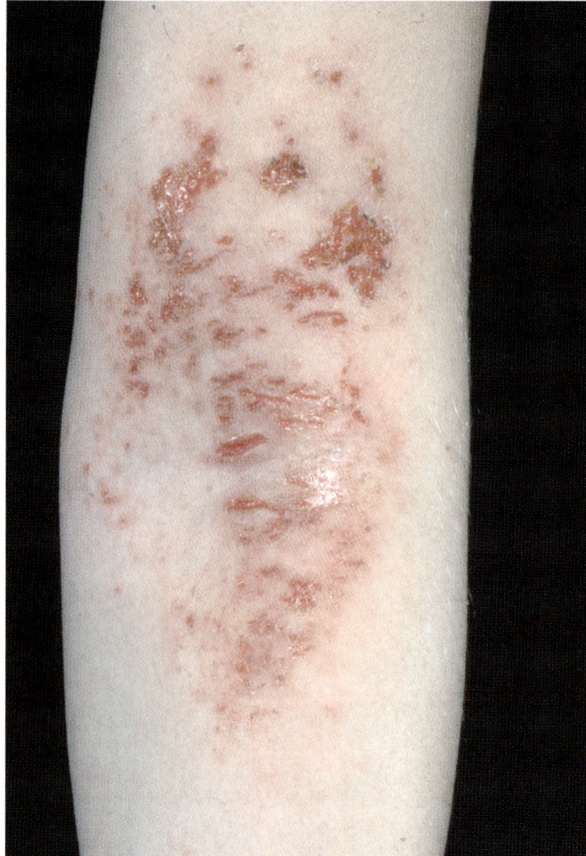

Figure 24.8. A typical flexural distribution of atopic dermatitis with secondary infection (note the yellowy exudate) as a result of continuous excoriation. (Courtesy of St. John's Institute of Dermatology, UK.)

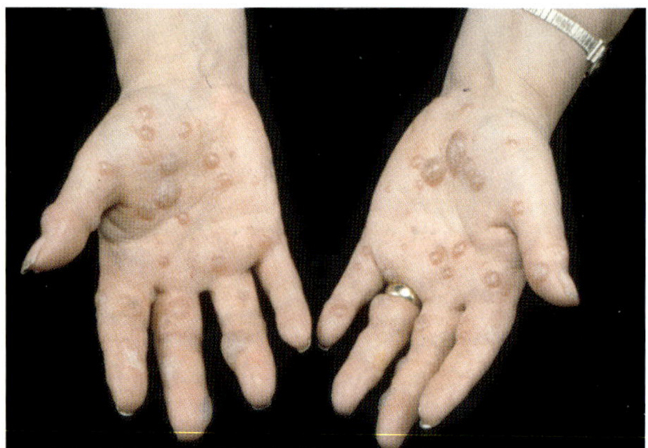

Figure 24.9. Acute pompholyx (dishydrotic eczema) of the palms. The tense bullae arise as a result of the edema occurring in the epidermis. (Courtesy of St. John's Institute of Dermatology, UK.)

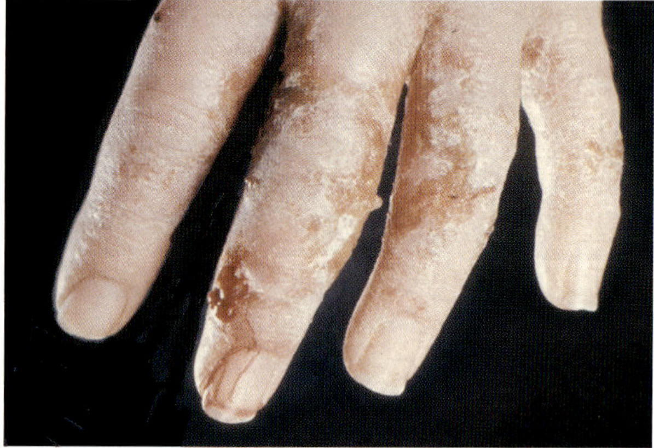

Figure 24.10. Infected hand eczema. The most likely organisms are Staphylococci and/or Streptococci. (Courtesy of St. John's Institute of Dermatology, UK.)

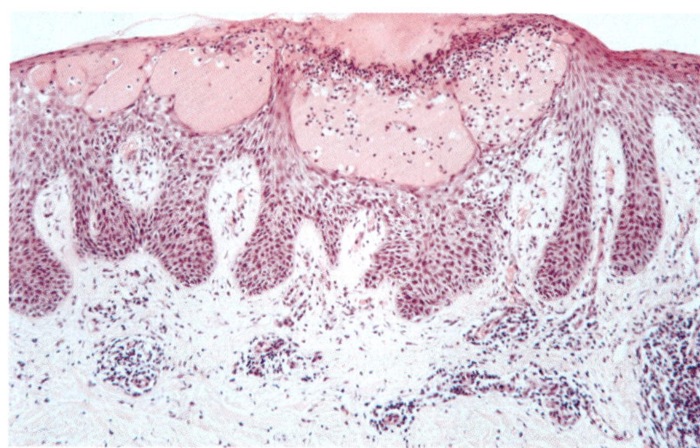

Figure 24.11. Acute atopic dermatitis showing intraepidermal spongiosis (edema) with fluid accumulation in macrovesicles and bullae. Exocytosis of lymphocytes produces intraepidermal vesicles. (Courtesy of St. John's Institute of Dermatology, UK.)

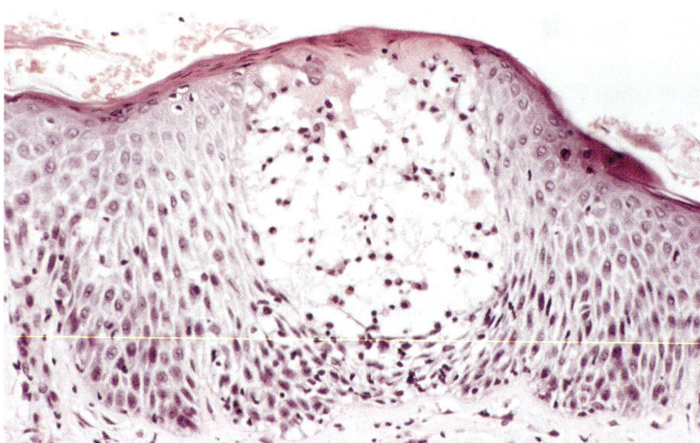

Figure 24.12. A spongiotic vesicle containing lymphocytes. (Courtesy of St. John's Institute of Dermatology.)

TABLE 24.1. UK working party diagnostic criteria for atopic dermatitis

Must have • itchy skin condition (or parental report of scratching or rubbing in a child)
AND
Three or more of the following: • History of involvement of skin creases (elbows, back of knees, front of ankles, around neck, cheeks in children under 10 years) • A personal history of atopy (asthmas, allergic rhinitis) or family history of atopy in first degree relative in children under 4 years of age) • History of generalized dry skin in last 1 year • Visible flexural eczema (or eczema involving cheeks/forehead and limbs in children under the age of 4 years) • Onset under the age of 2 years (not applied to children under 4 years of age)

REFERENCES

1. Nestle, F.O., Kaplan, D.H., Barker, J. 2009. Psoriasis. *N Engl J Med* **361**:496–509.
2. Liu, Y., Krueger, J.G., and Bowcock, A.M. 2007. Psoriasis: genetic associations and immune system changes. *Genes Immun* **8**:1–12.
3. Capon, F., Di Meglio, P., Szaub, J., et al. 2007. Sequence variants in the genes for the interleukin-23 receptor (IL23R) and its ligand (IL12B) confer protection against psoriasis. *Hum Genet* **122**(2):201–206.
4. Chang, J.C., Smith, L.R., Froning, K.J., et al. 1994. CD8+ T cells in psoriatic lesions preferentially use T-cell receptor V beta 3 and/or V beta 13.1 genes. *Proc Natl Acad Sci USA* **91**(20):9282–9286.
5. Gottlieb, A.B., Lifshitz, B., Fu, S.M., Staiano-Coico, L., Wang, C.Y., and Carter, D.M. 1986. Expression of HLA-DR molecules by keratinocytes, and presence of Langerhans cells in the dermal infiltrate of active psoriatic plaques. *J Exp Med* **164**(4):1013–1028.
6. Zheng, Y., Danilenko, D.M., Valdez, P., et al. 2007. Interleukin-22, a T(H)17 cytokine, mediates IL-23-induced dermal inflammation and acanthosis. *Nature* 445(7128):648–651.
7. Nestle, F.O., Turka, L.A., and Nickoloff, B.J. 1994. Characterization of dermal dendritic cells in psoriasis. Autostimulation of T lymphocytes and induction of Th1 type cytokines. *J Clin Invest* **94**(1):202–209.
8. Nestle, F.O., Conrad, C., Tun-Kyi, A., et al. 2005. Plasmacytoid predendritic cells initiate psoriasis through interferon-alpha production. *J Exp Med* 202(1):135–143.
9. Boyman, O., Hefti, H.P., Conrad, C., Nickoloff, B.J., Suter, M., and Nestle, F.O. 2004. Spontaneous development of psoriasis in a new animal model shows an essential role for resident T cells and tumor necrosis factor-{alpha}. *J Exp Med* **199**(5):731–736.
10. Lande, R., Gregorio, J., Facchinetti, V., et al. 2007. Plasmacytoid dendritic cells sense self-DNA coupled with antimicrobial peptide. *Nature* **449**(7162):564–569.
11. Williams, H., and Flohr, C. 2006. How epidemiology has challenged 3 prevailing concepts about atopic dermatitis. *J Allergy Clin Immunol* **118**:209–213.
12. Morar, N., Willis-Owen, S.A., Moffatt, M.F., and Cookson, W.O. 2006. The genetics of atopic dermatitis. *J Allergy Clin Immunol* **118**:24–34.
13. Bieber, T. 2008. Mechanisms of disease. Atopic dermatitis. *N Engl J Med* **358**:1483–1494.
14. Akdis, C.A., Akdis, M., Bieber, T., et al. 2006. Diagnosis and treatment of atopic dermatitis in children and adults: European Academy of Allergology and Clinical Immunology/American Academy of Allergy, Asthma and Immunology/PRACTALL Consensus Report. *J Allergy Clin Immunol* **118**:152–169. [Erratum, *J Allergy Clin Immunol*. 2006;**118**:724.]
15. William, H.C., Burney, P.G.J., Hay, R.J., et al. 1994. The UK Working Party's diagnostic criteria for atopic dermatitis I. Derivation of a minimum set of discriminators for atopic dermatitis. *Br J Dermatol* **131**: 383–396.

SUGGESTED READINGS

Bieber, T. 2008. Mechanisms of disease. Atopic dermatitis. *N Engl J Med* **358**:1483–1494.

Conrad, C., Boyman, O., Tonel, G., et al. 2007. Alpha 1 beta 1 integrin is crucial for accumulation of epidermal T cells and the development of psoriasis. *Nat Med* **13**(7): 836–842.

Kupper, T.S., and Fuhlbrigge, R.C. 2004. Immune surveillance in the skin: mechanisms and clinical consequences. *Nat Rev Immunol* **4**:211–222.

Lowes, M.A., Bowcock, A.M., and Krueger, J.G. 2007. Pathogenesis and therapy of psoriasis. *Nature* **445**(7130):866–873.

Mittermann, I., Aichberger, K.J., Bünder, R., Mothes, N., Renz, H., and Valenta, R. 2004. Autoimmunity and atopic dermatitis. *Curr Opin Allergy Clin Immunol* **4**:367–371.

Morar, N., Willis-Owen, S.A., Moffatt, M.F., and Cookson, W.O. 2006. The genetics of atopic dermatitis. *J Allergy Clin Immunol* **118**:24–34.

Nestle, F.O., Kaplan, D.H., Barker, J. 2009. Psoriasis. *N Engl J Med* **361**:496–509.

Nestle, F.O., di Meglio, P., Qin, J.Z., Nickoloff, B.J. 2009. Skin Immune Sentinels in Health and Disease. *Nat Rev Immunol* **9**. [Epub ahead of print].

25 Kidney Glomerulonephritis and Renal Ischemia

Jeremy S. Duffield and Joel M. Henderson

CASE 1

Presentation and Initial Assessment

A 64-year-old male with a history of diabetes mellitus type II, hypertension, hyperlipidemia, and known mild renal insufficiency (chronic kidney disease, stage 2) presented with fevers and chills, and worsening of kidney function (baseline plasma creatinine 1.5, increased to 1.8 mg/dL). His standing medications were modified and renal function was noted to return to baseline. One month later at follow up he reported fever, his renal function had deteriorated further with plasma creatinine level of 3.0 mg/dL, and he was found to be anemic. He received a blood transfusion and a course of antibiotics for presumed upper respiratory tract bacterial infection but one month later returned with lethargy, oliguria, and further deterioration in renal function with creatinine 4.4 mg/dL. At this point, he was referred to a nephrologist for further assessment.

At that assessment by a nephrologist, his family history was notable for thromboembolic disease, but he denied current upper or lower respiratory tract symptoms, rashes, joint pains, areas of numbness, black or red bowel movements. His medications at this presentation included Hydralazine Metoprolol Felodipine, Insulin subcutaneously, Ezetimibe, Glimepiride, Zolpidem, multi-vitamin qd, calcium, vitamin E, fish oil, and acetaminophen.

Clinical examination: weight 200 lbs, temperature 96.7F, pulse 63 beats per minute, blood pressure 160/80 mmHg, respiratory rate 20 per minute, peripheral blood oxygen saturation 97% on room air. He was a middle-aged man, pleasant, in no distress. Pupils and eye movements and oropharyngeal exams were normal. Neck exam was unremarkable, and the jugular venous pressure was low, there were no carotid artery bruits. The cardiac exam revealed regular rhythm and normal heart sounds. The lungs were normal to examination and the abdomen was soft without tenderness and no organomegaly was detected. Examination of the extremities, there was no edema or rashes, and his peripheral pulses were normal. His neurological exam found him to be alert and oriented, and examination of cranial and peripheral nerves was unremarkable.

Analysis of a urine specimen provided by the patient. Specific gravity of 1.010, pH 5, there was no glycosuria or ketonuria. However, there was 3+/4+ protein, large hemoglobin, and direct microscopy of sediment identified 5–9 white blood cells and 50–70 erythrocytes per high-powered field. Many of the erythrocytes were dysmorphic. There were many coarse granular casts, but no cellular casts.

Baseline laboratory findings on blood testing: sodium ions 128 mmol/L, potassium ions 4.8 mmol/L, chloride ions 99 mmol/L, bicarbonate ions 16, blood urea nitrogen 127 mg/dL, creatinine 7.14 µg/dL, glucose 398 mg/dL, total calcium 7.8 mg/dL. Complete blood count: white blood count 9,800/µL, hemoglobin 10 mg/dL, platelets 227,000/µL. The ESR was elevated at 93 mm/h.

Immunological assessment of plasma identified antinuclear antibodies present at 1:40 serum dilution; complement factor 3, 218 U/mL; and complement factor 4, and 38 U/mL. Blood cultures taken twice did not grow any organisms.

An echocardiogram showed no valvular vegetations and no pericarditis or left ventricle dysfunction. An ultrasound scan of the kidneys showed two normal sized kidneys with normal echogenicity and two small cysts. A chest X ray was performed which showed clear lung fields bilaterally.

Owing to anemia and evidence of gastrointestinal blood loss, the patient had an upper gastrointestinal endoscopy, which showed active duodenitis.

Because of the acute deterioration of renal function with an active urine sediment, the patient underwent a renal biopsy.

Renal Biopsy Findings

The biopsy consisted of an adequate sample of kidney cortex and medulla, and was examined by light, immunofluorescence, and electron microscopy. The findings are illustrated in Figures 25.1 through 25.4.

Forty-one glomeruli were present in the entire sample submitted; six of these glomeruli were globally sclerosed. Light microscopic examination revealed circumferential or segmental cellular crescents in 29 of the non-sclerosed glomeruli (Figures 25.1 and 25.2). Many of these cellular crescents included fibrinoid necrosis and karyorrhexis, fragmentation of Bowman's capsule, and active periglomerular inflammation (Figure 25.2). The fibrinoid necrosis and periglomerular fibrin deposition was particularly evident on immunofluorescence microscopy for fibrin (Figure 25.3). Many glomeruli also revealed an active glomerulitis, composed of mononuclear and polymorphonuclear inflammatory cells (Figure 25.2). Immunohistochemical staining indicated that most of these cells were CD3-positive T cells, although CD68-positive macrophages were also prominent. Periglomerular inflammation was also composed of CD3- and CD68-positive cells, along with few CD20-positive B cells (Figure 25.4).

The glomerular basement membranes in the nonlesional segments appeared slightly wrinkled but otherwise showed a normal appearance at the light microscopic level (Figures 25.1C, 25.2A, 25.2B, and 25.2F). There was mild mesangial expansion by PAS-positive material and mononuclear cells (presumably mesangial cells; Figures 25.1C, 25.2A, and 25.2B). Immunofluorescence microscopy revealed diffuse, strong linear deposition of IgG along the glomerular capillary walls (Figure 25.3). However, there was no evidence of discrete immune complex deposition in the glomeruli by immunofluorescence. This was confirmed by electron microscopy, as no electron dense deposits were seen along the peripheral glomerular capillary loops or within the mesangial matrix (Figure 25.3). Electron microscopy also showed changes of the glomerular endothelial cells indicative of mild endothelial injury, including segmental areas of swelling and loss of fenestrations (Figure 25.3). Glomerular visceral epithelial cells exhibited signs of injury, including foot process effacement, microvillous change, and vacuolization (Figure 25.3). Other ultrastructural findings included mild thickening of glomerular basement membranes, and mild mesangial expansion by matrix and cells (Figure 25.3).

In addition to the glomerular crescents, isolated acute inflammatory lesions were also seen in small arterial vessels. Two small arteries showed unequivocal fibrinoid necrosis of the wall (Figure 25.2C). Small interlobular arteries and arterioles revealed more widespread chronic damage, including moderate sclerosis of the media and hyaline accumulation in the intima (not illustrated). Focal C3 deposition was noted in the walls of small arteries and arterioles by immunofluorescence microscopy; however, these tended to be associated with areas of arterial and arteriolar sclerosis (not illustrated).

There was focal tubular atrophy and interstitial fibrosis apparent, involving approximately 25% of the sample (Figure 25.1). Patchy areas of interstitium were infiltrated by mononuclear inflammatory cells, neutrophils, and occasional eosinophils; these infiltrates were particularly prominent in the vicinity of active glomerular and vascular inflammatory lesions (Figures 25.1 and 25.2). Immunohistochemical staining demonstrated that the interstitial infiltrates were primarily composed of CD3-positive T cells and CD68-positive macrophages (Figure 25.4). Patchy areas of fibrin deposition were also noted in the interstitium by immunofluorescence microscopy (Figure 25.3). A few tubules showed features of acute tubular injury, including luminal distension and epithelial flattening (Figures 25.1 and 25.2).

Management and Follow Up

The patient had a clinical and pathological diagnosis of focal and necrotizing crescentic glomerulonephritis or rapidly progressive glomerulonephritis. The basement membranes showed linear IgG and C3 deposition characteristic of antiglomerular basement membrane disease. He had antibodies in plasma directed against the glomerular basement membrane measured at 179 ELISA units (normal < 20), consistent with a final diagnosis of anti-GBM disease. However, initial testing also indicated that he had antimyeloperoxidase (ANCA) antibodies at a titer of 1:160, but this test was not reproduced and may have been a false positive result due to the concomitant presence of antinuclear antibodies.

He was given IV methylprednisolone 1 g intravenously X3 then 60 mg prednisone orally, daily and was initiated on oral cyclophosphamide and plasmopheresis X 10 over 14 days. After 14 days of plasma exchange the anti-GBM ELISA had fallen to 38.

The patient's course of therapy was complicated by gastrointestinal bleeding from duodenal erosions that were detected by endoscopy. He required multiple transfusions of red blood cells. Owing to further deterioration of function, he was initiated on dialysis. Several weeks later, there was no recovery of renal function. He was maintained on chronic dialysis and was evaluated up for kidney transplantation.

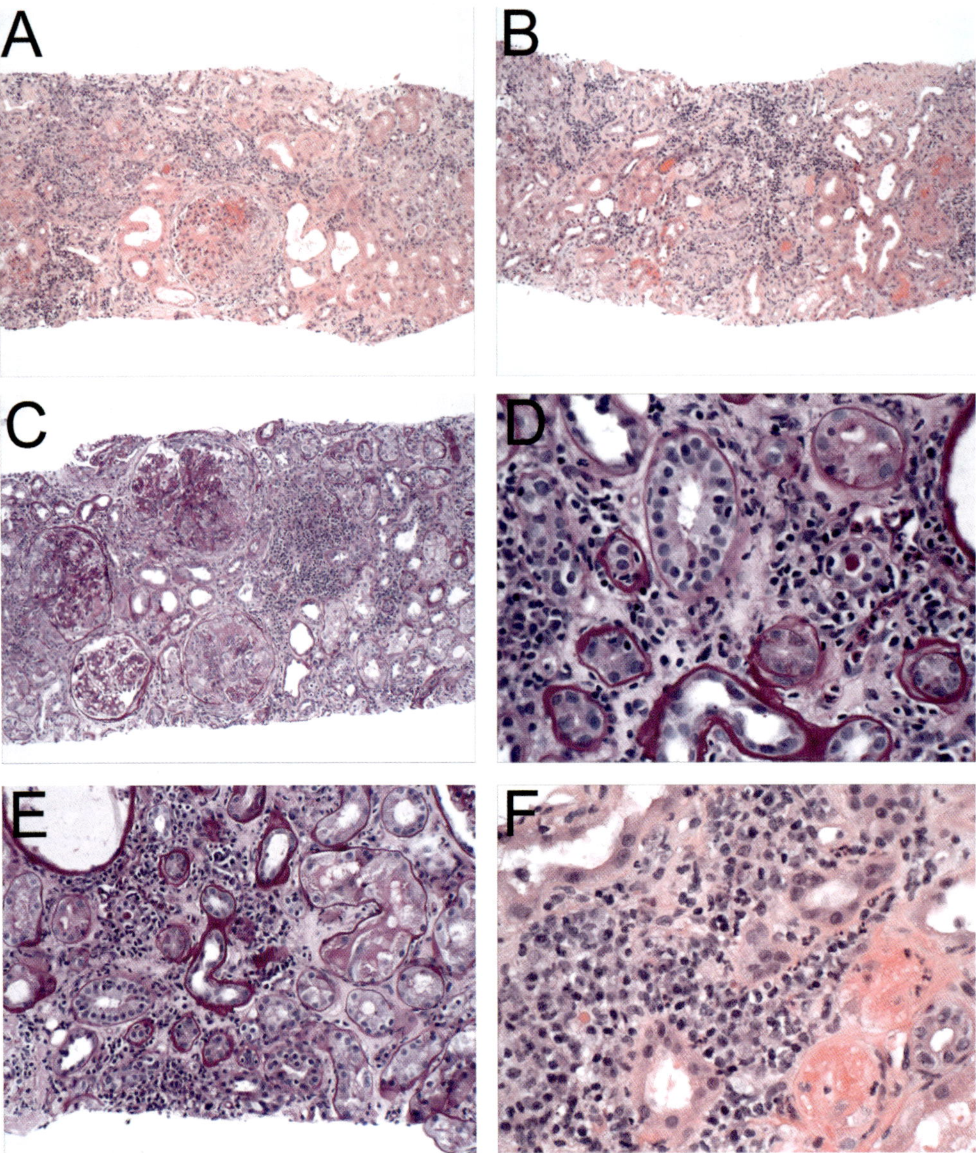

Figure 25.1. Kidney biopsy showing tubulointerstitial histopathology in Case 1. (A) This low power view shows an area of patchy infiltration of the cortical interstitium by acute and chronic inflammatory cells. There is also patchy tubular atrophy, characterized by shrinkage of the tubules, tubular basement membrane thickening and expansion of the intervening interstitium, often in association with the interstitial inflammation. Isolated tubules show features of acute tubular injury, characterized by luminal distension and epithelial flattening. A glomerulus shows a cellular crescent, with segmental fibrinoid necrosis. H&E, 100×. (B) Another region of the biopsy seen at low magnification shows similar changes to the region in A, with more extensive tubular injury. Some of the injured tubules contain red blood cells and/or necrotic debris in their lumens. H&E 100×. (C) Another region of the biopsy that shows tubulointerstitial changes similar to that seen in the regions represented in A and B, along with a cluster of glomeruli showing glomerular inflammatory lesions (cellular crescents with focal segmental fibrinoid necrosis). One glomerulus that does not show an inflammatory lesion (lower left) exhibits a slightly lobulated appearance (despite the lack of endocapillary infiltration) and mild expansion of the mesangial areas by matrix. PAS 100×. (D) High magnification view of tubulointerstitial inflammation, illustrating expansion of the interstitium by infiltrating inflammatory cells, primarily composed of mononuclear cells and occasional neutrophils and eosinophils. Lymphocytic invasion of the tubular epithelium (tubulitis) is apparent in both viable and atrophic tubules. A peritubular capillary at the lower right is occluded by neutrophils (capillaritis) and swollen epithelial cells are seen. PAS 400×. (E) High-magnification image illustrating tubulointerstitial inflammation. Several tubules show tubulitis. One tubule shows signs of epithelial injury, including epithelial swelling and vacuolization. PAS 200×. (F) An area of dense cortical inflammatory infiltration that has expanded the involved cortical interstitium. A variety of mononuclear inflammatory cells are apparent and are intermixed, including lymphocytes, plasma cells, and macrophages. Scattered neutrophils are also seen, and are concentrated at tubular basement membranes. One tubule with injured and focally necrotic epithelium (lower right) shows infiltration of the epithelial layer by neutrophils (tubulitis). H&E 400×.

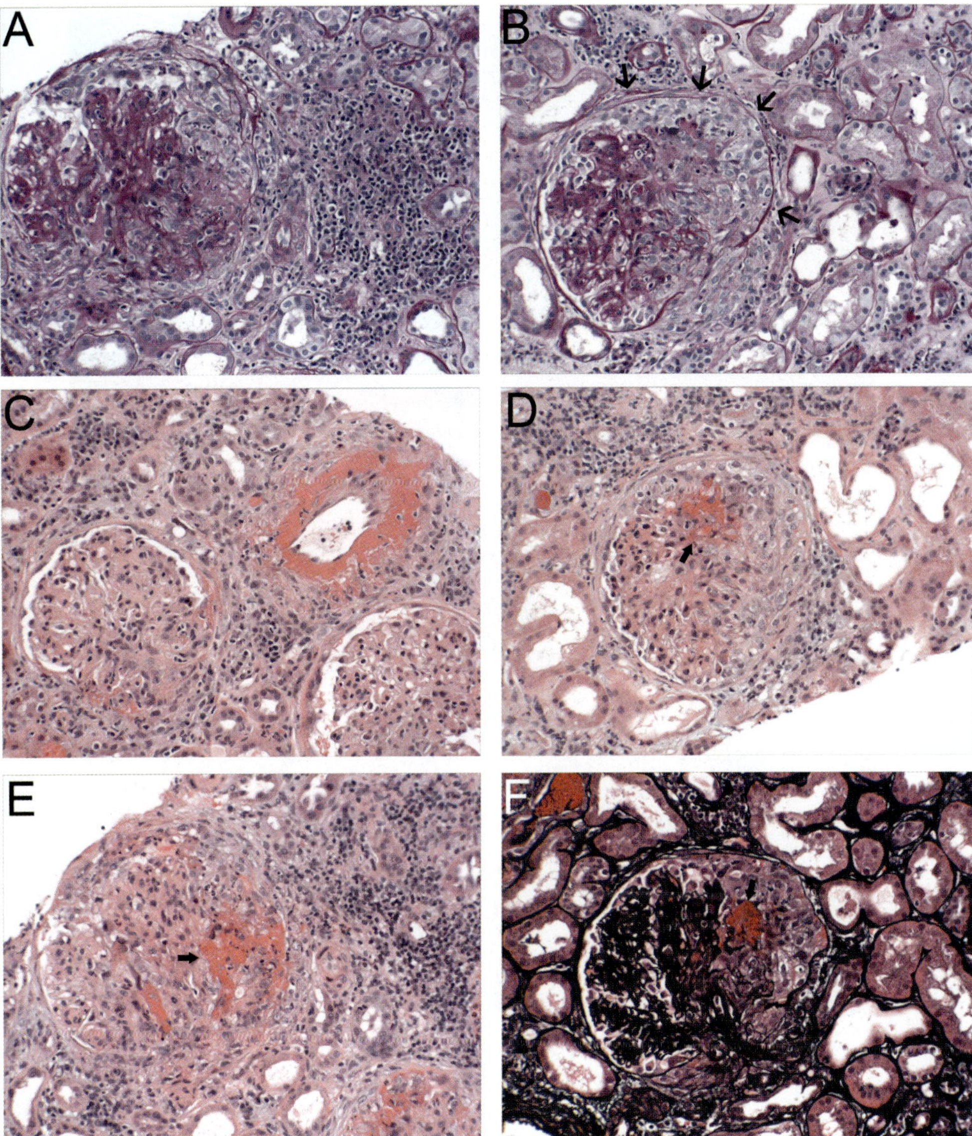

Figure 25.2. Glomerular histopathology in Case 1 kidney biopsy. (A,E) These PAS (A) and H&E-stained (E) serial sections demonstrate a glomerulus that is involved by a cellular crescent. The crescent is composed of proliferating epithelial cells, invading mononuclear and polymorphonuclear inflammatory cells, and apoptotic debris. The proliferating cells fill and expand Bowman's space, and penetrate the tuft between lobules. There is also a segmental area of fibrinoid necrosis involving the right side of the tuft, which appears deep red in the H&E-stained section (arrows) and also shows karyorrhexis or nuclear fragmentation. The tuft in the area of necrosis, including capillary loops and mesangium, is obliterated and fragmented. Bowman's capsule has also been obliterated adjacent to this area, best seen on the PAS-stained section. There is a prominent periglomerular mononuclear inflammatory infiltrate adjacent to the disrupted portion of Bowman's capsule, which is continuous with the crescent and fibrinoid necrosis. Scattered mononuclear and polymorphonuclear inflammatory cells are also observed throughout the intact portions of the tuft ("endocapillary inflammation"). Both images 200×. (B,D,F) Another glomerulus with an active cellular crescent (small arrows) and segmental fibrinoid necrosis, shown on PAS (B), H&E (arrow) (D) and silver-stained (F) sections (arrow). In this case, Bowman's capsule is less disrupted adjacent to the crescent and fibrinoid necrosis than that seen in (A) and (E). In association with this, the periglomerular inflammation, although present, is less pronounced, and fewer inflammatory cells are noted in the crescent and intact areas of the glomerular tuft. The silver stained section (F) highlights the disruption and fragmentation of capillary loops adjacent to the area of fibrinoid necrosis. Note the signs of injury in adjacent proximal tubules, including luminal distension, and epithelial flattening and irregularity. All images 200×. (C) A small artery in the kidney cortex shows transmural inflammation (arteritis), with invasion of the wall by mononuclear and polymorphonuclear inflammatory cells, apoptotic debris, and circumferential fibrinoid necrosis. The adjacent glomerulus on the left exhibits a segmental cellular crescent, with associated periglomerular inflammation and cell proliferation. Both glomeruli show endocapillary infiltration, predominantly by mononuclear inflammatory cells. H&E-200×.

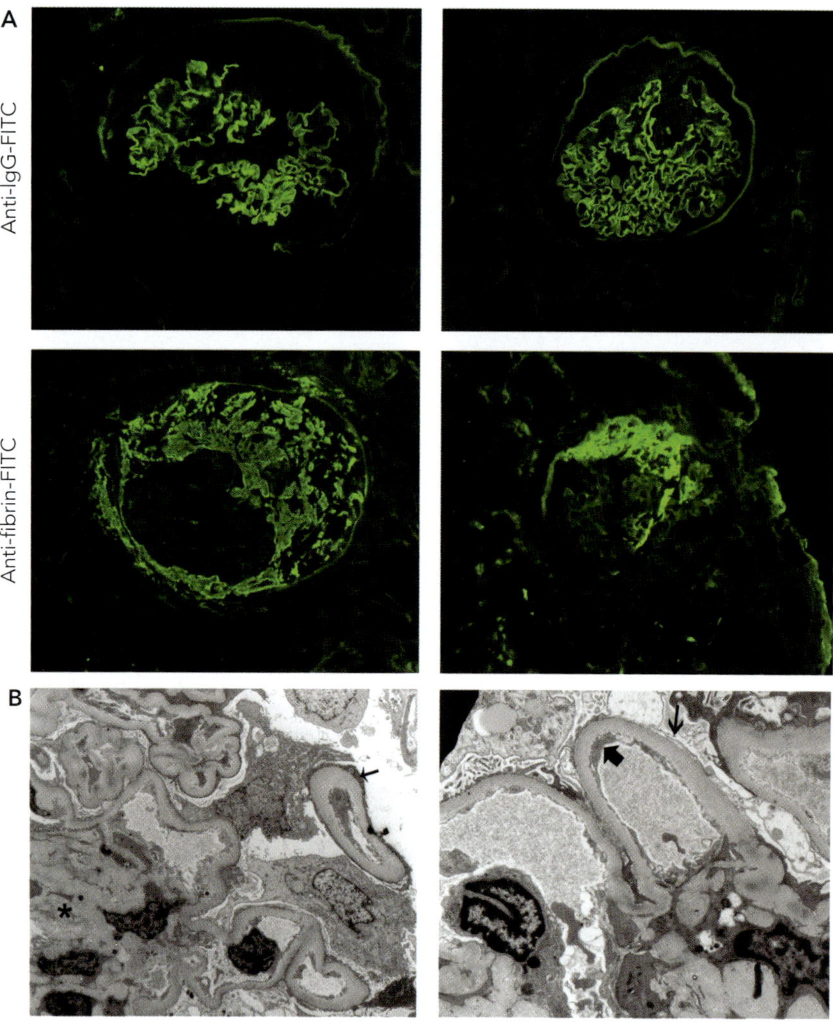

Figure 25.3. Indirect immunofluorescent microscopy of Case 1 kidney biopsy. (A) Top row: Images showing results of treatment with FITC-conjugated anti-IgG antibody. Examples of two glomeruli are shown, and both reveal strong linear staining of glomerular basement membranes, characteristic of anti-GBM antibody disease. Bottom row: Images showing results of treatment with FITC-conjugated anti-fibrin antibody. Same glomeruli as in top row; these images show strong staining of organized material that has accumulated in Bowman's space in association with infiltrating inflammatory cells and tissue necrosis, and corresponding to active cellular crescent formation. Areas of fibrin deposition are also apparent exterior to Bowman's capsule, in continuity with the inflammation involving the glomerulus, indicating associated periglomerular inflammation. Scattered fibrin deposition is also noted throughout the surrounding interstitium. All images 200×. (B) Electron micrographs of an intact glomerular segment uninvolved by glomerulitis (leukocytes in glomerulus) show expansion of the mesangial areas (*) and thickening of the glomerular basement membranes (small arrow), in the absence of electron dense deposits in these structures, indicating absence of immune complex deposits. There is evidence of mild endothelial injury, including segmental areas of swelling and loss of fenestrations (large arrow). Glomerular visceral epithelial cells (podocytes) also exhibit signs of injury, including segmental foot process effacement, microvillous change, and vacuolization (medium sized arrow) (left image 1800x, right image 5700x).

Discussion

This case represents a typical clinical presentation for a patient with rapidly progressive glomerulonephritis (RPGN). RPGN can be divided into three groups: (a) with immune complexes; (b) with linear IgG deposition in the basement membrane; and (c) with few immune deposits (pauci-immune), but usually with anti-neutrophil cytoplasmic antibodies detected in plasma. This patient had strong reactivity for IgG deposits in the GBM, and electron microscopy of the GBM showed it to be thickened but with no evidence of complexed deposits, in keeping with a diagnosis of anti-GBM disease. The presence in the blood of anti-GBM antibodies

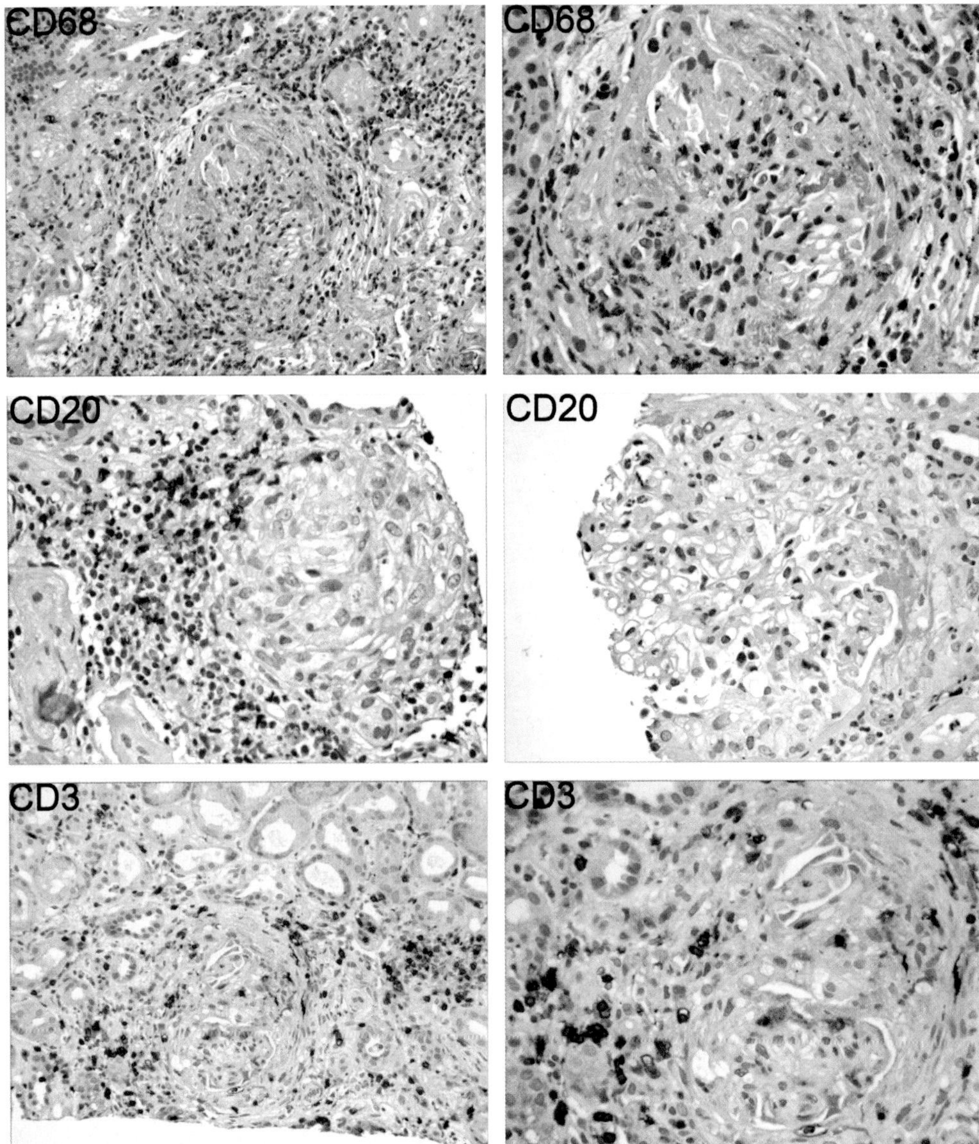

Figure 25.4. Immunohistochemical staining of Case 1 kidney biopsy with antibodies against leukocyte surface antigens (CD markers). All sections counterstained with hematoxylin. Top panels: At left, a section stained with antibodies against CD68 (a marker of macrophage-specific lysosomes) show numerous macrophages within areas of interstitial inflammation, and around a glomerulus involved by a segmental crescent (200×). At right, a higher magnification detail of the same glomerulus shows scattered macrophages within the intact portion of the tuft, and among proliferating cells in the crescent forming in Bowman's space (400×). Middle row: Sections stained with antibodies against CD20 (a marker of B-cell differentiation). Image at left shows few lymphocytes of B-cell lineage restricted to an area of periglomerular inflammation, surrounding a glomerulus involved by an active crescent. The glomerulus itself shows no B-cell infiltration (400×). Image at right shows a relatively intact glomerular tuft with segmental cellular crescent formation; again, no B cells are apparent in the tuft (400×). Bottom row: Sections stained with antibodies against CD3 (a marker of T-cell differentiation, the T-cell receptor). At left, scattered T cells are present in areas of interstitial and periglomerular inflammation surrounding a glomerulus with an active crescent, in a pattern similar to CD68-positive macrophages (200×). At right, a higher magnification view of the same glomerulus shows T cells within the developing crescent, but few are noted within the intact portions of the glomerular tuft (400×).

confirms the diagnosis. There were also reported to be ANCAs in the blood with a pANCA pattern of binding to fixed neutrophils as detected by immunfluorescence, usually suggestive of ANA-associated vasculitis that can re-present as a pauci-immune RPGN. Often anti-GBM disease and AAV occur in the same patient [1–7]. The diagnosis of RPGN was not made for several months during which the patient rapidly to assess kidney function. Unfortunately, the currently available tests to assess kidney function give a poor read-out when renal function is relatively preserved but give a much better read-out once there is substantial loss of kidney function. For this patient, by the time the diagnosis was made and treatment started, the severity of kidney disease was such that there was no recovery. This outcome was not unexpected from the series of patients with this particular diagnosis [8]. It is noteworthy that despite severe inflammation in the kidneys the patient exhibited mild, nonspecific symptoms, which is typical of diseases of the kidney. It behooves the physician assessing the patient to maintain a high degree of suspicion for kidney disease.

Examination of this patient's urine is extremely instructive since only severe glomerular inflammation (with a few exceptions) will produce the pattern of hematuria, proteinuria, leukocytes, and cellular casts in the urine sediment.

The biopsy findings show characteristic inflammation in the glomerulus with cellular destruction seen as fibrinoid necrosis and karryorrhexis. It is widely believed that neutrophils, monocytes, and cytotoxic T cells mediate destruction of glomerular cells, matrix, and membranes. It is noteworthy that few neutrophils were seen in these glomeruli and no B lymphocytes were seen in glomeruli but there were many monocyte/macrophages and T lymphocytes. In addition to leukocytes infiltration, native cells of the glomerulus undergo functional and phenotypic responses to injury. These included swelling migration, entry into cell cycle, deposition of pathological fibrous matrix which leads to scarring, and vacuolization, all of which may be a stress response to a hostile environment. Strikingly, in the glomerulus there is formation of a crescent of cells that can encase the glomerular tuft. This represents proliferation of parietal glomerular epithelial cells admixed with macrophages and T cells, and is strongly associated with severe disease and a poor prognosis for the kidney. In addition to glomerular changes, there are also notable tubular and interstitial changes. This patient had underlying chronic tubuloin terstitial damage which manifests as tubular atrophy chronic inflammation and fibrosis in the kidney interstitium, rarely is the glomerulus affected in isolation and any RPGN frequently is complicated by destructive inflammation involving the kidney interstitium.

CASE 2

Presentation and Initial Assessment

An 86-year-old man presented to his primary care physician with cough and shortness of breath with discomfort in the chest but no hemoptysis. His medical history included coronary artery disease and previous coronary artery bypass graft surgery hypertension, stroke, atrial fibrillation, type II diabetes mellitus, hyperlipidemia, and neoplasia involving the prostate gland for which he had received irradiation therapy, presented to his primary care physician with cough and shortness of breath with discomfort in the chest but no hemoptysis. A chest X ray in the emergency room showed a right lower zone consolidation. He was initially evaluated for pulmonary embolus but a V:Q scan that was negative, and Doppler scan of his legs, which revealed no evidence of deep venous thrombosis. In the emergency room, he was noted to have anemia with a hematocrit of 22% as well as abnormal renal function tests in routine clinical chemistry testing of blood. The creatinine was 3.8 mg/dL and the blood urea nitrogen was 52 mg/dL (a prior creatinine level was 1.4 mg/dL, but this resulted dated from). His primary care doctor was contacted and reported that the patient had experienced subacute deterioration of renal function tests over weeks to months, associated with increasing proteinuria which had not been detectable 2 years previously. A sonogram had indicated the kidneys to be smallish at 9 and 10 cm bipolar diameter, but that there had been no evidence of kidney obstruction.

There was no family history of kidney diseases.

His medications included Aspirin, Simvastatin, Casodex (bicalutamide), tamulosin, Sotalol, Digoxin, Furosemide, niacin, amitriptyline, nitroglycerin sublingually lanzoprozole.

Clinical examination: weight, 210 lbs; temperature, 97.3F; pulse, 60 per minute; blood pressure, 165/90 mmHg; respiratory rate, 22 breaths per minute; peripheral blood oxygen saturation, 96% on 3 L/min of oxygen delivered by nasal cannulae. He was an elderly man, pleasant, in no distress. Pupils and eye movements, and oropharyngeal exam, were normal. Neck exam was unremarkable, and the jugular venous pressure was elevated by 2 cm of water. There were no carotid artery bruits. The cardiac exam revealed regular rhythm and normal heart sounds except for a soft mid systolic murmur at the aortic area only. The right lung was dull to percussion in the right lower zone and there were coarse crackles heard on auscultation in that area. There was no definitive bronchial breathing. The left base was also somewhat dull but there were no focal signs. The abdomen was soft without tenderness and no organomegaly was detected. In the extremities, there was mild edema but no rashes, and his peripheral pulses were normal to examination. His joints were nontender with no overt arthritis. His neurological exam found him to be alert and oriented, and examination of cranial and peripheral nerves was unremarkable.

The patient provided a urine specimen. Analysis of this specimen revealed specific gravity of 1.007, pH 6, mild glycosuria, and no ketonuria. However, there was +++ protein, +++ hemoglobin, and direct microscopy of sediment identified 5–9 white blood cells and 50–70 erythrocytes per high-powered field. Many of the erythrocytes were dysmorphic. There were many coarse granular casts, but no cellular casts. Many leukocytes were seen and the urine dip tests revealed the presence of nitrites, indicative of urinary tract infection.

Baseline laboratory findings on blood testing revealed sodium ions 135 mmol/L, potassium ions 4.1 mmol/L, chloride ions 110 mmol/L, bicarbonate ions 22, blood urea nitrogen 72 mg/dL, creatinine 3.8 mg/dL, glucose 260 mg/dL and total calcium 8.1 mg/dL. A complete blood count revealed white blood count 6,700/μL, hemoglobin 11.0 mg/dL, and platelets 269,000/μL.

Immunological assessment was notable for anti MPO antibodies at a titer of 1:320. Antinuclear antibodies were negative.

Overnight culture of urine confirmed an *E. coli* urinary tract infection. Blood cultures were negative and there was no sputum available for culture.

He was initiated on azithromycin for possible diagnosis of pneumonia and as therapy for the urinary tract infection. The presence of right lower infiltrate, deteriorating renal function with increasing protenuria, and an active urinary sediment alerted the clinician to the possibility of a glomerulonephritis.

Serological studies for ANCA, ANA, dsDNA, anti-GBM antibodies, antibodies against hepatitis viruses and complement levels were all performed and a tuberculin test was placed to test for immune response to tuberculosis.

Renal Biopsy Findings

The ample biopsy consisted of kidney cortex and medulla, and included up to 40 glomeruli showing various stages of active inflammation and chronic damage. The findings are illustrated in Figure 25.5. Nine of the glomeruli showed active inflammation and crescent formation. The infiltrating cells consisted primarily of mononuclear inflammatory cells, with few polymorphonuclear cells. Three of these glomeruli also showed segmental areas of fibrinoid necrosis and karryorhectic debris. Several additional glomeruli exhibited fibrocellular (scarred) crescents, and eight glomeruli showed outright global sclerosis. Uninvolved glomeruli revealed only a mild degree of mesangial expansion, and glomerular basement membranes were of normal appearance and thickness.

The tubulointerstitium showed patchy areas of mononuclear interstitial inflammation; most of these infiltrates were localized within areas of interstitial scarring and fibrosis. Some of these areas also showed focal tubular atrophy. In other areas that were often adjacent to the areas of atrophy, tubules showed signs of acute injury, including luminal distension, epithelial flattening, and vacuolization.

Immunofluorescence microscopy showed moderate (2+/4+) linear deposition of IgG along glomerular capillary walls. Few glomeruli showed segmental mesangial IgM and C3 deposition. Scattered arteries and arterioles showed C3 deposition in their walls. Electron microscopy showed no electron dense deposits in the mesangium or capillary wall of intact glomeruli. Furthermore, the glomerular ultrastructure was preserved in intact glomeruli, including minimal evidence of segmental foot process effacement, only segmental loss of endothelial fenestrae, minimal expansion of mesangial areas, and normal glomerular basement membrane thickness.

Management and Follow Up

The presence of fibrinoid necrosis and karryorhectic debris, and fibrocellular crescents involving the glomeruli was consistent with a crescentic glomerulonephritis. The presence of linear IgG deposition along the glomerular basement membranes was suggestive of anti-GBM disease but the GBM thickness was normal and there were no immunocomplex deposits seen. The serological tests revealed no evidence for anti-GBM antibodies, but anti-MPO antibodies (ANCA) were elevated at 44 U/mL (< 6.0) when assessed by ELISA.

He was treated with oral prednisone 60 mg daily (reduced to 20 mg daily) and cyclophosphamide 75 mg daily, in addition to several other medications.

Creatinine level peaked at 5.2, and improved to 4.3. The estimated glomerular filtration rate was 12 mL/min, which is categorized as severe but stable chronic kidney disease. The patient's disease was complicated by hyperkalemia of 6.8 mEq/L, which had been managed with oral polysorbiol, but he continues to be stable and is living independently.

Discussion

This case represents a smoldering form of RPGN. It was not rapidly progressive, rather slowly progressive. Approximately 5%–10% of patients with glomerular inflammation with crescents have a smoldering course. This patient also had many other systemic diseases that can affect the kidney and lead to progressive loss of kidney function, including hypertension, atherosclerotic vascular disease affecting brain, heart, prostatic disease with malignancy, and infection. It was an astute nephrologist who was able to combine a series of symptoms and signs to make a diagnosis of RPGN and systemic vasculitis. Like Case 1, clinical vigilance was important in determining the outcome for this patient's kidneys. The patient had a lung consolidation and a complex of symptoms and signs that would have led many to a diagnosis of pneumonia, urinary tract infection, and acute or chronic kidney disease related to these infections. However, the presence of an atypical consolidation, systemic symptoms, deteriorating renal function, and urine microscopy testing showing

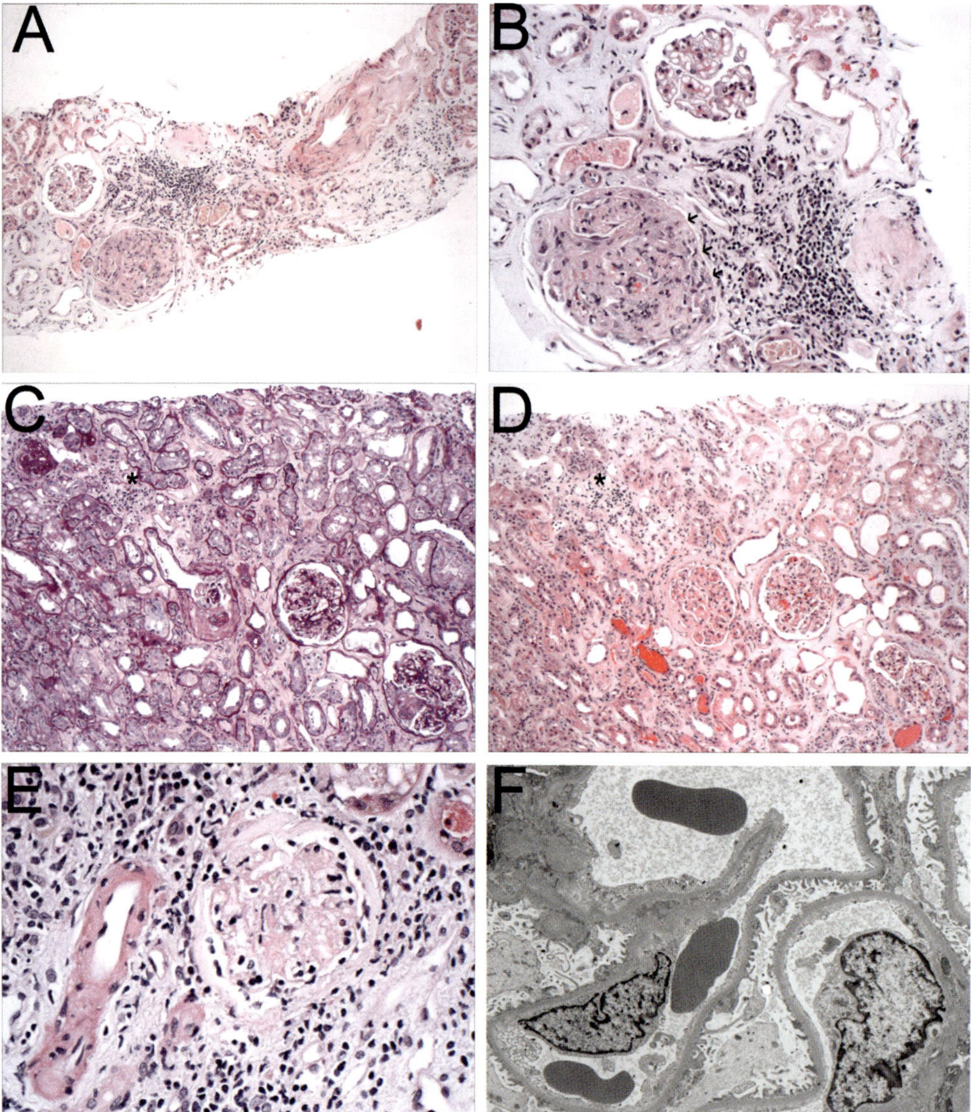

Figure 25.5. Histopathologic and ultrastructural findings in Case 2 kidney biopsy. (A) An area of chronic damage, with prominent interstitial expansion and fibrosis, associated with infiltration of the interstitium by mononuclear inflammatory cells. An artery at upper right shows severe arteriosclerosis and luminal narrowing. H&E 100×. (B) Higher magnification of group of three glomeruli in the same area, showing the spectrum of glomerular lesions in the biopsy. The glomerulus at lower left shows active crescent formation (arrows), with infiltration of the glomerular tuft by mononuclear and few polymorphonuclear inflammatory cells, as well as expansion of Bowman's space by proliferating epithelial cells and infiltrating inflammatory cells that penetrate the tuft. Note that capillary loops are not seen well in this glomerulus due to swelling of endothelial cells and the presence of leukocytes in the loops. A glomerulus at lower right is globally sclerosed. The glomerulus at top is lesion-free. A cluster of mononuclear inflammatory cells surround the sclerosed glomerulus. Red cell casts can be seen in tubules. Note also the increased matrix surrounding tubules within which there are spindle shaped cells H&E 200×. (C,D) Another region of the biopsy showing the spectrum of tubulointerstitial and glomerular damage. There is patchy interstitial expansion and fibrosis, associated with early tubular atrophy. Adjacent tubules show luminal expansion and epithelial flattening, indicative of focal acute tubular injury. There are inflammatory cells within the interstitium focally (asterisks) and more evenly distributed. Glomeruli show a variety of lesions, including a globally sclerosed glomerulus at upper left, a fibrocellular crescent second from left, an uninvolved glomerulus, and a glomerulus with a segmental cellular crescent at lower right. H&E, PAS 100×. (E) Another globally sclerosed glomerulus with surrounding chronic inflammation. Adjacent arterioles show moderate arteriolar sclerosis. H&E 400×. (F) Electron micrograph showing glomerular capillary loops with preserved architecture. There is minimal evidence of segmental foot process effacement, only segmental loss of endothelial fenestrae, minimal expansion of mesangial areas, and normal glomerular basement membrane thickness. No electron dense deposits are seen in the mesangium or capillary wall.

features suggestive of glomerular inflammation should always lead to a possible diagnosis of RPGN and systemic vasculitis.

The renal biopsy shows active glomerular inflammation with mononuclear cells that include lymphocytes and monocyte/macrophages and a few neutrophils. There is crescent formation in the involved glomeruli. Some glomeruli are healthy with no disease, and others have become fibrotic (sclerosed) and are no longer functioning. The basement membranes were not thickened and no antibodies were detected in glomeruli by immunofluorescent staining. These are the features of crescentic glomerulonephritis with no immune deposits. This pattern of disease is usually associated with the presence of circulating antineutrophil cytoplasm antibodies (ANCA), and is a form of autoimmune disease that frequently affects the kidneys. Indeed, testing confirmed anti-MPO ANCAs in blood, so he has an ANCA-associated vasculitis. This patient's disease centers on acute small vessel and capillary injury known as small vessel vasculitis, and is thought to be due to ANCA antibodies and autoreactive T cells directed at ANCAs leading to neutrophil activation on the endothelial surface of capillaries. This in turn leads to local capillary injury and secondary inflammation. The glomerulus, is frequently a focus for this initiating capillary injury due to its very high shear stresses. Similar to Case 1 we would expect to see many monocyte/macrophages and many T cells in the glomerular tuft and glomerular capillary loops by immunodetection. These can also be identified as mononuclear infiltrating cells on the H&E and PAS stained sections (Figure 25.5). In addition to the glomerular inflammation, there is severe medium vessel arteriosclerosis, similar to that which one would expect to see in the heart and brain of this patient. Around diseased blood vessels, there is significant fibrosis. However, none of these features is due to autoimmunity against the blood vessels (vasculitis); rather, they represent arteriosclerosis. The patient also had substantial chronic disease affecting the tubules and kidney interstitium. Arteriosclerotic vascular disease can result in kidney panrenchymal ischemia, which leads to chronic interstitial inflammation which we now know drives progressive interstitial kidney disease. However, ANCA vasculitis can involve any small vessels and capillaries, including the peritubular capillary network so it is possible that some of the interstitial fibrosis and tubular disease seen in the patient is due to ANCA associated Vasculitis in addition to arteriosclerosis and chronic ischemia.

CASE 3

Presentation and Initial Assessment

A 60-year-old man. His history included diabetes mellitus type II complicated by retinopathy which had required laser surgery, peripheral neuropathy, and hypertension of 10 years duration. He denied rashes, joint pains, or vomiting. In addition, he disclosed mild upper respiratory tract symptoms with bronchial secretions for which he had taken several 800 mg ibuprofen tablets to treat his symptoms. Notably he did not have hemoptysis or epistaxis. At systemic enquiry, he reported vague bilateral flank pains that were deep and not related to eating, bowel motions, musculoskeletal movement, or urination.

Further to the aforementioned complications of diabetes mellitus, he had a history of coronary artery disease, hyperlipidemia, and chronic kidney disease stage II, with a previously documented plasma creatinine level of 1.5 mg/dL. Further, he had documented mild restrictive lung disease of uncertain etiology, anxiety, prostatic hyperplasia, esophageal reflux disease, and erectile dysfunction.

At presentation his medications were aspirin, metformin, glyburide, Atenolol, Enalapril, Pioglitazone, Nifedipine, Omeprazole, Simvastatin, Fluticasone, and 2–3 days of ibuprofen 1,200 mg a day. He denied illicit drug use, tobacco or alcohol use, and he had no significant family history.

Upon clinical examination he was found to be in no distress, body temperature 99.2F, BP 170/101 upon initial assessment, subsequently 160/84 mmHg, heart rate 75 per minute, respiratory rate 20 per minute, and peripheral blood oxygen saturation of 95% while breathing room air. His head and neck examination was unremarkable and his jugular venous pressure was +3 to +4 cm water. Pulmonary exam revealed fine crackles in both lungs, in the lower zones left greater than right. Abdominal examination was unremarkable, except for mild epigastric tenderness. There was no peripheral rash or edema or joint tenderness, and peripheral neurological examination was unremarkable.

A Foley catheter had been placed in the bladder in the emergency room and he was oliguric (low urine volume).

Baseline laboratory findings on blood testing, his sodium ions were 139 mmol/L, potassium ions 6.3 mmol/L, chloride ions 102 mmol/L, bicarbonate ions 17, blood urea nitrogen 75 mg/dL, creatinine 10.6 μg/dL, glucose 161 mg/dL, total calcium 8.6 mg/dL, albumin 4.2, anion gap elevated at 20. Liver function tests normal. Complete blood count: white blood count 11,200/μL, hemoglobin 11.4 mg/dL, platelets 299,000/μL, PT-INR 1.0.

The patient provided a urine specimen. The centrifuged sediment was subjected to analysis. pH 7; however, there was 1+ protein, +++ hemoglobin, and direct microscopy of sediment identified many erythrocytes per high-powered field. Some of the erythrocytes were spiculated but none frankly dysmorphic. There were a few granular casts, but no cellular casts. Fractional excretion of sodium was 8%. Urine cultures were negative. Immunological assays were sent. An EKG revealed sinus rhythm but no other abnormities. Chest X-ray showed mild cardiomegaly, small left pleural effusion, no edema, and no pneumonia.

Sonogram of the kidneys found the urinary tract to be unremarkable and the kidneys were normal sized at 13.5, 12.1 cm longitudinally.

Urine sediment was stained with eosin to test for eosinophils – a sign of interstitial nephritis – but the test was negative.

Urine and blood were sent to the microbiology laboratory for culture, which the following day had grown enterococci sensitive to the antibiotic ampicillin. Blood cultures, however, were negative indicating any urine infection was localized.

Hyperkalemia was treated with intravenous calcium, intravenous dextrose with insulin and sodium polysorbitol, which binds intraluminal K^+ ions. In addition, he received intravenous volume resuscitation, for a component of hypovolemia. He was started on a fluroquinolone antibiotic to treat the urinary tract infection.

The patient had acute chronic renal insufficiency with hyperkalemia, acidosis, and a differential diagnosis that included prerenal disease and acute tubular necrosis, but also rapidly progressive glomerulonephritis. In addition, although less likely, acute interstitial nephritis and pyelonephrtitis had not been excluded.

However, he remained oliguric, and renal function tests indicated further deterioration. A renal biopsy was performed.

Renal Biopsy Findings

The kidney biopsy consisted of an adequate sample of cortex and medulla. Of 40 glomeruli in the sample, 16 exhibited global sclerosis. The remaining glomeruli were enlarged, and showed mesangial expansion by matrix and mesangial cells together with monocytes/macrophages. There was early mesangial nodule formation, indicative of early diabetic changes. The glomerular capillary wall appeared thickened on PAS stained sections (Figure 25.6A and 25.6B), and showed isolated double contours. The tubules showed extensive degenerative changes, including luminal distension and epithelial flattening and irregularity (Figure 25.6C and 25.6E). Several tubular epithelial cells exhibited striking epithelial vacuolization (Figure 25.6C and 25.6E). All of these features are indicative of acute tubular injury; overt necrosis was not seen. Focal tubular atrophy was noted in some areas, involving approximately 40% of the biopsy (Figure 25.6C, wide arrows). There was also a corresponding degree of interstitial fibrosis (Figure 25.6C, 25.6D, and 25.6F). A mild infiltrate of mononuclear inflammatory cells was present, mostly confined to areas of interstitial fibrosis (Figure 25.6C, 25.6D, and 25.6F). A dense perivenular infiltrate composed of mononuclear inflammatory cells was present in one area (Figure 25.6D, small arrows). Renal arteries showed prominent subintimal sclerosis and narrowing of the lumen (Figure 25.6D, wide arrows); arterioles also showed chronic changes including redundancy of the wall and hyalinosis (Figure 25.6D–25.6F, wide arrows). Immunofluorescence studies revealed no significant immune deposits. Tissue was unavailable for electron microscopy studies.

In summary, the predominant finding in the kidney biopsy was acute tubular injury. Focal interstitial inflammation was also noted in areas of tubular atrophy. These findings were present on a background of moderate chronic parenchymal damage, with features suggesting that diabetes and vascular disease (also likely exacerbated by diabetes) were responsible for the chronic damage. There was no evidence of active glomerulonephritis. The severe vascular changes leading to ischemia of the kidney would predispose to acute tubular injury.

Management and Follow Up

The patient's serological test results returned with no evidence of ANCAs, or ANA showed normal complement levels and no evidence of viral hepatitis or cryoglobulins.

On the second day after admission, the patient remained oliguric with plasma creatinine level increased further to 12. In the subsequent days, he had increasing urine volume and progressive recovery of function. He was managed without the need for dialysis, with careful adjustment of his plasma electrolytes and fluid balance, withdrawal of ibuprofen, Enalalpril, and metformin. Seven days later plasma creatinine was 2.8 mg/dL. He was discharged from the hospital. Instead of Enalapril, he was taking Labetalol, and for the infection, he was switched to 10 days of oral amoxicillin. In addition, he was receiving vitamin D and calcitriol. At a subsequent outpatient office follow-up, the plasma creatinine had stabilized at 1.5 mg/dL.

Discussion

The patient presented with rapid deterioration of renal function, a prodromal illness suggestive of vasculitis and urine sediment with features suggestive of glomerular inflammation. Fortunately, the patient had no glomerular inflammation. Rather he had a combination of diseases that can mimic RPGN but that may be no less benign to his kidneys in the ensuing years. He has severe arteriosclerosis of his medium and small vessels with consequent tubular and interstitial ischemia. In addition, he has features of diabetic nephropathy. The patient had a systemic illness, likely viral

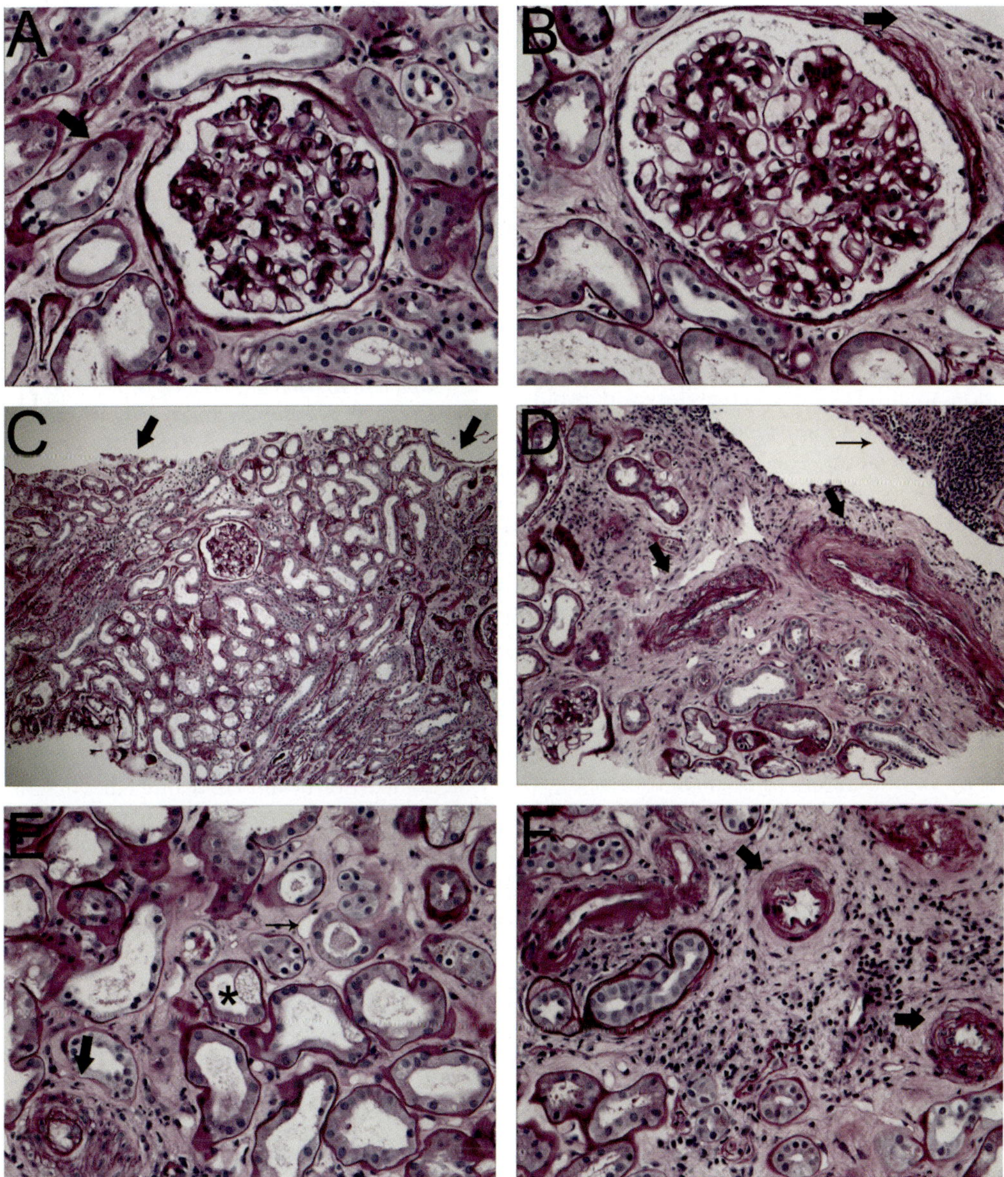

Figure 25.6. Histopathologic findings in Case 3 kidney biopsy. (A,B) High-magnification images of representative glomeruli. There is no evidence of active glomerulitis. However, there is diffuse mesangial expansion by matrix and mesangial cells and monocyte/macrophages (B arrow), and glomerular basement membranes and tubular basement membranes appear mildly thickened (arrow in A), both suggestive of early diabetic changes. PAS 400×. (C) Low power image showing spectrum of tubulointerstitial changes. Area at left (wide arrow) shows marked interstitial expansion and fibrosis, associated with mild mononuclear inflammatory infiltrates. Tubules in this area are relatively preserved. Tubules at midportion of image reveal luminal distension, epithelial flattening and irregularity, and focally prominent epithelial vacuolization (best seen at bottom of image), all indicative of acute tubular injury. Area at right (wide arrow) shows an area of focal tubular atrophy accompanied by interstitial fibrosis. PAS 100×. (D) Small arteries near the corticomedullary junction exhibit severe sclerosis (wide arrows). The artery at right shows severe thickening and fibrosis of the subintima, with associated narrowing of the lumen. An arteriole at lower left (adjacent to the glomerulus) shows redundancy of the wall and hyalinosis. A focus of perivenular inflammation, composed of a dense accumulation of mononuclear inflammatory cells reminiscent of a lymphoid aggregate, is also present (thin arrow). PAS 200×. (E) High-magnification image showing tubulointerstitial changes. Tubules throughout the image show luminal distension and epithelial flattening and irregularity, as well as focally prominent epithelial vacuolization (asterisk). Some tubules contain proteinaceous debris in their lumens. There is also diffuse interstitial fibrosis, with scattered mononuclear inflammatory cells. An arteriole (wide arrow) shows thickening and redundant layering of the wall ("onion-skinning") that is likely a result of chronic endothelial damage. PAS 400×. (F) Arterioles show moderate to focally severe sclerosis, with onion-skinning (wide arrows), narrowing of the lumen, and focal hyalinosis. Surrounding interstitium is expanded by fibrous tissue and infiltrated by mononuclear inflammatory cells. PAS staining magnification 400×.

infection, took cyclooxygenase-I (COX-I) inhibitors (nonsteroidal anti-inflammatory drugs) and became volume depleted due to poor oral intake, and likely decreased ability to concentrate urine. The combination of arteriosclerosis, volume depletion, and COX-I inhibitors can lead to hypoperfusion of the glomeruli and kidney interstitium and a resulting acute tubular necrosis. Acute tubular necrosis does not present with hematuria and proteinuria. The patient had proteinuria and hematuria due to the presence of diabetic nephropathy. Therefore the combination of tubular necrosis and diabetic nephropathy can mimic RPGN. With attention to fluids and electrolytes and avoidance of further COX-I inhibitors, the patient recovered kidney function. However, the patient will continue with chronically injured vessels that will likely lead to chronic progressive kidney disease. Although there was no evidence of severe inflammation in the glomeruli or interstitial compartment, monocytes/macrophages were nevertheless present in glomerlar tufts in mesangial areas. In areas of interstitial fibrosis and tubular disease there was a notable mononuclear infiltrate comprising monocytes/macrophages and lymphocytes. In addition, the severely diseased vessels were surrounded by inflammatory infiltrate. Increasingly, we have come to recognize that chronic kidney disease characterized by ischemia, fibrosis, and tubular atrophy is an inflammatory disease and leukocytes play direct roles in both promoting cell injury and also promoting fibrosis. The early stages of diabetic glomerulopathy are also characterized by leukocyte infiltrates where they also promote glomerular mesangial matrix deposition. Therefore, while this patient does not have autoimmunity or severe inflammatory vasculitis, he does nevertheless have a chronic inflammatory disease of his kidneys due to chronic vascular injury and diabetes mellitus. There are currently no specific anti-inflammatory therapies to counteract the deleterious role of leukocytes in this context.

REFERENCES

1. Peces, R., Rodriguez, M., Pobes, A., and Seco, M. 2000. Sequential development of pulmonary hemorrhage with MPO-ANCA complicating anti-glomerular basement membrane antibody-mediated glomerulonephritis. *Am J Kidney Dis* **35**:954–957.
2. Verburgh, C.A., Bruijn, J.A., Daha, M.R., and van Es, L.A. 1999. Sequential development of anti-GBM nephritis and ANCA-associated Pauci-immune glomerulonephritis. *Am J Kidney Dis* **34**:344–348.
3. Meisels, I.S., Stillman, I.E., and Kuhlik, A.B. 1998. Anti-glomerular basement membrane disease and dual positivity for antineutrophil cytoplasmic antibody in a patient with membranous nephropathy. *Am J Kidney Dis* **32**:646–648.
4. Endo, Y., Terada, M., Ohdaira, T., and Haraguchi, M. 1999. An autopsy case of Goodpasture's syndrome with P-ANCA and systemic vasculitis. *Nihon Kokyuki Gakkai Zasshi* **37**:55–60.
5. Inoue, A., Ino-oka, N., Konishi, K., Shindoh, Y., Suzuki, T., and Ono, Y. 1997. A case of Goodpasture's syndrome with myeloperoxidase specific anti-neutrophil cytoplasmic autoantibody (MPO-ANCA) during chronic interstitial pneumonia. *Nihon Kyobu Shikkan Gakkai Zasshi. Japanese Journal of Thoracic Diseases* **35**:1356–1362.
6. Neary, P., Kadlubowski, M., Thomson, D., Cavaye, M., and Cumming, A.D. Antiglomerular basement membrane disease with cANCA positivity without pulmonary involvement. *Nephrol Dial Transplant* 1996; **11**:693–695.
7. Jayne, D.R., Marshall, P.D., Jones, S.J., and Lockwood, C.M. 1990. Autoantibodies to GBM and neutrophil cytoplasm in rapidly progressive glomerulonephritis. *Kidney Int* **37**:965.
8. Levy, J.B., Hammad, T., Coulthart, A., Dougan, T., and Pusey, C.D. 2004. Clinical features and outcome of patients with both ANCA and anti-GBM antibodies. *Kidney Int* **66**(4):1535–1540.
9. Hellmark, T., Niles, J.L., Collins, A.B., et al. 1997. Comparison of anti-GBM antibodies in sera with to without ANCA. *J Am Soc Nephrol* **8**:376.
10. Weber, M.F., Andrassy, K., Pullig, O., et al. 1992. Antineutrophil-cytoplasmic antibodies and antiglomerular basement membrane antibodies in Goodpasture's syndrome and in Wegener's granulomatosis. *J Am Soc Nephrol* **2**:1227.
11. Geffriaud-Ricouard, C., No'l, L.H., Chauveau, D., et al. 1993. Clinical spectrum associated with antineutrophil cytoplasmic antibodies of defined antigen specificities in 98 selected patients. *Clin Nephrol* **39**:125.
12. Rutgers, A., Slot, M., van Paassen, P., van Breda Vriesman, P., Heeringa, P., and Tervaert, J.W. 2005. Coexistence of anti-glomerular basement membrane antibodies and myeloperoxidase-ANCAs in crescentic glomerulonephritis. *Am J Kidney Dis*. **46**(2):253–262. Review.
13. Lionaki, S., Jennette, J.C., and Falk, R.J. 2007. Anti-neutrophil cytoplasmic (ANCA) and anti-glomerular basement membrane (GBM) autoantibodies in necrotizing and crescentic glomerulonephritis. *Semin Immunopathol* **29**(4):459–474. Epub October 18, 2007. Review.

SUGGESTED READINGS

Brenner, B.M., and Rector, F.C. 2007. *Brenner and Rector's the Kidney*. Philadelphia, PA: Saunders.

Feehally, J., Floege, J., and Johnson, R.J. 2007. *Comprehensive Clinical Nephrology*. Philadelphia, PA: Mosby.

26 Inflammation in Cardiovascular Diseases

Kenneth K. Wu

INTRODUCTION

Inflammation has emerged as a key pathophysiological event in vascular diseases and the consequent cardiac and cerebral ischemic injury. There is ample evidence that inflammation is intimately involved in atherosclerosis. It mediates the initiation of atherosclerosis, promotes progression of the atherosclerotic lesions, and regulates atheromatous plaque stability [1,2]. There is also good evidence that inflammation plays a crucial role in ischemia-reperfusion cardiac and cerebral injury [3,4].

Inflammation is a complex process involving multiple cellular and molecular components. It is triggered by diverse proinflammatory mediators (PIM) which are generated directly and indirectly by microbial invasion, endotoxins, immune complexes, and cytokines. Vascular endothelium is subjected to proinflammatory insults, as it is in constant contact with circulating blood and along with it many environmental stressful factors. Fortunately, endothelium is endowed with potent anti-inflammatory molecules that confer resistance to damage by transient proinflammatory attacks. Once the insulting factors dissipate, endothelial cells return to its basal state. The mechanisms by which endothelial cells resist insults are likely to be very complex. One model is stress-coupled induction of anti-inflammatory and cytoprotective genes [5]. This mechanism allows for timely defense against transient insults. However, when insults by PIMs become persistent, this protective property wears out resulting in endothelial cell damage and functional defects and eventually endothelial apoptosis and necrosis. A number of uncontrolled disease states and/or risk factors are now well recognized to confer persistent insults and cause progressive endothelial damage: (1) uncontrolled hypertension causes constant blood flow changes creating mechanical and shear stresses that subject endothelial cells to constant stresses; (2) diabetes mellitus with high blood sugar damages vascular endothelial cells by glycation of functionally important vascular proteins; and (3) low-density lipoprotein (LDL) undergoes oxidation and the oxidized LDL causes endothelial damage. Chronic microbial invasion into blood and/or chronic release of endotoxins emerge as important causes of endothelial damage and vascular diseases. Immune complex and cytotoxic antibodies represent another group of factors that may cause vascular damage and vascular diseases. Control of the chronic diseases or risk factors is therefore important in preventing permanent endothelial damage.

RESPONSE OF ENDOTHELIUM TO PROINFLAMMATORY MEDIATORS

Stress-Coupled Anti-Inflammatory, and Cytoprotective Properties of Vascular Endothelium

Vascular endothelium comprises a single layer of cobblestone-like cells that express anti-inflammatory molecules on the cell surface and secrete vasoprotective and anti-inflammatory molecules into the circulating blood. These molecules appear to play major roles in conferring blood vessel resistance to diverse pro-inflammatory factors that invade the circulating blood and threat the integrity of vessel wall. The endothelial surface thrombomodulin (TM) and protein C receptor (EPCR) work in concert to provide anti-inflammatory and antithrombotic properties. TM anchors to endothelial cell surface with a single transmembrane domain [6,7]. The long extracellular region of TM contains terminal lectin-binding domain and six EGF repeat domains [8]. TM binds thrombin and protein C at the EGF domain where thrombin cleaves protein C to generate activated protein C (APC) [6,7]. APC detached from thrombomodulin degrades coagulation factors Va and VIIIa and thereby controls coagulation reactions and fibrin formation [8]. In addition to its anticoagulant activity, APC binds EPCR and mediates anti-inflammatory actions. APC has been shown to be effective in controlling sepsis-induced tissue damages and is clinically used in treating patients with septic shock.

Vascular endothelial cells produce several classes of compounds that control blood cell reactivity and blood cell interaction with vascular wall. Among the molecules, prostacyclin (PGI_2) and nitric oxide (NO) are the best characterized. PGI_2 is produced in endothelial cells via the catalytic actions of three enzymes: (1) phospholipase A_2 (PLA_2), (2) cyclooxygenase (COX), and (3) PGI_2 synthase (PGIS) [9]. Activation of vascular endothelial cells by shear stress, cytokines, lipopolysaccharides (LPS), or other agonists results in translocation of cytosolic PLA_2 to the outer membrane of endoplasmic reticulum (ER) and the contiguous perinuclear region. Activated $cPLA_2$ releases arachidonic acid (AA) from the phospholipids. AA is converted to PGH_2 by COX and PGH_2 is converted to PGI_2 by PGIS. Both COX isoforms (COX-1 and COX-2) are localized to the luminal membrane of ER via hydrophobic interactions while PGIS anchors to the external surface of ER by a single transmembrane domain. [10]. $cPLA_2$, COX, and PGIS appear to be physically and functionally coupled which facilitates transfer of intermediate metabolites. Both COX-1 and COX-2 are expressed in endothelial cells. There is increasing evidence to indicate that COX-2 is the predominant source of PGI_2 biosynthesis in humans [11,12]. PGI_2 is released into extracellular milieu where it inhibits platelet activation and aggregation, controls smooth muscle cell contraction, and suppresses monocyte activation, as well as inhibits monocyte interaction with vascular endothelium. Thus, in addition to being a thrombotic agent, PGI_2 possesses anti-inflammatory properties.

Nitric oxide is also constitutively produced by vascular endothelial cells under flow conditions. Shear stress activates endothelial nitric oxide synthase (eNOS also known as NOS III) which catalyzes the conversion of L-arginine to NO and L-citrulline. NO diffuses out of endothelial cells and acts on smooth muscle cells and platelets. NO inhibits platelet aggregation and acts synergistically with PGI_2 in controlling platelet and monocyte activation and monocyte interaction with vascular wall.

PGI_2 and NO also possess cytoprotective properties. It has been reported that PGI_2 protects against endothelial cell apoptosis by a peroxisome proliferator-activated receptor δ (PPARδ) pathway [13]. PGI_2-activated PPARδ enhances 14–3–3ε transcription and thereby increases 14–3–3ε protein expression. Increased 14–3–3ε levels enhance the capacity to bind and sequester Bad and thereby suppress Bad-mediated apoptosis via the mitochondrial pathway [13,14]. PGI_2 has also been reported to protect endothelial cell apoptosis by upregulating XIAP (X-linked inhibitor of apoptosis protein), which is a potent inhibitor of activated caspases and represents a major defense molecule against apoptosis [15]. NO confers antiapoptotic protection by nitration of Bcl-2 family proteins and via which it enhances the protective actions of Bcl-2 and thereby renders cells resistant to apoptosis.

The endothelial cells are adjusted to the proinflammatory insults by a coupling mechanism that switches COX-2 on and off. When endothelial cells are challenged by proinflammatory signals from shear and mechanical stresses, cytokines, endotoxins, and lipopolysaccharides (LPS), COX-2 expression is increased, resulting in the robust production of PGI_2. The mechanism by which diverse proinflammatory mediators stimulate COX-2 transcription has been characterized [16–18]. Several transactivators, notably AP-1 (C-Jun/C-Fos), C/EBPβ and NF-κB, are essential for COX-2 expression induced by LPS, cytokines, and growth factors [16–21]. These activators bind to respective motifs situated within 500 bp upstream of the transcription start site [22]. These functionally important transactivators have weak binding activity to COX-2 promoters at basal cellular state. The proinflammatory mediators increase their binding activities via kinase signaling pathways. Their binding activities are further enhanced by recruiting transcription co-activators, notably p300, which transmits the transcriptional activity to the transcription machinery [23,24]. Binding of transactivators as well as recruitment of p300 is reversible and is dissipated when the proinflammatory mediators level off. The stress-coupled COX-2 expression may also be regulated by posttranslational modification of COX-2. It has been reported that COX-2 has a short half-life, which is attributed to degradation via proteasome [25,26]. Cytokines such as interleukin-1β increases COX-2 stability thereby enhancing endothelial COX-2 proteins.

On the basis of in vitro studies, the proinflammatory mediators increase COX-2 expression in a time-dependent manner with maximal expression at 4–6 hours and return to baseline at 12–24 hours. The mechanism by which COX-2 expression is switched off has not been clearly elucidated but is thought to be linked to the clearance of the stress signals. Stress-coupled COX-2 expression in vascular endothelial cells results in amplified production of PGI_2 in a timely manner to defend against vascular damage by the insults from microbial infection, immune reactions, and physical forces. Once the transient insults dissipate, COX-2 levels return to the baseline within a finite period of time. This stress-coupled COX-2 expression and PGI_2 production is considered a key defense mechanism by which arterial wall is protected.

Alteration of Endothelial Phenotypes by Sustained Proinflammatory Stimulation

There is evidence from in vitro experiments and animal experiments that vascular endothelial cells chronically exposed to physical stresses and/or chemical stress signals eventually undergo functional alterations from being anti-inflammatory, antithrombotic, and protective to proinflammatory and prothrombotic. Thrombomodulin expression is downregulated at the transcriptional level by chronic exposure to PIMs leading to reduced surface thrombomodulin proteins [27]. Furthermore, endothelial surface thrombomodulin is abnormally cleaved during inflammation further reducing functional thrombomodulin [28,29]. In importance, chronic inflammatory signals trigger the endothelial cells to express adhesive molecules notably VCAM-1 (vascular cell adhesive molecule) and ICAM-1 (intercellular adhesion molecule-1). Expression of the adhesive molecules facilitates the interaction of leukocytes with endothelial surface via which the blood cells transmigrate into the vascular tissue to initiate vascular inflammation. This subject will be discussed in more detail in the next section.

Clinical studies have provided solid evidence for loss of ability of blood vessel to regulate vascular tone in chronic diseases including diabetes mellitus, hypertension, and other risk factors. A major cause of this is thought to be impairment of endothelial nitric oxide synthase (eNOS) activities. Nitric oxide producing capacity of eNOS is influenced by a number of factors. A major factor that suppresses NO availability is generation of superoxide. Superoxide reacts with NO to form peroxynitrite which is highly reactive and may contribute to endothelial damage. As NO is consumed to form peroxynitrite, the end result is reduction of NO and generation of peroxynitrite, creating a vasorestrictive and proinflammatory condition.

Proinflammatory mediators, especially the tumor necrosis factor (TNF) family proteins may induce endothelial apoptosis via the receptor-mediated apoptosis pathways. Furthermore, oxidants induced by PIMs may disrupt the mitochondrial membrane potential leading to release of proapoptotic factors such as cytochrome C from mitochondria to activate caspases [30]. Although vascular endothelial cells are endowed with a series of antiapoptotic molecules, sustained stimulation by PIMs may overwhelm the protective mechanisms and result in endothelial death. Exposure of the subendothelial matrix to blood cells and other constituents attracts platelets to the damaged vessel wall. Platelets undergo aggregation and release a myriad of factors that contribute to vascular inflammation.

BLOOD CELLS AND VASCULAR INFLAMMATION

Blood cells notably monocytes and lymphocytes are recognized to play critical roles in inducing vascular inflammation and promoting atherosclerosis and other types of vascular diseases [1,2]. Blood monocytes and lymphocytes are activated by proinflammatory factors present in circulating blood. The activated leukocytes produce and release proinflammatory cytokines that contribute to the inflammatory changes of endothelial cells with expression of adhesion molecules. Monocytes and lymphocytes interact with the adhesive molecules and transmigrate into the vascular tissue where they induce chronic vascular inflammatory changes. Blood platelets contribute to vascular inflammation by interacting with the subendothelial collagen and vWF after vascular endothelial denudation. Platelets release proinflammatory factors that induce smooth muscle cell migration to and proliferation at the intima. The role of neutrophils in vascular inflammation is unclear. However, neutrophils play a significant role in postischemic tissue inflammation and damage.

Activation of Blood Cells by PIMs

Circulating blood monocytes are highly reactive to PIMs such as LPS, endotoxins, cytokines, and immune mediators as well as CD-40 ligand (CD-40L) and P-selectin, which are expressed on platelets and endothelial cells. Stimulation of monocytes by the diverse factors results in the expression of proinflammatory cytokines and prostaglandins as well as adhesion molecules. LPS activates monocytes via interaction with specific receptors CD14 and Toll-like receptor 4. CD-40L acts by interaction with constitutively expressed CD-40 on monocyte surface. Proinflammatory cytokines and prostaglandins produced by activated monocytes are released into excellular milieu and acts in a paracrine manner to trigger local inflammation. Monocytes are attracted to the vascular wall by chemotactic factors such as MCP-1 where they roll on endothelial surface, adhere to endothelium, and migrate into the subendothelial zone. Once entering the vascular wall, monocytes undergo phenotypic changes to become macrophages, which express scavenger receptors and are the major cells that take up LDL. The LDL-loaded cells have a foamy appearances and therefore are called foam cells. Foam cells represent major building blocks of atherosclerosis. Vascular macrophages secrete cytokines to induce vascular inflammation that plays a critical role in the growth of atherosclerotic lesions. Interleukin-1β (IL-1β) and tumor necrosis factor-α (TNFα) are considered functionally important cytokines.

Under the stimulation of PIMs, circulating lymphocytes are activated, attracted to endothelium, and transmigrated into blood vessel wall. Biochemical mechanism by which lymphocytes enter into the vascular wall has not been as well investigated. However, it is now well recognized that lymphocytes enter into the vascular wall at early stage of atherogenesis and play critical roles in triggering and propagating atherosclerotic lesions.

Circulating platelets are essential components of hemostatic plugs and thrombi. Platelets do not interact with normal endothelium but are attracted to subendothelial regions when endothelium is severely damaged and becomes denuded. A single layer of platelets adhere to the subendothelium through interaction of platelet surface glycolprotein (GP) I_b–1X–V complex with von Willebrand factor (vWF) in the subendothelial tissue. The adhered platelets interact with subendothelial collagen via surface GP I_a-II_a and GP VI. Interaction with collagen results in platelet morphological changes, activation of signaling and metabolic pathways, and release of factors notably adenosine diphosphate (ADP) and thromboxane A_2 (TXA_2) leading to platelet aggregate formation. Platelets are also essential for coagulation and facilitate generation of thrombin, which is a powerful agonist of platelet activation. Thus, platelets possess coordinated programs for ensuring platelet aggregate and fibrin formations and are pivotal in thrombus generation. Despite its well-recognized role in thrombus formation on damaged vascular wall, especially ruptured atheromatous plaques, the involvement of platelets in vascular inflammation has not been as extensively investigated. Some recent reports suggest that activated platelets may adhere to endothelium stimulated by cytokines and transmigrate into the vessel. The results are based on in vitro experiments and should be further evaluated in vivo.

Activated platelets release biologically active factors from α granules. Several released factors such as P-selectin and CD-40L are attached to the outer surface of platelets. Monocyte surface constitutively expresses receptors for P-selectin and CD-40L. A number of studies have shown that binding of CD-40L to monocyte CD-40 has potent effects on monocyte activation and monocyte adhesion to vascular wall. Platelets form aggregates with monocytes probably through interactions between P-selectin and its receptors on monocyte surface. Monocytes may also be activated by CD-40L expressed on the cell surface of lymphocytes. Activated monocytes produce cytokines, which act on endothelial cells and other blood cells creating a positive loop of blood and vascular cell activation. These results indicate that PIMs trigger a series of blood cell activation processes resulting in cell–cell aggregation formation, amplification of PIM productions, and vascular endothelial activation. One of the intriguing questions is whether "low-grade" inflammatory signals from bacteria released from periodontic lesions or autoimmune products create a chronic blood cell activation state that poses increased risk of atherosclerosis and the consequent cardiac or cerebral tissue damage. As techniques for analyzing circulating blood cell activation are now available, it should be feasible to address this important issue by epidemiological and clinical studies.

Interaction of Blood Monocytic Cells with Endothelium

Activation of monocytes and lymphocytes leads to expression of integrins and lectins that recognize the adhesion molecules on the endothelial cell surface. Normal endothelium under laminar flow does not express adhesion molecules. Proinflammatory mediators such as cytokines, LPS, and oxidized LDL induce the expression of adhesion molecules notably vascular endothelial adhesion molecule (VCAM), intercellular adhesion molecule-1 (ICAM-1), and selectins. These adhesion molecules selectively interact with integrins and lectins expressed on activated monocytes or T lymphocytes. The activated monocytes and T lymphocytes roll and adhere to endothelium through sequential interactions with different classes of adhesion molecules. Mouse gene deletion experiments have revealed that VCAM-1 is essential for interaction of monocytic cells with endothelium and for early atherogenesis [31]. Chemotactic factors, notably monocyte chemotactic protein-1 (MCP-1), attract the adhered monocytes to migrate through endothelium and bind to surface chemokine receptors such as CCR-2 for MCP-1 [32].

Activated blood T lymphocytes adhere to endothelium by interacting with adhesion molecules such as VCAM-1 and transmigrate the endothelial cells into the subendothelium by the signals from chemokines including inducible protein 10 (IP-10), monokine induced by interferon-γ (Mig), and IFN-γ inducible T-cell α-chemoattractant (I-TAC) [1]. These chemoattractants bind to T lymphocyte surface chemokine receptors, CXCR-3 [33].

Mediation of Vascular Inflammation by Blood Cells

Transmigrated monocytes differentiate into tissue macrophages once localized to blood vessel wall. Monocytes/macrophages are capable of secreting PIMs in the vascular wall. Furthermore, they express scavenger receptors via which lipoprotein particles are engulfed. A major factor that activates macrophages is monocyte-colony stimulating factor (M-CSF) [34]. M-CSF enhances scavenger receptor expression and thereby increases lipoprotein uptake.

The lipoprotein-loaded macrophages have an appearance resembling foam cells. Under the stimulation of M-CSF and other factors such as CD-40L expressed on the surface of activated lymphocytes, macrophages and foam cells secrete myriad PIMs including cytokines, express cyclooxygenase-2 (COX-2), produce tissue factors and matrix metalloproteinases (MMPs), and generate reactive oxygen species (ROS). These powerful factors cross talk and trigger inflammation and tissue damage. They probably induce endothelial damage as well and initiate coagulation reactions via the tissue factor pathway. Thus, macrophages play a key role in inflammation, vascular damage, and lipid accumulation.

There is increasing evidence that T lymphocytes migrating into the subendothelial region initiate vascular inflammation by secreting PIMs and activating monocytes and macrophages. T lymphocytes interact with oxidized LDL and other antigens and functionally differentiate into TH1 and TH2 cells. TH1 cells secrete interferon-γ (IFN-γ), which induces the expression of CXC chemokines such as IP-10, Mig, and I-TAC and thereby attract T lymphocytes migration into the subendothelium. The TH cells produce cytokines that trigger inflammation and tissue damage. Importantly, activated T cells express CD-40L (known as CD154) which binds to CD-40 on monocyte surface and activates the expression in monocytes of tissue factors, MMPs and COX-2. These factors act in concert to induce inflammation, alter matrix, and cause coagulation activation. Interaction of CD-40 with CD-40L is considered crucial in all phases of atherosclerosis including plaque stability, and inhibition of CD-40 signaling was reported to be associated with improvement of atherosclerotic lesions.

Results form recent studies reveal that mast cells play a role in atherosclerosis. Similar to monocytes and T lymphocytes, mast cells are recruited into the subendothelial tissue by chemokines. Once arriving at the intima, mast cells secrete active compounds that contribute to inflammation and tissue damage.

In summary, PIMs in circulating blood activate endothelial cells and blood mononuclear cells (monocytes, T lymphocytes, and mast cells). The mononuclear cells adhere to endothelium and transmigrate into the vascular intima, where they are activated and secreted proinflammatory cytokines, prostaglandins, MMPs, and other proteolytic enzymes that create chronic inflammation and tissue damage. Furthermore, macrophages engulf lipoproteins to form foam cells. Foam cell infiltration at early stages of vascular inflammation gives rise to fatty streak appearance. As inflammation and tissue damage continue, the fatty streak appearance is replaced by raised atherosclerotic lesions, with eventual formation of atheromatous plaques.

INFLAMMATION IN ATHEROMA GROWTH AND PLAQUE FORMATION

Growth of Atheroma

Growth of the atheromatous lesions is geographically uneven and temporally discontinuous. Atheroma develops more rapidly at the sites of blood vessel bifurcation as a result of turbulent blood flow. Similarly, atheromatous plaques are more commonly noted at bifurcation sites. Growth of atheromatous lesions is dictated by the presence of abnormal blood flow, endotoxins, proinflammatory cytokines, prostaglandins, and growth factors. There is good evidence for regression of atheromatous lesions in experimental animals when dietary intake of high lipids is reduced. The periodic burst of growth is most likely contributed by invading microorganisms and endotoxins as well as immunological mediators.

Smooth muscle cell migration and proliferation contribute to progressive atheromatous lesions [35]. Smooth muscle cells normally are located in the media layer of blood vessels. The residential smooth muscle cells possess contractile properties, and have the primary function of contracting blood vessels. However, during the development of atherosclerosis, they are attracted to the intima by growth factors notably platelet-derived growth factor (PDGF). Once settled down in the intima, the smooth muscle cells alter phenotype. They lose contractile activities and acquire proliferative and secretory functions. They secrete glycoproteoglycans and other connective tissues. As a result of rapid proliferation, they increase the intimal thickness. The exact mechanism by which smooth muscle cells undergo drastic phenotypic changes is unclear.

Development of Atheromatous Plaque

Plaque formation is also geographically selective and temporally discontinuous. The exact factors that trigger plaque formation at selective spots in the atheromatous arterial wall have not been entirely identified. However, there is good evidence for locally active inflammation as a major cause of plaque formation and plaque composition. A mature plaque comprises a central core composed of necrotic tissues and fat infiltration surrounded by inflammatory cells. Atheromatous plaques are covered with a cap rich in cells and fibrous tissues, the so-called fibrous cap. This cap seals procoagulant and platelet activating materials within the vessel wall and prevents them from interaction with blood coagulation factors and platelets. The cap is composed primarily of smooth muscle cells and collagen secreted from smooth muscle cells. The cap structure and properties are determined by inflammation. Active infiltration of inflammatory cells in the cap results in cytokine and metalloproteinase production, which suppresses collagen production, induces smooth muscle cell apoptosis, and degrades preexisting fibrous tissues. The cap is replaced with inflammatory and necrotic tissues, and the fibrous cap becomes thin and vulnerable to rupture.

Plaque Rupture and Thrombus Formation

The fibrous cap prevents the plaque tissue factor (TF) and vWF from coming in contact with circulating platelets and coagulation factors. Once ruptured, vWF attracts platelets by binding to the GPIb/IX/V complex on platelet surfaces. Collagen binds platelets via two receptors: GPIa/IIa and GPVI [36]. Collagen activates platelets and induces membrane changes, which expose phosphatidylserine (PS) and phosphatidylethanolamine (PE). The membrane phospholipid changes facilitate coagulation reactions by providing surface for factor V and factor VIII binding. Coagulation is activated by binding of factor VIIa (activated factor VII) to cells expressing TF on their surface. Factor VIIa in circulating blood has weak catalytic activity but the catalytic activity is enhanced by three orders of magnitude when factor VIIa bind to TF. Factor VIIa cleaves and activates factor X and factor IX. Activated factor X (Factor Xa) and factor IX (factor IXa) are detached from factor VIIa–TF complex. Factor Xa binds to factor Va. The catalytic activity of factor Xa is greatly enhanced when it binds to factor Va, which is attracted to the surface of activated platelets as a result of the reorientation of the phospholipids. In the presence of calcium, factor Xa on platelet surface catalyzes the conversion of prothrombin (factor II) to thrombin. However, only a limited amount of thrombin is generated via VIIa/Xa pathway because factor Xa generation from this pathway is rapidly inhibited by a naturally occurring inhibitor, tissue factor pathway inhibitor (TFPI). Thrombin is generated in ample quantities by an alternative pathway: factor IXa. Factor IXa binds to factor VIIIa on activated platelet surface. Factor IXa catalyzes the conversion of factor X to factor Xa and factor Xa in turn converts prothrombin to thrombin on activated platelet surface. Thrombin catalyzes the formation of fibrin, which is consolidated through the cross-linking action of factor XIIIa. In parallel, platelets attracted to the ruptured surface form aggregates. Platelet aggregation is initially triggered by collagen and vWF, amplified by ADP and TXA_2 and further amplified by thrombin. As the disrupted surface is enriched in TF-bearing macrophages and smooth muscle cells, thrombus formation composed of platelet aggregates enmeshed by fibrin fibriles could develop at a very rapid rate and may occlude the entire artery in a short period of time. This creates catastrophic clinical outcome such as sudden death and massive myocardial infarction. However, probably more commonly, plaque rupture may cause small fissures on which small thrombi are formed. The small sizes of thrombi have no immediate impact on blood supply to tissues and therefore do not cause clinical problems. However, they may become organized and embedded into the atheromatous mass, which contributes to the growth of atheromas and progression of vascular narrowing. Severe coronary arterial narrowing results in reduced blood flow and hypoxia especially during exercises causing clinical manifestations of angina pectoris. Thrombi formed in the carotid arteries tend to be detached and lodge at distal arterial trees to cause acute arterial occlusion. The most common site of occlusion is middle cerebral artery. The artery-to-artery embolization causes catastrophic ischemia and cerebral infarction. Small thrombi may be organized and contribute to arterial narrowing. As the narrowing increases, blood flow diminished and ischemia symptoms may develop (transient ischemic attacks and "minor stoke").

INFLAMMATION AND ISCHEMIA-REPERFUSION INJURY

Common Inflammatory Response to Tissue Ischemia

Extensive investigations using ischemia-reperfusion (I/R) animal models have established inflammation as a key event in tissue damage following vascular occlusion. There is a common sequence of events that occurs with all types of I/R tissue injury including myocardial and cerebral tissue damage following coronary artery and middle cerebral artery occlusion, respectively [37,38]. The inflammatory response to I/R is illustrated in Figure 26.1. Arterial occlusion results in reduced oxygen tension, ATP depletion, and necrotic cell death. It is generally thought that dying cells trigger inflammation through generation of reactive oxygen species (ROS), complement activation, upregulation of Toll-like receptors (TLR) and activation of transcriptional programs via transactivators such as NF-κB. Transcriptional activation by proinflammatory transactivators results in expression of proinflammatory cytokines, chemokines, and adhesive molecules on microvascular endothelial cells. Neutrophils in the microvascular circulation roll, adhere, and transmigrate the endothelium by interacting with E-selectin, P-selectin, and ICAM-1, which are expressed on vascular endothelial cells under the inflammatory stimulation. Neutrophils accumulated at the ischemic tissues induce acute inflammatory changes by generation of ROS, notably superoxide ions via NADPH oxidase and the production of proteolytic enzymes. It is estimated that neutrophils transmigrate and accumulate at the

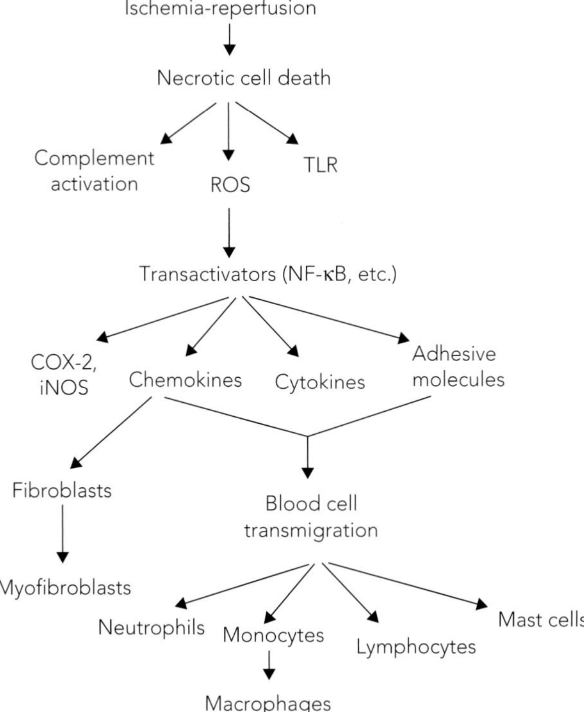

Figure 26.1. A simplified scheme illustrating the factors involved in triggering blood and tissue cell migration and activation. Abbreviations: ROS, reactive oxygen species; TLR, Toll-like receptor; COX-2, cyclooxygenase-2; iNOS, inducible nitric oxide synthase.

ischemic heart or brain tissues within a few hours after ischemia. Neutrophil numbers reduce within 24–48 hours. Neutrophil accumulation in microvasculature may occlude the microvascular circulation and enhance ischemic injury. Transmigration of monocytes into the ischemic tissues occurs later than neutrophils but monocytes accumulate for a longer period of time and phenotypically convert to macrophages. Monocytes are attracted to the ischemic tissues by chemokines, notably MCP-1 and IL-8 which interact with their receptors on monocytes. Monocytes/macrophages elicit chronic inflammatory damage primarily by producing and releasing proinflammatory cytokines. Under inflammatory stimulation, monocytes express cyclooxygenase-2 (COX-2) and PGE synthase (mPGES-1) resulting in the production of proinflammatory PGE_2.

Among cytokines and growth factors implicated in ischemic injury, IL-1β, IL-6, and tumor necrosis factor-α (TNF-α) are extensively investigated and their roles are better documented. These cytokines induce tissue damage by stimulating proinflammatory molecules such as PGE_2 derived from COX-2 and nitric oxide (NO) via inducible nitric oxide synthase (iNOS). They induce the expression of metalloproteinases (MMP) which alter the extracellular matrix.

Tissue-Specific Inflammatory Features

Analysis of postischemic inflammation in the heart and brain tissues reveals tissue-specific differences. In the postischemic heart tissues, mast cells, and fibroblasts contribute to fibrosis. Mast cells are residential cells in normal heart tissues. Following ischemia, they increase in numbers and undergo degranulation, which releases active compounds such as histamine that contributes to inflammation. Furthermore, mast cells activation was considered to contribute to fibrosis [38]. The exact roles that mast cells play remains to be elucidated. Another major cell type that infiltrates postischemic heart tissues is fibroblasts. Fibroblasts migrate to ischemic tissues where they transdifferentiate into myofibroblasts. Myofibroblasts secrete collagen and other matrix proteins that constitute the major supporting tissues of fibrosis.

The postischemic inflammatory responses are considered a defense mechanism to repair tissue damage caused by reactive oxidants and other factors. However, the inflammatory response often proceeds beyond control and causes tissue damage. As a tissue repairment process, the inflammatory process elicits angiogenesis via expression of vascular endothelial growth factor (VEGF) and cell proliferation to offset the cell loss due to necrosis and apoptosis. Transdifferentiation and proliferation of myofibroblasts represent an important step in tissue repairment [39,40]. However, myofibroblasts cause dire consequences such as excessive fibrosis that impair the cardiac function. Inflammation also contribute to myocyte apoptosis resulting in cardiac dilatation and vascular remodeling.

In contrast, inflammation-associated fibrosis is not a major feature in postischemia brain damage. Instead, inflammatory mediators in ischemic brain tissues induce glial cell activation. Of the three major types of glial cells in the brain, GFAP-positive astrocytes, residential glial cells, are activated and become hypertrophic a few hours after ischemia [37]. Approximately 24 hours after ischemia, microglial cells are activated. Another major change following brain ischemia is degradation of basal lamina, an extension of the extracellular matrix of microvasculature. Microvascular lamina functions not only as a barrier but also as a structure to anchor microvascular endothelial cells and astrocytes. Anchoring of astrocyte end-foot to the microvasculature via basal lamina has been shown to be functionally important in regulating microvascular blood flow. Thus, loss of basal lamina results in microvascular dysfunction and erythrocyte leakage. Disruption of basal lamina is attributed at least in part to rapid generation of MMP-2. Another mechanism by which endothelial cells and astrocytes become detached from the basal lamina is downregulation of several integrins such as $\alpha_1\beta_1$ and $\alpha_3\beta_1$ on endothelial cells and $\alpha_1\beta_1$ and $\alpha_6\beta_4$ on astrocyte surfaces. These integrins interact with molecules in the basal lamina and mediate anchoring endothelial cells and astrocytes to the matrix. Thus, astrogliosis, microglial activation, and microvascular dysfunction are major consequences of postischemic inflammation. These cellular changes contribute to neuronal apoptosis after ischemia. There is no apparent matrix deposition in the infracted brain as contrast to fibrosis being a key event in myocardial infarcts.

Postischemia inflammation in other organs such as kidneys and lungs share with the heart and brain common inflammation features but is likely to possess tissue-specific features because of mobilization and activation of specific residential inflammatory cells. The tissue-specific features of those organs have not been as extensively investigated as heart or brain.

Postischemic Transcription of Prolnflammatory Genes

The common features of postischemia inflammation could be attributed to the activation of a transcription program that mediates the expression of proinflammatory genes. The factors that trigger the initiation of the transcriptional activation have not been completely identified. Recent studies indicate that during the acute phase of inflammation, a major triggering factor is ROS. Both H_2O_2 and superoxide ions are shown to activate the expression of proinflammatory genes. However, there are reports that ROS suppress expression of genes such as COX-2 and iNOS. Other more powerful factors may be responsible for activation of transcription of proinflammatory chemokines and adhesive molecules that recruit blood cells into the ischemic tissues as well as activation of cytokine transcription that leads to amplified expression of proinflammatory genes. Cytokines notably IL-1β and IL-6 and TNF-α stimulate monocyte transcriptional programs that result in overexpression of cytokines that serve to propagate the expression of proinflammatory genes as well as expression of COX-2, iNOS, and MMPs. Therefore, the inflammation cascade may be considered to have an initiation phase and a propagation phase. The propagation phase amplifies the proinflammatory gene expression by forming a positive regulatory loop. It induces the expression of key inflammatory mediators such as eicosanoids, NO, and MMPs (Figure 26.2).

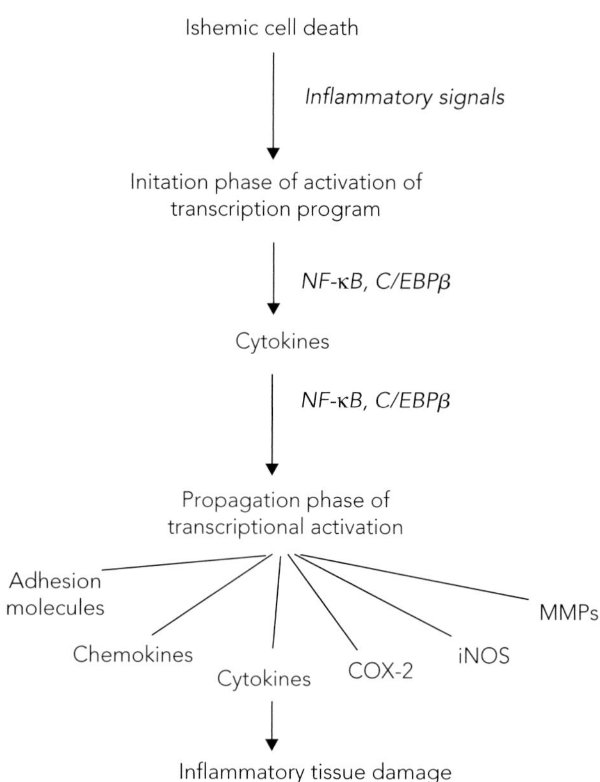

Figure 26.2. Hypothetic scheme of induction of inflammatory tissue damage by ischemic cell death. The inflammatory response induced by ischemic cell death is divided into initiation phase and propagation phase of transcriptional stimulation mediated by transcription factors such as NF-κB and C/EBPβ. The scheme is a simplified view of a much more complex process.

TRANSCRIPTIONAL REGULATION OF PROINFLAMMATORY GENES

The transcriptional program that initiates the expression of proinflammatory genes is a key event that sustains inflammatory responses in blood vessel diseases such as atherosclerosis and in postischemic tissue damages such as myocardial infarction. The promoter regions of a majority of proinflammatory genes have characteristics of inducible genes which harbor TATA motif and a series of *cis*-activating elements at the 5'-flanking promoter region. The promoter is activated by binding of transactivators to their respective binding sites, which recruit co-activators such as p300. p300 interacts with the transcription machinery to initiate transcription. p300 possesses histone acetyltransferase activities that induce open chromatin structures to facilitate binding of transactivators [41–43]. Although transactivators that bind and activate promoter may differ among the proinflammatory genes, several transactivators are considered to be involved in the activation of diverse proinflammatory genes and thus are considered to occupy a central position in inducing inflammation.

NF-κB

NF-κB comprises two families of proteins: (1) the Rel family, which is composed of C-Rel (p75), RelA (p65), and RelB (p68); and (2) NF-κB family which is composed of NF-κB1 (p105/p50) and NF-κB2 (p100/p52) [44]. The two families of proteins form heterodimers or homodimers which bind to κB *cis*-acting elements. Both families of proteins are localized to cytoplasm in resting cells. The Rel proteins are sequestered in cytoplasm through binding to inhibitors, called IκBs, while NF-κB proteins are retained in cytoplasm because they contain intramolecular IκB-like inhibitory domains [44]. Upon stimulation by proinflammatory signals, Rel is dissociated from IκB by IκB degradation via ubiquitin-proteasome. IκB degradation involves phosphorylation of IκB by IκB kinases (IKK). Phosphorylated IκB is targeted by ubiquitin enzymes for polyubiquinition and the polyubiquinized IκB is degraded in proteasome. IκB degradation releases Rel. Formation of free NF-κB1 (p50) or NF-κB2 (p52) is mediated by a different mechanism. p50 and p52 exist in cytoplasm of resting cells as p105 and p100 proteins, respectively. The C-terminal region of p100 or p105 contains IκB-like structure. Upon activation, the C-terminal region is removed by proteolytic enzymes. Free Rel and NF-κB form dimers, which are translocated into nucleus, where they bind to IκB motifs and activate the transcription of proinflammatory genes. The most common isoform of NF-κB in proinflammatory gene expression is p65 (RelA) /p50. C-Rel (p75)/p50 is also frequently involved in proinflammatory gene activation. Among NF-κBs, p65 possesses binding sites for p300 and recruits p300. Furthermore, p300 acetylates p65 and increases p65 activity [45,46]. NF-κB binds to the promoter regions of diverse cytokines, adhesive molecules as well as COX-2 and iNOS [23,46] and is critical for induction or upregulation of the expression of diverse proinflammatory genes. NF-κB has been implicated in mediating inflammatory diseases including postischemic inflammation and tissue damage.

C/EBP

CCAAT/enhanced binding proteins (C/EBP) comprise six members of basic leucine zipper transcription factors [47,48]. They are divided into two subgroups based on sequence homology: one group comprises C/EBPα, β, and δ and the other, C/EBPγ, ε, and δ. The β isoform, also known as nuclear factor IL6 (NF-IL6) drives the promoter activities of IL-6 and myriad proinflammatory genes. C/EBPβ in resting cells has weak DNA binding activities because its activity is inhibited by an intramolecular inhibitory element [18]. C/EBPβ in cells stimulated by proinflammatory mediators is phosphorylated which results in releasing the inhibitory element and exposure of the kinase active sites. Several kinases such as p90 ribosomal S6 kinases (RSK), calmodulin-dependent kinase IV (CamK-IV), protein kinase A (PKA), and extracellular regulated protein kinase (ERK) were reported to phorylate C/ERPβ [49–53]. We have shown that phosphorylation of C/EBPβ by RSK is crucial for C/EBPβ binding to COX-2 promoter and C/EBPβ-mediated COX-2 transcription.

C/EBPβ binding is also regulated at the translational level. C/EBPβ mRNA contains several translation start codons that are used as a start site in ribosomal translation to generate a series of truncated isoforms [21]. In human cells, the full length of C/EBPβ is 46 kDa. Two major truncated forms, a 41 kDa that is also known as LAP (liver-enriched transcription activating protein) and a 16 kDa, which is known as LIP (liver-enriched transcription inhibitory protein), regulate the expression of COX-2 and other proinflammatory genes, respectively. LIP functions as a dominant mutant of C/EBPβ, which inhibits binding of full-length C/EBPβ and 41 kDa isoform to the C/EBP cognitive sites on the promoter of proinflammatory genes. Like NF-κB, C/EBPβ is considered to mediate postischemic inflammation and tissue damage.

ANTI-INFLAMMATORY THERAPEUTIC STRATEGIES

Despite the prominent roles of inflammation in vascular diseases and postischemic tissue damage, there is no effective anti-inflammatory treatment for atherosclerosis, myocardial infarction, ischemic stroke, or other postischemic tissue damage. One of the reasons for the failure is that inflammatory responses are intended for protecting tissues and for tissue repairment. It is the uncontrolled inflammation that contributes to vascular, cardiac, or brain tissue damage and is the target for therapy. As the currently available anti-inflammatory drugs do not distinguish between "physiological" and "pathologic" inflammation, they do not have clear beneficial effects on reducing atherosclerosis and/or inflammation-mediated tissue injury. In fact, the nonsteroidal anti-inflammatory drugs (NSAIDs) and selective COX-2 inhibitors (COXIBs) are associated with cardiovascular complications [54]. One of the explanations for the complications is that COX-2 and proinflammatory cytokines such as TNF-α have Yin (proinflammatory) and Yang (protective) functions. In this regard, COX-2 has been extensively investigated and will be discussed in more detail to illustrate this important point.

Paradoxical Effects of COX-2 Inhibitors

COX-2 expression is highly induced by cytokines, endotoxins, and the growth factor in inflammatory and neoplastic cells [27]. The expressed COX-2 proteins are coupled to PGE synthases (PGES), notably the microsomal PGES-1 (mPGES-1) isoform, and catalyzes robust production of proinflammatory prostaglandins such as PGE_2, COX-2, and mPGES-1 are overexpressed in atherosclerosis, postischemic brain tissues, and cardiac tissues [55]. COX-2/mPGES-1 is considered an important mediator of inflammatory joint diseases such as osteoarthritis. Selective COX-2 inhibitors (COXIBs) were documented to be effective in controlling arthritic inflammation and relieving inflammatory symptoms and were widely used by patients in treating joint diseases. It was subsequently reported that one of the potent COXIB drugs, roficoxib (Vioxx) was significantly associated with increased risk of myocardial infarction [56]. Additional analysis reveal that nonselective COX-2 inhibitors including diverse classes of NSAIDs were also associated with increased risk of myocardial infarction. Those reports raised major concerns about the safety of COXIBs and NSAIDs and led to withdrawal of Vioxx from the market [54].

The reasons that COX-2 inhibitors increase risk of myocardial infarction are not entirely clear but several studies have pointed to the disruption of the protective effects of vascular COX-2. Vascular endothelial cells express both isoforms of COX (COX-1 and COX-2). COX-2 is shown to be the predominant isoform for prostacyclin (PGI_2) synthesis [11,12]. Although COX-2 level is very low at resting endothelial cells in culture, endothelial COX-2 expression in vivo is stimulated by shear stress. It is functionally coupled to PGI_2 synthase (PGIS) and catalyzes the production of PGI_2. PGI_2 has several vascular and tissue protective properties including inhibition of platelet aggregation, protection of endothelial cells against apoptosis, and prevents vasoconstriction. Thus, COX-2/PGIS pathway is critical in maintaining vascular integrity.

COX-2 has also been shown to mediate vascular inflammation in atherosclerosis. There is evidence to suggest COX-2 expression in the infiltrating monocytes of atherosclerotic lesions is enhanced [57]. The consequence of COX-2 overexpression and the metabolic shift is the overproduction of PGE_2. COX-2 expression was reported to be increased in cerebral and cardiac ischemic tissues following ischemia-reperfusion. COX-2 overexpression is considered to be a key player in inflammation in ischemic tissues [4,46]. It thus is somewhat paradoxical that clinical use of COXIBs and NSAIDs is associated with increased risk of myocardial infarction (MI).

A prerequisite for MI is rupture of the atheromatous plaque, which causes rapid platelet aggregation and coagulation cascade activation resulting in thrombus formation and acute occlusion of coronary arteries. It may be speculated that COX-2 may be involved in protection against plaque rupture via the PGI synthase pathway. Inhibition of COX-2 results in loss of the protective PGI_2 thereby increasing the risk of plaque rupture. This thesis should be investigated.

Transcriptional Control as Therapeutic Strategy

The Yin-Yang functions of proinflammatory mediators are not limited to COX-2. It has been suggested that TNF-α may possess protective action against tissue damage besides its well-recognized proinflammatory action. Since complete elimination of those dual-activity mediators is unlikely to achieve optimal therapeutic effects as it may cause undesirable complications, and since induction of proinflammatory gene expressions requires multiple transcription activators, it may be possible to suppress partially the expression of multiple proinflammatory genes such as COX-2, IL-1β, IL-6, and TNF-α by blocking the binding of a selected transactivator such as NF-κB or C/EBPβ [58]. This possibility is supported by the findings that salicylates (aspirin and sodium salicylate) inhibit COX-2 transcriptional activation by proinflammatory mediators by blocking C/EBPβ binding to COX-2 promoter [20]. The reported results further show that salicylates target RSK and thereby suppress C/EBPβ phosphorylation and transcriptional activity [59]. By suppressing C/EBPβ binding without affecting NF-κB, C-Jun/C-Fos, or CREB, COX-2 expression in response to stimuli is reducing by ~50%. Salicylate is a natural plant-signaling molecule and is enriched in plant sources such as willow bark. Willow bark extracts were reported to possess anti-inflammatory activities, which were attributed to salicylates. Sodium salicylate was as effective as aspirin in controlling pain and inflammation and yet it has very weak inhibitory actions against COX-2 or COX-1 catalytic activity. Since salicylate suppresses COX-2 transcriptional activity by inhibiting RSK-mediated C/EBPβ binding and transactivation, it is possible that the anti-inflammatory action of salicylate may be attributed to partial transcriptional inhibition of COX-2 through selective suppression of C/EBPβ binding. It is to be noted that the action of salicylate is demonstrated in in vitro cellular studies. Its anti-inflammatory actions in vascular and cardiac as well as cerebral tissues should be investigated in animals and humans. It is of interest that other ingredients of plants such as resveratrol and catechins are active in controlling COX-2 transcription in response to stimuli by proinflammatory mediators. Thus, the natural products may represent an important source of developing new transcription-based anti-inflammatory therapy of atherosclerosis and ischemia-reperfusion tissue damage.

REFERENCES

1. Ross, R.A., Monteith, P.R., and McAdam, J.G. 1993. Case report: polymorphous haemangioendothelioma, a rare

cause of persistent lymphadenopathy. *J R Nav Med Serv* **79**:80–82.
2. Libby, P. 2002. Inflammation in atherosclerosis. *Nature* **420**:868–874.
3. Frangogiannis, N.G., Smith, C.W., and Entman, M.L. 2002. The inflammatory response in myocardial infarction. *Cardiovasc Res* **53**:31–47.
4. del Zoppo, G., Ginis, I., Hallenbeck, J.M., Iadecola, C., Wang, X., and Feuerstein, G.Z. 2000. Inflammation and stroke: putative role for cytokines, adhesion molecules and iNOS in brain response to ischemia. *Brain Pathol* **10**:95–112.
5. Wu, K.K. 1998. Injury-coupled induction of endothelial eNOS and COX-2 genes: a paradigm for thromboresistant gene therapy. *Proc Assoc Am Physicians* **110**:163–170.
6. Dittman, W.A., and Majerus, P.W. 1990. Structure and function of thrombomodulin: a natural anticoagulant. *Blood* **75**:329–336.
7. Sadler, J.E. 1997. Thrombomodulin structure and function. *Thromb Haemost* **78**:392–395.
8. Esmon, C.T. 1995. Thrombomodulin as a model of molecular mechanisms that modulate protease specificity and function at the vessel surface. *FASEB J* **9**:946–955.
9. Wu, K.K., and Liou, J.Y. 2005. Cellular and molecular biology of prostacyclin synthase. *Biochem Biophys Res Commun* **338**:45–52.
10. Smith, W.L., Garavito, R.M., and DeWitt, D.L. 1996. Prostaglandin endoperoxide H synthases (cyclooxygenases)-1 and -2. *J Biol Chem* **271**:33157–33160.
11. McAdam, B.F., Catella-Lawson, F., Mardini, I.A., Kapoor, S., Lawson, J.A., and FitzGerald, G.A. 1999. Systemic biosynthesis of prostacyclin by cyclooxygenase (COX)-2: the human pharmacology of a selective inhibitor of COX-2. *Proc Natl Acad Sci USA* **96**:272–277.
12. Belton, O., Byrne, D., Kearney, D., Leahy, A., and Fitzgerald, D.J. 2000. Cyclooxygenase-1 and -2-dependent prostacyclin formation in patients with atherosclerosis. *Circulation* **102**:840–845.
13. Liou, J.Y., Lee, S., Ghelani, D., Matijevic Aleksic, N., and Wu, K.K. 2006. Protection of endothelial survival by peroxisome proliferator-activated receptor-delta mediated 14–3–3 upregulation. *Arterioscler Thromb Vasc Biol* **26**:1481–1487.
14. Liou, J.Y., Ghelani, D., Yeh, S., and Wu, K.K. 2007. Nonsteroidal anti-inflammatory drugs induce colorectal cancer cell apoptosis by suppressing 14–3–3epsilon. *Cancer Res* **67**:3185–3191.
15. Liou, J.Y., Ellent, D.P., Lee, S., et al. 2007. Cyclooxygenase-2-derived prostaglandin e2 protects mouse embryonic stem cells from apoptosis. *Stem Cells* **25**:1096–1103.
16. Herschman, H.R. 1996. Prostaglandin synthase 2. *Biochim Biophys Acta* **1299**:125–140.
17. Wu, K.K. 1995. Inducible cyclooxygenase and nitric oxide synthase. *Adv Pharmacol* **33**:179–207.
18. Wu, K.K., Liou, J.Y., and Cieslik, K. 2005. Transcriptional Control of COX-2 via C/EBPbeta. *Arterioscler Thromb Vasc Biol* **25**:679–685.
19. Schroer, K., Zhu, Y., Saunders, M.A., et al. 2002. Obligatory role of cyclic adenosine monophosphate response element in cyclooxygenase-2 promoter induction and feedback regulation by inflammatory mediators. *Circulation* **105**:2760–2765.
20. Saunders, M.A., Sansores-Garcia, L., Gilroy, D.W., and Wu, K.K. 2001. Selective suppression of CCAAT/enhancer-binding protein beta binding and cyclooxygenase-2 promoter activity by sodium salicylate in quiescent human fibroblasts. *J Biol Chem* **276**:18897–18904.
21. Zhu, Y., Saunders, M.A., Yeh, H., Deng, W.G., and Wu, K.K. 2002. Dynamic regulation of cyclooxygenase-2 promoter activity by isoforms of CCAAT/enhancer-binding proteins. *J Biol Chem* **277**:6923–6928.
22. Tazawa, R., Xu, X.M., Wu, K.K., and Wang, L.H. 1994. Characterization of the genomic structure, chromosomal location and promoter of human prostaglandin H synthase-2 gene. *Biochem Biophys Res Commun* **203**:190–199.
23. Deng, W.G., Zhu, Y., and Wu, K.K. 2003. Up-regulation of p300 binding and p50 acetylation in tumor necrosis factor-alpha-induced cyclooxygenase-2 promoter activation. *J Biol Chem* **278**:4770–4777.
24. Deng, W.G., Zhu, Y., and Wu, K.K. 2004. Role of p300 and PCAF in regulating cyclooxygenase-2 promoter activation by inflammatory mediators. *Blood* **103**:2135–2142.
25. Rockwell, P., Yuan, H., Magnusson, R., Figueiredo-Pereira, M.E. 2000. Proteasome inhibition in neuronal cells induces a proinflammatory response manifested by upregulation of cyclooxygenase-2, its accumulation as ubiquitin conjugates, and production of the prostaglandin PGE(2). *Arch Biochem Biophys* **374**:325–333.
26. Mbonye, U.R., Wada, M., Rieke, C.J., Tang, H.Y., Dewitt, D.L., and Smith, W.L. 2006. The 19-amino acid cassette of cyclooxygenase-2 mediates entry of the protein into the endoplasmic reticulum-associated degradation system. *J Biol Chem* **281**:35770–35778.
27. Moore, K.L., Andreoli, S.P., Esmon, N.L., Esmon, C.T., and Bang, N.U. 1987. Endotoxin enhances tissue factor and suppresses thrombomodulin expression of human vascular endothelium *in vitro*. *J Clin Invest* **79**:124–130.
28. Ishii, H., and Majerus, P.W. 1985. Thrombomodulin is present in human plasma and urine. *J Clin Invest* **76**:2178–2181.
29. Salomaa, V., Matei, C., Aleksic, N., et al. 1999. Soluble thrombomodulin as a predictor of incident coronary heart disease and symptomless carotid artery atherosclerosis in the Atherosclerosis Risk in Communities (ARIC) Study: a case-cohort study. *Lancet* **353**:1729–1734.
30. Suen, D.F., Norris, K.L., and Youle, R.J. 2008. Mitochondrial dynamics and apoptosis. *Genes Dev* **22**: 1577–1590.
31. Cybulsky, M.I., Iiyama, K., Li, H., et al. 2001. A major role for VCAM-1, but not ICAM-1, in early atherosclerosis. *J Clin Invest* **107**:1255–1262.
32. Boring, L., Gosling, J., Cleary, M., and Charo, I.F. 1998. Decreased lesion formation in CCR2-/- mice reveals a role for chemokines in the initiation of atherosclerosis. *Nature* **394**:894–897.
33. Mach, F., Sauty, A., Iarossi, A.S., et al. 1999. Differential expression of three T lymphocyte-activating CXC chemokines by human atheroma-associated cells. *J Clin Invest* **104**:1041–1050.
34. Clinton, S.K., Underwood, R., Hayes, L., Sherman, M.L., Kufe, D.W., and Libby, P. 1992. Macrophage colony-stimulating factor gene expression in vascular cells and in experimental and human atherosclerosis. *Am J Pathol* **140**: 301–316.
35. Doran, A.C., Meller, N., and McNamara, C.A. 2008. Role of smooth muscle cells in the initiation and early progression of atherosclerosis. *Arterioscler Thromb Vasc Biol* **28**:812–819.
36. Wu, K.K., and Matijevic-Aleksic, N. 2005. Molecular aspects of thrombosis and antithrombotic drugs. *Crit Rev Clin Lab Sci* **42**:249–277.
37. Dirnagl, U., Iadecola, C., and Moskowitz, M.A. 1999. Pathobiology of ischemic stroke: an integrated view. *Trends Neurosci* **22**:391–397.

38. Frangogiannis, N.G. 2008. The immune system and cardiac repair. *Pharmacol Res.* **58**(2):88–111.
39. Willems, I.E., Havenith, M.G., De Mey, J.G., and Daemen, M.J. 1994. The alpha-smooth muscle actin-positive cells in healing human myocardial scars. *Am J Pathol* **145**:868–875.
40. Tomasek, J.J., Gabbiani, G., Hinz, B., Chaponnier, C., and Brown, R.A. 2002. Myofibroblasts and mechano-regulation of connective tissue remodelling. *Nat Rev Mol Cell Biol* **3**:349–363.
41. Ogryzko, V.V., Schiltz, R.L., Russanova, V., Howard, B.H., and Nakatani, Y. 1996. The transcriptional coactivators p300 and CBP are histone acetyltransferases. *Cell* **87**:953–959.
42. Boyes, J., Byfield, P., Nakatani, Y., and Ogryzko, V. 1998. Regulation of activity of the transcription factor GATA-1 by acetylation. *Nature* **396**:594–598.
43. Furia, B., Deng, L., Wu, K., et al. 2002. Enhancement of nuclear factor-kappa B acetylation by coactivator p300 and HIV-1 Tat proteins. *J Biol Chem* **277**:4973–4980.
44. Karin, M. 1999. The beginning of the end: IkappaB kinase (IKK) and NF-kappaB activation. *J Biol Chem* **274**:27339–27342.
45. Chen, L.F., Mu, Y., and Greene, W.C. 2002. Acetylation of RelA at discrete sites regulates distinct nuclear functions of NF-kappaB. *EMBO J* **21**:6539–6548.
46. Deng, W.G., and Wu, K.K. 2003. Regulation of inducible nitric oxide synthase expression by p300 and p50 acetylation. *J Immunol* **171**:6581–6588.
47. Wedel, A., and Ziegler-Heitbrock, H.W. 1995. The C/EBP family of transcription factors. *Immunobiology* **193**:171–185.
48. Akira, S., and Kishimoto, T. 1997. NF-IL6 and NF-kappa B in cytokine gene regulation. *Adv Immunol* **65**:1–46.
49. Nakajima, T., Kinoshita, S., Sasagawa, T., et al. 1993. Phosphorylation at threonine-235 by a ras-dependent mitogen-activated protein kinase cascade is essential for transcription factor NF-IL6. *Proc Natl Acad Sci USA* **90**:2207–2211.
50. Trautwein, C., Caelles, C., van der Geer, P., Hunter, T., Karin, M., and Chojkier, M. 1993. Transactivation by NF-IL6/LAP is enhanced by phosphorylation of its activation domain. *Nature* **364**:544–547.
51. Buck, M., Poli, V., van der Geer, P., Chojkier, M., and Hunter, T. 1999. Phosphorylation of rat serine 105 or mouse threonine 217 in C/EBP beta is required for hepatocyte proliferation induced by TGF alpha. *Mol Cell* **4**:1087–1092.
52. Wegner, M., Cao, Z., and Rosenfeld, M.G. 1992. Calcium-regulated phosphorylation within the leucine zipper of C/EBP beta. *Science* **256**:370–373.
53. Dlaska, M., and Weiss, G. 1999. Central role of transcription factor NF-IL6 for cytokine and iron-mediated regulation of murine inducible nitric oxide synthase expression. *J Immunol* **162**:6171–6177.
54. Fitzgerald, G.A. 2004. Coxibs and cardiovascular disease. *N Engl J Med* **351**:1709–1711.
55. Cipollone, F., Prontera, C., Pini, B., et al. 2001. Overexpression of functionally coupled cyclooxygenase-2 and prostaglandin E synthase in symptomatic atherosclerotic plaques as a basis of prostaglandin E(2)-dependent plaque instability. *Circulation* **104**:921–927.
56. Mukherjee, D., Nissen, S.E., and Topol, E.J. 2001. Risk of cardiovascular events associated with selective COX-2 inhibitors. *JAMA* **286**:954–959.
57. Schonbeck, U., Sukhova, G.K., Graber, P., Coulter, S., and Libby, P. 1999. Augmented expression of cyclooxygenase-2 in human atherosclerotic lesions. *Am J Pathol* **155**:1281–1291.
58. Wu, K.K. 2006. Transcription-based COX-2 inhibition: a therapeutic strategy. *Thromb Haemost* **96**:417–422.
59. Cieslik, K., Zhu, Y., and Wu, K.K. 2002. Salicylate suppresses macrophage nitric-oxide synthase-2 and cyclo-oxygenase-2 expression by inhibiting CCAAT/enhancer-binding protein-beta binding via a common signaling pathway. *J Biol Chem* **277**:49304–49310.

SUGGESTED READINGS

Libby, P. 2002. Inflammation in atherosclerosis. *Nature* **420**:868–874.

Liou, J.Y., Lee, S., Ghelani, D., Matijevic-Aleksic, N., and Wu, K.K. 2006. Protection of endothelial survival by peroxisome proliferator-activated receptor-delta mediated 14–3–3 upregulation. *Arterioscler Thromb Vasc Biol* **26**:1481–1487.

Wu, K.K. 2006. Transcription-based COX-2 inhibition: a therapeutic strategy. *Thromb Haemost* **96**:417–422.

Wu, K.K., Liou, J.Y., and Cieslik, K. 2005. Transcriptional control of COX-2 via C/EBPbeta. *Arterioscler Thromb Vasc Biol* **25**:679–685.

27 Models of Acute Inflammation – Air-Pouch, Peritonitis, and Ischemia-Reperfusion

André L.F. Sampaio, Neil Dufton, and Mauro Perretti

IMPORTANCE OF STUDYING CELL TRAFFICKING IN VIVO

The innate response has long been compartmentalized into several facets traditionally termed redness, heat, pain, and edema. Inflammatory insults induce release of a plethora of tightly regulated intra- and extracellular mediators, rapidly modulating the local microenvironment. The initial "humoral" response is largely a nongenomic reaction conducted by constitutive proteins and metabolic products released by resident cells (depending on the site of insult), for example, histamine, tumor necrosis factor alpha (TNF-α), and metabolites of the arachidonic acid cascade. These nonspecific or "classical" signals concurrently increasing vascular dilation, permeability, and blood flow to allow the exudation of fluid and protein. Consequentially, a variety of systemic mediators, cytokines, chemokine, and centrally acting molecules are produced to orchestrate cellular infiltration, local activation of blood-borne cells, and dispose of the inflammogen; ultimately, resolution of inflammation occurs in a time-dependent and space-regulated fashion, resulting in restoration of tissue integrity. When these symptoms persist, through deregulation or repeated insult, chronic inflammatory conditions (such as rheumatoid arthritis [RA]) can occur resulting in loss of function locally as well as effecting systemic pathology.

Among this multitude of molecular and cellular processes, migration of blood white cells to the site of inflammation is central to the overall response orchestrated by the host. Cellular migration is a complex process employing both the innate and adaptive arms of the immune system to confront and resolve inflammatory insult. Circulating leukocytes exist as transiently activated cells, constantly interacting with vascular walls via interaction of adhesion molecules (see Chapter 17). In an inflammatory response nonspecific "danger signals" are released often as a result of pattern recognition, complement activation, cellular necrosis, and so on leading to the classic cellular cascade of activation, chemotaxis, adhesion, and transmigration.

This chapter is aimed at illustrating experimental models suitable to dissect the properties of mediators responsible for leukocyte activation and migration. A number of inflammatory models have been developed to investigate the intricate nature of the inflammatory process with both theoretical and clinical relevance. Here we will discuss three models ranging from a relatively simplistic, air-pouch model of cell migration (we often refer to it as in vivo chemotaxis) to the complex "classical" processes observed following peritonitis (inflammation provoked by an external stimulus) and ischemia-reperfusion (inflammation from within).

AIR-POUCH MODEL

The air-pouch model is inexpensive and not technically demanding, requiring two injections of air into the dorsal intrascapular region to generate a discrete pouch. The nature of the air-pouch makes it particularly suitable for use in rodents.

The air-pouch was first developed as an in vivo model representative of an inflamed synovium. Subcutaneous injection of air (day 0 and day 3) leads to the formation of a *lining tissue*, a very thin layer of 2–3 cells where a combination of macrophage- and fibroblast-like resident cells coexists. Application of acute stimuli leads to a predominant influx into the air-pouch of polymorphonuclear leukocytes, whereas large doses of inflammogens or nondissolving xenobiotics produce a predominant infiltration of mononuclear (chronic) cells (Figure 27.1). When proper stimuli are used and the inflammatory reaction becomes chronic, the air-pouch lining becomes progressively thicker and shows clearest resemblance to

Day 0 and 3 inject 2.5 mL of air by intradermal (i.d) injection into the dorsal interscapular region

By day 6 the air-pouch lining shows features similar to synovium. These features can be exacerbated with repeated injections of air and extended time courses up to 30 days

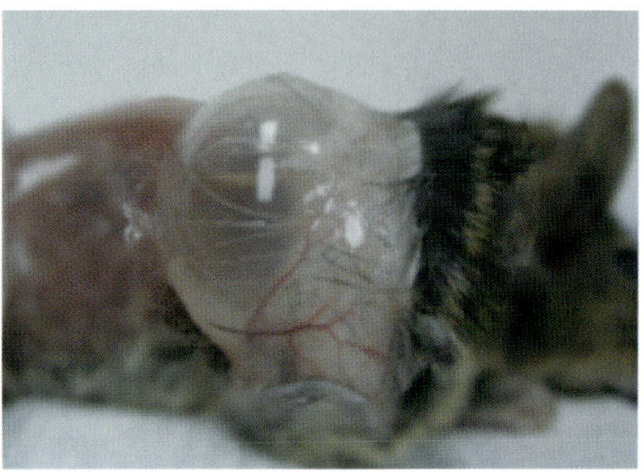

Characteristic macroscopic view of a 6-day air pouch membrane displaying the developed mirovasculature

Figure 27.1. How to make an air-pouch (schematic) and how a *real* one is. Top panel: representation of the procedure required to produce an air-pouch; two injections of air on day 0 and day 3 will form a pouch (in essence an in vivo *test tube*). Bottom panel: a photograph of the microvasculature present in the lining tissue (inside "skin") of the air-pouch, following inflammation with 1 mg zymosan (tissue sample collected at 4 h post-zymosan injection) (*note: air-pouch was reinflated to highlight its structure and vasculature*).

articular synovium around day 14, before becoming more fibrous toward day 30. These features offer both a simplistic and robust microenvironment that allows great flexibility for experimental design.

Cytokine and Chemokine Stimulation

The air-pouch can be useful when translating in vitro observations as a primary in vivo screen particularly effective when comparing stimulus-dependent responses in an acute or chronic setting. The simplistic nature of the air-pouch model has allowed it to be widely adapted to incorporate a variety of specific and nonspecific stimuli (Table 27.1). Initial models of leukocyte recruitment instigated by classical inflammatory stimuli were refined in the early 1990s by the improvements of recombinant technology, allowing administration of sufficient doses of cytokines and chemokines to produce trafficking of white blood cells from the vessels into the pouch volume.

Interleukin 1β (IL-1β) was the first cytokine to be introduced locally into the air-pouch as a potent neutrophil chemoattractant: it produces a rapid acute response, occurring within 2–6 hours, and the air-pouch is now a well-characterized model of neutrophil recruitment. Further studies have highlighted IL-8 and TNF-α as other stimulants able to inducing a similar primary response with a propensity for neutrophil accumulation within 6 hours. The rapid, robust, and reproducible responses produced by these cytokines can allow dissection of molecular mechanisms in in vivo settings, for instance determining the impact of downstream mediators and adhesion molecules.

In some cases, for example, CC chemokines, the air-pouch is not suitable to produce a marked response. Application of human recombinant eotaxin failed to

TABLE 27.1. Inflammatory agents that can trigger specific or mixed leukocyte accumulation into the air-pouch model

Stimulus	Target leukocyte	Dose/ pouch	To learn more read the following
IL-1β	Neutrophil	1–100 ng	Perretti et al., *Agents Actions*, **38**, C64, 1993
TNF-α	Neutrophil	0.1–10 ng	Tessier et al., *J Immunol*, **159**, 3595, 1997
IL-8	Neutrophil	0.01–10 μg	Perretti et al., *Br J Pharmacol*, **112**, 801, 1994
Eotaxin	Eosinophil	200–800 ng	Das et al., *J Leukoc Biol*, **64**, 156, 1998
Zymosan	Neutrophils Monocyte Lymphocyte	0.5%–2%	Dawson et al., *Agents Actions*, **38**, 255, 1983
Carrageenan	Neutrophils Monocyte Lymphocyte	0.5%–2%	Sedgwick et al., *Agents Actions*, **18**, 429, 1986
Granulomatous: Croton oil	Neutrophils Monocyte Lymphocyte	0.1%–1% in complete Freund's adjuvant	Ventrelli et al., *Boll Soc Ital Biol Sper*, **68**, 641, 1992

induce accumulation of eosinophils; a response was obtained when circulating eosinophils were increased by immunizing mice to an antigen (ovalbumin) and when an excess of mast cells was applied to the air-pouch before eotaxin (Table 27.1).

Protocols

The protocol outlined refers to volumes and treatments optimized for a mouse model; however, the methodology is applicable to rats, though proportions do vary between species.

Air-Pouch Formation

- Age and weight matched mice should be between 6 and 10 weeks old or >22 g in weight for this procedure. Males are often the most applicable sex for inflammatory models.
- Induction of the air-pouch is undertaken with light anesthetic (e.g., isoflurane), inserting a 25-gauge needle – connected to a 2.5 mL syringe – in the dorsal intrascapular region, delivering subcutaneously a volume of 2.5–3 mL air.
- It is particularly important to generate a discrete pouch, care should be taken to prevent dispersion of the air toward the head and forearms by directing the air injected with the syringe toward the dorsum with the thumb and finger.
- The pouch is reinflated on day 3 in a similar manner. Ensure the second injection is within the same pouch created on day 0. It is imperative that a secondary and distinct air-pouch does not form, as this will invalidate the model.
- The pouch is ready for experimentation from day 6; as described earlier, by this time there is formation of a lining layer (below the inner side of the skin raised by creation of the air-pouch) which expresses some characteristics similar to those of a synovial structure.

Induction of Inflammation

- An inflammatory reaction can be induced by local injection of a specific chemoattractant or adjuvant of choice depending on the complexity of the response interested to generate. Doses should be calculated for a volume of 0.5 mL inflammogen per each air-pouch, following the practical dose ranges reported in Table 27.1.
- Inject 0.5 mL inflammogen solution into the air-pouch in animals under light anesthesia, using a 1-mL syringe and a 21-gauge needle.

Washing the Air-Pouch

- Cells that migrate to the air-pouch can be recovered by washing the air-pouch with lavage buffer (see recipe in Supporting Methods).
- Kill the animals using a humane procedure, carbon dioxide is preferred to avoid any local tissue damage.
- Inject 2-mL lavage buffer into the top of the pouch with a syringe and 20-gauge needle. Gently massage the pouch before recovering the lavage fluid back into the syringe. If the air pouch has been correctly induced it is possible to completely recover the entire 2-mL of buffer injected.
- Place the samples into polypropylene tubes and keep on ice.
- In some cases it may be indicated to open the air-pouch producing a small incision with a scissor, inject 2-mL of lavage fluid, and then recover the

washing fluid with a 3-mL plastic Pasteur pipette. In these cases, the tissue of the air-pouch can be collected and used for further analyses (e.g., histology or immunohistochemistry of the lining layers).

Analysis

There are three main quantitative analytical methods that are advisable following inflammatory cell harvest:

1. Cell counting using nuclear staining with Turk's solution allows differential counts to be assessed using Neubauer chamber.
2. Flow cytometry analysis following staining of cells with specific monoclonal antibodies conjugated to fluorescent molecules. For example, Gr1 stains for neutrophils and F4/80 stains for monocytes and macrophages.
3. Finally, cell pellets and supernatants can be separated by centrifugation and used for molecular-based as well as biochemical analyses.

We will now elaborate on method 1 where total or differential cell counts are obtained. It is important to consider a suitable dilution of the lavage fluids collected in relation to the strength of the cell trafficking response elicited. In most cases, mixing 0.1 mL of cell exudate with 0.4 or 0.9 mL of Turk's solution in a 1.5-mL Eppendorf tube is recommended. Samples must be vortexed to ensure homogeneous cellular suspension prior to dilution with Turk's. Then, vortex diluted samples before placing 10 μL into the Neubauer chamber, counting each sample in duplicate.

The Neubauer chamber calculates the number of cells (n) that would be present in 1 mL multiplied by a factor of 10^4. The sum obtained can then be multiplied by five or ten, depending on the dilution factor used for Turk's staining, to give the number of cells/mL in the exudate.

To express data as the total number of cells per air-pouch and hence per mouse, the sum obtained earlier must be multiplied by the volume used to wash the air-pouch. These calculations are applicable to individual cell populations counted since, as stated earlier, Turk's allows to distinguish between polymorphonuclear leukocytes and mononuclear cells.

In summary,

Number of cells per mL of lavage fluids recovered from the washed air-pouch:

$$\text{Cells/mL} = (n \times 10^4) \times 5$$

Number of cells migrated into the air-pouch in response to the inflammogen applied:

$$\text{Total cells/pouch} = (\text{Cells/mL}) \times 2$$

MODEL OF EXPERIMENTAL PERITONITIS

Peritonitis is a clinical feature observed in patients undergoing long-term peritoneal dialysis or in some postsurgery complications after peritoneal surgery. Experimental peritonitis can help the understanding of the mechanisms of inflammation in that particular cavity; in some cases, the local inflammatory reaction in the peritoneum can rapidly disseminate and can lead to a systemic syndrome associated with high morbidity.

Experimental peritonitis is widely used as an inflammatory model for drug screening and testing. Depending on the triggering agent used, specific inflammatory mediators, enzymes, or receptors can contribute to the leukocyte recruitment process, facilitating the evaluation of the effect of compounds of interest. More recently, this self-resolving model of inflammation has been used to study the mechanisms and molecules that contribute for the resolution of inflammation.

Rodents are usually the species of choice for peritonitis induction; however, the use of larger animals has been reported. The peritoneal cavity, as with the air-pouch, is suitable model for injection of different flogistic agents capable of mediating the recruitment of specific leukocyte subsets (Table 27.1). However, the peritoneal cavity is probably best employed studying multifaceted inflammation using nonspecific inflammogens. For this reason, we will focus on zymosan-induced peritonitis, which resembles the opportunistic infection of the peritoneal cavity produced by microorganisms in immunocompromised patients.

Zymosan Peritonitis

Zymosan is prepared from bakers' yeast (*Saccharomyces cerevisiae*) and is used to trigger acute inflammation. Zymosan particles are opsonized upon contact with blood proteins and activate the complement cascade via the classical pathway that will generate the C5a fragment that will contribute for acute neutrophil accumulation. Opsonized zymosan particles are rapidly phagocytosed by resident macrophages, via the C3 receptor (Mac-1; CD11b/CD18) and Dectin, triggering activation and consequent cytokine release, especially CXCL1 (or KC) and IL-1β, contributing to the first wave of neutrophil accumulation.

The inflammatory response observed after peritoneal injection of zymosan, with doses ranging from 0.1 to 10 mg per cavity, induces a large influx of blood-borne leukocytes, such as monocytes, monocyte-derived macrophages, and neutrophils, masterminding the innate immune response. Figure 27.2 presents profiles of cell influx as measured across a 48-hour time course.

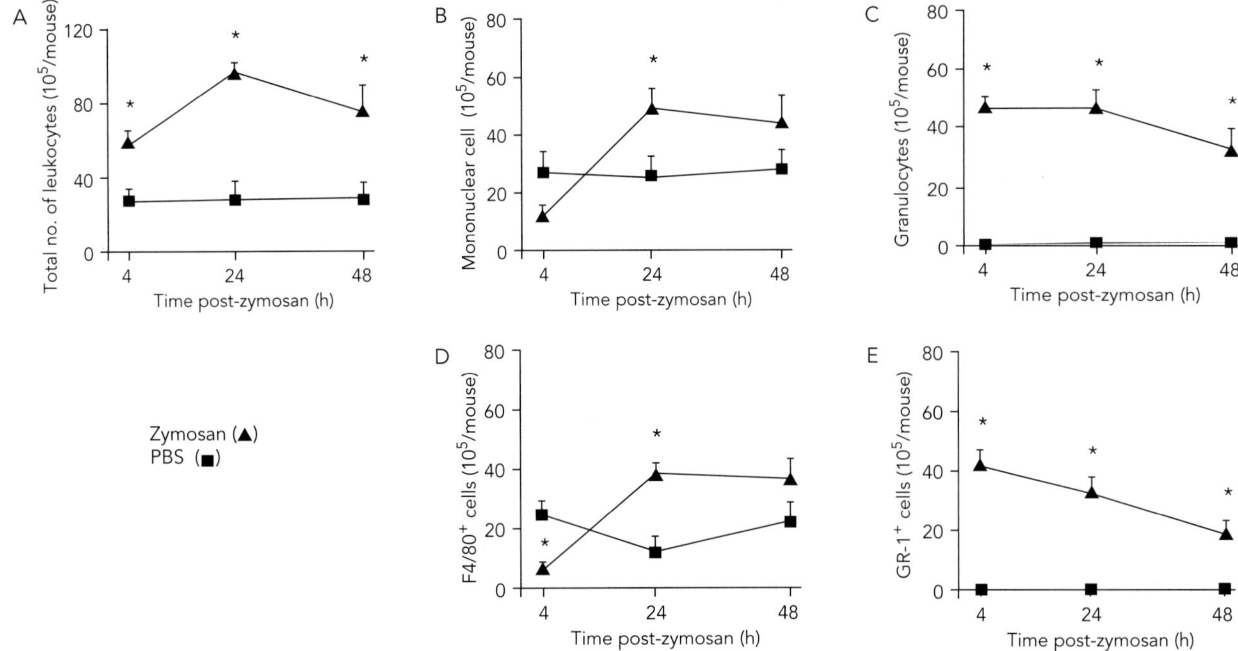

Figure 27.2. Time-course of zymosan peritonitis. Representative time course of white blood cell recruitment into the peritoneal cavity inflamed with zymosan. Mice (six per group) received an intraperitoneal injection of 1 mg zymosan (▲) or 0.5 mL PBS (■) at time 0, and humanely killed at the reported time point: leukocyte accumulation was evaluated as described in the text. * $p < 0.05$ when compared to vehicle-injected (PBS) mice.

Preparation of Zymosan

Zymosan is a component of yeast wall, and it is insoluble. Originally zymosan was prepared from live bakers' yeast by simply boiling it, followed by several washes. Ready prepared and dried zymosan can be bought from different commercial suppliers; however, the boiling step (30 minutes, using a large flask and a small volume of water, e.g., 100 mg zymosan in 200 mL of water in a 2.5 L flask), and the multiple wash steps (again, use large amounts of water and recover the insoluble zymosan by centrifugation at 1,000 rpm for 20 minutes at 4°C) are still necessary. Zymosan preparations (stored as aliquots at –20°C) obtained after boiling and washing will produce responses that are more consistent across time.

Induction of Peritonitis

Zymosan injection will provoke peritonitis in both mice and rats; commonly doses of 1 mg per mouse cavity and 2–3 mg per rat cavity are given. On the day of the experiment, remove an aliquot of zymosan from the freezer and leave to defrost at room temperature. Resuspend to a concentration suitable for injecting a volume of 0.5–1 mL per mouse cavity or 1–5 mL per rat cavity.

Zymosan particles can decant in the tube were the suspension was prepared or in the syringe that is used for the injection; hence, it is important to keep a homogeneous suspension by shaking (simple, steady inversion of the tube three or four times) when filling the syringe prior to injection. A useful tip is to fill the syringe with volume sufficient for two animals to minimize sedimentation of zymosan in the syringe.

Inject the lower part of the abdominal cavity, taking care to avoid injuring the liver to minimize a potential hemorrhage that will interfere with the evaluation of leukocyte recruitment.

Harvesting of Recruited Cells

After a determined time period, animals are killed and the skin is separated from the abdomen wall with the help of forceps via a careful incision; ensure the abdominal wall is not perforated. After the abdomen is exposed up to 2–3 or 6–10 mL of ice-cold PBS, for mice or rats, respectively, is injected with the help of a syringe with 21-gauge needle. The abdomen is gently massaged to facilitate the complete washing of the cavity and recovery of recruited leukocytes. It is highly advisable to use heparin, EDTA, or the combination of both to avoid leukocytes to aggregate (see wash buffer recipe). The recovery of the wash fluid can be performed via an incision and by the use of a plastic Pasteur pipette. If sterile conditions are required, lavage fluid can be recovered by using a syringe without opening the cavity, as described earlier for aseptic washing of the air-pouch. For best practice, access to cell culture hood will ensure the most likely sterile conditions.

Migrated cells can be determined by light microscopy after staining in Turk's, with differential cell counts (distinguishing between polymorphonuclear and mononuclear cells) also attained by light microscopy analyses. For more detailed identification of the migrated cell populations, peritoneal cells harvested from the cavities can be stained with specific antibodies and analyzed by flow cytometry. Figure 27.2D and 27.2E show profiles of influx of F4/80+ cells – for monocytes/macrophages – and GR-1+ cells – for granulocytes. The latter marker can also pick up a monocyte population with intermediate level of expression, so it is advisable to quantify Gr1+ high cells as granulocytes.

Finally, zymosan peritonitis can also be used in combination with intravital microscopy to observe the dynamics of leukocyte rolling, adhesion, and transmigration into the mesentery tissue.

Summary of Protocol for Zymosan Peritonitis in the Mouse

- Injection of zymosan (1 mg/cavity), in a final volume of 500 μL pyrogen-free PBS, with control animals receiving an identical volume of PBS.
- At the desired time point for analysis, use a humane procedure to terminate the animals.
- Expose of the abdominal cavity via an incision in the skin, without opening the cavity.
- Detach the skin from the abdominal wall with the help of forceps.
- Injection of 3 mL of ice-cold wash buffer with a syringe equipped with a 21-gauge needle.
- Perform a gentle massage of the abdomen (for 10–20 seconds).
- Holding the abdominal wall with forceps, make a small incision and carefully insert a Pasteur pipette for lavage fluid collection.
- Transfer the collected fluid to a 15-mL tube on ice.
- Repeat the collection step until the fluid has been completely harvested. Note that usually the volume recovered is not greater than 2.8 mL.
- With a help of a 5-mL syringe, the volume of the fluid collected can be accurately measured.
- Take a 100-μL aliquot and dilute in 900 μL of Turk's solution for total leukocyte count in a Neubauer chamber (see recipe and procedures).
- The remaining wash fluid can be centrifuged for collection of the cell-free lavage fluid for the determination of soluble mediators and/or cells can be used for cellular analysis by flow cytometry, western blotting, RNA isolation, or other second step technique.

ISCHEMIA-REPERFUSION (I/R)

The effects of deprivation of blood supply in tissues have been the aim of studies from several years. Cardiovascular diseases that are caused by ischemic episodes followed by reperfusion are characterized by high morbidity and mortality in developed countries. The ischemia/reperfusion (I/R) induced microvascular dysfunction has been described in most organs and is recognized as a problem in several medical and surgical procedures, including organ transplantation, angioplasty, and cardiopulmonary bypass. Recently, some light has been put on this field by studies on the pre- and postconditioning of tissues to reperfusion, opening new venues for the prevention of deleterious effects of reperfusion so that emerging new questions can be answered.

Ischemia-reperfusion injury of the intestine is a significant problem in abdominal aortic aneurism surgery, small bowel transplantation, cardiopulmonary bypass, and neonatal necrotizing enterocolitis. The cellular components of the intestine are very susceptible to episodes of ischemia and further reperfusion results in increased damage to the mucosa.

Ischemia-Reperfusion: Inflammation from Within

The postischemic inflammatory reaction activated in a reperfused vessel is observed not only within that specific microvascular bed, but can also be observed at remote sites as a consequence of systemic activation, that is, release of inflammatory mediators in the circulation. These remote and systemic responses can lead to the development of systemic inflammatory response syndrome (SIRS) or the multiple organ dysfunction syndrome (MODS).

The I/R injury is associated with the generation of a cascade of multiple cytokines/chemokines, other inflammatory mediators, and expression of adhesion molecules on endothelial and stromal cells. In this context, increased levels of TNF-α in the early phase of I/R, play an important role in the formation of the injury and contribute to neutrophil accumulation; when high levels of systemic TNF-α are generated, it can provoke lethality as seen in some models of intestinal I/R.

This crucial control of I/R by an endogenous anti-inflammatory pathway has been suggested in other models of injury including stroke and heart infarct. In fact, it is clear that, at variance from inflammatory reactions provoked by xenobiotics and other external agents (e.g., infections) where a prompt, robust, effective, and time-dependent response is life-saving, I/R activates the microcirculation provoking a *response from within*. There must be molecular sensors of ischemia, or of I/R, that alert the body to activate

Mesenteric microcirculation as observed in sham operated mice

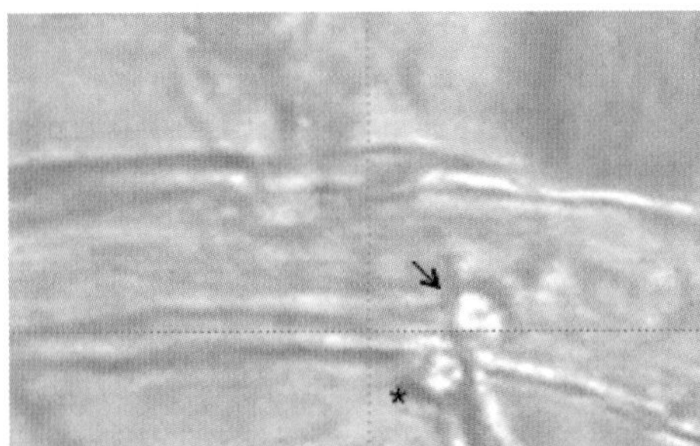

Mesenteric microcirculation as observed after 25 min ischemia and 90 min reperfusion

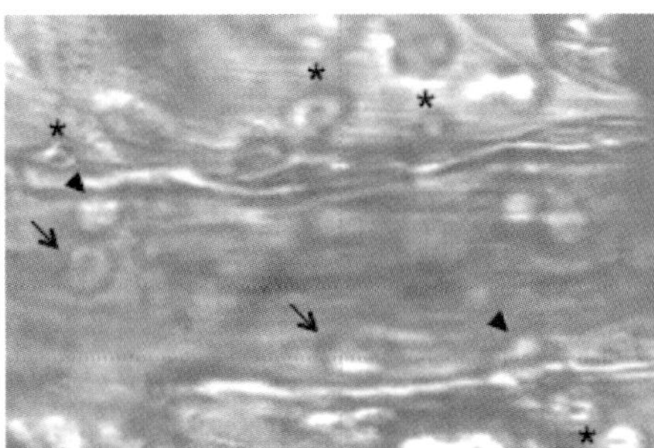

Figure 27.3. Pictures of sham (noninflamed) and postreperfusion (inflamed) vasculature. Representative images of intravital microscopy showing postcapillary venules of the mesenteric microcirculation. Left panel: Sham-operated mouse, with arrow indicating a rolling cell, and asterisk an emigrated one. Right panel: Inflammation of the microvessel after a 25-minute ischemia (achieved by closure of the superior mesenteric artery) and 90-minute reperfusion. Signs of cell/endothelial interactions are evident, with many rolling (some indicated with an arrow), adherent (indicated with an arrowhead), and emigrated (asterisk) white blood cells. Note the couple of leukocytes squeezing through the endothelial wall (top section, asterisk to the right and asterisk to the left).

rapidly the endogenous anti-inflammatory arm of the response: in models of I/R injury, pharmacological activation of endogenous anti-inflammatory mechanisms by the means of annexin-1, melanocortins peptides and lipoxins exert a potent protective role, providing strong rationale for exploiting these anti-inflammatory and tissue-protective pathways for novel pharmacological intervention and therapeutic development, likely identifying the receptors responsible for bringing about these protective properties.

Cell/Endothelium Interaction in Reperfusion

During I/R injury the inflammatory cytokine cascade is activated; this activates the vessel endothelium as well as the circulating leukocytes so that the processes of cell rolling, adhesion, and emigration can be triggered, leading to tissue infiltration by extravasated neutrophils. However, this leukocyte accumulation and activation is thought to account for some of the signs of injury left in the tissue.

A reliable protocol to produce significant neutrophil accumulation in the mesenteric microvasculature can be observed after 35 minutes of ischemia followed by 90 minutes of reperfusion. Intravital microscopy analysis of the mesenteric microvasculature demonstrates an increased extent of white blood cells interacting with the vessel, hence rolling and adhering, before extravasating. Figure 27.3 illustrates cell interaction in the mesenteric microcirculation after I/R, as compared to sham-operated animals. Chapter 17 illustrates some of the molecular and cellular events that must occur to promote and sustain these intercellular interactions.

Protocol for I/R of the Mouse Mesentery

- Age and weight matched mice should be between 2–4 weeks old or >15 g in weight for this procedure. Males are often the most applicable sex for inflammatory models.
- Mice were anesthetized intraperitoneally (i.p.) with a mixture of 7.5 mg/kg xylazine and 150 mg/kg ketamine hydrochloride.
- The abdominal cavity is opened by a small incision and the gut with the mesenteric vascular bed exteriorized.
- The mesenteric artery is occluded by the use of a surgical micro clip (micro-aneurysm clip Harvard Apparatus; Kent, UK). Alternatively, the artery can be occluded using a silk surgical thread.
- Sham-operated animals will have the mesentery exteriorized but the artery clamping step will be skipped.
- The gut should be placed back inside the cavity and a piece of gauze soaked in saline (0.9% NaCl) can be placed over the incision to avoid the gut from drying.
- During this procedure the levels of anesthesia have to be checked and, if necessary, a boost has to be administered.
- After the ischemia period (35 minutes), the clip is removed and reperfusion is allowed for a required time (90 minutes in the example shown in Figure 27.3).

(Alternative combinations of ischemia and reperfusion can be used; in some protocols, ischemia can be as short as 15 minutes and reperfusion as long as 24 hours. In the protocols were a longer period of reperfusion

occurs, the surgical interventions have to be modified to enable animal recovery.)

- At the given time-point of reperfusion, animals are killed; tissue and blood samples collected for cytokine/chemokine determinations, myeloperoxidase (MPO) analysis (*see supporting material*), western blotting for protein expression, PCR and other techniques for gene expression.
- Though it may vary the aim of the project, the best practice is to infuse the mouse with at least 30 mL of PBS, via a butterfly in the heart and an incision in the vena cava, to flush the tissue and remove any trace of blood in the mesenteric vasculature, therefore analyzing the parameters described earlier and other biochemical markers in clean tissue samples without the "interference" of residual blood and leukocytes.

SUPPORTING INFORMATION

Differential Cell Counting

Turks staining allows distinctions between leukocytes in view of their nucleus morphology. Neutrophils are easily identifiable by their polymorphonuclear (PMN) structure. Monocytes/macrophages are apparent by their kidney-shaped nucleus and larger cytoplasm. Mast cells will occasionally be observed in cellular exudates as fully stained large cells, very deep blue-purple in coloration, and highly granular. Finally, lymphocytes are distinguished by their spherical nucleus and small cytoplasm.

Samples should be vortexed prior to dilution for differential counts. It is important to consider a suitable dilution of cell exudates depending on the extent of the inflammatory response. As an example, a 1:10 dilution is often a good starting point, hence 0.1 mL of cell exudate are mixed with 0.9 mL Turk's solution in a 1.5 mL Eppendorff tube. Vortex diluted samples before placing 10 μL into the Neubauer chamber.

Measurement of Myeloperoxidase Activity

Leukocyte MPO activity is assessed by measuring the H_2O_2 dependent oxidation of 3,3',5,5'-tetramethylbenzidine (TMB). Mesenteric (or any other) tissue samples from sham and I/R animals are homogenized in ice-cold PBS containing 0.5% hexadacyl trimethylammonium bromide (HTAB) and MOPS (10 mM). The homogenate is centrifuged at 13,000×g for 5 minutes. In a 96-well plate, 20 μL homogenate supernatant is plated together with 160 μL of tetramethylbenzidine (TMB; 0.5 mg/mL) and 20 μL of H_2O_2 (0.1 mM). Optical density is read at 620 nm and assessed against a standard curve constructed with human recombinant MPO (0.031–1 U/mL) or known numbers of PMN pellets (10^4–10^6 cells). Therefore, readings for the unknown samples are interpolated onto the standard curve; the obtained values (e.g., U/mL or cells equivalent/mL) would need to be normalized accordingly to protein content (use Bradford assay). The final data are then expressed as units of MPO per mg of protein.

Note: Samples homogenized in ice-cold PBS containing 0.5% hexadacyl trimethylammonium bromide (HTAB) and MOPS (10 mM) are suitable for cytokine/chemokine determination using standard ELISA methods.

Carboxymethylcellulose Preparation

When the air-pouch model is employed for cytokines and chemokines, it is advisable to use carboxymethylcellulose (CMC) as solute. At low concentrations, CMC provokes a mild inflammatory reaction that would help magnifying the effect of the cytokine. Therefore, for a 0.5% w/v CMC solution, sprinkle CMC onto a beaker (again, use 200 mL beaker for 50-mL solution) containing sterile pyrogen-free PBS under constant stirring at room temperature. Within few minutes, after the addition of CMC, under stirring, a clear solution is obtained. To ensure that no lumps are present, the CMC solution is centrifuged at 400×g for 10 minutes. This solution can be kept on ice for addition of inflammatory stimuli, and injection to animals.

Washing Buffer

The lavage buffer used to harvest cells from air-pouches or peritoneal cavities consists of sterile PBS supplemented with 3 mM EDTA and 25 U/mL heparin to aid detachment of cells adhering to the lining tissues.

Turk's Solution

Turk's solution consists of 0.1% crystal violet dissolved in 3% acetic acid prepared in distilled water. The acidic solution will lyse erythrocytes, facilitating the counting of white blood cells.

SUGGESTED READINGS

Ajuebor, M.N., Das, A.M., Virag, L., Flower, R.J., Szabó, C., and Perretti, M. 1999. Role of resident peritoneal macrophages and mast cells in chemokine production and neutrophil migration in acute inflammation: evidence for an inhibitory loop involving endogenous IL-10. *J Immunol* **162**:1685–1691.

Bannenberg, G.L., Chiang, N., Ariel, A., et al. 2005. Molecular circuits of resolution: formation and actions of resolvins and protectins. *J Immunol* **174**:4345–4355.

Granger, D.N. 1999. Ischemia-reperfusion: mechanisms of microvascular dysfunction and the influence of risk factors for cardiovascular disease. *Microcirculation* **6**:167–178.

Edwards, J.C., Sedgwick, A.D., and Willoughby, D.A. 1981. The formation of a structure with the features of synovial lining by subcutaneous injection of air: an *in vivo* tissue culture system. *J Pathol* **134**:147–156.

Isaji, M., and Naito, J. 1992. Comparative studies on inflammatory reactions induced by non-immunological and immunological stimuli in an air pouch and in a carboxymethyl cellulose (CMC)-induced inflammatory pouch. *Int J Exp Pathol* **73**:231–239.

Lundy, S.R., Dowling, R.L., Stevens, T.M., Kerr, J.S., Mackin, W.M., and Gans, K.R. 1990. Kinetics of phospholipase A2, arachidonic acid, and eicosanoid appearance in mouse zymosan peritonitis. *J Immunol* **144**:2671–2677.

Perretti, M., and Flower, R.J. 1993. Modulation of IL-1-induced neutrophil migration by dexamethasone and lipocortin 1. *J Immunol* **150**:992–999.

Perretti, M., Harris, J.G., and Flower, R.J. 1994. A role for endogenous histamine in interleukin-8-induced neutrophil infiltration into mouse air-pouch: investigation of the modulatory action of systemic and local dexamethasone. *Br J Pharmacol* **112**:801–808.

Sedgwick, A.D., and Lees, P. 1986. Studies of eicosanoid production in the air pouch model of synovial inflammation. *Agents Actions* **18**:429–438.

Sedgwick, A.D., Moore, A.R., Al-Duaij, A.Y., Edwards, J.C., and Willoughby, D.A. 1985. The immune response to pertussis in the 6-day air pouch: a model of chronic synovitis. *Br J Exp Pathol* **66**(4):455–464.

Willoughby, D.A., Sedgwick, A.D., Giroud, J.P., Al-Duaij, A.Y., and de Brito, F. 1986. The use of the air pouch to study experimental synovitis and cartilage breakdown. *Biomed Pharmacother* **40**:45–49.

28A Experimental Models of Glomerulonephritis

Aidan Ryan, Denise M. Sadlier, and Catherine Godson

INTRODUCTION

In the human kidney, some 1 million glomeruli filter 180 liters of plasma daily, allowing passage of low-molecular-weight products while restricting the passage of albumin and larger macromolecules. The resulting urine is extensively modified in the renal tubular system, with changes to both composition and volume necessary to maintain extracellular volume and homeostasis. As a consequence of this, immune complexes formed in the circulation are delivered at a high rate to the intraglomerular capillary bed and trapping occurs primarily in the mesangium and or on the subendothelial surface of the capillary wall. In contrast to in vitro models, which are somewhat limited to assessing isolated cell, antibody, and antigen function, or indeed, human biopsy specimens which give a snapshot at a particular clinical stage, in vivo animal models can outline how structure and function changes with initiation, progression, and potential regression within affected organs and the various cellular and humoral factors involved in disease progression. The nephron is the functioning unit of the kidney and the glomerulus is a branching network of capillaries responsible for plasma filtration and the initial step in urine formation. This chapter will review the use of experimental animal models in delineating the pathogenesis of glomerulonephritis (GN). GN at its basic definition is the term used to describe an inflammatory process involving the glomeruli characterized morphologically by an influx of leucocytes and cellular proliferation often accompanied by glomerular capillary wall abnormalities. Patients with GN frequently develop chronic kidney disease and although the exact percentage varies depending on sample population, it represents a significant fraction of the etiology of end-stage kidney disease (ESKD). Although the lesions at the capillary give rise to the early clinical manifestation, it is the downstream effect of this disease process (namely nephron replacement with tubulointerstitial fibrosis) that is the main determinant of the decline in glomerular filtration rate (GFR) and development of ESKD. Initially, we will describe the functional structure of the glomerulus, and the various types of animal models of GN commonly in use together with some of the limitations associated with them. We will subsequently describe how these experimental models have helped further understanding of GN as a disease process, under the broad headings of the mediators involved, structural change in the glomerulus, and animal models used to reflect specific types of glomerular pathology.

NORMAL GLOMERULAR STRUCTURE AND FUNCTION

Developments in the area of glomerular biology have shed light on the functional role of the various structural components of the glomerular barrier and how abnormalities in these structures may result in disease. The glomerulus is a branching network of capillaries originating at the afferent arteriole and draining into the efferent arteriole. Lining this capillary bed, on the inner surface of the glomerular basement membrane (GBM) is a thin layer of fenestrated (diameter 70–100 nm) endothelial cells. External to this, covering the urinary surface of the GBM, are the podocytes/epithelial cells with foot processes with intervening slit diaphragms. The GBM between these two layers consists of a dense central layer (the lamina densa) and the thinner peripheral layers (lamina interna and rara) that consist of type IV collagen, laminin, fibronectin, enactin, and proteoglycans such as heparin sulfate. (Figure 28A.1 shows a section of normal rat kidney.) The glomerular barrier depends on the following factors: the GBM, the endothelium, the epithelium (podocytes), a strong negative charge (due to the presence of glycoproteins in the endothelium), net glomerular

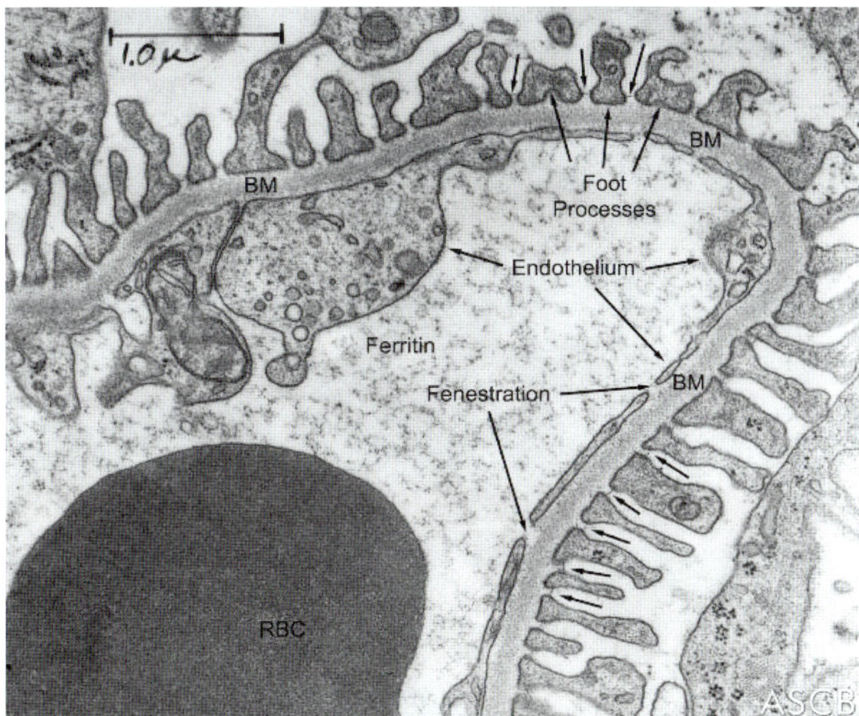

Figure 28A.1. Structure of the rat glomerulus. Peripheral area of normal rat kidney. The capillary wall is composed of three distinct layers: the endothelium, its fenestra, the basement membrane, and foot processes of the renal epithelium. In a number of places (arrows), a slit membrane can be seen bridging the narrow gap between foot processes. The vascular lumen contains a red blood cell and ferritin granules. (©2006 MG Farquhar. All rights reserved. Reprinted under license from The American Society for Cell Biology. Farquhar, M.G. Glomerular capillary of normal kidney. ASCB Image & video Library. August 2006; SFC-4. Available at http://cellimages.ascb.org.)

filtration pressure, and molecule size (less than 50 Å) [1]. Insight into the role played by each component has been greatly enhanced by models that allow selective alteration of the glomerular structure; however, it is worth noting the importance of not only structural defects but also changes in glomerular hemodynamics that contribute to altered permeability [2]. The glomerular barrier is highly size- and charge-selective, in that it markedly restricts the passage of large anionic proteins. The exact location of this barrier function is still the subject of debate but most would agree that the above structures function as a dynamic integrated unit with damage to any portion leading to severe proteinuria [3]. The dynamic nature of this interaction has been shown in experiments using murine models of vascular endothelial growth factor-A (VEGF-A) knockdown and knockout models. These models have shown that podocyte-derived VEGF is necessary for endothelial cell integrity and mesangial cell differentiation [4]. A consideration of normal glomerular anatomy provides a useful insight into how various types of GN are classified pathologically and present clinically. Diseases affecting subendothelial and mesangial cells are readily accessible to circulating inflammatory cells, and cause severe damage, with endothelial cell injury, activation of procoagulant cascades, fibrin deposition, and exudative lesions leading to hematuria, proteinuria. Subepithelial diseases, however, are typified by flattening of the foot processes that engage the basement membrane, resulting in glomerular barrier disruption and proteinuria. Due to location, subepithelial diseases are separated by the GBM from the circulation, and are characterized by complement-mediated podocyte injury with a lack of inflammatory reaction [5].

OVERVIEW OF EXPERIMENTAL ANIMAL MODELS

Mammalian kidney development has helped to elucidate the general concepts of epithelial–mesenchymal interactions, inductive signaling, and epithelial cell polarization. Cell culture studies using nephrons from various species have demonstrated a similar genetic expression of patterns that can be related to time scales in morphogenesis [6]. Thus animal studies provide critical information that aid our understanding of renal development and offer insights into the pathogenesis of human renal disease although the kidneys of rats and mice are smaller, are unipapillate, have fewer urinary tubules, smaller urinary space but similar

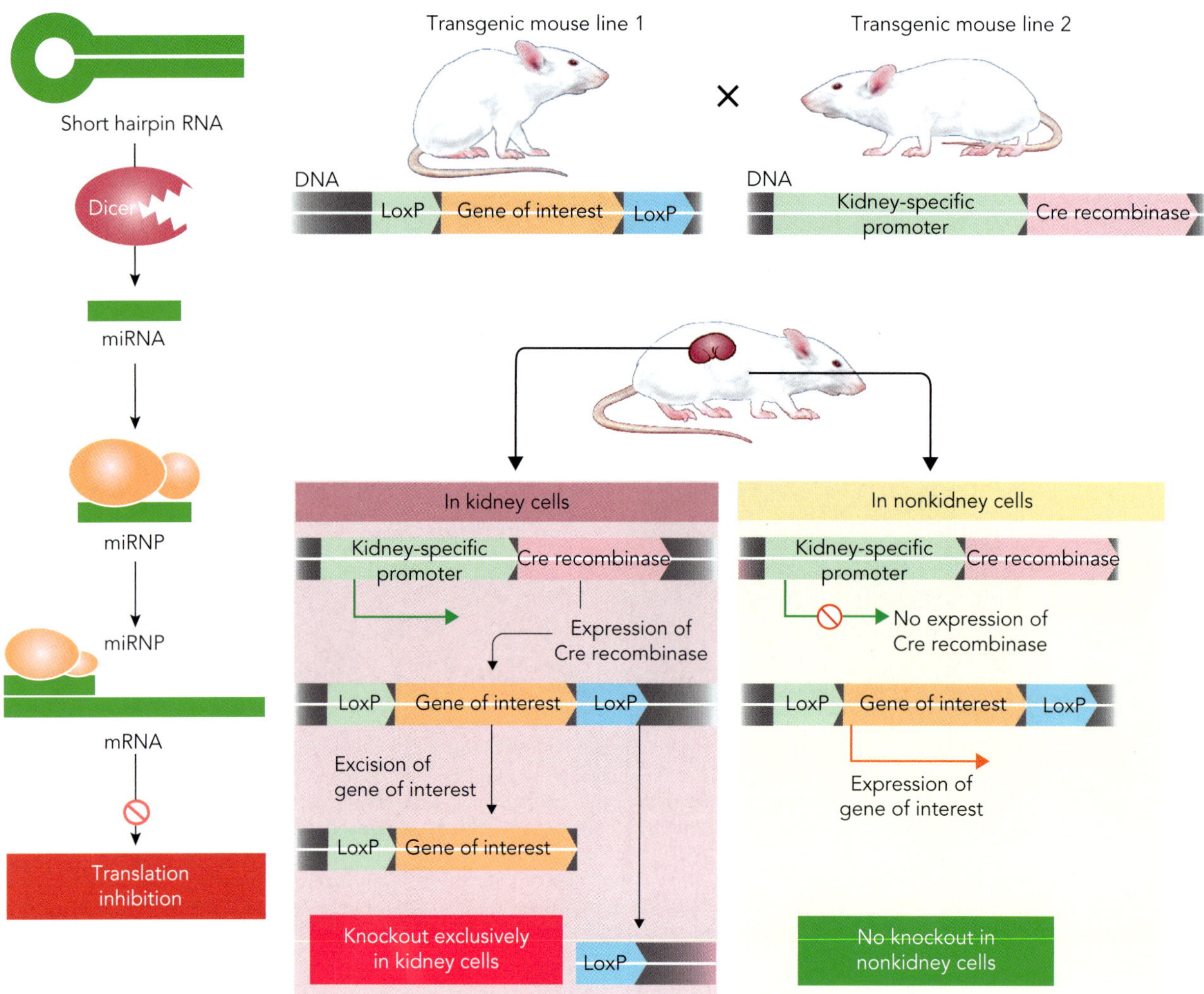

Figure 28A.2. *(continued).*

pelvis and similar medullary-cortex volume ratios [7]. The most commonly used species for experimental animal models of GN are the rat and the mouse. Use of *Xenopus*, chick embryos, and *Drosophila* have enabled assessment of isolated gene function, with identification of human renal disease genes and elucidation of pathways of renal development that have remained conserved through evolution [8,9]. The rat offers ease of handling due to size, lower maintenance requirements, and as a consequence has been a preferred choice for surgical and physiological procedures. However, the choice of model and ability to control function at the genetic level is far greater with mice rather than rats. The development in the early 1980s of transgenic mice along with subsequent embryonic stem cells has allowed a more accurate assessment of gene function through knockout, knockin, and site-specific recombinases, which allow both spatio and temporal-specific control over gene expression. Use of this technology makes it possible to assess the effect of certain genes by removing the gene completely or transiently altering its expression and assessing the effects of point mutations, respectively [10] (see Figure 28A.2).

More recently, technology has developed which allows for control of expression at the level of RNA using short-hairpin RNA (shRNA) and micro-RNA (mi-RNA) that allow interference with translation (see Figure 28A.2), affording even greater flexibility than use of gene recombinases, in that they offer the ability to reduce gene expression by varying degrees. This may overcome certain adverse consequences of gene knockout models. It is likely that this technology will complement genetic recombination rather than replace it, since it is presently not possible to induce point mutation, or completely turn off gene expression. It remains unknown whether this effect is permanent [11]. Other strategies that attenuate gene expression by interfering with cytosolic mRNA (message RNA) or translated protein include DNA enzymes, antisense oligonucleotides, decoys, ribozymes, aptamers, and

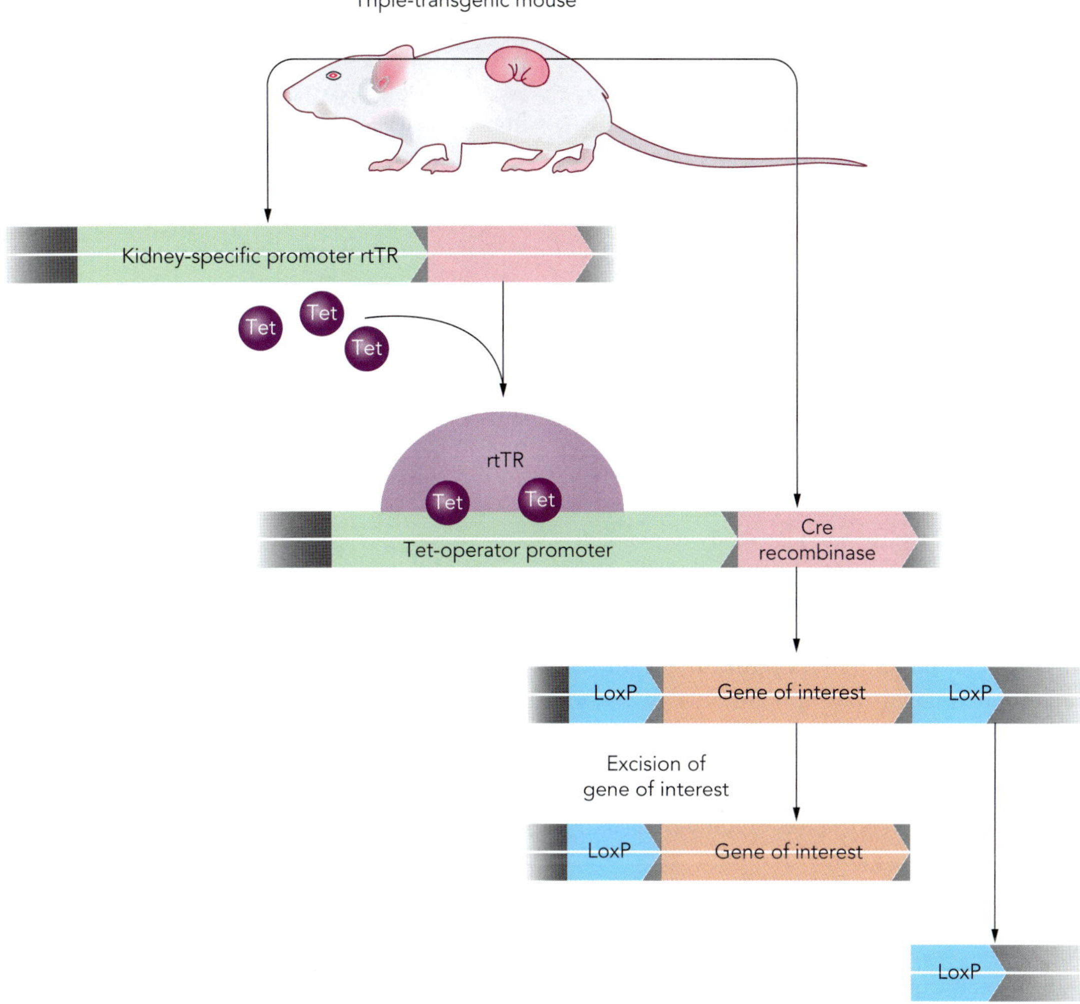

Figure 28A.2. Overview of methods for interfering with gene expression, including RNAi. (Used with permission and adapted from Gawlik A, Quaggin SE. *Physiology* 19:245–252, 2004.)

alternative splicing and these are reviewed elsewhere [12,13].

From the point of view of recreating the exact pathophysiological conditions that mimic human pathologies, some obstacles remain in both of the above species [14]. However, the strength of the animal experimental model lies in the delineation of the pathogenesis subsequent to this process, which we will now discuss further.

ELEMENTS INVOLVED

The pathogenesis of GN is complex and multifactorial and is not completely understood. Despite this complexity, it is apparent that the histopathological changes that occur in different types of GN have overlapping similarities in the elements involved such as leukocyte infiltration, complement activation, antibody deposition, changes in resident glomerular cells in terms of number size and phenotype, and in the composition of the extracellular matrix. A hallmark of GN is the presence of leucocytes, particularly macrophages and T lymphocytes, within both the glomerulus and the interstitium. Recruitment of leucocytes from the circulation depends on chemokines and increased renal injury has been shown to arise with increasing levels of the chemokine monocyte chemotactic protein 1 [15]. Activated glomerular macrophages produce a variety of proinflammatory mediators, promote fibrin and extracellular matrix deposition, cause local tissue damage, and as we will see further on, induce proinflammatory responses of resident renal cells. It is, however, clear that this macrophage population is heterogeneous, two distinct subpopulations designated M1 and M2 have been proposed to distinguish the classical and alternative activation states, respectively. The former following the traditional cytotoxic, proinflammatory route and the latter involved in increased processes that are associated with promoting the resolution of inflammation [16]. The clinical significance and potential for phenotypic switching between these two subtypes is a focus of intense investigation. A central role for T cells in GN has been shown in many experimental models, mediated by cytokine secretion, direct cytotoxic effects, activating macrophages, B cell, and antibody induction [17]. Recent experimentation has shown that not only is T-cell activation important but that differential activation of T helper cell subsets may help explain the variety of patterns of injury in GN [18], with Th1 subset associated with crescentic GN and Th2 subset with a membranous pattern. Within the kidney itself, both cell types undergo activation mediated by proinflammatory cytokines, secreted infiltrating, and resident renal cells. Experimental studies blocking these cytokines such as IL-1, TNF-α, and IFN-γ have shown some protection [19] from injury mediated by direct cell damage such as apoptosis and indirect through antibody-dependent cell-meditated cytotoxicity. The next step in the process of leukocyte infiltration is mediated by adhesion molecules of which there are three major classes, selectins, integrins, and the immunoglobulin superfamily. Experiments have shown that upregulation of these molecules is associated with increased leukocyte infiltration of the glomerulus and proteinuria in certain GN models. One such example showed that the differential effects of different subclasses of antiglomerular basement antibodies on glomerular infiltration and proteinuria in Wistar Kyoto rats was due to levels of lymphocyte function associated antigen-1 (LFA-1) and intercellular adhesion molecule (ICAM-1), to which the above were directly proportionate [20]. Neutrophils (PMN) are a major protective effector of the innate immune system, they are short-lived and respond to signals from injured tissues and the presence of common foreign proteins through Toll-like receptors (TLRs). In GN, they typically participate in association with other leukocytes such as macrophages and T cells as part of an integrated immune response. As might be predicted, chemokine and adhesion molecules associated with PMN invasion rise and peak during the early stages of the disease process. Most models that have studied the role of PMN in GN relate to antineutrophil cytoplasmic antibody (ANCA)-associated crescentic GN (see later). Experimental evidence including models involving depletion and transfer of PMN has shown the critical role of PMN [21]. Intravital microscopy has confirmed that ANCA can induce neutrophil localization and degranulation in glomeruli in vivo [22]. The complement pathway plays an important role in the development of GN and consists approximately of 30 plasma and membrane-bound proteins, consisting of three different pathways (classical, lectin, and alternative) that converge in the activation of the central complement molecule C3. Complement has been known for some time to play an important role in lupus nephritis (LN), but in a paradoxical manner and the use of accurate mouse models has shed some light as to the mechanisms behind this paradox. Through the use of genetically altered mice and recombinant protein inhibitors it has been demonstrated that proteins C1q and C4 are protective whereas complement activation later in the pathways is deleterious [23]. Antibody-mediated cell damage occurs via several mechanisms which include complement activation, antibody-dependent cell-mediated cytotoxicity, and leukocyte activation via Fc-gamma. At the center of antibody production is the B cell, which plays an important role in antigen presentation, control of T-cell migration and activation, and production of regulatory cytokines. Animal models of GN that have shown the importance of antibody-mediated tissue damage have included the NZB/W model of LN, whereby use of mice deficient in Fc-gamma receptors are resistant to the development of GN [24] and evidence for the pathological role of the ANCA in small vessel vasculitis by upregulation of endothelial cell expression of adhesion molecules [25]. As outlined earlier, animal models have provided insight into the functions of glomerular structures. This evidence suggests that far from being bystanders intrinsic renal cells actively participate in the inflammatory process in GN. A classic experiment in this context has investigated the source of the proinflammatory cytokine TNF-α in crescentic GN: TNF-α chimeric mice were created by transplanting normal wild-type bone marrow into irradiated TNF-α-deficient recipients and vice versa. These results showed that those mice that only had bone marrow–derived TNF-α had significantly decreased crescents and better serum creatinine, indicating that the intrinsic renal cells are the major source of TNF-α contributing to inflammatory injury in crescentic GN [26]. Other sources of evidence showing the importance of resident renal cells in the pathogenesis of GN have demonstrated both proliferation and activation of resident fibroblasts and transition of epithelial cells to a mesenchymal-like phenotype, coupled to matrix accumulation leading to the development of fibrosis that may result in ESKD [27–29]. The role of TLRs have been shown to be an essential component of innate immunity providing defense against microorganisms but also leading to the induction of signals that control the activation of adaptive responses including autoimmune responses and allorecognition. In connection with this, recent experiments on progression of LN in MRLlpr/lpr mice have shown that TLR9 via CpG oligonucleotide stimulation leads to an increase in anti-DNA autoantibodies, proteinuria, progression from mild to crescentic GN and increased interstitial fibrosis [30], and inhibition of TLR7 and TLR9 in lupus-prone mice has shown decrease in autoantibodies, proteinuria, and decreased GN [31]. Providing further evidence of the importance in overlap of the innate and the adaptive immune system, TLR4 has been shown in mouse models of cryoglobinemic membranoproliferative glomerulonephritis (MPGN) to be constitutively expressed by podocytes and is upregulated in MPGN where it may modulate expression of proinflammatory chemokines that result in deposition of immune complexes [32].

STRUCTURAL CHANGES

Insight into the role of the various components of the glomerular barrier has come from murine models with selective podocyte damage and deletion of several molecules including α-3 chain of type IV collagen and nephrin [33–35]. Podocyte damage has been implicated as having a central role in the development of sclerosis in most forms of glomerular disease [36,37]. The sequence of anatomic events resulting from podocyte injury has been described in multiple animal model systems [38]. The results of podocyte damage include: increased intraglomerular pressure, denuded GBM areas, adhesion to Bowman's capsule, focal segmental and global glomerular sclerosis that is associated with misdirected filtration into the interstitial compartment contributing to interstital fibrosis and injury. The response of the podocyte to injury is to undergo hypertrophy and to a limited extent, cell division followed by apoptosis and podocyte depletion. Depending on the stage of glomerular development and associated environmental factors, podocyte dysfunction, injury, or loss can result in a broad spectrum of clinical syndromes, which include congenital nephrotic syndrome of the Finnish type, immune and inflammatory GN [39]. The major components of the GBM are type IV collagen, laminin, enactin, and sulfated proteoglycans [40] with the former consisting of α-3, 4, and 5 chains. Creation of a mouse with an α-3 chain knockout results in severe GBM defects with intact podocyte processes and by the fifth week, these mice develop proteinuria. This α-3 knockout is a model for autosomal recessive Alport's syndrome in which a genetic mutation results in the COL4A3 gene leads to thin basement membrane nephropathy [41]. Nephrin is a major structural component of the slit diaphragm connecting podocyte foot processes in the glomerular wall [42] and it is encoded by the NPHS1 gene. Mutations in this gene have been found to be responsible for the rare genetic disorder termed congenital nephrotic syndrome of the Finnish type. Mouse models that have targeted nephrin that result in slit-diaphragm defects have shown that massive proteinuria may be induced without obvious podocyte effacement or endothelial abnormalities [43]. Another protein that binds to nephrin is podocin, which is encoded by NPHS2 and podocin-null mice have been shown to develop severe proteinuria and die a few days after birth from renal failure [44]. It is believed that podocin may act to stabilize nephrin and facilitate cell signaling. The glomerular endothelium is highly specialized in order to facilitate the formation of a protein-free ultrafiltrate, the function of which is highly dependent on fenestrae to maintain hydraulic conductivity and a glycocalyx that contributes toward a charge and size selective barrier [45]. Disruption of this barrier has been shown in several animal studies to result in proteinuria but its importance in GN rests not alone with its structural properties but also with endothelial damage facilitating the development of a proinflammatory state [46]. This effect has been shown to occur not only through increased leukocyte recruitment but also through upregulation of adhesion molecules and complement activation, which as we will discuss later, play a key role in GN pathogenesis. Although the initial disease manifestation in GN arises as a consequence of glomerular involvement, it is the onset of tubulointerstitial fibrosis that leads to a decline in GFR and thus the onset of chronic kidney disease. Few studies have dealt with direct investigation of the pathways involved in how a nephron undergoes irreversible degeneration during a specific kidney disease. Progression of kidney disease arises as a result of nephron loss, which leads to compensatory mechanisms including glomerular hypertension, hyperfiltration, hypertrophy, and an increase in proteinuric glomerular leaking leads to an increase in tubulointerstitial inflammation, which leads to further tissue damage and fibrosis [47]. A correlation exists between the severity of glomerular injury and the level of excreted protein and the rate of disease progression; however, this may be further accounted by glomerular damage that directly encroaches upon the tubule leading to nephron degeneration [48]. Renal fibrosis is characterized by an exaggerated accumulation of fibrous collagen, by an increased incidence of interstitial cells in the renal interstitium and by the emergence of myofibroblast. The source of these fibroblasts has been in a variety of accelerated models of renal fibrosis such as 5/6 nephrectomy, unilateral ureteric obstruction model which have shown contributions from bone marrow, resident fibroblasts, and EMT [49]. It can be appreciated therefore that with progressive chronic kidney disease, with degeneration of the nephron mass, that the interstitial damage will dominate over the glomerular damage score due to the decrease in glomeruli number. More critical still is the fact that there is now evidence to show that decline in renal function in chronic kidney disease correlates best with remaining number of nephrons [50].

SPECIFIC DISEASE MODELS

In this section, we will outline the more commonly used animal models of specific types of GN and for

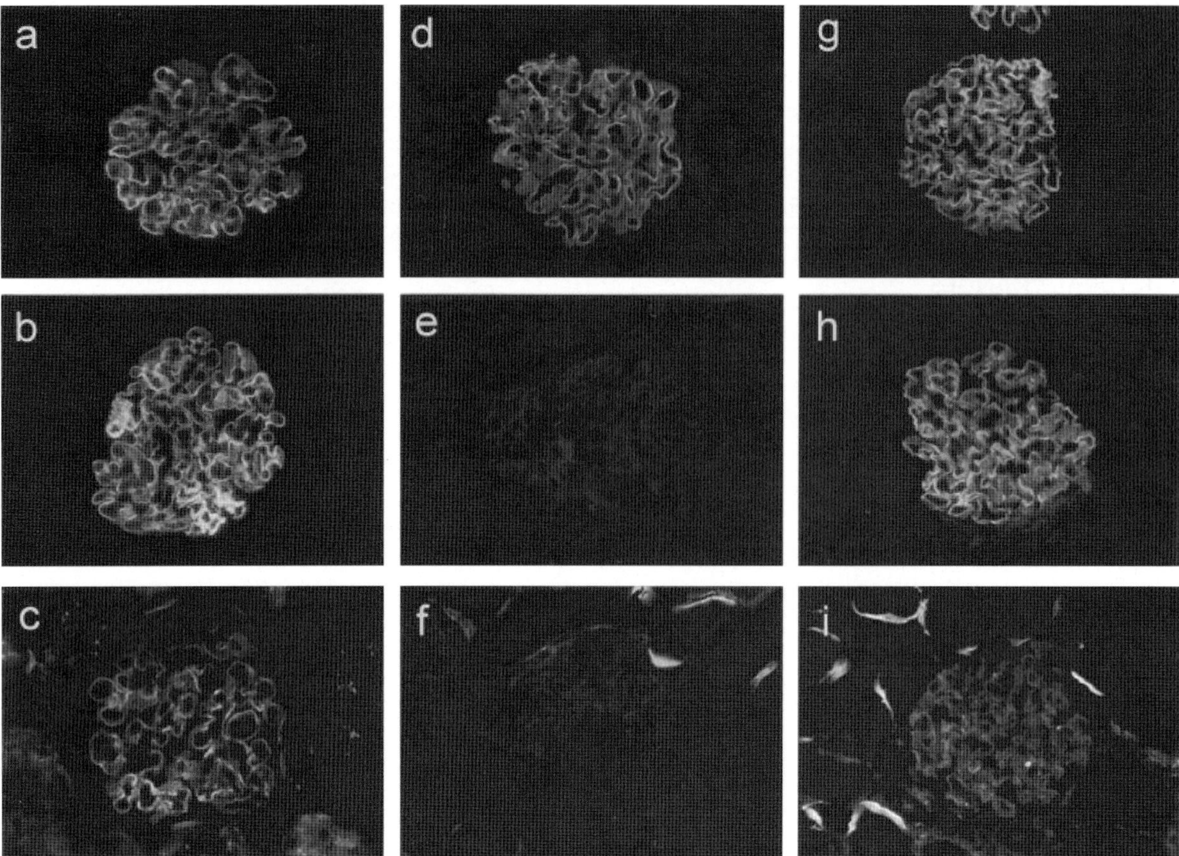

Figure 28A.3. Immunofluorescence staining of a rat model of anti-GBM disease. Immunofluorescence staining of glomerular deposition of humoral immune reactants in rat anti-GBM disease. Anti-GBM disease induced in a control animal shows (a) strong linear deposition of sheep anti-GBM antibody, (b) linear deposition of rat IgG, and (c) patchy linear deposition of rat C3. Anti-GBM disease induced in a cyclophosphamide (CyPh) treated animal shows (d) strong linear deposition of sheep anti-GBM antibody, (e) absence of rat IgG, and (f) absence of rat C3. Anti-GBM disease induced in a CyPh treated animal given passive transfer of rat anti-sheep IgG serum shows (g) strong linear deposition of sheep anti-GBM antibody, (h) linear deposition of rat IgG, and (i) patchy linear deposition of rat C3. Original magnification, 400×. (Adapted by permission from Macmillan Publishers Ltd: Kidney International: Ikezumi, Y., et al. 2003. Adoptive transfer studies demonstrate the macrophages can induce proteinuria and mesangial cell proliferation. *Kidney Int* 63:83–95.)

the sake of clarity we have decided to restrict our discussion in this section to rapidly progressive glomerulonephritis (RPGN), where accurate models of human diseases have been fully characterized. A comprehensive table outlines other types of GN and the most frequently used animal models. RPGN is a clinical syndrome characterized by a sudden decline in kidney function often accompanied by oliguria or anuria with features of GN including dysmorhic erythrocyturia, glomerular proteinuria, and crescents on renal biopsy which may result in ESKD if untreated. Several conditions can lead to this syndrome. These conditions are classified according to biopsy findings [51]: (1) anti-GBM antibodies; (2) granular immune complex deposition in the kidney; and (3) minimal or Pauci-immune deposition. Goodpasture syndrome consists of a triad of proliferative crescentic GN, pulmonary hemorrhage, and the presence of anti-GBM antibodies [52]. The classic anti-GBM model (termed experimental autoimmune GN) involves immunization with complete Freund's adjuvant and the antigen, which is the non-collagenous domain of the α-3 chain of type IV collagen [53]. This leads to crescentic GN in association with deposition and circulating anti-GBM antibodies (Figure 28A.3). The most susceptible strain is the Wistar-Kyoto rat in which crescentic GN is invariably reproduced using the above conditions [54]. Work undertaken using this model has shown that in addition to the disease being transferred using antibodies; it can also be transferred with T cells derived from immunized rats [55]. This emphasizes the importance of both humoral and cellular mediated processes but also provides a rationale for therapies that target both. The disease has also been induced in mice using a similar process in C57B/6 mice; however, the disease is less severe and less reproducible, but still shows the same dependence on both cellular and humeral processes [56,57].

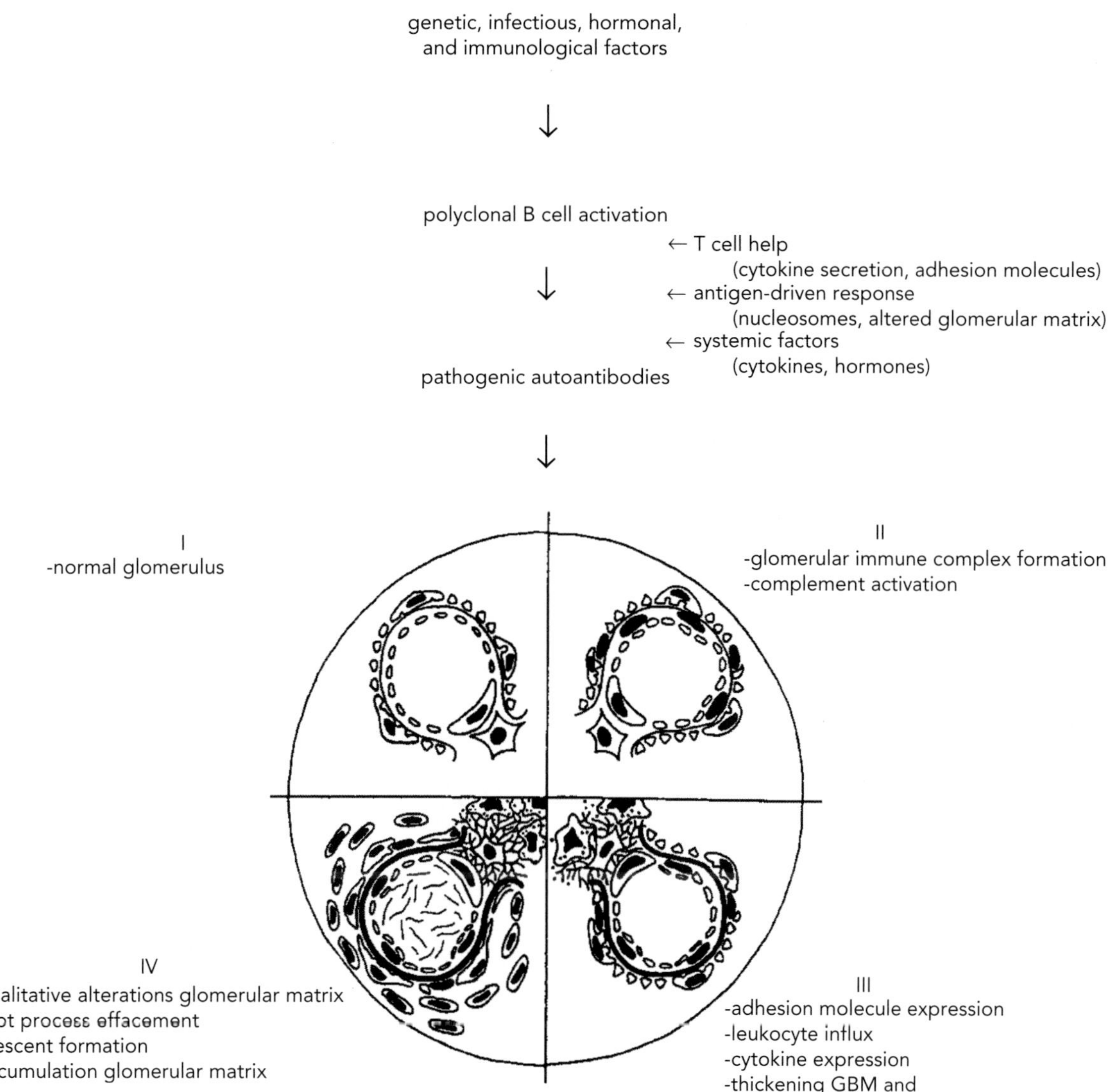

Figure 28A.4. Schematic representation of a hypothetical pathway for the development of experimental lupus nephritis.

SLE is a clinical syndrome with protean manifestations that are defined by the American College of Rheumatology [58] and its renal manifestations are defined by the WHO/ISN (World Health Organization/International Society of Nephrology) classification of biopsy findings [59]. Clinically, the disease may present with a spectrum from mild microscopic hematuria or proteinuria to nephrotic syndrome to progressive renal impairment and requirement for renal replacement therapy. With regard to experimental models, both rat and murine strains exist that develop a background consistent with systemic illness in humans and a reproducible model of lupus nephritis. Traditionally, the classical LN model was induced using a nephrotoxic nephritis, which was formed by colocalization of alloantibody raised against kidney antigens and the host response to the foreign antibody. This model has been used in both rats (WKY) and mice but had more reproducibility in the rat [60,61]. Developments in transgenic mice, however, have allowed creation of spontaneous models that have further insights into the pathogenesis of LN.

The most commonly used murine strains for investigation have included MRL/Mpj, MRL/MpJ-Ipr, NZB hybrids, and BXSB animals [62]. Results from experimentation on these transgenic mice have resulted in the ability to evaluate the role of candidate gene loci in disease susceptibility, adhesion molecule dysregulation in disease initiation and progression. Gene susceptibility loci that have been located in mouse and human models have included C-reactive protein/serum albumin P component, DNASE1, C1q, Fc-gamma R2A/2A, and PDCD1 [63,64]. Many of the products of these genes have been shown to be involved in clearing apoptotic cells. Knockout mouse models affecting these genes have been shown to develop spontaneous

TABLE 28A.1. Reliable experimental models of glomerulonephritis

GN type	Species	Model name	Reference
Membranous nephropathy	Rat	Heymann nephritis	[66]
Membrano-proliferative	Rat Mouse	Anti-Thy-1 nephritis, Factor-H deficient	[67,68]
Alport's syndrome	Mouse	Col4a3–/–	[69]
IgA nephropathy	Mouse	ddY	[70]
Focal Segmental Global Sclerosis	Rat Mouse	Remanent kidney model, puromycin aminonucleoside, and adriamycin neph-ropathy. Transgenic models affect-ing genes related to podocyte function	[71,72]
Human immuno-deficiency virus	Mouse	Transgenic podocin/VPR	[73]

SLE and GN. The importance of the role of adhesion molecules in GN and in promoting inflammatory cell infiltration has been outlined previously and strategies that have targeted adhesion molecules in LN, particularly MRL/Fas, have included monoclonal antibodies against adhesion epitopes, soluble adhesion molecule fragments, antisense oligonucleotides to inhibit synthesis, and promotion of anti-inflammatory networks. The effects of these strategies on ICAM, β2-integrin, LFA-1 and membrane activated complex-1 (MAC-1) has lead to improvements in serum and histological markers of severity of LN [62].

Pauci-immune GN occurs usually as part of a systemic small-vessel vasculitis or may occur as an isolated renal lesion, characterized histologically by focal necrotizing crescentic GN, with little or no glomerular staining for immunoglobulin on immunofluorescence microscopy. Before the development of a satisfactory transgenic model for P-ANCA (perinuclear-ANCA) vasculitis, the most established model involved immunization of Brown Norway rats with mercuric chloride; however, this did not reflect the clinical pattern of disease [45]. A more representative model involves adoptive transfer of splenocytes from MPO–/– (myeloperoxidase) mice immunized with MPO into RAG–/– mice which induced vasculitis and pauci-immune crescentic GN. Subsequent experimentation showing the pathological role of anti-MPO ANCA was demonstrated with transfer of antibody alone from the MPO immunized animals into wild-type animals [25] and a similar successful disease model has been shown in the WKY rat immunized with human MPO.

SUMMARY

In this brief overview of the use of experimental animal models in outlining the pathogenesis of GN, we provided an introduction to the types of animal models, followed by an evaluation of how these models have been used to show the structural changes in glomerular biology, the mediators involved and their use in specific types of GN. It is clear that compared with biopsy specimens and in vitro models animal models offer an insight from the early stage of disease initiation, pathogenesis through to potential regression. However, owing to the time commitments necessary with creating an accurate animal model it is clear that choice of the specific type of model in terms of species, size, genetic background, how reproducible the disease in that particular species are critical considerations. This overview outlines the importance of changes in renal structure and the role of the mediators involved in the pathogenesis of GN. The approach taken describes how use of animal models have provided insight into RPGN initiation and progression and where possible we have listed models of other types of GN that are the most accurate and reproducible that are currently available (see Table 28A.1).

REFERENCES

1. Haraldsson, B., Nystrom, J., and Denn, W.M. 2008. Properties of the glomerular barrier and mechanisms of proteinuria. *Physiol Rev* **88**:451–487.
2. D'Amico, G., and Bazzi, C. 2003. Pathophysiology of proteinuria. *Kidney Int* **63**:908–925.
3. Pavenstadt, H., Kriz, W., and Kretzler, M. 2003. Cell biology of the glomerular podocyte. *Physiol Rev* **83**:253–307.
4. Eremina, V., Cui, S., Gerber, H., et al. 2006. Vascular Endothelial Growth Factor A signaling in podocyte-endothelial compartment is required for mesangial cell migration and survival. *J Am Soc Nephrol* **17**:724–735.
5. Nangaku, M., and Couser, W.G. 2005. Mechanisms of immune-deposit formation and the mediation of immune renal injury. *Clin Exp Nephrol* **9**:183–191.
6. Horster, M.F., Braun, G.S., and Huber, S.M. 1999. Embryonic renal epithelia: induction, nephrogenesis and cell differentiation. *Physiol Rev* **79**:1157–1191.
7. Khan, S.R. 1997. Animal models of kidney stone formation: an analysis. *World J Urol* **15**:236–243.
8. Dressler, G.R. 2006. The cellular basis of kidney development. *Annu Rev Cell Dev Biol* **22**:509–529.

28B Glomerulonephritis and Ischemia Reperfusion injury

Jagdeep Obhrai and Jeremy S. Duffield

INTRODUCTION

The kidney receives disproportionately high blood flow, Blood flows to each renal filtering unit, known as the nephron, through a specialized vascular bed called the glomerulus, which is by design uniquely leaky to solutes and plasma proteins (Figure 28B.1). Human glomeruli filter 144 liters of plasma filtrate a day from the bloodstream (Figure 28B.2). Disruption of the glomerulus interferes with blood flow to the downstream segments of the nephron due to the presence of a second capillary network deriving from the glomerular capillaries, frequently culminating in ischemia of the nephron, with consequent chronic inflammatory responses to ischemic injury of parenchymal cells. Furthermore, the unique regulation of vascular tone in the kidney paradoxically can worsen ischemic insults by triggering vasoconstriction. Physiologically, compromised blood flow in the medulla of the kidney is a prerequisite for effective reabsorption of ≈142 liters a day of filtered solute back into the blood compartment, but renders the high-energy-requiring components of the medulla especially susceptible to ischemic injury.

The glomerulus is particularly prone to inflammatory injury, perhaps due to the tendency of immune complexes and other abnormal proteins to become lodged in the specialized filtering basement membranes, which then trigger immune responses not seen in other regions of the vasculature. Rarely autoimmunity directed at unique proteins within the specialized glomerular filtering membranes also triggers stereotyped responses in the glomerulus, which we now recognize as immune cell and parenchymal cell responses to injury. Like other organs, the kidney has a remarkable capacity to repair and regenerate itself. Whereas inflammation is a hallmark of progressive disease in the kidney, inflammation is also strikingly associated with normal repair. However, the role of inflammation in repair has received relatively little attention. We will focus on the tremendous advances that have been made in understanding the initiation and progression of inflammatory diseases focused on the glomerulus of the kidney; we will also look at the role of inflammation in normal repair that follows injury, and point to where lessons from understanding the role of inflammation in repair might help direct inflammation in a reparative direction in other diseases where progressive damage is overriding.

GLOMERULONEPHRITIS

The microscopically beautiful histological patterns of inflammation in the glomerulus combined with a paucity of understanding of the pathogenesis, has long fascinated Pathologists, Immunologists and Nephrologists. The evolution of therapies for kidney disease have mirrored those developed for oncological and hematological diseases indicating commonality of disease mechanisms. The study of injury in the glomerulus has centered around four major animal models: spontaneous nephritis in rodent models of systemic lupus erythematosus (SLE), a disease of systemic autoimmunity; induced nephritis using antibodies directed at unique molecules in the glomerular filtering membranes (glomerular basement membrane [GBM]), referred to as nephrotoxic nephritis (NTN); autoimmunity against the GBM; and injury directed at the mesenchymal cells of the glomerulus, called mesangial cells. More recently, two additional models have been developed to study deposition of circulating immune complexes in the glomerulus and to study a form of capillary injury or vasculitis of the glomerulus, which lacks immune complexes, but has circulating antibodies directed against neutrophil and monocyte (Mo) proteins. The patterns of inflammation and injury in glomerulonephritis range from very mild to severe, and are stereotyped responses to injury that can be identified in many different diseases. Mild

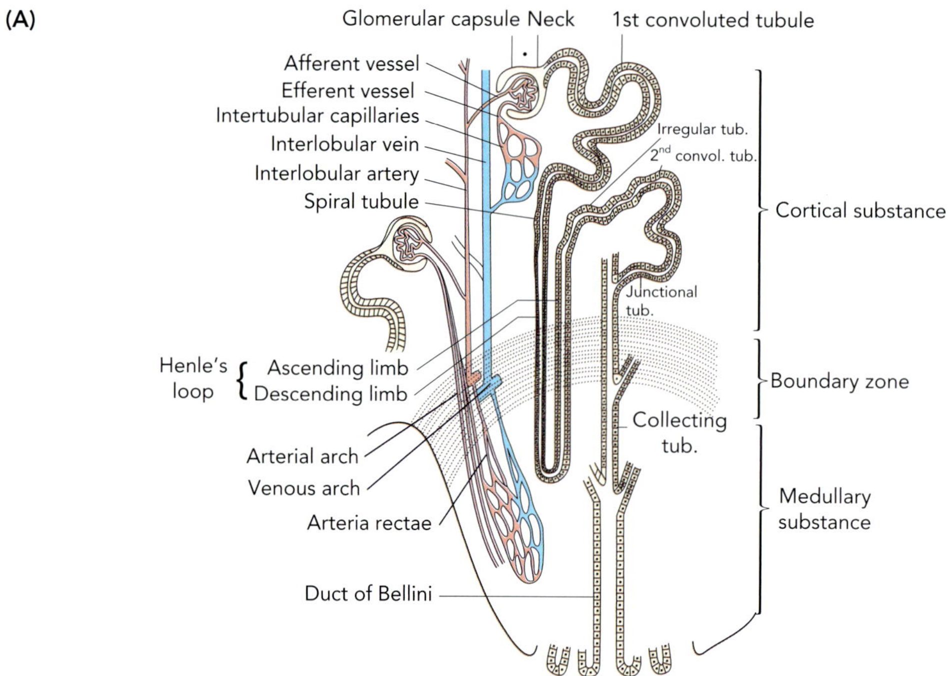

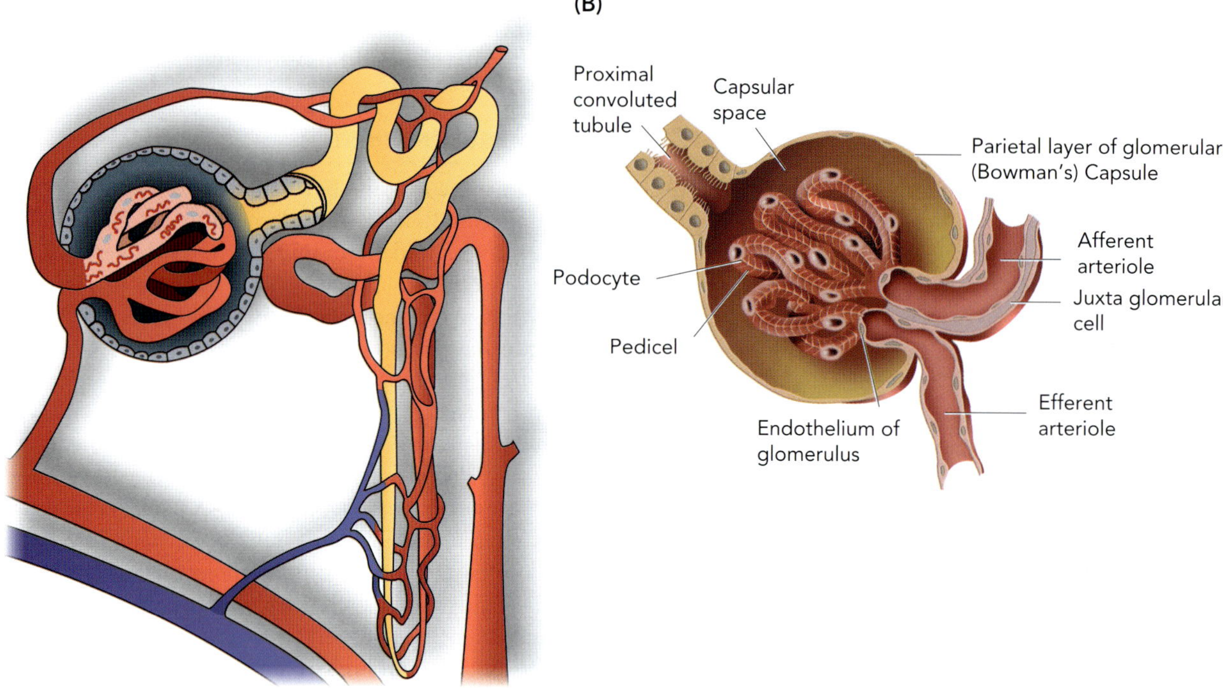

Figure 28B.1. Anatomy of the Nephron, the functional unit of the kidney, and its blood supply. (A) Diagram indicating arteriolar flow to the glomerulus, the component that filters plasma filtrate through fenestrate endothelia and a specialized endothelial basement membrane that is leaky to plasma filtrate. The glomerular capillaries recombine to form the postglomerular (efferent) arteriole, which then divides to form a second peritubular capillary network. This second network is not only the sole blood supply for the nephron but is also specialized to permit reabsorption of solutes reabsorbed from the glomerular filtrate by the tubule of the nephron. (B) High-power image of the nephron (left) and glomerulus (right) showing its vital components, and emphasising the specialized microvasculature of the kidney from the glomerulus to the peritubular capillary plexus.

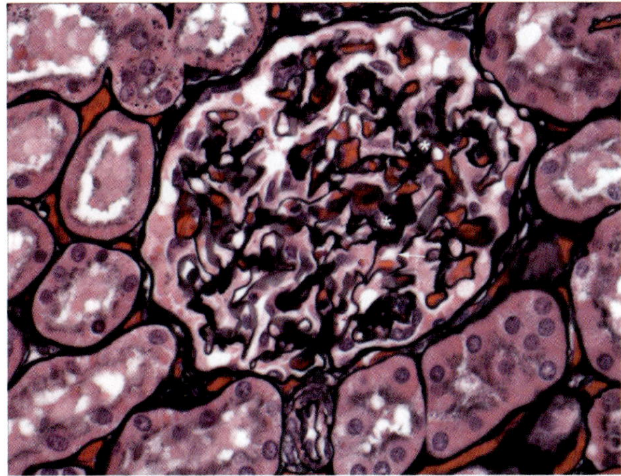

Figure 28B.2. Normal human glomerulus. Jones's Silver stain of normal human glomerulus showing glomerular basement membranes (GBM) (black), capillaries containing erythrocytes (red) and glomerular endothelial cells (arrow). Mesangial areas are also identified (*). Many podocytes can be seen attached to the GBM. The GBM is a major site of immunologically mediated leukocytes and complement activation.

inflammation is characterized by proliferation of mesenchymal mesangial cells, increased mesangial matrix, and inflammatory cells in the mesangium. More severe disease characterized by capillary loop injury with swollen endothelial cells, denuded endothelium, adherent intracapillary inflammatory cells, and changes to the highly specialized pericyte-like epithelial cells, called podocytes. The most severe disease is characterized by focal areas of necrosis or apoptosis of mesangial, endothelial cells or podocytes, associated with leukocytes, and laying down of abnormal matrix called sclerosis. In addition, podocytes are also lost, capillary loops may thrombose, GBM may be broken and direct leak of plasma into the filtering space (Bowman's) is associated with proliferation and inflammation of the overlying parietal epithelial cells, and is known as a crescent. Frequently surrounding crescents, there is periglomerular inflammation and fibrosis.

MODELS OF RENAL SYSTEMIC LUPUS ERYTHEMATOSUS

SLE is a complex systemic autoimmune disease that can affect multiple organs. Most patients with SLE demonstrate some renal abnormalities and overt disease impacts patient survival [1]. Defects in innate and adaptive immune mechanism underlie this disease. The development of autoantibodies directed at nuclear antigens is a central feature of the lupus syndrome. These antibodies target nucleic acids and their associated proteins and comprise the immune complexes that are detected in patients with SLE and mice with SLE-like nephritis [2]. Many mouse models of SLE have been developed, and many that develop autoantibodies also generate immune complex deposition. This review focuses on the few models that have severe renal disease as measured by significant changes in glomerular architecture or biochemical evidence of renal disease as part of the phenotype. These models have been critical to our understanding of the genetics and pathogenesis of the syndrome and have served as platforms of discovery for diagnostics and therapeutics.

TABLE 28B.1. Models for SLE-related renal disease

Spontaneous SLE-like strains	New Zealand strains MRL/Mp and related strains BXSB
Genetically modified SLE-like strains	
T cell activation	CD45-deficiency [2a] Gadd45a-deficiency [2b]
B cell activation	Lyn-deficiency [2c] BAFF transgenic [2d] CD40L transgenic [2e]
Antigen presenting cells	Dendritic cells [2f]
Complement	C1q-deficiency [71] Serum amyloid P-deficiency [2g]
Cytokines	Type I interferons [2h] Interferon γ and IL-4 [2i]

Disease Models

Renal disease in lupus is mostly studied using mice that [1] spontaneously develop disease, or [2] have been genetically designed to develop disease by breeding (e.g., congenic strains), or by targeted genetic manipulation. Drug-induced models are also available but are less commonly employed [3,4] (Table 28B.1). The most important distinction between human SLE and murine models is that, in humans, renal disease frequently varies with time, whereas all of the mouse models of SLE show progression of disease with aging of the mouse.

Among spontaneous SLE-like models, three have frequently been employed: New Zealand hybrids (NZ), MLR, and BXSB. All three strains develop a complex disease pattern, with renal and extrarenal disease. The histopathology of the renal disease varies among these strains of SLE-prone mice, but all three develop glomerulonephritis (GN) and tubulointerstitial abnormalities. These abnormalities are associated with antinuclear antibodies, anti-double-stranded (ds)DNA antibodies, and immune complexes [5].

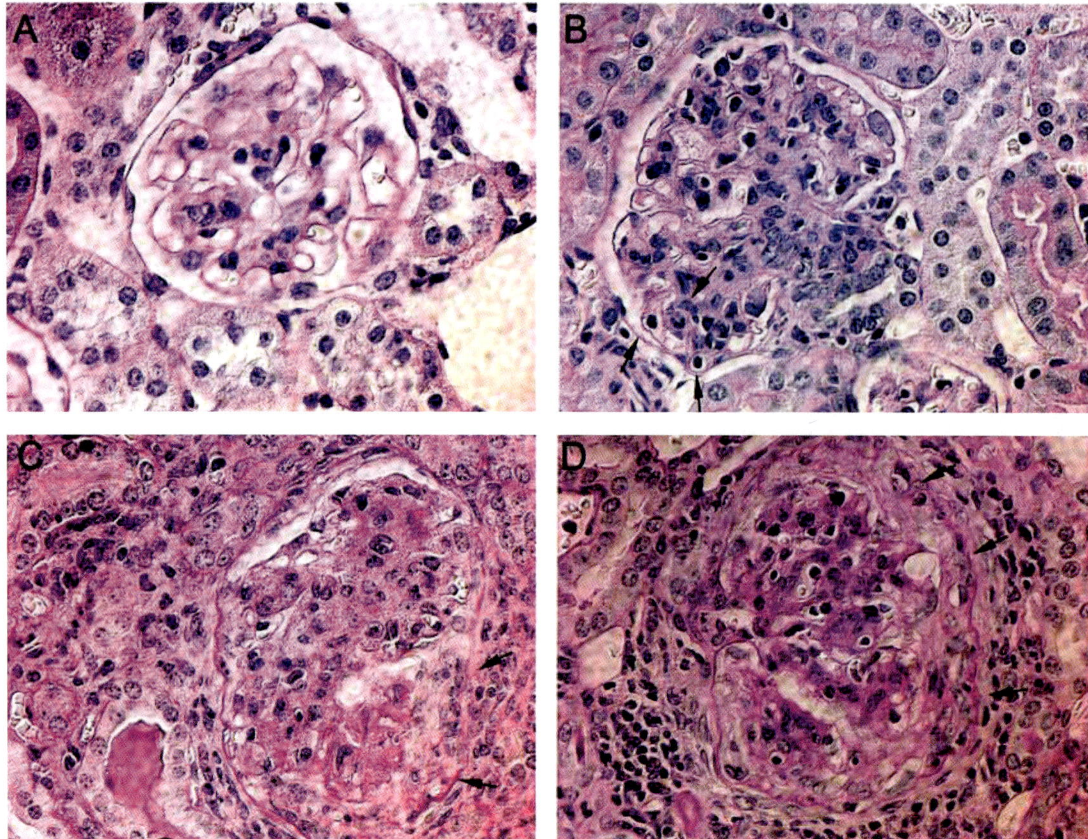

Figure 28B.3. Characteristic glomerular lesions of lupus nephritis in the $MRL_{lpr/lpr}$ mouse. (A) Normal glomerulus showing capillary loops, basement membrane and in addition to podocytes and endothelial cells, supporting mesangial cells are present (B) Intermediate disease showing mesangial proliferation. There are monocytes in the mesangial area and mononuclear cells in capillary loops (arrows) (C-D) Severe disease showing mesangial proliferative disease, loss of capillary loops due to swelling of endothelial cells, andadhesion of leukocytes. In addition there are prominent firbrocelluar crescents primarily due to proliferation of parietal epithelial cells, the hallmark of severe disease.

NZ Hybrids

NZ hybrids were the first spontaneous SLE models and may best reflect human SLE. NZ Bielschowsky black mice (NZ/Bl or NZB) develop an interstitial nephritis, B-cell activation, and antierythrocyte autoantibodies [6]. Older NZB mice develop proteinuria (selective plasma protein leak into the urine) due to a proliferative mesangial glomerulonpehritis [7]. NZ White (NZW) mice develop autoantibodies but are largely without overt disease. The cross between NZB and NZW mice, BWF1, has features associated with human SLE including a female predominance of disease, hemolytic anemia, and proteinuric renal disease by four months of age. These mice have anti-dsDNA antibodies and ~1/3 have antierythrocyte antibodies. The renal disease in the BWF1 mice is a chronic, obliterative GN with mesangial deposits, basement membrane thickening, and cellular proliferation and tubulointerstitial infiltrate composed of B cells, T cells, and CD11c+ myeloid cells [8,9]. By immunofluorescence, mesangial and basement membrane deposition of immune complexes comprising IgG, IgM, and complement (C3) are seen. Extraglomerular deposition of immune complexes develops as the disease progresses, and death from renal disease occurs by 1 year of age [5].

MRL Models

Murphy's recombinant large mice (MRL/Mp) develop a systemic autoimmunity that is accelerated when it is accompanied by Fas-deficiency (lpr). MRL/*lpr* mice develop rapid, severe autoimmunity with defined stages [10]. Antibodies to small ribonucleoproteins, ribosomes, IgG (rheumatoid factor), and cryoglobulins (precipitating immune complexes (ICs) are seen. In addition to the kidney, salivary glands, joints, and skin are affected. Striking lymphadenopathy and the accumulation of immune cells, which are not features of human SLE, also develop as MRL/*lpr* mice age and develop disease [5]. Both male and female MRL/*lpr* mice develop severe disease (Figure 28B3). Renal disease occurs more rapidly than in NZ models but is associated with less proteinuria. Histologically, GN in MRL/*lpr* mice is classically diffuse and proliferative but can also have features of membranous GN, crescent formation, and a mononuclear and T-cell infiltrate. Immune complexes are seen in glomerular capillary walls and mesangium. Tubulointerstitial inflammation with T cells and mononuclear cells is also observed [11].

BXSB Models

Inflammatory disease in the BXSB model is linked to a Y-chromosome accelerator (*Yaa*) and therefore is manifest in males [12]. Lymphadenopathy and splenomegaly are present but less pronounced than in MRL strains and proteinuria is similar to the MRL mice. Unique to SLE-prone strains, BXSB mice manifest a monocytosis [13]. ANA titers in these mice are lower than either the NZ or MRL models. Antibodies to ssDNA and dsDNA and erythrocytes also develop. Renal disease in this model is manifested as an acute, proliferative GN with loss of capillary loop architecture, membranous thickening, and mesangial expansion.

Congenic Strains and Genetics of SLE Nephritis

To understand the contribution of genetic factors to the development of SLE, investigators have selectively inbred strains with spontaneous development of SLE-like clinical features – particularly from crosses of NZ and BXSB strains – and transferred them to SLE-resistant strains (e.g., C57BL/6 or C57BL/10). Using this strategy, studies suggest that disease susceptibility in the mouse models is dependent on the presence of multiple susceptibility loci (epistasis).

Genes within the MHC class II locus in mice were among the first to be associated with the development of SLE and parallel data from human studies linking HLA-DR2 and DR3 with development of disease. In the BWF1 NZ hybrid model, two SLE susceptibility genes are located in the H2 region, and MHC can regulate the development of kidney disease [14–16]. Some MHC II haplotypes promote the pathogenic response in the kidney whereas others may suppress autoimmunity (i.e., presence of autoantibodies) [17]. In BXSB mice, MHC is also linked to the development of renal disease [18].

NZ hybrid (NM2410) mice have been used to identify multiple disease promoting and suppressing loci successfully. These loci, known as *Sle1*, *Sle2*, and *Sle3*, are present on chromosomes 1, 4, and 7, respectively, and each one contains many potential disease-promoting genes [14,19,20]. C57BL/6 (B6) congenics with one of these loci (e.g., *Sle1*), despite having autoantibody production, do not develop profound disease in the kidney. However, B6.*Sle1*/*Sle2* and B6.*Sle1*/*Sle3* bicongenics both develop severe GN and early mortality, whereas the *Sle2*/*Sle3* combination is spared. The *Sle1* locus is syntenic with the human SLE susceptibility loci containing genes involved in the complement cascade and Fcγ-receptors. It contains genes for B-cell survival signals (SLAM), regulatory T cells, complement receptors (CR2), and Fcγ-receptors, and regulates the development of autoantibody and nephritis [21–24]. Genes within *Sle2* affect B-cell responses while *Sle3* genes affect T-cell activity. The fact that all three *Sle* disease susceptibility loci are derived from NWZ strain that does not have autoimmunity indicates that, in addition to disease-susceptibility loci, there must also be disease-suppressor loci [15,25].

Another NZ hybrid strain (NZM2328) has been useful in understanding how genetics impact development of end-organ injury. In this stain, the *Cngz1* locus (chr. 1) is associated with GN while *Adnz1* (chr. 4) is associated with anti-dsDNA production. Replacing the *Adnz1* locus with intervals from wild-type mice eliminated anti-dsDNA antibody but did not affect the development of GN. Understanding how genes in the *Cngz1* locus impact disease will increase our understanding of how end-organ gene expression affects development of immune-mediated disease [26].

From BXSB and NZB hybrid congenics, additional important disease susceptibility loci have been uncovered. *Bxs3*, which is linked to autoantibody production and renal disease, maps to the same interval as *Nba2*, which was isolated from NZB mice, and *Sle1* [27]. *Bsx3* is syntenic to the human locus that encodes PD-1 [28]. *Nba2* is a large locus containing candidate genes associated with the Type I interferon response [29,30].

Pathogenesis of Kidney Disease in SLE

Genetic studies of patients with SLE and in autoimmune-prone mice indicate that an interplay of innate and adaptive immunity is responsible for the development of renal disease in SLE. One fundamental question in lupus nephritis has been whether kidney disease is due to autoimmunity against the kidney or whether kidney disease is due to a "bystander" effect of the overactivated immune system damaging the kidney by IC deposition in the glomerulus and an innate immune response to those complexes. The studies presented later indicate that both of these mechanisms of injury contribute to disease progression.

Central to the hypothesis that innate and adaptive immune responses are at play is the B-cell response. B cells produce autoantibody that complexes with nuclear antigens. These immune complexes (IC), which are composed of apoptotic debris, and are usually cleared by the innate immune system. However, when generated in excess, the IC's are deposited in the vasculature, especially the GBM of the kidney. The

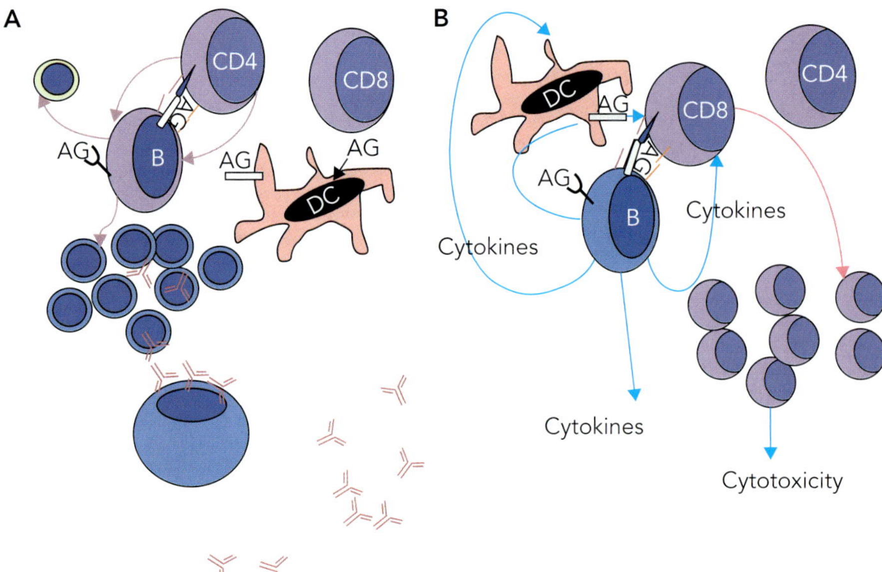

Figure 28B.4. Diagram indicating the role of B cells in autoimmunity. (A) Classical view of B cells recognizing antigen via the B-cell receptor, presenting peptides on MHC class II to CD4 T cells that provide paracrine signals necessary to trigger proliferation of B cells, synthesis of antibody, and subsequent differentiation in long-lived plasma cells. (B) Current view of B cells recognizing antigen via the B-cell receptor, presenting peptides on MHC class II, and delivering paracrine signals to CD8 T cells triggering their differentiation to cytotoxic T cells directed against the presented peptides. B cells additionally provide paracrine signals to dendritic cells promoting successful presentation of antigen to the lymphocytes, and the B cell generated proinflammatory cytokines that have systemic biological effects.

consequence of IC deposition in the glomerulus is central to the injury that ensues. However in addition to a primary role of B cells in auto-antibody formation and therefore IC deposition, data from multiple models indicate that B cells are capable of injuring the kidney (indirectly) even in the absence of antibody generation. An increasing role for B cells in antigen presentation has been appreciated particularly in the initiation of autoimmune responses (Figure 28B.4).

The induction of kidney disease in SLE requires signaling through B- and T-cell receptors and germ-line–encoded receptors (Fc-receptors, Toll-like receptors (TLRs), and complement receptors) that are expressed on innate and adaptive immune cells. Binding of immune complexes via these receptors can either provoke or dampen: (a) autoantibody production and (b) expression of cytokines and other effector molecules that mediate disease (Figure 28B.5). Evidence for the importance of these receptors derives from genetic studies of autoimmune-prone mice that have isolated loci containing receptor genes and targeted genetic manipulation that illustrates a causative role for these receptors in the development of disease.

Cells of the Immune Response

B cells are required for the development of SLE renal disease in mice [31], and signaling downstream of the B-cell receptor is abnormal in SLE model strains [32]. Autoantibody precedes the development of disease, and elimination of B cells in SLE-prone mice eliminates the renal disease [33]. Autoantibody in SLE targets soluble, extracellular nucleic acids and nuclear-associated proteins. Immune complexes bind the B-cell receptor and germ-line–encoded receptors (complement receptors, Fc-receptors, and TLRs – discussed later) on B cells in lymphoid organs. Deposition in the glomerulus has been associated with pathogenesis [34,35].

However, multiple models indicate that antibody production can be dissociated from renal disease in SLE. In the BWF1 strain, Fc-Receptor γ chain–deficient mice (lacking functional FcRI, FcRIII, and FcγRIV) generate autoantibody and immune complexes but do not develop nephritis. However, MRL/*lpr* mice that have B cells genetically engineered to lack the capacity to secrete soluble antibody still develop nephritis, albeit at a lower level [11,26,36]. B cells have been shown to be potent antigen-presenting cells in this disease context, secrete cytokines that promotes disease, and are able to induce the development of cells that mediate cell-mediated autoimmunity against the kidney (and other organs) [37–40].

The role of T cells in the development of SLE kidney disease is accepted from work in all three major mice models [41–44], but the mechanisms by which T cells exert their affects are not completely understood.

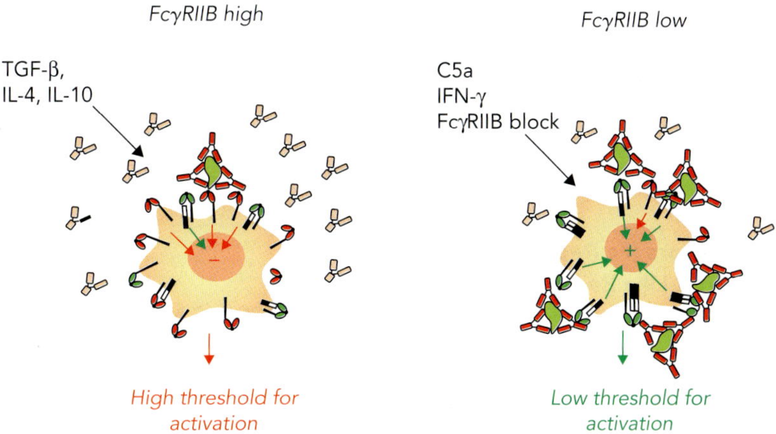

Figure 28B.5. Diagram of proposed model by which monocytes become activated by immune complexes. When FcγRIIb is expressed at low levels and activating FcγRs are present at high levels there is binding predominantly to the activating FcγRs, resulting in activation. However, when the converse is true, FcγRIIb primarily binds immune complexes and does not result in activation of leukocytes. The cytokines TGF-β, IL-4, and IL-10 all promote high levels of FcγRIIb expression.

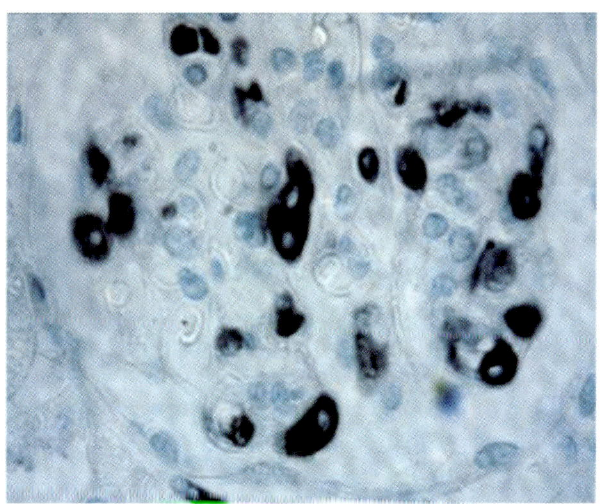

Figure 28B.6. Leukocytes in the glomerular capillaries. ED1 positive monocytes (black) in the capillary loops of a rat glomerulus 24 hours after injection of nephrotoxic serum.

In addition to promoting autoantibody production [2], there is data that αβT cells mediate direct nephrotoxicity. T-cell infiltrates are seen in active nephritis in the MRL/lpr and NZ2328 models, and MHC is upregulated in injured kidneys in the NZ2328 [45,46]. In addition, T-cell-directed therapies mitigate disease without obvious impact on B cells [9]. T cells with regulatory properties also impact the development of renal disease in SLE models but data among different models has been inconsistent. The role of CD4+CD25+FoxP3+ regulatory T cells in murine SLE models is controversial with different models yielding differing results regarding the number and suppressive capacity of these cells [47–50]. Similarly, NK T cells exacerbate disease in the BWF1 model but suppress disease in MRL/*lpr* mice [51,52]. γδ T cells are also reported to suppress disease on the MLR/*lpr* background [53].

Several lines of evidence support a role for dendritic cells (DCs) in the development of renal disease in lupus-prone mouse strains. Monocyte-lineage cells accumulate in the kidneys of BWF1 and MRL/*lpr* mice as they develop nephritis (Figure 28B.6). DCs and macrophages express TLRs, FcRs, and complement receptors that bind immune complexes. Macrophages (Mϕs) are adept at phagocytosis of immune complexes and are therefore important cells in the resolution and remodeling of the glomerulus and both monocytes and differentiated Mϕs are identified in the injured glomerulus in large numbers (Figure 28B.6). However, Mϕs and Monocytes are also activated, particularly in the presence of inflammatory cytokines such as IFN-γ, by danger molecules released from injured cells, and by ligation of FcγRs by IgGFc domains within ICs to liberate proinflammatory cytokines and nitrogen and oxygen radicals [54,55]. Monocytes and Mϕs are well recognized as effectors of cellular injury, and most studies point to the predominant role of Mϕs in lupus nephritis models as leukocytes that promote progression of disease and inflammation. Neutrophils are also frequently seen in the lupus nephritis models and are also activated by ICs via FcγRs, and generate similar but frequently more potent oxygen and nitrogen radicals in addition to the release of enzymes that destroy basement membranes interstitial matrices and cells.

Plasmacytoid DCs express large amounts of Type I interferons (IFN-α and -β), a critical cytokine in the development of SLE in humans and mice [30,56,57].

The induction and resolution of renal disease is associated with infiltrating macrophage [55,58]. DCs also secrete Type I IFNs and other critical B-cell differentiation factors (BAFF and APRIL) integral to the development of lupus-like syndromes in mice and SLE in humans [56,57,59,60].

Germ-Line–Encoded Receptors (Fc-Receptors, Complement Receptors, Toll-Like Receptors, and Chemokine Receptors)

Binding Fc-receptors to their ligands facilitates DC maturation and B-cell differentiation. Both activating and inhibitory Fc-receptors are associated with autoimmunity in multiple mouse models of SLE [61]. In the BWF1 strain, the absence of activating receptors requiring FcR γ-chain (FcγRI, FcγRIII, FcγRIV) protects against nephritis and absence of inhibitory receptor FcγRIIb can promote renal disease. In both cases, hematopoietic cell expression of the Fc-receptors is critical for their effects (Figure 28B.6) [36,62]. Like Fc-receptors, TLRs are expressed on both B cells and innate immune cells and can blunt or exacerbate disease [63,64]. B cells and DC can respond to immune complexes via TLRs, and autoantibody production and severity of renal disease is modulated by these receptors. TLR7 binds ssRNA, activates B cells, and regulates the development of autoantibody and renal disease. Signaling through TLR9, which recognizes CpG DNA, modulates the development of specific autoantibody production, but surprisingly, blunts renal disease [65–68].

Absence of early complement proteins C1 through C4, have long been recognized to predispose to SLE in humans [69], and mice deficient in these proteins develop spontaneous lupus disease with nephritis, underscoring the importance of complement components in preventing immune cell activation by immune complexes [70,71]. Conversely, blocking C5, a late complement protein, in BWF1 mice results in decreased renal disease, consistent with established roles for the membrane attack complex C5–C9 in parenchymal cell injury [72]. These findings, along with studies showing that deficiencies in other proteins involved in the processing and safe clearance of apoptotic debris can result in autoimmunity, indicate that innate immunity, including complement factors, functions primarily in a is protective role against the development of SLE, through the scavenging of apoptotic debris [73], but clearly these safe mechanisms of clearance become readily overwhelmed.

Type I Interferons (IFNα-β)

IFNα-β expression is increased in patients with SLE and in multiple models of SLE [24,57]. In the B6.*lpr* and NZB models, disruption of signaling through the IFN receptor attenuates the development of renal disease, and in the BWF1 model, IFNαβ exacerbates disease [29,74,75]. It should be noted that MLR/lpr mice do not share the requirement of IFNαβ signaling for the development of disease. Rather, in this strain, IFNγ is particularly important for expression of disease [76].

NEPHROTOXIC NEPHRITIS

Disease Models

The nephrotoxic nephritis (NTN) model of kidney disease was originally developed in rats in 1939 using repeated IV injections of rabbit anti-rat kidney antiserum to induce a glomerulonephritis. The ensuing disease is a crescentic glomerulonephritis (CGN) with many similarities to the immune-complex–associated glomerulonephritides in humans [77] (Figure 28B.7). The nephrotoxic antigen is prepared by mincing of rat kidney, and separation of a soluble fraction. The NTN model has become the most widely used model of glomerulonephritis, and although the method for preparing kidney antigen has been refined, it still remains a "dirty" antigen preparation that contains many antigens from many cells and extracellular matrix proteins. The nephrotoxic serum is frequently cleared of antierythrocyte antibodies to prevent hemolysis but not anti-endothelial antibodies, rendering endothelial injury a significant feature of this model. The model has been used in dogs, rabbits, and other large animals, but in the past 10–15 years it has been refined for use in mice. There are several important differences between the NTN model in mice compared with other rodents. There is greater heterogeneity, and greater capillary thrombos is frequently seen in mice whereas focal necrosis is less strikingly seen [78–80].

Pathogenesis

Each batch of nephrotoxic serum is different from the last, with resultant disease which has differing contributions from innate immunity, adaptive immunity, and the coagulation cascade, so comparison of studies from one laboratory with another is often not feasible [81,82]. Nevertheless, in well-controlled studies this model has provided great insight into disease mechanisms that are applicable to human disease. There are typically two phases to the disease. The first involves planted antigens against glomerular basement membrane (GBM), and likely endothelial cells (± other glomerular cells) complex, which promotes antibody-directed cell cytotoxicity (ADCC) by monocytes and neutrophils. The second phase includes formation of the late complement complexes MAC as well as early complement activation, chemokine

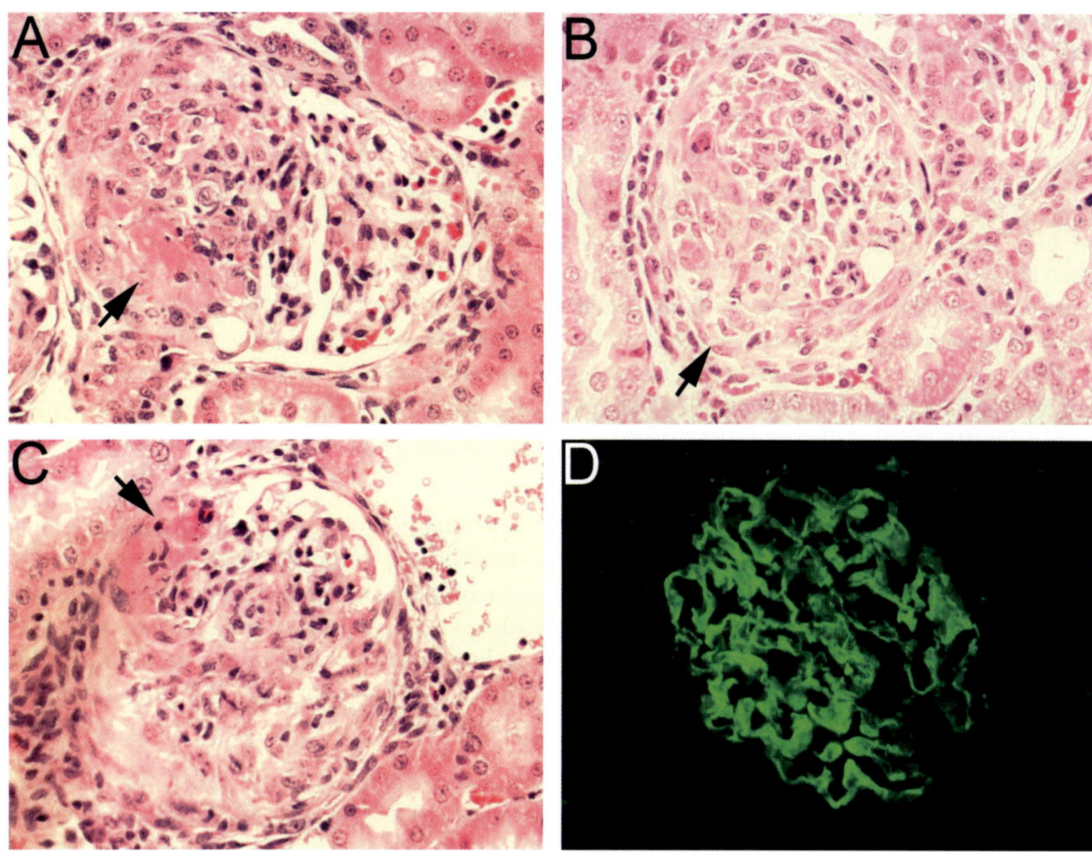

Figure 28B.7. Anti-GBM disease in Wistar Kyoto rats immunized with synthesized rat Col4A3 protein. (A) Not focal proliferative disease with an area of fibrinoid necrosis (arrow – fibrin protein in the mesangial area shows as intense pink) and is associated with nuclear debris or apoptotic bodies known as karyorexis. (B) Glomerulus showing proliferative change and a crescent (arrow). (C) Glomerulus showing a cellular crescent but also another area of fibrinoid necrosis (arrow). (D) Immunoflurescent image detecting IgG deposition. Note linear deposition along the glomerular basement membrane. (Reproduction courtesy of Dr. John Reynolds, Imperial College London.)

release, and release of proinflammatory lipid mediators, and injured endothelia release cytokines and chemokines [83–89]. Neutrophil adhesion, margination, and degranulation ensue. There is early protein leak across the GBM into the urine (proteinuria) and often renal dysfunction. There may be significant capillary thrombosis at this point [78–80,90]. In addition, there is clearance of ICs and endothelial proliferation [91]. In some models, this initial phase 'burns out' and resolves leaving varying degrees of focal sclerosis in the glomerulus [92]. However, in many examples, a secondary phase develops which involves both cellular and antibody-mediated immunity against the planted antigen, and leukocyte recruitment in response to cellular injury [93,94]. The possibility that neoantigens are exposed by the initial injury and an autoimmune cell–mediated immunity develops has been suggested but there is no data to support this hypothesis. This secondary phase or adaptive immunity phase results in focal glomerular injury with mesangial and endothelial necrosis and further capillary injury, which promotes glomerular crescent formation, the histological hallmark of severe glomerular injury (Figures 28B.3 and 28B.7) (see earlier). As in humans, the glomerular disease in rodents is associated with varying degrees of interstitial disease [95,96]. In mice, this is particularly prominent. The mechanisms of this component of the disease are less clearly understood, but include both immune-mediated injury to tubular basement membrane, direct injury to tubular capillaries, and also secondary ischemic injury, which promotes innate inflammatory responses.

One adaptation to the model that has evolved is pre-immunization of the recipient animal, subcutaneously with IgG corresponding to the administered serum. This accelerates the development of the later adaptive immune response since planted antibodies are instantly recognized by host antibodies and reactive T and B cells. Studies from different laboratories using different nephrotoxic sera report that the early phase is characterized by neutrophil influx, and complement activation and that the secondary cell-mediated phase also has features of a host-initiated humoral response [97]. The immune response to the planted antibodies in the glomerulus, and the subsequent cell-mediated and humoral phases of response to glomerular injury have been the focus of much research, and yielded the greatest advances in our understanding of pathogenesis.

Genetics

Marked genetic variation in the presentation of disease has been appreciated in mice and rats. Disease in C57BL6 inbred mice which have a Th1 bias was noted to be associated with a prominent cell-mediated response to immune complex deposition in the GBM, whereas Balb/c inbred mice with a more Th2 bias had a greater humoral response [98]. The development of congenic strains to study this disease is not well established. However, by hypothesis driven studies, using mice with mutations in specific inflammatory cytokines, important roles for IFNγ, interleukins, other cytokines including TNF-α, lymphocyte and monocyte chemokines, and leukocyte cell adhesion molecules have become established in disease initiation and progression [81,99–102]. In rats where the disease models tend to be more consistent and bear greatest similarities to human CGN, the Wistar Kyoto (WKY) inbred rat has greatest susceptibility to initiation and progression of NTN following a single injection of NTN serum. Mutations in a region of chromosome 13 designated *Crgn1* were identified in the inhibitory receptor FcγRII expressed by B cells and Mϕs, and the activating receptor FcγRIII. Mutations in FcγRIII were associated with overactivated Mϕs when stimulated by immune complexes, implicating FcγRIII as an important disease susceptibility gene [103].

In keeping with an important role for FcγRs in NTN progression, mutation of FcR γ-chain, which is necessary for cell surface expression of FcγRI, III, and IV in rodents, prevents the development of NTN in mice. Mutation of the inhibitory receptor FcγRII is associated with more severe disease, and greater development of antibodies against the planted antibody [104–106]. FcγRII on B cells is an important regulator of differentiation to plasma cells and therefore regulates antibody production. It is likely that the targeting FcγRII on B cells will be a future target in antibody-associated CGN. However, FcγRII is also expressed by inflammatory Mϕs and its importance in disease progression is less well explored. These studies strongly implicate neutrophils and monocytes in glomerulonephritis.

Immune Cells

B Cells

There is little evidence that B cells play a prominent role in the initiation or progression of NTN disease in mice. B cells are necessary for the development of a humoral response to planted antigen in the glomerulus, but the presence of autologous antibody during the autologous phase, at least in mice, has been shown to be less important in disease progression than initially thought [107].

T Cells

Conflicting reports about the role of T cells in the NTN model have been reported. This may reflect the different NTN sera used or the species [108–112]. Nevertheless, the consensus is that T cells are important activators of Mo/Mϕs, and that this may be one mechanism through which T cells act. In one group of studies, CD8 effector T cells and CD4 cells were important disease effectors, possibly through cytokine-mediated activation of Mo/Mϕs in the glomerulus, akin to a delayed-type hypersensitivity reaction [110,113]. In other studies, both αβT cells and γδT cells promoted glomerular injury following planted antigen by mechanisms independent of humoral responses [107]. The authors of these studies conclude that, in addition to direct T-cell cytotoxicity, T-cell-mediated monocyte activation by IFNγ or GM-CSF may be important in monocyte activation in the glomerulus. There is no evidence for the development of autoreactive T-cell clones in these disease models. Importantly, no studies have provided clear evidence for any role for NK lymphoid cells as disease effectors.

Myeloid Cells

Many studies have identified the importance of neutrophils (PMNs) in the initiation of glomerular injury [83,85,114,115]. PMNs bind to complexed planted antigens and through integrin adhesion molecule–dependent mechanisms bind to glomeular capillaries or diapedese and induce injury by liberation of cytokines and oxygen and nitrogen radicals. The role of monocytes and their tissue siblings, macrophages in this process has also been suggested by several studies [97,114–124]. Mo/Mϕs have FcγRs, may be activated by IC ligation, and have phagocytic and radical generating capacities similar to PMNs, although their recruitment to the glomerulus normally follows behind the rapid and early PMN recruitment. At least in the early phase of disease, one role for Mo/Mϕs may be in IC clearance rather than injury. However, in mutant mice with reduced Mo/Mϕ, mice with conditionally ablated numbers Mϕs, or mice mutated for chemokine receptors necessary for the recruitment of Mo/Mϕ all point to a central role for Mo/Mϕ in the perpetuation of glomerular injury and also the progression of the secondary inflammation in the cortex and medulla that ensues, particularly in the later progressive phase of glomerular disease. The role of T cells in Mo/Mϕ activation in the glomerulus has been suggested from

TABLE 28B.2. Fcγ receptors in humans and rodents

Human FcγR	Equivalent rodent FcγR	Consequence of crosslinking	IgG subclass specificity
FcγRI	FcγRI	Possible activation	IgG1
FcγRIIb	FcγRII	Possible inhibition	IgG1, IgG2a, IgG2b
FcγRIIa	FcγRIII	Possible activation	IgG1
FcγRIIIA	FcγRIV	Possible activation	IgG2a, IgG2b

some studies but it is unclear whether T cells play any role in Mϕ-mediated disease in the kidney interstitium [110,113]. More likely is that Danger Associated Molecular Patterns (DAMPs) released from injured parenchymal cells are important molecular factors in myeloid cell activation.

Germ-Line–Encoded Receptors

Both genetic studies in rats and gene-targeted studies in mice have demonstrated an important role for FcγRs in development and progression of NTN. Mice lacking FcγRI, III, and IV have markedly attenuated disease. More recent studies of using blocking antibodies against FcγRIV and also mice with combined mutations in FcγRI and FcγRIII point to activation of Mo/Mϕs and neutrophils specifically FcγRIV as a major factor in disease development [104,105,125]. FcgRIV selectively binds IgG2a and IgG2b. However, genetic studies in rats suggest FcγRIII on Mϕs is a key determinant of disease progression. FcγRIII has selectivity for IgG1 [103]. The importance of the inhibitory receptor FcγRIIb in limiting disease is reported, and given the absence of a role for B cells in the heterologous phase of this disease it is likely that FcγRII expressed by Mϕs, and possibly neutrophils, is an important component receptor in regulating Mϕ activation in the progression of NTN [105,125]. Importantly, mouse monocytes express FcγRIIb at low levels only; rather, it is induced in macrophages so the relative inhibitory and activator signals from FcγRs in monocytes may be an oversimplification (Figure 28B.6 and Table 28B.2).

Integrins play key roles in the adhesion and diapedesis of inflammatory cells and possibly contribute to leukocyte activation. Much work around the early events in the NTN models indicate that without neutrophil adhesion and TNF-α liberation disease does not ensue [115,126]. CD11b/CD18 (complement receptor 3), VLA4, and ICAM-1 have been implicated in these early adhesion/activation events [83]. The role of late complement components (C5b-9) and the generation of the MAC are key factors in initiation of renal injury [86,127,128]. In contrast, the *absence* of early complement components such as C1q can not only result in spontaneous nephritis due to impaired clearance of ICs and apoptotic cells, but is also associated with exacerbation of the NTN model, implicating early component complement receptors in the safe removal of ICs via specific receptors such as $\alpha_1\beta_2$ integrin that are widely expressed [71,125]. It has long been known that coadministration of bacterial lipopolysaccharide (LPS), a ligand for TLRs 2, 4, and CD14, can exacerbate injury in the NTN model. While TLR2 or 4 ligation can profoundly activate myeloid leukocytes, there has been surprisingly little literature on the role of TLRs in the progression of NTN, suggesting they may not play a prominent role [129]. One possible explanation for this is that co-ligation of TLR4 and FcγRs on Mϕs induces a profoundly anti-inflammatory cell phenotype, another possible explanation is that TLRs are ligated less by endogenous ligands [130–132]. Chemokine receptors on leukocytes and release of their ligands (chemokines) are important events in initiation and perpetuation of NTN disease. Monocyte chemokines, MIP-2 and its receptor CXCR2, MCP-1 and its receptor CCR2 contribute to Mo/Mϕ recruitment and injury, and Rantes and its receptor CCR1, and IP10 and its receptor CXCR3 have been reported to play important roles in T-cell recruitment and activation in this model [102,133–135].

Cytokines

In addition to the many reports indicating roles for TNF-α in disease initiation and perpetuation, many interleukin family members have been implicated in leukocyte activation and disease perpetuation [136–142].

ANTI-GBM NEPHRITIS AND ALPORT'S MOUSE

Disease Models

An uncommon form of severe CGN in humans known as Goodpasture's disease or anti-GBM disease was discovered in the 1980s to be due to autoimmunity against the GBM, specifically the noncollagenous domain (NC1) of the α3 chain of collagen IV (Col4A3), whose expression is restricted to specialized basement membrane of the glomerulus, lung, and cochlear in humans. The disease is characterized by autoantibody and complement deposition linearly along the GBM and by severe inflammatory injury of the glomerulus similar to that seen in NTN models. Mutations in the Col4A3 gene also lead to an autosomal-recessive inherited glomerulonephritis called Alport's syndrome, due to primary abnormalities of GBM. In addition, Alport's syndrome may be due to mutations in Col4A4 and Col4A5, which also have highly restricted expression and cause abnormalities of GBM.

Although the NTN model shares certain similarities to anti-GBM disease, a cleaner model of autoimmunity specifically against the NC1-domain of Col4A3 has been developed in rats and mice. There are three models: (a) rodents are repeatedly immunized subcutaneously with purified bovine α3(IV)NC1 fragments, or (b) purified rodent α3(IV)NC1 immunogenic peptides synthesized in vitro, or (c) more crudely and possibly most successfully rodents are immunized subcutaneously with rodent glomerular total collagen which has been collagenase digested [143]. The models have worked most successfully in disease-susceptible WKY rats, and have worked less reproducibly in mice, and appear to be notably strain dependent [144]. Immunization with synthesized rodent α3(IV)NC1 is reported to be most effective in outbred CD1 mice or DBA1 mice [145]. Nevertheless, specific autoimmunity against α3(IV) collagen has been achieved and induces glomerular disease in rodents resulting in CGN with stereotyped hypercellularity in the glomerulus, inflammatory cells, capillary injury, and crescent formation due to proliferation of the surrounding parietal epithelial cells Figure 28B.7 [146]. The models using rodent total glomerular collagen or purified bovine α3(IV) most readily produce disease in rats and mice but the autoimmune response may not be against α3(IV)NC1 [147].

A mutant mouse lacking Col3A4 was generated and develops progressive glomerulonephritis similar to humans with Alport's over an 8-week period. Why abnormalities of the GBM should lead to severe stereotyped inflammatory disease of the glomerulus is unclear, but it is likely that in this structure where there is high shear forces and high flux of plasma proteins, that dysfunction of basement membrane leads to abnormalities of shear and flow and accumulation of plasma proteins in filtering glomerular capillaries which subsequently leads to injury of endothelial cells and the overlying specialized pericyte cells, the podocytes. This injury recruits inflammatory cells that enhance injury and loss of pericytes from the GBM causes plasma protein leak and injury to parietal epithelial cells.

Pathogenesis

In the models of anti-GBM nephritis, there is direct injury from deposition of antibodies along the GBM, which, similar to the models described earlier, can induce innate immune cell activation. Antibody deposition may occur without immune cell activation [148]. There is, however, a clear component which is cell-mediated autoimmunity directed against the GBM, and occurs independently of antibody deposition [146,149].

Genetics

Goodpasture's disease in humans is associated with MHC class II alleles, positively associated with DR15 and DR4, and negatively associated with DR7 and DR1 [150–152]. In a comprehensive study of strain susceptibility to glomerulonephritis in mice to immunity induced by purified bovine α3(IV)NC1 fragments, only mice with MHC class II haplotypes H-2^s, H-2^b, and H-2^d developed CGN [146]. Other strains developed no disease despite deposition of antibody in the GBM [146]. These studies confirmed that passive transfer of antibodies was able to trigger low-level protein leak from the kidney but insufficient to trigger full disease. In comprehensive studies of autoimmunity against purified rat glomerular collagens, there was much greater disease susceptibility in WKY rats compared with Lewis rats, despite the strains having similar MHC class II haplotypes [153]. This susceptibility to disease comprised several components: (a) greater deposition of antibody in the glomerulus; (b) greater specificity of antibody; (c) a failure to clear antibodies from glomeruli after passive transfer; but in addition, (d) despite equivalent antibody deposition initially after transfer of antibodies, disease ensued only in the WKY rats, implicating differential responses by inflammatory cells. Passive transfer of antibodies was able to trigger inflammatory disease in the glomeruli of WKY rats leading to a CGN and glomerulosclerosis, confirming that anti-GBM antibodies alone are sufficient to cause disease in genetically susceptible animals [153]. Using the purer model of autoimmunity against rat α3(IV) NC1 domain in WKY rats compared with Lewis rats, only the WKY rats developed any disease, and the Lewis rats failed to generate antibodies to the pure domain.

Immune Cells

T Cells

The anti-bovine GBM model was dependent on the presence of the T-cell receptor. In addition to the necessity of T cells for the generation of antibody, nephritogenic effector T cells were postulated. In the rat model of this disease, antibodies against CD8 effectively block the progression of disease even after it is established, pointing to a major role for effector nephritogenic autoimmune T cells in disease progression [149]. However, autoreactive CD4 T cells are present in human disease and the importance of costimulatory molecules (see later) in the development of experimental disease suggest CD4 T cells are important in autoantibody production.

Myeloid Cells

The models of anti-GBM disease are characterized by glomerular macrophages. The fact that the WKY rat has recently been associated with overactive Mϕs due to mutations in FcγRIII suggests that Mϕs may play an important role in clearance of ICs from the glomerulus but that their activation in this process may be important in disease progression [103]. In the Alport's mouse lacking Col4A3, two studies implicate Mϕs in the progression of disease and the development of interstitial fibrosis [96].

Germ-Line–Encoded Receptors

Successful use of these models has been restricted to only a few centers. The difficulties in establishing a reliable model in mice has limited the progress of research in this area. However, T-cell costimulatory molecules play nonredundant roles in the induction of disease [154,155], and the inhibitory receptor FcγRIIb in mice enhances susceptibility of mice to the development of anti-GBM disease. The relative roles of B-cell overactivation and macrophage overactivation remain to be elucidated [156].

MESANGIOPROLIFERATIVE GLOMERULONEPHRITIS AND THE ANTI-THY1.1 RAT MODEL

Milder forms of human glomerulonephritis are often characterized by proliferation of mesangial cells, the mesenchymal cells of the glomerulus that perform structural roles in health but become matrix forming cells in disease (Figure 28B.8) [157]. In more severe glomerulonephritis, there is also mesangial proliferation, which is often eclipsed by other glomerular events or excessive mesangial cell death. Mesangial proliferation has been studied in great detail in the Thy1.1 model of kidney disease in rat due to serendipitous finding that a thymic T-cell antigen Thy1.1 is also exclusively expressed by mesangial cells in normal rat and antibodies against this antigen mediate an antibody and complement-mediated mesangial cell lysis [158]. The glomerular lysis is followed by intense innate immune responses to injury, and intense proliferation of surviving mesangial cells which in some rat strains then resolves over several weeks, but in others leads to focal, then global glomerulosclerosis [159]. Nevertheless, this is one of the few models in the kidney in which there is successful repair following injury and it has enabled the study of repair mechanisms and the study of the role of inflammation in repair of the glomeurlus. Congenic studies suggest that gene expression by parnechymal cells of the kidney rather than immune cells dictates the capacity of the glomerulus to repair itself, but these factors remain obscure

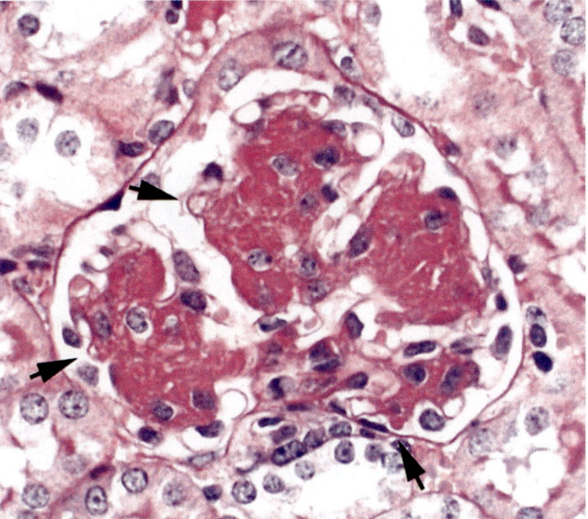

Figure 28B.8. PAS stained glomerular cross section from mouse overexpressing thymic stromal lymphopoietin (TSLP) showing membranoproliferative glomerulonephritis with prominent mesangial expansion due to the deposition of immune complexes and immune complex deposition in capillary loops (arrows). (Reproduction courtesy of Dr. Charles Alpers, University of Washington, Seattle, WA.)

[159]. The early phase of expansion of mesangial cells and matrix has furthered the study of mesenchymal cell matrix deposition and the response of mesenchymal cells to injury [160–163]. Unfortunately, there has been no equivalent model in mice, limiting the study of genetic mutation of candidate genes. There are several mesangiolysis models in mice including Habu Snake Venom and antimesangial serum models, but these have been less reproducible and result in far less mesangial proliferation than seen in the rat [164].

Prominent advances from this model include the dissection of the roles of platelet-derived growth factors (PDGFs) and their receptors in mesenchymal mesangial cell proliferation, and the role of apoptosis and macrophages in the resolution of glomerular inflammation [162,163,165,166]. Monocytes are a prominent cell type in the mesangium of the glomerulus throughout the repair phase. Selective depletion of Mo limited mesangial cell proliferation and mesangial cell matrix accumulation suggesting one role for Mo/Mϕ in acute injury is the regulation of myofibroblast cells, and that paradoxically monocytes may play an important role in recovery from inflammation [167–169].

MODELS OF ANTINEUTROPHIL CYTOPLASMIC ANTIBODY (ANCA)-ASSOCIATED VASCULITIS

A common form of severe CGN in humans originally described as pauci-immune glomerulonephritis due to a paucity (though not absence) of antibody deposition in the injured glomeruli was discovered in the 1980s to be associated with the production of autoantibodies,

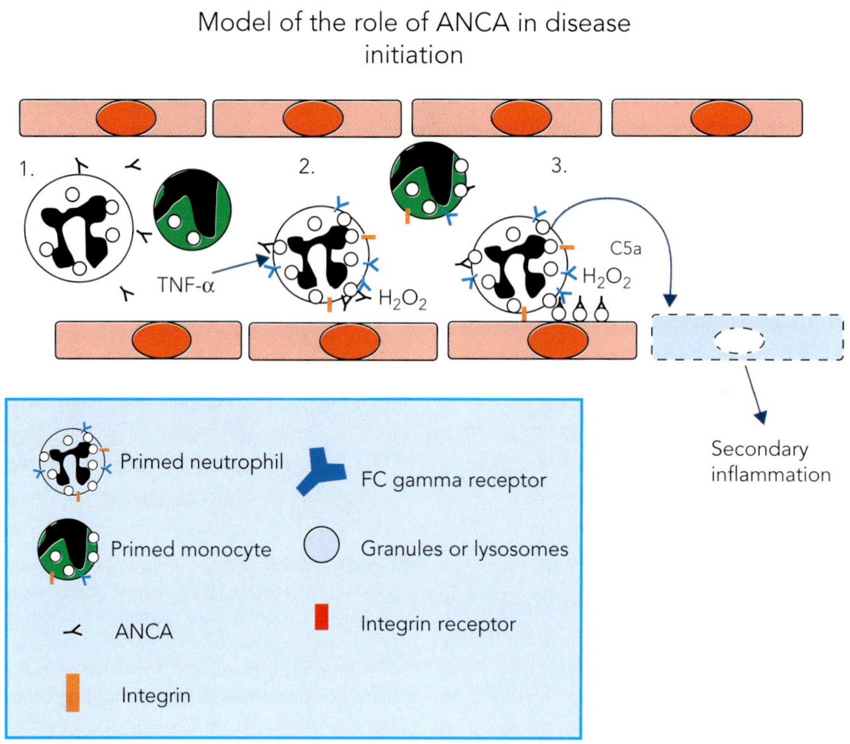

Figure 28B.9. Model of mechanism by which ANCAs initiate injury to endothelial cells of small vessels. Cytokine exposed neutrophils (and possibly monocytes) expose autoantigens on the plasma membrane and release low level of autoantigen which adheres to the endothelial surface. ANCAs in the circulation bind exposed autoantigens on leukocyte plasma membrane and endothelium triggering micro-immune complex formation and leukocyte activation via Fcg receptors and late complement component activation. Activated neutrophils release toxic products and radicals that promote local injury.

ANCAs, against lysosome and granule enzymes found predominantly in neutrophils and monocytes/macrophages [170,171]. The disease may be associated with systemic manifestations which are characterized by arteriole, venule, and capillary inflammation, known as vasculitis (Figures 28B.8 and 28B.9). However, the glomerular capillaries are a major site of injury microvascular in many cases, possibly reflecting the unusual hemodynamics, and stresses in the glomerulus whose endothelial cells are uniquely fenestrated. The major antigens against which antibodies have been generated are myeloperoxidase and proteinase 3, enzymes important in neutrophil respiratory burst and also elastin degradation. The link, however, between these autoantibodies found in the circulation and inflammation in the glomerulus which is characterized by absence of glomerular antibody deposition or IC deposition, has perplexed the scientific community. Are circulating autoantibodies just a marker of disease or are they pathogenic? Two new mouse models of autoimmunity against mouse or human myeloperoxidase (MPO) have been developed successfully, and merit our attention [172,173]. In addition, recent study of humoral alloimmunity following kidney transplantation has indicated that antibodies and ICs can be deposited in capillaries but may not be seen at a later date by immunodetection methods due to their complete removal. Nevertheless, these "invisible" ICs can be pivotal in inciting innate injurious cellular responses by neutrophils and monocytes leading to capillary injury [174,175]. It is therefore possible that in ANCA associated CGN although IC are not visible, they have previously deposited at a low level which is nevertheless, sufficient to trigger inflammatory disease.

One of the new anti-MPO antibody models in mice has demonstrated that, similar to the passive transfer of anti-GBM antibodies described earlier [147], passive transfer of anti-MPO antibodies in mice can induce necrotizing CGN and also systemic vasculitis (Figure 28B.10) [172]. The reported disease is milder than if splenocytes are transferred, indicating that cell-mediated immunity is also an important component of disease. Interestingly, in rats, and another mouse model of anti-MPO ANCA vasculitis, passive transfer of antibodies alone was insufficient to induce disease, rather to induce disease a second stimulus is required that generated mild glomerular injury [173,176,177]. Whether these differences are attributable to strain

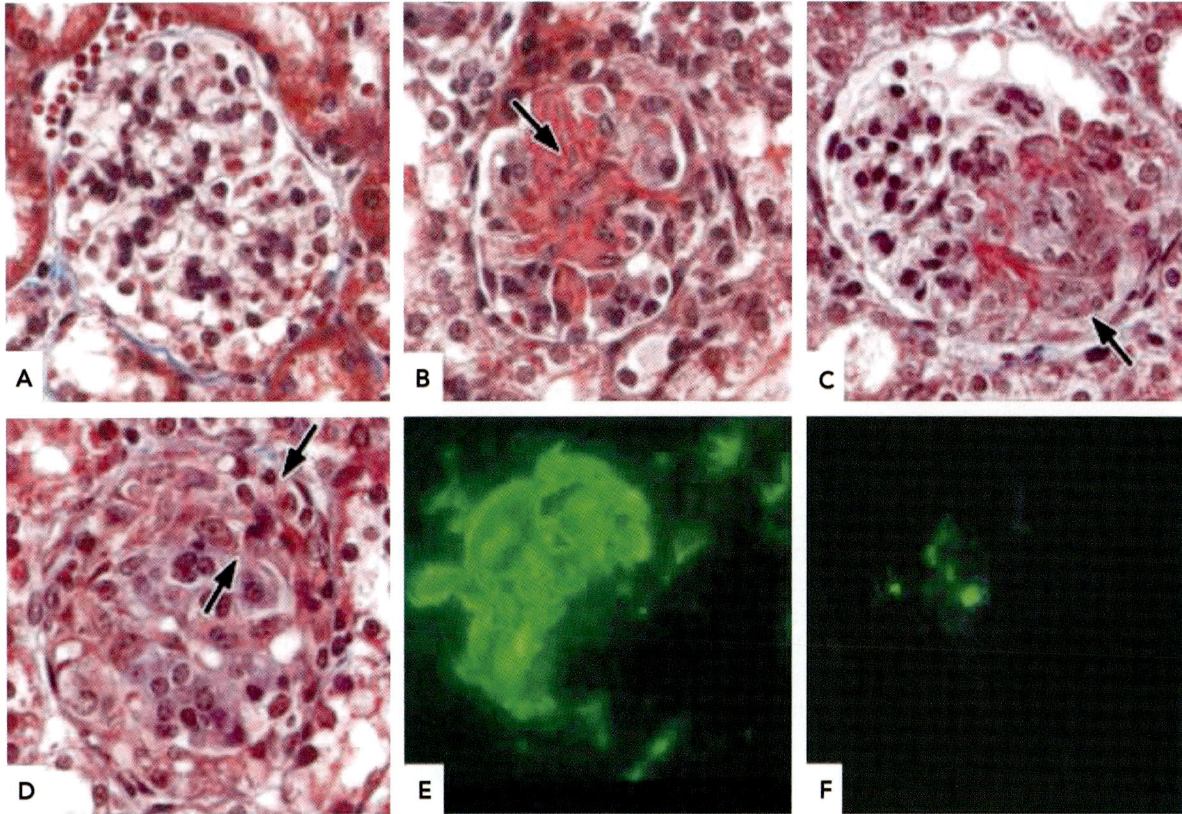

Figure 28B.10. Induction of severe glomerular injury following passive transfer of anti-MPO antibodies to mice. (A) Normal glomerulus, (B) areas of fibrinoid necrosis, and (C) cellular crescent (C and D). (E) Immunodetection of fibrin and (F) low level deposition of IgM antibodies. (Reproduction courtesy of Jennette/Falk JCI 2003.)

variations or environmental factors or the IgG isotype has not been elucidated. Nevertheless, as in anti-GBM disease, antibodies alone are sufficient to induce the full spectrum of disease in the glomerulus. What are the mechanisms by which antibodies cause disease? Primed neutrophils expose MPO on the surface and can bind ANCAs. Complexed ANCAs on the neutrophil can trigger activation by FcγR-mediated PMN activation [178]. In addition, local margination of neutrophils in glomeruli, or glomerular endothelial perturbation will lead to release of MPO in the glomerular capillary on the endothelial surface resulting in ANCA binding and local IC formation which will recruit neutrophils and monocytes, and may result in IC-mediated activation of myeloid leucocytes by FcγRs (Figure 28B.9) [178]. Thus although the disease in the glomeulus is characterized as pauci-immune, it is thought that disease initiation requires local IC formation that has long gone by the time it is viewed by the pathologist [177]. As in other disease models, activation of late complement components by local ICs recruits neutrophils and may result in membrane attack complex (MAC) formation, thus triggering secondary capillary injury, and recruitment of other leukocytes. Complexed antibodies on endothelia may activate Mo/Mφs that are in the process of clearing the immune complexes, setting up an inflammatory cascade that results in the stereotyped patterns of injury described earlier.

Although there is an unequivocal role for ANCAs in disease initiation, one of the anti-MPO antibody mouse models can result in the full spectrum of disease without the presence of B cells, suggesting that the disease can occur independently of antibodies [173]. In these studies, T cells were reported to be important effectors of glomerular and vascular disease. Although there are differences between these models, it appears that both an autoimmune-cell–mediated response to neutrophil antigens as well as antibodies can initiate and propogate disease. Studies using these models have not been verified by independent research groups and further study using these models is warranted.

CRYOGLOBULINEMIC MOUSE

Like SLE nephritis, there are many other diseases of the glomerulus characterized by IC deposition in the GBM. Some of these diseases are characterized predominantly by circulating ICs; others are characterized by local IC formation. Unlike SLE there is no obvious autoimmunity. Several models of immune complex disease have been used with variable traction including serum sickness models and systemic immunization against apoferritin [89]. A new model in mice closely mimicking human disease induced by hepatitis viruses, known as membranoproliferative GN, in which IC in the circulation may precipitate in cool temperatures (cryoprecipitate) has been developed. Transgenic (tg) mice overexpressing thymic stromal lymphopoietin (TSLP), an epithelial cytokine that induces dendritic cell–mediated CD4+ T-cell and B-cell responses, develop polyclonal mixed cryoglobulinemia (Type III) with renal disease closely resembling human cryoglobulinemic membranoproliferative glomerulonephritis, present within 8–12 weeks (Figure 28B.8). This new model is still in evolution and is strain dependent. Nevertheless, several important facets of the response of the glomerulus to ICs have been elucidated. First, the disease is modified by Fcγ receptors. In the absence of FcγRII, there is amelioration of disease and a reduction in IC formation in the circulation. B-cell FcγRII expression is strongly implicated in this outcome but the role of FcγRII on Mo/Mϕs is less clear and will require further study. Second, conditional depletion of Mo/Mϕs confirms that inflammatory macrophages are important effectors of glomerular injury in the glomerulus [179–181].

Ischemia Reperfusion Injury

Ischemia reperfusion injury (IRI) to the kidney is frequently a major or minor component of many presentations of kidney disease in humans. Compromised renal blood flow due to (a) systemic factors (shock, infection, dehydration, systemic inflammatory response syndrome); (b) arteriosclerosis which compromises vascular autoregulation as seen in chronic diseases including hypertension and diabetes mellitus; and (c) pharmacological agents that disrupt vascular autoregulation, all contribute to IRI. In addition, many inflammatory diseases of the glomerulus will lead to secondary ischemia of other areas of the kidney by virtue of the postglomerular capillaries [182]. IRI is a major component of the both the short-term and long-term kidney dysfunctions seen after renal allograft transplantation [183]. IRI is therefore both the cause of renal disease and the result of renal injury. Inflammatory factors impact the development of IRI disease by a variety of mechanisms and hold the promise of being able to mitigate disease.

Until 15 years ago, it was thought that IRI was a disease of the kidney parenchyma but studies identified inflammatory leukocytes associated with parenchymal injury, initially neutrophils but subsequently many cellular components of the innate and adaptive immune systems. Inflammation has reluctantly revealed itself to be critical to the pathogenesis of acute kidney IRI. One facet of acute IRI is that, following a severe insult, in the absence of continued or iterative insults, the kidney shows remarkable capacity for regeneration and remodeling, and functional recovery. On the other hand, however, chronic IRI is increasingly recognized as part of a final common pathway of inflammatory kidney injury that is collectively termed chronic kidney disease. There is growing recognition that inflammatory cells, in addition to promoting injury, may play a part in the recovery from renal IRI. Understanding the function of these cells in healing has become an active area of investigation.

The kidney has specialized vascular network necessary for generation of urine, and this makes the organ particularly susceptible to decreased perfusion (Figure 28B.1: renal anatomy). In order for reabsorption of electrolytes and water to occur from glomerular filtrate back into the circulation (>98% by volume), a high interstitial salt concentration must be maintained within the normal medulla. This necessitates low blood flow and therefore "physiological ischemia." Within the kidney, different regions are particularly sensitive to decreased blood flow. The S3 segment of the proximal tubule is highly metabolic yet lies in the medulla and unlike other areas of the tubule lacks anerobic respiration [184]. This segment is highly susceptible to IRI. Following acute injury, surviving or healthy areas of kidney compensate for loss of renal mass. Sometimes, this is accomplished by renal hypertrophy, but frequently this process is compromised and the compensation is by increased filtration within each glomerulus, known as glomerular hypertension. However, the consequence of this overwork is chronic injury to the glomerulus, and segmental scarring which itself leads to chronic ischemia and inflammation [185]. Pharmacological agents and diseases that chronically injure medium sized and small vessels impact negatively on this process.

The recognition that inflammation impacts the development of renal IRI syndromes has created new avenues for diagnostic and therapeutic discovery. Indeed, emerging evidence indicates inflammatory markers are predictive of clinical renal injury and that suppression of innate immune responses may blunt the development of the syndrome.

IRI Models

The classic model for acute kidney IRI is designed to mimic clinical syndromes of abrupt, global loss of renal blood flow with preservation of overall blood flow. This model allows the study of renal IRI in isolation from other factors and is directly relevant to clinical situations such as cardiac surgery, with cross-clamping of the aorta, and severe hypovolemia with collapse of systemic perfusion.

Renal pedicles, unilaterally or bilaterally, are clamped for a predetermined period of time followed

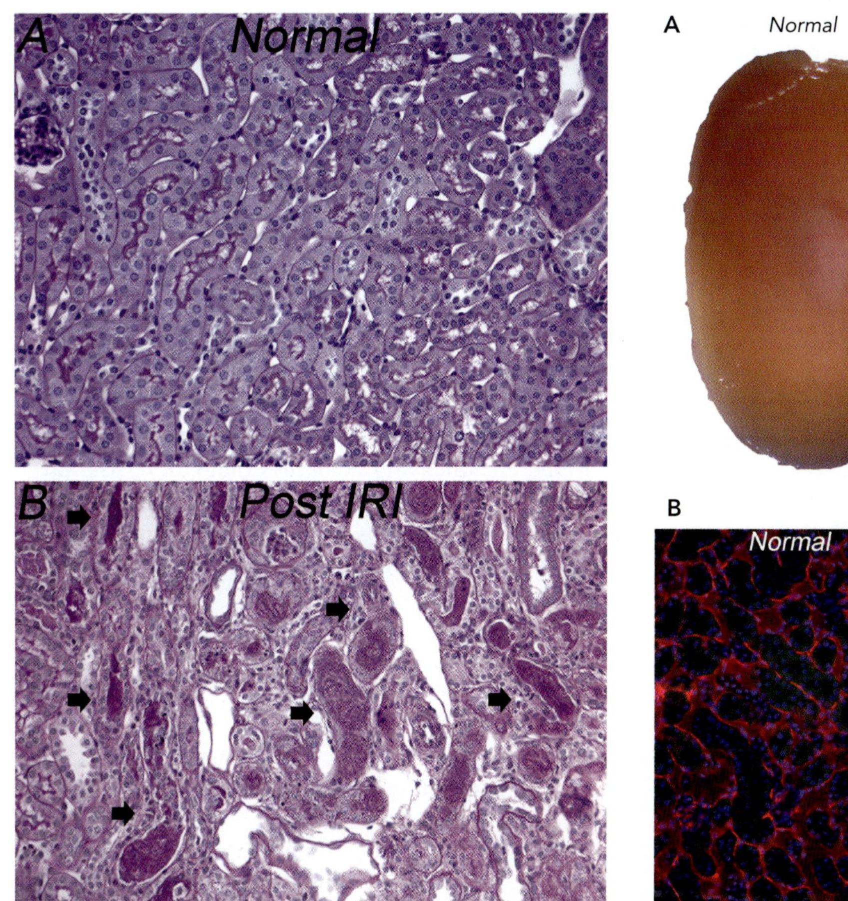

Figure 28B.11. PAS stained photomicrographs of normal mouse kidney cortex and d2 following kidney IRI. Note sever tubule injury with intratubular debris and casts (arrows).

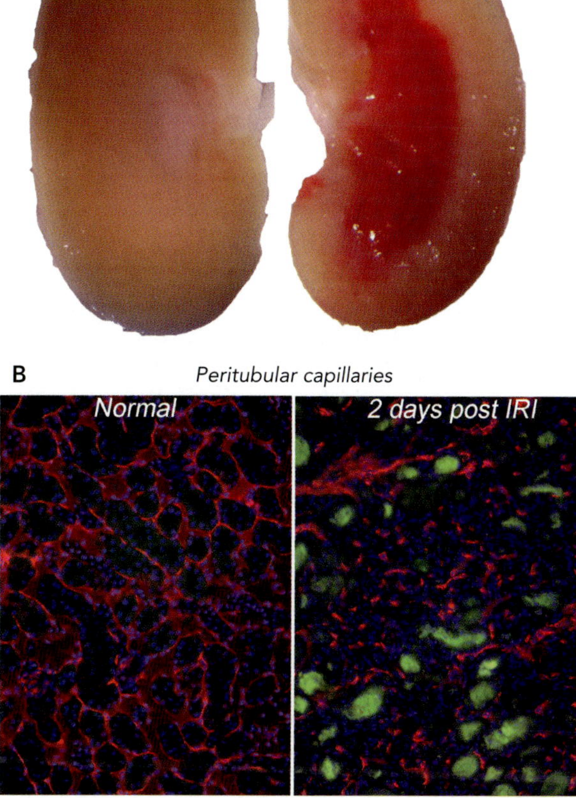

Figure 28B.12. Vascular changes in the kidney following IRI. (A) whole mouse kidneys 48 h following IRI. Note that the post IRI kidney has very poor blood flow in the medulla, detected by the presence of erythrocytes (red) in peritubular capillaries). (B) Florescence micrograph of the peritubular capillaries (CD31-red) in normal mouse kidney cortex and in kidney cortex 5 d following IRI.

by release. The development of disease in this model follows an established pattern of severe functional compromise peaking at 24–48 hours followed by progressive recovery over the ensuing two weeks. The model is temperature-dependent, and differing degrees of injury can easily be provoked. The unilateral clamping model is used to study of the impact of kidney ischemia on the ipsilateral kidney or on the animal overall. In this model, the contralateral kidney is functional, meaning there are only small (< 50%) changes in renal function. However, the contralateral kidney is not normal as it experiences increased blood flow that results from the induction of the injury. To study the impact of ischemia and renal failure on the animal overall, both renal pedicles can be clamped and severity of injury can be determined by measuring an increase in the concentration of molecules that are cleared by the kidney (i.e., serum creatinine and urea nitrogen).

Acute IRI kidney is characterized by epithelial cell death and an intense inflammatory response (Figure 28B.11). Surviving epithelial cells assume mesenchymal characteristics following injury and migrate along denuded tubular basement membrane and in the process repair the basement membrane. The delicate peritubular capillary network (Figure 28B.12) is also compromised by IRI and may result in long-term interstitial ischemia, present after functional recovery. The tempo of inflammatory cell recruitment after injury is similar to that observed in other organs acutely injured by nonimmune mechanisms. Early recruitment of neutrophils (6–36 hours after injury) is followed by intense recruitment of monocyte-lineage cells (beginning 36–48 hours) and also T lymphocytes, and these latter inflammatory cell populations in fact increase during resolution of injury and repair of the kidney. In addition to leukocytes, antibody, predominantly IgM, complement components and circulating pentraxin proteins are widely deposited in the kidney following IRI. Whereas parenchymal cells in glomeruar disease have been shown to play important roles in regeneration and remodeling after injury, it has been appreciated that injured epithelial cells induce phagocytic genes, acquire a phagocytic phenotype-like macrophages, and contribute to epithelial tubule repair by clearing cellular debris and apoptotic cells from within the tubule lumen allowing successful recanalization of the lumen and flow of glomerular filtrate [186].

The bilateral clamp injury results in a precipitous loss of renal function as determined by increased blood concentration of creatinine and urea. The rise in these molecules starts within hours of clamping, peaks at 24h to 48h and does not return to normal until almost 2 weeks after injury. A significant diuresis is seen one week after injury and remains marginally elevated for months afterward. More subtle defects in kidney function such as decreases in concentrating ability are also evident long after the initial insult. Proteinuria also developed 4 months after IRI. Histologically, the injury has profound, long-lasting effects. Though tubular morphology by 2 months after IRI is relatively normal, interstitial fibrosis develops several months post-injury and vascular architecture exhibits rarefaction [187,188]. This loss of microvasculature after acute injury is thought to link acute injury to long-term, chronic compromise in renal function [184].

Several other models of ischemic injury have been used. The Goldblatt Kidney is a model of chronic renal ischemia in which the lumen of the renal artery is reduced to restrict blood flow. In this model, renal ischemia leads to hypertension and chronic inflammation of the kidney interstitium [189]. Another model of repetitive ischemia has been developed in the heart. A ligature is placed around a coronary artery and brought to the surface. Daily application of tension to this ligature from the surface induces transient occlusion and therefore a repeated IRI [190].

Pathogenesis

There are several components to the pathogenesis in the model of acute IRI. There is the initial hypoxia induced by clamping. This necessitates anaerobic respiration by cells of the kidney, and those epithelial cells of the S3 segment of the proximal tubule are most compromised by acute lack of oxygen and nutrients. Following release of the clamp, a myriad of events occur. Among them are release of toxic compounds, activation of the coagulation cascade, the generation of intravascular thrombosis, generation of proinflammatory cytokines and chemokines by endogenous cells, and the fixation of complement. Peritubular vessels undergo a paradoxical vasoconstriction that exacerbates the hypoxia and nutrient starvation (Figure 28B.12). There is recruitment of leukocytes, in the early phase after reperfusion, predominantly neutrophils, and release of oxygen radicals and cytokines by recruited immune cells. At 48 hours after injury there is already tremendous proliferation of surviving cells and, at this point, there is a peak of functional compromise. Thereafter recovery, repair, and remodeling of the injured organ become its overriding features. Functional recovery is complete by 14 days, but many components of repair continue thereafter.

Immune Cells

Innate

Neutrophils accumulate in the kidney early after IRI, but their role in the development of the syndrome is debatable. In general, leukocytes do not incite phlogistic responses to mild injury or during phagocytosis. However, neutrophils are packed with preformed enzymes and cytokines that can wreak havoc on parenchymal cells if released uncontrollably. The injured endothelium is an important focus for activation of leukocytes, including neutrophils. Endothelial cells upregulate integrins including VCAM and ICAM-1. Systemic administration of antibodies directed against the neutrophil adhesion receptor ICAM-1 and genetic ICAM-1 mutation both reduced leukocyte recruitment, and ameliorated injury during early phases of disease. Antibodies directed against TNF-α or IL-1β deriving predominantly from neutrophils, but also by injured panrenchmal cells, have both been reported to attenuate early injury suggesting that neutrophils may exacerbate the parenchymal injury (so call debriding/sterilizing function) [191–194]. However, 24 hours after injury, neutrophils may also generate potent antiinflammatory compounds that limit this period of inflammatory injury [187] (Figure 28B.13).

Monocyte-Lineage Cells

Although monocytes and macrophages appear to be major effectors in the progressive glomerulonephritides, their primary task is phagocytosis and they have the capacity to both exacerbate injury (debridement and sterilization) and promote repair by mechanisms that involve clearance of cellular and matrix debris, regulation of the cell cycle of neighboring cells, and liberation of reparative and antiinflammatory cytokines [169]. Studies to date implicate early recruited monocyte/macrophages to IRI kidney with a similar role to neutrophils, that is, in exacerbation of injury [195]. However, the number of monocytes in the initial 24 hours after injury in the kidney is small and further studies that rigorously distinguish monocytes from neutrophils are required to clarify the role of early monocyte recruitment to the kidney [195]. In the days following IRI injury, when there is intense repair and regeneration, inflammatory macrophages are the predominant immune cell in the kidney. At this stage they appear to switch function to a predominantly reparative capacity [196,197].

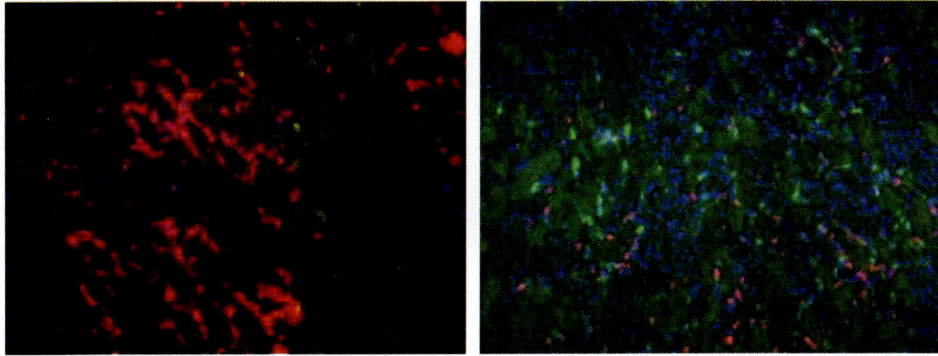

Figure 28B.13. Neutrophils and monocytes in the after IRI kidney. (A) CD11b+ leukocytes (red) surrounding injured tubules 48 hours following IRI of the kidney (B). Kidney macrophages (green) and neutrophils (red) 48 hours following injury.

Adaptive

Multiple mouse models lacking components of the adaptive immune system have been used to study the role of this arm of immunity in regulating renal IRI. A consensus from this work has not emerged. Work in mice that lack T cells (*nu/nu* mice) and B cells (μ-MT mice) implicates these both B cells and CD4+ T cells as important mediators of the IRI [198,199]. RAG-1-deficient mice, which lack T and B cells, did not confirm these data, and interestingly following IRI, CD4+ T cells, but not B cells are prominent tissue leukocytes [200]. B cells, however, generate antibodies and have other important functions that may act systemically rather than locally in the kidney. It may be that the other differences in the models (e.g., other cell types or the presence of natural antibody) can explain the variability in results reported [198]. NKT cells, which are additionally deficient in RAG-deficient mice but are present in small numbers early after injury, have recently been reported to exacerbate renal IRI [201]. Further studies are required to clarify the roles of these distinct cell types in injury or repair of the kidney.

Humoral Responses

The role of immunoglobulins in the immune response to injury is unclear. IRI studies on kidneys of RAG-1$^{-/-}$ mice that lack T and B cells report that disease progresses and recovers similarly regardless of the presence of these immune cells [200]. However, IRI injury in the gut requires B cells [202]. Natural IgM antibodies opsonize dead cells and have been found to be widely deposited in the after IRI organs including heart, skeletal muscle, gut, limb, and also kidney [202–206]. Deposition of IgM, possibly also directed at neoantigens exposed by injury, can lead to complement activation via the mannan binding lectin pathway of activation and promote injury [207]. However, much complementary activation in the post-IRI kidney is reported to occur via the alternative pathway. Complement activation may assist with nonphlogistc removal dead cells, but C5a and C5b-9 have also been shown to contribute to renal injury independent from their role as neutrophil chemoattraction, possibly by direct induction of apoptosis of parenchymal cells [208–212]. Overall, activation of the alternative complement pathway in renal IRI contributes significantly to exacerbation of injury during the inflammatory response [208–212].

Soluble Endogenous Factors Stimulating IRI

A small number of studies have implicated tissue factors in the progression and resolution of IRI disease, including, IFN-γ, α-melanocyte stimulating hormone, selectins, heparin binding-EGF, and the bone morphogenetic proteins [199,213,214]. Importantly, the injured epithelial cells of the proximal tubule and also the highly metabolic components of the loop of Henle generate both proinflammatory and chemotactic cytokines, including TNF-α, MCP-1, IL-1β, IL-6, IL-8, TGF-β, and RANTES [215]. Furthermore, T-cell costimulatory ligands, CD40L, B7-1, B7-2 may be induced by injured epithelial cells rendering the injured epithelial cell with the potential to contribute to local activation of T lymphocytes [183,216]. Another important biological active compound, adenosine, is released from injured cells and has gained increasing attention. Adenosine has potent anti-inflammatory function by virtue of its ability to bind adenosine receptors on leukocytes. There are four adenosine receptors, with differing cell specificity. Adenosine agonists are anti-inflammatory and this effect is due to ligation of leukocyte receptors [183,216]. Some controversies exist about the mechanism by which adenosine and adenosine agonists promote resolution of injury, but this avenue of therapy remains promising.

One interesting caveat is that the mechanisms by which inflammatory monocytes switch to become anti-inflammatory leukocytes is obscure. One hypothesis has been the process of phagocytosis of apoptotic cells as an anti-inflammatory stimulus, but investigations of leukocyte biology implicate adenosine release from injured tissues in this biological switch. Further studies in this area are merited [217].

Lipid Mediators

The role of lipid mediators in regulating the inflammatory response following IRI has also gained increasing attention. Many small molecular lipid mediators are generated during injury. Some of these such as leukotrienes and prostaglandins exert proinflammatory responses. However, several more recently identified lipids function as natural mediators of resolution of inflammation and their benefits can be exploited by exogenous administration in concentrations higher than those naturally achieved. Among these compounds are a group of compounds known as eicosanoids that are generated by the activity of the enzyme lipoxygenase. Active products, lipoxin-A4, 15-epi-lipoxin-A4, or their synthetic analogues, have been described in several models of IRI injury in the kidney and elsewhere reduce injury and promote resolution [218–221]. These compounds act on inflammatory cells promoting less activation and promoting greater phagocytocytic capacity. Two other groups of bioactive lipid mediators that are derived from oxygenation the omega-3 fatty acid, docosahexaenoic acid (DHA) called resolvins and protectins have been described [221]. The compounds were initially identified as generated by neutrophils following interactions with endothelial cells but it is likely that parenchymal cells can also generate these compounds. Their actions are widespread but include inhibiting activation of other inflammatory cells. In models of inflammation that resolve, protectins and resolvins are induced following the initial wave of injury and inflammation and therefore peak during the resolution phase of injury. The after IRI kidney generates these compounds in moderate quantities and this can be enhanced by systemic administration of the parent lipid, DHA. Systemic administration of resolvins and protectins, however, markedly attenuate renal injury [187].

REFERENCES

1. Cameron, J.S. 1999. Lupus nephritis. *J Am Soc Nephrol* **10**:413–424.
2. Shlomchik, M.J., Craft, J.E., and Mamula, M.J. 2001. From T to B and back again: positive feedback in systemic autoimmune disease. *Nat Rev Immunol* **1**: 147–153.

2a. Majeti, R., Xu, Z., Parslow, T.G., Olson, J.L., Daikh, D.I., Killeen, N., Weiss, A. 2000. An inactivating point mutation in the inhibitory wedge of CD45 causes lymphoproliferation and autoimmunity. *Cell* **103**:1059–1070.

2b. Salvador, J.M., Hollander, M.C., Nguyen, A.T., Kopp, J.B., Barisoni, L., Moore, J.K., Ashwell, J.D., Fornace, A.J. 2002. Mice lacking the p53-effector gene Gadd45a develop a lupus-like syndrome. *Immunity* **16**:499–508.

2c. Hibbs, M.L., Tarlinton, D.M., Armes, J., Grail, D., Hodgson, G., Maglitto, R., Stacker, S.A., Dunn, A.R. 1995. Multiple defects in the immune system of Lyn-deficient mice, culminating in autoimmune disease. *Cell* **83**:301–311.

2d. Mackay, F., Woodcock, S.A., Lawton, P., Ambrose, C., Baetscher, M., Schneider, P., Tschopp, J., Browning, J.L. 1999. Mice transgenic for BAFF develop lymphocytic disorders along with autoimmune manifestations. *J Exp Med* **190**:1697–1710.

2e. Higuchi, T., Aiba, Y., Nomura, T., Matsuda, J., Mochida, K., Suzuki, M., Kikutani, H., Honjo, T., Nishioka, K., Tsubata, T. 2002. Cutting edge: ectopic expression of CD40 ligand on B cells induces lupus-like autoimmune disease. *J Immunol* **168**:9.

2f. Zhu, J., Liu, X., Xie, C., Yan, M., Yu, Y., Sobel, E.S., Wakeland, E.K., Mohan, C. 2005. T cell hyperactivity in lupus as a consequence of hyperstimulatory antigen-presenting cells. *J Clin Invest* **115**:1869–1878.

2g. Bickerstaff, M.C., Botto, M., Hutchinson, W.L., Herbert, J., Tennent, G.A., Bybee, A., Mitchell, D.A., Cook, H.T., Butler, P.J., Walport, M.J., Pepys, M.B. 1999. Serum amyloid P component controls chromatin degradation and prevents antinuclear autoimmunity. *Nat Med* **5**: 694–697.

2h. Santiago-Raber, M.L., Baccala, R., Haraldsson, K.M., Choubey, D., Stewart, T.A., Kono, D.H. 2003. Theofilopoulos AN: Type-I interferon receptor deficiency reduces lupus-like disease in NZB mice. *J Exp Med* **197**:777–788.

2i. Peng, S.L., Moslehi, J., Craft, J. 1997. Roles of interferon-gamma and interleukin-4 in murine lupus. *J Clin Invest* **99**:1936.

3. Satoh, M., Kumar, A., Kanwar, Y.S., and Reeves, W.H. 1995. Anti-nuclear antibody production and immune-complex glomerulonephritis in BALB/c mice treated with pristane. *Proc Natl Acad Sci USA* **92**:10934–10938.
4. Mendlovic, S., Brocke, S., Shoenfeld, Y., et al. 1988. Induction of a systemic lupus erythematosus-like disease in mice by a common human anti-DNA idiotype. *Proc Natl Acad Sci USA* **85**:2260–2264.
5. Andrews, B.S., Eisenberg, R.A., Theofilopoulos, A.N., et al. 1978. Spontaneous murine lupus-like syndromes. Clinical and immunopathological manifestations in several strains. *J Exp Med* **148**:1198–1215.
6. Borchers, A., Ansari, A.A., Hsu, T., Kono, D.H., and Gershwin, M.E. 2000. The pathogenesis of autoimmunity in New Zealand mice. *Semin Arthritis Rheum* **29**:385–399.
7. Mellors, R.C. 1966. Autoimmune disease in NZB/BL mice. 3. Induction of membranous glomerulonephritis in young mice by the transplantation of spleen cells from old mice. *J Exp Med* **123**:1025–1034.
8. Hurd, E.R., and Ziff, M. 1978. Association of interstitial nephritis with tubule cell injury and proliferation in NZB/NZW mice. *Clin Exp Immunol* **32**:1–11.
9. Schiffer, L., Sinha, J., Wang, X., et al. 2003. Short term administration of costimulatory blockade and cyclophosphamide induces remission of systemic lupus erythematosus nephritis in NZB/W F1 mice by a mechanism downstream of renal immune complex deposition. *J Immunol* **171**:489–497.

10. Reilly, C.M., and Gilkeson, G.S. 2002. Use of genetic knockouts to modulate disease expression in a murine model of lupus, MRL/lpr mice. *Immunol Res* **25**:143–153.
11. Chan, O.T., Hannum, L.G., Haberman, A.M., Madaio, M.P., and Shlomchik, M.J. 1999. A novel mouse with B cells but lacking serum antibody reveals an antibody-independent role for B cells in murine lupus. *J Exp Med* **189**:1639–1648.
12. Murphy, E.D., and Roths, J.B. 1979. A Y chromosome associated factor in strain BXSB producing accelerated autoimmunity and lymphoproliferation. *Arthritis Rheum* **22**:1188–1194.
13. Wofsy, D., Kerger, C.E., and Seaman, W.E. 1984. Monocytosis in the BXSB model for systemic lupus erythematosus. *J Exp Med* **159**:629–634.
14. Mohan, C., Yu, Y., Morel, L., Yang, P., and Wakeland, E.K. 1999. Genetic dissection of Sle pathogenesis: Sle3 on murine chromosome 7 impacts T cell activation, differentiation, and cell death. *J Immunol* **162**:6492–6502.
15. Subramanian, S., Yim, Y.S., Liu, K., Tus, K., Zhou, X.J., and Wakeland, E.K. 2005. Epistatic suppression of systemic lupus erythematosus: fine mapping of Sles1 to less than 1 mb. *J Immunol* **175**:1062–1072.
16. Vyse, T.J., Halterman, R.K., Rozzo, S.J., Izui, S., and Kotzin, B.L. 1999. Control of separate pathogenic autoantibody responses marks MHC gene contributions to murine lupus. *Proc Natl Acad Sci USA* **96**:8098–8103.
17. Zhang, D., Fujio, K., Jiang, Y., et al. 2004. Dissection of the role of MHC class II A and E genes in autoimmune susceptibility in murine lupus models with intragenic recombination. *Proc Natl Acad Sci USA* **101**:13838–13843.
18. Merino, R., Fossati, L., Lacour, M., Lemoine, R., Higaki, M., and Izui, S. 1992. H-2-linked control of the Yaa gene-induced acceleration of lupus-like autoimmune disease in BXSB mice. *Eur J Immunol* **22**:295–299.
19. Mohan, C., Morel, L., Yang, P., et al. 1999. Genetic dissection of lupus pathogenesis: a recipe for nephrophilic autoantibodies. *J Clin Invest* **103**:1685–1695.
20. Sobel, E.S., Mohan, C., Morel, L., Schiffenbauer, J., and Wakeland, E.K. 1999. Genetic dissection of SLE pathogenesis: adoptive transfer of Sle1 mediates the loss of tolerance by bone marrow-derived B cells. *J Immunol* **162**:2415–2421.
21. Morel, L., Blenman, K.R., Croker, B.P., and Wakeland, E.K. 2001. The major murine systemic lupus erythematosus susceptibility locus, Sle1, is a cluster of functionally related genes. *Proc Natl Acad Sci USA* **98**:1787–1792.
22. Kumar, K.R., Li, L., Yan, M., et al. 2006. Regulation of B cell tolerance by the lupus susceptibility gene Ly108. *Science* **312**:1665–1669.
23. Giles, B.M., Tchepeleva, S.N., Kachinski, J.J., et al. 2007. Augmentation of NZB autoimmune phenotypes by the Sle1c murine lupus susceptibility interval. *J Immunol* **178**:4667–4675.
24. Namjou, B., Kilpatrick, J., and Harley, J.B. 2007. Genetics of clinical expression in SLE. *Autoimmunity* **40**:602–612.
25. Morel, L., Tian, X.H., Croker, B.P., and Wakeland, E.K. 1999. Epistatic modifiers of autoimmunity in a murine model of lupus nephritis. *Immunity* **11**:131–139.
26. Waters, S.T., McDuffie, M., Bagavant, H., et al. 2004. Breaking tolerance to double stranded DNA, nucleosome, and other nuclear antigens is not required for the pathogenesis of lupus glomerulonephritis. *J Exp Med* **199**:255–264.
27. Haywood, M.E., Rogers, N.J., Rose, S.J., et al. 2004. Dissection of BXSB lupus phenotype using mice congenic for chromosome 1 demonstrates that separate intervals direct different aspects of disease. *J Immunol* **173**:4277–4285.
28. Prokunina, L., Castillejo-Lopez, C., Oberg, F., et al. 2002. A regulatory polymorphism in PDCD1 is associated with susceptibility to systemic lupus erythematosus in humans. *Nat Genet* **32**:666–669.
29. Santiago-Raber, M.L., Baccala, R., Haraldsson, K.M., et al. 2003. Type-I interferon receptor deficiency reduces lupus-like disease in NZB mice. *J Exp Med* **197**:777–788.
30. Rozzo, S.J., Allard, J.D., Choubey, D., et al. 2001. Evidence for an interferon-inducible gene, Ifi202, in the susceptibility to systemic lupus. *Immunity* **15**:435–443.
31. Shlomchik, M.J., Madaio, M.P., Ni, D., Trounstein, M., and Huszar, D. 1994. The role of B cells in lpr/lpr-induced autoimmunity. *J Exp Med* **180**:1295–1306.
32. Shlomchik, M.J. 2008. Sites and stages of autoreactive B cell activation and regulation. *Immunity* **28**:18–28.
33. Klinman, D.M. 1990. Polyclonal B cell activation in lupus-prone mice precedes and predicts the development of autoimmune disease. *J Clin Invest* **86**:1249–1254.
34. Madaio, M.P., Carlson, J., Cataldo, J., Ucci, A., Migliorini, P., and Pankewycz, O. 1987. Murine monoclonal anti-DNA antibodies bind directly to glomerular antigens and form immune deposits. *J Immunol* **138**:2883–2889.
35. Ehrenstein, M.R., Katz, D.R., Griffiths, M.H., et al. 1995. Human IgG anti-DNA antibodies deposit in kidneys and induce proteinuria in SCID mice. *Kidney Int* **48**:705–711.
36. Clynes, R., Dumitru, C., and Ravetch, J.V. 1998. Uncoupling of immune complex formation and kidney damage in autoimmune glomerulonephritis. *Science* **279**:1052–1054.
37. Yan, J., Harvey, B.P., Gee, R.J., Shlomchik, M.J., and Mamula, M.J. 2006. B cells drive early T cell autoimmunity in vivo prior to dendritic cell-mediated autoantigen presentation. *J Immunol* **177**:4481–4487.
38. Harris, D.P., Haynes, L., Sayles, P.C., et al. 2000. Reciprocal regulation of polarized cytokine production by effector B and T cells. *Nat Immunol* **1**:475–482.
39. Yin, Z., Bahtiyar, G., Zhang, N., et al. 2002. IL-10 regulates murine lupus. *J Immunol* **169**:2148–2155.
40. Ishida, H., Muchamuel, T., Sakaguchi, S., Andrade, S., Menon, S., and Howard, M. 1994. Continuous administration of anti-interleukin 10 antibodies delays onset of autoimmunity in NZB/W F1 mice. *J Exp Med* **179**:305–310.
41. Wofsy, D., Ledbetter, J.A., Hendler, P.L., and Seaman, W.E. 1985. Treatment of murine lupus with monoclonal anti-T cell antibody. *J Immunol* **134**:852–857.
42. Wofsy, D., and Seaman, W.E. 1985. Successful treatment of autoimmunity in NZB/NZW F1 mice with monoclonal antibody to L3T4. *J Exp Med* **161**:378–391.
43. Lawson, B.R., Koundouris, S.I., Barnhouse, M., et al. 2001. The role of alpha beta+ T cells and homeostatic T cell proliferation in Y-chromosome-associated murine lupus. *J Immunol* **167**:2354–2360.
44. Peng, S.L., Madaio, M.P., Hughes, D.P., et al. 1996. Murine lupus in the absence of alpha beta T cells. *J Immunol* **156**:4041–4049.
45. Anderson, B.E., McNiff, J., Yan, J., et al. 2003. Memory CD4+ T cells do not induce graft-versus-host disease. *J Clin Invest* **112**:101–108.
46. Bagavant, H., Deshmukh, U.S., Wang, H., Ly, T., and Fu, S.M. 2006. Role for nephritogenic T cells in lupus glomerulonephritis: progression to renal failure is accompanied

by T cell activation and expansion in regional lymph nodes. *J Immunol* **177**:8258–8265.
47. Scalapino, K.J., Tang, Q., Bluestone, J.A., Bonyhadi, M.L., and Daikh, D.I. 2006. Suppression of disease in New Zealand Black/New Zealand White lupus-prone mice by adoptive transfer of ex vivo expanded regulatory T cells. *J Immunol* **177**:1451–1459.
48. Bagavant, H., and Tung, K.S. 2005. Failure of CD25+ T cells from lupus-prone mice to suppress lupus glomerulonephritis and sialoadenitis. *J Immunol* **175**:944–950.
49. Monk, C.R., Spachidou, M., Rovis, F., et al. 2005. MRL/Mp CD4+,CD25- T cells show reduced sensitivity to suppression by CD4+,CD25+ regulatory T cells in vitro: a novel defect of T cell regulation in systemic lupus erythematosus. *Arthritis Rheum* **52**:1180–1184.
50. Cuda, C.M., Wan, S., Sobel, E.S., Croker, B.P., and Morel, L. 2007. Murine lupus susceptibility locus Sle1a controls regulatory T cell number and function through multiple mechanisms. *J Immunol* **179**:7439–7447.
51. Zeng, D., Liu, Y., Sidobre, S., Kronenberg, M., and Strober, S. 2003. Activation of natural killer T cells in NZB/W mice induces Th1-type immune responses exacerbating lupus. *J Clin Invest* **112**:1211–1222.
52. Yang, J.Q., Saxena, V., Xu, H., Van Kaer, L., Wang, C.R., and Singh, R.R. 2003. Repeated alpha-galactosylceramide administration results in expansion of NK T cells and alleviates inflammatory dermatitis in MRL-lpr/lpr mice. *J Immunol* **171**:4439–4446.
53. Peng, S.L., Madaio, M.P., Hayday, A.C., and Craft, J. 1996. Propagation and regulation of systemic autoimmunity by gammadelta T cells. *J Immunol* **157**:5689–5698.
54. Bergtold, A., Gavhane, A., D'Agati, V., Madaio, M., and Clynes, R. 2006. FcR-bearing myeloid cells are responsible for triggering murine lupus nephritis. *J Immunol* **177**:7287–7295.
55. Schiffer, L., Bethunaickan, R., Ramanujam, M., et al. 2008. Activated renal macrophages are markers of disease onset and disease remission in lupus nephritis. *J Immunol* **180**:1938–1947.
56. Banchereau, J., and Pascual, V. 2006. Type I interferon in systemic lupus erythematosus and other autoimmune diseases. *Immunity* **25**:383–392.
57. Graham, R.R., Kozyrev, S.V., Baechler, E.C., et al. 2006. A common haplotype of interferon regulatory factor 5 (IRF5) regulates splicing and expression and is associated with increased risk of systemic lupus erythematosus. *Nat Genet* **38**:550–555.
58. Bloom, R.D., Florquin, S., Singer, G.G., Brennan, D.C., and Kelley, V.R. 1993. Colony stimulating factor-1 in the induction of lupus nephritis. *Kidney Int* **43**: 1000–1009.
59. Banchereau, J., Pascual, V., and Palucka, A.K. 2004. Autoimmunity through cytokine-induced dendritic cell activation. *Immunity* **20**:539–550.
60. Groom, J.R., Fletcher, C.A., Walters, S.N., et al. 2007. BAFF and MyD88 signals promote a lupuslike disease independent of T cells. *J Exp Med* **204**:1959–1971.
61. Lin, Q., Xiu, Y., Jiang, Y., et al. 2006. Genetic dissection of the effects of stimulatory and inhibitory IgG Fc receptors on murine lupus. *J Immunol* **177**:1646–1654.
62. Bolland, S., and Ravetch, J.V. 2000. Spontaneous autoimmune disease in Fc(gamma)RIIB-deficient mice results from strain-specific epistasis. *Immunity* **13**:277–285.
63. Pisitkun, P., Deane, J.A., Difilippantonio, M.J., Tarasenko, T., Satterthwaite, A.B., and Bolland, S. 2006. Autoreactive B cell responses to RNA-related antigens due to TLR7 gene duplication. *Science* **312**:1669–1672.
64. Christensen, S.R., Kashgarian, M., Alexopoulou, L., Flavell, R.A., Akira, S., and Shlomchik, M.J. 2005. Toll-like receptor 9 controls anti-DNA autoantibody production in murine lupus. *J Exp Med* **202**:321–331.
65. Lartigue, A., Courville, P., Auquit, I., et al. 2006. Role of TLR9 in anti-nucleosome and anti-DNA antibody production in lpr mutation-induced murine lupus. *J Immunol* **177**:1349–1354.
66. Christensen, S.R., Shupe, J., Nickerson, K., Kashgarian, M., Flavell, R.A., and Shlomchik, M.J. 2006. Toll-like receptor 7 and TLR9 dictate autoantibody specificity and have opposing inflammatory and regulatory roles in a murine model of lupus. *Immunity* **25**:417–428.
67. Viglianti, G.A., Lau, C.M., Hanley, T.M., Miko, B.A., Shlomchik, M.J., and Marshak-Rothstein, A. 2003. Activation of autoreactive B cells by CpG dsDNA. *Immunity* **19**:837–847.
68. Lau, C.M., Broughton, C., Tabor, A.S., et al. 2005. RNA-associated autoantigens activate B cells by combined B cell antigen receptor/Toll-like receptor 7 engagement. *J Exp Med* **202**:1171–1177.
69. Walport, M.J. 2001. Complement. Second of two parts. *N Engl J Med* **344**:1140–1144.
70. Paul, E., Pozdnyakova, O.O., Mitchell, E., and Carroll, M.C. 2002. Anti-DNA autoreactivity in C4-deficient mice. *Eur J Immunol* **32**:2672–2679.
71. Botto, M., Dell'Agnola, C., Bygrave, A.E., et al. 1998. Homozygous C1q deficiency causes glomerulonephritis associated with multiple apoptotic bodies. *Nat Genet* **19**:56–59.
72. Wang, Y., Hu, Q., Madri, J.A., Rollins, S.A., Chodera, A., and Matis, L.A. 1996. Amelioration of lupus-like autoimmune disease in NZB/WF1 mice after treatment with a blocking monoclonal antibody specific for complement component C5. *Proc Natl Acad Sci USA* **93**:8563–8568.
73. Carroll, M.C. 2004. A protective role for innate immunity in systemic lupus erythematosus. *Nat Rev Immunol* **4**:825–831.
74. Braun, D., Geraldes, P., and Demengeot, J. 2003. Type I Interferon controls the onset and severity of autoimmune manifestations in lpr mice. *J Autoimmun* **20**: 15–25.
75. Mathian, A., Weinberg, A., Gallegos, M., Banchereau, J., and Koutouzov, S. 2005. IFN-alpha induces early lethal lupus in preautoimmune (New Zealand Black x New Zealand White) F1 but not in BALB/c mice. *J Immunol* **174**:2499–2506.
76. Balomenos, D., Rumold, R., and Theofilopoulos, A.N. 1998. Interferon-gamma is required for lupus-like disease and lymphoaccumulation in MRL-lpr mice. *J Clin Invest* **101**:364–371.
77. Smadel, J.E., and Swift, H.F. 1939. The effect of prolonged administration of sulfanilamide on rats with nephrotoxic nephritis. *J Clin Invest* **18**:757–762.
78. Cunningham, M.A., Kitching, A.R., Tipping, P.G., and Holdsworth, S.R. 2004. Fibrin independent proinflammatory effects of tissue factor in experimental crescentic glomerulonephritis. *Kidney Int* **66**:647–654.
79. Kitching, A.R., Kong, Y.Z., Huang, X.R., et al. 2003. Plasminogen activator inhibitor-1 is a significant determinant of renal injury in experimental crescentic glomerulonephritis. *J Am Soc Nephrol* **14**:1487–1495.
80. Kitching, A.R., Holdsworth, S.R., Ploplis, V.A., et al. 1997. Plasminogen and plasminogen activators protect against renal injury in crescentic glomerulonephritis. *J Exp Med* **185**:963–968.

81. Kitching, A.R., Holdsworth, S.R., and Tipping, P.G. 1999. IFN-gamma mediates crescent formation and cell-mediated immune injury in murine glomerulonephritis. *J Am Soc Nephrol* **10**:752–759.
82. Ring, G.H., Dai, Z., Saleem, S., Baddoura, F.K., and Lakkis, F.G. 1999. Increased susceptibility to immunologically mediated glomerulonephritis in IFN-gamma-deficient mice. *J Immunol* **163**:2243–2248.
83. Mulligan, M.S., Johnson, K.J., Todd, R.F., 3rd, et al. 1993. Requirements for leukocyte adhesion molecules in nephrotoxic nephritis. *J Clin Invest* **91**:577–587.
84. Wu, X., Pippin, J., and Lefkowith, J.B. 1993. Attenuation of immune-mediated glomerulonephritis with an anti-CD11b monoclonal antibody. *Am J Physiol* **264**: F715–721.
85. Wu, X., Pippin, J., and Lefkowith, J.B. 1993. Platelets and neutrophils are critical to the enhanced glomerular arachidonate metabolism in acute nephrotoxic nephritis in rats. *J Clin Invest* **91**:766–773.
86. Hughes, J., Nangaku, M., Alpers, C.E., Shankland, S.J., Couser, W.G., and Johnson, R.J. 2000. C5b-9 membrane attack complex mediates endothelial cell apoptosis in experimental glomerulonephritis. *Am J Physiol Renal Physiol* **278**:F747–F757.
87. Savige, J.A., Dash, A.C., and Rees, A.J. 1989. Exaggerated glomerular albuminuria after cobra venom factor in anti-glomerular basement membrane disease. *Nephron* **52**:29–35.
88. Hebert, M.J., Takano, T., Papayianni, A., et al. 1998. Acute nephrotoxic serum nephritis in complement knockout mice: relative roles of the classical and alternate pathways in neutrophil recruitment and proteinuria. *Nephrol Dial Transplant* **13**:2799–2803.
89. Anders, H.J., Vielhauer, V., Kretzler, M., et al. 2001. Chemokine and chemokine receptor expression during initiation and resolution of immune complex glomerulonephritis. *J Am Soc Nephrol* **12**:919–931.
90. Nangaku, M., Alpers, C.E., Pippin, J., et al. 1997. Renal microvascular injury induced by antibody to glomerular endothelial cells is mediated by C5b-9. *Kidney Int* **52**:1570–1578.
91. Iruela-Arispe, L., Gordon, K., Hugo, C., et al. 1995. Participation of glomerular endothelial cells in the capillary repair of glomerulonephritis. *Am J Pathol* **147**:1715–1727.
92. Holdsworth, S.R., and Bellomo, R. 1984. Differential effects of steroids on leukocyte-mediated glomerulonephritis in the rabbit. *Kidney Int* **26**:162–169.
93. Karkar, A.M., Tam, F.W., Proudfoot, A.E., Meager, A., and Rees, A.J. 1993. Modulation of antibody-mediated glomerular injury in vivo by interleukin-6. *Kidney Int* **44**:967–973.
94. Tipping, P.G., Worthington, L.A., and Holdsworth, S.R. 1987. Quantitation and characterization of glomerular procoagulant activity in experimental glomerulonephritis. *Lab Invest* **56**:155–159.
95. Duffield, J.S., Erwig, L.P., Wei, X., Liew, F.Y., Rees, A.J., and Savill, J.S. 2000. Activated macrophages direct apoptosis and suppress mitosis of mesangial cells. *J Immunol* **164**:2110–2119.
96. Rodgers, K.D., Rao, V., Meehan, D.T., et al. 2003. Monocytes may promote myofibroblast accumulation and apoptosis in Alport's renal fibrosis. *Kidney Int* **63**:1338–1355.
97. Neugarten, J., Feith, G.W., Assmann, K.J., Shan, Z., Stanley, E.R., and Schlondorff, D. 1995. Role of macrophages and colony-stimulating factor-1 in murine anti-glomerular basement membrane glomerulonephritis. *J Am Soc Nephrol* **5**:1903–1909.
98. Huang, X.R., Holdsworth, S.R., and Tipping, P.G. 1997. Th2 responses induce humorally mediated injury in experimental anti-glomerular basement membrane glomerulonephritis. *J Am Soc Nephrol* **8**:1101–1108.
99. Huang, X.R., Kitching, A.R., Tipping, P.G., and Holdsworth, S.R. 2000. Interleukin-10 inhibits macrophage-induced glomerular injury. *J Am Soc Nephrol* **11**:262–269.
100. Kitching, A.R., Tipping, P.G., and Holdsworth, S.R. 1999. IL-12 directs severe renal injury, crescent formation and Th1 responses in murine glomerulonephritis. *Eur J Immunol* **29**:1–10.
101. Kitching, A.R., Tipping, P.G., Mutch, D.A., Huang, X.R., and Holdsworth, S.R. 1998. Interleukin-4 deficiency enhances Th1 responses and crescentic glomerulonephritis in mice. *Kidney Int* **53**:112–118.
102. Segerer, S., Banas, B., Wornle, M., et al. 2004. CXCR3 is involved in tubulointerstitial injury in human glomerulonephritis. *Am J Pathol* **164**:635–649.
103. Aitman, T.J., Dong, R., Vyse, T.J., et al. 2006. Copy number polymorphism in Fcgr3 predisposes to glomerulonephritis in rats and humans. *Nature* **439**:851–855.
104. Wakayama, H., Hasegawa, Y., Kawabe, T., et al. 2000. Abolition of anti-glomerular basement membrane antibody-mediated glomerulonephritis in FcRgamma-deficient mice. *Eur J Immunol* **30**:1182–1190.
105. Suzuki, Y., Shirato, I., Okumura, K., et al. 1998. Distinct contribution of Fc receptors and angiotensin II-dependent pathways in anti-GBM glomerulonephritis. *Kidney Int* **54**:1166–1174.
106. Park, S.Y., Ueda, S., Ohno, H., et al. 1998. Resistance of Fc receptor- deficient mice to fatal glomerulonephritis. *J Clin Invest* **102**:1229–1238.
107. Rosenkranz, A.R., Knight, S., Sethi, S., Alexander, S.I., Cotran, R.S., and Mayadas, T.N. 2000. Regulatory interactions of alphabeta and gammadelta T cells in glomerulonephritis. *Kidney Int* **58**:1055–1066.
108. Tipping, P.G., and Holdsworth, S.R. 2006. T cells in crescentic glomerulonephritis. *J Am Soc Nephrol* **17**:1253–1263.
109. Li, S., Holdsworth, S.R., and Tipping, P.G. 2000. B7.1 and B7.2 co-stimulatory molecules regulate crescentic glomerulonephritis. *Eur J Immunol* **30**:1394–1401.
110. Huang, X.R., Tipping, P.G., Apostolopoulos, J., et al. 1997. Mechanisms of T cell-induced glomerular injury in anti-glomerular basement membrane (GBM) glomerulonephritis in rats. *Clin Exp Immunol* **109**:134–142.
111. Kuroda, T., Kawasaki, K., Oite, T., Arakawa, M., and Shimizu, F. 1994. Nephrotoxic serum nephritis in nude rats: the role of cell-mediated immunity. *Nephron* **68**:360–365.
112. Kusuyama, Y., Nishihara, T., and Saito, K. 1981. Nephrotoxic nephritis in nude mice. *Clin Exp Immunol* **46**:20–26.
113. Huang, X.R., Holdsworth, S.R., and Tipping, P.G. 1994. Evidence for delayed-type hypersensitivity mechanisms in glomerular crescent formation. *Kidney Int* **46**:69–78.
114. Tam, F.W., Karkar, A.M., Smith, J., et al. 1996. Differential expression of macrophage inflammatory protein-2 and monocyte chemoattractant protein-1 in experimental glomerulonephritis. *Kidney Int* **49**: 715–721.
115. Tang, T., Rosenkranz, A., Assmann, K.J., et al. 1997. A role for Mac-1 (CDIIb/CD18) in immune complex-stimulated neutrophil function in vivo: Mac-1 deficiency abrogates

sustained Fcgamma receptor-dependent neutrophil adhesion and complement-dependent proteinuria in acute glomerulonephritis. *J Exp Med* **186**:1853–1863.
116. Wilson, H.M., Chettibi, S., Jobin, C., Walbaum, D., Rees, A.J., and Kluth, D.C. 2005. Inhibition of macrophage nuclear factor-kappaB leads to a dominant anti-inflammatory phenotype that attenuates glomerular inflammation in vivo. *Am J Pathol* **167**:27–37.
117. Duffield, J.S., Tipping, P.G., Kipari, T., et al. 2005. Conditional ablation of macrophages halts progression of crescentic glomerulonephritis. *Am J Pathol* **167**:1207–1219.
118. Erwig, L.P., Kluth, D.C., and Rees, A.J. 2003. Macrophage heterogeneity in renal inflammation. *Nephrol Dial Transplant* **18**:1962–1965.
119. Segerer, S., Cui, Y., Hudkins, K.L., et al. 2000. Expression of the chemokine monocyte chemoattractant protein-1 and its receptor chemokine receptor 2 in human crescentic glomerulonephritis. *J Am Soc Nephrol* **11**:2231–2242.
120. Erwig, L.P., and Rees, A.J. 1999. Macrophage activation and programming and its role for macrophage function in glomerular inflammation. *Kidney Blood Press Res* **22**:21–25.
121. Karkar, A.M., Smith, J., Tam, F.W., Pusey, C.D., and Rees, A.J. 1997. Abrogation of glomerular injury in nephrotoxic nephritis by continuous infusion of interleukin-6. *Kidney Int* **52**:1313–1320.
122. Tipping, P.G., and Holdsworth, S.R. 1988. Isolation and characterization of glomerular macrophages in experimental glomerulonephritis. *Immunol Cell Biol* **66** (Pt 2):147–151.
123. Holdsworth, S.R., Tipping, P.G., Hooke, D.H., and Atkins, R.C. 1985. Role of the macrophage in immunologically induced glomerulonephritis. *Contrib Nephrol* **45**:105–114.
124. Holdsworth, S.R., and Tipping, P.G. 1985. Macrophage-induced glomerular fibrin deposition in experimental glomerulonephritis in the rabbit. *J Clin Invest* **76**:1367–1374.
125. Kaneko, Y., Nimmerjahn, F., Madaio, M.P., and Ravetch, J.V. 2006. Pathology and protection in nephrotoxic nephritis is determined by selective engagement of specific Fc receptors. *J Exp Med* **203**:789–797.
126. Coxon, A., Cullere, X., Knight, S., et al. 2001. Fc gamma RIII mediates neutrophil recruitment to immune complexes. a mechanism for neutrophil accumulation in immune-mediated inflammation. *Immunity* **14**:693–704.
127. Pruchno, C.J., Burns, M.M., Schulze, M., et al. 1991. Urinary excretion of the C5b-9 membrane attack complex of complement is a marker of immune disease activity in autologous immune complex nephritis. *Am J Pathol* **138**:203–211.
128. Van Zyl Smit, R., Rees, A.J., and Peters, D.K. 1983. Factors affecting severity of injury during nephrotoxic nephritis in rabbits. *Clin Exp Immunol* **54**:366–372.
129. Brown, H.J., Sacks, S.H., and Robson, M.G. 2006. Toll-like receptor 2 agonists exacerbate accelerated nephrotoxic nephritis. *J Am Soc Nephrol* **17**:1931–1939.
130. Obhrai, J., and Goldstein, D.R. 2006. The role of toll-like receptors in solid organ transplantation. *Transplantation* **81**:497–502.
131. Shishido, T., Nozaki, N., Yamaguchi, S., et al. 2003. Toll-like receptor-2 modulates ventricular remodeling after myocardial infarction. *Circulation* **108**:2905–2910.
132. Gerber, J.S., and Mosser, D.M. 2001. Reversing lipopolysaccharide toxicity by ligating the macrophage Fc gamma receptors. *J Immunol* **166**:6861–6868.
133. Topham, P.S., Csizmadia, V., Soler, D., et al. 1999. Lack of chemokine receptor CCR1 enhances Th1 responses and glomerular injury during nephrotoxic nephritis. *J Clin Invest* **104**:1549–1557.
134. Tesch, G.H., Schwarting, A., Kinoshita, K., Lan, H.Y., Rollins, B.J., and Kelley, V.R. 1999. Monocyte chemoattractant protein-1 promotes macrophage-mediated tubular injury, but not glomerular injury, in nephrotoxic serum nephritis. *J Clin Invest* **103**:73–80.
135. Anders, H.J., Frink, M., Linde, Y., et al. 2003. CC chemokine ligand 5/RANTES chemokine antagonists aggravate glomerulonephritis despite reduction of glomerular leukocyte infiltration. *J Immunol* **170**:5658–5666.
136. Kitching, A.R., Turner, A.L., Wilson, G.R., Edgtton, K.L., Tipping, P.G., and Holdsworth, S.R. 2004. Endogenous IL-13 limits humoral responses and injury in experimental glomerulonephritis but does not regulate Th1 cell-mediated crescentic glomerulonephritis. *J Am Soc Nephrol* **15**:2373–2382.
137. Kitching, A.R., Katerelos, M., Mudge, S.J., Tipping, P.G., Power, D.A., and Holdsworth, S.R. 2002. Interleukin-10 inhibits experimental mesangial proliferative glomerulonephritis. *Clin Exp Immunol* **128**:36–43.
138. Kitching, A.R., Tipping, P.G., Huang, X.R., Mutch, D.A., and Holdsworth, S.R. 1997. Interleukin-4 and interleukin-10 attenuate established crescentic glomerulonephritis in mice. *Kidney Int* **52**:52–59.
139. Tam, F.W., Smith, J., Karkar, A.M., Pusey, C.D., and Rees, A.J. 1997. Interleukin-4 ameliorates experimental glomerulonephritis and up-regulates glomerular gene expression of IL-1 decoy receptor. *Kidney Int* **52**:1224–1231.
140. Eitner, F., Westerhuis, R., Burg, M., et al. 1997. Role of interleukin-6 in mediating mesangial cell proliferation and matrix production in vivo. *Kidney Int* **51**:69–78.
141. Schwarting, A., Tesch, G., Kinoshita, K., Maron, R., Weiner, H.L., and Kelley, V.R. 1999. IL-12 drives IFN-gamma-dependent autoimmune kidney disease in MRL-Fas(lpr) mice. *J Immunol* **163**:6884–6891.
142. Sugiyama, M., Kinoshita, K., Kishimoto, K., et al. 2008. Deletion of IL-18 receptor ameliorates renal injury in bovine serum albumin-induced glomerulonephritis. *Clin Immunol* **128**:103–108.
143. Reynolds, J., Mavromatidis, K., Cashman, S.J., Evans, D.J., and Pusey, C.D. 1998. Experimental autoimmune glomerulonephritis (EAG) induced by homologous and heterologous glomerular basement membrane in two substrains of Wistar-Kyoto rat. *Nephrol Dial Transplant* **13**:44–52.
144. Kalluri, R., Meyers, K., Mogyorosi, A., Madaio, M.P., and Neilson, E.G. 1997. Goodpasture syndrome involving overlap with Wegener's granulomatosis and anti-glomerular basement membrane disease. *J Am Soc Nephrol* **8**:1795–1800.
145. Hopfer, H., Maron, R., Butzmann, U., Helmchen, U., Weiner, H.L., and Kalluri, R. 2003. The importance of cell-mediated immunity in the course and severity of autoimmune anti-glomerular basement membrane disease in mice. *FASEB J* **17**:860–868.
146. Kalluri, R., Danoff, T.M., Okada, H., and Neilson, E.G. 1997. Susceptibility to anti-glomerular basement membrane disease and Goodpasture syndrome is linked to MHC class II genes and the emergence of T cell-mediated immunity in mice. *J Clin Invest* **100**:2263–2275.
147. Reynolds, J., Albouainain, A., Duda, M.A., Evans, D.J., and Pusey, C.D. 2006. Strain susceptibility to active induction and passive transfer of experimental

autoimmune glomerulonephritis in the rat. *Nephrol Dial Transplant* **21**:3398–3408.
148. Stevenson, A., Yaqoob, M., Mason, H., Pai, P., and Bell, G.M. 1995. Biochemical markers of basement membrane disturbances and occupational exposure to hydrocarbons and mixed solvents. *QJM* **88**:23–28.
149. Reynolds, J., Norgan, V.A., Bhambra, U., Smith, J., Cook, H.T., and Pusey, C.D. 2002. Anti-CD8 monoclonal antibody therapy is effective in the prevention and treatment of experimental autoimmune glomerulonephritis. *J Am Soc Nephrol* **13**:359–369.
150. Fisher, M., Pusey, C.D., Vaughan, R.W., and Rees, A.J. 1997. Susceptibility to anti-glomerular basement membrane disease is strongly associated with HLA-DRB1 genes. *Kidney Int* **51**:222–229.
151. Phelps, R.G., and Rees, A.J. 1999. The HLA complex in Goodpasture's disease: a model for analyzing susceptibility to autoimmunity. *Kidney Int* **56**:1638–1653.
152. Rees, A.J., Peters, D.K., Compston, D.A., and Batchelor, J.R. 1978. Strong association between HLA-DRW2 and antibody-mediated Goodpasture's syndrome. *Lancet* **1**:966–968.
153. Reynolds, J., Cook, P.R., Ryan, J.J., et al. 2002. Segregation of experimental autoimmune glomerulonephritis as a complex genetic trait and exclusion of Col4a3 as a candidate gene. *Exp Nephrol* **10**:402–407.
154. Reynolds, J., Khan, S.B., Allen, A.R., Benjamin, C.D., and Pusey, C.D. 2004. Blockade of the CD154-CD40 costimulatory pathway prevents the development of experimental autoimmune glomerulonephritis. *Kidney Int* **66**:1444–1452.
155. Reynolds, J., Tam, F.W., Chandraker, A., et al. 2000. CD28-B7 blockade prevents the development of experimental autoimmune glomerulonephritis. *J Clin Invest* **105**:643–651.
156. Nakamura, A., Yuasa, T., Ujike, A., et al. 2000. Fcgamma receptor IIB-deficient mice develop Goodpasture's syndrome upon immunization with type IV collagen: a novel murine model for autoimmune glomerular basement membrane disease. *J Exp Med* **191**:899–906.
157. Johnson, R.J., Iida, H., Alpers, C.E., et al. 1991. Expression of smooth muscle cell phenotype by rat mesangial cells in immune complex nephritis. Alpha-smooth muscle actin is a marker of mesangial cell proliferation. *J Clin Invest* **87**:847–858.
158. Brandt, J., Pippin, J., Schulze, M., et al. 1996. Role of the complement membrane attack complex (C5b-9) in mediating experimental mesangioproliferative glomerulonephritis. *Kidney Int* **49**:335–343.
159. Aben, J.A., Hoogervorst, D.A., Paul, L.C., et al. 2003. Genes expressed by the kidney, but not by bone marrow-derived cells, underlie the genetic predisposition to progressive glomerulosclerosis after mesangial injury. *J Am Soc Nephrol* **14**:2264–2270.
160. Floege, J., Johnson, R.J., Alpers, C.E., et al. 1993. Visceral glomerular epithelial cells can proliferate in vivo and synthesize platelet-derived growth factor B-chain. *Am J Pathol* **142**:637–650.
161. Floege, J., Eng, E., Lindner, V., et al. 1992. Rat glomerular mesangial cells synthesize basic fibroblast growth factor. Release, upregulated synthesis, and mitogenicity in mesangial proliferative glomerulonephritis. *J Clin Invest* **90**:2362–2369.
162. Yoshimura, A., Gordon, K., Alpers, C.E., et al. 1991. Demonstration of PDGF B-chain mRNA in glomeruli in mesangial proliferative nephritis by in situ hybridization. *Kidney Int* **40**:470–476.
163. Floege, J., Johnson, R.J., Gordon, K., et al. 1991. Increased synthesis of extracellular matrix in mesangial proliferative nephritis. *Kidney Int* **40**:477–488.
164. Yo, Y., Braun, M.C., Barisoni, L., et al. 2003. Anti-mouse mesangial cell serum induces acute glomerulonephropathy in mice. *Nephron Exp Nephrol* **93**:e92–106.
165. Floege, J., Eng, E., Young, B.A., et al. 1993. Infusion of platelet-derived growth factor or basic fibroblast growth factor induces selective glomerular mesangial cell proliferation and matrix accumulation in rats. *J Clin Invest* **92**:2952–2962.
166. Baker, A.J., Mooney, A., Hughes, J., Lombardi, D., Johnson, R.J., and Savill, J. 1994. Mesangial cell apoptosis: the major mechanism for resolution of glomerular hypercellularity in experimental mesangial proliferative nephritis. *J Clin Invest* **94**:2105–2116.
167. Westerhuis, R., van Straaten, S.C., van Dixhoorn, M.G., et al. 2000. Distinctive roles of neutrophils and monocytes in anti-thy-1 nephritis. *Am J Pathol* **156**:303–310.
168. Chen, Y.M., Hu-Tsai, M.I., Lin, S.L., Tsai, T.J., and Hsieh, B.S. 2003. Expression of CX3CL1/fractalkine by mesangial cells in vitro and in acute anti-Thy1 glomerulonephritis in rats. *Nephrol Dial Transplant* **18**:2505–2514.
169. Duffield, J.S. 2003. The inflammatory macrophage: a story of Jekyll and Hyde. *Clin Sci (Lond)* **104**:27–38.
170. Goldschmeding, R., van der Schoot, C.E., ten Bokkel Huinink, D., et al. 1989. Wegener's granulomatosis autoantibodies identify a novel diisopropylfluorophosphate-binding protein in the lysosomes of normal human neutrophils. *J Clin Invest* **84**:1577–1587.
171. Jennette, J.C., Wilkman, A.S., and Falk, R.J. 1989. Anti-neutrophil cytoplasmic autoantibody-associated glomerulonephritis and vasculitis. *Am J Pathol* **135**:921–930.
172. Xiao, H., Heeringa, P., Hu, P., et al. 2002. Antineutrophil cytoplasmic autoantibodies specific for myeloperoxidase cause glomerulonephritis and vasculitis in mice. *J Clin Invest* **110**:955–963.
173. Ruth, A.J., Kitching, A.R., Kwan, R.Y., et al. 2006. Anti-neutrophil cytoplasmic antibodies and effector CD4+ cells play nonredundant roles in anti-myeloperoxidase crescentic glomerulonephritis. *J Am Soc Nephrol* **17**:1940–1949.
174. Mauiyyedi, S., Pelle, P.D., Saidman, S., et al. 2001. Chronic humoral rejection: identification of antibody-mediated chronic renal allograft rejection by C4d deposits in peritubular capillaries. *J Am Soc Nephrol* **12**:574–582.
175. Collins, A.B., Schneeberger, E.E., Pascual, M.A., et al. 1999. Complement activation in acute humoral renal allograft rejection: diagnostic significance of C4d deposits in peritubular capillaries. *J Am Soc Nephrol* **10**:2208–2214.
176. Brouwer, E., Huitema, M.G., Klok, P.A., et al. 1993. Antimyeloperoxidase-associated proliferative glomerulonephritis: an animal model. *J Exp Med* **177**:905–914.
177. Yang, J.J., Jennette, J.C., and Falk, R.J. 1994. Immune complex glomerulonephritis is induced in rats immunized with heterologous myeloperoxidase. *Clin Exp Immunol* **97**:466–473.
178. Ben-Smith, A., Dove, S.K., Martin, A., Wakelam, M.J., and Savage, C.O. 2001. Antineutrophil cytoplasm autoantibodies from patients with systemic vasculitis activate neutrophils through distinct signaling cascades: comparison with conventional Fcgamma receptor ligation. *Blood* **98**:1448–1455.
179. Kowalewska, J., Muhlfeld, A.S., Hudkins, K.L., et al. 2007. Thymic stromal lymphopoietin transgenic mice develop cryoglobulinemia and hepatitis with

similarities to human hepatitis C liver disease. *Am J Pathol* **170**:981–989.
180. Muhlfeld, A.S., Segerer, S., Hudkins, K., et al. 2003. Deletion of the fcgamma receptor IIb in thymic stromal lymphopoietin transgenic mice aggravates membranoproliferative glomerulonephritis. *Am J Pathol* **163**:1127–1136.
181. Guo, S., Wietecha, T.A., Hudkins, K., et al. 2008. Macrophages are essential contributors to kidney injury in murine cryoglobiulinemia-associated membranoproliferative glomerulonephritis. *J Am Soc Nephrol* **19**:Abstract in press.
182. Brouwer, E., Klok, P.A., Huitema, M.G., Weening, J.J., and Kallenberg, C.G. 1995. Renal ischemia/reperfusion injury contributes to renal damage in experimental anti-myeloperoxidase-associated proliferative glomerulonephritis. *Kidney Int* **47**:1121–1129.
183. Chandraker, A., Takada, M., Nadeau, K.C., Peach, R., Tilney, N.L., and Sayegh, M.H. 1997. CD28-b7 blockade in organ dysfunction secondary to cold ischemia/reperfusion injury. *Kidney Int* **52**:1678–1684.
184. Duffield, J.S., and Bonventre, J.V. 2004. Acute renal failure from Bench to Bedside: Chapter 43. In *Chronic Kidney Disease, Dialysis and Transplant*. Pereira (ed), pp. 765–786. Philadelphia: WB Saunders Co.
185. Dworkin, L.D., Hostetter, T.H., Rennke, H.G., and Brenner, B.M. 1984. Hemodynamic basis for glomerular injury in rats with desoxycorticosterone-salt hypertension. *J Clin Invest* **73**:1448–1461.
186. Ichimura, T., Asseldonk, E.J., Humphreys, B.D., Gunaratnam, L., Duffield, J.S., and Bonventre, J.V. 2008. Kidney injury molecule-1 is a phosphatidylserine receptor that confers a phagocytic phenotype on epithelial cells. *J Clin Invest* **118**:1657–1668.
187. Duffield, J.S., Hong, S., Vaidya, V.S., et al. 2006. Resolvin D series and protectin D1 mitigate acute kidney injury. *J Immunol* **177**:5902–5911.
188. Basile, D.P., Donohoe, D., Roethe, K., and Osborn, J.L. 2001. Renal ischemic injury results in permanent damage to peritubular capillaries and influences long-term function. *Am J Physiol Renal Physiol* **281**:F887–F899.
189. Johnson, R.J., Rodriguez-Iturbe, B., Schreiner, G.F., and Herrera-Acosta, J. 2002. Hypertension: a microvascular and tubulointerstitial disease. *J Hypertens Suppl* **20**:S1–S7.
190. Haudek, S.B., Xia, Y., Huebener, P., et al. 2006. Bone marrow-derived fibroblast precursors mediate ischemic cardiomyopathy in mice. *Proc Natl Acad Sci USA* **103**:18284–18289.
191. Rabb, H., Mendiola, C.C., Dietz, J., et al. 1994. Role of CD11a and CD11b in ischemic acute renal failure in rats. *Am J Physiol* **267**:F1052–F1058.
192. Linas, S.L., Whittenburg, D., Parsons, P.E., and Repine, J.E. 1995. Ischemia increases neutrophil retention and worsens acute renal failure: role of oxygen metabolites and ICAM 1. *Kidney Int* **48**:1584–1591.
193. Kelly, K.J., Williams, W.W., Jr., Colvin, R.B., and Bonventre, J.V. 1994. Antibody to intercellular adhesion molecule 1 protects the kidney against ischemic injury. *Proc Natl Acad Sci USA* **91**:812–816.
194. Kelly, K.J., Williams, W.W., Jr., Colvin, R.B., et al. 1996. Intercellular adhesion molecule-1-deficient mice are protected against ischemic renal injury. *J Clin Invest* **97**:1056–1063.
195. Day, Y.J., Huang, L., Ye, H., Linden, J., and Okusa, M.D. 2005. Renal ischemia-reperfusion injury and adenosine 2A receptor-mediated tissue protection: role of macrophages. *Am J Physiol Renal Physiol* **288**:F722–F731.
196. Arnold, L., Henry, A., Poron, F., et al. 2007. Inflammatory monocytes recruited after skeletal muscle injury switch into antiinflammatory macrophages to support myogenesis. *J Exp Med* **204**:1057–1069.
197. Lin, S.L., Nowlin, B.T., Nathans, J., Lang, R.A., and Duffield, J.S. 2008. Macrophage-delivered Canonical Wnt Signaling is a mediator of Tissue Repair in the Injured Kidney. *J Am Soc Nephrol* **19**:Abstract in press.
198. Burne-Taney, M.J., Ascon, D.B., Daniels, F., Racusen, L., Baldwin, W., and Rabb, H. 2003. B cell deficiency confers protection from renal ischemia reperfusion injury. *J Immunol* **171**:3210–3215.
199. Burne, M.J., Daniels, F., El Ghandour, A., et al. 2001. Identification of the CD4(+) T cell as a major pathogenic factor in ischemic acute renal failure. *J Clin Invest* **108**:1283–1290.
200. Park, P., Haas, M., Cunningham, P.N., Bao, L., Alexander, J.J., and Quigg, R.J. 2002. Injury in renal ischemia-reperfusion is independent from immunoglobulins and T lymphocytes. *Am J Physiol Renal Physiol* **282**:F352–F357.
201. Li, L., Huang, L., Sung, S.S., et al. 2007. NKT cell activation mediates neutrophil IFN-gamma production and renal ischemia-reperfusion injury. *J Immunol* **178**:5899–5911.
202. Zhang, M., Austen, W.G., Jr., Chiu, I., et al. 2004. Identification of a specific self-reactive IgM antibody that initiates intestinal ischemia/reperfusion injury. *Proc Natl Acad Sci USA* **101**:3886–3891.
203. Zhang, M., Alicot, E.M., and Carroll, M.C. 2008. Human natural IgM can induce ischemia/reperfusion injury in a murine intestinal model. *Mol Immunol* **45**:4036–4039.
204. Zhang, M., Michael, L.H., Grosjean, S.A., Kelly, R.A., Carroll, M.C., and Entman, M.L. 2006. The role of natural IgM in myocardial ischemia-reperfusion injury. *J Mol Cell Cardiol* **41**:62–67.
205. Chan, R.K., Verna, N., Afnan, J., et al. 2006. Attenuation of skeletal muscle reperfusion injury with intravenous 12 amino acid peptides that bind to pathogenic IgM. *Surgery* **139**:236–243.
206. Austen, W.G., Jr., Zhang, M., Chan, R., et al. 2004. Murine hindlimb reperfusion injury can be initiated by a self-reactive monoclonal IgM. *Surgery* **136**:401–406.
207. Zhang, M., Takahashi, K., Alicot, E.M., et al. 2006. Activation of the lectin pathway by natural IgM in a model of ischemia/reperfusion injury. *J Immunol* **177**:4727–4734.
208. Arumugam, T.V., Shiels, I.A., Strachan, A.J., Abbenante, G., Fairlie, D.P., and Taylor, S.M. 2003. A small molecule C5a receptor antagonist protects kidneys from ischemia/reperfusion injury in rats. *Kidney Int* **63**:134–142.
209. de Vries, B., Kohl, J., Leclercq, W.K., et al. 2003. Complement factor C5a mediates renal ischemia-reperfusion injury independent from neutrophils. *J Immunol* **170**:3883–3889.
210. De Vries, B., Matthijsen, R.A., Wolfs, T.G., Van Bijnen, A.A., Heeringa, P., and Buurman, W.A. 2003. Inhibition of complement factor C5 protects against renal ischemia-reperfusion injury: inhibition of late apoptosis and inflammation. *Transplantation* **75**:375–382.
211. Thurman, J.M., Ljubanovic, D., Edelstein, C.L., Gilkeson, G.S., and Holers, V.M. 2003. Lack of a functional alternative complement pathway ameliorates ischemic acute renal failure in mice. *J Immunol* **170**:1517–1523.
212. Zhou, W., Farrar, C.A., Abe, K., et al. 2000. Predominant role for C5b-9 in renal ischemia/reperfusion injury. *J Clin Invest* **105**:1363–1371.

213. Rabb, H., Ramirez, G., Saba, S.R., et al. 1996. Renal ischemic-reperfusion injury in L-selectin-deficient mice. *Am J Physiol* **271**:F408–F413.

214. Chiao, H., Kohda, Y., McLeroy, P., Craig, L., Linas, S., and Star, R.A. 1998. Alpha-melanocyte-stimulating hormone inhibits renal injury in the absence of neutrophils. *Kidney Int* **54**:765–774.

215. Bonventre, J.V. 2003. Molecular response to cytotoxic injury: role of inflammation, MAP kinases, and endoplasmic reticulum stress response. *Semin Nephrol* **23**:439–448.

216. Takada, M., Chandraker, A., Nadeau, K.C., Sayegh, M.H., and Tilney, N.L. 1997. The role of the B7 costimulatory pathway in experimental cold ischemia/reperfusion injury. *J Clin Invest* **100**:1199–1203.

217. Macedo, L., Pinhal-Enfield, G., Alshits, V., Elson, G., Cronstein, B.N., and Leibovich, S.J. 2007. Wound healing is impaired in MyD88-deficient mice: a role for MyD88 in the regulation of wound healing by adenosine A2A receptors. *Am J Pathol* **171**:1774–1788.

218. Godson, C., Mitchell, S., Harvey, K., Petasis, N.A., Hogg, N., and Brady, H.R. 2000. Cutting edge: lipoxins rapidly stimulate nonphlogistic phagocytosis of apoptotic neutrophils by monocyte-derived macrophages. *J Immunol* **164**:1663–1667.

219. Leonard, M.O., Hannan, K., Burne, M.J., et al. 2002. 15-Epi-16-(para-fluorophenoxy)-lipoxin A(4)-methyl ester, a synthetic analogue of 15-epi-lipoxin A(4), is protective in experimental ischemic acute renal failure. *J Am Soc Nephrol* **13**:1657–1662.

220. Goh, J., Godson, C., Brady, H.R., and Macmathuna, P. 2003. Lipoxins: pro-resolution lipid mediators in intestinal inflammation. *Gastroenterology* **124**: 1043–1054.

221. Serhan, C.N., and Chiang, N. 2002. Lipid-derived mediators in endogenous anti-inflammation and resolution: lipoxins and aspirin-triggered 15-epi-lipoxins. *Scientific World Journal* **2**:169–204.

29 Asthma

Bruce D. Levy

INTRODUCTION

Asthma is a disease of chronic airway inflammation. This condition is prevalent worldwide and accounts for significant morbidity, excess mortality, and substantial health care expenditures [1,2]. Asthma is clinically defined by three characteristics, namely reversible airflow obstruction, airway hyperresponsiveness, and airway inflammation [1]. There is no cure for asthma, but many therapies have been developed to lessen the burden of the disease. In light of the need for additional therapeutics to treat and ultimately cure asthma, several animal experimental models have been developed to perform preclinical investigation of asthma pathogenesis and novel therapeutics. Simply stated, asthma is only a human disease. None of the current animal models entirely recapitulates asthma [3], but they have proven very useful in the investigation of asthma traits. In this chapter, the most common animal models of asthma and their features will be described with particular attention to the airway inflammatory responses.

ASTHMA PATHOBIOLOGY

Asthma has a complex pathogenesis and can be considered a clinical syndrome of intermittent dyspnea, wheezing, chest tightness, and/or cough. In most subjects, airway inflammation is present [1]. The inflammatory cell infiltrate is enriched with eosinophils, T lymphocytes and, in some cases, neutrophils, especially in the setting of asthma exacerbations. This complex, chronic airway inflammation is likely initiated and driven by signals from sentinel cells in the airway, including airway epithelia and dendritic cells, responding to provocative stimuli. A broad range of respirable agonists can perpetuate or exacerbate asthmatic airway inflammation. Asthma has been clinically characterized as extrinsic or intrinsic. Extrinsic asthma results from exposure to antigens to which the patient is specifically allergic. This inflammatory response is principally driven by TH2 cytokines and is associated with elevated levels of serum IgE. It is most common in children and young adults. Of interest, the population-attributable risk for atopy on human asthma is less than 50% [3]. Intrinsic asthma has a more complex phenotype and is less commonly related to an allergic diathesis. Subjects with intrinsic asthma have respiratory symptoms that can be caused by one or more of several nonallergic airway provocative stimuli, including inhalation of chemicals such as cigarette smoke or cleaning agents, ingestion of aspirin, upper or lower respiratory tract infection, stress, loud talking or laughter, exercise, cold air, food preservatives, and several other factors. Some asthmatic individuals are characterized as having mixed asthma in which either allergic or nonallergic stimuli can trigger symptoms.

ANIMAL MODELS AS A RESEARCH TOOL FOR ASTHMA

The first animal models of asthma were developed over 100 years ago (reviewed in reference [4]). For a complex phenotype of inflammatory lung disease in asthma, an in vivo model of the disease is an ideal tool for investigating immunological processes at the cellular, molecular, and genetic levels. While in vitro systems utilizing cells obtained from asthmatic individuals can be useful for modeling aspects of the disease in translational research, the in vitro culture systems are devoid of the complex interactions between cell types within the lung and the circulation, the physical processes of respiration and humoral, hormonal, and autacoid signals that are present in vivo. Animal models have not been developed to fully resemble human asthma, but they are quite useful for investigation of asthma traits. There are several examples of the identification of pivotal mechanisms for asthma pathobiology via animal models. Recognition of the importance of TH2 cytokines, including interleukin-5 and interleukin-13, in the development of asthma emerged from animal models [5,6]. These in vivo systems have also proved useful in drug discovery, and several current asthma therapies were successfully tested in preclinical studies in both small and large animal models of asthma. Because there are also examples of pathogenetic findings and drug candidates effective in animal testing that have failed in clinical trials for human asthma [7,8], it is critical to remember that these in vivo experimental systems are only models. Relevance of the findings in animals for asthma can only be established through rigorous human clinical investigation.

MURINE MODELS OF ASTHMA

Advantages

Murine experimental models of asthma are ideal for investigating allergic airway responses. In addition to practical concerns, such as relatively low costs and ease of housing, mice are a non-endangered species that afford several distinct advantages as an animal model [9]. There is now a detailed understanding of mouse genetics that can be manipulated by gene deletion and insertion technology to assay the influence of signaling pathways on airway inflammation and hyperresponsiveness. Numerous mouse-specific probes for cellular and molecular effectors in allergy and inflammation are available. There are several strains of inbred mice that vary in the degree of airway inflammation and hyperresponsiveness that result from allergen sensitization and challenge [10,11]. On the other hand, mice also have limitations for modeling asthma. Unlike humans, they do not spontaneously develop allergy or asthma [12], so they are most useful for modeling asthma traits rather than the entire asthma phenotype [13,14].

Asthma Traits

Allergic Airway Inflammation

To model allergic airway inflammation in mice, animals are first sensitized to an allergen and then challenged by respiratory tract exposure to the same allergen [4,11,15,16]. In the most common models, an allergen is combined with an adjuvant, such as aluminum hydroxide, to initiate a strong TH2 phenotype. The most common allergen in use is chicken ovalbumin (OVA), but many other allergens have been utilized in animal models of asthma. Many investigators have chosen to use allergens that are more relevant than OVA to human asthma, such as ragweed, cockroach, cat (FelD1), and dust mite (DerP1) allergens. To induce airway inflammation, the allergen is delivered to the respiratory tract via aerosol. The resulting inflammatory cell infiltrate, predominantly eosinophils and T lymphocytes, is prominent in medium to small airways as well as alveoli (Figure 29.1). There is also perivascular inflammation. Many of these features are distinct from human asthma in which the airway inflammation is centered predominantly around medium to small airways and generally spares the alveolar spaces [1,3]. Some investigators have chosen to both sensitize and challenge animals via the airway, as this leads to a more robust peribronchial inflammatory infiltration in which mast cells play a more integral role in the response to allergen [17,18].

Airway Hyperresponsiveness

Airways constrict in responsive to provocative stimuli. A rapid change in airway caliber can lead to difficulty breathing, cough, wheezing, or chest tightness. These are the cardinal symptoms of asthma and, not surprisingly, most of the morbidity associated with asthma stems from a susceptibility in the airways of asthmatics to provocation by stimuli that in healthy subjects are well-tolerated. While some airways responsiveness can be attributed to specific allergens, asthmatic airways will also display hyperresponsiveness to nonspecific stimuli, including cigarette smoke, perfumes, and airborne pollutants. Other triggers for airway hyperresponsiveness include pharmacological stimuli, such as methacholine and histamine, and physical stimuli, such as cold, dry air.

Chronic airway hyperresponsiveness is fundamental to asthma, but has yet to be successfully replicated in experimental animal models of asthma. Of interest, antigen-induced responses often increase both airway inflammation and hyperresponsiveness [12]. When an asthmatic individual has an allergen-initiated late phase response (*vide infra*), their airway responsiveness will be increased for the next few days or weeks. This acute to subacute allergen-induced airways hyperresponsiveness can develop in several animal models of asthma and has been used by many investigators to study the mechanisms and regulation of this phenotype. A typical provocative agent for mice is methacholine that can be administered via inhalation when animals are intubated and sedated on a ventilator circuit. While it is not possible to have mice perform a forced expiratory maneuver, the methacholine-initiated changes in measures of airway physiology, such as lung resistance (R_L), can be monitored in the ventilated animals during tidal breathing. Mice given increasing doses of methacholine will have marked increases in R_L (Figure 29.2). The dose–response relationship or the effective dose of methacholine leading to a 200% increase in

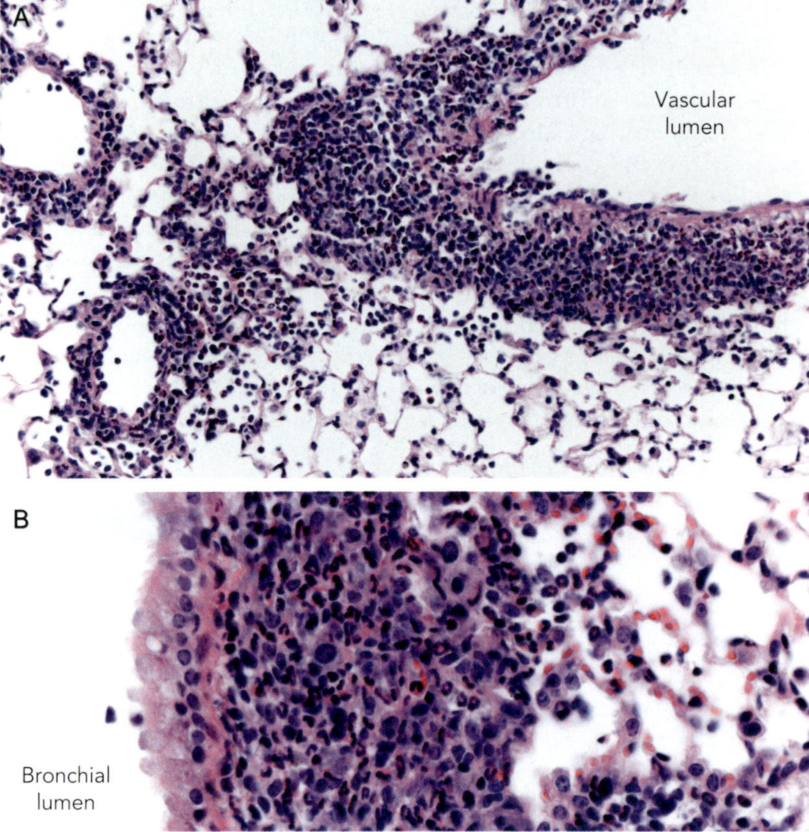

Figure 29.1. Histopathology of allergic airway inflammation in mice. After sensitization by intraperitoneal injection with chicken OVA and aluminum hydroxide, mice were exposed 20 minutes a day to inhaled OVA for 4 consecutive days. After 24 hours, the last airway allergen challenge, mice were killed and lungs prepared for histology. Representative lung tissue sections (magnifications: ×200 (A), ×400 (B), were obtained from fixed, paraffin-embedded lung tissue, prepared and stained with hematoxylin and eosin. Bronchial and vascular lumens are indicated.

lung resistance (aka ED_{200}) can be compared between mice that are genetically distinct or have been exposed to different experimental interventions.

Example of a Protocol

The most common murine model employs chicken OVA as an allergen for systemic sensitization and then proceeds to expose sensitized animals to an OVA aerosol to direct the allergic inflammation to the airways. Aluminum hydroxide (alum) is used as an adjuvant to generate a strong TH2 immune response with the production of antigen-specific IgE. These adaptive immune responses differ by mouse age, sex, and strain, so it is important to control carefully for these variables. An example of a typical experimental protocol [19] follows:

a. To initiate the model (Protocol Day 0), OVA (grade III, 50 µg) and aluminum hydroxide (2 mg) are mixed and the solution (200 µL) is injected intraperitoneally. For control animals, alum without OVA in 0.9% saline is injected.
b. The allergic reaction is boosted with a second injection of antigen (OVA/alum 50 µg/2 mg; intraperitoneally) one week later (Protocol Day 7).
c. Allergen challenge is performed on protocol day 14. The animals are individually exposed to an aerosol of 6% OVA in 0.9% saline for 20 minutes (Figure 29.3). This is repeated daily for a total of four consecutive days (Protocol days 14, 15, 16, and 17). Control animals are exposed similarly except to saline alone without OVA. It is best to utilize separate nebulization systems for OVA and saline controls as even small amounts of antigen can induce a response in control animals.
d. On protocol day 18 (i.e., 24 hours after the last allergen challenge), the lung's immune response can be characterized by either bronchoalveolar lavage, histology or testing for airway hyperresponsiveness (AHR). Because AHR testing can lead to mediator release with increases in lung resistance, it is best to use separate animals for AHR testing and assessment of lung inflammation (BAL or histology).

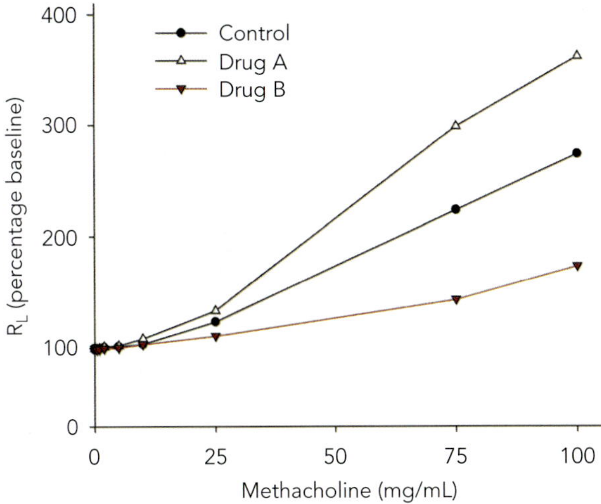

Figure 29.2. Measurement of airway hyperresponsiveness in ventilated mice. Illustration of a representative dose–response relationship for methacholine-induced changes in lung resistance (R_L) in OVA sensitized and airway challenged mice that were exposed to drug A (upward white triangle), drug B (downward red triangle), or a vehicle control (black circle). Drug A led to a relative increase in R_L compared to the control animals at doses of methacholine greater than 10 mg/mL, thus increasing airway hyperresponsiveness. Drug B had the opposite actions, leading to a dampening of airway responsiveness to methacholine.

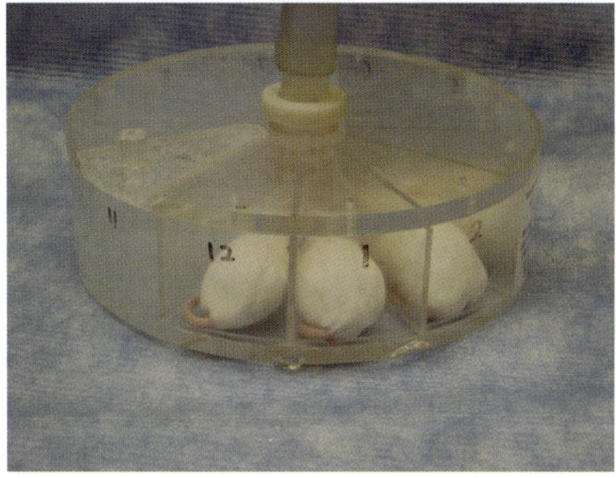

Figure 29.3. Exposure of mice to inhaled allergen. Two views of a nebulization apparatus in which animals are placed into individual chambers. Once loaded, allergen is nebulized through the central port for equal distribution to each animal. Separate devices (including tubing and nebulizers) are used for allergen and 0.9% saline controls to avoid cross contamination.

Factors Influencing Host Response to Antigen

a. Nature of allergen – Allergen sensitization induces IgE production and a TH2 type immune response. Chicken OVA is most commonly used because it is simple to control the diet and environment of inbred animals so that the investigator can be certain the animals have not been previously exposed to chicken OVA. In addition to OVA, allergic responses can be initiated in mice with several other antigens, including shared human antigens such as ragweed, FelD1, and DerP1 [16].
b. Adjuvant – The use of adjuvants can augment or skew the immune response depending on the nature of the adjuvant and its concurrent administration with allergen. Examples of common adjuvants include aluminum hydroxide that provides a preferential TH2 immune response and Freund's adjuvant that leads to a TH1 type immune response [4].
c. Route and mode of sensitization – In addition to systemic sensitization, mice can also be sensitized directly via the upper or lower respiratory tract. Distinct from the more common methods of initial systemic sensitization in which allergen is given intraperitoneally or subcutaneously, if the allergen is first provided via the respiratory tract prior to intraperitoneal injection then the systemic TH2 phenotype is dampened with little or no IgE generated, and there is attenuation of eosinophil recruitment to the lung after subsequent aerosol antigen challenge [20].
d. Genetics – There are clear genetic influences on both allergic airway inflammation and AHR. Mice strains can differ significantly in these traits [10]. For example, both traits can be studied in BALB/c mice, yet these animals easily develop significant eosinophilia but are relatively more resistant to methacholine-induced AHR. The opposite is true for the C57Bl6 strain.

Early versus Late Response

Many human asthmatic individuals experience both an early and late response to inhaled allergen. The early-phase response lasts approximately 30–60 minutes and is principally characterized by bronchospasm from released mediators that trigger airway smooth muscle constriction. Roughly half of individuals

TABLE 29.1. Comparison between the structure of mouse and human respiratory systems

Parameter	Human	Mouse
Respiratory rate	15–20 breaths per minute	~150 breaths per minute
Tidal volume	500 mL	0.150 mL
Functional residual capacity	2.4–3.0 L (~35–45 μL/g body wt)	~14–17 μL/g body weight (more compliant chest wall)
Airway size and branching	>20 branching airways with gas exchanging respiratory bronchioles	Wider relative to body size with 6–8 branching airways with gas exchanging respiratory bronchioles
Airway smooth muscle	Present in several generation bronchi	Only present in first generation bronchi
Submucosal glands	Abundant in large and medium-sized airways	Only in trachea
Testing for AHR	Forced expiratory maneuver	Intubated, sedated on a ventilator, and measured during tidal breathing
Mediator responses	Bronchoconstriction to LTC_4/D_4, histamine, and neurokinins	Bronchoconstriction to cholinergic agonists and serotonin. No response to LTC_4/D_4, histamine, and neurokinins
Housing conditions	Dirty environment with exposures to indoor/outdoor allergens and irritants	Clean, often pathogen-free barrier facility

experiencing an early-phase response will also experience a late-phase response. The bronchoconstriction of the late-phase response is more often multifactorial with contributions from inflammatory cells that have entered the lung in response to the allergen challenge, as well as the consequent airway smooth muscle contraction, mucus, and edema resulting from the release of their provocative mediators. Distinct from agonist (e.g., methacholine) induced airway hyperresponsiveness, a limitation of the mouse model is that it appears that these responses either do not occur or are minimal in mice.

Measuring Airway HyperResponsiveness

The respiratory system of the mouse is quite distinct from humans (Table 29.1), but humans and mice share susceptibility for the development of airways hyperresponsiveness [14]. Increased airway reactivity is a diagnostic hallmark of asthma. This asthma trait can be measured in animals by either invasive or noninvasive methods. The most common methods for mice use methacholine to initiate airway constriction that increases airflow obstruction. Noninvasive measurement of lung function utilizes barometric plethysmography of unrestrained animals, while invasive measures of lung resistance are performed on sedated, tracheotomized, and mechanically ventilated mice.

Plethysmography is performed by placing spontaneously breathing animals into a whole-body plethysmograph [15]. Mice are then exposed to inhaled methacholine or 0.9% saline as a control. Methacholine is a nonspecific irritant and is also useful to measure AHR in humans. With plethysmography, changes in airway function are measured and expressed as an enhanced pause (Penh). This parameter is empirically derived and intended to reflect changes in the inspiratory and expiratory box flow waveforms in conjunction with early and late expiratory box flows. The principal advantage of this method is that the Penh measurement can be repeatedly performed on living, spontaneously breathing animals. The disadvantage is that Penh is a dimensionless parameter and not a direct measure of lung physiology. Although Penh appears to correlate with lung resistance, this disadvantage has limited its use by many investigators.

A more direct method of measurement of changes in airway physiology is performed in sedated and mechanically ventilated animals (Figure 29.4) [21]. With this technique, changes in volume relative to elastic recoil pressure changes between end inspiration and end expiration are used to determine lung resistance and dynamic compliance. The mice are tracheotomized and ventilated via a cannula that is connected to a manifold that features an in-line aerosol nebulizer to provide methacholine or saline alone. Lung resistance is a measure of resistance to airflow and is determined by flow, airway size, and lung tissue resistance (Figure 29.2). The murine chest wall is so compliant that its relative contribution is negligible (Table 29.1). Because the mouse chest wall is substantially more compliant than the human chest wall,

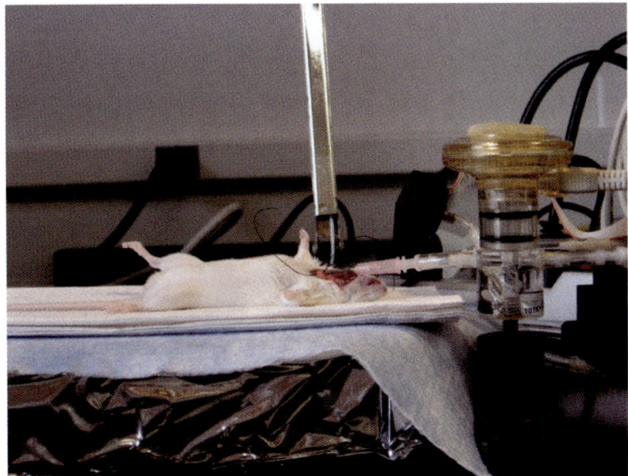

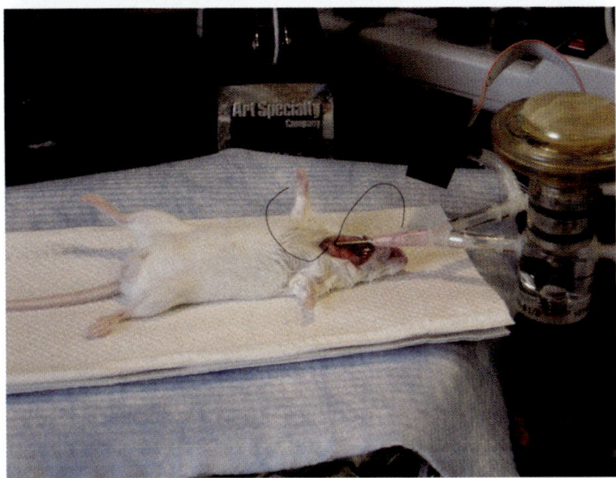

Figure 29.4. Measurement of methacholine-induced airway responsiveness. Two views of an anesthetized mouse (~20 g) intubated with a 19-G tracheal tube and ventilated (150 breaths per minute, 200 μL tidal volume). The neck has been opened and the skin and neck muscles dissected to expose the trachea. After measurement of baseline values, saline or methacholine is delivered by in-line nebulizer (right) into the inhalation port.

the functional residual capacity is much lower and the respiratory rate much higher. Lung tissue resistance is higher at lower respiratory rates, but mice breathe at rates greater than 120 per minute. Thus, when measured at baseline in ventilated animals, total lung resistance (airway resistance plus lung tissue resistance) will principally reflect airway resistance with little contribution from lung tissue resistance. Methacholine administration can affect both airway and lung tissue resistance, so in challenged animals it cannot be assumed that total lung resistance is primarily a reflection of airway constriction. Of interest, the route of methacholine administration can be an important influence of these parameters. Intravenous methacholine will constrict large and small airways and aerosolized methacholine will impact both airway and lung tissue resistance.

Modeling Chronic Airway Inflammation

Because asthma is a disease of chronic inflammatory changes with remodeling of the airway structure, several models have been developed to better reflect these pathological changes in mice. Of interest, there have been recent reports of murine models that repeatedly expose the animals to allergen for extended periods of time (e.g., 75 days) leading to the development of both allergic airway inflammation and airway structural changes, including goblet cell hyperplasia and deposition of collagen beneath airway epithelial cells and in the interstitium [22]. The timing of allergen exposure is in a chaotic pattern because the regular and repeated exposure of mice to allergen will lead to tolerance. While longer-term models of allergen challenge may provide a pathologic airway that more closely resembles the human condition, there are still insurmountable challenges in the differences in airway structure and function (see Table 29.1). Still these more chronic experimental models hold the promise of better understanding pathophysiological factors during longer-term airway inflammation and interactions between epithelial and inflammatory cells in the process of airway fibrosis.

Modeling Resolution

In most instances, the airway inflammation of asthma does not resolve completely. However, in health, inhalation of potential allergens or provocative stimuli leads to an acute inflammatory response that is self-limited. In asthmatic individuals, avoidance of allergen or provocative stimuli can passively decrease inflammation to restore airway homeostasis. However, a growing body of evidence supports a signaling network that actively promotes the resolution of inflammation in tissue catabasis [23]. Several classes of natural anti-inflammatory chemical mediators have been identified that are produced at sites of inflammation, including the lung. Lipoxins, protectin D1, and resolvin E1 can all prevent the development allergic airway responses in a murine model of asthma [19,21,24,25]. Because the clinical presentation of asthma is after the disease has already developed, more recent investigations have focused on the natural factors that promote resolution of allergic airway responses and mechanisms that counter these protective signals to perpetuate inflammation and maladaptive airway responses. Investigation of the resolution phase of allergic airway responses during the week after cessation of allergen exposure has recently uncovered important roles for IL-17 in the persistence of airway inflammation [24]. Inhibition of IL-17 production appears to be a common point of regulation for lipoxins, resolvin E1, and protectin D1 that can also accelerate the pace of resolution of allergic airway

TABLE 29.2. Airway inflammation and hyperresponsiveness – comparison of animal models to human asthma

Parameter	Human	Mouse	Rat	Guinea pig	Dog	Sheep	Monkey
Natural allergy	Yes	No	No	No	Yes	Yes	Yes
Airway inflammation – after allergen sensitization and challenge	Yes	Yes	Yes	Yes	Yes	Yes	Yes
Airway hyperresponsiveness	Yes	Yes	Yes	Yes	Yes	Yes	Yes
Airway response to allergen – early phase	Yes	Maybe	Yes	Yes	Yes	Yes	Yes
Airway response to allergen – late phase	Yes	No	Yes	Yes	Yes	Yes	Yes

responses [24,25]. Of interest, the anti-inflammatory and proresolving mediators lipoxin A_4 and protectin D1 are generated in lower amounts during severe and uncontrolled human asthma, reflecting a deficiency in these conditions in the natural protective mechanisms for airway homeostasis [25,26].

ADDITIONAL ANIMAL MODELS

Although mice are now most commonly used for investigation of asthma pathogenesis, several other animal models are in use. A few of the more common species and models are described in this section and reviewed in Table 29.2.

Rat

Rats are a popular species for models of asthma (reviewed in [4]). Similar to mice, it is easy to measure cellular and biological markers of allergic inflammatory disease in blood and bronchoalveolar lavage, and airway hyperresponsiveness can be measured. The most common strain in use is the Brown Norway rat. At present, the availability of reagents and ease of gene insertion or deletion is more limited for rats than for mice. Typical protocols involve systemic sensitization to any of a wide range of antigens, including OVA. Once sensitized, Brown Norway rats display increased airway hyperresponsiveness and acute bronchoconstriction to inhaled specific allergen. Similar to human asthma and unlike mice, sensitized rats experience both an early and late phase bronchoconstricting response to allergen. Also unlike mice, the early and late phase responses in the rat are dependent on cysteinyl leukotrienes [27].

Guinea Pig

Scientists have used guinea pigs in the investigation of asthma pathobiology for over a century [4]. Guinea pig models of asthma were crucial to the testing of two of the most common classes of asthma medications, namely β_2-receptor agonists and corticosteroids. These animals are easily sensitized to antigens and when aerosol challenged, guinea pigs will develop significant eosinophilia and airway hyperresponsiveness. The two principal allergens used with guinea pigs are *Ascaris suum* and OVA. Isolated guinea pig airways responses are very similar to human airways, with a few notable exceptions (e.g., leukotrienes). Allergen sensitized and challenged guinea pigs develop both early and late phase responses with an early phase response that can be so strong that pretreatment with antihistamines are necessary to prevent death. Guinea pigs display high baseline levels of eosinophils. In addition to IgE, IgG1 also plays important roles in allergic responses in these animals. Relative to rodents, the utility of guinea pigs in asthma models is limited by increased variability in responsiveness, many fewer reagents, and inbred strains, plus a lack of available genetic technology.

Dogs

More recently, dogs have served as a very useful preclinical model of asthma [28]. These animals will spontaneously develop allergic responses with atopic dermatitis, rather than asthma, as the natural expression of this allergic diathesis. The dog airways are proportionally larger than most mammals and may be less prone to bronchoconstriction. However, allergic reactions of both the upper and lower respiratory tracts can be elicited by allergen sensitization and aerosol challenge. These responses can be amplified by selectively breeding dogs that spontaneously generate high levels of IgE. Shortly after birth, puppies are systemically sensitized to a relevant human allergen, such as ragweed pollen, at regular intervals. After they mature, the sensitized dogs are aerosol challenged and then systemic and airway markers of inflammatory responses determined. Airway hyperresponsiveness can also be characterized after aerosol challenge, but repeated challenge can lead to AHR that can last months prior to resolution. While this species is quite useful for preclinical testing, the maintenance of colonies of allergic dogs is expensive and labor intensive.

Sheep

Like dogs, sheep can spontaneously develop allergy [4,12]. Of interest, when sensitized animals are allergen challenged by aerosol, the airway inflammatory infiltrate contains both eosinophils and neutrophils. Similar to humans, allergen challenge will lead to airway hyperresponsiveness as well as both early and late phase responses. Mediators recovered from BAL fluids are quite similar when compared to humans, namely leukotrienes, histamine, prostaglandins D_2 and E_2, and kinins. The ease of induction of human-like asthmatic traits in these animals must be balanced against the high cost and labor involved in tending to their care.

Non-Human Primates

Rhesus, squirrel, and cynomolgus monkeys will develop a natural sensitivity to *Ascaris suum* antigen that can be useful in challenge testing [12]. Airway responses to allergen challenge are quite similar to humans with an immediate bronchoconstriction, increase in airways resistance, and change to a more rapid shallow breathing pattern. Colonies of cynomolgus monkeys with both early and late phase responses have been bred and are currently used as preclinical asthma models.

SUMMARY AND CONCLUSIONS

In summary, there are several animal models of asthma traits, but no convincing experimental model that replicates human asthma. Experimental models of allergic airway inflammation and hyperresponsiveness have uncovered many cellular, molecular, and genetic features of the pathogenesis and pathophysiology of asthma. Mouse models of asthma are useful for investigations of airway inflammation, epithelial mucus, and airway hyperresponsiveness, but larger animal models are preferred for investigation of early and late phase responses to allergens. No matter the preclinical animal model chosen for study, it is essential to perform clinical translational research in human subjects with asthma to establish the relevance of experimental findings in animals. While much has been learned from these models that governs the early onset and upstroke of airway inflammation, there is still much to learn from future study of the mechanisms that lead to chronic inflammation in disease and the mechanisms that propel the lung toward resolution in health.

KEY POINTS

1. Animal models are useful experimental tools for preclinical investigation into the underlying mechanisms and potential treatment of select asthma traits, namely allergic airway inflammation and airway hyperresponsiveness.
2. Murine models of asthma have several advantages over other species, including relatively low cost, ease of housing, availability of probes, status as a non-endangered species, and importantly there is now a detailed understanding of mouse genetics that can be manipulated by gene deletion and insertion technology to assay the influence of specific genes on airway inflammation and hyperresponsiveness.
3. It is important to understand limitations of murine models of asthma. There are structural differences between murine and human respiratory systems, and unlike humans and several larger animals, mice do not spontaneously develop allergy.
4. Methods have been developed to monitor the degree of allergic airway responses in animals. Airway inflammation can be determined by bronchoalveolar lavage and histology. Airway hyperresponsiveness can be measured by either noninvasive (Penh) or invasive (R_L) means.

REFERENCES

1. Busse, W.W., and Lemanske, R.F., Jr. 2001. Asthma. *N Engl J Med* **344**:350–362.
2. Serra-Batlles, J., Plaza, V., Morejon, E., Comella, A., and Brugues, J. 1998. Costs of asthma according to the degree of severity. *Eur Respir J* **12**:1322–1326.
3. Wenzel, S., and Holgate, S.T. 2006. The mouse trap: it still yields few answers in asthma. *Am J Respir Crit Care Med* **174**:1173–1176; discussion 1176–1178.
4. Zosky, G.R., and Sly, P.D. 2007. Animal models of asthma. *Clin Exp Allergy* **37**:973–988.
5. Hamelmann, E., Takeda, K., Schwarze, J., Vella, A.T., Irvin, C.G., and Gelfand, E.W. 1999. Development of eosinophilic airway inflammation and airway hyperresponsiveness requires interleukin-5 but not immunoglobulin E or B lymphocytes. *Am J Respir Cell Mol Biol* **21**:480–489.
6. Wills-Karp, M., Luyimbazi, J., Xu, X., et al. 1998. Interleukin-13: central mediator of allergic asthma. *Science* **282**:2258–2261.
7. Bryan, S.A., O'Connor, B.J., Matti, S., et al. 2000. Effects of recombinant human interleukin-12 on eosinophils, airway hyper-responsiveness, and the late asthmatic response. *Lancet* **356**:2149–2153.
8. Gavett, S.H., O'Hearn, D.J., Li, X., Huang, S.K., Finkelman, F.D., and Wills-Karp, M. 1995. Interleukin 12 inhibits antigen-induced airway hyperresponsiveness, inflammation, and Th2 cytokine expression in mice. *J Exp Med* **182**:1527–1536.
9. Shapiro, S.D. 2006. Animal models of asthma: pro: allergic avoidance of animal (model[s]) is not an option. *Am J Respir Crit Care Med* **174**:1171–1173.
10. Drazen, J.M., Takebayashi, T., Long, N.C., De Sanctis, G.T., and Shore, S.A. 1999. Animal models of asthma and chronic bronchitis. *Clin Exp Allergy* 29(Suppl 2):37–47.
11. Kips, J.C., Anderson, G.P., Fredberg, J.J., et al. 2003. Murine models of asthma. *Eur Respir J* **22**:374–382.

12. Abraham, W.M., and Baugh, L.E. 1995. Animal models of asthma. In *Asthma and Rhinitis,* W.W. Busse and S.T. Holgate (eds), pp. 961–977. Boston: Blackwell Scientific Publications.
13. Finkelman, F.D., and Wills-Karp, M. 2008. Usefulness and optimization of mouse models of allergic airway disease. *J Allergy Clin Immunol* **121**:603–606.
14. Irvin, C.G. 2008. Using the mouse to model asthma: the cup is half full and then some. *Clin Exp Allergy* **38**:701–703.
15. Corry, D.B., and Irvin, C.G. 2006. Promise and pitfalls in animal-based asthma research: building a better mouse-trap. *Immunol Res* **35**:279–294.
16. Pichavant, M., Goya, S., Hamelmann, E., Gelfand, E.W., and Umetsu, D.T. 2007. Animal models of airway sensitization. *Curr Protoc Immunol* Chapter 15, Unit 15; 18.
17. Takeda, K., Hamelmann, E., Joetham, A., et al. 1997. Development of eosinophilic airway inflammation and airway hyperresponsiveness in mast cell-deficient mice. *J Exp Med* **186**:449–454.
18. Taube, C., Miyahara, N., Ott, V., et al. 2006. The leukotriene B4 receptor (BLT1) is required for effector CD8+ T cell-mediated, mast cell-dependent airway hyperresponsiveness. *J Immunol* **176**:3157–3164.
19. Levy, B.D., De Sanctis, G.T., Devchand, P.R., et al. 2002. Multi-pronged inhibition of airway hyper-responsiveness and inflammation by lipoxin A(4). *Nat Med* **8**:1018–1023.
20. Holt, P.G., Batty, J.E., and Turner, K.J. 1981. Inhibition of specific IgE responses in mice by pre-exposure to inhaled antigen. *Immunology* **42**:409–417.
21. Levy, B.D., Lukacs, N.W., Berlin, A.A., et al. 2007. Lipoxin A4 stable analogs reduce allergic airway responses via mechanisms distinct from CysLT1 receptor antagonism. *FASEB J* **21**:3877–3884.
22. Henderson, W.R., Jr., Tang, L.O., Chu, S.J., et al. 2002. A role for cysteinyl leukotrienes in airway remodeling in a mouse asthma model. *Am J Respir Crit Care Med* **165**:108–116.
23. Serhan, C.N. 2007. Resolution phase of inflammation: novel endogenous anti-inflammatory and proresolving lipid mediators and pathways. *Annu Rev Immunol* **25**:101–137.
24. Haworth, O., Cernadas, M., Yang, R., Serhan, C., and Levy, B. 2008. Resolvin E1 regulates interleukin 23, interferon-gamma and lipoxin A4 to promote the resolution of allergic airway inflammation. *Nat Immunol* **9**:873–879.
25. Levy, B.D., Kohli, P., Gotlinger, K., et al. 2007. Protectin D1 is generated in asthma and dampens airway inflammation and hyperresponsiveness. *J Immunol* **178**:496–502.
26. Levy, B.D., Bonnans, C., Silverman, E.S., et al. 2005. Diminished lipoxin biosynthesis in severe asthma. *Am J Respir Crit Care Med* **172**:824–830.
27. Sapienza, S., Eidelman, D.H., Renzi, P.M., and Martin, J.G. 1990. Role of leukotriene D4 in the early and late pulmonary responses of rats to allergen challenge. *Am Rev Respir Dis* **142**:353–358.
28. Barrett, E.G., Rudolph, K., Bowen, L.E., and Bice, D.E. 2003. Parental allergic status influences the risk of developing allergic sensitization and an asthmatic-like phenotype in canine offspring. *Immunology* **110**:493–500.

SUGGESTED READINGS

Busse, W.W., and Lemanske, R.F., Jr. 2001. Asthma. *N Engl J Med* **344**:350–362.

Corry, D.B., and Irvin, C.G. 2006. Promise and pitfalls in animal-based asthma research: building a better mouse-trap. *Immunol Res* **35**:279–294.

Finkelman, F.D., and Wills-Karp, M. 2008. Usefulness and optimization of mouse models of allergic airway disease. *J Allergy Clin Immunol* **121**:603–606.

Kips, J.C., Anderson, G.P., Fredberg, J.J., et al. 2003. Murine models of asthma. *Eur Respir J* **22**:374–382.

Pichavant, M., Goya, S., Hamelmann, E., Gelfand, E.W., and Umetsu, D.T. 2007. Animal models of airway sensitization. *Curr Protoc Immunol* Chapter 15, Unit 15; 18.

Zosky, G.R., and Sly, P.D. 2007. Animal models of asthma. *Clin Exp Allergy* **37**:973–988.

30 Animal Models of Rheumatoid Arthritis

H.B. Patel, B. Dawson, F. Humby, M. Blades, C. Pitzalis, M. Burnet, and M. Seed

RHEUMATOID ARTHRITIS

Rheumatoid arthritis (RA) is an extremely painful, debilitating, and destructive inflammatory disease of diarthrodial joints. It affects between 0.5% and 1% of the world's population, with women having a threefold prevalence. Using traditional therapy, the majority will develop moderate disability at 2 years, with 40% being unable to work at 5 years. The disease has a high cost in pain, disability, and deformity. Morbidity is high, with a reduced lifespan. The advent of new treatment paradigms has meant that disability has been reduced, orthopedic surgeries have a much reduced rheumatoid joint replacement case-load, and whilst still severe, the aim of therapy is to preserve normal lifestyle and work patterns.

This is achieved by the use of low-dose methotrexate, leflunomide, or sulphasalazine, followed by biologic therapies, mainly anti-tumor necrosis factor (anti-TNF-α), anti-B cell (anti-CD20), recombinant human IL-1 receptor antagonist (rhIL-1ra), or anti-interleukin-6 (anti-IL-6). There remains a severe problem that despite these regimes a significant proportion (up to 40% for anti-TNF-α) do not respond. In addition, these treatments are expensive in their own right, and heavy on clinical resources for administration and monitoring. There is thus a continuing requirement for the development of improved therapeutics through drug discovery and further development of current therapeutics and their targets. A detailed understanding of the pathogenic mechanisms of RA are also required to fulfill these aims. Animal models of rheumatic disease continue to play a significant role in this process.

Rheumatoid arthritis comprises an inflammatory granulomatous tissue termed pannus that invades diarthrodial joints eroding bone and cartilage as it does so. This results in profound deformity. If left to run its course, joints become totally devoid of cartilage, with severely eroded bone, and ultimately ankylosis. After initiation, pannus develops from the cartilage – pannus junction with angiogenesis, to erode into the bone and cartilage, filling the joint space. The synovial lining thickens from 1–2 cells to 6 or more, with an active recruitment of neutrophils, monocytes, and T- and B-lymphocytes (Figure 30.1). Pannus forms villus projections that grow into the synovial space. The synovial fluid contains neutrophils and macrophages, with some fibrin and tissue fragments. The granulomatous tissue has a large monocyte lineage component, possesses mast cells, fibroblasts, plasma cells, and dendritic cells, and can become organized into regions of T cells or B cells to form germinal centers that can be found in approximately 10% of cases (Figure 30.1).

At the synovial interface with cartilage and bone, the destructive nature of the disease is played. Erosion is initiated at the point of ligament insertion at the periosteum, where cartilage, synovium, and bone meet. The recruitment of monocytes to the area, coupled with RANKL and M-CSF signaling, differentiate into osteoclast precursors and thank osteoclasts. These resorb bone to such an extent that the erosion penetrates the subchondral bone to finally reach into the marrow. Inflammation is then initiated in the bone marrow that again feeds the synovial inflammation to ultimately result in the marginal erosion characteristic of the disease. Osteoblasts are either scarce or suppressed, otherwise they have the potential to reverse this process. Living bone is degraded by multinucleate osteoclasts.

Cartilage is eroded by proteinases sourced from synoviocytes, synovial fibroblasts, neutrophils, and the chondrocytes themselves from within the cartilage, which are responding to cytokines such as IL-1. Aggrecanases degrade proteoglycan to reveal collagen to be degraded by MMP3. The counter-regulatory metalloproteinase inhibitory pathways, comprising the tissue inhibitors of metalloproteinase (TIMPs) appear not to be upregulated, and thus ineffective.

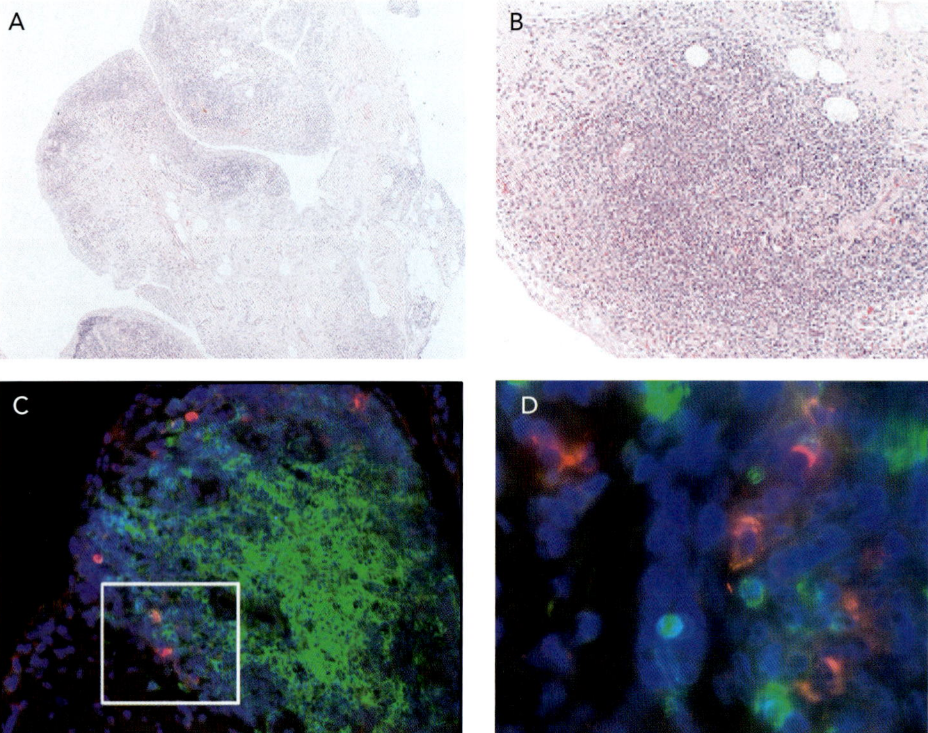

Figure 30.1. Rheumatoid synovium. (A) General overview of an RA synovial villus showing intense infiltration and synovial lining layer proliferation (H&E). (B) Higher power view of a typical lymphoid aggregate from the same patient (H&E). (C) and (D) B cells (stained with anti-CD20, green) producing anticitrullinated protein antibodies (stained with citrullinated fibrin, red) surrounding a B-cell aggregate. (A and B) M.Blades & B Hands Biobank PEAK (MRC, UK) ID = 038/09, Knee, Grade 3 RA Disease. (C and D) F. Humby et al., *PLoS Med.* 2009 Jan 13;6(1):e1. Reproduced with kind permission from the publisher.

RA can be split into sero-positive and sero-negative forms, with the production of rheumatoid factor, an IgM antibody to IgG, anticitrullinated peptide IgG antibodies (ACAP) and the stress protein immunoglobulin binding protein (BiP). Both ACAP and anti-BiP pre-date the clinical manifestations of rheumatoid arthritis.

The development of this reaction has a multifactorial genetic component, many of the mutations appearing to indicate a role of the T cell in much of the rheumatoid population [1]. Mutations have been found in the self epitope (SE) alleles of the MHC class II complex of the HLA-DR used by CD4 helper cells, as well as the MHC class II molecule DR3, the inhibitory CTLA-4 T-cell ligand for accessory cell CD80/CD86, and protein tyrosine phosphatase N22 (PTPN22) that functions to suppress T-cell responsiveness. Other non-T-cell variations include peptidyl arginine deiminase type IV (PAD) involved in protein citrullination at sites of inflammation to which an autoimmune antibody response is raised, and migration inhibitory factor. The prevalence of the HLA-DR4 SE gene modifications has given rise to the shared epitope hypothesis. However, a single shared epitope has not been found and it should be noted that twin concordance is only 15%.

Numerous factors have been proposed through the reactivity of T cells or antibodies from some patients to amongst others, peptidoglycans, collagen, and Epstein–Barr virus, and more recently citrullinated peptides, each being shown to be arthritogenic or enhance arthritis through animal modeling. These all point to the central role of not only the T cell, but also the B cell. Anti-CD20 anti-B-cell therapy is highly successful [2]. One consideration has been that at the initiation of RA disease there is a restricted antigen, and as chronicity develops, epitope drift occurs with the disease becoming self-sustaining. Peptides citrullinated at sites of inflammation through the action of PAD, such as the lungs of smokers, are considered to provide an explanation for the strongest environmental link to RA. Rheumatoid pannus contains citrullinated peptides (Figure 30.1). In addition it synthesizes RF and citrullinated antibodies (Figure 30.1), and when transplanted into immunodeficient SCID mice is remarkably robust, retaining its architecture, cytokine synthetic profile, and capacity to produce these antibodies for several weeks.

Figure 30.2 illustrates in a simplified form how these cells interact. Rheumatoid T cells express CD69

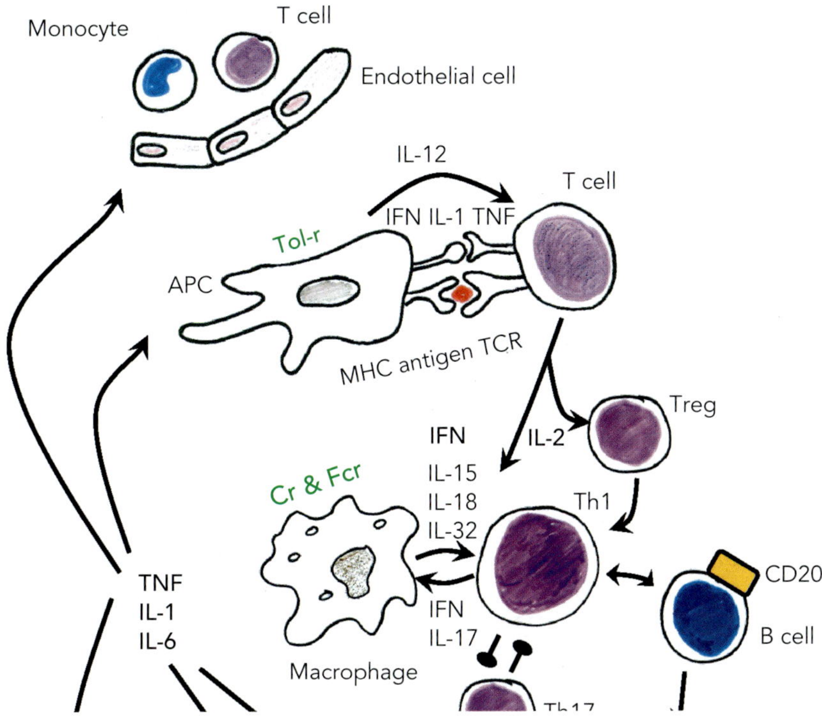

Figure 30.2. Immune processes in the rheumatoid synovium, and the destruction of cartilage and bone. These processes, either in their entirety, or in parts, are modeled in animals to produce inflammatory erosive joint disease. Central to this process is the presentation of an antigen by antigen presenting cells (APC) to T cells to produce a Th1 and Th17 response. The production of IFN-γ aggravates a pathogenic co-operation between Th1 cells and macrophages, enhanced further by ligation of complement receptors (Cr) and Fc receptors (FcR) to produce central proinflammatory cytokines, especially TNF, IL-6, and IL-1. These in turn mediate, either directly or through synoviocytes and RANKL, the stimulation of osteoclast differentiation and bone resorption, as well cartilage destruction through chondrocyte stimulation and synovial cartilage erosion.

indicating activation, as well as expression of HLA-DR, but they are often reported to be anergic to antigen. They are important for the hyper-expression of TNF-α seen in RA synovium ex vivo. Anti-CTLA-4, anti-CD4, and soluble CTLA-4 all have a degree of efficacy in RA, with the sCTLA-4 being judged as potent as anti-TNF-α therapy. Regulatory T cells (Tregs, CD4+ CD25+ cells) are also present in rheumatoid synovium, but appear unresponsive, and express low FoxP3, which is required for their function. Anti-TNF-α therapy restores Treg function. T-helper 17 (Th17) cells secrete IL-17, which has pleiotropic proinflammatory actions on a variety of cells found within the synovium, including stimulating the destruction of bone and cartilage, and the synthesis of a wide variety of proinflammatory cytokines.

The role of the B cell in the synthesis of autoantibodies and thus RA is evidenced by the profound activity of anti-CD20 therapy. CD20 is expressed on pre-B cells and mature B cells alone, and thus host–antibody responses to pathogens remains unaffected, whilst RF and ACAP antibodies are depleted.

The presence of lymphoid cells in the synovium ensures a proinflammatory network of cytokines and chemokines is dominant in this disease [3]. Disaggregation and reaggregation experiments show that T cells are important in co-operating with monocyte/macrophages to produce a cascade of cytokines with macrophage-derived TNF-α playing a central role [4]. Processes affected by TNF-α include chondrocyte-induced cartilage destruction, osteoclast activation, HLA-DR expression, endothelial cell expression of adhesion molecules, chemokine synthesis, and induction of the acute phase response. IL-1, IL-6, IL-15, IL-23, Il-17, IL-22 are all considered important targets in their own right. In addition, the chemokines, which are involved in the recruitment of cells into pannus, contribute a final layer.

MODELS OF INFLAMMATORY ARTHRITIS

The role of animal models in elucidating and testing the hypotheses that have lead to this current level of understanding cannot be understated. Such models rely on the same processes illustrated in Figure 30.2, but may utilize in antigens common with RA, such as collagen, or peptidoglycans, muramyl dipeptide, but also novel antigens such as heterogeneous nuclear ribonucleoprotein A2 (hnRNP-A2) and other unidentified antigens that may (or may be absent) in the mineral oil arthritides for example. The dependency on the T or B cell can differ between models, for example, collagen-based arthritides possess strong T cell as well as IgG antibody dependency being transferrable through primed T cells, or serum, from diseased animals. However, the rat Freund's adjuvant arthritis is unresponsive to anti-B-cell therapies, and is transferable through primed T cells and not serum. The facility to be able to transfer disease by sera has enabled the arthritogenicity of rheumatoid autoantibodies to be demonstrated, and knockout/knockin mice for a wide variety of cytokines, chemokines, cellular receptors, signaling pathways remain important tools in the dissection of polyarticular joint disease.

Since the induction of an autoimmune response to antigen is a common trigger, Figure 30.2 can be taken to summarize the induction processes of most of the common models. It should be noted that the type of immune response is important. The induction of inflammation into the joint need not necessarily lead to a sustainable destructive arthritis. Table 30.1 shows the effect of a variety of inflammatory stimuli on the degree of inflammation, bone erosion, and cartilage damage over 5 days in rats. Each of these consists of inflammation as a consequence of the activation of a different arm of the immune response, namely type of hypersensitivity as classified by Gel and Coombs. Each type may be induced in rodents using a combination of three basic ingredients: animal strain, antigen, and adjuvant. Adjuvants are agents added to antigens that enhance the translocation of the antigen to the lymph node. They also enhance retention at the site of injection to prolong delivery of the antigen, induce local reactions involving mast cells and Th cells to enhance chemokine synthesis, induce local cytokine synthesis and systemic cytokines to increase the circulating population of immune cells, and stimulate the innate response through accessory cell pattern recognition receptors (Toll receptors). The latter serves to direct the immune response toward Th1 cell–mediated immunity and IgG synthesis, diverting it from Th1 IgE-dominated allergic responses.

1. *Strain of animal.* Different strains of mice and rats react to antigenic stimuli in different ways according to inbred defects in their immune systems. For example, the Balb/c inbred mouse strain is particularly responsive in responding with Th2 allergies (Type II hypersensitivities), but is relatively unresponsive to Th1 (type IV hypersensitivity). The c57/black mouse is the opposite. The Lewis rat has a good response to cell-mediated immunity (Th1), whilst the Wistar can be used for Th2 (allergy) models. Some strains of Wistar are susceptible to Th1 stimuli.

TABLE 30.1. The relationship between inflammatory stimuli, join inflammation, bone loss, and cartilage proteoglycan (GAG) loss in Wistar CFHB rats injected into one knee and followed for up to 7 days

	Joint inflammation	Bone loss	GAG loss
Carageenan	+++	0	0
RP Arthus	++	0	– –
Mtb	+++	+++	+++
Zymosan	++	++	+
mBSA	+++	+++	++

RP, reverse passive; M.tb., heat killed mycobacterium tuberculosis; mBSA, methylated bovine serum albumin antigen-induced arthritis. + = worse; – = improved; o = no change.

2. *Antigen.* The antigen used is important, so ovalbumin is used for allergic (Th2) reactions, whilst keyhole limpet antigen, methylated bovine serum albumin (mBSA), or mycobacterial antigens are used for the cell-mediated (Th1) immunities. In addition, cartilage components such as collagen or peptidoglycans can be used to create autoimmunity to host collagen and proteoglycans within the joints.
3. *Adjuvant.* The use of alum containing adjuvants diverts the immune system toward Th2 allergic responses, and mycobacterial and similar adjuvants divert the response to the Th1 pathway through its stimulation of macrophage or antigen-presenting cell Toll receptors and IL-12 synthesis.

There is a sensitization step to an antigen, during which antigen-presenting cells process the antigen and present it (or them) to T cells to produce cell-mediated immunity. T cells also co-operate with B cells to produce an IgG humoral response. The fixation of complement by these antibodies, and their binding to Fc receptors (FcγR) that can regulate neutrophil as well as macrophage function, can be a key event in disease initiation, and joint inflammation occurs thereafter. Since individual animals may succumb to disease at different times, the initiation of the disease may be synchronized by a boost injection comprising either the antigen in adjuvant, or a toll receptor agonist such as lipopolysaccharide (LPS). A comparison of the sensitization protocols is given in Table 30.2.

ARTHRITIS ASSESSMENT

The polyarthritis models can be assessed in a variety of ways. Since these are systemic diseases, visual scoring systems have been developed that relate to the pattern of inflammation that is presented for that particular model. This can include scoring the severity of inflammation in each paw and digit, as in the

TABLE 30.2. Summary of the most common sensitization and challenge protocols for autoimmune polyarthritis and monoarticular models of inflammatory arthritis

Arthritis	Species sex strain	Antigen/adjuvant	Boost	Challenge
Adjuvant arthritis (AA)	Rat M/F Lewis	Heat-killed Mycobacterium *tuberculosis*/FIA	None	Autoimmune
Collagen-induced arthritis (active)	Rat F Lewis or DA	Chicken or bovine collagen-II/FIA	None	Autoimmune
Collagen arthritis (active)	Mouse M dba/1	s.c./i.d. chicken or bovine collagen/ FCA	Optional synchronization with LPS or collagen/ Freund's incomplete adjuvant	Autoimmune
	Mouse M c57bl/6	s.c./i.d. chick collagen/enhanced FCA		Autoimmune (relapsing)
Collagen arthritis (Passive)	Mouse most strains (e.g., balb/c, c57bl6)	None	None	i.v. anti-CII antibodies
K/BxN (passive)	Most strains (e.g., balb/c, c57bl6)	None	None	i.v. anti-GI antibodies
Pristane	Rat M/F Lewis or DA	s.c. pristane	None	Autoimmune (?)
Pristane	Mouse M CBA/Igb or DBA/101aHsd	i.p. pristane	Optional i.p. boost	Autoimmune (?)
Avridine	Rat M/F Lewis or DA	s.c. avridine (CP-20961, lipoidamine)	None	Autoimmune (?)
Antigen-induced	Rat M/F Lewis	s.c. mBSA/ FCA	mBSA/FIA or mBSA/FCA	i.a. mBSA
	Mouse M/F C57bl/6	s.c. mBSA/FCA	mBSA/FIA or mBSA/FCA or Bordatella pertussis	i.a. mBSA
SCW (active)	Rat F Lewis	i.p. SCW /no adjuvant	None	Autoimmune
SCW (monoarticular)	Rat F Lewis	None	None	i.a. SCW
	Mouse M c57/bl6	None	None	i.a. SCW chronic erosive form: 4× i.a. injections
Zymosan	Rat/mouse	None	None	i.a. zymosan
Carageenan	Rat/mouse	None	None	i.a. carrageenan
FCA	Rat/mouse	None	None	i.a. FCA

FCA, Freund's complete adjuvant (mineral oil + heat-killed ***Mycobacterium tuberculosis***); FIA, Freund's incomplete adjuvant (mineral oil); SCW, streptococcal cell wall component peptidoglycans-polysaccharide (PGPS); CII = type-2 collagen; mBSA, methylated bovine serum albumin; M, male; F, female; LPS, lipopolysaccharide; anti-GI, anti-glucose-6-phosphate isomerase.
Genetic models which develop spontaneous arthritis include K/BxN, NZB/NZW, HuTNF Tg, TNF gene mutation in AUUUA motif, Tristetraprolin −/−, IL-1RA −/−.

case of collagen arthritis, or in the case of rat adjuvant arthritides, extra-articular signs such as ear and nose lesions, as well as ribbed tail can be included. The patterns of arthritis can vary quite widely between sources of antigen and strain, so are usually developed with the particular method in mind. Examples are given in Table 30.3. These end points are quite distinct from direct measures of joint inflammation through swelling which assess the size of the hind paws, through either water displacement (mercury plethysmometery though highly accurate, is extremely hazardous and should never be undertaken), or by using calipers to measure dorsoventral thickness. Looking at the pharmacology of the models, it will be clear that the erosion of joint structures can be dislocated from the severity of the inflammation, hence it cannot be inferred that an inhibition of arthritis is necessarily accompanied by an inhibition of joint erosion.

The most rigorous assessment (e.g., see Table 30.4) of joint pathology is by microscopy. A variety

TABLE 30.3. Examples of clinical scoring systems for polyarthritis models rheumatoid arthritis (see text for description of models)

Rat adjuvant arthritis	0–16: 0–3 for lesions of each ear, paw (×4), and tail, or local variations thereof. 0 (no arthritis), 1 (one arthritic joint of the paw), 2 (two arthritic joints of the paw), 3 (more than three arthritic joints of the paw), 4 (whole paw joints involved).
Rat CIA	0–8: 0–4 for each hind paw (0, no erythema or swelling; 1, isolated ankle swelling; 2, swelling/erythema of ankle and proximal ½ of tarsal joints; 3, swelling/erythema of ankle and all tarsal joints up to metatarsalphalangeal joints; 4, swelling/erythema of entire paw including digits). Or: 0–4 for erythema swelling for each of the 4 paws. Or: 0–3 for erythema and swelling. Max = 6.
Mouse CIA	0–8: Paw involvement =0–2, +1 for ankylosis. Or: 0–22: each digit =1, palm and sole = 1. Or: 0–12: for each paw b 0, normal; 1, erythema/oedema; 2, visible joint distortion; 3, ankylosis on flexion. Or: 0–16: for each paw 1, mild erythema; 2, mild swelling and erythema; 3, gross swelling and erythema; 4, gross deformity and inability to use limb.
Mouse and rat pristane	0–28: 1 point for each swollen or red toe, 1 point for midfoot digit and knuckle, and 5 points for a swollen ankle.
K/BxN	0–4: 1 point for each affected limb; 0.5 for a limb with only mild swelling/redness or only a few digits affected.

CIA, collagen-induced arthritis; K/BxN, arthritis developed by KRN/ MHC class II molecule A^{g7} (K/BxN mice) or anti-glucose-6-phosphate isomerase antiserum.

of scoring methods have been developed, each specific to the type of arthritis being assessed, since the pathology of these lesions differ. Whilst rigorous pathological examination can be considered definitive, utilizing cell counts at different defined levels through the tissues, scoring systems permit the more rapid assessment of samples, and the comparison between groups of large numbers of animals. Scoring of cartilage damage can take into account physical erosion, but in those models where frank erosion is not seen, the loss of matrix, namely proteoglycan, can be visualized by the loss of proteoglycan-binding stains such as Safronin-O or Toluidine Blue. Bone erosion can be visualized and scored by X-ray, but the use of microcomputed tomography X-ray now allows the assessment of multiple joints within one limb (see Figure 30.3). Erosions in murine collagen arthritis for example start in the metacarpo-phalyngeal joints, with erosion of the ankles and talus later. In adjuvant arthritis the disease is focused on the ankle joint. Algorithms can calculate the surface area of entire bones, and surface roughness, such that statistically relevant data can be extrapolated.

Sampling of tissues for gene or cytokine analysis is one the prime advantages for the use of animal models. In models of arthritis, sampling serum synovial tissues, spleens, lymph nodes, synovial washouts, or even whole joints, provides very useful pathobiological information. The source of sample has direct bearing on the output that will be obtained. For example in a variety of polyarthritis models during the chronic phase, synovial fluid washouts or sera may not demonstrate certain cytokines, for example TNF-α, but immunohistology may demonstrate high expression within the lesion. Thus through intelligent use of tissues high quality and relevant information can be gleaned.

HUSBANDRY

The models described in this chapter are severe and often painful, especially the polyarthritic diseases. Good animal husbandry is very important, and if walking becomes difficult, soft food, or the placement of food in the cage as opposed in hoppers is required, as well as gel-based liquid delivery systems. Special note should be taken to make sure animals can move to food. Environmental stress can interfere with the incidence and severity of the diseases, and should be strictly controlled. Animals do lose body mass as the disease gets more severe which reflects both cachexia and reduced feeding. In adjuvant arthritis the loss of body mass reflects a cachexia through altered liver glucose handling as well as reduced feeding. Body mass can serve as a good index of animal well-being when giving anti-rheumatic drugs. Body mass can thus be used as a global indicator of general well-being, and any increase in its loss with a therapeutic is cause for concern and should be monitored closely. Steroids induce weight loss through glucocorticoid side effects. Nonsteroidal antiinflammatory drugs can improve body mass, as can low-dose methotrexate, for example, through improvement in well being and reduced disease.

TABLE 30.4. Scoring systems for histopathology and joint erosion

Rat adjuvant arthritis	0–9: 0–2 for synovial hyperplasia, bone erosion, and inflammatory infiltrates with leukocytes (0 = absent, 1 = mild, 2 = severe). Or: Synovial infiltrate, including monocyte/macrophages, lymphocytes, and polymorphonuclear cells, and blood vessels, sum of counts in three ×1,000 microscopic fields. Or: 0–3 for cartilage and bone separately. Cartilage (toluidine blue proteoglycan staining) 0 = fully stained cartilage, 1 = destained cartilage, 2 = destained cartilage with synovial cells invasion, and 3 = complete loss of cartilage. Bone erosion: 0 = normal, 1 = mild loss of cortical bone at few sites, 2 = moderate loss of cortical and trabecular bone, and 3 = marked loss of bone at many sites.
Rat CIA	0–20: Talus joint, 0–5 for: Cartilage (fibrillations/erosions/clefts, chondrocyte necrosis, proteoglycan loss), subchondral bone resorption, osteophytes, and inflammation. 0 = no change, 1 = minimal, 2 = slight, 3 = moderate, 4 = severe, 5 = massive). Or: 0–3 for each of cells in synovial fluid (0 = no cells, 1 = a few cells, 2 = joint cavity partly filled, 3 = joint cavity totally filled); synovitis (0 = healthy, 1 = mild, 2 = substantial, 3 = severe thickening), cartilage (0 = normal, 1 = minor loss of cartilage surface, 2 = clear loss, 3 = cartilage almost absent from whole joint); and bone (0 = normal, 1 = minor signs of destruction, 2 = up to 30% destruction, 3 ≥ 30% destruction).
Mouse CIA	0–3 for each of inflammatory infiltrate, cartilage surface erosions, and cartilage proteoglycan depletion.
Mouse antigen-induced arthritis	0–20: Score 0–5 for: Inflammation (soft tissue inflammation, synovitis, cellular infiltration, angiogenesis); pannus (hypertrophic synovial tissue with tight adhesion to cartilage); cartilage damage (loss of matrix and destruction on both condylar surfaces); bone erosion (area and depth of subchondral bone damage).
K/BxN Arthritis	0–18: Talus and/or midfoot. Inflammation: (1) Minimal infiltration of inflammatory cells and/or mild edema; (2) mild infiltration; (3) moderate infiltration; (4) marked infiltration; and (5) severe infiltration. Bone erosion: (1) Small areas of resorption, not readily apparent on low magnification, in trabecular or cortical bone; (2) more numerous areas of resorption, not readily apparent on low magnification, in trabecular or cortical bone; (3) obvious resorption of trabecular and cortical bone, without full thickness defects in the cortex; loss of some trabeculae; lesions apparent on low magnification; (4) full thickness defects in the cortical bone and marked trabecular bone loss, without distortion of the profile of the remaining cortical surface; and (5) full thickness defects in the cortical bone and marked trabecular bone loss, with distortion of the profile of the remaining cortical surface. Cartilage damage away from pannus: (1) Minimal to mild loss of cartilage with no obvious chondrocyte loss or collagen disruption; (2) mild loss of cartilage with mild (superficial) chondrocyte loss and/or collagen disruption; (3) moderate loss of cartilage with moderate multifocal (depth to middle zone) chondrocyte loss and/or collagen disruption; (4) marked loss of cartilage with marked multifocal (depth to deep zone) chondrocyte loss and/or collagen disruption; and (5) severe diffuse loss of cartilage with severe multifocal (depth to tidemark) chondrocyte loss and/or collagen disruption. Cartilage damage adjacent to pannus: (1) Pannus formation with superficial cartilage destruction; (2) pannus formation with moderate cartilage destruction (depth to the middle zone); and (3) pannus formation with marked cartilage destruction (depth to the tidemark).

RAT ADJUVANT ARTHRITIS

Adjuvant arthritis takes its name from the agent that induces it, Freund's complete adjuvant (FCA). This is 85% mineral oil (Bayol F) and 15% emulsifier (Arlacel-A), complete with finely ground heat killed mycobacterial fragments. This is commonly *Mycobacterium tuberculosis* (*M.tb.*) strain H37ra, or *Mycobacterium butyricum* (*M.bt.*). This is administered at the base of the tail. Old protocols utilized injections into a hind footpad, however this has been largely discontinued owing to animal husbandry concerns. This did permit the dual assessment of systemic and local disease mechanisms. Tracking of the mineral oil and mycobacterial particles shows that they distribute quickly throughout the body, and accumulate in lymph tissues. A powerful systemic invasive inflammatory periarthritis is induced that involves not only the joints but also the formation of granulomas within meninges, skin bone marrow and eyes, as well as in the connective tissue. The animals lose body mass. The predominant feature is a profound and crippling inflammation of the hind paws. The joints of the ankle and paws become seriously eroded, but the knees are less involved. The disease is accompanied by splenomegaly, increased gut permeability and liver involvement, such that the absorption and metabolism of drugs can be deranged. The microstructure of hepatocytes is deranged, and

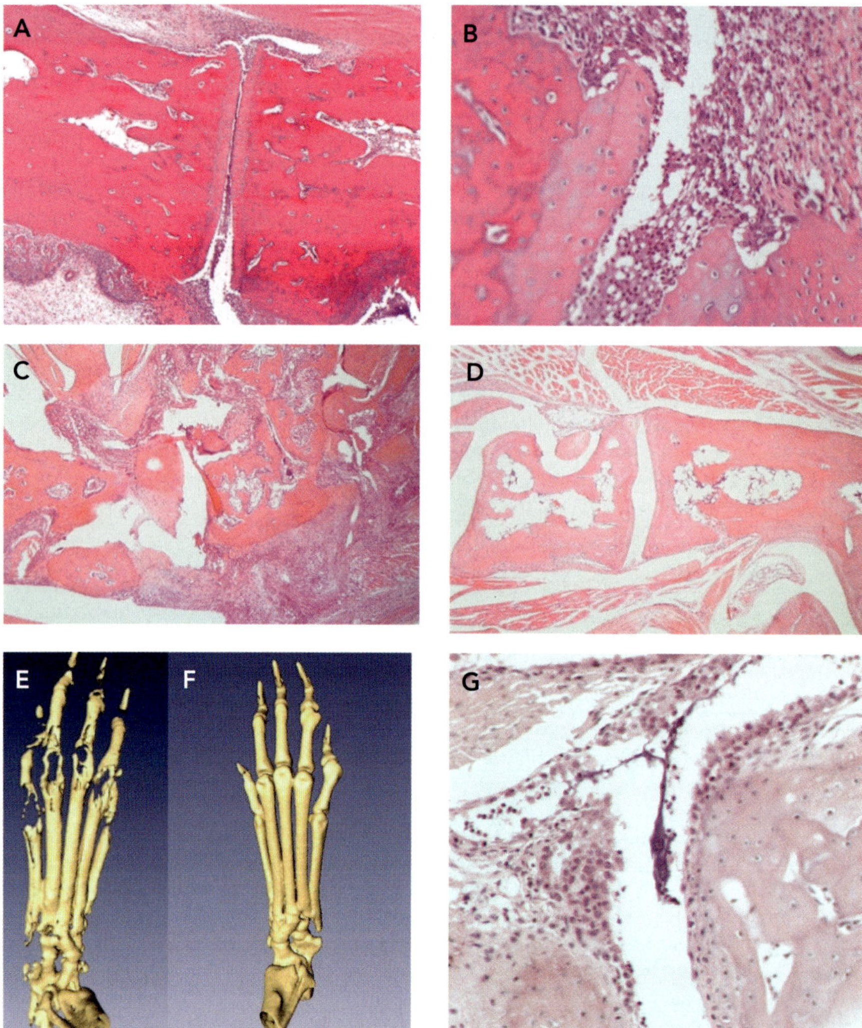

Figure 30.3. Joint destruction in collagen and K/BxN arthritis. (A) Midfoot of a rat 20 days after sensitization to bovine type-2 collagen in Freund's complete adjuvant. Cells have infiltrated the joint and are eroding the joint margins. (B) High power showing the firm adhesion of inflammatory cells to the entire surface of the cartilage, eroding pits into it, as well as invading bone. (C) Mouse ankle joint 35 days after sensitization to bovine type-2 collagen showing complete destruction of the joint architecture, cartilage, and bone. (D) Mouse ankle joint 35 days after sensitization to bovine type-2 collagen, but treated with the anti-inflammatory steroid dexamethasone at 0.1 mg/kg daily from the day of boost at day 21. Note the complete protection against joint destruction. For comparison micro-CT isosurface plots of a mouse collagen arthritic paw, and (E) after treatment with dexamethasone (F). The initiation of stifle joint destruction induced by the intravenous injection of K/BxN anti-GI antibody into a C57b/6 mouse. Note the synovitis, and the firm adhesion of inflammatory cells to the cartilage surface (G).

bile flow reduced. Glucose responsiveness is attenuated and ureogenesis mechanisms altered, and P450 and b5 and the activities of NADH-b5 reductase, NADPH-ferrihaemoprotein reductase, P450 mixed function oxidase, FAD-monooxygenase, and several enzymes involved in conjugation are reduced. P450 has been reported to be as low as 16% of normal, N-demthylase 7% and NADPH oxidase 43%. The metabolism of haem is raised, and may explain the changes in P450 levels. Thus, drug metabolism can sometimes be deranged. Plasma albumin is reduced, and plasma glycoproteins raised, which again may affect the pharmacokinetics of plasma protein bound drugs. Spleens display splenitis, lymphoid hyperplasia, giant cell reaction, granulomas, necrosis, and abscess formation. Thus, one common endpoint for assessing the relief of systemic disease is a measure of spleen mass.

It is an auto-immune disease, since it can be passed to nonarthritic irradiated recipients through T cell transfer from diseased animals. These T cells express

CD4 and the α,β-TCR (T-cell receptor). Pan T-cell depletion reduces the onset of the disease, as does CD4 cell depletion. Susceptibility is conferred by genes within the MHC class II region. Adjuvant arthritis is not transferred by serum, but serum from adjuvant arthritis rats can induce a flare in rats with cyclophosphamide attenuated disease, whilst serum from the cyclophosphamide animals does not. Strain is important, the Lewis and DA rats being the favored strains, some Wistar strains are susceptible, but Fisher F344 rats are moderately resistant. Resistance is genetically dominant, unlike collagen arthritis where susceptibility is dominant. The MHC phenotype is weaker than that for the rat collagen arthritis, but exerts a strong influence on the severity of disease. Thus, non-MHC phenotypes have a significant contribution to play. The MHC gene is *Aia1* found on chromosome 20. Non-MHC QTLs include *Aia2* and *Aia3* on chromosome 4. Conserved synteny among rats, mice and humans, suggests that *Aia1, Aia2,* and *Aia3* contain candidate genes for several autoimmune diseases including diabetes (*Aia3*:IDDM9), systemic lupus erythematosus, inflammatory bowel disease, asthma/atopy, multiple sclerosis, and RA (*Aia3:*RA2).

Histologically, the predominant inflammatory cell is the neutrophil, which is accompanied by destruction of the synovial lining. At initiation, fibrin is deposited long the bone shafts, followed by fibrin deposition in the joints. The striking feature is a strong perisoteal inflammation comprising monocytic cells, that progress to invade the joints. I-a (MHC) expressing cells line the cartilage, and lymphoid I-a expressing antigen presenting cells are diffusely scattered in the deeper inflamed tissue, some with dendritic appearance. However, immunoglobulin-containing cells may be low in number or absent. Large quantities of CD4 T cells are seen adjacent to pannus, but not next to cartilage, and there is a moderate scattering of CD8 cells. Bone erosion is both osteoclastic and due to the inflammatory tissue. Osteoblastic production of new perisoteal bone can be seen, with osteoclastic erosion of old bone. Osteoclasts erode through original and new bone, and marrow spaces are excavated in new bone. Articular cartilage remains relatively untouched, though dead cartilage can be seen associated with neutrophils or macrophages. Bone is ultimately eroded away from under the cartilage leaving it supported by fibrocartilage. At late stages, fibrosis occurs with ankylosis. Lymphocyte aggregates and plasma cell infiltrates are not seen. With a polyarthritis, there is special focus on destruction of the ankle joint.

Numerous candidates for the endogenous antigen involved in adjuvant arthritis have been proposed. Muramyl dipeptide induces adjuvant disease similar to adjuvant arthritis, and antibodies to type-II collagen (CII) are expressed. Heat shock protein-65 (HSP65) is considered to be a major endogenous antigen, with anti HSP65 antibodies cross reacting with cartilage proteoglycans being expressed as the disease develops. Susceptible rats appear not to be fully tolerant to rat HSP65. HSP65 pretreatment induces disease tolerance. Originally T cells specific for the sequence 180–188 of mycobacterial heat shock protein suggested this was the epitope, being capable of inducing disease, and induce tolerance through the modulation of IFNγ and IL-17 down regulation. Detailed epitope mapping reveals a range of epitopes on rat HSP65 within the C-terminal between 418 and 535, to which draining lymph node cells from diseased rats respond well. These epitopes appear to be disease cryptic, and T-cell clones specific for these epitopes can confer tolerance. This appears to be broken by processed bacterial HPS65. There thus appear to be a dual role for HSP65-restricted T-cell clones as a result of diversification between reactivity to self and bacterial HS65 peptides. T-cell responsiveness is involved in the development of the disease while tolerance to self HSP65 is reasserted during the chronic phase to terminate the disease. Synovial T cells include both effector and suppressor subtypes.

DRUGS

Adjuvant arthritis is notoriously susceptible to cyclo-oxygenase (COX) inhibition by nonsteroidal anti-inflammatory drugs (NSAIDs), which inhibit not only the inflammation, but also joint destruction. This model was used as a screen for NSAIDs, so there is little need to list them all, with aspirin and indomethacin being the archetypes with ED50s at around 100 mg/kg and 0.2–2 mg/kg respectively. In fact, NSAIDs are so effective, treated animals can gain weight at a similar rate to normal, nondiseased controls, unlike those treated with steroids which lose weight due to the trans-activation side effects. The rat acute phase protein, α_1GP, is suppressed, as are antibody titers, but they do not do so in RA. The effect of NSAIDs is supposed to be due to the inhibition of prostaglandin E_2, which is involved in osteoclast differentiation and activation. COX-2 is expressed by the inflammatory synovial cells and chondrocytes. Its inhibition by antisense inhibits both the inflammation and joint erosion. Pharmacological separation of the two Cox enzymes shows that the inhibition of Cox-1 by SC-58560 is ineffective at inhibiting bone erosion, but the Cox-2 inhibitor celecoxib is effective.

Antiinflammatory steroids also suppress every disease marker, but also induce further weight loss and cause thymic, splenic, and adrenal involution. If treatment is stopped, a rapid rebound of the disease occurs. Of the disease-modifying drugs (DMARDs), cyclophosphamide, chlorambucil, and methotrexate

TABLE 30.5. Characteristics and properties of the major rodent models of rheumatoid arthritis

	RA	Avridine/ pristane	AA	RCIA	MCIA	SCW	AIA	K/BxN (passive)	Irritant (zymo-san or carra-geenan)
MHC	++	++	++	++	++	++	++	–	–
Sex	F > M	Rat M/F mouse	F = M	F > M		F > M	F = M	F = M	M = F
		M							
RF	++	+ (pristane)	–	–	–		–	–	–
ACPA	++				++/–	–			–
Anti-CII	+/–	–	–	++	++	–	–		–
Chronicity	Decades	Rat < 1 month Mouse 3 months	<1 month	<1 month	Months	Months	Months	Months	1 week
Flare	++	+ (pristane)	–	–	–(autol CII+)	+	+ (induced)		
Ankylosis	When burnt out		++	++	++				
Antigen	Epitope spread	?	Bacterial	CII	CII	Bacterial	mBSA	GI	
Polyarthritis	Symmetrical	Y	Y	Y	Y	Y	Mono	Y	N
Synovitis	++	++	++	++	++	++	++	++	++
Cartilage erosion	++	++	+/–	++	++	+/–	++	++	+ Zym – Car
Ir to cartilage	+/?	–	–	++	++	–	++	++	–
Bone	++	++	++	++	++	++	++	++	–
Neutrophil	+ (SF)	++	++	++	+	++	+	++	–
Monocyte	++	++	++	++	++	++	++	–	–
CD4+ T	++	++	++	++	++	++	++	–	–
Th17 T	++		+		++	++			–
B cells	Synov Aggregates	–/?	–	+	+ Diffuse	–	++		–
Ab Isotype	IgG/IgM	–	–	IgG2	IgG2	–	IgG2	IgG1	–
Cytokines	++	++	++	++	++	++	++	++	
Methotrexate		M + R+	++	++	++	++	++		+
Sulphasalazine	++/–			+/–					
Leflunomide	++/–		++		++				
Anti-TNF	++/–	M+/–	++	++/–	++	+		++	
Anti-IL-1	++		++	++	++		++	++	
Anti-IL-6	++				++				
Anti-CD20	++	Nr			++		++		
Extra-articular	++	–	++						
Non-RA features			Non RA extra-articular features	Periostitis	Periostitis	Periostitis	Osteophytes chondro-phytes		

Key: RA = Rheumatoid Arthritis; AA = Adjuvant Arthritis; RCIA = Rat Collagen Arthritis; MCIA = Mouse Collagen Arthritis; SCW = Streptococcal Cell Wall Arthritis; AIA = Antigen-induced Arthritis. F = Female; M = Male; Ir = Immune response; Ab = antibody; Zym = Zymosan; Car = Carrageenan. For all others see text.

are effective in reducing clinical score when given prophylactically. Azathioprine has efficacy close to its toxicity. Methotrexate is active at low doses (<0.03 mg/kg), as it is the clinic. All of these drugs also inhibit the PPD skin test for cellular responses to the antigen. Responses to aurothiomalate vary, depending on the strain of rat used and route of administration. Gold also has a narrow therapeutic window, so evidence of reduced disease coupled with loss of body mass may be considered to be a toxic action. Auranofin, the oral pro-drug is more reliably effective, and can increase body mass. Chloroquine and hydroxyl-chloroquine are generally ineffective, as is D-penicillamine and salazopyrine. Cytotoxic drugs as a group appear to reduce the severity of secondary lesions, leaving established disease unaffected. Levamisole may exacerbate established disease. Leflunomide, if given before disease becomes established – that is, before 12 days after sensitization – prevents adjuvant disease. It also normalizes adjuvant lymph node cell responses without affecting lymph nodes from untreated rats. This conforms to its actions on proliferating T cells, unlike cyclophosphamide, which inhibits the lymphocyte responses of both diseased and control T cells.

Referring to Table 30.5, of the more modern therapies, adjuvant arthritis responds to leflunomide in the acute and chronic phases with a reduction in joint pathology as well as acute phase proteins and circulating antibodies. It responds to anti-TNFα therapy, through the soluble TNFα- receptor, via anti-TNFα therapy, or through IL-1 receptor antagonists. The soluble TNF receptor and IL-1ra are effective when administered over the first 7 days by continuous infusion osmotic minipump, pulse therapy by subcutaneous injection being ineffective. The two therapies are synergistic and reduce bone erosion as well as inflammation. These illustrate the care that has to be taken to take into regard the exposure and pharmacokinetics to ensure good drug coverage when using biologics in rodent models, and not least the consideration of species cross-reactivity or the lack of it.

Adjuvant arthritis remains a popular model for evaluating modern drug targets. To illustrate the use of adjuvant arthritis in modern drug discovery, 38-MAP-kinase inhibitors inhibit both TNFα synthesis and arthritis, while adenosine A3 agonists can inhibit bone erosion. Anti-IL-1 therapy inhibits the erosive nature of the disease more than the inflammatory endpoints. IL-4 administration reduces severe disease, but IL-10 is only effective against mild disease. Both reduce neutrophil activation in situ as does the rat neutrophil chemotactic factor IL-8 homologue. IFNγ only has effects on inflammation and bone erosion rIFNγ suppressed the secondary phase from day 18 onward, as well as the splenomegaly and erythrocyte sedimentation rate, an indirect measure of the acute phase response. Both joint inflammation and joint erosion are inhibited by an antagonistic IL-17-IgG1 fusion protein.

RAT COLLAGEN ARTHRITIS

This model, discovered accidentally by Trentham and colleagues during [5] attempts to generate anti-type II collagen (CII), was the first collagen-derived arthritis and was followed by the mouse version some years later. Whilst similar to adjuvant arthritis described above, it is milder and often preferred for this reason. CII comprises 80%–90% of the collagen in hyaline cartilage. The arthritis is induced by the sensitization of susceptible rat strains (Inbred strains: DA, LEW, BB-DR, not F344 nor BN strains; outbred: some Wistar strains are also susceptible, but by no means many of them) to native heterologous collagen emulsified in a mineral oil, namely Freund's incomplete adjuvant (FIA). An unusual feature of this model is that the inclusion of Freund's adjuvant complete (FCA) with heat killed mycobacterium tuberculosis (M.tb.) elicits a lower severity of disease, and may lead to a mild adjuvant arthritis. Native collagen is important, denatured is ineffective, as are extraarticular collagen such as CI. CII is thus kept in its native form by careful dissolution in cold 0.1M acetic acid during its preparation. The inoculum can be injected intradermally in several places on the lower back, or at the base of the tail. Old protocols utilized sensitization in the footpad, but this has been discontinued owing to animal husbandry concerns. Some protocols utilize a boost, but arthritis develops between 11 and 14 days, reaching a maximum at around 20 days. A boost at 20 days can be given to establish a long-term erosive disease.

Like adjuvant arthritis, it appears that fibrin deposition and a periostitis occurs before inflammation within the joint space. Fibrin is deposited over the fat pad, soft tissues, and cartilage, followed by synovial hyperplasia, and infiltration by neutrophils and monocytes. The periostitis and intense mononuclear accumulation is associated with areas of ligament attachment, which then moves into the synovial space leading to invasion of the joint space, with an influx of CD90+ CD4+ Ly2- (CD8-) Ia (MHC class II) expressing T cells (with little evidence of suppressor T cells) and subsequent destruction of cartilage and bone (Figure 30.3). The Cartilage surface is the predominant focus for cellular attack. The pattern of erosions differs to adjuvant arthritis, being less severe, but involving both the ankle and midfoot.

The transfer from CIA rats of anti-CII IgG antibodies to give equivalent circulating levels of antibody in the recipient induces a transient arthritis (7 days) only, whilst the antibody itself has a circulating half life of <48 hours, without pannus formation. Sera from nonresponder rats is ineffective. It could be argued that such arthritis could be due to complement fixation, but in these cases, the anti-CII did not bind CIq. The transfer of T cells on the other hand into nude (athymic) rats, induces an erosive chronic (21-day) arthritis without inducing anti-CII IgG. Depletion of TCRα/β on a prophylactic basis inhibits disease, unlike rat adjuvant arthritis, whilst established disease is resistant. Rat CIA is susceptible to tolerization to collagen peptides, created by oral administration of CII collagen. Incidence is reduced by approximately 20%, and delays onset by 50%. This results in reduced anti-CII-IgG2 (a Th1 product), whilst increasing anti-CII IgG1 (a Th2 product) with lower expression of MHC-II and fewer T cells.

The genetic basis of rat CIA centers on the rat MHC Class II (Class Ia in the mouse) RT1 complex. LEW ($RT1^{l}$) and the Wistar strain diabetes resistant Biobank $RT1^{u}$ (DR-BB) rats are susceptible, but LEW. B3 ($RT1^{n/n}$) and diabetes prone BB (DP-BB) $RT1^{u}$ and PVG.$RT1^{u}$ hooded rats are not. Unlike adjuvant arthritis, susceptibility, as opposed to resistance, is dominant in rat CIA. PVG.RT^{U} rats do exhibit a mild disease detectable by histology, but the erosive disease of the LEW rats is considerably greater, and accompanied by large changes in the draining popliteal lymph nodes. The DR-BB rat CIA exhibits a bidirectional bone eroding disease more akin to the human disease than in other strains, involving erosion via inflammation at the subchondral bone, as well as from the cartilage. These rats express both RT1.D and RT1.B homologues of human HLA-DR and DQ respectively. The sequence within the antigen presenting cleft of the MHC-II molecule bears a close resemblance to those found in HLA-Dr1 and HLA-DR4 RA patients (Table 30.6). Analysis of the collagen epitopes reveals that these rats immunized to human, bovine, or chicken CII possess IgG2a that recognizes the human arthritogenic cyanogen bromide collagen fragment CB11, and more mildly, the CB9 arthritogenic fragments. Epitope scanning of CB11 shows the immunodominant region to be in the 37–45 amino acids. However, immunization to this antigen creates an antibody response that cross-reacts with denatured bovine CII but not rat CII, and with no arthritis. Therefore, whilst CIA anti-CII antibodies may cross-react with CB11 fragments of a variety of species, these may not be the endogenous epitopes that initiate the disease in the host animal. In addition, there are other non-RT genetic influences controlling T-cell receptor, Ig subtypes, or complement components that are thought to play a role.

Approximately 80% of RA patients express antibodies specific for citrullinated peptides. As detailed for mouse collagen arthritis, proteins citrullinated at arginine can be detected in rat CIA synovial, as is the enzyme responsible, peptidyl arginine deiminase-4

TABLE 30.6. An example of commonality between Rodent and human arthritic haplotypes.

	β Chain residue
	69 HVRIII 75 79
Rat: DR BB/Wor-UTM RT1.Dβ(u)	E R R R A A V D T Y C
Human HLA DRβ	
HLA Dr1-Dw1	E Q R R A A V D T Y C
HLA DR4-Dw4	E Q K R A A V D T Y C
HLA DR4-Dw14	E Q R R A A V D T Y C
HLA DR4-Dw15	E Q R R A A V D T Y C
HLA DRw10	E R R R A A V D T Y C

The Biobreeding (DR BB/WorUTM) rat HVRIII RT1.Dβ(u) allele compared to the HRVIII for human HLA DRβ associated with RA
Source: Adapted from Watson WC et al. *J Exp Med*., 1990;172:1331–1339.

(PAD4) from day 21 after the onset of disease. Citrulline can be seen in damaged cartilage, as well as infiltrating cells, and a major portion in extracellular deposits, perhaps citrullinated fibrin. B-cell tolerance to an autologous antigen, rat serum albumin can be broken by citrullination. Citrullination of CII increases its antigenicity in rats of a strain that has mild susceptibility to CII arthritis (LEW.1AV1 strain).

DRUGS

Rat collagen arthritis, like all animal arthritides is highly sensitive to antiinflammatory steroids, with dexamethasone for example having an ED50 of 0.01 mg/kg. Prednisolone also reduces inflammation, antibody titer and disease incidence. Betamethasone also inhibits fibrinogen deposition, bone and cartilage erosion, and as a marker of joint destruction, cartilage oligomatrix protein (COMP). NSAIDs reduce inflammation, but have little effect on the erosion of bone and cartilage. In one published experiment, indomethacin had little effect on the arthritis, with no action on antibody titer, and reduced disease incidence. Gold therapy, D-penicillamine, and levamisole had no effect, and may even increase anti-CII titer. Cyclophosphamide reduces inflammation, antibody titer, and incidence. In the established lesion, dosing from day 21, indomethacin again reduced inflammation, but not antibody titer. Levamisole and D-penicillamine both potentiated antibody production, as did Gold therapy. Cyclophosphamide was effective across the board. X-ray analysis revealed that levamisole increased bone erosions, whilst D-penicillamine, indomethacin, and prednisolone reduced erosions.

It is clear that collagen arthritis is not an exact replica of human disease, but more akin to a polychondritis. Rat CIA appears to be less used for assessing modern drug targets, the mouse being preferred. However, it is sensitive to low-dose methotrexate, and both sTNFR and IL-1ra inhibit rat CIA, and inhibit bone resorption, unlike adjuvant arthritis where IL-1ra is ineffective against bone erosion. P38 MAP kinase plays an important role in the signaling for cytokine synthesis. Prophylactic p38 inhibition is accompanied by reduced paw inflammation, body mass loss, and bone erosion, accompanied by reduced serum IL-1β and TNFα as well as ankle IL-1β, and reduced CD4, CD8a cells within the bone marrow. RANKL stimulated osteoclast differentiation was also inhibited in vitro. This profile could be considered a perfect introduction to a successful drug in the clinic. However, to date p38 inhibitors, whilst effective, have not proven to be as efficacious as anti-TNFα or anti-CD20 in the RA clinic when assessed by modern clinical response criteria.

MOUSE COLLAGEN ARTHRITIS

Mouse CIA is based on the rat model, but developed further due to the use of this species in gene technology [6]. As in the rat, this is a polyarthritic response, clinical signs start with inflammation in the digits, that can recover and relapse, but builds so that the majority of digits are affected, and the pads visibly inflamed. All four paws are affected, but different sources and batches of collagen, coupled with different strains, result in different patterns of disease. It is initiated in susceptible strains [7], namely the DBA1/J and NFR/N mice (haplotype H-2q), and the C57.B10RIII (H-2r), with better responses seen in males. Increasing the concentration of mycobacterium within the sensitizing emulsion can increase the "take" in this strain. Heterologous collagen is used, usually either bovine or chick type-II collagen, but H-2q mice will respond to mouse CII (50% incidence as opposed to 80%–100% for DBA/1). The H-2r haplotypes do not respond to mouse or chicken collagen. In addition, as in the rat, the type and form of the collagen is important. CI does not induce the arthritis, and neither does denatured CII. Minor collagens of cartilage such as CIX can also be used. CII results in the generation of CD4 T helper cell responses, anti-collagen IgG generation with the clinical signs of the disease correlating with appearance of the antibodies in the circulation. The IgG antibody titer is important for the severity of the disease. Both IgG2a and IgG2b are expressed, with the former having a greater complement fixing potential. Mice that are nonresponders usually have a low antibody titer.

This reaction is transferable from arthritic mice by either the transfer of T cells or the transfer of serum into naive mice [8]. The Th-1 dependency of this collagen arthritis is proven through the expression of Th1 cytokines by lymphoid organs, and the influx of Th1 T cells after the acute neutrophilic phase. Indeed, sensitization to collagen using Freund's incomplete adjuvant is only successful in inducing disease if supplemented with a boost of IL-12 showing the importance of diversion to the Th1 pathway since IL-12 is the costimulatory signal secreted by dendritic cells during antigen presentation to T cells. However, if nonsusceptible strains are subjected to this protocol, a Th1 response to collagen-II is raised, but without a well-developed antibody response or arthritis. Boosting with lipopolysaccharide heightens response through the induction of IL-12 synthesis.

This model thus comprises both type-IV hypersensitivity to collagen type-II, as well as an immune complex disease.

Disease can be transferred through lymph node, spleen, or T cells. Whilst this is reportedly difficult to achieve, requiring concentrated numbers of cells, it shows the T-cell component is important. CD4 T-cell cell depletion abrogates the established disease. For the passive transfer through antibodies, the level of disease is also low unless a mixture of antibodies to multiple collagen antigens is used. Commercial preparations contain up to four different clones of antibodies, and have been developed to induce a rapid and severe arthritis. In addition, anti-CII antibodies isolated from an RA patient expressing high titers induced CIA in mice, albeit low (25% incidence). This passive disease is not solely due to the presence of antibody, since in the chronic phase, it can in turn be transferred by auto-reactive T cells, which well illustrates the interdependence of the two systems.

Pregnancy in RA induces remission. CIA induced in pregnant C57.B10 mice has a reduced incidence, anti-CII antibodies, and severity, and followed post partum with an exacerbated response. "Take" in female mice is low, and ovulation and fertility reduced such that pregnancy does not occur.

ANTIBODY EPITOPES IN MOUSE COLLAGEN ARTHRITIS AND RA

As previously mentioned, a common feature of RA is the presence of RF and anti-citrullinated peptide antibodies (ACPAs). CIA mice do not express rheumatoid factor. However, epitope analysis of 253 antigens, both citrullinated and native, shows that CIA DBA/1 mice generate low level antibodies to two citrullinated epitopes by day 21, which then doubles by day 31, and spreads to 13 epitopes by day 55. At day 21 the antibodies react to CII, CV, HCgp39 (another purported RA antigen), and glucose-6-phosphate isomerise (the antigen in K/BxN induced arthritis). Citrullination of CII enhances its antigenicity, and transgenic mice with the human HLA-DRB1 shared epitope have an exaggerated CD4 T-cell response when exposed to peptides that are citrullinated at the shared epitope-specific recognition sites. It is thought that peptidyl arginine deiminase (PAD) expressed at sites of inflammation citrullinates arginine residues on proteins to enhance autoantigenicity in susceptible individuals. Smoking induces PAD in the damaged inflamed lung, which is thought to enhance anticitrulline responses that then cross-react with joint citrullinated peptides when the joint space is breached. Mouse CIA may be unique amongst the mouse models in exhibiting ACPAs, using ELISA most other models appear negative for ACPAs. However, this may reflect the sensitivity of the assay, mouse collagen ACPAs being detected by epitope arrays.

PASSIVE COLLAGEN ANTIBODY-INDUCED ARTHRITIS

In the case of serum-transferred arthritis [8], termed collagen antibody-induced arthritis (CAIA), the strain of the recipient mouse is less important than the active arthritis, with even Balb/c mice being responsive. In this event, the arthritis is initiated without the sensitization phase, with signs appearing within days (Figure 30.4), and is very severe. This antibody-induced arthritis can be induced in T and B cell deficient mice, though the effect is less severe. The administration of anti-CTLA-4 antibody, which attenuates T-cell APC interactions, does reduce CAIA if administered during the chronic phase. Thus, there is evidence that autoantibodies induce a secondary T-cell reaction that further exacerbates the disease. Neutrophils are also important; their depletion reduces the severity of the symptoms. IgG FcγR engagement is very important, knockout mice for the FcRγ chain being resistant to disease. FcγRIII knockout mice have reduced severity, whilst absence of the inhibitory form, FcγRIIb, exacerbates CAIA disease (but not in balb/c mice). Complement is also a factor, C3 depletion with cobra venom factor reduces the response, as does C5 knockout. C5a is a highly powerful neutrophil chemoattractant and is high levels are seen in RA synovial fluid.

HISTOPATHOLOGY OF CIA

In the ankle, the antibodies bind cartilage, fix complement, and initiate the recruitment and activation of neutrophils via Fcγ receptors. Neutrophil adhesion to the cartilage at this early phase appears important. This neutrophilic layer remains to strip the cartilage of its smooth surface, to form deep erosions. An intense interstitial neutrophilic infiltration is seen into soft tissues. The synoviocytes at the point of insertion proliferate, and in a manner similar to RA, pannus growth with bone and cartilage erosion is initiated. The ankle joints and knees are both affected. The attack is concentrated against the cartilage, and starts early with the attachment of neutrophils with roughening and erosion into the cartilage surface. As the arthritis matures, the synovium comprises large numbers of monocytes and lymphocytes after about two weeks. The bone marrow becomes inflammatory, but this is not as pronounced as that seen with adjuvant arthritis discussed above. Bone is eroded by osteoclasts, and erosions can result in the coalescence of the inflammatory tissue and bone marrow. At later stages cartilage erosion and pannus formation at the margins is seen and eventually the total destruction of cartilage. In very chronic cases, new hyaline cartilage can be seen, but often being eroded by neutrophils. New perisoteal bone growth occurs, and in these late stages of the disease (>45 days) ankylosis occurs as is seen in end stage RA. At this stage the synovium posses immune cells, macrophages, and is fibrotic. In the classical CIA, neither lymphocyte nor plasma cell aggregates are seen.

The knee lesions differ in that they do not show loss of the surface of hyaline cartilage, but show a proliferative synovitis with pannus, with erosion of the bone and cartilage at the margins and underneath the pannus.

ROLE OF THE MHC–TCR SYSTEM IN MOUSE CIA

The H-2 haplotypes, representing the immune response (Ir) system, in the mouse equates to the human MHC class II system. The mouse class II gene complex comprises two loci, I-A and I-E, each of which code for one alpha and one beta chain polypeptide to form one class II molecule. Thus the arthritis susceptible H-2^q strains code for I-A, that equates to the RA susceptibility haplotype HLA-DR [7]. However, it should be noted that DBA mice do not possess a functional I-E, and thus do not have a functional equivalent of HLA-DR, but HLA-DQ. The mutations in H-2^q, namely at positions 85, 86, 88, and 89, occur in the antigen presenting cleft of the I-A^q molecule, in a similar part of the molecule RA susceptibility variations are seen in the HLA-DQ molecule. The H-2^p, H-2^{w5}, H-2^b, H-2^s, H-2^d, H-2^{w17}, and H-2^{w3} haplotypes are resistant. The stronger penetration of the arthritis in the H-2^q and H-2^r haplotypes to heterologous CII appears to correlate with their stronger cross-reactive autoantibody response with heterologous compared to mouse collagen. However, whilst the H-2^q haplotype does respond with a CIA of sorts, the NFR/N (haplotype H-2^q) strain does this in the absence of a mouse-anti-mouse CII antibody response. Other genes also exert an influence, and chromosome-17 on

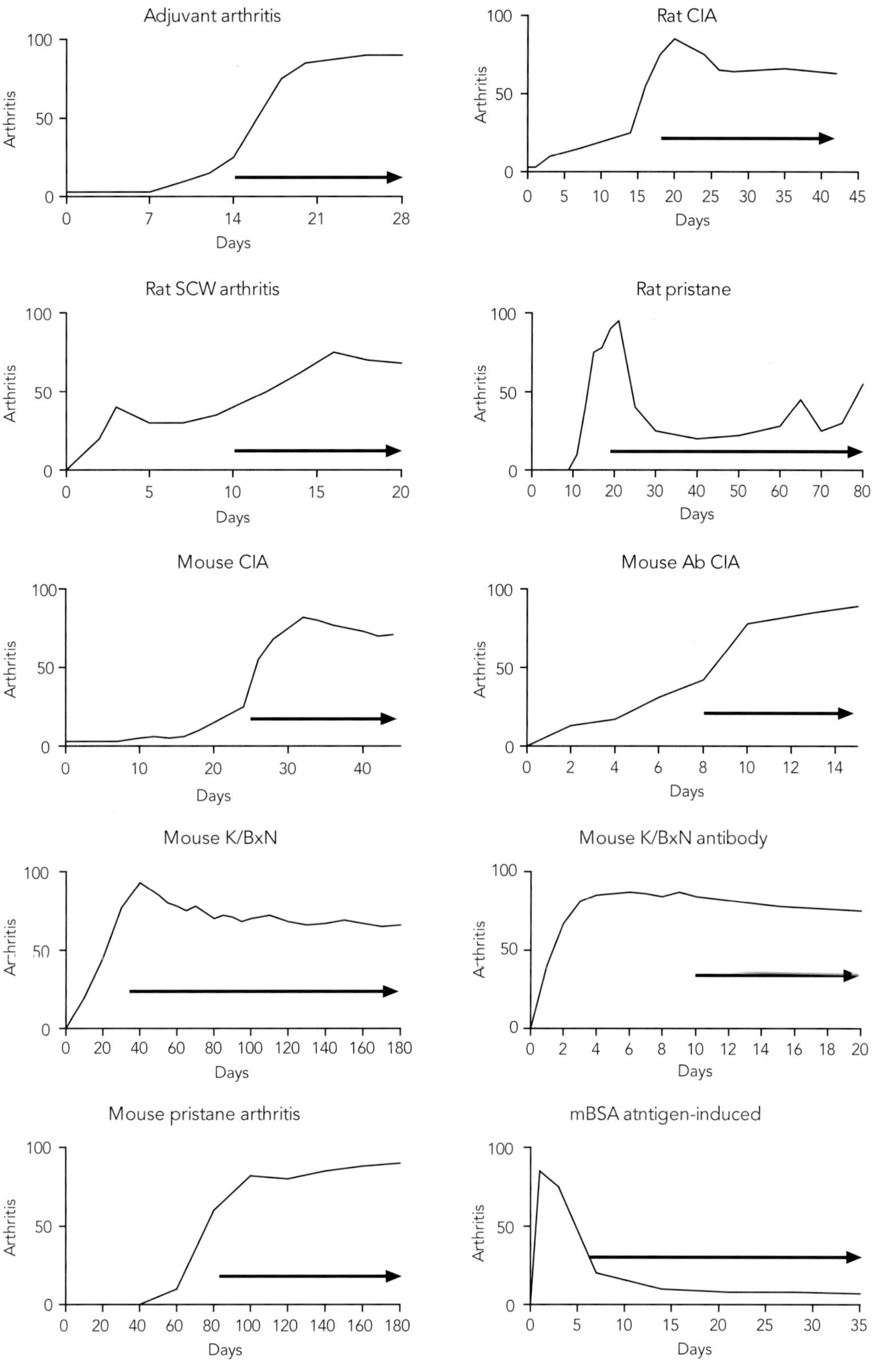

Figure 30.4. The time courses of inflammation in arbitrary units and joint erosion (arrow) in a variety of mouse and rat polyarthritis models. Inflammation can be assessed either by a visual clinical score (Tables 30.3 and 30.4), or by direct measures such a plethysmometry or paw thickness with calipers. When assessing a drug for its actions on the sensitization process, it is dosed from day 0. If the action on established disease is required, it is given after the disease has initiated in the experiment, or at a pre-defined day. An alternative is to dose on an individual animal basis, often after two or more successive days of a positive clinical score, and takes into account variability in the dates of onset between animals, especially in non-synchronised murine CIA.

which the MHC-II genes reside, such as TNFα, MHC class I, heat shock proteins, complement, and mechanisms involved in antigen processing. H-2^q mice lacking C5 for example produce autologous antibodies, but do not induce disease.

The immuno-reactive region of CII bound in the MHC-II molecule has been narrowed to positions 256–270 ($CII_{256-270}$). The minimal motif that can be generated from this is 260–267, glycosylated at K^{264}. The I-A^p molecule of the H-2^p mice that do not respond with CIA, bind this peptide at a much lower level than I-A^q. Human HLA-DR4 molecules bind the $CII_{260-273}$ peptide, and mice expressing DR4 respond with CIA, and can respond to a CII peptide from the same region. HLA-DR1 transgenic mice are susceptible to CIA, and respond to $CII_{259-273}$.

T cells from CIA mice respond to this molecule, the glycosylation of the K_{264} site with a variety of sugars potentiates T-cell responsiveness in a wide variety of CII restricted T cells, with the galactosylated form being the most potent. The origins of the glycosylated–CII would depend on the posttranslational modifications carried out by cartilage chondrocytes. However, it appears that lymph node and spleen cells from these mice respond with an efficiency 100 times greater than the glycosylated form. Finally, the glycosylated form of the peptide binds human HLA-DR1.

USAGE

Being the species it is, the mouse has been used for the investigation of a prodigious variety of endogenous molecules, biologics, and small drug molecules. The c57 strain are the most useful since many transgenic modifications are carried out in this strain, and in order to respond, these often require a heavier M.tb. load in the FCA to induce a response. The other major strain used in transgenics, the 129 (haplotype H-2bc) is not susceptible to CIA.

Taking into account the major features of Figure 30.2, transgenics and biologics have demonstrated the role of many of these in the polyarthritis. The mouse CIA was instrumental in proving that TNFα was central to the rheumatic response. TNFα$^{-/-}$ mice do not respond to the arthritis, and TNFα receptor$^{-/-}$ mice have a lower incidence and milder disease. The use of anti-TNFα regimes reveals a more complex role, in that its suppression can increase arthritogenic Th1 and T17 in lymph nodes, whilst inhibiting their influx into joints. It does appear that once arthritis is established, inhibiting the TNF system has less effect. Variability in response between research groups may arise from the slow progression of the disease through the joints, having an action on each joint at different times (for example see Table 30.7). IL-1 is very important, anti-IL-1 therapy being very effective, especially when cartilage degradation is taken into account. IL-1 converting enzyme (ICE) $^{-/-}$ mice have a reduced response, and introducing IL-1 receptor antagonist protein (IL-1ra, IRAP) expressing vectors into the synovium of one joint has efficacy in both that and the contralateral joint.

From the central role of macrophage-derived IL-12 in the stimulation of IFNγ and the development of the Th1 response, it would be expected that ablation would suppress mouse CIA. IL-12 administration exacerbates disease, and inhibition does occur in IL-12$^{-/-}$ mice. Anti-IL-12 given at sensitization and onset (day 21) also inhibits the disease. However, when anti-IL-12 is given in established disease it has a markedly reduced action, and withdrawal exacerbates disease. Late treatment with antil-IL-12 leads to disease enhancement through a paradoxical enhancement of enhanced IL-10 expression. Indeed IL-10, a Th1 inhibiting product of the Th2 system, reduces CIA. IL-10 neutralization exacerbates established disease. IL-4, the other Th2 cytokine is rarely seen in RA; it has little action alone, but is profoundly effective when combined with IL-10. The antisera to both of these Th2 cytokines have effects consistent to this data. When administered to CIA mice IL-23, another member of the IL-12 family produced by macrophages, stimulates the appearance of Th17 cells. Th17 cells and their product IL-17 play a significant role in the development of erosive CIA. IL-23$^{-/-}$ mice do not respond with CIA, they do not suffer from the bone erosion and are absent in Th17 T cells. IL-12$^{-/-}$ mice produce more Th17 cells. It thus appears that end stage disease is controlled differently to the mid and early stages, with IL-23 promoting and IL-12 protecting against CIA.

B-cell depletion [9], highly effective in seropositive RA, when given prior to sensitization, delays disease onset, autoantibody production, and reduces the severity of inflammation and joint destruction. However, the effects are attenuated by the reappearance of B cells.

Regarding bone destruction, receptor activator of nuclear factor B ligand (RANKL) is a central factor for osteoclast differentiation. It is antagonized by a natural receptor antagonist, osteoprotegerin (OPG). OPG$^{-/-}$ CIA mice have reduced osteoclast numbers and correlated with this, bone erosion. RANKL mediates the erosive actions of IL-17, with anti-RANKL treatment inhibiting this. Anti-RANKL also reduces osteoclast numbers and bone erosion in mouse CIA.

The predominance of the neutrophil in CIA is in marked difference to RA, as is the pattern of neutrophil erosion of cartilage. The role of the FcR for IgG, which are expressed on macrophages and neutrophils, and complement is in open debate. C5-deficient mice do not develop clinical signs of disease, or erosion, but still possess cell-mediated and antibody-mediated

TABLE 30.7. Examples of the action of biologic therapies on murine collagen-induced arthritis

Treatment	P or T	Incidence	Time onset	CII cell response dth	CII ab	Erosion	Inflammation
Anti-B cell	P		S	S	S	S	S
Anti-Ia	P	S	S	NS	Transient		
	T	NS	NS	NS	NS		
antiCD4	P	S	S		S		
	T	<S	<S		NS		
Anti-CD40L	P		S		S		
Anti-CTLA-4	P			S	S		S
rhIL1RA (High Dose)	P	S	S			S	S
	T					S	S
Anti-IL2r	P	S	S		S		
Anti-IL6	P	S			NS		S
IFNγ	T	S	S	S	S		S
Anti-IFNγ	P	NS	NS		NS	S	S
	T						increase
Anti-TNFα	P	S	S		S	NS/S	S
@3wk	T				NS	NS/S	S
@2mo	T				NS	NS	S/NS
TNFsR	T	S				S/NS	S
Anti-IL1	P					S	S
@3wk	T					S	S
@2mo	T					S	S

(T, therapeutic dosing regime, P, prohylactictic dosing regime; S, significant, NS, not significant). There are often large differences in responses to biologics between authors, and so these are not definitive. Avidity, biodistribution, and dose play major parts in negative responses, rhIL1Ra for example requires high dose, osmotic pump infusion, since leaving 2% of IL-1 free is sufficient to maintain disease. Thus there many reports of the in-activity of this treatment. Results for anti TNF therapy are especially variable. Prophylactic dosing with anti-TNF is effective, but therapeutic is not, since TNF is involved in the early stages of the synovitis. However, each joint as it succumbs to arthritis can be considered to have early stage disease no matter how far into the process. Thus sometimes therapeutic dosing in studies where the disease onset is not synchronized (by stimulating with lipopolysaccharide for example) may show some efficacy.

responses. $Fc\gamma R^{-/-}$ mice are protected against CIA. However, subsets of FcR have stimulatory or suppressive roles, and the complement binding activities of the anti-CII IgG isotypes differ in their binding of both complement and FcR subtypes. The interactions between these systems and the end result remain an active area of research.

DRUGS

Compared to adjuvant arthritis, the mouse CIA presented a refreshing change. Table 30.8 summarizes the effect of a variety of anti-rheumatic agents in CIA. Essential differences can be seen, depending on whether drugs are given as prophylactic or therapeutic treatments. Overall, the antiinflammatory steroids are the most effective. This may be interpreted to mean that it is easier to effect change in the early phase, especially if the drug has actions on leukocyte infiltration at this stage. Thus, this model responds to a certain extent to NSAIDs, etodolac, for example, if given prophylactically delays onset and incidence slightly, with no effect on anti-CII. Given therapeutically it is inactive. In general, cyclooxygenase-2 inhibitors have only marginal actions, exemplified by drugs such as rofecoxib, celecoxib, deracoxib, and SC58236. Piroxicam, a traditional NSAID, in comparative studies appears more potent against the clinical score. Combining

TABLE 30.8. Examples of responses of the mouse collagen-induced model of arthritis to anti-rheumatic drugs

Treatment		Incidence	Time onset	CII cell response dth	CII ab	Erosion	Inflammation
Aspirin						NS	
Etodolac	P		S		NS	NS (ank S)	NS
Benoxaprofen	P	S (slight)					S/NS
Indomethacin	P	S	NS		NS		S
Naproxen	P	NS					NS
D-Penicillamine	P	NS					Increase
Auranofin	P	NS					Increase
Gold salt	P				NS		NS
Levamisole	P				NS		NS
Cyclophosphamide	P	S	S		S		S
	T	S					S
Prednisolone	P	S					S
	T	S					S

P, prophylactic treatment; T, therapeutic; S, significant; NS, not significant; ank = ankylosis. Note that in some instances well-known antirheumatic agents may potentiate disease in this model. This is as a consequence of the specific nature of the disease mechanism that does not completely mimic that of human disease. This is not uncommon in other arthritis models.

Piroxicam with an inhibitor of 5-lipoxygenase results in a potentiation of efficacy. In this instance, the effect on erosive disease is unknown.

With respect to DMARDs, the model is insensitive to aurothiomalate, chloroquine, colchicines, and levamisole, but responds well to cyclophosphamide and low-dose methotrexate. The sensitivity and delicate balance between helper and suppressor T cell subsets can be illustrated by the markedly contradictory effects of cyclosporine, a suppressor of T-cell IL-2 synthesis and thus proliferation. If given at initiation administration it prevents CIA, but if given later can have a variety of effects dependent on the laboratory reporting the data, or the time of administration. This can vary from being suppressant, ineffective to frankly pro-arthritic when given to established disease. This is duplicated by the cyclosporine partial mimic, FK506.

Modern drug targets such as p38 MAPkinase and adenosine A3 are responsive in the mouse collagen arthritis. A novel area is the use of short inhibitory RNA (siRNA). Delivery of siRNA has been problematic, but a variety of delivery systems are effective, for example lipoplexes. siRNA to IL-6, IL-1, and IL-18 administered weekly are effective against both joint inflammation and cartilage and bone erosion in both prophylactic and therapeutic settings. The combination of all three reduced all features significantly when compared to anti-TNFα siRNA alone.

This model has been used to a great extent in the investigation biologial therapeutics [10]. Table 30–7 gives some idea as to the use to which the model has been put in the field of biologics. It is important that a full appreciation of the pharmacodynamic properties of the biologic involved is known. Different agents used by different research groups, aimed at the same target, can provide very different results. This is well illustrated in the literature. As described before the time of dosing can have a very important bearing on the response. The table exemplifies these variabilities in response seen with interleukin-1 receptor antagonist (rhIL-1RA, or IRAP), as well as the anti-TNF therapies [4,11,12] and interferon-gamma.

K/BxN ARTHRITIS

The group of Mathis and Benoist fortuitously discovered that T-cell receptor transgenic mice recognizing an epitope of bovine RNase (KRN) crossed with non-obese diabetic (NOD) MHC class II A^{g7} mice generated offspring (K/BxN mice) that develop a spontaneous arthritis resembling human rheumatoid arthritis [13]. The most recent of the major arthritis models discovered, the knowledge from previous models described above, and the application of transgenic, biologic, and analytical technologies has meant that much has been learned within a short period of time. The onset of arthritis is observed at 3–4 weeks of age presenting clinically as a chronic, progressive, polyarthritic disease with a distal to proximal gradient of severity. The transgenic TCR recognizes residues 282 to

294 of glucose-6-phosphate isomerase (GPI) as the antigen presented by NOD antigen presenting cells in the context of MHC class II molecule I-A^{g7} [14]. This molecule is also recognized by antibodies generated in the murine CIA. It was this ubiquitously expressed glycolytic enzyme that was identified as the crucial driver for T-lymphocyte initiation and B-lymphocyte production of arthritogenic immunoglobulins, which lead to disease in the offspring. Early T-lymphocyte requirement is determined by the fact that administration of anti-CD4 neutralizing antibodies prevent disease but only if given at least 5 days before the onset of the disease. Furthermore, KRNxNOD mice without B-lymphocytes do not develop arthritis, confirming the necessity of this cell type for initiation. Unusually it seems this model may be more Th2 than Th1. Though not full Th2, the expression of IL-4 rather than IL-12 is required for full disease. It should be noted that IL-4 is absent from RA synovium.

K/BxN SERUM TRANSFER MODEL

The major problem with the transgenic model has been the limited availability of the mice. However, what has been most useful is that serum from KRNxNOD offspring can transfer disease to a wide panel of inbred-strains and gene-disrupted mice with the same onset of arthritic features being observed in almost 100% of mice in a fashion similar to that seen with the transfer of arthritis by anti-CII antibodies [8]. Whilst a single i.p. injection of antibodies confers a transient arthritis related to the level of residual circulating antibodies, multiple administrations confer an erosive arthritis similar to that seen in the KRNxNOD mice. The standard protocol is to dose i.p. on days 0 and 2. The antibodies appear to be of the IgG1 (Th2) subclass, with indications that they are closely related and thus derived from a narrow range of B cells. Like the passive CII model, except for the fact that CIA antibodies are of the Th1 IgG2a and IgG2b class, mixtures of antibodies derived from serum are required, though an arthritis-inducing monoclonal antibody has been developed.

This model thus allows in-depth analysis of the effector phase of arthritis where T- and B-lymphocytes seem to play a less crucial role in the pathophysiology; an observation confirmed by injection of purified anti-GPI immunoglobulins into $RAG^{-/-}$ mice (absent of mature T- and B-lymphocytes) with the same onset of arthritic characteristics as K/BxN mice. Thus, the arthritogenic anti-GPI immunoglobulins are essential for the cascade of events discussed below that ultimately lead to synovitis and joint destruction.

Anti-GPI immune complexes and complement protein C3 are found to colocalize on the surface of articular joints, a space deficient of complement regulatory/inhibitory proteins. A specific interaction of the GPI enzyme active site and cartilage proteoglycans is thought to result in its concentration at the cartilage surface since glycosaminoglycan degradation with ABC or β-glucuronidase prevents GPI binding to cartilage. The localization of the antigen at the cartilage surface, like collagen, is an ideal situation for initiation of the alternative pathway of complement, which requires a surface for its activation. In fact, Factor B- and not C4-deficient mice (early effector proteins of the alternative and classical pathway of complement, respectively) are protected from K/BxN serum induced arthritis highlighting the role of the alternative complement pathway in this model. C5aR, a downstream component of the complement pathway with chemoattractant properties, is also implicated in the pathophysiology. As in mouse CIA, there is a need for FcγRIII, as mice deficient of this receptor have a delayed disease onset with reduced ankle swelling and joint erosion compared to wild-types. FcγRIII is expressed on mast cells, neutrophils, and macrophages, cells all highlighted to play a role in the pathophysiology of K/BxN serum induced arthritis. Mast cells do play a crucial role in this model. Using W/W^v and Sl/Sl^d mice (mice deficient of mast cells), the absence of mast cells protects them from disease. After restoration with wild-type mast cells the mice revert to disease susceptibility. Furthermore, immune complex activated mast cells are the source of IL-1 that contributes to the arthritic profile observed. The role of the macrophage is demonstrated through selective depletion with intra-articular clodronate liposomes to deplete synovial macrophages. No subsequent signs of disease are seen, confirmed by the absence of pannus formation, cartilage destruction, and bone erosion. Neutrophils are another contributing factor in K/BxN arthritis and are found in synovial joint spaces and fluid within 48 hour of disease induction. The neutrophils are involved in both initiation and progression of disease as neutropenic mice rendered by treatment with a neutrophil depleting antibody are completely resistant to disease even when neutrophils are depleted up to 5 days after serum transfer. Collectively, the aforementioned innate immune mediators and inflammatory cells are the source of proinflammatory molecules that orchestrate a complex network of occurrences leading to the chronic inflammation and joint destruction observed in K/BxN serum transferred arthritis.

HISTOPATHOLOGY

The active K/BxN arthritis is initiated by the deposition of fibrin, followed by synovial edema and subsynovial cellular infiltration and angiogenesis. Weeks later, an intense inflammatory synovitis occurs with hyperplastic synovitis. Evidence of fibrosis appears at this stage, with some evidence of lymphocyte aggregates. The fibrinoid material is present during the chronic phase and concomitant with the ingress of neutrophils into the joint space. Pannus occurs with degradation of bone and juxtaposed cartilage. Months later the joints are fibrotic, with no evidence of cartilage, but bone with remodeling and chaotic presentation. This affects the joints of all limbs except the hip. The sacroiliac joint is not involved and the spine sparingly. The histopathology of the passive serum transfer model is less well described as the active model, and it is reported the features in general are the same as the active arthritis. There is a significant synovial inflammation with pannus formation, cartilage erosion accompanied by adhesion of granulocytes to the cartilage surface, similar to that seen with CIA and antigen-induced arthritis (Figure 30.3G) is eroded by pannus. There is bone erosion that is dependent on RANKL and cartilage degradation accompanied by a neutrophilic synovial fluid. Cartilage distant from the pannus loses proteoglycan staining with toluidine blue. One difference between the active and passive forms of the disease is that in the passive there is heterogeneity in the joints involved. Some adjacent joints can have either severe infiltration, or no detectable disease.

USAGE

Anti-IL-1 receptor antibodies reduce disease, and IL-$1^{-/-}$ mice do not respond to serum-induced arthritis. The same occurs for TNF$\alpha^{-/-}$ mice, but disease escape can occur in these mice. IL-6 deficiency is ineffective. The variability in response with the TNF$\alpha^{-/-}$ mice may be environmentally linked, but this is solely due to the absence of any convincing genetic evidence thus far. In some instances, anti-TNFα therapy is ineffective in the K/BxN mouse, although anti-TNFα is effective in preventing SCID mice from the passive transfer of arthritis through DBA/1 splenocytes.

The angiogenesis featured in the histological presentation of the disease indicates that inhibition of the potent angiogenesis stimulant, vascular endothelial growth factor (VEGF) may be effective. Indeed, anti-VEGF therapy is effective in reducing, if not abrogating the erosive disease, as is selective VEGFR1 kinase inhibition. Suppression of the alternative VEGFR2 receptor is effective.

Heme-oxygenase, an enzyme critical for the scavenging of reactive-oxygen species, through the synthesis of bilirubin and biliverdin, reduces disease if induced pharmacologically with protoporphyrins. In addition, the inhibition of the cell cycle with cyclin-D inhibitors is also effective.

The use of this model is thus discovering quite different mechanisms to those hitherto seen with the other major forms of experimental arthritis, not least because of its reliance on a Th2 bias. How these mechanisms relate to clinical disease is so far speculative and could be very revealing.

ANTIGEN-INDUCED ARTHRITIS

Originally described in the rabbit using bovine serum albumin, this model is now used in the rat [15] and more commonly in the mouse. Mice of the H-2^b haplotype (such as the c57bl/6) are sensitized with an emulsion of M.tb. and antigen, methylated bovine serum albumin (mBSA), either in the flanks or in the base of the tail [16]. Subsequent subcutaneous challenge with mBSA induces a classical type-IV delayed type hypersensitivity Jones-Mote reaction, complete with an acute IgG2 dependent Arthus stage, with neutrophils, followed by the cell-mediated stage characterized by the influx of lymphocytes and monocytes.

mBSA challenge in the stifle joint induces a vigorous monoarticular arthritis that peaks at around 7 days and then wanes. Whilst anti-mBSA IgG2 antibodies are induced, susceptibility of animals to i.a. challenge is low after serum transfer, but high after T cell transfer. It may be that, like CIA and K/BxN, altering the dose, the antibody mixture, or the administration schedule could reveal an antibody dependent mBSA arthritis. In any event, the induction of arthritis is thus T cell dependent, in keeping with the type IV hypersensitivity protocol. There is an early activation of activation of macrophages and Th1 and Th2 cell populations over the next 6 days. The Th responses differ depending on the maturity of the arthritis, with Th2 responses in the inguinal draining lymph nodes, and later overlapping Th1-like/Th2-like peaks in the spleen. There is chronic elevation of synovial IL-1beta mRNA and spleen IL-6 mRNA. B cell activation is not considered to be involved in the generation of the hypersensitivity response of this model. CD4+ cell depletion inhibits both the acute reaction, and severity of the chronic disease. Antigen-induced arthritis is transferred into SCID mice by CD4+ cells and not CD8+ cells.

Sensitization to bovine serum albumin, as opposed to mBSA, elicits inflammation but not a chronic erosive form. The specific arthritic reaction induced by mBSA is thought to be due to its poly-cationic charged nature, which results in its retention within the joint, as well as hyaline cartilage, binding to the negatively charged elements within [17]. The induction of inflammation, or trypsinisation of cartilage, reduces the binding of mBSA to the cartilage. When injected into naive and immune mice, mBSA is found within the exudate, synovial lining cells, hyaline cartilage, menisci, and tendons. The joint structures in nonimmune mice are retained, with mBSA being detectible 28 days later. In immune mice, mBSA becomes depleted from these structures in areas associated with the advance of the granulomatous inflammation, and underlying chondrocyte death. IgG deposition within the cartilage is seen from the second day of injection, increases to day 7 and then remains. mBSA is retained chronically in the femoral and tibial margins, but becomes depleted in proportion to the degree of inflammation.

The reaction is accompanied by fibrin deposition and synovial lining hyperplasia, and the inflammation is initially neutrophilic. This develops into a monocytic and lymphocytic reaction accompanied by synovial lining hyperplasia, soft tissue inflammation and angiogenesis, infrapatellar fat pad inflammation, accompanied by loss of cartilage proteoglycan (staining for Safronon-O or Toluidine Blue), suppression of proteoglycan synthesis, periostitis, and bone erosion. The suppression of proteoglycan synthesis is due to IL-1, being inhibited by IL-1 receptor antagonist protein (IRAP), and not anti-TNF nor anti-IL-6. However, if mice are boosted with *Bordatella pertussis*, a more chronic reaction is induced which is accompanied by a monocytic pannus, fibrosis and the severe erosion of joints. Joint erosion is related to the boosted circulating IgG2 anti-mBSA antibodies. The erosion follows immune complex fixation at the cartilage surface and is dependent of the presence of macrophages. Patellar osteophyte and chondrophyte formation is a feature of this model (Figure 30.5).

This model is Th1 dependent, with an accumulation of mBSA-specific IFNγ-secreting T cells. This is in keeping with the sensitization protocol for other skin hypersensitivity models, and the pattern of anti-mBSA IgG reflects this with IgG2b being predominant, and IgG1 much less so (IgG2b=IgG3>IgG3=IgM=IgA>>IgG 1). Despite the poor IgG2a response reported by some authors for the collagen induced arthritis in C57bl/6 mice, and this model, circulating IgG2a has been reported in antigen-induced arthritis. IgG binds mast cells, macrophages, and neutrophils via Fcγ receptors. Mice lacking FcγRI and FcγRIII demonstrate equivalent synovitis, but lacking both have reduced synovial inflammation. FcγRI is essential for joint destruction. FcγRII is inhibitory, and its absence exacerbates exudative inflammation and cartilage erosion. IL-1 is expressed in the synovium, and anti-IL-1 inhibits reduces the joint swelling, and neutrophil influx after 4 days, but not earlier. IL-6, and TNF are expressed in the acute phase, IL-4 expression being reduced. Indeed, anti-IL-4 exacerbates disease.

RA is characterized by phases of disease flare. This can be mimicked in the antigen-induced arthritis. During the chronic phase, if mBSA is injected intravenously, an exudative flare reaction in the affected joint is induced, that is abrogated by T-cell depletion, and not complement depletion, confirming the T cell dependence of the immune compartment (as opposed to erosive compartment) of the disease. This is characterized by fibrin deposition and an influx of neutrophils, over and above the previously monocyte dominated pathology. 9 days later, the monocytic dominance is restored. In addition, re-challenge into the knee around one month after primary challenge elicits the formation of lymphoid aggregates at 5–9 months similar to those found in RA. Many of these possess discrete T and B cell zones. Lymphoid structures are dependent on cellular responses to chemokines, and not only are they reduced in CCR7 or CXCR5 deficiency, cellular and antibody responses are reduced along with the arthritis. CXCR5$^{-/-}$ mice possess reduced erosive disease.

USAGE

The pharmacology of this model illustrates the importance of measuring cartilage erosion in addition to inflammation [18]. When dosed therapeutically, indomethacin for example inhibits inflammation to a greater degree than it does erosion, in fact erosion could be considered to be accelerated. Antiinflammatory steroids inhibit both, and the first line antirheumatic drugs sulphasalazine and methotrexate also inhibit erosion and inflammation, as does the historical antirheumatic DMARD, azathioprine. D-penicillamine is largely ineffective in this model.

IL-6 knock out mice develop less severe arthritis, with reduced TNFα and IL-1β synthesis and Safronin-O staining. sTNFR reduces the inflammation associated with the chronic phase when administered on day 6. The suppression of CCRL7 delays the inflammation, but has no effect on joint erosion, whilst CXCR5 deletion inhibits both, accompanied by the reduced formation of ectopic germinal centers.

The relative mechanistic simplicity, and the monoarticular nature of this model makes it very suitable for dissecting mechanisms, especially using transgenic mice for effects on the generation of the pure Th1 immune response, and the determination of the erosive potential of the disease. For example, the initiation of T cell responses is dependent on antigen presenting cell CD80 and CD86, and the inhibition of their function by anti-CD80 and anti-CD86 inhibits the arthritis. PAR-2$^{-/-}$ mice show reduced Th1 responses and arthritis. However, K/BxN, zymosan, and FCA arthritis do not, these being Th2 and innate in immune nature. This shows a selective effect on the generation of the Th1 response. Suppressor T cells are important in regulating the disease, depletion of CD25-expressing cells before the arthritis leads to an exacerbation of arthritis, as indicated by reduced joint inflammation and erosion. Transfer of CD4+CD25+ cells at the time of challenge decreases the severity, but does not abrogate the disease completely. No changes in the systemic mBSA-specific immune response occurs, so the effect appears to be synovial in nature. In vivo migration studies do show an accumulation of CD4+CD25+ cells into the inflamed joint as compared with CD4+CD25- cells. However, Treg cells appear unable to affect acute or chronic inflammation once it is established.

The model has been especially useful in investigating the role of TIMPS and metalloproteinases. The inhibition of MMPS is considered an important therapeutic goal to prevent the destructive aspects of disease. However, their role in the inflammatory process may be more important. The deficiency of TIMP-3 (inhibits MMPs, aggrecanases, and TNFα converting enzyme) exacerbated the acute inflammation and circulating TNFα after immunization. Aggrecan mediated cartilage destruction appeared unaffected, illustrating its importance in the inflammatory process, differentiating roles in inflammation and destruction. PCR analysis shows that during the acute phase, MMP3 is indeed enhanced, but MMPS 8 and 9 do not differ greatly from baseline, despite Affychip gene analysis showing an enhancement. The direct assay of enzyme activity shows that MMP9 (macrophage gelatinase) is greatly enhanced during the chronic macrophage dominated erosive phase. The neutrophlilic nature of the acute disease enables neutrophil selective therapies to be investigated, such as leukotriene B4 synthesis inhibition, or CXCR1 and CXCR2.

The role of annexin-1 (formerly lipocortin) in the severity of inflammatory arthritis has been demonstrated using this model. A significant exacerbation of arthritis is observed in Anx-1($^{-/-}$) mice compared with wild-type (WT) mice. This is associated with increased mRNA expression of synovial IL-1β, TNFα, IL-6, and macrophage migration inhibitory factor. Dexamethasone prevents synovitis and bone damage in an annexin-1 dependent fashion, as well as reducing circulating anti-mBSA IgG. This is since annexin-1$^{-/-}$ mice do not respond to dexamethasone treatment.

RAT ANTIGEN-INDUCED ARTHRITIS

In the rat, a similar reaction can be initiated through sensitization to mBSA in FCA. A chronic erosive arthritis, like in mice, can be conferred by an mBSA/FCA boost. The larger stifle joint of the rat enables easier measurement of joint inflammation, as well as the assay of cartilage components, and bone loss [19]. Algesia in the affected joint can also be measured through observation [20] as well as using more quantitative paw pressure plate technology.

The acute phase is C5a dependent [20]. On intraarticular challenge cytokine expression is elevated in the synovial membrane with hours [21]. IL-1β, IL-6, TNFα, IL-2, and IFNγ are elevated, as are IL-4 and IL-10 at day 1. The invading pannus is initiated early by day 3, through cell recruitment. Conversion to the chronic phase is associated with a reduction in all cytokines expressed in the synovial membrane, except IL-1β. At day 14, there is a significant neutrophilic inflammation and by day 28, both neutrophils and macrophages can be seen lining the cartilage, accompanying its destruction and severe synovial fibrosis. At day 49, the tissue lining the cartilage comprises macrophages and fibrocytes. Cartilage loses proteoglycan in the early phases, accompanied by MMP in the synovial fluid. Osteophyte and chondrophyte formation accompanies the erosion of cartilage in the chronic phases, especially circumscribing the patella (Figure 30.5). The reaction is, like the mouse, transferable by activated mBSA-specific lymphocytes. Circulating antibodies to mBSA are induced, as are antibodies to proteoglycan and type-I and type-II collagen. The reaction is inhibited by cyclosporine, steroids, the NSAIDs indomethacin, ibuprofen, and the DMARD methotrexate, but not by D-penicillamine. However, detailed analysis reveals that whilst Piroxicam is antiinflammatory during the acute phase, its administration during the chronic phase exacerbates the destructive reaction. This matches acute data found in other acute arthritidies in the rat such as the M.tb. monoarticular arthritis (Piroxicam and naproxen, but not diclofenac or tiaprofenic acid) [22], and zymosan (all NSAIDs tested), and antigen-induced arthritis in the rabbit (indomethacin). The size of the rat joint means that the assay of synovial contents is easier, IL-6 and TNFα for example, are raised, and reduced by anti-C5 complement therapy, as is circulating TNFα.

HUMAN: MOUSE RA SYNOVIUM XENOGRAFTS

The severe combined immunodeficient (SCID) mouse is used in the field of cancer research to investigate human tumors. The immune deficient status of the SCID mice can be taken advantage of to graft pieces of human rheumatoid synovium and investigate their function in vivo without the accompanying problems associated with ex vivo organ culture, such as artificial growth medium, and tissue necrosis through poor oxygen perfusion and nutrient diffusion. After grafting, the human synovial vasculature anastamoses with the mouse via a fibrous capsule and the tissue becomes perfused with the host blood [23]. The grafts remain stable for over a month and often longer, before slowly starting to lose their architecture. Some grafts appear to recruit murine monocytes or T cells over time to a minor degree, whilst others do not. Immuno-fluorescence assays of human cell markers in the main do not detect human cells in host mouse

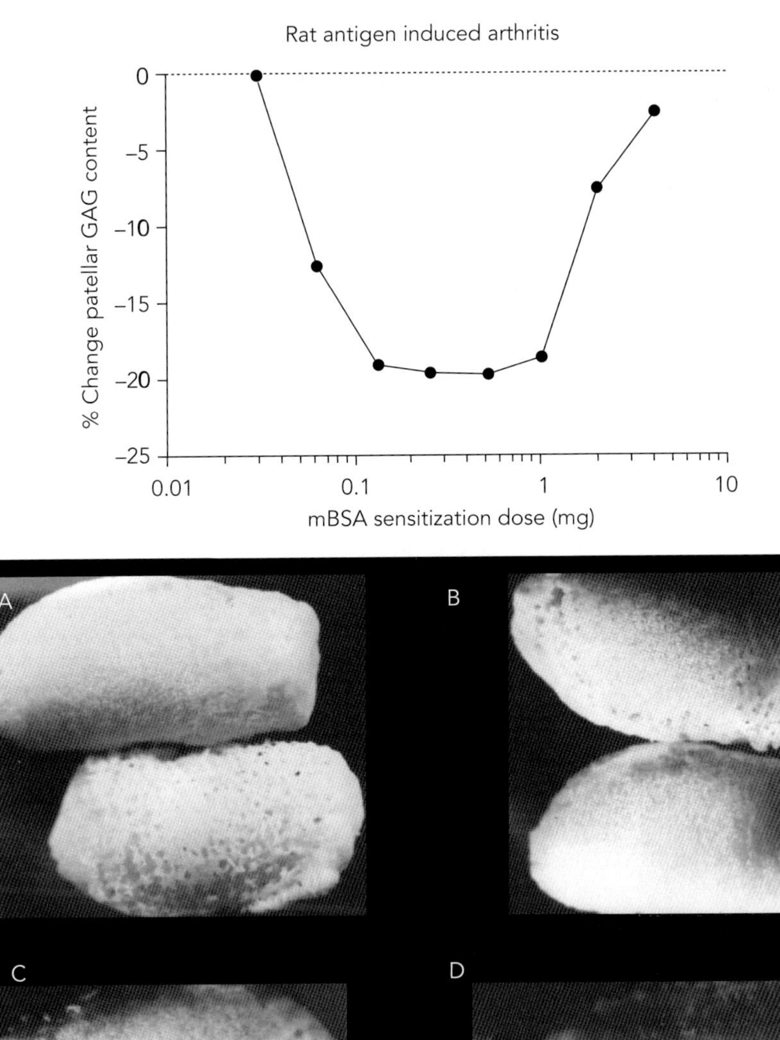

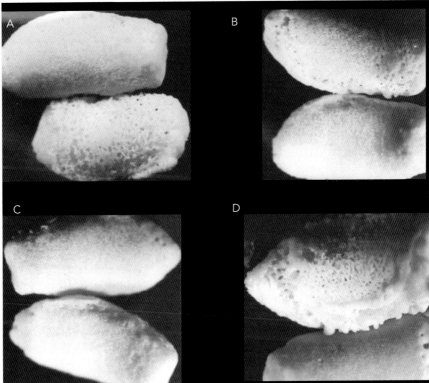

Figure 30.5. Rat antigen-induced arthritis. Rats were sensitized with methylated bovine serum albumin in an emulsion with Freund's complete adjuvant at the base of the tail and challenged into one knee with mBSA two weeks later. Graph: The importance of establishing the sensitization dose of an antigen. As the dose is raised, the degradation of patellar cartilage proteoglycan matrix (GAG) is increased. As the dose is increased further, tolerance is induced and challenge no longer has a destructive effect on the cartilage. (A) Patellar bone erosion in a 4 day antigen arthritis joint, note the bone resorption in the affected joint compared to the contralateral control (upper bone). (B) A representative rat treated with leflunomide (control lower bone), and (C) cyclosporine (control upper bone). (D) The chronic fibrotic phase of antigen-induced arthritis the bone recovers with chondrophyte and osteophyte formation, day 55. (control lower bone)

lymphoid organs after implantation. Some report that grafts become depleted of T-lymphocytes, which could indicate turnover of T cells in the diseased tissue, and others that CD3 cells remain after engraftment. At 4 weeks cells still express Ki-67 or PCNA, markers of cell division, and the grafts continue to release human IgM, IgG, RF, and ACPAs, and IL6 for example, all of which can be detected in the circulating blood. PCNA is associated with the proliferating $\alpha v\beta 3$ angiogenic blood vessels, and proliferating fibroblasts. CD68 macrophages start with a concentration at the synovial lining layer, which becomes less demarcated and distribution more diffuse. CD3 cells can congregate around blood vessels. Fibroblasts become denser with time, and the tissue become more densely packed and organized, unlike the looser less organized tissue at implantation. Engrafted pannus at four weeks contains cells expressing a variety of markers such as those indicating helper cells (CD4), fewer T-suppressor cells (CD8 subsets), and B cells (CD20). B-cell aggregates although smaller and less organized have also been demonstrated up to 4 weeks post-transplantation. Cytokines, IL-6 and TNFα for example, are synthesized though IL-6 synthesis can wane unless stimulated by a further stimulus or recruitment of T cells. Enzymes indicating the graft's potential to degrade bone and cartilage, such as tartrate-resistant acid phosphatise (TRAP) positive cells (osteoclasts), and matrix metalloproteinases (MMPs) including MMP1, 3, and 9 are also expressed. T/B cell segregation has also been demonstrated within rheumatoid lymphoid aggregates, with CD4+ T cells predominantly surrounding the B-cell aggregate and CD8+ T cells located mainly within the perifollicular area. The macrophage marker CD68 is seen scattered widely through the tissues. MMP1 expressing cells appear to be at the margins, and MMP9 more evenly distributed, but more expressed in pannus tissue. It should be noted that synovial biopsies can be differentiated into those obtained from the synovial villus, and those from pannus juxtaposed to cartilage, and may have different characteristics.

USAGE

The model can be used in a variety of ways (see Table 30.9). Not only drugs but also human-specific biologics can be assessed on these tissues, without heed to species cross-reactivity.

Once the vascular anastamoses are established (3 weeks) human lymphocytes can be injected intravenous. and their accumulation within the graft assessed either through labeling with a tracer (e.g., technetium, indium) or via histology using human-specific cell markers, with assessment at a variety of times after infusion from hours to several days. Isolated blood lymphocytes, monocytes, or T cells can be studied, and U937 cells, a human myeloid precursor cell line, can also be used [24]. Care should be taken if specific RA tissue homing is being investigated, since it can be shown that human cells can traffic to other inflammatory tissues as well, showing certain systems are a general feature of the tissue inflammatory state. To control for cell trafficking under noninflamed conditions, human skin can be transplanted as a control in the same mice. Effective treatments preventing cell recruitment include anti-CD11a (ICAM-1), and αEβ7 (E-cadherin).

TABLE 30.9. The use of the Human: SCID (hu/SCID) mouse rheumatoid synovium xenograft model
Determination of key molecular mediators initiating and maintaining rheumatoid synovitis Mechanism determination and manipulation of (auto) antibody production within synovial tissue
Human lymphocyte migration into the graft: adhesion molecules, chemokines
Human lymphocyte subsets, cloned, or transgenic cell function
Erosive capacity and mechanisms
Human cytokine and chemokine synthesis and actions
Human rheumatoid gene expression under resting and investigational conditions
Assessment of synovial effects of human specie-specific biologics
Determination of the translational utility of experimental therapeutics in a human in vivo system
Comparisons between OA, RA, and other human inflammatory conditions
Drug delivery to human RA synovium
Determination of synovium-specific markers using in vivo phage display

Adhesion Molecule expression changes over time. VCAM-1 and E-Selectin expression decreases with time, whilst ICAM-1 is maintained. The injection of TNFα (1000U) into the graft significantly upregulates ICAM-1, and this is accompanied by an anti-ICAM-1 inhibitable enhancement of human mononuclear cell recruitment into the grafts after i.v. infusion. This effect is seen both with IL-1 and TNF. E-Selectin and VCAM-1 are unaffected. Increasing the dose of TNFα to 2000U, and giving IL-1β as opposed to IL-1α, can result in a heightened expression of E-selectin in addition to VCAM-1 and ICAM-1, but not MHC-II. The rarity of RA synovial biopsies means that detailed dose response curves are a luxury, so differential concentration dependent expression profiles are difficult to justify. However, it is tempting to speculate that adhesion molecule expression profiles may vary according to the concentration of TNFα. The antiinflammatory cytokine IL-10 reduces ICAM-1 expression, and its partner IL-4 a mild increase in ICAM-1. IL-10 and IL-4 reduce human peripheral blood mononuclear cell recruitment by the grafts. These are significant observations, establishing the importance of ICAM-1 in the recruitment of mononuclear cells into human RA synovium.

This system can thus be used to assess the efficacy of human specific biologics. One endpoint can be tissue size, anti-human IL-6 receptor antibodies reduce graft size by half, whilst Auranofin (an oral disease modifying anti-rheumatic drug) treatment does not. Anti-IL-6 reduces inflammatory cell involvement, and the grafts are replaced by fibrous and adipose tissue. MMP1 and MMP9 expression is markedly reduced. This would indicate that the erosive potential of the tissues would be reduced. The erosiveness of the RA synovium can be assessed either by the determination of metalloproteinase expression or by implantating the grafts in juxtaposition with cartilage, or other matrices such as dentine. Synoviocytes can be seen to adhere and erode into the matrices. In the case of human cartilage co-implants, tissue can be seen to burrow into the matrix creating furrows and tunnels. The implants can be graded according to their erosive capacity, and will depend on the donor tissues. Normal donor tissue is not erosive. The cell types associated with this erosion appears to be fibroblastic (type B synoviocytes), with some macrophages (type A) with little contribution by lymphocytes. The administration of TNFα accelerates erosion, whilst IL-10 and IL-4 reduce it.

Selective depletion and reconstitution experiments have also provided very useful information. The depletion of T cells with anti-CD2 for example results in not only an almost total suppression of TNFα, IL-1β, IFNγ, and IL-15 mRNA expression, but also the expression of MMPs. In addition to this, CD68 staining cells also disappear. The administration of autologous CD4 T cells restores TNFα synthesis, stimulates IFNγ synthesis and maintains CD68 cells. IFNγ will replace the role of CD4 T cells in anti-CD2 treated mice. Noncytotoxic CD8 suppressor T-cells (lacking CD40L, the cytotoxic molecules perforin and Granzyme-A) derived from autologous RA synovial have the opposite effect. These cells are seen closely associated with lymphoid aggregates and their selective depletion results in disaggregation of the ectopic germinal centers, an inability to retain dendritic cells, and reduced IgG synthesis by the grafts. The depletion of B cells from synovial tissues containing B-cell aggregates, results in dissociation of the aggregates, loss of CD4 T-cells, and reduced inflammatory cells, IFNγ and IL-1B. Micro-dissection of follicle derived CD4 cells followed by expansion and injection stimulates graft IFNγ, TNFα, and IL-1β synthesis.

This also enables the investigation of disease subsets, with HLA-DRB1 matched cells and tissues being required for responses, and mismatched combinations are unresponsive. Comparison between follicle-containing grafts, and nonfollicle grafts show that nonfollicle-bearing grafts do not respond to CD4 cells from follicle bearing samples.

The capacity of this model to operate independently of new inflammatory cell influx has also allowed the recent demonstration that ectopic germinal centre like structures within RA synovial tissue are independently functional, need not rely on continual cellular recruitment, and support local autoantibody production [25]. SCID mice transplanted with aggregate RA synovial tissue expressing activation induced cytidine deaminase (AID), the enzyme required for the production of class switched high affinity antibodies and CD21Long isoform (a specific marker of follicular dendritic cells) are critically associated with the detection of human anti-CCP antibodies within mouse sera. This association demonstrates for the first time that disease specific autoantibodies are actively manufactured at the primary site of inflammation, the synovial membrane and furthermore that manufacture occurs within synovial tissue containing functional B-cell aggregates.

The active angiogenesis within the grafts enables research into this area. The grafts express vascular endothelial growth factor (VEGF) and angiopoitein-1, both potent angiogenic factors. The administration of antiangiogenic factors, such as endostatin and TNP-470 (aka AGM-1470), reduce blood vessel density, and the retention of inflammatory cell infiltrates. Cell apoptosis is also induced within the grafts.

Prospective mining of novel targets in the synovial structures is also enabled through this model. This is exemplified by the screening of bacteriophage for adhesion to synovial structures in vivo in the transplanted mice. This approach has resulted in the identification of novel and site-specific molecular structures within RA synovium that can be selectively ligated by novel peptides, and have the potential for drug targeting. Genetic manipulation of cells derived from RA synovial can be investigated, one such example are RA synovial fibroblasts, whereby manipulation of MMP, cytokine, cell signaling pathway, expression has been manipulated to provide detailed knowledge of the function of these cells in cartilage erosion in vivo.

OTHER MODELS

There are a large variety of other models of inflammatory arthritis that have been used to dissect immune and erosive processes within joints. A wide variety of inflammatory substances and cytokines can be administered intraarticularly. In general inflammation can be associated with the acute loss of cartilage proteoglycans, followed by their recovery. The injection of streptococcal cell wall (SCW) into the ankle joint in rats is one such model [26], or knee joints in mice. In the rat, this develops an intraarticular inflammation that is not overtly erosive, but in the mice cartilage erosion occurs when given repeatedly [27]. This acute arthritis is susceptible to anti-TNF therapy, namely Etanercept. An interesting observation is that this dependence on TNF is converted to a Th17 dependence when flares are induced by repeated intraarticular injection of SCW. An alternative i.a. model, heat killed mycobacterium tuberculosis, a granulomatous reaction is initiated that induces proteoglycan loss in rats. This is inhibited by DMARDs, such as methotrexate, azathioprine, and cyclophosphamide, but exacerbated by some NSAIDs. Zymosan induces an acute monoarthritis, and again NSAIDs may exacerbate the loss of proteoglycan whilst reducing the inflammation. Interleukin-1 in a variety of species induces a mild inflammation, and acute proteoglycan loss. Unlike those models described previously in this chapter, these acute arthritides are not notable for their destructive capacity.

A systemic polyarthritis can be induced in female Lewis rats by SCW given intraperitoneally. F344 rats are not susceptible, even though they have the same haplotype as the Lewis. This has been taken to infer MHC independence of the disease; however, it appears that F344 rats are tolerized to the SCW bacterial antigens through their gut flora since germ-free animals are susceptible to the disease [28]. In addition, resistance to disease can be transferred by F344 haematopeitic cells to irradiated Lewis rats, and vice versa susceptibility is T-suppressor cells are important for the lack of F344 responsiveness [29]. There is, unusually for a polyarthritis model, an acute phase peaking at 3 days, followed by a chronic erosive phase from day 10 to 28. Erosion is associated with the development of a pannus, and erosion of bone by multinucleate cells, namely osteoclasts. The primary phase is not T cell dependent, but the secondary phase is, these cells recognizing SCW antigens as well as HSP65. Spleen cell transferable tolerance induced by HSP65 prevents disease induction. IL-12 is an important factor in the development of the primary phase, preparing the chronic Th1 dependent arthritis. Anti-SCW antibodies are minimal and not involved in the disease. Splenectomy prevents the secondary phase. Whilst the least severe of the polyarthritis models, it responds to methotrexate, and is used primarily for revealing mechanisms in disease flares. In the c57bl/6 mouse, a chronic arthritis is induced after i.p. injection of SCW. Flares induced by lipopolysaccharide or gut flora products, may indicate a role of these in flares associated with RA. The chronic phase is resistant to anti-TNF therapy, and cartilage proteoglycan loss is IL-1 dependent. It has a pharmacology similar to that of RA.

Another series of polyarthritides are induced by mineral oils, such as squalene, avridine, and pristane [30]. There is a syndrome in man of mineral oil induced arthritis. Freund's incomplete adjuvant, that is Freund's complete adjuvant without the heat killed mycobacterium tuberculosis, whilst it does not induce arthritis in Lewis rats, does do so in DA (Dark Agouti) rats. There is thus a different mechanism operating between the two. These models are T cell transferable, are dependent on antigen-presenting cells, and are erosive with pristane-induced arthritis in DA rats being the most severe. The arthritic capacity of the injected oil disperses rapidly (within 15 minutes) of injection, and the oils disseminate rapidly, with specific concentration in the lymph nodes. This leads to an acute phase reaction, synovial hyperplasia, and cytokine expression. Disease develops at around 9 days, with pristane-induced arthritis being a relapsing form. Anti-αβTCR prevents and abrogates disease, indicating TCR dependency. In the absence of any identifiable auto-antigens, it has been hypothesized that mineral oil distorts the T-cell populations such that self-reactive T cells predominate. Joints are immune privileged, and exposure of novel auto-antigens may induce a joint selective response. Genetic studies have identified a negative association with Rt1n and RT1h, that is, these haplotypes suppress or abrogate disease, but there are several quantitative trait loci involved in susceptibility and severity, but the evidence of MHC dependency has been thwarted until recently by the lack of an identifiable antigen.

Epitope screening of common RA associated auto-antigens in pristane-induced arthritis identified heterogeneous nuclear ribonucleoprotein A2 (hnRNP-A2) [31]. IgG and IgM to are raised a week before disease onset, and sustained during the acute phase of the arthritis. CD4+ cells from spleens react to this protein with IFNg synthesis, and not IL-4, indicating a Th1 response, and hnRNP-A2 over expressed in pristane treated rats prior to arthritis expression. It may be possible then other mineral oil models may include either this or other antigens as yet to be discovered.

SUMMARY

A synthesis of the mechanisms that can induce arthritis in rodents highlights that a cohort of the major mechanisms of the innate and adaptive immune responses are involved in the generation of erosive inflammatory disease. It also follows that these also are involved rheumatoid arthritis. However, none of the models alone accurately reproduce the human condition. Some, like the monoarticular models, are not intended to do so, but are intended to investigate different molecular and cellular arms of the immune system as applied to the arthritic response. It is clear that the MHC, T cells, autoantibodies, B cells, antigen-presenting cells, fibroblasts, cytokines, and the innate system are all involved, and can separately be manipulated to modify arthritis. An overarching hypothesis as to whether each of the multiple systems plays a greater or more central role than the other in human disease has yet to be reached. The advantage of the animal models has been that they can be investigated from the point of induction, whilst human RA has historically presented as a chronic disease. The difficulty in finding a single RA auto-antigen probably reflects this, the temporal distance from the point of induction with concomitant epitope spreading providing a confused picture. The realization of the importance of early and aggressive treatment of RA has lent some credence to the lessons that animal models have been showing, and the availability of early RA clinics to researchers is revealing much more than has been gleaned from the mature disease resulting from years of DMARD therapy.

Each of the key processes of immunity can be found in RA, however, it is the erosive nature of the disease that as yet is not in whole explained by disease modeling. The major differences within the arthritis models are between the rat and the mouse. The murine systems have a high involvement of the antibody as well as the T-cell-dependent responses, whilst in the rat, apart from the collagen arthritis, antibodies play a lesser role. The most significant advance is the huSCID xenotransplant model, which provides a unique bridge to human disease in a wide variety of ways (Table 30.9). No model is in itself correct. Experience shows that those research programs that utilize more than one model of arthritis are more successful in translating novel therapeutics into the rheumatoid arthritis clinic. Rather than being the fact that positive data from two or more systems is more likely to lead to more relevant data, it may reflect a more diligent approach to the problem.

REFERENCES

1. Cope, A.P. 2008. T cells in rheumatoid arthritis. *Arthritis Res Ther* **10**(Suppl 1):S1.
2. Silverman, G.J., and Boyle, D.L. 2008. Understanding the mechanistic basis in rheumatoid arthritis for clinical response to anti-CD20 therapy: the B-cell roadblock hypothesis. *Immunol Rev* **223**:175–185.
3. Moissec, P. 2008. Dynamic interactions between T cells and dendritic cells and their derived cytokines/chemokines in the rheumatoid synovium. *Arthritis Res Ther* **10**(Suppl 1):S2.
4. Brennan, F.M., and McInnes, I.B. 2008. Evidence that cytokines play a role in rheumatoid arthritis. *J Clin Invest* **118**:3533–3546.
5. Trentham, D.E., Townes, A.S., and Kang, A.H. 1977. Autoimmunity to type II collagen an experimental model of arthritis. *J Exp Med* **146**(3):857–868.
6. Luross, J.A., and Williams, N.A. 2001. The genetic and immunopathological processes underlying collagen-induced arthritis. *Immunology* **103**(4):407–416.
7. Holmdahl, R.H., Bockermann, R., Backlund, J., and Yamada, H. 2002. The molecular pathogenesis of collagen-induced arthritis in mice – a model for rheumatoid arthritis. *Ageing Res Rev* **1**:135–147.
8. Nandakumar, K.S., and Holmdahl, R. 2006. Antibody-induced arthritis: disease mechanisms and genes involved at the effector phase of arthritis. *Arthritis Res Ther* **8**:223.
9. Yanaba, K., Hamaguchi, Y., Venturi, G.M., Steeber, D.A., St. Clair, E.W., and Tedder, T.F. 2007. B cell depletion delays collagen-induced arthritis in mice: arthritis induction requires synergy between humoral and cell-mediated immunity. *J Immunol.* **179**:1369–1380.
10. Wooley, P.H. 2004. Immunotherapy in collagen-induced arthritis: past, present, and future. *Am J Med Sci* **327**(4):217–226.
11. Joosten, L.A., Helsen, M.M., Saxne, T., van De Loo, F.A., Heinegard, D., and van Den Berg, W.B. 1999. IL-1 alpha beta blockade prevents cartilage and bone destruction in murine type II collagen-induced arthritis, whereas TNF-alpha blockade only ameliorates joint inflammation. *J Immunol* **163**(9):5049–5055.
12. Joosten, L.A., Helsen, M.M., van de Loo, F.A., and van den Berg, W.B. 2008. Anticytokine treatment of established type II collagen-induced arthritis in DBA/1 mice: a comparative study using anti-TNFalpha, anti-IL-1alpha/beta and IL-1Ra. *Arthritis Rheum.* **58**(2 Suppl):S110–S122.
13. Kouskoff, V., Korganow, A.S., Duchatelle, V., Degott, C., Benoist, C., and Mathis, D. 1996. Organ-specific disease provoked by systemic autoimmunity. *Cell* **87**(5):811–822.
14. Kamradt, T., and Schubert, D. 2005. The role and clinical implications of G6PI in experimental models of rheumatoid arthritis. *Arthritis Res Ther* **7**(1):20–28.
15. Griffiths, R.J. 1992. Characterisation and pharmacological sensitivity of antigen arthritis induced by methylated bovine serum albumin in the rat. *Inflamm Res* **35**(1–2):88–95.
16. Staite, N.D., Cutshaw, L.G., Dunn, C.J. 1989. The effects of different sensitisation protocols on arthritic responses in antigen-induced arthritis. *Inflamm Res* **27**:338–340.
17. Van Den Berg, W., Van Beusekom, H.J., Van De Putte, L.B.A., Zwarts, W.A., and Van Der Sluis, M. 1982. Antigen handling in antigen-induced arthritis in mice an autoradiographic and immunofluorescence study using whole joint sections. *Am J Pathol* **108**:9–16.
18. Crossley, M.J., Spowage, M., and Hunneyball, I.M. 1987. Studies on the effects of pharmacological agents on antigen-induced arthritis in BALB/c mice. *Drugs Exp Clin Res* **13**(5):273–277.
19. Seed, M.P. 2003. The assessment of inflammation, cartilage matrix and bone loss in experimental mono-articular arthritis in the rat. *Methods Mol Biol* **225**:161–174.
20. Woodruff, T.M., Strachan, A.J., Dryburgh, N., et al. 2002. Antiarthritic activity of an orally active c5a receptor antagonist against antigen-induced monarticular arthritis in the rat. *Arthritis Rheum* **46**:2476–2485.
21. Pohlers, D., Siegling, A., Buchner, E., et al. 2005. Expression of cytokine mRNA and protein in joints and lymphoid organs during the course of rat antigen-induced arthritis. *Arthritis Res Ther* **7**:R445–R457.
22. Seed, M.P., and Gardner, C.R. 2005. The modulation of intra-articular inflammation, cartilage matrix and bone loss in mono-articular arthritis induced by heat-killed Myobacterium tuberculosis. *Inflammopharmacology* **12**(5–6):551–567.
23. Davis, L.S., Sackler, M., Brezinschek, R.I., et al. 2002. Inflammation, immune reactivity, and angiogenesis in a severe combined immunodeficiency model of rheumatoid arthritis. *Am J Pathol* **160**:357–367.
24. Wahid, S., Blades, M.C., De Lord, D., et al. 2000. Tumour necrosis factor-alpha (TNF-a) enhances lymphocyte migration into rheumatoid synovial tissue transplanted into severe combined immunodeficient (SCID) mice. *Clin Exp Immunol* **122**:133–142.
25. Humby, F., Bombardieri, M., Manzo, A., et al. 2009. Ectopic lymphoid structures support ongoing production of class switched autoantibodies in rheumatoid synovium. *PLoS Med* **6**(1):e1.

26. Schopf, L., Anderson, K., and Jaffe, B.D. 2006. Rat models of arthritis: similarities, differences, advantages and disadvantages in the identification of novel therapeutics. In: *In vivo Models of Inflammation. Progress in Inflammation Research*. Parnham, M.J. (ed.), 2nd Ed., Vol 1., pp. 1–34. Basel: Birkhauser.
27. Koenders, M.I., Kolls, J.K., Oppers-Walgreen, B., et al. 2005. Interleukin-17 receptor deficiency results in impaired synovial expression of interleukin-1 and matrix metalloproteinases 3, 9, and 13 and prevents cartilage destruction during chronic reactivated streptococcal cell wall-induced arthritis. *Arthritis Rheum* **52**:3239–3247.
28. van den Broek, M.F. 1989. Streptococcal cell wall-induced polyarthritis in the rat. Mechanisms for chronicity and regulation of susceptibility. *APMIS* **97**:861–878.
29. van Bruggen, M.C., van den Broek, M.F., and van den Berg, W.B. 1991. Streptococcal cell wall-induced arthritis and adjuvant arthritis in F344 – Lewis and in Lewis – F344 bone marrow chimeras. *Cell Immunol* **136**(2):278–290.
30. Holmdahl, R., Lorentzen, J.C., Lu, S., et al. 2001. Arthritis induced in rats with non-immunogenic adjuvants as models for rheumatoid arthritis. *Immunol Rev* **184**:184–202.
31. Hoffmann, M.H., Tuncel, J., Skriner, K., et al. 2007. The rheumatoid arthritis-associated autoantigen hnRNP-A2 (RA33) is a major stimulator of autoimmunity in rats with pristane-induced arthritis. *J Immunol* **179**:7568–7576.

SUGGESTED READINGS

Andersson, A.K., Li, C., and Brennan, F.M. 2008. Recent developments in the immunobiology of rheumatoid arthritis. *Arthritis Res Ther* **10**(2):204. Epub March 14, 2008.

Asquith, D.L., Miller, A.M., McInnes, I.B., and Liew, F.Y. 2009. Animal models of rheumatoid arthritis. *Eur J Immunol.* **39**(8):2040–2044.

Billingham, M.E.J. 1983. Models of arthritis and the search for anti-arthritic drugs. *Pharmacol Ther* **21**:389–428.

Hegen, M., Keith, J.C., Mary, C., and Nickerson-Nutter, C. 2008. Utility of animal models for identification of potential therapeutics for Rheumatoid Arthritis. *Ann Rheum Dis* **67**:1505–1515.

Henderson, B., Edwards, J.C.W., and Pettipher, E.R. (eds.) 1995. *Mechanisms and Models in Rheumatoid Arthritis*. London: Academic Press.

Joosten, L.A.B., and van den Berg, W. 2006. Murine collagen-induced arthritis. In: *In Vivo Models of Inflammation. Progress in Inflammation Research*. Parnham, M.J. (ed.), 2nd Ed., Vol 1, pp. 1–34. Basel: Birkhauser.

Kannan, K., Ortmann, R.A., and Kimpel, D. 2005. Animal Models of rheumatoid arthritis and their relevance to human disease. *Pathophysiology* **12**:167–181.

Paska, W., McDonald, K.J., and Crofy, M. 1986. Studies on type-II collagen induced arthritis in mice. *Inflamm Res* **18**:413–420.

Rioja, I., Bush, K.A., Buckton, J.B., Dickson, M.C., and Life, P.F. 2004. Joint cytokine quantification in two rodent arthritis models: kinetics of expression, correlation of mRNA and protein levels and response to prednisolone treatment. *Clin Exp Immunol* **137**:65–73.

Schett, G., Stach, C., Zwerina, J., Voll, R., and Manger, B. 2008. How do anti-rheumatic drugs protect joints from damage in rheumatoid arthritis. *Arthritis Rheum* **58**:2936–2948.

Seed, M.P. 2009. Development of disease-modifying treatments for rheumatoid arthritis. *Pharm J* **282**:191–192.

Seed, M.P., and Walsh, D. (eds.) *Angiogenesis in Inflammation: Mechanisms and Clinical Correlates*. Progress in Inflammation Research. Basel: Birkhauser.

Stevenson, C.S., Marshall, L.A., and Morgan, D.W. (Eds.) 2006. *In Vivo Models of Inflammation*, 2nd Ed., Vol 1. *Progress in Inflammation Research*. Basel: Birkhauser.

Stevenson, C.S., Marshall, L.A., and Morgan, D.W. (Eds.) 2006. *In Vivo Models of Inflammation*, Vol 2. *Progress in Inflammation Research*. Basel: Birkhauser.

Tak, P-P. (Ed.) 2009. *New Therapeutic Targets in Rheumatoid Arthritis. Progress in Inflammation Research*. Basel: Birkhauser.

Van den Berg, W.B., and Miossec, P. (Eds.) 2004. *Cytokines and Joint Injury. Progress in Inflammation Research*. Basel: Birkhauser.

Van den Berg, W., van Lent, P.L., Joosten, L.A., Abdollahi-Roodsaz, S., and Koenders, M.I. 2007. Amplifying elements of arthritis and joint destruction. *Ann Rheum Dis* **666**(s3):iii45–iii48.

Wooley, P. 1991. Animal models of arthritis. *Curr Opin Rheumatol* **3**:407–420.

Wooley, P. 2008. Immunotherapy in collagen-induced arthritis: past, present, and future. *Am J Med Sci*. **327**:217–226.

31 Ocular Inflammation Models

Karsten Gronert

INTRODUCTION

The eye is one of the primary sense organs. This unique organ is composed of highly specialized tissues: an outgrowth of the central nervous system, the retina, on the posterior end and a highly modified transparent skin, the cornea, on the anterior end (Figure 31.1). In the simplest terms, the primary function of the eye is to convert light energy into nerve action potentials. This essential and evolutionarily conserved function directly depends on maintaining (1) the refractive properties of the cornea, lens, aqueous humor, and vitreous humor; (2) formation and reabsorption of fluids that keep intraocular pressure in the fluid-filled eyeball constant; and (3) the function of retinal neurons. The eye, as an organ that faces the external environment, provides a unique opportunity to directly access highly specialized and diverse tissues such as the immune-privileged avascular cornea and the conjunctiva, a mucosal tissue that directs host defense and immune response of the ocular surface. Moreover, since light has to pass through the eye, physiological processes and cells in the anterior (cornea), interior (lens, uvea), and posterior (neural and vascular retinal layers) tissues of the eye (Figure 31.1) can be studied by noninvasive procedures in animals and humans [1–7].

The visual axis is a delicate sensory organ that has to maintain an avascular and transparent state in order to preserve ocular function. Hence, in the eye, especially in the cornea and retina, the threat of inflammation is incompatible with good vision. However, it acutely straddles inflammatory and immunogenic threats as the ocular surface is continuously exposed to sheering stress of eyelid motion and prolonged periods of hypoxia associated with sleep and environmental pathogens and irritants; the retina is vulnerable to blood-borne pathogens, oxidative stress, and inflammatory mediators, even though it is isolated by a well-developed blood tissue barrier. Therefore, the eye contains highly developed circuits to control inflammation, maintain host defense, and promote ordered and rapid wound healing, as loss of sight is a powerful negative selecting force. The immune response in the eye, even though it shares many features with immunity in other tissues, is atypical. The eye like other tissues that are extremely vulnerable to inflammation-induced tissue injury, such as the brain, pregnant uterus, and testes, or tissues that are chronically challenged with excessive irritants or pathogens such as the oral cavity have developed multiple adaptations to minimize collateral tissue injury by restraining immune and inflammatory responses [2,5,8–10].

Pioneering and seminal work by Streilein and colleagues [2,5,8,11] has demonstrated that these atypical immune/inflammatory responses are not due to immunological ignorance but are highly evolved and conserved adaptations that lead to privileged immune and repair responses in selected tissue, such as the anterior chamber of the eye. A striking feature of the eye is the presence of avascular transparent tissues, namely the cornea and lens, and its unique status as an immune-privileged site that accepts, sometimes indefinitely, allogeneic grafts without immune rejections. This highly adapted immune response has been extensively studied in the cornea and termed anterior chamber–associated immune deviation (ACAID). It involves complex regulatory circuits that connect the eye, thymus, spleen, and sympathetic nervous system [2,5,12]. A body of work has developed, over a 30-year period, elegant eye models to study inflammatory/immune responses, wound healing, heme-angiogenesis, and lymph-angiogenesis [13–22]. As in most tissues, execution of these fundamental and self-resolving processes in the eye are tightly regulated by interdependent circuits that release or generate specific and temporally defined arrays of mediators. Formation or bioactions of distinct classes of these mediators have been studied in animal models of ocular inflammation, and

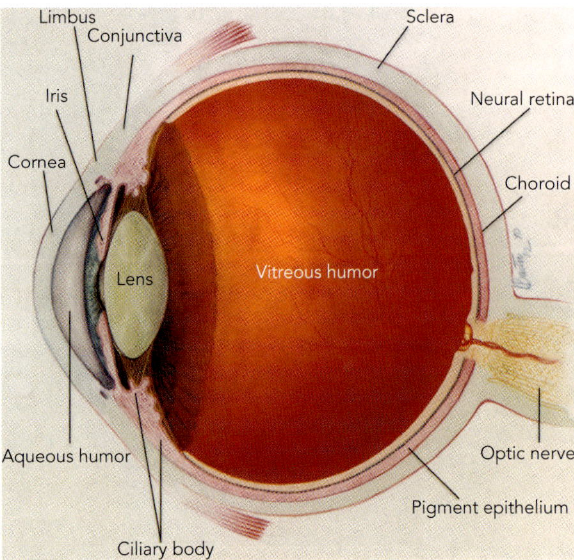

Figure 31.1. Artist rendition of the human eye (National Eye Institute, National Institutes of Health, www.nei.nih.gov).

these include (1) protein mediators such as cytokine/chemokines, growth factors, and angiogenic factors [21,23–28]; (2) lipid autacoids such as eicosanoids, docosanoids, and platelet-activating factors [17,29–34]; (3) adhesion molecules such as selectins and integrins [23,35–40]; (4) Toll receptors [41,42]; and (5) complement [43]. Temporal and localized formation of these chemical mediators initiates mobilization and regulates activation of effector cells of ocular inflammation, such as neutrophils (PMN), macrophages, mast cells, and lymphocytes, and orchestrates the eventual resolution of the inflammatory event. Several animal species have been utilized as models to gain insights into the pathogenesis and molecular mechanisms of ocular inflammatory diseases. Owing to the inherent complexity of inflammatory/reparative responses and the fact that in vivo models mimic clinical features of ocular inflammatory diseases, animal models have provided fundamental advances in the discovery and validation of novel therapeutic targets. Owing to sequencing of the mouse genome and genetic engineering, which allows the targeted deletion or insertion of genes, mice have become a favorite model to elucidate and establish the role of specific receptors, chemical mediators, adhesion molecules, and enzymes in the inflammatory, immune, and reparative response (Tables 31.1 and 31.2).

CORNEA

The cornea (Figure 31.1) is the transparent and avascular anterior tissue of the eye [3,6,7]. It is a relatively simple tissue, principally composed of five to six layers of epithelial cells that, as in all mucosal tissues, forms a critical barrier to the external environment and together with the tear film maintains the optical properties of the cornea. The epithelium of the cornea is continuous with the epithelium of the adjacent conjunctiva, a vascularized and lymphatic mucosal tissue that is critical to the ocular immune response and an important source of tear mucin. The anatomical characteristics of the cornea are largely due to the collagen-rich stroma, which accounts for 90% of the cornea and contains keratocytes (corneal fibroblasts) and nerve fibers. The cornea is one of the most innervated tissues containing 300–400 times higher density of nerve endings than the skin. A single layer of endothelial cells at the posterior surface of the cornea, which faces the aqueous humor of the eye, dynamically regulates water content of the cornea. In humans and primates, these corneal endothelial cells do not proliferate; hence loss of corneal endothelial cells is irreversible. Even though the cornea is avascular, blood constituents are essential for corneal metabolism, inflammatory responses, and wound healing. The cornea is bordered by the limbus region, which contains the corneal epithelial stem cells and a vascular arcade that is derived from both the ophthalmic and carotid arteries. It is the vasculature of the limbus region that is the primary source of leukocyte for the cornea. However, the healthy cornea contains no PMN, macrophages, or lymphoid cells. Even though recent evidence suggests that a small heterogeneous population of langerhans or dendritic cells resides in the central cornea [44], especially in mice, most antigen-presenting cells are restricted to the periphery in the healthy cornea. Several mouse models of corneal inflammation (keratitis) have been developed to study leukocyte trafficking, host defense, wound healing, and inflammatory neovascularization. These models range from tissue injury to infection, which result in acute self-resolving or chronic nonresolving inflammatory/reparative responses. The simple and unique anatomy of the cornea has provided opportunities to investigate the contribution of resident or recruited effector cells to basic inflammatory/reparative responses.

Epithelial Injury Models

A well-established animal model for inducing a mild and self-resolving inflammatory/reparative response in the cornea is mechanical abrasion, which removes epithelial cells from a defined area of the cornea [17,36,45–51]. The most common methods employ a diamond blade that is used in human refractive surgery or a spinning metal burr device (Figure 31.2) that is used to remove foreign objects from human corneas. This mechanical abrasion removes the epithelium while leaving the underlying stroma intact, thereby creating a reproducible, uniform, and quantifiable epithelial injury. This epithelial injury induces

TABLE 31.1. Common strains, genetic knockout, transgenic, or mutant mice that are used as models of corneal inflammation

Mouse	Engineered gene	Inflammation model	Effect/phenotype
CD18$^{-/-}$	Adhesion molecule, PMNβ2 integrin	Keratitis: epithelial injury	↓Wound healing, PMN
P-selectin$^{-/-}$	Adhesion molecule, platelet (P-CAM)	Keratitis: epithelial injury	↓Wound healing, PMN
E-selectin$^{-/-}$	Adhesion molecule, endothelial cell (E-CAM)	Keratitis: epithelial injury	↓Wound healing, PMN
E- and P-Selectin$^{-/-}$ double knockout	Adhesion molecules	Keratitis: epithelial injury	↓Wound healing, PMN
12/15-LOX$^{-/-}$	Leukocyte-type 12/15-lipoxygenase (Alox15)	Keratitis: epithelial injury Cornea: neovascularization	↓Wound healing ↑PMN, angiogenesis
HO-2$^{-/-}$	Heme-oxygenase-2	Keratitis: epithelial injury Cornea: neovascularization	↑Inflammation, ↓healing ↑Angiogenesis/inflammation
Lum$^{-/-}$	Proteoglycan lumican	Keratitis: stromal wounds with infection	↓Wound healing, apoptosis ↓Inflammation, leukocytes
MRL strain	None	Keratitis: alkaline burn	↓PMN, inflammation, opacity ↑Wound healing
BALB/C strain	None	Keratitis: microbial	No corneal perforation
C57/BL6 strain	None	Keratitis: microbial	Corneal perforation
IL-12$^{-/-}$ BL6 strain	Cytokine, interleukin-12	Keratitis: microbial	↓IFN-γ, ↑perforation
IL-6$^{-/-}$ BL6 strain	Cytokine, interleukin-6	Keratitis: viral	↓Opacity, chemokines
β_2-Microglobulin$^{-/-}$ BL6 strain	CD8+ T-cell knockout	Keratitis: microbial	No perforation
ICE$^{-/-}$	Caspase 1 knockout	Keratitis: microbial	↓Il-1β, PMN
TLR2$^{-/-}$	Toll receptor 2	Keratitis: TLR2 agonist	↓Edema, PMN
TLR9$^{-/-}$	Toll receptor 9	Keratitis: TLR9 agonist	↓Edema, PMN, chemokines
MyD88$^{-/-}$	TLR adaptor protein	Keratitis: microbial	↓Edema, PMN, chemokines
C3H/HeJ strain	TLR4 point mutation, TLR4 knockout	Keratitis: microbial	↓Chemokines, PMN, edema
TSP$^{-/-}$	Thrombospondin (1 and/or 2)	Cornea: neovascularization	↑Angiogenesis
VEGF-A$^{188/188}$	Vascular endothelial growth factor-A, noVEGF-A^{120} or VEGF-A^{164}	Cornea: neovascularization	↓Angiogenesis (heme/lymph)
VEGF-A$^{164/164}$	No VEGF-A^{120}/VEGF-A^{188}	Cornea: neovascularization	↓Angiogenesis (heme/lymph)
NOD strain MRL/lpr strain MRL/MpJ-fas+/fas+	Non-obese diabetic mouse Lymphoproliferation mutant	Dry eye	Chronic infiltration of CD4+ lymphocytes into lacrimal glands

a mild inflammatory/reparative response characterized by complete wound healing within 4–7 days, no neovascularization, and eventual resolution of inflammation (Figure 31.2). The degree of inflammation, edema, and leukocyte infiltration correlates directly with the size of the epithelial injury. Wound healing is monitored and quantified by fluorescein staining of the epithelial defect in anesthetized animals (Figure 31.2) and immunohistology methods are well developed for the cornea. Despite the small size of the tissue, especially in mice, advanced analytical methods such as fluorescence-based kinetic PCR (real-time PCR), mass-spectrometry–based lipidomics, and cytokine/chemokines arrays have been employed to assess inflammatory pathways in the cornea (Figure 31.3).

The epithelial abrasion model addresses multiple aspects of acute and self-resolving inflammation and wound repair that are relevant to the pathology of corneal injuries such as laceration, abrasion, and refractive surgery as well as general mucosal epithelial injuries. Genetic knockout mice (Table 31.1) for the adhesion molecules CD18, P-selectin, and E-selectin [36,37,51];

TABLE 31.2. Common strains, genetic knockout, transgenic or mutant mice that are used as models of uveal or retinal inflammation.

Mouse	Engineered gene	Inflammation model	Effect/phenotype
IL-17$^{-/-}$	Interleukin 17	Uveitis: uveoretinitis	↓Autoimmune uveoretinitis
CD4+ STAT3$^{-/-}$	STAT3 (cytokine signaling) deletion in CD4+ lymphocyte	Uveitis: uveoretinitis	Resistant to autoimmune uveoretinitis
Osteopontin$^{-/-}$	Secreted phosphoprotein 1	Uveitis: uveoretinitis	↓Autoimmune uveoretinitis
IFNγ$^{-/-}$	Interferon γ	Uveitis: uveoretinitis	↑Autoimmune uveoretinitis
Fyn$^{-/-}$	Tyrosine kinase in lymphocyte differentiation	Uveitis: uveoretinitis	↑Autoimmune uveoretinitis
BLT1$^{-/-}$	Leukotriene B_4 receptor 1	Uveitis: uveoretinitis	↓Autoimmune uveoretinitis
AIRE$^{-/-}$	Autoimmune regulator	Uveitis: uveoretinitis	Spontaneous uveoretinitis
MC5r$^{-/-}$	Immunosuppressive peptide (melanocortin) receptor	Uveitis: uveoretinitis	↑Autoimmune uveoretinitis
HLA-A29	Major histocompatibility allele HLA-A29 transgenic	Uveitis: uveoretinitis	Spontaneous uveoretinitis
BALB/C strain	None	Uveitis: microbial	Minimal anterior uveitis
C57/BL6 strain	None	Uveitis: microbial	Significant anterior uveitis
C3aR$^{-/-}$ C5aR$^{-/-}$	Receptors for complement component C3a and C5a	Retinopathy: laser injury	↓Choroidal neovascularization ↓Leukocyte recruitment
SOD1$^{-/-}$	Superoxide dismutase 1	Retinopathy: age-induced	↑Choroidal neovascularization ↑Drusen (protein/lipid deposit)
Ccl-2$^{-/-}$ Ccl-2r$^{-/-}$	Chemokine, MCP1 MCP1 receptor, Ccl-2r	Retinopathy: age-induced	↑Choroidal neovascularization ↑Drusen (protein/lipid deposit)
CX3CR1$^{-/-}$	CX3C chemokine receptor1	Retinopathy: laser injury	↑Choroidal neovascularization ↑Microglial cell accumulation
S1p2$^{-/-}$	Sphingosine 1-phosphate receptor 2	Retinopathy of prematurity	↓Macrophage, COX-2 ↓Neovascularization
MIP-1α$^{-/-}$	Macrophage inhibitory protein (MIP-1α)	Allergic conjunctivitis	↓Mast cell degranulation ↓Edema
IL-12$^{-/-}$	Interleukin-12	Allergic conjunctivitis	No immediate hypersensitivity reaction
IFNγ$^{-/-}$	Interferon γ	Allergic conjunctivitis	↑Conjunctivitis

inflammatory mediator pathways 15-lipoxygenase and heme-oxygenase-2 [49,50,52] and proteoglycans such as lumican [53]; as well as antibody depletion of PMN suggest that inflammatory responses are intimately linked to normal wound healing in this model.

References 50 and 51 about epithelial abrasion models that employ knockout mice, molecular biology, immunohistology, or lipid autacoid analyses as experimental tools.

Chemical Burn

Chemical burns are a severe and common corneal injury in humans [7,54]. In order to develop effective and safe treatment options, alkali burn of the cornea by sodium hydroxide or chemical cauterization by silver nitrate were established as animal models to delineate the pathogenesis and sequelae of this severe, chemically induced keratitis. These chemical burns result in pronounced corneal injury that destroys the epithelial layer and penetrates the anterior stroma [46,54–57]. Corneal edema and leukocyte infiltration are observed within a few hours, and neovascularization follows within days. Depending on the concentration of the chemical and area that is exposed, the injury is mild to severe. A standard method is the application of filter paper disks, which have been soaked in 0.1–1.0 N sodium hydroxide, to the cornea for ~30–60 seconds. In a mild alkali burn (0.1–0.25 N NaOH), the epithelium regenerates and corneal blood vessels regress generally within 10 days, whereas severe alkali burn (1 N sodium hydroxide) results in exacerbated and prolonged inflammatory responses, marked by massive PMN infiltration by 1

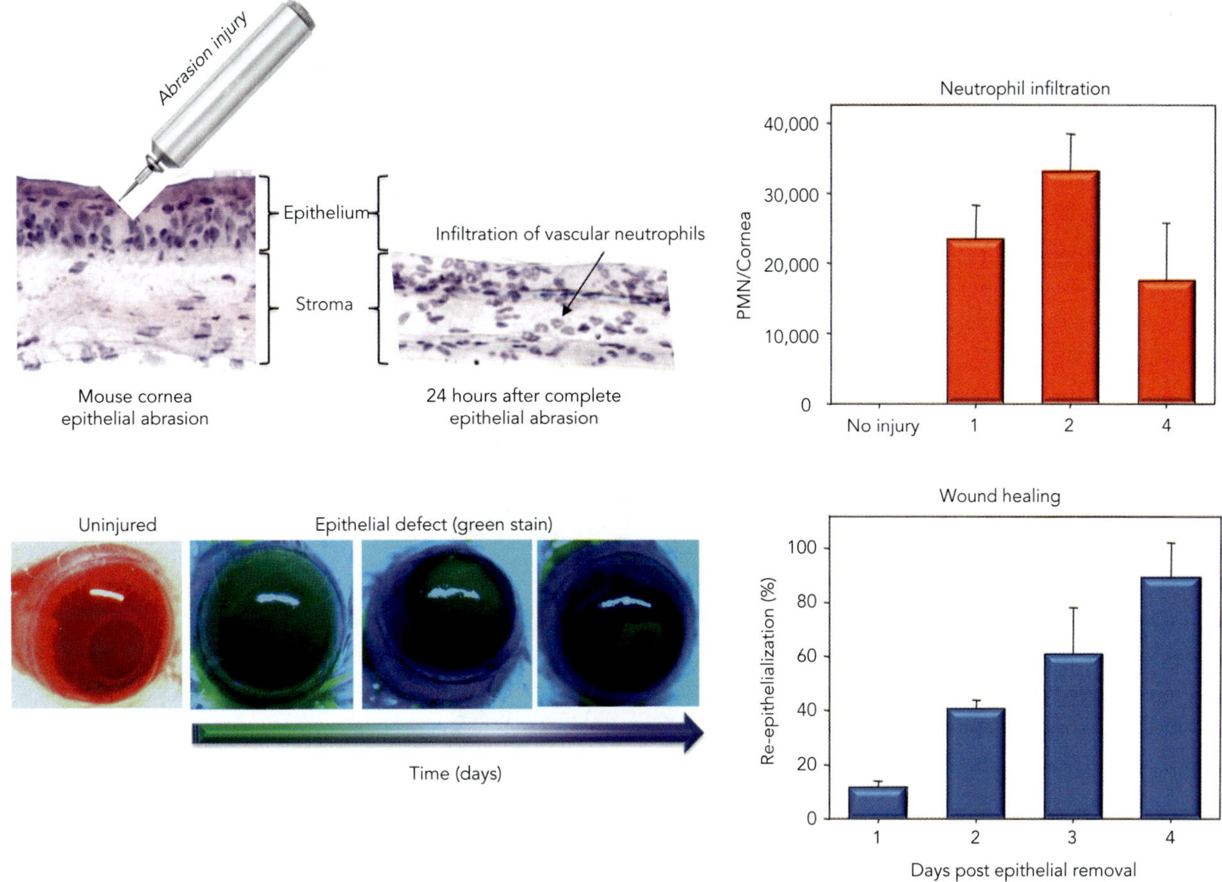

Figure 31.2. Epithelial abrasion-induced keratitis. Illustrated is the removal of the corneal epithelium by an Algerbrush in mice. Histology (hematoxylin/eosin staining) shows infiltrating PMN in a mouse cornea after epithelial injury. PMN infiltration was quantified by measuring myeloperoxidase activity (a specific marker enzyme for PMN). Epithelial defect was quantified by staining of the stroma with fluorescein (green). Total green staining area (epithelial defect) was quantified by image analyses and is a direct measure of wound healing in the cornea.

week and appearance of lymphocytes by 3 weeks. In severe alkali burns, neovascularization is persistent, and, even though it is rapidly initiated, there is only partial wound healing and the critical morphology of the cornea is not restored and often associated with formation of perforating ulcers.

Exacerbated inflammation is a key feature of the pathology of chemical burns and depletion of PMN in vivo by treatment with granulocyte-specific GR-1 antibody greatly improves corneal wound healing in mice. The inbred mouse strain MRL has a unique phenotype (Table 31.1) of accelerated wound healing in the alkali burn model, which is associated with a reduced inflammatory response in the cornea, which may provide insight into the genetic basis of controlled leukocyte activation [58]. Alkali burn, as a model of severe corneal injury, has been used to study persistent leukocyte infiltration, inflammatory neovascularization, recurrent epithelial erosion, ulceration, edema, and failed wound healing. Advanced molecular methods such as gene microarrays [56] and real-time PCR [58] have been utilized to define the response to corneal trauma at the genetic level.

Reference 58 discusses alkaline burn–induced injury, which uses inbred mouse strains and molecular biology approaches to define an intrinsic phenotype of accelerated wound healing and reduced inflammation.

Microbial Keratitis

Inflammation is an essential feature and protective response of ocular host defense [6,7]. Models of microbial keratitis have been developed to understand the mechanism and dynamics of leukocyte migration and the role of inflammatory mediators (e.g., cytokines/chemokines) and Toll receptors in the ocular response to infection [21,23,42,59]. The cornea is highly resistant against bacterial and viral infection and subsequent inflammation due to the effective epithelial barrier, tear film, and rapid microbial/viral clearance. Hence, most keratitis models apply bacteria, viruses, or lipopolysaccharide (LPS, endotoxin) topically after

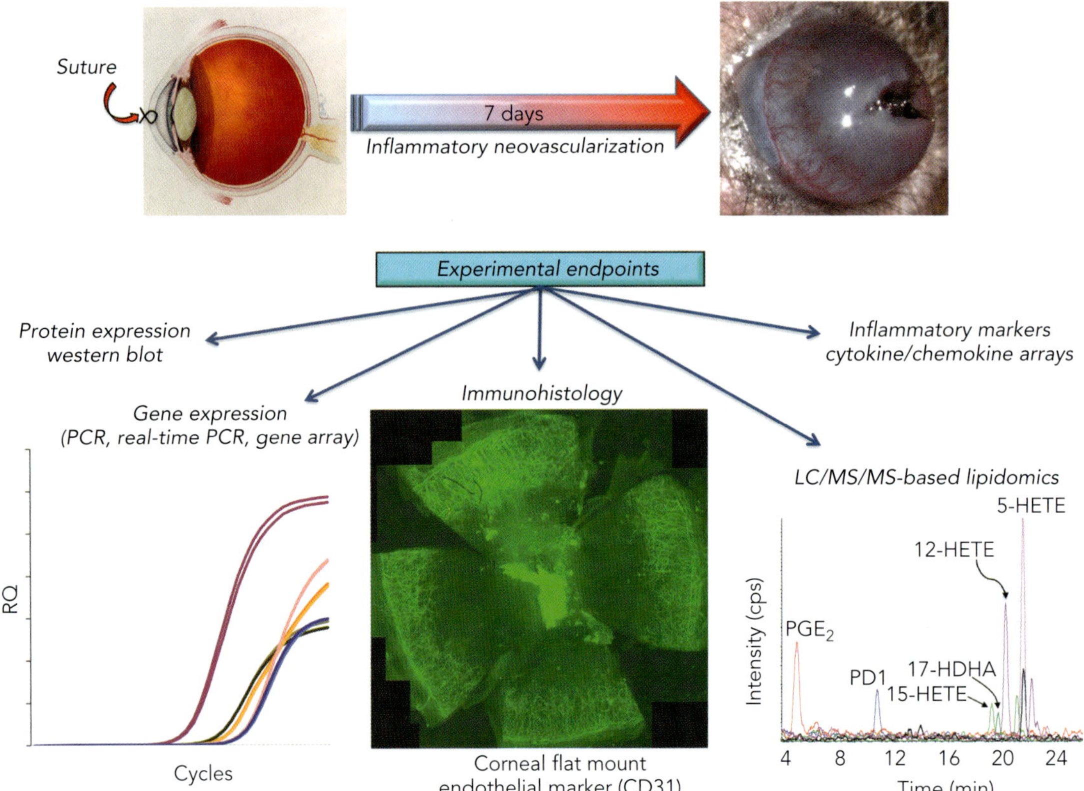

Figure 31.3. Suture-induced inflammatory neovascularization. Illustrated is the placement of a central corneal suture, which induced inflammatory neovascularization within 7 days in mouse eyes. Several advanced molecular analytical and immunohistological methods have been developed to assess inflammatory and angiogenic pathways in this model. Shown are mass-spectrometry–based analyses of lipid autacoids (eicosanoids, docosanoids), immunohistochemistry for a specific marker of vascular endothelial cells (CD31), and real-time PCR analyses of genes that regulate angiogenesis and inflammation.

epithelial abrasion, by direct injection into the stroma or by placement of contaminated contact lenses on the cornea [13,21,59–62]. Microbial keratitis induces a robust innate immune response that is characterized by rapid infiltration of PMN, followed by macrophages. Depending on the bacterial load and mouse strain, microbial keratitis is associated with scarring, persistent PMN, opacity, and corneal perforation. A frequent microbial choice for infection with life bacteria or sterile infection with LPS is *Pseudomonas aeruginosa*, which causes widespread opportunistic infections, notably in the lungs of patients with cystic fibrosis and in the eye due to extended contact lens usage [6,21,59].

Two inbred mouse strains (C57/BL6 and Balb/c) provide an elegant model to investigate the role of T helper cells (T_H) and pathway-specific cytokines, which are associated with directing immune and inflammatory responses in microbial keratitis [21,63]. The adaptive immune response in Balb/c mice is predominantly a humoral (T_H2) response, while C57/BL6 mice have a predominant cell-mediated (T_H1) response. Hence, the sequelae of microbial keratitis exhibit a strain-specific phenotype, which in Balb/c mice is characterized by a milder response with no corneal perforation, while infected corneas of C57BL/6 mice perforate. Several mice with targeted genetic deletions (knockout mice, Table 31.1) have been developed to study the role of cytokines (IL-12, IL-6, ICE [IL-1β knockdown], β_2-microglobulin [CD8 T-cell knockout] [21,59,62]), Toll receptors [TLR2, TLR3, C3H/HeJ (TLR4 mutant), TLR9 [21,60,61]), and toll-receptor signaling (MyD88 [60,61]). An elegant method for monitoring in vivo trafficking of bone marrow–derived leukocytes [13] has been developed. This method uses irradiated mice, which have received bone marrow transplants from green fluorescent protein transgenic mice (EGFP mice) to visualize the dynamic migration and recruitment of the bone marrow–derived leukocytes (fluorescent green) in the infected cornea.

Reference 63 discusses the marked differences in inflammatory phenotypes of inbred mouse strains in response to bacterial infections and Reference 13 discusses about the unique advantage of the eye to monitor migration of leukocytes in an acute inflammatory response.

Inflammatory Neovascularization

Formation of blood vessel from existing vessels is termed angiogenesis and eventual formation of functional new microvasculature is termed neovascularization. This process is a fundamental feature of wound healing, chronic inflammation, tumor growth, and a protective response to increase blood flow in ischemic tissue [6]. The cornea, due to its normal avascular state (it lacks blood and lymph vessels), has been used as a principle tissue to study angiogenic processes and neovascularization [7,64,65]. Corneal injury and inflammation, despite numerous mechanisms that keep the cornea avascular, can induce a robust angiogenic response [66–72]. Heme-angiogenesis and lymph-angiogenesis are primarily associated with inflammation and leukocyte activation/recruitment and are considered protective responses to deliver inflammatory or immune effector cells to the avascular cornea. In several models of corneal inflammation, such as microbial keratitis, chemical burn, and cauterization, neovascularization is part of the pathogenesis but the degree of the angiogenic response depends on many parameters that are difficult to control and quantify in these experimental models.

Suture-induced corneal neovascularization is an established mouse model to delineate molecular mechanisms, role of specific cell types, and action of chemical mediators in the pathological formation of blood vessels (heme-angiogenesis) and lymph vessels (lymph-angiogenesis) [66–73]. A large suture is placed intrastromally at the center (Figure 31.3) of the cornea, or, for a more uniform response, three sutures are spaced equally along a 2-mm diameter around the central cornea. Placement of the sutures evokes a robust and sustained inflammatory response. Unlike in acute ocular inflammatory response, where the initial influx of PMN is followed by macrophages around day 2 and PMN begin to resolve by day 7, in chronic suture–induced injury the predominant leukocyte-type, even 7 days after suture placements, are PMN. Neovascularization and lymph-angiogenesis can be observed by day 2 but peaks by 7 days, and persists as long as the sutures remain in the cornea. Removal of the sutures will result in significant regression of blood and lymph vessels within weeks and eventual complete regression within 4–8 months [73]. Immunohistological methods are well developed for the morphometric analyses of pathogenic heme-angiogenesis and lymph-angiogenesis [66–70,73]. In addition, advanced molecular analytical methods (Figure 31.3) such as mass spectrometry (LC/MS/MS)-based lipidomics, fluorescence-based kinetic PCR (QPCR/real-time PCR), and chemokines/cytokine arrays have been employed to assess the formation of inflammatory and angiogenic mediators [66]. The important role of resident corneal pathways, which inhibit angiogenesis and/or inflammation, have been investigated (Table 31.1) in thrombospondin (TSP-1$^{-/-}$, TSP-2$^{-/-}$, TSP1,2$^{-/-}$ [70]), 15-lipoxygenase (12/15-LOX$^{-/-}$), and heme-oxygenase-2 (HO-2$^{-/-}$ [66]) knockout mice. The function of different isoforms of vascular endothelial growth factor-A (VEGF-A), has been investigated in VEGF-A$^{164/164}$ and VEGF-A$^{188/188}$ transgenic mice [68]. VEGF-A is the principle angiogenic mediator and therapeutic target of VEGF neutralizing antibodies or fusion proteins, which bind and neutralize VEGF-A and are a new clinical approach for the treatment of age-related macular degeneration and colon cancer.

References 68, 71, and 73 discuss how suture-induced lymph- and heme-angiogenesis are a quantifiable and dynamic model and the role of leukocytes in this inflammatory response.

Corneal Transplantation (Penetrating Keratoplasty)

Penetrating keratoplasty, or corneal grafting, is the most successful and common tissue transplantation, a fact that has been attributed to the immune-privileged environment of the cornea, which, sometimes indefinitely, tolerates foreign antigens [7,74]. The corneal allograft, in general, possesses immunological privilege; however, this immune privilege is compromised in inflamed or vascularized recipient corneas leading to inevitable graft failure due to immune rejection. The cornea is an ideal model to study transplant rejection because high-risk transplantation can be induced experimentally by activating the immune/inflammatory system in the recipient cornea tissue prior to keratoplasty. The availability of highly inbred mouse strains for syngeneic or allogeneic transplantation have been an important model to define immune privilege of normal corneal allografts and genetic knockout mice have been important to define the role of mediators and pathways that lead to high-risk transplantations [75–79]. A common method for inducing an inflamed and immune-compromised recipient cornea has been suture-induced inflammatory neovascularization (section "Inflammatory neovascularization," Figure 31.3). On the other hand, epithelial abrasion of the graft cornea (section "Epithelial injury models"), which protects against immune rejection of the graft, has established the immunogenicity of the graft epithelium [76,80–84].

Reference 75 discusses knockout mice and molecular biology methods and the role of chemokines in corneal transplant immune rejection.

Keratoconjunctivitis Sicca (Dry Eye)

Unstable or deficient tear film leads to a common ocular disease called dry eye syndrome or *keratoconjunctivitis sicca*. This ocular surface disease is characterized by epithelial defects, and irritation and loss of epithelial barrier function. It most commonly affects middle-aged or elderly women and is diagnostic of Sjorgen's syndrome (autoimmune disease of salivary and lacrimal glands) [7,85]. Dry eye syndrome is a multifactorial disease associated with chronic infiltration of T helper lymphocytes (CD4+) into the central glands for tear film formation (lacrimal glands) and ocular surface inflammation.

Animal models have been developed to mimic part of the pathology of human dry eye syndrome. Multiple approaches [15,16,86–88] have been employed to induce experimental dry eye syndrome: (1) pharmacological agents, surgery, or desiccating environmental chambers to disrupt the tear film and induce ocular irritation; and (2) mutant and transgenic mouse strains that develop lacrimal gland inflammation (dacryoadenitis). Several mutant and transgenic mice (Table 31.1) such as nonobese diabetic mice (NOD), MRL/MpJ-fas+/fas+ or MRL/MpJ-fas^{lpr}/fas^{lpr}, NZB/NZW F1, and IQI/Jic mice develop chronic infiltration of CD4+ lymphocytes into lacrimal glands [15,89]. However, the phenotype of abnormal lymphocytes in these mice is also associated with other significant systemic pathologies. Moreover, dacryoadenitis or surgical removal of lacrimal glands does not necessarily result in ocular surface disease (i.e., dry eye syndrome) due to compensatory tear formation by other glands. Hence, pharmacological approaches have been developed to block neural control of tear film formation by antimuscarinic agents (anticholinergic), such as scopolamine, in conjunction with exposure to constant high airflow in order to disrupt the tear film. Recent advances in this method eliminate the need for cholinergic blockage by placing mice in a controlled-environmental chamber that provide continuous airflow and low humidity, thereby effectively disrupting the tear film [86]. The model of evaporative dry eye mimics many aspects of the epithelial defects of dry eye and eventually leads to ocular surface inflammation [15,85,86]. However, unlike human dry eye syndrome, this epithelial injury model does not have an immune component or is associated with progressive lymphocyte infiltration of the lacrimal glands. Standard methods for measuring tear production (Schirmer Test), immunohistochemistry, histology, and the assessment of established inflammatory markers have been widely used to assess epithelial defect and immune/inflammatory processes in all the dry eye models.

Reference 15 provides a critical review of animal models of dry eye syndrome and the use of transgenic or knockout mice.

UVEA

The uveal tract or middle lymphovascular tunic of the eye is located between the cornea/sclera and retina and includes the iris, ciliary body in the anterior segment, and the choroid in the posterior segment (Figure 31.1). Uveitis or intraocular inflammation of the eye is a recurrent autoimmune disease of the eye with wide-ranging etiology, which can lead to temporary or permanent loss of vision and glaucoma [6,19,20,22]. The most common form of uveitis is anterior uveitis with inflammation in either the iris or ciliary body. These inflammatory disorders are initially acute, mild in symptoms, and, in general, last no longer than 3 months; however, it is the recurrence of acute uveitis that can eventually lead to vision loss. The most common animal models are endotoxin-induced uveitis and autoimmune uveitis, which exhibit many of the key features of human uveitis, namely dilation of vessels, macromolecule leakage, edema, and infiltration of leukocytes and lymphocytes.

Autoimmune Anterior Uveitis

Immunization with soluble bovine retinal extracts, uveal melanin or collagen induces organ-specific autoimmune disease in a several mammalian species, characterized by progressive infiltration of CD4+ T lymphocytes of the ciliary body and iris [19,20,22]. Even though PMN are present, lymphocytes are the predominant inflammatory cells, and regression of inflammation is slow. Even after 4 weeks, mononuclear cells persist in the iris and ciliary body. The degree and extent of uveal inflammation is antigen-dependent and can induce uveitis, uveoretinitis, and/or panuveitis (see section "Uveoretinitis"). Hence, not all antigens, especially soluble retinal proteins, induce clinical features of the most common form of human uveitis (anterior uveitis). Immunization with a specific ocular collagen (type I collagen α-chain) induces a more localized anterior uveitis and models the clinical features of human idiopathic acute anterior uveitis. However, collagen-induced uveitis, unlike other experimental autoimmune uveitis, is not associated with recurrent flare-up of inflammation, a key feature of the human disease, unless the animal is reexposed to the antigen [19].

Reference 19 reviews clinical aspects of human anterior uveitis and assesses available animal models of immune- and endotoxin-induced anterior uveitis.

Endotoxin-Induced Anterior Uveitis

Endotoxin- or LPS-induced uveitis is a popular model of acute and rapidly resolving anterior uveitis, which mimics clinical features of Gram-negative bacterial infections [19,22,90,91]. Uveitis is induced in rodents, rabbits, or guinea pigs by either local or systemic injection of LPS. Unlike autoimmune uveitis, which is mediated by mononuclear cells, this inflammation is characterized by massive infiltration of PMN into the ciliary body and iris by 8–12 hours and infiltration of the aqueous humor, which peaks by 24 hours. However, by 48 hours, this acute inflammatory response begins to resolve and the eyes appear normal 1 week after an LPS challenge. In mice, responses to LPS are highly strain-specific (see section "Microbial keratitis," Table 31.1) and in C3H/HeN and C57/B6 mice, LPS induces robust to mild uveitis while in Balb/c and 129/J LPS-induced uveitis is minimal.

Reference 90 is a historical reference for the serendipitous discovery that injection of bacteria at remote tissue sites induces endotoxin-induced uveitis in rats.

Uveoretinitis

Experimental uveoretinitis is the prototype autoimmune disease model that was developed by Wacker and colleagues in 1965 [20,92]. Immunization with soluble retinal antigens induces a CD4+ T_H1 driven inflammatory disease that affects the retina and choroid (vascular tissue between the retina and sclera, Figure 31.1). The autoimmune-mediated inflammation of the retina and surrounding structures in mouse and rat models exhibits many clinical characteristics of human autoimmune uveoretinitis. These animal models have been critical in advancing our understanding of immune modulators and effector cells in this potentially debilitating inflammatory disorder. In general, the antigen-induced inflammatory response results in leukocyte infiltration of the vitreous humor, choroid, or subretinal leukocyte infiltration, and inflammation of the retinal vessels and edema. To induce uveoretinitis requires activation of the innate immune response; hence, a standard protocol is subcutaneous injection of human or bovine interstitial retinal binding proteins and *Mycobacterium tuberculosis* in Freund's Adjuvant (immune potentiator). This model of induced uveitis is a CD4+ T cell–mediated disease and knockout mice (Table 31.2) for cytokine pathways (IL-17 [93]), CD4+ STAT3 [94], osteopontin [95,96], IFN-γ [97,98], tyrosine kinase fyn [98], leukotriene pathways (leukotriene B_4 receptor, 5-lipoxygenase [99]), immunosuppressive neuropeptide pathways (melanocortin 5 receptor [100]) have been essential to establish the role of lymphocyte and associated pathways in uveoretinitis.

However, human uveoretinitis is often a spontaneous disease with no apparent inciting stimulus, while experimental uveoretinitis requires inoculation with retinal antigen and activation of Toll receptors. Thus, insights into the pathogenesis of human uveoretinitis from these models have limitations. A few immunologically or genetically modified animal models (Table 31.2) have been produced to model spontaneous occurring uveitis [20]: (1) grafting rat embryonic thymus into the immune-deficient athymic mice (Nude, nu/nu) [101]; (2) deletion of the AIRE protein in mice (AIRE $^{-/-}$), which regulates peripheral antigen presentation in thymus [102]; (3) using transgenic mice that express the class I major histocompatibility gene HLA-A29 [103]; and (4) transgenic expression of foreign antigen in mice under control of a retinal promoter and crossing these mice with antigen-specific T cell transgenic mice or adaptive transfer of the antigen-specific T cells from uveitic mice [20].

Reference 97 discusses genetic knockout mice and molecular biology methods and about the role of leukocytes, mononuclear cells, and chemokines in uveoretinitis and Reference 102 discusses about genetic deletion of an autoimmune regulator induces spontaneous uveoretinitis.

RETINA

The retina is the innermost tunic of the eye and surrounds the vitreous humor (Figure 31.1). The highly complex neural tissue develops from the diencephalon and accounts for 30% of the total sensory input to the brain [1,6,104]. The retina is the only tissue of the central nervous system that can be observed and whose function and microanatomy can be studied without invasive procedures [104]. The neuro-retina is composed of three general layers: (1) the ganglion cells, which are adjacent to the vitreous humor; (2) intermediate neurons; and (3) the posterior layer of photoreceptors, which are intimately associated with the pigmented epithelial cells [1]. The subretinal space, like the cornea, is an immune-privileged site and the retinal pigmented epithelial (RPE) cells have an important role in maintaining the immune privilege [105]. RPE are highly specialized and are essential to maintain and phagocytose the photoreceptors and form the retinal-blood barrier. The RPE are tightly associated with the choroid (vascular tunic of the eye situated between the retina and sclera), which as the most vascularized tissue in the body has the highest blood flow [104] and provides vital nutrients and oxygen to the RPE and photoreceptors (per weight, the retina is the most oxygen-consuming tissue in the body). Hence, the choroidal circulation is significantly affected by diseases that have a vascular pathogenesis; these include diabetes, hypertension, leukemia, and arteriosclerosis (e.g., diabetic retinopathy is a leading cause of blindness with 12,000–24,000 cases/year in the USA; Source: National Eye Institute, 2008, www.nei.nih.gov). The etiology of retinopathy is multifactorial and increasing evidence suggests that inflammation contributes to the pathogenesis [1,104,106,107].

Retinopathy

Animal models have provided important insights into the mechanism of choroidal and retinal neovascularization [104,108]. Neovascularization is a key feature of age-related macular degeneration (AMD), which is the leading cause of blindness in individuals over 60 years of age, and retinopathy of prematurity (ROP), a common blinding disease in premature infants. A standard model for AMD is focal laser injury (laser coagulation) to the RPE and supporting membrane (Bruch's membrane), which induces choroidal neovascularization (CNV). The role of inflammatory pathways in CNV and other clinical features of AMD have been established by genetic deletion (Table 31.2) of receptors in the complement cascade (C3aR and C5aR [109]), superoxide dismutase (SOD1 [110]), and chemokines and/or chemokine receptors (MCP1 and CCR2 [111], CX3CR1 [112]).

Reference 104 reviews clinical aspects of age-related macular degeneration and the advances animal models have provided in understanding this leading cause of blindness in humans.

Retinopathy of Prematurity

During the final stages of vascular development, which occurs in humans right before birth, the retina is very sensitive to sudden changes in oxygen tension or levels of angiogenic factors. A premature increase in environmental oxygen tension due to a premature birth initially leads to regression of retinal blood vessels and is followed by a second phase of pathogenic hypoxia-induced retinal neovascularization, which is a major cause of blindness in newborns [113]. As in other proliferative retinopathies, inflammatory pathways contribute to the pathogenic neovascularization [106]. The development of the retinal vasculature in several animal species (rodents, cats, and dogs), unlike humans, occurs postnatal and resembles the immature retinal vasculature development stage of premature infants. A mouse model has been developed, which exposes neonatal mice to 75% oxygen for 5 days after which they are returned to normal room air. This model exhibits cardinal features of human retinopathy of prematurity and certain aspects of diabetic retinopathy, namely pathogenic blood vessel growth. Inflammation is a distinct component of the pathogenesis of this hypoxia-induced neovascularization as evidenced by studies that demonstrate inflammatory mediator expression [106,113,114] in the retina or used sphingosine 1-phosphate receptor knockout mice [115]. Sphingosine 1-phosphate is a pleiotropic lipid mediator that regulates cell proliferation, migration, and differentiation. Mice deficient in the sphingosine 1-phosphate receptors (Table 31.2) demonstrate a phenotype of decreased macrophage accumulation, decreased cyclooxygenase-2 expression, and reduced pathological vascularization in this retinopathy model. Standard molecular biology methods, immunohistology, LC/MS/MS-based lipidomics, fundus photography, and fluorescein angiography are well developed to investigate pathways and mediators of inflammation and angiogenesis [106,113–115].

References 106 and 115 discuss immunohistology, dietary manipulation, molecular biology, mass-spectrometry–based lipidomics analyses, and knockout mice and the role of inflammatory mediators in retinopathy.

CONJUNCTIVA

The conjunctiva is a mucous membrane that covers the exposed anterior surface of the sclera and forms a membrane-lined sac that stretches from the corneo-scleral limbus to the muco-cutaneous junction on the eyelids (Figure 31.1). This tissue is the tear film reservoir, a major source of secreted mucin and antimicrobial peptides [6,7,14]. Like most mucosal tissue, it contains a large number of mast cells and is the primary immune tissue of the anterior surface of the eye. Conjunctivitis is defined as either an allergic or nonallergic form. However, the most common form of conjunctivitis in human and in experimental animal models is allergic inflammation, which is predominantly driven by mast cells and lymphocytes with the cardinal symptoms of itchy eyes and vascular congestion of the conjunctiva.

Conjunctivitis

Allergic or immune-driven diseases of other mucosal tissues have been studied in detail (e.g., asthma, inflammatory bowel diseases) and animal models have provided important advances in understanding the pathophysiology and immunology. However, studies and animal models for conjunctivitis are limited [14]. In general, most animal models have employed active immunization protocols using primarily ovalbumin or ragweed as an antigen. A variety of protocols have been established for different antigens and time periods of antigen sensitization and challenge that are animal species and strain specific [14]. The animal model of conjunctivitis was first developed in guinea pigs and has been established in both rats and mice. As with most immediate type I hypersensitivity reactions, experimental conjunctivitis is an immunoglobulin (IgE antibody) driven response that depends on activation of T_H2 type helper T cells and mast cells. The critical role of T_H2 cells and mast cells in experimental antigen–induced conjunctivitis has been confirmed in knockout mice (Table 31.2) for cytokines that regulate their activity, namely IL-12 [116], IFNγ [116], and MIP-1α [117].

Reference 14 reviews the development and experimental protocols of animal models of conjunctivitis.

IN VITRO MODELS OF INFLAMMATION

Due to the accessibility of the eye for direct experimental manipulation, cell lines or isolated cells are not as extensively used as models of ocular inflammation as is the case for other tissues and organs, such as the gastrointestinal tract, blood, and central nervous system. Several immortalized or transformed lines of corneal epithelial (CE) and retinal pigment epithelial (RPE) are available from both human and murine sources and are routinely used to investigate molecular mechanisms that regulate cell proliferation, inflammatory pathways, and apoptosis.

REFERENCES

1. Odgen, T.E., Hinton, D.R., and Schachat, A.P. 2004. *Retina*. Philadelphia: Mosby.
2. Streilein, J.W. 1995. Unraveling immune privilege. *Science* **270**:1158–1159.
3. Kaplan, H.J. 2007. Anatomy and function of the eye. *Chem Immunol Allergy* **92**:4–10.
4. Niederkorn, J.Y. 2006. See no evil, hear no evil, do no evil: the lessons of immune privilege. *Nat Immunol* **7**:354–359.
5. Streilein, J.W. 2003. Ocular immune privilege: therapeutic opportunities from an experiment of nature. *Nat Rev Immunol* **3**:879–889.
6. Kumar, V., Fausto, N., and Abbas, A. 2004. *Robbins and Cotran Pathological Basis of Disease*. Philadelphia: W.B. Saunders Company.
7. Krachmer, J.H., Mannis, M.J., and Holland, E.J. 2005. *Cornea*. Philadelphia: Elsevier Mosby.
8. Streilein, J.W. 2003. Ocular immune privilege: the eye takes a dim but practical view of immunity and inflammation. *J Leukoc Biol* **74**:179–185.
9. Kaplan, H.J., and Niederkorn, J.Y. 2007. Regional immunity and immune privilege. *Chem Immunol Allergy* **92**:11–26.
10. Niederkorn, J.Y., and Kaplan, H.J. 2007. Rationale for immune response and the eye. *Chem Immunol Allergy* **92**:1–3.
11. Niederkorn, J.Y., and Kaplan, H.J. 2007. *Immune Response and the Eye*. New York: Karger.
12. Niederkorn, J.Y. 2007. The induction of anterior chamber-associated immune deviation. *Chem Immunol Allergy* **92**:27–35.
13. Carlson, E.C., Drazba, J., Yang, X., and Perez, V.L. 2006. Visualization and characterization of inflammatory cell recruitment and migration through the corneal stroma in endotoxin-induced keratitis. *Invest Ophthalmol Vis Sci* **47**:241–248.
14. Groneberg, D.A., Bielory, L., Fischer, A., Bonini, S., and Wahn, U. 2003. Animal models of allergic and inflammatory conjunctivitis. *Allergy* **58**:1101–1113.
15. Barabino, S., and Dana, M.R. 2004. Animal models of dry eye: a critical assessment of opportunities and limitations. *Invest Ophthalmol Vis Sci* **45**:1641–1646.
16. Barabino, S., Rolando, M., Chen, L., and Dana, M.R. 2007. Exposure to a dry environment induces strain-specific responses in mice. *Exp Eye Res* **84**:973–977.
17. Gronert, K. 2005. Lipoxins in the eye and their role in wound healing. *Prostaglandins Leukot Essent Fatty Acids* **73**:221–229.
18. Chakravarti, S. 2001. The cornea through the eyes of knockout mice. *Exp Eye Res* **73**:411–419.
19. Bora, N.S., and Kaplan, H.J. 2007. Intraocular diseases – anterior uveitis. *Chem Immunol Allergy* **92**:213–220.
20. Forrester, J.V. 2007. Intermediate and posterior uveitis. *Chem Immunol Allergy* **92**:228–243.
21. Hazlett, L.D. 2007. Bacterial infections of the cornea (Pseudomonas aeruginosa). *Chem Immunol Allergy* **92**:185–194.
22. Smith, J.R., Hart, P.H., and Williams, K.A. 1998. Basic pathogenic mechanisms operating in experimental models of acute anterior uveitis. *Immunol Cell Biol* **76**:497–512.
23. McDermott, A.M., Perez, V., Huang, A.J., et al. 2005. Pathways of corneal and ocular surface inflammation: a perspective from the cullen symposium. *Ocul Surf* **3**:S131–S138.
24. Chang, J.H., Gabison, E.E., Kato, T., and Azar, D.T. 2001. Corneal neovascularization. *Curr Opin Ophthalmol* **12**:242–249.
25. Calder, V.L. 2005. Cytokine profiles in conjunctival allergy and inflammation. *Ocul Surf* **3**:S142–S144.
26. Nour, M., and Chodosh, J. 2005. Chemokine signaling pathways in corneal fibroblasts. *Ocul Surf* **3**:S149–S151.
27. Vallochi, A.L., Commodaro, A.G., Schwartzman, J.P., Belfort, R., Jr., and Rizzo, L.V. 2007. The role of cytokines in the regulation of ocular autoimmune inflammation. *Cytokine Growth Factor Rev* **18**:135–141.
28. Planck, S.R., Rich, L.F., Ansel, J.C., Huang, X.N., and Rosenbaum, J.T. 1997. Trauma and alkali burns induce distinct patterns of cytokine gene expression in the rat cornea. *Ocul Immunol Inflamm* **5**:95–100.
29. Bito, L.Z. 1986. Prostaglandins and other eicosanoids: their ocular transport, pharmacokinetics, and therapeutic effects. *Trans Ophthalmol Soc U K* **105**(Pt 2):162–170.
30. Camras, C.B., Bito, L.Z., and Eakins, K.E. 1977. Reduction of intraocular pressure by prostaglandins applied topically to the eyes of conscious rabbits. *Invest Ophthalmol Vis Sci* **16**:1125–1134.
31. Bazan, H., and Ottino, P. 2002. The role of platelet-activating factor in the corneal response to injury. *Prog Retin Eye Res* **21**:449–464.
32. Bazan, H.E. 2005. Cellular and molecular events in corneal wound healing: significance of lipid signalling. *Exp Eye Res* **80**:453–463.
33. Bazan, N.G. 2006. Cell survival matters: docosahexaenoic acid signaling, neuroprotection and photoreceptors. *Trends Neurosci* **29**:263–271.
34. Gronert, K. 2008. Lipid autacoids in inflammation and injury responses: a matter of privilege. *Mol Interv* **8**:28–35.
35. Burns, A.R., Li, Z., and Smith, C.W. 2005. Neutrophil migration in the wounded cornea: the role of the keratocyte. *Ocul Surf* **3**:S173–S176.
36. Li, Z., Burns, A.R., and Smith, C.W. 2006. Two waves of neutrophil emigration in response to corneal epithelial abrasion: distinct adhesion molecule requirements. *Invest Ophthalmol Vis Sci* **47**:1947–1955.
37. Petrescu, M.S., Larry, C.L., Bowden, R.A., et al. 2007. Neutrophil interactions with keratocytes during corneal epithelial wound healing: a role for CD18 integrins. *Invest Ophthalmol Vis Sci* **48**:5023–5029.
38. Becker, M.D., Garman, K., Whitcup, S.M., Planck, S.R., and Rosenbaum, J.T. 2001. Inhibition of leukocyte sticking and infiltration, but not rolling, by antibodies to

ICAM-1 and LFA-1 in murine endotoxin-induced uveitis. *Invest Ophthalmol Vis Sci* **42**:2563–2566.
39. Planck, S.R., Han, Y.B., Park, J.M., O'Rourke, L., Gutierrez-Ramos, J.C., and Rosenbaum, J.T. 1998. The effect of genetic deficiency of adhesion molecules on the course of endotoxin-induced uveitis. *Curr Eye Res* **17**:941–946.
40. Smith, J.R., Subbarao, K., Franc, D.T., Haribabu, B., and Rosenbaum, J.T. 2004. Susceptibility to endotoxin induced uveitis is not reduced in mice deficient in BLT1, the high affinity leukotriene B4 receptor. *Br J Ophthalmol* **88**:273–275.
41. Johnson, A., and Pearlman, E. 2005. Toll-like receptors in the cornea. *Ocul Surf* **3**:S187–S189.
42. Yu, F.S., and Hazlett, L.D. 2006. Toll-like receptors and the eye. *Invest Ophthalmol Vis Sci* **47**:1255–1263.
43. Jha, P., Bora, P.S., and Bora, N.S. 2007. The role of complement system in ocular diseases including uveitis and macular degeneration. *Mol Immunol* **44**:3901–3908.
44. Hamrah, P., and Dana, M.R. 2007. Corneal antigen-presenting cells. *Chem Immunol Allergy* **92**:58–70.
45. Cao, Z., Said, N., Amin, S., et al. 2002. Galectins-3 and -7, but not galectin-1, play a role in re-epithelialization of wounds. *J Biol Chem* **277**:42299–42305.
46. Carlson, E.C., Wang, I.J., Liu, C.Y., Brannan, P., Kao, C.W., and Kao, W.W. 2003. Altered KSPG expression by keratocytes following corneal injury. *Mol Vis* **9**:615–623.
47. Mohan, R., Chintala, S.K., Jung, J.C., et al. 2002. Matrix metalloproteinase gelatinase B (MMP-9) coordinates and effects epithelial regeneration. *J Biol Chem* **277**:2065–2072.
48. Sharma, G.-D., He, J., and Bazan, H.E.P. 2003. p38 and ERK1/2 Coordinate Cellular Migration and Proliferation in Epithelial Wound Healing: evidence of cross-talk activation between map kinase cascades. *J Biol Chem* **278**:21989–21997.
49. Seta, F., Bellner, L., Rezzani, R., et al. 2006. Heme Oxygenase-2 Is a Critical Determinant for Execution of an Acute Inflammatory and Reparative Response. *Am J Pathol* **169**:1612–1623.
50. Gronert, K., Maheshwari, N., Khan, N., Hassan, I.R., Dunn, M., and Laniado Schwartzman, M. 2005. A role for the mouse 12/15-lipoxygenase pathway in promoting epithelial wound healing and host defense. *J Biol Chem* **280**:15267–15278.
51. Li, Z., Rumbaut, R.E., Burns, A.R., and Smith, C.W. 2006. Platelet response to corneal abrasion is necessary for acute inflammation and efficient re-epithelialization. *Invest Ophthalmol Vis Sci* **47**:4794–4802.
52. Biteman, B., Hassan, I.R., Walker, E., et al. 2007. Interdependence of lipoxin A4 and heme-oxygenase in counter-regulating inflammation during corneal wound healing. *FASEB J* **21**:2257–2266.
53. Vij, N., Roberts, L., Joyce, S., and Chakravarti, S. 2005. Lumican regulates corneal inflammatory responses by modulating Fas-Fas ligand signaling. *Invest Ophthalmol Vis Sci* **46**:88–95.
54. Reim, M., Kottek, A., and Schrage, N. 1997. The cornea surface and wound healing. *Prog Retin Eye Res* **16**:183–225.
55. Yoshida, S., Yoshida, A., Matsui, H., Takada, Y., and Ishibashi, T. 2003. Involvement of macrophage chemotactic protein-1 and interleukin-1beta during inflammatory but not basic fibroblast growth factor-dependent neovascularization in the mouse cornea. *Lab Invest* **83**:927–938.
56. Fang, Y., Choi, D., Searles, R.P., and Mathers, W.D. 2005. A time course microarray study of gene expression in the mouse lacrimal gland after acute corneal trauma. *Invest Ophthalmol Vis Sci* **46**:461–469.
57. Ogawa, S., Yoshida, S., Ono, M., et al. 1999. Induction of macrophage inflammatory protein-1alpha and vascular endothelial growth factor during inflammatory neovascularization in the mouse cornea. *Angiogenesis* **3**:327–334.
58. Ueno, M., Lyons, B.L., Burzenski, L.M., et al. 2005. Accelerated wound healing of alkali-burned corneas in MRL mice is associated with a reduced inflammatory signature. *Invest Ophthalmol Vis Sci* **46**:4097–4106.
59. Hazlett, L.D. 2005. Inflammatory response to Pseudomonas aeruginosa keratitis. *Ocul Surf* **3**:S139–141.
60. Johnson, A.C., Li, X., and Pearlman, E. 2008. MyD88 functions as a negative regulator of TLR3/TRIF-induced corneal inflammation by inhibiting activation of c-Jun N-terminal kinase. *J Biol Chem* **283**:3988–3996.
61. Johnson, A.C., Heinzel, F.P., Diaconu, E., et al. 2005. Activation of toll-like receptor (TLR)2, TLR4, and TLR9 in the mammalian cornea induces MyD88-dependent corneal inflammation. *Invest Ophthalmol Vis Sci* **46**:589–595.
62. Fenton, R.R., Molesworth-Kenyon, S., Oakes, J.E., and Lausch, R.N. 2002. Linkage of IL-6 with neutrophil chemoattractant expression in virus-induced ocular inflammation. *Invest Ophthalmol Vis Sci* **43**:737–743.
63. McClellan, S.A., Huang, X., Barrett, R.P., van Rooijen, N., and Hazlett, L.D. 2003. Macrophages restrict Pseudomonas aeruginosa growth, regulate polymorphonuclear neutrophil influx, and balance pro- and anti-inflammatory cytokines in BALB/c mice. *J Immunol* **170**:5219–5227.
64. Singh, N., Jani, P.D., Suthar, T., Amin, S., and Ambati, B.K. 2006. Flt-1 intraceptor induces the unfolded protein response, apoptotic factors, and regression of murine injury-induced corneal neovascularization. *Invest Ophthalmol Vis Sci* **47**:4787–4793.
65. Ambati, B.K., Nozaki, M., Singh, N., et al. 2006. Corneal avascularity is due to soluble VEGF receptor-1. *Nature* **443**:993–997.
66. Bellner, L., Vitto, M., Patil, K.A., Dunn, M.W., Regan, R., and Laniado-Schwartzman, M. 2008. Exacerbated corneal inflammation and neovascularization in the HO-2 null mice is ameliorated by biliverdin. *Exp Eye Res* **87**(3):268–278.
67. Chen, L., Hamrah, P., Cursiefen, C., et al. 2007. Vascular endothelial growth factor receptor-3 mediates induction of corneal alloimmunity. 2004. *Ocul Immunol Inflamm* **15**:275–278.
68. Cursiefen, C., Chen, L., Borges, L.P., et al. 2004. VEGF-A stimulates lymphangiogenesis and hemangiogenesis in inflammatory neovascularization via macrophage recruitment. *J Clin Invest* **113**:1040–1050.
69. Cursiefen, C., Chen, L., Saint-Geniez, M., et al. 2006. Nonvascular VEGF receptor 3 expression by corneal epithelium maintains avascularity and vision. *Proc Natl Acad Sci USA* **103**:11405–11410.
70. Cursiefen, C., Masli, S., Ng, T.F., et al. 2004. Roles of thrombospondin-1 and -2 in regulating corneal and iris angiogenesis. *Invest Ophthalmol Vis Sci* **45**:1117–1124.
71. Maruyama, K., Ii, M., Cursiefen, C., et al. 2005. Inflammation-induced lymphangiogenesis in the cornea arises from CD11b-positive macrophages. *J Clin Invest* **115**:2363–2372.
72. Streilein, J.W., Bradley, D., Sano, Y., and Sonoda, Y. 1996. Immunosuppressive properties of tissues obtained from

eyes with experimentally manipulated corneas. *Invest Ophthalmol Vis Sci* **37**:413–424.

73. Cursiefen, C., Maruyama, K., Jackson, D.G., Streilein, J.W., and Kruse, F.E. 2006. Time course of angiogenesis and lymphangiogenesis after brief corneal inflammation. *Cornea* **25**:443–447.
74. Hori, J., and Niederkorn, J.Y. 2007. Immunogenicity and immune privilege of corneal allografts. *Chem Immunol Allergy* **92**:290–299.
75. Hamrah, P., Yamagami, S., Liu, Y., et al. 2007. Deletion of the chemokine receptor CCR1 prolongs corneal allograft survival. *Invest Ophthalmol Vis Sci* **48**:1228–1236.
76. Streilein, J.W. 2003. New thoughts on the immunology of corneal transplantation. *Eye* **17**:943–948.
77. Chen, L., Huq, S., Gardner, H., de Fougerolles, A.R., Barabino, S., and Dana, M.R. 2007. Very late antigen 1 blockade markedly promotes survival of corneal allografts. *Arch Ophthalmol* **125**:783–788.
78. Shen, L., Jin, Y., Freeman, G.J., Sharpe, A.H., and Dana, M.R. 2007. The function of donor versus recipient programmed death-ligand 1 in corneal allograft survival. *J Immunol* **179**:3672–3679.
79. Hargrave, S.L., Hay, C., Mellon, J., Mayhew, E., and Niederkorn, J.Y. 2004. Fate of MHC-matched corneal allografts in Th1-deficient hosts. *Invest Ophthalmol Vis Sci* **45**:1188–1193.
80. Yamagami, S., Dana, M.R., and Tsuru, T. 2002. Draining lymph nodes play an essential role in alloimmunity generated in response to high-risk corneal transplantation. *Cornea* **21**:405–409.
81. Yamagami, S., and Dana, M.R. 2001. The critical role of lymph nodes in corneal alloimmunization and graft rejection. *Invest Ophthalmol Vis Sci* **42**:1293–1298.
82. Cursiefen, C., Cao, J., Chen, L., et al. 2004. Inhibition of hemangiogenesis and lymphangiogenesis after normal-risk corneal transplantation by neutralizing VEGF promotes graft survival. *Invest Ophthalmol Vis Sci* **45**:2666–2673.
83. Yamagami, S., Hamrah, P., Zhang, Q., Liu, Y., Huq, S., and Dana, M.R. 2005. Early ocular chemokine gene expression and leukocyte infiltration after high-risk corneal transplantation. *Mol Vis* **11**:632–640.
84. Amescua, G., Collings, F., Sidani, A., et al. 2008. Effect of CXCL-1/KC production in high risk vascularized corneal allografts on T cell recruitment and graft rejection. *Transplantation* **85**:615–625.
85. Barabino, S., and Dana, M.R. 2007. Dry eye syndromes. *Chem Immunol Allergy* **92**:176–184.
86. Barabino, S., Shen, L., Chen, L., Rashid, S., Rolando, M., and Dana, M.R.. 2005. The controlled-environment chamber: a new mouse model of dry eye. *Invest Ophthalmol Vis Sci* **46**:2766–2771.
87. De Paiva, C.S., Corrales, R.M., Villarreal, A.L., et al. 2006. Corticosteroid and doxycycline suppress MMP-9 and inflammatory cytokine expression, MAPK activation in the corneal epithelium in experimental dry eye. *Exp Eye Res* **83**:526–535.
88. Li, S., Nikulina, K., DeVoss, J., et al. 2008. Small proline-rich protein 1B (SPRR1B) is a biomarker for squamous metaplasia in dry eye disease. *Invest Ophthalmol Vis Sci* **49**:34–41.
89. van Blokland, S.C., and Versnel, M.A. 2002. Pathogenesis of Sjogren's syndrome: characteristics of different mouse models for autoimmune exocrinopathy. *Clin Immunol* **103**:111–124.
90. Rosenbaum, J.T., McDevitt, H.O., Guss, R.B., and Egbert, P.R. 1980. Endotoxin-induced uveitis in rats as a model for human disease. *Nature* **286**:611–613.
91. Shen, D.F., Buggage, R.R., Eng, H.C., and Chan, C.C. 2000. Cytokine gene expression in different strains of mice with endotoxin-induced uveitis (EIU). *Ocul Immunol Inflamm* **8**:221–225.
92. Wacker, W.B. 1991. Proctor Lecture. Experimental allergic uveitis. Investigations of retinal autoimmunity and the immunopathologic responses evoked. *Invest Ophthalmol Vis Sci* **32**:3119–3128.
93. Yoshimura, T., Sonoda, K.H., Miyazaki, Y., et al. 2008. Differential roles for IFN-gamma and IL-17 in experimental autoimmune uveoretinitis. *Int Immunol* **20**:209–214.
94. Liu, X., Lee, Y.S., Yu, C.R., and Egwuagu, C.E. 2008. Loss of STAT3 in CD4+ T cells prevents development of experimental autoimmune diseases. *J Immunol* **180**:6070–6076.
95. Hikita, S.T., Vistica, B.P., Jones, H.R., et al. 2006. Osteopontin is proinflammatory in experimental autoimmune uveitis. *Invest Ophthalmol Vis Sci* **47**:4435–4443.
96. Kitamura, M., Iwabuchi, K., Kitaichi, N., et al. 2007. Osteopontin aggravates experimental autoimmune uveoretinitis in mice. *J Immunol* **178**:6567–6572.
97. Su, S.B., Grajewski, R.S., Luger, D., et al. 2007. Altered chemokine profile associated with exacerbated autoimmune pathology under conditions of genetic interferon-gamma deficiency. *Invest Ophthalmol Vis Sci* **48**:4616–4625.
98. Fukushima, A., Yamaguchi, T., Ishida, W., Fukata, K., Udaka, K., and Ueno, H. 2005. Mice lacking the IFN-gamma receptor or fyn develop severe experimental autoimmune uveoretinitis characterized by different immune responses. *Immunogenetics* **57**:337–343.
99. Liao, T., Ke, Y., Shao, W.H., et al. 2006. Blockade of the interaction of leukotriene b4 with its receptor prevents development of autoimmune uveitis. *Invest Ophthalmol Vis Sci* **47**:1543–1549.
100. Taylor, A.W., Kitaichi, N., and Biros, D. 2006. Melanocortin 5 receptor and ocular immunity. *Cell Mol Biol (Noisy-le-grand)* **52**.53–59.
101. Ichikawa, T., Taguchi, O., Takahashi, T., et al. 1991. Spontaneous development of autoimmune uveoretinitis in nude mice following reconstitution with embryonic rat thymus. *Clin Exp Immunol* **86**:112–117.
102. Anderson, M.S., Venanzi, E.S., Klein, L., et al. 2002. Projection of an immunological self shadow within the thymus by the aire protein. *Science* **298**:1395–1401.
103. Szpak, Y., Vieville, J.C., Tabary, T., et al. 2001. Spontaneous retinopathy in HLA-A29 transgenic mice. *Proc Natl Acad Sci USA* **98**:2572–2576.
104. Rattner, A., and Nathans, J. 2006. Macular degeneration: recent advances and therapeutic opportunities. *Nat Rev Neurosci* **7**:860–872.
105. Zamiri, P., Sugita, S., and Streilein, J.W. 2007. Immunosuppressive properties of the pigmented epithelial cells and the subretinal space. *Chem Immunol Allergy* **92**:86–93.
106. Connor, K.M., SanGiovanni, J.P., Lofqvist, C., et al. 2007. Increased dietary intake of omega-3-polyunsaturated fatty acids reduces pathological retinal angiogenesis. *Nat Med* **13**:868–873.
107. Noel, A., Jost, M., Lambert, V., Lecomte, J., and Rakic, J.M. 2007. Anti-angiogenic therapy of exudative age-related macular degeneration: current progress and emerging concepts. *Trends Mol Med* **13**:345–352.
108. Uemura, A., Kusuhara, S., Katsuta, H., and Nishikawa, S. 2006. Angiogenesis in the mouse retina: a model

system for experimental manipulation. *Exp Cell Res* **312**:676–683.

109. Nozaki, M., Raisler, B.J., Sakurai, E., et al. 2006. Drusen complement components C3a and C5a promote choroidal neovascularization. *Proc Natl Acad Sci USA* **103**:2328–2333.
110. Imamura, Y., Noda, S., Hashizume, K., et al. 2006. Drusen, choroidal neovascularization, and retinal pigment epithelium dysfunction in SOD1-deficient mice: a model of age-related macular degeneration. *Proc Natl Acad Sci USA* **103**:11282–11287.
111. Ambati, J., Anand, A., Fernandez, S., et al. 2003. An animal model of age-related macular degeneration in senescent Ccl-2- or Ccr-2-deficient mice. *Nat Med* **9**:1390–1397.
112. Truman, L.A., Ford, C.A., Pasikowska, M., et al. 2008. CX3CL1/fractalkine is released from apoptotic lymphocytes to stimulate macrophage chemotaxis. *Blood* **112**(13):5026–5036.
113. Chen, J., and Smith, L.E. 2007. Retinopathy of prematurity. *Angiogenesis* **10**:133–140.
114. Smith, L.E. 2003. Pathogenesis of retinopathy of prematurity. *Semin Neonatol* **8**:469–473.
115. Skoura, A., Sanchez, T., Claffey, K., Mandala, S.M., Proia, R.L., and Hla, T. 2007. Essential role of sphingosine 1-phosphate receptor 2 in pathological angiogenesis of the mouse retina. *J Clin Invest* **117**:2506–2516.
116. Magone, M.T., Whitcup, S.M., Fukushima, A., Chan, C.C., Silver, P.B., and Rizzo, L.V. 2000. The role of IL-12 in the induction of late-phase cellular infiltration in a murine model of allergic conjunctivitis. *J Allergy Clin Immunol* **105**:299–308.
117. Miyazaki, D., Nakamura, T., Toda, M., Cheung-Chau, K.W., Richardson, R.M., and Ono, S.J. 2005. Macrophage inflammatory protein-1alpha as a costimulatory signal for mast cell-mediated immediate hypersensitivity reactions. *J Clin Invest* **115**:434–442.

SUGGESTED READINGS

Barabino, S., and Dana, M.R. 2004. Animal models of dry eye: a critical assessment of opportunities and limitations. *Invest Ophthalmol Vis Sci* **45**:1641–1646.

Bora, N.S., and Kaplan, H.J. 2007. Intraocular diseases – anterior uveitis. *Chem Immunol Allergy* **92**:213–220.

Groneberg, D.A., Bielory, L., Fischer, A., Bonini, S., and Wahn, U. 2003. Animal models of allergic and inflammatory conjunctivitis. *Allergy* **58**:1101–1113.

McDermott, A.M., Perez, V., Huang, A.J., et al. 2005. Pathways of corneal and ocular surface inflammation: a perspective from the cullen symposium. *Ocul Surf* **3**:S131–S138.

Rattner, A., and Nathans, J. 2006. Macular degeneration: recent advances and therapeutic opportunities. *Nat Rev Neurosci* **7**:860–872.

Streilein, J.W. 2003. Ocular immune privilege: therapeutic opportunities from an experiment of nature. *Nat Rev Immunol* **3**:879–889.

32 Atherosclerosis in Experimental Animal Models

Aksam Merched and Lawrence Chan

The term *atheroma*, derived from Greek and meaning "porridge," was first proposed by Albrecht von Haller in 1755 to label the degenerative process observed in the intima of arteries. London surgeon Joseph Hodgson (1788–1869) published in 1815 his *Treatise on the Diseases of Arteries and Veins* in which he claimed that inflammation was the underlying cause of atheromatous arteries. Atherosclerosis is now widely appreciated as an inflammatory disease involving the vascular wall. Histologically, the lipid-laden foam cells of the fatty streak, which characterize the plaque at an early stage, are derived from macrophages. In time, the lipid/necrotic core is covered with fibrous tissue composed mainly of α-actin positive smooth muscle cells, and thus forms the fibrolipid plaque. Large amounts of T lymphocytes are found surrounding the plaque and in the fibrous cap, pointing to a role for the body's immune system in atherosclerosis.

Advanced complex atheromata that set the stage for overt clinical events in atherosclerosis are preceded by less complex lesions. The factors that enable some lesions to progress while others regress remain unclear. It is clear, however, that lack of regression is associated with persistent inflammation in the vascular wall. Most studies to date rely heavily on animal models to define mechanistic pathways [1,2].

THE USE OF ANIMAL MODELS IN ATHEROSCLEROSIS RESEARCH

The ideal animal model of cardiovascular disease should mimic the human subject metabolically and pathophysiologically, be large enough to permit physiological and metabolic studies, and develop end-stage disease comparable to that in humans. Given the complex multifactorial nature of cardiovascular disease, no one species fulfills all these criteria.

Until recently, the most popular animal models for atherosclerotic research had been restricted to relatively large animals, mainly rabbits, and to a lesser extent, pigs and nonhuman primates, which have provided invaluable insight in the disease process. However, investigations using large animal models have many drawbacks, mainly related to cost and difficulty of maintenance and handling of the colonies. With the advent of genetic engineering, transgenic mouse models have supplemented the classical dietary cholesterol-induced disease models such as the cholesterol-fed hamster, rabbit, pig, and monkey [3].

THE MOUSE AS A MODEL OF ATHEROSCLEROSIS

Previously thought to be atherosclerosis-resistant, in the past two decades, mice have become the mainstay of atherosclerosis research. Paigen et al. first demonstrated a huge variation in atherosclerosis susceptibility in 10 different inbred mouse strains, when they were feeding a proatherogenic diet. The most susceptible strains, for example, C57BL/6, required a diet high in cholesterol and the bile salt sodium cholate to develop measurable lesions.

More recently, the ability to genetically modify mice to overexpress, or knock out, or knockdown expression of specific genes has greatly facilitated the definition of pathways in the atherogenic process. All current mouse models of atherosclerosis rely heavily on perturbations of lipoprotein metabolism or cellular processes through dietary or genetic manipulations.

Use of mice in atherosclerosis research offers many advantages, including ease of handling, short generation time, and large litter size. Another important attribute, exploited in many biological applications, is the extensive genetic information available on the numerous inbred strains. Mice as a model also provide a convenient animal system for bone marrow transplantation experiments. This technique is now used extensively in atherosclerosis research as it provides valuable insight into the role of bone marrow–derived hematopoietic cells on the process.

A major breakthrough in atherosclerosis research was the creation of two mouse dyslipidemic models, apolipoprotein E (apoE)- and LDL receptor (LDLR)-deficient mice, by homologous recombination techniques. In contrast to the prior models, mice lacking functional apoE or LDLR genes develop extensive arterial lesions that progress from foam cell-rich fatty streaks to fibroproliferative plaques with lipid/necrotic cores, typical of the spectrum of human lesions. These two popular mouse models both harbor defects in lipid metabolism and severe dyslipidemias that underlie atherosclerosis development.

Use of LDLR-deficient and apoE-deficient mice has allowed the dissection of nonlipid factors that influence the severity and character of lesions. Various genetic modifications have been introduced to generate phenotypes that resemble human atherosclerotic lesions, and numerous additional mouse models have been created based on these two models [4–7].

ApoE KNOCKOUT

The most widely used mouse model is the apoE-deficient mouse strain, which spontaneously develops lesions analogous to human atherosclerotic lesions. Therefore, the availability of the *apoE-/-* mouse can be considered an important landmark in atherosclerosis research. ApoE is an important ligand for the uptake of lipoproteins by different receptors in the LDLR gene family. It has systemic effects on plasma lipoproteins but also it is involved locally within the arterial wall where it is synthesized by monocytes/macrophages in cholesterol homeostasis and inflammation. Mice deficient in apoE (apoE–/–) were first produced in 1992 in different laboratories. Lesions in the apoE-deficient mouse, as in humans, tend to develop at vascular branch points and progress from foam cell fatty streaks to the fibroproliferative stage with well-defined fibrous caps and lipid-containing necrotic cores, although the establishment of ruptured plaque in apoE-deficient mice was not without controversy. These mice develop spontaneous plaques when they are on regular chow, but one can induce accelerated and more severe atherosclerotic plaques in many vascular beds by putting them on a high-fat diet. Lesion progression occurs at a greatly accelerated pace as compared with that in humans or large animal models; this is a major advantage for experiments on atherosclerosis.

A shortcoming of the *apoE–/–* mouse is that the lipoprotein profile is very different to that in human patients. In these mice, the plasma cholesterol is carried almost exclusively in very low-density lipoprotein (VLDL) and intermediate density lipoprotein (IDL), rather than in LDL, as in humans.

A common experimental design is to breed mice carrying "candidate genes" suspected of being involved in either protecting against or promoting atherogenesis to *apoE–/–* background. Investigation into these cross-bred mice can often reveal some novel roles of the candidate(s) in question in atherosclerosis development [8].

LDL RECEPTOR DEFICIENT MICE

Deficiency of LDLR is the underlying cause of familial hypercholesterolemia (FH) in humans. The LDLR–/– mouse is a model for human FH. LDLR–/– mice were created by targeting the genes of embryonic stem cells. By eliminating the functional gene for LDLR, the mouse model displays a mildly elevated cholesterol level with a lipoprotein profile simulating that of the human plasma lipoprotein profile (where cholesterol is mainly confined to the LDL fraction). For example, when they are fed a 10% fat and cholesterol-enriched diet, they develop a twofold increase in total cholesterol levels due to high LDL and VLDL levels. LDLR–/– mice do develop mild but significant atherosclerosis while on a regular chow. However, they develop large atherosclerotic lesions in their aorta when they are fed diets enriched in saturated fat without cholate. The lesions found in these mice are of simple morphology, consisting predominantly of lipid-laden macrophages. Some of the features of advanced lesions, including necrotic cores and calcification, are generated only after prolonged feeding of high-fat diets.

As in the case of models involving *apoE–/–*, *LDLR–/–* mice have been a useful model to study the role of candidate genes in atherosclerosis, although to a lesser extent than *apoE–/–* mice. They have also served as an excellent host for bone marrow transplantation experiments and have been used extensively for this purpose [9,10].

OTHER MOUSE MODELS

More recently, apoE and LDLR double-knockout (apoE/LDLr-DKO) mice have been created, representing a new mouse model that develops severe hyperlipidaemia and atherosclerosis. On a regular chow diet, the progression of atherosclerosis is more marked in apoE/LDLR-DKO mice than in mice deficient for apoE alone, making the double knockout mouse a suitable model for atherosclerosis without having to feed the animals an atherogenic diet. Other reported mouse models include a *APOE3Leiden* mouse model that was found to be less susceptible to atherosclerosis than the *apoE–/–* mouse. Other genetic manipulations led to *ApoBEC-1–/– x LDLr–/–*, *ApoB100 transgenic x LDL–/–*, and *ApoCII transgenic x LDLR–/–* mouse models [10].

INFLAMMATORY MODULATORS OF ATHEROSCLEROSIS

Russell Ross developed his popular "response to injury" hypothesis of atherogenesis in the mid-1970s, postulating that atherosclerotic lesions arise as a result of focal injury to the arterial endothelium, followed by adherence and aggregation of platelets. Ross revisited his "response to injury" theory in 1986, considering that "subtle endothelial injury" was the *primum movens* in atherosclerosis, and published in 1999 an insightful review entitled "Atherosclerosis: a chronic inflammatory disease." The view that atherosclerosis is indeed a chronic inflammatory disease initiated by monocyte/lymphocyte adhesion to activated endothelial cells is now widely accepted and substantiated by experimental and clinical observations.

After recruitment to vascular injury sites, the infiltrated monocytes differentiate into macrophages, which take up lipids and form lipid-laden foam cells in the unique microenvironment of the vascular intima. These facets provide the pathological basis for early atherosclerosis, referred to as fatty streak lesions. Under the influence of various genetic and environmental risk factors, the early fatty streak lesion progresses into a complex lesion, typically characterized by a lipid-rich core covered by a fibrous cap, and a large number of activated inflammatory cells, particularly macrophages and T cells.

With the recognition of the inflammatory nature of the atherosclerotic process, increasing attention has been devoted to the role of cytokines and inflammatory mediators. Thus, a large number of mouse models have been generated by genetic manipulations of the immune system. Atherosclerosis progression is affected in mice lacking functional immune cells [11].

Macrophages

Activated macrophages influence the function of vascular cells by synthesizing and releasing a variety of bioactive molecules, such as proinflammatory cytokines, chemokines, growth factors, reactive oxygen species, and eicosanoids. Macrophages, by expressing different cytokines and costimulatory factors and by presenting antigens, also shape the T-cell response in atherosclerosis. Thus, macrophages are positioned, by the nature of their potent innate immune signaling, to play central roles in vascular inflammation. Numerous investigations have targeted macrophages and macrophage-specific functions in mice.

By way of example, we will consider two bioactive molecules, monocyte chemoattractant protein-1 (MCP-1) and interferon γ (IFN-γ), which are important in normal macrophage function. MCP-1 plays a major role in recruitment of macrophages. Deletion of the MCP-1 gene has been shown to protect against monocyte recruitment in mice overexpressing apolipoprotein B. MCP-1-deficient LDLR–/– mice also showed reduced macrophage recruitment, suggesting that the role of MCP-1 in atherosclerosis is to attract CCR2-bearing monocytes into the vessel wall. On the other hand, overexpression of MCP-1 generates mice with more extensive atherosclerotic lesions. These studies provide a proof of principle for the role of MCP-1 in atherosclerotic lesion development. IFN-γ is known to stimulate the expression of proatherogenic molecules such as intracellular adhesion molecule 1, and decrease antiatherogenic effects by upregulating the expression of lipoprotein receptors on macrophages. Mice with a combined deficiency of IFN-γ and apoE exhibit a substantial reduction in lipid accumulation in arteries, presumably resulting from an increase of atheroprotective phospholipid/apoA-IV-rich particles [12].

Lymphocytes

The involvement of T and B lymphocytes in atherosclerosis occurs in a complex way. Mice deficient in either recombinase-activating gene 1 or 2 (RAG1 or RAG2) do not produce functional T or B cells, owing to a defect in V(D)J recombination. The net effect of a deficiency in both T and B cells is a 40%–80% reduction in atherosclerotic lesion development, as observed in apoE–/– or LDLr–/– mice crossed into a recombination activating gene (Rag)-deficient background or crossed with severe combined immunodeficiency (SCID) mice. Interestingly, the effect may vary according to the site of the lesion, immunodeficiency being protective in the aortic root but not in the thoracic and abdominal aorta or in the brachiocephalic trunk.

CD4+ and CD8+ cell depletion reduces fatty streak formation in C57BL/6 mice, indicating that T cells aggravate fatty streak formation. Transfer of CD4+ cells from immunocompetent apoE–/– mice to immunodeficient apoE–/– × SCID–/– mice leads to drastic increase in atherosclerotic lesions, indicating a proatherogenic role for T cells. Furthermore, MHC-I- or MHC-II-deficient mice show larger lesions compared with normal mice.

On the other hand, B cells appear to exert a protective effect. Induction of humoral immunity by immunization of hypercholesterolemic apoE–/– mice with oxLDL reduces lesion size in association with the production of high levels of IgM type anti-oxLDL antibodies, probably from B1 cells. Interestingly, the IgM type anti-oxLDL antibodies recognize similar oxidation-specific epitopes on apoptotic cells and are structurally and functionally identical to classic "natural" anti-phosphotidylcholine antibodies that provide protection against pneumococcal infection. Immunization of LDLr–/– mice with *Streptococcus pneumoniae* induces high circulating levels of oxLDL-specific T15 IgM, indicating molecular mimicry between epitopes of oxLDL and *S. pneumoniae*; it leads to a reduction in the extent of atherosclerosis, confirming a protective role of this humoral immune response in murine cholesterol-induced atherosclerosis.

Splenectomy induces increase in atherosclerosis in cholesterol-fed apoE–/– mice, an effect that is abrogated by the transfer of purified B cells from the spleen of atherosclerotic apoE–/– (but not from the spleen of nonatherosclerotic mice), suggesting a protective immunity provided by splenic B cells that have been "educated" by prior in vivo exposure to atherosclerotic antigens.

Over the past few years, investigations into the autoimmune basis of atherosclerosis have focused on a population of T cells with regulatory properties, called regulatory T cells (or Treg cells), which actively suppress immune activation and maintain immune homeostasis. Most, if not all, naturally occurring Treg cells are CD4+ single-positive cells that constitutively express the CD25 molecule. They are produced in the normal thymus where unique interactions between their TCRs and self-peptide/MHC complexes expressed on the thymic stromal cells are required for their development. Costimulatory signals mediated by engagement of CD28 by CD80/CD86 (B7) are essential for the development and homeostasis of Treg cells. Mice deficient in CD28 or B7 molecules lack Treg cells and are at increased risk of autoimmune diabetes. Even though IL-2 is a vital cytokine for Treg, recent studies suggest that expression of the forkhead transcription factor Foxp3, irrespective of CD25 expression or MHC restriction, defines the naturally occurring Treg cell lineage. Besides the role of IL-2 in the Treg cell development and maintenance, two immunosuppressive cytokines, TGF-β and IL-10, have been shown to mediate, at least in part, Treg function in vivo.

The development of these regulatory cells may be promoted in vitro and in vivo by a specific set of antigens and under particular conditions. Results from two studies suggest that mucosal administration of heat shock protein 65 antigen reduces plaque development. However, the mechanisms behind this effect are not fully understood.

Antigen-specific clones of Treg cells have also been shown to induce both antigen specific and nonspecific bystander immune suppression in vitro, and when introduced in vivo.

Administration of ovalbumin-specific Treg cells to apoE–/– mice was associated with a significant reduction in atherosclerotic plaque development and a marked reduction in the relative accumulation of inflammatory cells. Atherosclerosis in apoE–/–RAG-2–/– mice is exacerbated after transfer of splenocytes with Treg deficiency (from CD28- or B7-deficient mice) compared with the transfer of wild-type splenocytes, a process that is abrogated after the reconstitution of a normal CD4+CD25+ Treg cell compartment, suggesting that innate or acquired impairment of natural Treg cell function may promote atherosclerosis [13].

Dendritic Cells

Dendritic cells (DCs) are specialized antigen-presenting cells (APCs) that are potent stimulators of both T and B cell–mediated immune responses. DCs have been identified in atherosclerotic plaques and may cluster with T cells within the lesions. DCs showed impaired migratory function in hypercholesterolemic mice due to inhibitory signals generated by platelet-activating factor (PAF) and oxLDL. Whether these abnormal migratory properties directly affect the atherosclerotic process is still unknown. DCs exhibit considerable plasticity. The balance between proinflammatory and anti-inflammatory signals in the local microenvironment greatly affect DC maturation and distinct subsets of DCs elicit distinct T-helper responses.

In addition to targeting cell plaque component, many other studies have focused on specific cytokines, cytokine receptors, adhesion molecules in apoE or LDLR mouse background. Below, we have included an exciting ever-growing area of bioactive lipid mediators with a summary of the latest research in the field [13].

Bioactive Lipid Mediators and Resolution of Inflammation

The popular view that all lipid mediators are proinflammatory arises largely from the finding that nonsteroidal anti-inflammatory drugs block the biosynthesis of prostaglandins. Of interest, one class of arachidonic acid–derived mediators, the lipoxins generated from arachidonic acid (AA) and aspirin-triggered lipoxins, were the first mediators recognized to have both endogenous anti-inflammatory and proresolving actions, indicating that not all lipid mediators are "bad guys" in controlling inflammation and in impeding resolution. In contained sites of inflammation, lipoxins are temporally dissociated from other proinflammatory mediators such as the prostaglandins and leukotrienes that are biosynthesized in the initial steps of the acute inflammatory response. Two new families of compounds identified within the resolution phase are called resolvins (Rv) and protectins generated from omega-3 polyunsaturated fatty acids (eicosapentaenoic acid and docosahexaenoic acid).

The resolution of inflammation and the return of tissues to homeostasis were widely held as passive events until the characterization of these novel biochemical pathways and lipid-derived mediators that

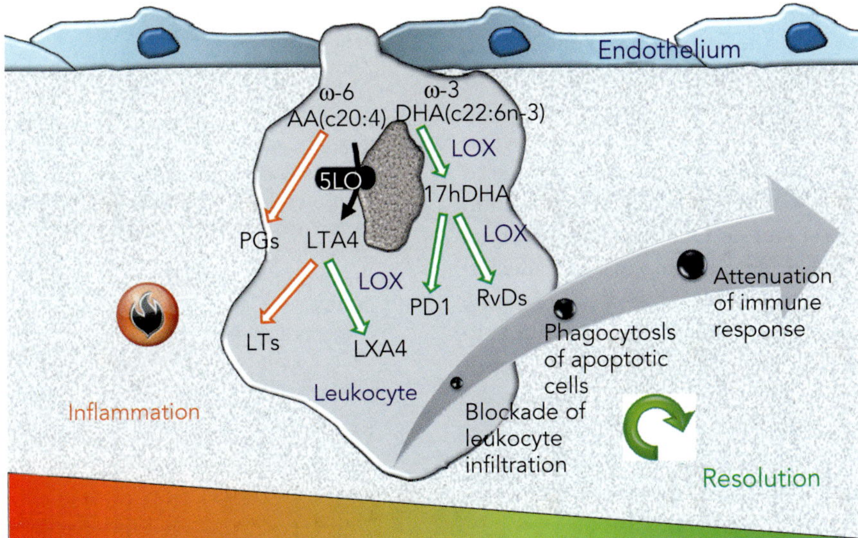

Figure 32.1. Model of atherosclerosis as a nonresolving form of vascular inflammation. The essential ω-6 polyunsaturated fatty acid (PUFA) arachidonic acid (AA) is released from phospholipids in cells by the action of cytosolic phospholipase A2. The free AA is sequestered at the nuclear envelope and brought into contact with 5-LO by an accessory protein named 5-LO-activating protein (FLAP). After specific enzymatic steps, AA is converted into different families of mediators: prostaglandins (PGs) and leukotrienes (LTs), which are mostly proinflammatory molecules, and lipoxins such as LXA4, a stop-inflammation mediator. The essential ω-3 PUFA DHA is converted to two novel mediators, RvD1 and PD1, which promote resolution. We propose a model in which deficiency of macrophage 12/15LO leads to a deficiency in proresolving end products, RvD1 and PD1 as well as LXA4, locally at the site of the ongoing inflammation, crippling multiple proresolving functions, leaving the proinflammatory milieu unabated, and fueling atherosclerosis progression. LOX, lipoxygenase.

are actively turned on in resolution and that possess potent anti-inflammatory and proresolving actions. Given the potent actions of lipoxins, resolvins, and protectins in models of human disease, deficiencies in resolution pathways may contribute to many diseases and offer exciting new potential for therapeutic control via resolution.

The production of these specific lipid-derived mediators is initiated by lipoxygenases. Experiments with animals and humans support a proinflammatory role for the 5-lipoxygenase system. Recent biologic and genetic findings implicate the 5-LO pathway in atherosclerosis. Mehrabian et al. reported that heterozygotes for the 5-LO gene on the LDLr–/– background had considerably reduced aortic lesions, compared with the advanced lesions observed in 5-LO–/–LDLr–/– mice, despite equivalent hypercholesterolemia. The 5-LO pathway also promotes pathogenesis of hyperlipidemia-dependent aortic aneurysm.

In contrast, results from animal experiments show a range of responses with the 12/15-lipoxygenase pathways in atherosclerosis. To date, the only two clinical epidemiology human studies both support an antiatherogenic role for 12/15-lipoxygenase downstream actions. Very recently, we tested the hypothesis that atherosclerosis results from a failure in the resolution of local inflammation by analyzing apolipoprotein E–deficient mice with (i) global leukocyte 12/15-lipoxygenase deficiency, (ii) normal enzyme expression, or (iii) macrophage-specific 12/15-lipoxygenase overexpression. Results from these tests indicate that 12/15-lipoxygenase expression protects mice against atherosclerosis via its role in the local biosynthesis of lipid mediators including lipoxin A_4, resolvin D1, and protectin D1 (Figure 32.1). These mediators exert potent agonist actions on macrophages and vascular endothelial cells that can control the magnitude of the local inflammatory response. Taken together, these findings suggest that a failure of local endogenous resolution mechanisms may underlie the unremitting inflammation that fuels atherosclerosis [14,15].

CONCLUSION

This chapter reviews the major mouse models used in atherosclerosis research. The development and application of apoE–/– and LDLR–/– mice, as well as other genetic mouse models, have enabled significant advances to be made in cardiovascular research. These targeted mouse models have played an important role and will continue to be an integral part in atherosclerosis research.

A major concern is how closely these models simulate human disease. Murine lesions have not, as yet, progressed to the stage of plaque rupture, with the exception of events seen in a small proportion of atherosclerotic mice. However, the striking similarities in the morphology of lesions formed in mice compared with specific stages of the human disease indicate that these models are good simulations that will continue to contribute to our understanding of the mechanisms involved in the atherogenic process. With the wide availability of mouse models, the revolution in genomics and functional genomics has the prospect of uncovering many more genes that are important in the atherogenic process.

REFERENCES

1. Tedgui, A., and Mallat, Z. 2006. Cytokines in atherosclerosis: pathogenic and regulatory pathways. *Physiol Rev* **86**:515–581.
2. Hansson, G.K. 2005. Inflammation, atherosclerosis, and coronary artery disease. *N Engl J Med* **352**:1685–1695.
3. Russell, J.C., and Proctor, S.D. 2006. Small animal models of cardiovascular disease: tools for the study of the roles of metabolic syndrome, dyslipidemia, and atherosclerosis. *Cardiovasc Pathol* **15**:318–330.
4. Daugherty, A. 2002. Mouse models of atherosclerosis. *Am J Med Sci* **323**:3–10.
5. Jawien, J., Nastalek, P., and Korbut, R. 2004. Mouse models of experimental atherosclerosis. *J Physiol Pharmacol* **55**:503–517.
6. Linton, M.F., Atkinson, J.B., and Fazio, S. 1995. Prevention of atherosclerosis in apolipoprotein E-deficient mice by bone marrow transplantation. *Science* **267**: 1034–1037.
7. Wouters, K., Shiri-Sverdlov, R., van Gorp, P.J., van Bilsen, M., and Hofker, M.H. 2005. Understanding hyperlipidemia and atherosclerosis: lessons from genetically modified apoe and ldlr mice. *Clin Chem Lab Med* **43**:470–479.
8. Zhang, S.H., Reddick, R.L., Piedrahita, J.A., and Maeda, N. 1992. Spontaneous hypercholesterolemia and arterial lesions in mice lacking apolipoprotein E. *Science* **258**:468–471.
9. Ishibashi, S., Brown, M.S., Goldstein, J.L., Gerard, R.D., Hammer, R.E., and Herz, J. 1993. Hypercholesterolemia in low-density lipoprotein receptor knockout mice and its reversal by adenovirus-mediated gene delivery. *J Clin Invest* **92**:883–893.
10. de Winther, M.P., and Heeringa, P. 2003. Bone marrow transplantations to study gene function in hematopoietic cells. *Methods Mol Biol* **209**:281–292.
11. Ross, R. 1999. Atherosclerosis – an inflammatory disease. *N Engl J Med* **340**:115–126.
12. Yan, Z.Q., and Hansson, G.K. 2007. Innate immunity, macrophage activation, and atherosclerosis. *Immunol Rev* **219**:187–203.
13. Mallat, Z., Gojova, A., Brun, V., et al. 2003. Induction of a regulatory T cell type 1 response reduces the development of atherosclerosis in apolipoprotein E-knockout mice. *Circulation* **108**:1232–1237.
14. Serhan, C.N., Yacoubian, S., and Yang, R. 2008. Anti-inflammatory and proresolving lipid mediators. *Annu Rev Pathol* **3**:279–312.
15. Merched, A., Ko, K., Gotlinger, K., Serhan, C.N., and Chan, L. 2008. Atherosclerosis: Evidence for impairment of resolution of vascular inflammation governed by specific lipid mediators. *FASEB J* **22**(10):3595–3606. doi:10.1096/fj.08–112201.

SUGGESTED READINGS

Hansson, G.K., Robertson, A.L., and Soderberg-Naucler, C. 2006. Inflammation and atherosclerosis. *Annu Rev Pathol Mech Dis* **1**:297–329.

Merched, A.J., Ko, K., Gotlinger, K.H., Serhan, C.N., and Chan, L. 2008. Atherosclerosis: evidence for impairment of resolution of vascular inflammation governed by specific lipid mediators. *FASEB J* **22**:3595–3606.

33 Oral Inflammation and Periodontitis

Alpdogan Kantarci, Hatice Hasturk, and Thomas E. Van Dyke

INTRODUCTION

The oral cavity is a complex organ system composed of salivary glands, tongue, tonsils, and teeth. The tissues of the oral cavity are either mineralized-hard (e.g., enamel, dentin, cementum, and bone) or soft tissues (e.g., mucosa, periodontal ligament, and gingiva), which altogether maintain a complex system of function and esthetics. The oral cavity is the entrance to and a major component of the gastrointestinal tract as the first site in the body to break down the food due to its masticatory function; it is also the gateway for respiration. Oral systems are crucial for proper phonation; the alignment of the teeth affects how words are pronounced. Facial esthetics is also directly associated with the shapes of the teeth and soft tissues surrounding them. With this unique blend of components and functions, the health and the disease of the oral cavity, therefore, present an important area of research as well as a challenge to the maintenance of general health. As in many diseases common to humankind, oral pathologies are associated with alterations in tissue homeostasis. Oral pathological conditions and diseases have been recognized as important health problems since the dawn of the early civilizations. For example, golden toothpicks found in Mesopotamia have indicated that Sumerians were practicing oral hygiene as early as 3000 B.C. Various herbal medications were used by Babylonians and Assyrians to "treat" periodontal problems. Egyptians, Chinese, and Indians all have written documentation of treating dental and periodontal inflammation, ulcerations, and abscesses. Greeks and Islamic physicians have also contributed to the understanding of oral problems as major health issues to be treated. As in other medical fields, dentistry has been elevated to a scientific and systematic understanding level in the eighteenth century. Pierre Fauchard published *"The Surgeon Dentist"* (1728) and covered all aspects of oral diseases and their treatments. John Riggs, an American dentist, pointed out the inflammatory nature of the periodontal disease in the nineteenth century. Dental science was closely affected by the discovery of "germs" by Louis Pasteur and Robert Koch at the end of the nineteenth century as well as new methodologies such as x-ray radiographs by Wilhelm Roentgen. "The Microorganisms of the Mouth" was published by Miller in 1890 and marked the introduction of modern microbiology into dentistry [1]. Today, we know that the oral cavity is highly populated with more than 400 known and cultivable bacteria [2–4] and presents a constant battleground for the host defense systems and the invader microorganism species. The understanding of oral pathologies is closely linked to elucidating the mechanisms of inflammatory changes in the oral tissues. Since the tissues of the oral cavity are unique in many structural, biological, and physiological aspects, this chapter will first briefly present the tissues of the oral cavity in "health" and then focus on the inflammatory processes that underlie the pathological conditions. The most complex of the oral tissues is the periodontium, including both the hard and the soft tissues in a functional system; therefore, the diseases of the periodontal tissues will be used as a model for presentation of specific mechanisms of inflammation.

BIOLOGY OF TISSUES OF ORAL CAVITY AND PERIODONTIUM

The tissues of the oral cavity interact with each other and the rest of the mammalian body in a complex and integrated way. There are few areas of the body where biology meets function and esthetics as in the oral cavity. One of the main functions of oral tissues is mastication. Teeth are specialized structures, where each group has a different role. The tooth is the only hard tissue complex that erupts outside the soft tissues of the mammalian body. It is inserted into the specialized alveolar bone of the jaws. Alveolar bone is covered by layers of soft tissue composed of connective tissue and epithelia.

The oral cavity is lined with a highly specialized mucosa with distinct functional and structural characteristics. Oral epithelium is the continuation of the mucosa, which also lines the tongue and the palate. Outside the oral cavity, oral mucosa is continuous with the skin of the lips. Structurally, it is possible to distinguish the oral mucosa into three different areas:

1. Masticatory mucosa: The mucosa that covers the gingiva and the hard palate and is directly involved in the mastication process together with the teeth.
2. Tongue mucosa: The mucosa that covers the dorsum of the tongue.
3. Lining mucosa: The remaining mucosal surfaces of the oral cavity.

Saliva is a unique secretory fluid that is produced by the highly specialized salivary glands. There are two types of salivary glands: major and minor glands. The major role of saliva is to provide physical lubrication to soft and hard tissues of oral cavity. In addition to its cleansing and humidifying effect, saliva also provides an important protection in defensive mechanisms of the oral cavity. It is a rich source of sIgA and molecules such as histatins [5]. The gingival crevicular fluid, on the other hand, is an exudate that is secreted by the inflammatory cells and the epithelium into the crevice between the teeth and the soft tissues. Its volume increases proportionally with inflammation. In addition to its defensive functions, gingival crevicular fluid is also used to assess the inflammatory changes and the stage of inflammation during periodontal disease activity.

Dental Structures and Periodontium

A tooth consists of several compartments with distinct characteristics. These structures are lost because of periodontal inflammation (Figure 33.1). *Enamel* is the layer of the teeth that covers the external surfaces outside the gingiva and extends into the oral cavity. The main function of the enamel is to protect the other compartments of the teeth and resist mechanical and chemical forces. *Dentin* lies just below the enamel in the crown of the tooth and extends throughout the root. This is a continuous tissue with parallel tubuli and its main role is to connect the pulp of the teeth to the layers that cover the surfaces (enamel and cementum). The *cementum* layer can be considered a continuation of the enamel on the root surface. Cementum is attached to the alveolar bone by the fibers of the periodontal ligament. The *pulp* of the tooth is the source of vitality, and contains blood vessels and nerve endings through which the teeth are connected to the circulation and the nervous system of the body.

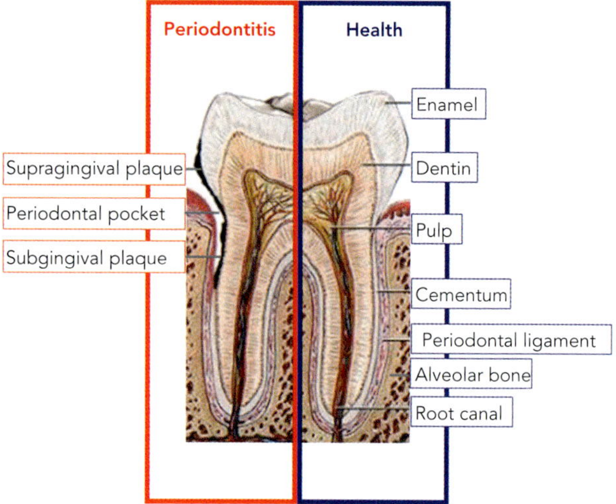

Figure 33.1. Dental and periodontal tissues in health and disease. The tooth is attached to the alveolar bone via periodontal ligament. The fibers of the periodontal ligament are inserted into the root cementum of the tooth. The surface of the tooth is covered with enamel. Dentin separates the pulp from the enamel and cementum. When periodontal inflammation reaches to the alveolar bone, bone resorption takes place and periodontal pocket is formed. Exposed root surfaces in periodontitis are covered with bacterial biofilm.

The periodontium is the organ complex that consists of the cementum of the root surfaces, the periodontal ligament, the alveolar bone, and the gingiva. The main function of the periodontium is to "attach" the tooth to the alveolar bone. Through this attachment, the teeth serve as the main masticatory organ. The periodontium, which consists of the hard tissues of the teeth and the alveolar bone and the soft tissues of the gingiva and the periodontal ligament, supports the integrity of the oral cavity. A complete dentition with a healthy periodontium is, therefore, essential to the homeostasis of all the organs in the oral cavity and the rest of the gastrointestinal tract. Dental and periodontal structures form a developmental, biologic, and functional unit and undergo changes throughout life. Although age itself is not a major determinant of these changes, it has a cumulative effect on dental health. In addition to life-related events, functional alterations and oral environment as well as genetic factors continuously affect the biology of the dental and periodontal structures.

Root Cementum

The cementum is a mineralized tissue covering the root surfaces of the teeth. It is a hard tissue similar to bone, and contains collagen fibers embedded in an organic matrix and a mineral content, which is mainly hydroxyapatite (65%). The principal fibers between the root cementum and the alveolar bone are called Sharpey's fibers and are the direct continuation of the Sharpey's fibers in the periodontal ligament. On the other hand, the cementum does not have any vascularization or nerves. Root cementum also does not show any alternating deposition or resorption layers. In contrast, there is a continuous increase of the thickness of the cementum throughout life due to deposition of new layers. Among the many unique functions of the root cementum, attaching the periodontal ligament fibers, which originate from the alveolar bone to the root, is the most important. This is a critical function during the repair process following the inflammatory damage.

Alveolar Bone

The alveolar bone is the part of the maxilla and mandible where teeth are inserted. Bundle bone of the alveolus is also referred to as "alveolar bone proper" and harbors the teeth. This part of the alveolar bone consists of the "alveolar socket" lined with a thin plate of cortex facing the root surfaces and a thick cortical bone under the gingiva. Alveolar bone proper is continuous with the alveolar process of the jaws and forms the basal bone plate. The area between the alveolar socket and the compact jawbone walls are lined by cancellous bone, which is also known as the spongiosa. The cancellous bone contains bone trabeculae. The shape of the trabeculae of the alveolar bone is directly affected by the occlusal forces. The development of the alveolar bone is parallel to the tooth eruption. The bone is attached to the tooth root cementum through the periodontal ligament and Sharpey's fibers and provides a solid support to the teeth. Apart from providing the major support for the teeth in the oral cavity, the main function of the alveolar bone is to distribute forces generated during mastication and occlusion. There are several types of cells in the alveolar bone. The cell that is responsible for bone formation is the osteoblast and for resorption, the osteoclast. There are also osteocytes, which are pyknotic osteoblasts reside in between the layers of mature bone with a major function of communicating through the canaliculi of the bone. As in other bone types elsewhere in the body, the alveolar bone is covered with the periosteum. The alveolar bone is consistently renewed throughout the life by resorption and apposition processes. These events are regulated by many functional demands such as mastication or tooth movement as well as pathological events. A typical response of the alveolar bone to the inflammatory injury is the increased bone resorption stimulated by the osteoclastic activity.

Gingiva

The gingiva is a part of the masticatory mucosa and covers the alveolar bone proper. The gingiva also surrounds the cervical portion of the teeth. There are two different compartments of the gingiva: (1) epithelium, and (2) connective tissue. Gingival epithelium may be differentiated as *oral epithelium*, which is the continuation of the mucosal surfaces and faces the oral cavity; *oral sulcular epithelium*, which faces the tooth and does not get into contact with the dental surfaces; and *junctional epithelium*, which provides a hemidesmosomal contact between the tooth and the gingiva. The oral epithelium is attached to the underlying connective tissue with indentations inserted and extending into the connective tissue. These ridge-like structures are called *rete pegs* and are present only at the oral epithelium. In healthy and "normal" sulcular and junctional epithelia, rete pegs are not observed; the basal membrane delineates epithelial connection to the underlying connective tissue. Rete peg formation in sulcular epithelium and the extension of these structures deeper into the connective tissue underlying the oral epithelium is an important histopathological distinction between the healthy and inflamed gingiva. The oral epithelium is a keratinized, stratified, and squamous epithelium. There are four distinct layers of the oral epithelium from connective tissue to the surface: (1) *stratum basale*, (2) *stratum spinosum*, (3) *stratum granulosum*, and (4) *stratum corneum*. These layers are delineated based on the degree to which the keratin-producing cells are differentiated. Sulcular and junctional epithelia lack the keratinization and therefore the *stratum corneum*. The cells of the *stratum basale* of the epithelium are cylindrical or cuboidal. This layer is in contact with the basement membrane, which separates the epithelium from the connective tissue and the cells of the *stratum basale* can undergo mitotic cell division. This is an important feature through which renewal of the epithelial layer takes place.

The predominant tissue of the gingiva is the connective tissue, which is also known as the *lamina propria*. The connective tissue is rich with fibers of collagen; fibroblasts are the main cell type of the gingival connective tissue; and the gingival blood and nerve supply is in the connective tissue. Apart from these structural components, gingival connective tissue is also the site of inflammation when periodontal health is lost. Inflamed gingiva harbors numerous inflammatory cells among the extracellular matrix components released by the fibroblasts. The gingiva is continuous with the alveolar mucosa, which is a form of lining mucosa and it is not considered to be involved in mastication directly. An easily recognizable border known as the mucogingival junction makes the distinction between the gingival epithelium and the alveolar mucosa. Except for the palate where mucogingival junction does not exist due to the coverage of both the alveolar bone proper and the alveolar process by the masticatory mucosa, this structure is important during the extension of the inflammatory changes beyond the gingiva and forms a border.

Periodontal Ligament

The periodontal ligament is a unique structure found in between the teeth and the alveolar bone. There is no other part of the human body where periodontal ligament is found. It is a highly vascular connective tissue, which surrounds the roots of the teeth and connects the root cementum to the alveolar bone socket wall. It is also continuous with the connective tissue of the gingiva and contains collagen fiber bundles that are highly specialized with various directions as a result of the functional and structural properties of the dentition. In addition to the structural components and the cells, the periodontal ligament is rich with blood vessels originating from the alveolar bone and the connective tissue of the gingiva providing the vascularization for the teeth through the pulp. Another important function of the periodontal ligament is to provide the nerve supply with receptors for occlusal forces, which allows the teeth respond to the impact of mastication and trauma. Thus, the periodontal ligament acts as a suspension for the teeth in alveolar socket creating a mechanical force distribution and playing an extremely critical role for the homeostasis of teeth, alveolar bone, and the other organs of the oral cavity. In an inflamed periodontium, the periodontal ligament recedes apically due to the bone loss and the teeth become mobile in the alveolar socket losing their capability to resist the forces of occlusion with the antagonistic teeth.

ETIOLOGICAL FACTORS OF ORAL INFLAMMATION

Periodontal inflammation presents a complex pathogenesis (Figure 33.2). The main etiological factor is the microorganism-induced dental biofilm. The host response to the microbes and their products or virulence factors determine the impact of tissue destruction. Major risk factors to this line of inflammatory process can be genetic or environmental (e.g., smoking).

Microbial Factors

With its diverse areas and functional properties, the oral cavity also presents one of the most complex microbial ecologies in the human body. There are more than

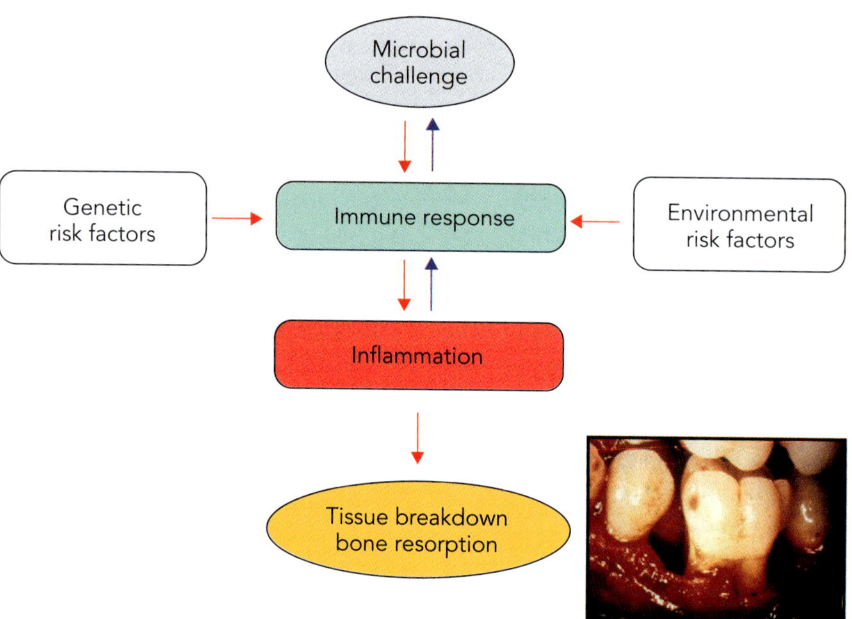

Figure 33.2. Pathogenesis of periodontal inflammation. The microbial infection precedes the activation of the host response and periodontal inflammation. There is a two-way interaction between the host and the microbial biofilm; increased inflammation as a result of activated immune response leads to further colonization of root surfaces by the microorganisms advancing the microbial front apically along the root while recruiting more immune cells to the lesion. Genetic and environmental risk factors determine the predisposition and severity of the inflammatory response in periodontal tissues. Because of periodontal inflammation, connective tissue breaks down and alveolar bone resorbs leading to the loss of teeth without any signs of decay.

400 bacteria species reported to date [2,4,6], and any individual may harbor 150 or more different species [4]. The oral cavity is the entrance to the pharynx and is named as the "oropharynx"; there is a dynamic equilibrium between the adhesion and removal of microorganisms through mastication, swallowing, breathing, tongue movement, washout by saliva, and motion of the ciliae. Survival of the microorganisms is dependent on this dynamic interaction. The formation of the bacterial biofilm on the soft and hard tissues of the oral cavity is the direct result of the adherence of the various bacterial species in a microecological system.

The oral cavity has several different ecosystems: hard surfaces (teeth, implants, and restorations), periodontal pocket, buccal and palatal soft tissue surfaces, tongue dorsum, and tonsils. These ecosystems serve as niches for biofilm formation and their physical and functional characteristics determine the type and the complexity of the biofilm. While most species, with the exception of spirochetes, have the capacity to colonize all these surfaces, bacterial adhesion to epithelial surfaces as well as the aerobic status of the niches determines the accumulation of the pathogenic species, which elicit different levels of inflammatory response.

The hard surfaces of the oral cavity present a unique example for microbial colonization in the human body. Since the teeth are the only vital hard structures of mammalians exposed to the environment, they provide a hard, nonshedding surface and allow the continuous and extensive accumulation of the biofilm. The teeth also present a unique ectodermal interruption where there is a unique seal between the external environment and the internal parts of the body. Therefore, in contrast to anywhere else in the human body, the hard tissue biofilm in the oral cavity is the major reason for the formation of local caries. Meanwhile, the interaction of the bacterial species in this complex ecosystem with the host creates a substantial inflammatory impact by eliciting the immune response at both the local and the systemic levels, which forms the basis of the generalized response to the periodontal and peri-implant inflammation. Based on this knowledge, teeth are considered one of the primary habitats of bacteria throughout the body and as a "port of entry" for many pathogens [7]. While cleaning of their surfaces or extraction of the teeth eradicates the caries-causing bacteria, the periodontal pathogens are not completely eliminated and the "sterility" of the oral cavity is virtually impossible. This suggests the importance of the host in creating a suitable environment for the bacteria to colonize and survive in the oral cavity.

The hard tissues are the major sites for biofilm formation in the oral cavity and the source for bacterial growth. Studies have also demonstrated that various species could adhere to and colonize the soft tissues of the mouth [8,9]. This is the result of the direct bacterial adhesion to the epithelial cells. While the same mechanism applies to all other soft tissues of the human body, the high turnover of the epithelium in the oral cavity due to masticatory forces and washout effects provided by the saliva and gingival crevicular fluid prevents massive accumulation of microorganisms. However, these natural cleansing mechanisms are insufficient once pockets form around the teeth and implants where a more anaerobic environment is created and the integrity of the epithelial lining in the sulcus (which is a nonkeratinized epithelium) is disrupted. Therefore, the weakest link in the oral cavity for infection is the periodontal or peri-implant area and the extended impact of periodontal inflammation, therefore, can impact systemic health due to elicited immune responses.

The biofilm forms on the hard tissues of the teeth or the implant surfaces after the salivary proteins accumulate and create a "sticky" protective layer, which is known as the pellicle. The pellicle formation is the first step in the adhesion of the bacterial species [10]. If left uncleaned, the accumulation of the bacteria on the pellicle and further expansion of the film layer occurs. This biofilm is called "dental plaque." Dental plaque can be studied as supra- and subgingival plaque based on its localization on the tooth/implant surfaces. In a healthy oral cavity with no sign of periodontal destruction and pocket formation, dental plaque is supragingival. Accumulation of supragingival plaque results in a local inflammatory response called gingivitis where the inflammation is restricted to the soft tissues of the periodontium, namely the gingival epithelium and connective tissue. The major species that are associated with the supragingival plaque formation are Gram-positive rods and cocci such as *Streptococcus mitis*, *Streptococcus sanguis*, *Actinomyces naeslundii*, *Actinomyces viscosus*, and various *Eubacterium* species. It has been demonstrated that more plaque accumulation will lead to increased severity of gingivitis where the plaque advances the subgingival environment due to the relative deepening of the pocket as a result of the increased swelling in the gingival margin. During this progression, the plaque composition also shifts to a more Gram-negative and rod-dominated plaque [4]. While it is still unclear how and if the gingivitis progresses into periodontitis, all periodontitis lesions are preceded by gingivitis. The distinction between the periodontitis and the gingivitis is the extent of the lesion and the destruction of the hard tissues (alveolar bone), in addition to the gingival inflammation. Periodontitis is associated with subgingival plaque accumulation where the bacteria adhere to the root surfaces and the sulcus epithelium apically. The major species at this stage of the disease are *Streptococcus oralis*, *Streptococcus intermedius*, *Peptostreptococcus micros*, *Porphyromonas gingivalis*, *Prevotella intermedia*, *Tannerella forsythia*, and *Fusobacterium nucleatum*.

The environmental conditions of the subgingival region are different from those above the gingival margin. The crevice between the tooth surfaces and the gingiva is filled and washed by the gingival crevicular fluid, which also supplies nutrients for the bacteria. The redox potential of the subgingival region is also lower than the rest of the oral cavity. This feature provides an anaerobic environment where more pathogenic bacteria can colonize. It is not clear if the anaerobic subgingival environment is a consequence of the bacterial activity, but this habitat is certainly a beneficial one for the progression of periodontal infection. Since strictly anaerobic bacteria cannot survive outside the pocket, subgingival colonization of the periodontium by these pathogens is secondary to the initiation of the infection, and there is accumulated evidence suggesting that the host response may provide a favorable environment for such a population of strictly anaerobic bacteria to grow below the gingival margin, signifying a shift from gingivitis to periodontitis.

As mentioned earlier, the dental plaque is considered as a "biofilm." Similar to the other parts of the body, the biofilm refers to an organized structure rather than a random accumulation of the microorganisms [11]. Microcolonies of different species of bacteria as well as other microorganisms (e.g., viruses) are nonrandomly situated in a matrix, which is mainly polysaccharide in structure. The lower layers are denser and the microorganisms are firmly bound. A looser layer covers the base of the biofilm. This layer is more irregular and mixed with the fluids of the oral cavity as it becomes more superficial. The nutrients penetrate the biofilm through diffusion. The oxygen content of the deeper layer is substantially lower compared to the surface, facilitating an abundance of anaerobic species as the biofilm becomes thicker. In addition to the polysaccharides, the matrix of the dental biofilm is rich with proteins, glycoproteins, and lipids as well as inorganic compounds such as calcium, phosphorus, and trace amounts of sodium, potassium, and fluoride. The source of the organic components is the host and bacterial cells and fluids such as saliva and crevicular fluid. The inorganic components derive mainly from saliva and gingival crevicular fluid. While saliva provides most of the compounds for the supragingival plaque, the serum transudate of crevicular fluid is the source for the subgingival minerals.

Host-Related Factors

Owing to its highly specialized structure and function, and in response to extremely high levels of bacteria even during good health, the oral cavity is a unique immune system. The dynamics of this immune response determine the progression and the impact of the inflammatory changes in oral tissues. As in the other parts of the mammalian body, the oral cavity has both innate (non-specific) and acquired (specific) immune mechanisms to respond to microbial agents. Polymorphonuclear granulocytes (= neutrophils, PMN) and mononuclear phagocytes form the first line of defense as the cells of immune system in oral tissues. These cells are accompanied by humoral components, such as complement proteins, which are directly involved in killing bacteria and facilitate their phagocytosis by PMN and mononuclear phagocytes. In addition to the phagocytic cells, lymphocytes play a role in the advancement of periodontal lesions into chronicity. Similar to other pathologies of inflammatory nature, oral inflammatory diseases are characterized by an abundance of lymphocytes and plasma cells, as well as antibodies against various bacterial species.

In addition to these cellular and humoral components of the immune system, the epithelia of mucosal surfaces in the oral cavity form a strong barrier against invasion by bacteria, viruses, mechanical trauma, and various injuries. The highly specialized oral epithelium also has a high turnover capacity and it regulates the recognition of various bacterial products through specific receptors (e.g., Toll-like receptors, TLR). While these receptors are not unique to the oral epithelium or epithelial cells in general, pathogen recognition through the interaction of these receptors with the surface ligands on the bacteria initiate a series of immune mechanisms carried out by the activity of the cellular components of the oral immune defense system. In addition to the recognition of pathogenic organisms, oral epithelial cells are also directly involved in defense mechanisms through the release of antimicrobial peptides and proteins (e.g., BPI, LL-37). Finally, the oral epithelium regulates the adherence of the immune cells and their chemotactic migration.

Resident cells of the oral tissues are the first cellular structures exposed to pathogen-associated molecular patterns after the epithelial barrier is crossed with these structures. TLR are the major pattern-recognition receptors identified in various oral tissues for pathogen-associated molecular patterns. These patterns or ligands for TLR include the lipopolysaccharides, outer membrane proteins, and fimbriae from various bacterial species. While most TLR are expressed in oral tissues, TLR-2, TLR-4, and TLR-9 have been linked to functional properties in ligand-specific activation of the oral defense mechanisms [12,13]. TLR-ligand interaction results in the activation of other surface molecules and signaling events (e.g., MyD88) leading to the release of proinflammatory cytokines such as TNF-α and IL-1β. Most of the work on TLR-mediated recognition of ligands of oral bacteria origin have focused on and identified pathways of regulation in macrophages. In addition to the macrophages, neutrophils and fibroblasts of the oral connective tissues and periodontal ligament also express TLR and can also contribute to the progression of inflammation by IL-1, IL-6, and IL-8 release when stimulated by bacterial ligands.

Macrophages, which are usually found at low levels in healthy oral tissues, are directly activated in inflammation. Apart from presenting antigen to the lymphocytes and advancing the inflammatory response to chronicity, oral macrophages contribute to the inflammation by releasing proinflammatory cytokines in response to bacterial or inflammatory activators, thus generating an acute-phase response. Acute-phase response, in turn, generates a systemic reaction and increases many key inflammatory proteins such as C-reactive protein (CRP). CRP levels in patients with periodontitis are increased, similar to the levels of those with cardiovascular diseases [14–16]. Periodontitis also leads to significant and substantial increases in proinflammatory cytokine levels (e.g., TNF-α, IL-1) in gingival crevicular fluid. Proinflammatory cytokines are correlated with increased clinical attachment loss around the affected teeth and can be used as indicators of periodontal inflammatory activity. Macrophages also take part in the clearance of the apoptotic or necrotic cellular debris and direct clearance of bacteria through phagocytosis. Another major function of oral macrophages is to regulate the osteoclastic activity. This function is crucial for the destruction of oral tissues during the inflammatory response through the release of RANKL, IL-1, PGE_2, and TNF-α. In addition to the resident macrophages, monocytes from the peripheral blood are recruited to the site of inflammation in oral tissues. These migrating cells can be differentiated into macrophages or osteoclasts through the action of bacterial (e.g., LPS) or inflammatory (e.g., GM-CSF) ligands. Activated macrophages, in turn, generate a lymphocytic response through the activation of T cells.

Neutrophils play a pivotal role in oral inflammation [17]. These cells are normally not residential to the oral tissues and are recruited from peripheral blood upon injury. Until recently, defective neutrophil functions were believed to predispose a person to infection. However, there is a growing body of data implying that this presumption may not be valid. On the basis of recent findings, the neutrophil abnormalities in periodontal inflammation (specifically in localized aggressive periodontitis) are the result of a chronic hyperactivated or primed state of the neutrophil. The excess activity and release of toxic products from the neutrophils is responsible, in part, for the tissue destruction in chronic periodontal inflammation. Peripheral blood neutrophils from individuals with localized aggressive periodontitis subjects exhibit increased adhesion, reduced chemotaxis, reduced surface glycoprotein GP110, abnormal phagocytosis and killing, and increased superoxide generation both in resting and stimulated cells. Several signal transduction anomalies are linked to the functional alterations in neutrophils. For example, the later phase of the calcium response associated with membrane channel activation and an influx of extracellular calcium is compromised, correlating strongly with defective chemotaxis. Total calcium-dependent PKC activity of neutrophils is lower. Neutrophils from subjects with localized aggressive periodontitis exhibit a marked increase in diacylglycerol (DAG) in both stimulated and unstimulated cells, and this increase is associated with a pronounced decrease in DAG kinase activity, all of which suggesting that the cells exist in a "primed" or pre-activated state.

The final stage of the oral and periodontal immune response is characterized by the recruitment and differentiation of the lymphocytes. T cells are usually recruited through the action of macrophages and their biological mediators while B cells can be directly stimulated with pathogen recognition molecular patterns. Their main function is to generate antigen-specific antibodies after being differentiated into plasma cells in oral tissues.

Genetic Factors

Early studies on the pathogenesis of oral inflammation presented the paradigm that the process was exclusively the result of the microbial infection and that inflammatory pathologies of the oral cavity are due to the lack of appropriate removal of such infectious agents. These concepts were fulfilled with findings that once the etiological factors were removed through oral hygiene practices, "most" of the disease is "reversed" or "prevented." While hygiene is extremely important for the prevention of disease in the oral cavity, similar to the other parts of the body, the natural history of disease progression is not the same in every individual and in different populations [18]. The susceptibility to disease varies on an individual basis, affecting the incidence and progression of oral diseases. The infectious organisms are necessary to initiate the process; however, the severity of the disease is determined by the host's genetic factors, which regulate the defense mechanisms.

The theory that familial heritage may influence disease patterns and that there may be susceptibility in certain groups is not a new concept. A complex disease such as periodontitis and oral inflammatory conditions linked to genetic markers on the other hand have not been an easy field of research in which to identify specific traits. With the development of new tools to map and identify various genetic factors, which determine the natural course of inflammation in individuals and populations, substantial progress has been made over the past decade. These studies demonstrate the complexity of the problem while some patterns have been associated with inflammatory conditions.

Studies on the genetic basis of oral inflammation can be exemplified by aggressive periodontitis. Aggressive periodontitis shows familial aggregation [19]. In some populations, an autosomal dominant transmission has been shown, while within the same or other populations it has been demonstrated that the transmission of aggressive periodontitis is autosomal recessive, demonstrating the population specificity of the disease. Because of their role in human immune responses and already established association with many other autoimmune disorders, human leukocyte antigens (HLA) have been considered risk markers for aggressive periodontitis. Studies have demonstrated that class I and II HLA located on chromosome 6 co-clustered with important inflammatory markers such as TNF-α and abnormal HLA coding could account for inflammatory conditions including periodontitis. Similar to aggregation studies, the results suggest that population-specific HLA distribution may predispose individuals differently to aggressive forms of periodontitis. Another line of evidence that aggressive periodontitis could be linked to genetic basis comes from linkage analyses. These studies have demonstrated that the gene for aggressive periodontitis may be co-localized on chromosome 4 (4q11–13) with *Dentinogenesis imperfecta* [20] or by itself on chromosome 1 (1q25) [21]. Such studies have been recent, and chromosomal locations need to be narrowed to the region of interest. Aggressive periodontitis can also be the oral manifestation of various genetic diseases such as leukocyte adhesion deficiency, Chediak–Higashi syndrome, and Ehlers–Danlos syndrome (Types IV and VIII). One of the best well-characterized genetic disorders linked to aggressive periodontitis is the Papillon–Lefèvre syndrome, which affects the periodontal tissue integrity in both primary and secondary dentition and is caused by mutations in cathepsin C gene located on chromosome 11 [22].

Many genetic factors play a role in the homeostasis of oral tissues. A recent review has shown that epigenetics is becoming an emerging field in inflammation [23]. Gene tolerization because of transient silencing of genes and priming for activation by genetic regulation may lead to the prevention of pathologies. Oral inflammation and its association to epigenetics is an emerging field, while very little has been known about how the dynamic equilibrium regulates this interaction.

PATHOBIOLOGY OF ORAL INFLAMMATION

The lesions of the oral, and in particular, periodontal tissues are chronic inflammatory lesions. The most common form of oral inflammation is gingivitis, which refers to the inflammation of the gingiva, limited to the soft tissues of epithelium and connective with no destruction in the alveolar bone or the periodontal ligament [24]. The clinical symptoms of gingivitis are reflective of typical characteristics of the inflammation with erythema, edema, swelling, and bleeding. Different stages of gingivitis have been defined as the "initial," "early," and "established" lesions, which reflect histopathological observations at different levels of gingival inflammation [25]. In essence, these changes are the result of vascular inflammation and cellular infiltration leading to the destruction of the extracellular matrix and loss of collagen in the gingiva. Most importantly, a distinct shift in the composition of the microbial biofilm occurs in parallel with these histopathological changes in the gingiva [26]. The plaque eventually becomes more dominated by Gram-negative anaerobic species as the gingivitis evolves from the initial lesion into an established inflammatory lesion. This observation from the experimental gingivitis in humans and models in animals led to the paradigm that altered microbial flora is the direct inducer of the progression of inflammatory changes in the gingival tissues, eventually leading to the development of periodontitis. In addition to the plaque-associated forms of gingivitis, gingival inflammation may also be the result of hormonal changes (e.g., puberty or menstruation), medication use, systemic diseases, or fibrotic side effects of certain drugs. These forms of gingival inflammation are histopathologically similar to the plaque-associated forms and resemble other types of inflammatory changes elsewhere in the body.

Advanced inflammation in periodontal tissues affects deeper components of the periodontium. The most common form of such changes is the chronic periodontitis, which involves the periodontal ligament destruction and the loss of alveolar bone in addition to the gingival inflammation. This is the form of oral inflammation, which eventually leads to the increased mobility of the teeth and their loss. The inflammatory lesion in periodontitis is similar to advanced gingivitis dominated by vascular changes and cellular infiltration with the abundance of plasma cells. Therefore, it was originally called an "advanced" lesion [25]. Although all periodontitis lesions are the extension of gingival inflammation to the alveolar bone and the attachment apparatus, not all gingivitis cases progress into periodontitis. Therefore, the linearity of disease progression from gingivitis to periodontitis is questionable. The microflora changes in the dental biofilm however, are substantial as seen in the later stages of gingivitis compared to the initial lesion.

In addition to the loss of alveolar bone, periodontitis is characterized by clinical attachment loss between the teeth and the alveolar bone and pathological pocket formation as a result of the extension of the sulcus around the teeth. The periodontal pocket is usually filled with extensive accumulation of subgingival plaque and calculus. The clinical pattern of periodontitis is characterized by phases of active and passive progression of inflammation. During the active phases, the bleeding is common and progression of the inflammation is rapid while the passive phases represent chronicity with slower or moderate advancement of the lesion. The limiting factor in periodontitis progression is the apical extent of the sulcular epithelium, which becomes the "pocket epithelium."

Chronic periodontitis is immunologically characterized by complement activation through alternative pathway and the formation of antigen-specific antibodies against microbes and their components. The destruction of the periodontal tissues is associated with increased matrix metalloproteinase (MMP) activity. This family of enzymes is released by multiple cellular sources such as the neutrophils and macrophages. Some periodontopathogen microbes also have the capacity to modulate the host-mediated MMP release. Other soluble factors released from cells of the oral tissues during the inflammation include proteases, cytokines, and prostaglandins. Proteases and MMPs break down the collagen structure. As the inflammatory response progresses, the epithelial cells apically migrate along the root surface and form the "periodontal pocket." The migration of the epithelium is accompanied by the infiltration of leukocyte-dense inflammation, further activating the osteoclastic activity and bone resorption. As the destructive pattern continues, the accumulation of subgingival plaque increases and biofilm widens on the exposed root surfaces, which further propagate the destructive lesion. Eventually, the periodontal lesion results in the loss of the teeth.

Chronic periodontitis may also be the result of systemic factors (e.g., diabetes) or viral pathogens (e.g., HIV). These nonbacteria-associated forms of chronic periodontitis usually present a more aggressive progression with the contribution of non-dental–related factors. For example, HIV-periodontitis is the net outcome of the reduced host resistance to bacteria. In general, oral lesions are common in HIV-infected subjects. Even before presenting the full-blown clinical symptoms of AIDS, most individuals infected with HIV show symptoms of oral lesions such as necrotizing ulcerative gingivitis and periodontitis, ulcerations, secondary viral infection-associated lesions (e.g., *Candida albicans*, *Herpes simplex*), oral hairy leukoplakia, oral non-Hodgkin's lymphoma, and Kaposi's sarcoma. Highly active antiretroviral therapy has substantially reduced the incidence of inflammatory lesions; however, the availability of costly medications is not universal and there have been recent reports that the virus may be gaining resistance to the treatment protocols [27]. HIV-gingivitis is a persistent, linear, and erythematous gingivitis observed in HIV-positive individuals. This type of gingival inflammation is called linear gingival erythema (LGE) and may not respond to conventional treatment. The microflora in LGE resembles that of periodontitis rather than non-HIV gingivitis. LGE lesions may extend into the alveolar mucosal and may undergo spontaneous remission. A possible role for the *Candida* species has been proposed. Necrotizing forms of gingivitis or periodontitis may be an important clinical finding in HIV-positive subjects. These lesions typically present with ulcers interdentally and may extend into the alveolar bone, exposing the bone and leading to necrosis. Necrotizing ulcerative lesions of the gingival, periodontal, and other oral tissues are extremely painful and severely destructive. These conditions are usually identical to *Cancrum oris* (noma) and may be associated with severe immunodeficiency of suppressed CD4-positive helper T lymphocytes. The incidence and severity of chronic periodontitis are increased in HIV-infected subjects. The mechanism of the link between the virus and periodontal destruction, however, is not clear and confounding factors such as reduced oral hygiene have not been ruled out.

Aggressive periodontitis is the second-most-abundant form of periodontal destruction, with an onset at early ages and rapid progression of the lesions. When certain teeth are affected by the aggressive periodontitis (incisors and first molars), it is referred to as "localized aggressive periodontitis (LAP)." If there are no specific localization of the aggressive periodontitis, and the lesions impact all areas of dentition, this form is referred to as "generalized aggressive periodontitis (GAP)." The linearity between LAP and GAP is not clear and LAP is linked to multiple genetic and immunologic defects. For example, LAP is considered to be a "neutrophil-mediated tissue injury" and presents with primed and hyperactivated neutrophil phenotype [17]. Several signaling defects have been associated with the phenotypic changes in LAP neutrophils and genetic predisposition has been well established in certain populations. While retaining its multi-microbial etiology, LAP is also strongly associated with the presence of *Aggregatibacter actinomycetemcomitans*. In this sense, LAP represents a different disease phenotype than the GAP since such a distinctive microbial flora has not been associated with GAP.

In addition to these forms of gingivitis and periodontitis, necrotizing forms of periodontal diseases are examples of oral inflammatory conditions. These lesions are rapidly developing lesions affecting specific locations of dentition. Necrotizing lesions may be limited to gingivitis or may involve the alveolar bone (necrotizing periodontitis). Necrotizing forms of gingival inflammation are characterized by ulcers, pain, and bleeding as well as fever and lymphadenopathy. Together with the abscesses, these lesions are the only forms of periodontal inflammation that lead to the progression of painful oral lesions of soft tissues. In addition to the pain, abscesses of periodontal tissues are accompanied by pus formation. While *Fusobacterium* species are associated with necrotizing lesions, overall, the same anaerobic bacterial species are common in these forms of "acute" periodontal diseases.

Apart from these most common inflammatory lesions of the oral cavity involving the periodontal tissues, pulpal inflammation as a result of caries-forming bacteria affects the vitality of the teeth. These lesions, collectively referred to as the endodontic inflammation, occur in a relatively closed environment within the pulp chamber. Since the only natural opening to the pulp chamber is the apex of the tooth, any inflammatory lesion advancing from the hard tissues of the teeth (dentin, enamel) rapidly leads to an acute inflammation and eventually to the loss of the vitality of the teeth. Left untreated, pulpal inflammatory conditions progress through the apices of the teeth and result in peri-apical abscesses. These lesions are dominated with inflammatory cellular and extracellular exudate and are considered to be dental emergencies.

RESOLUTION OF ORAL INFLAMMATION

Oral diseases are inflammatory processes in which microbial etiological factors induce a series of host responses that mediate an inflammatory cascade of events in an attempt to protect and heal the periodontal tissues. The progression of periodontal disease and its commonality with other systemic disorders such as cardiovascular disease and diabetes is based on the inflammatory events underlying the pathogenesis. Recent work has shown that in addition to the "on" signals that initiate the inflammatory events, periodontal tissues are capable of generating "off" or "stop" signals as checkpoint controls in inflammation (Figure 33.3) [28]. These control mechanisms are specific resolving cellular and biochemical circuits that have evolved to activate resolution, thus limiting the uncontrolled dissemination of inflammation. Several common inflammatory pathways of pathogenesis in chronic periodontitis have already been defined where neutrophils play an important role as an inducer in conveying the inflammatory processes. These neutrophil-mediated pathways, which could very well be shared by the other phagocytes (such as monocytes)

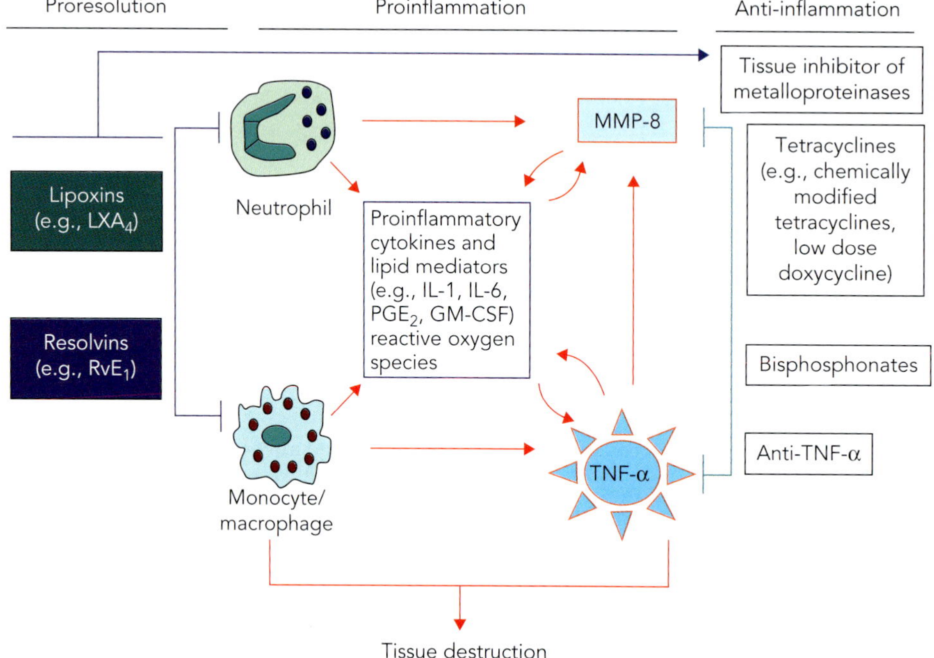

Figure 33.3. Oral inflammation and resolution. Resolution of inflammation differs from "suppression" or "blockage" of inflammation in the sense that it is an active and natural process through which the body restores its homeostasis. "Anti-inflammatory" strategies to stop the oral inflammation utilize several inhibitors of inflammatory processes at various stages of inflammatory response resulting in side effects due to the blockage of resolution circuits. (Reproduced from [28] with permission).

and cells of the adaptive immune system, could be used to control and manage inflammatory processes. Likewise, findings from LAP as a disease model where neutrophil responses are genetically altered show that this form of periodontitis represents a valuable model in which neutrophil-mediated tissue destruction follows similar routes. In addition to the traditional pro- and anti-inflammatory mechanisms, novel endogenous lipid mediators (lipoxins and resolvins) represent a "pro-resolution" phase in limiting the severity and duration of inflammation. Blockage of proinflammatory pathways could be accomplished not only by immune modulation therapy (e.g., anti-TNF-α), bisphosphonates, tetracyclines (e.g., chemically modified tetracyclines, low-dose doxycycline), or activation of TIMPs, all of which have side effects or short-comings, but also through shutting down of the cellular and molecular circuits endogenously by lipoxins (e.g., lipoxin A_4) or resolvins (e.g., resolvin E_1). Proresolution has a net advantage over the traditional methods by suppressing the excessive cellular activity [29].

In an attempt to identify the role of lipid-based resolution agonists in periodontal inflammation, transgenic rabbits overexpressing 15-lipoxygenase-type 1 enzyme and their response to inflammatory challenge were studied [30]. Skin challenges with either LTB_4 or IL-8 showed that 15-LO transgenic rabbits give markedly reduced neutrophil recruitment and plasma leakage at dermal sites with LTB_4, compared to responses in wild-type rabbits. Neutrophils from transgenic rabbits also exhibited a dramatic reduction in LTB_4-stimulated granular mobilization that was not observed with peptide agonists. In 15-LO transgenic rabbits, *Porphyromonas gingivalis*–induced periodontitis and leukocyte-mediated bone destruction were markedly reduced and local inflammation was not observed compared to wild-type rabbits. Since enhanced lipoxin production was associated with an increased anti-inflammatory status of 15-LO transgenic rabbits, a stable analog of LXA_4 was applied to gingival crevice subject to periodontitis. Topical application of the 15-epi-16-phenoxy-parafluoro-lipoxin-A_4 stable analog (ATLa) dramatically reduced leukocyte infiltration, and "prevented" bone loss as well as tissue inflammation monitored by radiological and histologic parameters. Together these results indicated that overexpression of 15-LO type I and lipoxin A_4 is associated with dampened neutrophil-mediated tissue degradation and bone loss and suggested that enhanced endogenous anti-inflammation is an active process in periodontitis. The enhanced anti-inflammation status suggested that 15-LO autacoids such as lipoxins could be targets in diseases where inflammation and bone destruction are pathogenic features as in periodontitis and arthritis.

Porphyromonas gingivalis–induced periodontitis and leukocyte-mediated bone destruction were used to test the resolution efficacy of resolvin E1, an example of bioactive products of omega-3 fatty acid transformation circuits initiated by aspirin treatments that counter proinflammatory signals [31]. Resolvin E1 specifically binds to human neutrophils at a site that is functionally distinct from the LX receptor. Topical application of resolvin E1 in rabbit periodontitis dramatically prevented the inflammation-induced tissue and bone loss associated with periodontitis similar to the impact of topical LXA_4 (Figure 33.4).

Following this observation, the potency of resolution agonists in "treatment" of inflammatory diseases was tested. *Porphyromonas gingivalis*–induced periodontitis and leukocyte-mediated bone destruction was again used as the model since this model presented a disease profile similar to human periodontitis, initiated by the bacteria and advanced by the host's inflammatory mechanisms. The treatment of established periodontitis using resolvin E1 as a monotherapy in rabbits was compared to structurally related lipids PGE_2 and LTB_4. PGE_2 and LTB_4 each aggravated the development of periodontitis and worsened the severity of disease. Promotion of resolution of inflammation as a therapeutic target with resolvin E1 resulted in complete restoration of the local lesion, and reduction in the systemic inflammatory markers CRP and IL-1β. The results for the first time demonstrated that tissue homeostasis can be fully restored, the lost bone can be fully "regenerated," the periodontal attachment between the teeth and the alveolar bone can be completely recovered, gingival inflammation can be completely resolved, and the mobility of the teeth can be "cured" by using the topical nanomolar range resolvin E1 (Figure 33.5) [32]. This critical finding is important beyond the dental and periodontal applications due to the powerful demonstration that a very complex tissue such as periodontium with both hard and soft tissues can be fully regenerated through the agonists of "resolution of inflammation," which direct the inflammatory process toward restoration of tissue homeostasis.

ANIMAL MODELS FOR PERIODONTAL INFLAMMATION

As in other fields of biological research, clinical questions are key to understanding the pathological processes in the oral cavity. Animal models are necessary to test in vivo mechanisms underlying the oral pathologies and to prove cause-and-effect relationships. Animal models are also used to identify the pathways of mechanism of oral pathologies and inflammation, and to test the potential of novel therapeutics. Animal models also answer the limitations of human studies.

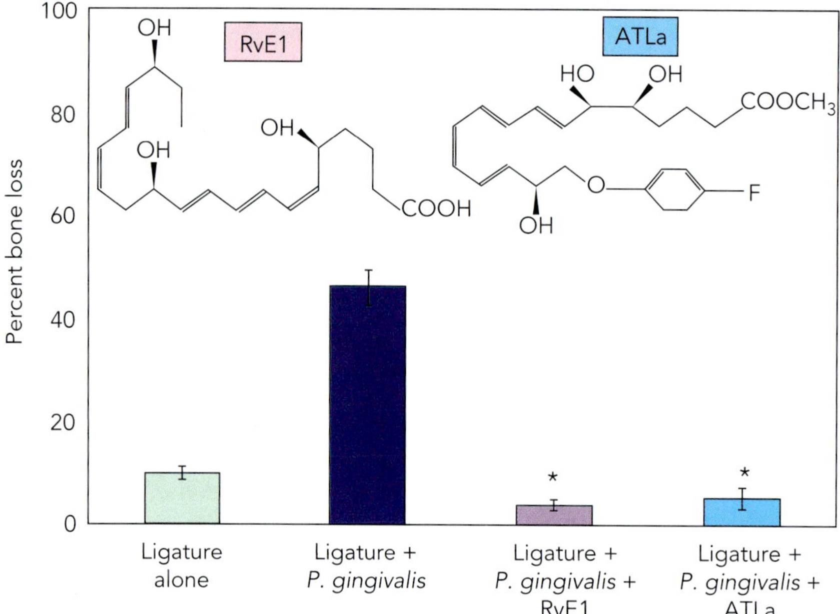

Figure 33.4. Prevention of periodontal inflammation. Pathogen (*Porphyromonas gingivalis*)-induced experimental periodontitis can be prevented by topical application of resolution agonists (e.g., LXA_4, RvE1) in rabbit model. (Reproduced from [28] with permission).

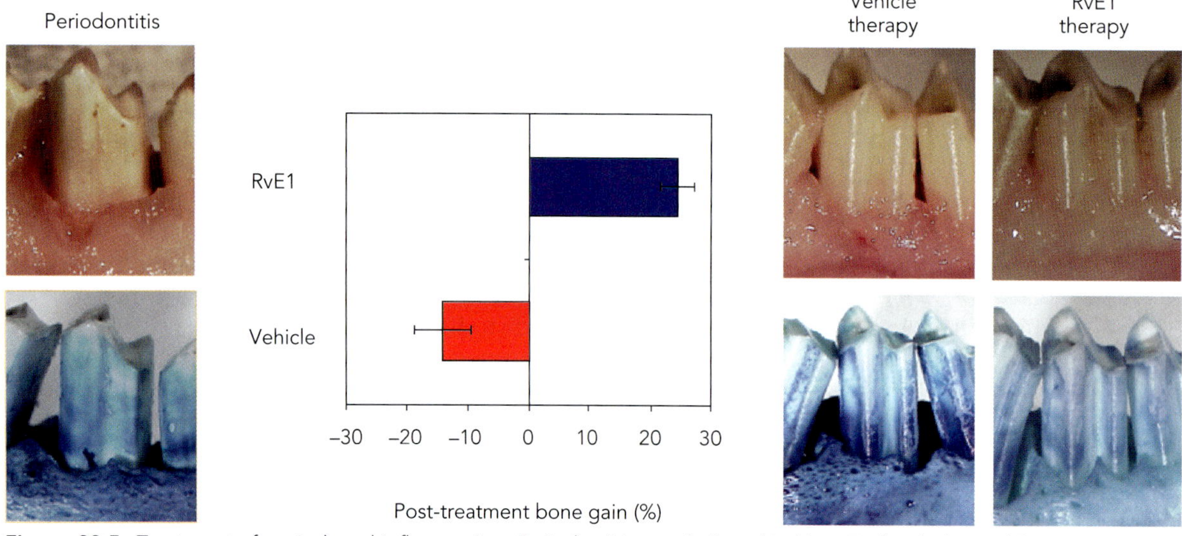

Figure 33.5. Treatment of periodontal inflammation. Periodontitis was induced in New Zealand white rabbits with 3–0 silk ligature and topical application of the human periodontal pathogen *Porphyromonas gingivalis* for 6 weeks with all characteristics of human periodontitis including soft and hard tissue destruction. Following 6 weeks of treatment, topical application of ~4 μg of RvE1 three times per week resulted in restoration of tissues to health, while the periodontal disease progressed in the vehicle-therapy group. Analyses of direct measurements on defleshed jaws revealed that RvE1 restored the lost bone (~95%), while the vehicle-alone treated group showed further bone loss, respectively.

Similar to all the other animal models of human disease, animal models of oral inflammation and periodontitis are far from perfect and there exists no single animal model completely applicable to all oral and periodontal diseases. For example, in the case of experimentally inducing periodontitis in animals using human bacteria, there is still controversy about which bacteria actually initiate the events that lead to irreversible breakdown of periodontal tissues in humans [3,33]. There is only indirect evidence that the defined set of "periodontopathogen" bacteria causes disease. Similarly, it is not known whether they represent important constituents in a polymicrobial infection, and if the infection is polymicrobial what other bacteria participate. The choice of bacteria has important implications as does the decision to study a single or polymicrobial infection. Therefore, it is difficult to reproduce the human disease in an animal while this represents still an important transition between benchtop in vitro experimentation and human clinical trials for many newly developed therapeutic agents.

As demonstrated in the experiments summarized earlier, the rabbit presents an excellent model to study the inflammatory changes in human periodontitis. The original studies on rabbits have focused on the atherosclerotic changes and cardiovascular inflammatory changes. Since the animal proved to be a suitable model to evaluate the activation of inflammatory cascades, periodontal inflammation has been induced to test its availability to human periodontal disease research [34]. The rabbit microflora is not pathogenic as demonstrated by the lack of periodontitis development when only ligatures are applied. This distinct feature, in addition to the comparably larger size of the rabbit than the rodents, presents it as a model to induce the periodontitis as seen in humans by using the pathogenic bacteria of human origin. The results from the studies on rabbit experimental periodontitis demonstrated that the rabbit is not a sterile animal. The species of the rabbit oral microflora cross-react with DNA from human bacteria and with the introduction of human pathogen (e.g., *Porphyromonas gingivalis*) the non-pathogenic species increase in the biofilm on the rabbit teeth. When the inflammation is resolved and periodontal tissue architecture is restored, the externally introduced human pathogen is cleared and the rabbit oral microflora is reverted to the baseline levels of detection [32]. This observation not only confirms the complexity of the human periodontitis, but also demonstrates the interaction between the microbial structures and the host inflammatory processes in the development of the diseases.

While small animals are suitable to study the inflammatory processes, research on alveolar bone regeneration around the teeth requires larger animals for testing various compounds in combination with resolution agonists. No studies have been performed to this end yet. Historically, dogs have been extensively used to study the bone regeneration in periodontal treatment. This model provided considerable data on how newly developed grafting materials as well as other regenerative strategies such as barrier membranes, root surface conditioning, mechanical treatment, antibiotics, and the combination of such techniques would result in restoration of the lost periodontal architecture. Periodontal defects were either "created" by mechanical means (e.g., surgical burs) or through the introduction of various bacterial species. Dogs, therefore, provided a valuable model for studying both the acute and chronic effects of periodontitis and the findings were immediately transferable to humans. While the animal findings were not always reproducible in humans, the data obtained through studies on dogs showed that the "periodontal regeneration" is feasible. The suitability of dog as a model to study the periodontal regeneration is hindered by the reduced availability of dogs for biological research over the recent decade.

A swine model of periodontal regeneration has also been developed. This model represents a prospective system where similar inflammation pathways to human disease processes can be combined in a large enough animal to study the impact of resolution of inflammation on the regenerative processes in periodontium. For example, ligature-induced inflammation can be reproduced around the dental implants in pig jaws [35]. Preliminary work has shown that while the alveolar bone architecture is considerably different in swine with larger marrow spaces and a more fragile cortical plate than human jaws, the bone deposition dynamics and the attachment apparatus of periodontium are comparable to humans (Figure 33.6). Indeed, to better understand the structure and function of the junctional epithelium, early studies on pig model, an omnivore whose dietary and dental development and occlusion patterns are similar to the human condition, and which, because of its size, is more readily amenable to experimental manipulation have focused on the anatomical aspects [36]. These findings were also compared to the well-described epithelium in the rat. These studies demonstrated that the porcine junctional epithelium has predictable morphological and biochemical features, which established the pig as an advantageous model to study the basic and clinical biology of this unique epithelium. Later studies from other centers also confirmed the suitability of the pig model to study periodontal regeneration. Specifically, stem-cell–mediated tissue engineering research has been the focus of such studies [37,38].

SUMMARY

As in other tissues of the mammalian body, inflammation is an active process in the tissues of oral cavity. The oral cavity and thus the inflammatory response of its tissues present unique challenges and mechanisms due to the complex integration of both soft and hard tissue compartments and fluids. Understanding the concepts of immune response to the microbial factors and the impact of genetic determinants of susceptibility to disease is crucial for identifying the pathways of inflammation. Since oral inflammation presents a risk for systemic health and follows similar pathways of inflammatory response elsewhere in the body, preventing and treating the inflammatory lesions of the oral tissues is also important for the maintenance

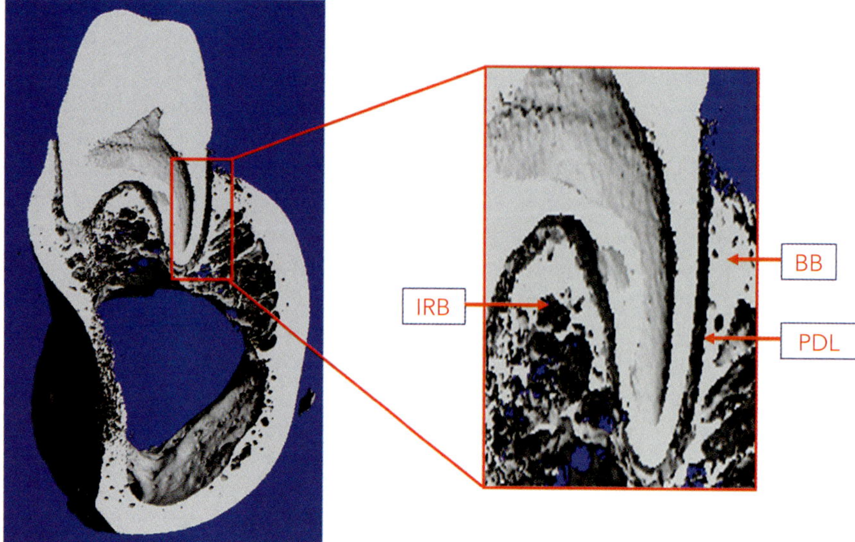

Figure 33.6. Pig model to study periodontal regeneration. Microcomputerized tomographical (micro-CT) imaging of the pig molar demonstrates the similarity of the swine model to human dentition. IRB, interradicular bone; BB, buccal bone; PDL, periodontal ligament.

of the homeostasis of the whole human body. For example, periodontal inflammation is one of the major complications of diabetes [39]. Likewise, cardiovascular diseases are associated with periodontal inflammation and the list of systemic diseases linked to oral inflammatory lesions is growing. In vitro experimentation and in vivo models together with high-throughput analyses are in place and have already provided substantial information on how the oral inflammation is initiated and progresses. Future studies will be focused on the utilization of this information for development of therapeutic strategies to resolve the oral inflammation as well as their contribution to the systemic diseases.

REFERENCES

1. Shklar, G., and Carranza, F.A. 2006. The historical background of periodontology. In *Carranza's Clinical Periodontology*, Newman, M., Takei, H.H., Klokkevold, P., and Carranza, F.A. (eds.), pp. 1–10. St. Louis, MO: Elsevier.
2. Paster, B.J., Olsen, I., Aas, J., and Dewhirst, F.E. 2006. The breadth of bacterial diversity in the human periodontal pocket and other oral sites. *Periodontol 2000* **42**:80–87.
3. Pihlstrom, B.L., Michalowicz, B.S., and Johnson, N.W. 2005. Periodontal diseases. *Lancet* **366**(9499):1809–1820.
4. Socransky, S.S., and Haffajee, A.D. 2005. Periodontal microbial ecology. *Periodontol 2000* **38**:135–187.
5. Li, J., Helmerhorst, E.J., Troxler, R.F., and Oppenheim, F.G. 2004. Identification of in vivo pellicle constituents by analysis of serum immune responses. *J Dent Res* **83**(1):60–64.
6. Haffajee, A.D., and Socransky, S.S. 2005. Microbiology of periodontal diseases: introduction. *Periodontol 2000* **38**:9–12.
7. Danser, M.M., Timmerman, M.F., van Winkelhoff, A.J., and van der Velden, U. 1996. The effect of periodontal treatment on periodontal bacteria on the oral mucous membranes. *J Periodontol* **67**(5):478–485.
8. Mager, D.L., Ximenez-Fyvie, L.A., Haffajee, A.D., and Socransky, S.S. 2003. Distribution of selected bacterial species on intraoral surfaces. *J Clin Periodontol* **30**(7):644–654.
9. Sachdeo, A., Haffajee, A.D., and Socransky, S.S. 2008. Biofilms in the edentulous oral cavity. *J Prosthodont* **17**(5):348–356.
10. Ronstrom, A., Edwardsson, S., and Attstrom, R. 1977. Streptococcus sanguis and streptococcus salivarius in early plaque formation on plastic films. *J Periodontal Res* **12**(5):331–339.
11. Costerton, J.W., Stewart, P.S., and Greenberg, E.P. 1999. Bacterial biofilms: a common cause of persistent infections. *Science* **284**(5418):1318–1322.
12. Hajishengallis, G., Sojar, H., Genco, R.J., and Denardin, E. 2004. Intracellular signaling and cytokine induction upon interactions of Porphyromonas gingivalis fimbriae with pattern-recognition receptors. *Immunol Invest* **33**(2):157–172.
13. Yumoto, H., Chou, H.-H., Takahashi, Y., Davey, M., Gibson III, F.C., and Genco, C.A. 2005. Sensitization of human aortic endothelial cells to lipopolysaccharide via regulation of Toll-like receptor 4 by bacterial fimbria-dependent invasion. *Infect Immun* **73**(12):8050–8059.
14. Buhlin, K., Gustafsson, A., Pockley, A.G., Frostegård, J., and Klinge, B. 2003. Risk factors for cardiovascular disease in patients with periodontitis. *Eur Heart J* **24**(23):2099–2107.
15. Franek, E., Blaschyk, R., Kolonko, A., et al. 2006. Chronic periodontitis in hemodialysis patients with chronic kidney disease is associated with elevated serum C-reactive protein concentration and greater intima-media thickness of the carotid artery. *J Nephrol* **19**(3):346–351.
16. Iwamoto, Y., et al., 2003. Antimicrobial periodontal treatment decreases serum C-reactive protein, tumor

necrosis factor-alpha, but not adiponectin levels in patients with chronic periodontitis. *J Periodontol* **74**(8):1231–1236.

17. Kantarci, A., Oyaizu, K., and Van Dyke, T.E. 2003. Neutrophil-mediated tissue injury in periodontal disease pathogenesis: findings from localized aggressive periodontitis. *J Periodontol* **74**(1):66–75.
18. Löe, H., Anerud, A., Boysen, H., and Morrison, E. 1986. Natural history of periodontal disease in man. Rapid, moderate and no loss of attachment in Sri Lankan laborers 14 to 46 years of age. *J Clin Periodontol* **13**(5):431–445.
19. Benjamin, S.D., and Baer, P.N. 1967. Familial patterns of advanced alveolar bone loss in adolescence (periodontosis). *Periodontics* **5**(2):82–88.
20. Boughman, J.A., Halloran, S.L., Roulston, D., et al., 1986. An autosomal-dominant form of juvenile periodontitis: its localization to chromosome 4 and linkage to dentinogenesis imperfecta and Gc. *J Craniofac Genet Dev Biol* **6**(4):341–350.
21. Li, Y., Xu, L., Hasturk, H., Kantarci, A., DePalma, S.R., and Van Dyke, T.E. 2004. Localized aggressive periodontitis is linked to human chromosome 1q25. *Hum Genet* **114**(3):291–297.
22. Hart, T.C., Harta, P.S., Bowden, D.W., et al. 1999. Mutations of the cathepsin C gene are responsible for Papillon-Lefevre syndrome. *J Med Genet* **36**(12):881–887.
23. Wilson, A.G. 2008. Epigenetic regulation of gene expression in the inflammatory response and relevance to common diseases. *J Periodontol* **79**(8 Suppl):1514–1519.
24. Mariotti, A. 1999. Dental plaque-induced gingival diseases. *Ann Periodontol* **4**(1):7–19.
25. Page, R.C., and Schroeder, H.E. 1976. Pathogenesis of inflammatory periodontal disease. A summary of current work. *Lab Invest* **34**(3):235–249.
26. Loesche, W.J., and Syed, S.A. 1978. Bacteriology of human experimental gingivitis: effect of plaque and gingivitis score. *Infect Immun* **21**(3):830–839.
27. Cohen, J. 2002. Therapies. Confronting the limits of success. *Science* **296**(5577):2320–2324.
28. Kantarci, A., Hasturk, H., and Van Dyke, T.E. 2006. Host-mediated resolution of inflammation in periodontal diseases. *Periodontol 2000* **40**:144–163.
29. Serhan, C.N., Chiang, N., and Van Dyke, T.E. 2008. Resolving inflammation: dual anti-inflammatory and pro-resolution lipid mediators. *Nat Rev Immunol* **8**(5):349–361.
30. Serhan, C.N., Jain, A., Marleau, S., et al. 2003. Reduced inflammation and tissue damage in transgenic rabbits overexpressing 15-lipoxygenase and endogenous anti-inflammatory lipid mediators. *J Immunol* **171**(12):6856–6865.
31. Hasturk, H., Kantarci, A., Ohira, T., et al. 2006. RvE1 protects from local inflammation and osteoclast-mediated bone destruction in periodontitis. *FASEB J* **20**(2):401–403.
32. Hasturk, H., Kantarci, A., Goguet-Surmenian, E., et al., 2007. Resolvin E1 regulates inflammation at the cellular and tissue level and restores tissue homeostasis in vivo. *J Immunol* **179**(10):7021–7029.
33. Sakamoto, M., Umeda, M., and Benno, Y. 2005. Molecular analysis of human oral microbiota. *J Periodontal Res* **40**(3):277–285.
34. Jain, A., Batista, E.L., Jr., Serhan, C., Stahl, G.L., and Van Dyke, T.E. 2003. Role for periodontitis in the progression of lipid deposition in an animal model. *Infect Immun* **71**(10):6012–6018.
35. Hickey, J.S., O'Neal, R.B., Scheidt, M.J., Strong, S.L., Turgeon, D., and Van Dyke, T.E. 1991. Microbiologic characterization of ligature-induced peri-implantitis in the microswine model. *J Periodontol* **62**(9):548–553.
36. Marks, S.C., Jr., McKee, M.D., Zalzal, S., and Nanci, A. 1994. The epithelial attachment and the dental junctional epithelium: ultrastructural features in porcine molars. *Anat Rec* **238**(1):1–14.
37. Sonoyama, W., Liu, Y., Fang, D., et al., 2006. Mesenchymal stem cell-mediated functional tooth regeneration in swine. *PLoS ONE* **1**:e79.
38. Liu, Y., Zheng, Y., Ding, G., et al., 2008. Periodontal ligament stem cell-mediated treatment for periodontitis in miniature swine. *Stem Cells* **26**(4):1065–1073.
39. Oliver, R.C., Brown, L.J., and Loe, H. 1998. Periodontal diseases in the United States population. *J Periodontol* **69**(2):269–278.

SUGGESTED READINGS

Kantarci, A., Hasturk, H., and Van Dyke, T.E. 2006. Host-mediated resolution of inflammation in periodontal diseases. *Periodontol* 2000 **40**:144–163.

Serhan, C.N., Chiang, N., and Van Dyke, T.E. 2008. Resolving inflammation: dual anti-inflammatory and pro-resolution lipid mediators. *Nat Rev Immunol* **8**(5):349–361.

34 Pathogens and Inflammation

Julio Aliberti

Here, we discuss the modulation of inflammation upon pathogen invasion, including the new pathway triggered by lipoxin involved in the repression of immune response during infection, as well as this mechanism from the perspective of the pathogen, which pirates the host's lipoxygenase machinery to its own advantage as a probable immune-escape mechanism.

INNATE IMMUNITY AND PATHOGENS

It is well known that a series of pattern-recognition receptors is involved in the recognition of different microbial pathogens and induction of the innate response. Such receptors recognize distinct biochemical patterns of molecules displayed by the invading pathogen. The repertoire of innate immune receptors is very broad and includes several classes of germ-line–encoded proteins such as Toll-like receptors (TLRs), scavenger receptors, and C-type lectins. This wide array of recognition molecules allows the host to detect a variety of microbial molecules including carbohydrates, lipids, nucleic acids, and proteins [1,2]. Distinct TLR ligands provide distinct activation status and cytokine production patterns for antigen-presenting cells (APCs), resulting in the induction of differential immune responses. Thus, TLRs are critical molecules to induce not only inflammatory responses but also fine-tune adaptive immune responses depending on invading pathogens [3]. TLRs activation can upregulate costimulatory molecules on APC, thus enhancing the activation of adaptive T-cell responses. It has been reported that the activation of TLRs on DCs leads to the induction of IL-6, which downregulates the function of CD4+CD25+ T-regulatory cells, further stimulating the adaptive T-cell response. Finally, different TLRs can have different patterns of functional responses, depending on the adaptor molecules with which they associate.

During infection, that is, antiviral responses, TLRs activated by viral ligands lead to activation of the TNF receptor-inhibitory factor (TRIF), interferon-regulated factor 3 (IRF3), specific signaling pathways, resulting in the release of IFN-β and an antiviral gene program [4]. The result of this program is the inhibition of viral replication. Interestingly, some viruses encode proteins that inhibit TLR signaling, thereby escaping this response.

Gram-positive bacteria contain a variety of ligands that activate the innate system (i.e., lipoteichoic acids, teichoic acids, and peptidoglycans). These cell-wall–associated ligands are able to activate TLR2 to induce the release of cytokines, and these ligands also activate C-reactive protein, which accelerates the clearance of the pathogen [5]. Finally, these ligands activate a platelet-activating factor receptor, which triggers endocytosis. The resultant inflammatory responses can help eliminate the pathogen, but they also trigger damage to host tissues.

Dendritic cells (DCs) are key APC that also have a broad functional role in mediating innate immune responses and influencing adaptive immunity. DCs can be divided into functional subsets, and these subsets can be defined according to cell-surface markers, and play specific roles in host defense to infection. During viral infection, plasmacytoid DCs produce type I IFNs, which contribute to the rapid control of infection. CD8+ DCs are involved in priming cytolytic T lymphocytes either by direct presentation of viral antigen or by cross-priming. A third subset, CD11b+ DCs, is involved in priming CD4+ helper T cells [6].

Upon parasitic infection, DCs use TLR as well as other receptors to recognize and respond to microbial pathogens [7]. DC-specific ICAM-3-grabbing nonintegrin (DC-SIGN) is able to recognize carbohydrate motifs on various bacteria and parasites, and is regulated by cytokines, such that the local environment can regulate DC interaction with microbial pathogens. It is well established that DCs have the ability to dictate the development of Th1 versus Th2 responses in response to parasitic infection [7].

During experimental models of infection (i.e., *Toxoplasma gondii*, *Mycobacterium tuberculosis*, and *Trypanosoma cruzi* among others), an NK cell-activating cytokine, IL-12, is essential for triggering IFN-γ-dependent immune responses to microbes; for example, IL-12-deficient animals are extremely susceptible to *T. gondii* infection [8]. Several cell populations are known to produce IL-12 in vitro and in vivo, including B cells, macrophages, neutrophils, and DCs [9]. Nevertheless, DCs, which are an abundant source of IL-12 in vivo, seem to be the most relevant cell population for the development of a parasite-specific type 1 immune response. Reise Sousa and colleagues observed that splenic mouse CD8α+ DCs produce IL-12 in response to *T. gondii* in the absence of costimulatory signals [10]. So, DCs can either activate the immune system by recognition of parasite-derived molecules or can potentially harbor initial replication of the intracellular parasites.

ADAPTATIVE IMMUNE RESPONSES TO INFECTION

In 1986, two types of murine helper T-cell clones were described (Th1 and Th2) that have had enormous effect on the way exploration of immune responses to pathogen invasion is understood [11]. Initially, the paradigm was adopted and it seemed as though immune responses to various classes of infecting organisms could be neatly classified as either eliciting a Th1 or a Th2 response and that immunity to an organism required one of these specific responses for clearance of infection. Then it became apparent that in many cases a balance of these two types of T-cell responses was actually required to eliminate infection. Currently, it is beginning to emerge that, in fact, the paradigm may not be nearly as clear-cut as was initially thought.

The Th1/Th2 dogma can be simply described in the following way: CD4+ T cells have been categorized into two subsets (Th1 and Th2) according to the profile of cytokines they produce. Generally, intracellular infections rely on a Th1-type response to resolve infection and extracellular parasites (mainly helminths) elicit a Th2 type response. Prototype Th1 cytokines include IL-12, IFN-γ, and TNF. These cytokines are proinflammatory and evoke cell-mediated immune responses and these cytokines are typically needed to resolve bacterial, viral, and protozoal infections. Th2 cytokines include IL-4, IL-5, and IL-10. These cytokines are anti-inflammatory and typically evoke humoral immune responses and Th2 responses are classically described in immune responses to helminthes (i.e., schistosomiasis).

The cytokine released by APC into the immunologic synapse, mainly by DCs, is emerging as key and hence the biology of this phenotype. The early production of IL-12 in the immunological synapse between APC and Th0 T cell has been shown to prime Th1 responses in response to intracellular microbial infections and the secretion of IL-10 leads to Th2 T-cell responses.

The infection in which this paradigm of immunity is best studied and most easily understood is *Leishmania*. It is well known that resolution of a murine *Leishmania* infection depends on an early Th1 response and that susceptibility to disease is associated with early Th2 domination. DCs from susceptible mouse strains are an early source of the Th2 biasing cytokine IL-4. *Leishmania amazonensis* amastigotes and metacyclics are able to enter and activate DCs of both susceptible and resistant mouse strains. DCs from susceptible BALB/c mice, however, fail to produce CD40-induced IL-12, but rather produce IL-4. Transfer of these BALB/c DCs into syngeneic amastigote-exposed mice results in a significantly higher parasite-specific Th2 response than that seen in resistant mouse strains. DCs taken from IL-4−/− mice indicate that this Th2 biasing is at least partially due to amastigote-carrying DCs. Thus host susceptibility, at least in part, may be due to the Th2 biasing effect of infected DCs. Similar findings have been demonstrated for several infectious agents including *Toxoplasma gondii*, *Trypanosoma cruzi*, and *Cryptosporidium parvum*.

The need for both T-cell phenotypes acting sequentially has been recognized in *Plasmodium chabaudi* AS infection of the resistant C57BL/6 and susceptible AJ mouse strains. It has become clear that early Th1 responses cause a rapid drop in parasitemia and a later Th2 response elicits an antibody response that finally clears parasitemia. Several studies have shown an important role for early IL-12 production and DCs in *Plasmodium* and *Toxoplasma* infections. In a mouse model of this disease, it has been shown that the spleen, mainly DCs, is the primary source of IL-12. Resistant C57BL/6 mice show an early Th1 response in the spleen as shown by increased levels of TNF-α and IFN-γ mRNA expression and low levels of IL-4 expression in splenic lymphocytes. TNF-α mRNA levels correlate with cytokine concentrations in blood. In addition, anti-TNF antibody given early in the infection renders resistant mice susceptible. The susceptibility of AJ mice however is correlated with lower levels of early Th1 cytokine mRNA expression in the spleen and also later on in the disease around the time of death. Hence, it is apparent that the timing and the site of Th1 cytokine release is crucial in outcome. In resistant mice, if the early Th1 response is not followed by a Th2 response, effective antiparasite antibodies are not generated and the infection persists.

The critical role of IFN-γ production by CD4+ T cells in controlling *M. tuberculosis* was recognized early [12–15]. IFN-γ knock-out mice are highly susceptible to *M. tuberculosis* infection, with deficient production of reactive nitrogen intermediates (key mycobactericidal effectors) and defective granuloma formation [16]. The high mortality observed in these mice is associated with the lack of generation of Th1 cells – nevertheless, no Th2 response is "uncovered" in these mice. Indeed, IL-4-deficient mice have no detectable difference in resistance to *M. tuberculosis* infection, suggesting that disease does not derive from a Th1/Th2 ratio imbalance, such as is observed in *Leishmania major* infection models [16].

Given the central role played by IL-12 in driving IFN-γ responses, it is not surprising that IL-12 has been shown to be essential for protection against *M. tuberculosis* infection in mice [12,15]. IL-12p40-deficient mice are highly susceptible to this infection, exhibiting high levels of mycobacterial dissemination and proliferation, organ failure, and high mortality [17]. Interestingly, IL-12p35-deficient mice, although slightly more resistant than their IL-12p40-knockout counterparts, still exhibit high susceptibility to *M. tuberculosis* infection, thus indicating that IL-12 has a dominant role in mediating resistance to *M. tuberculosis* infection [17]. However, other cytokines that share the IL-12p40 subunit, such as IL-23, may also contribute to resistance to this pathogen [17].

In humans, strong evidence for the importance of IFN-γ and IL-12 in host defense against mycobacterial infection has been described [18,19]. Thus, it is well established that the protective immune response to *M. tuberculosis* is dependent on the host's ability to initiate Th1 cellular responses.

TNF is another key proinflammatory mediator that contributes to the development of resistance to *M. tuberculosis*. TNF- or TNFR1-deficient mice, as well as mice treated with neutralizing anti-TNF mAb, exhibit high levels of mycobacteria growth in the lungs and peripheral organs, along with high mortality [16]. More interestingly, during chronic disease, neutralization of TNF causes mortality by disrupting granuloma structure [16].

MODULATION OF IMMUNE RESPONSE AND PREVENTION OF PATHOLOGY DURING INFECTION: ROLE OF CYTOKINES

The simple way of understanding the Th1/Th2 paradigm of immune response to protozoan infections, namely, early Th1 cytokines clear the infection and later Th2 cytokines prevent serious ongoing inflammation, may well be an over-simplification of the truth. Cytokine interactions are complex and timing as well as specific location of release are just two of the factors that may impact strongly on the effect of a particular cytokine. The beneficial and detrimental view of Th1 versus Th2 cytokines is less and less a view held onto as a dogma useful in the explanation of immunopathogenesis of disease.

Several studies have shown a role for classical Th2 cytokines in an early appropriate disease protecting immune response against intracellular protozoal infections. IL-4, when present during initial activation of DCs, can instruct them to produce IL-12 and promote a Th1 response thus protecting normally susceptible BALB/c mice against *L. major* infection. If IL-4 is present later during the period of T-cell priming, a Th2 response develops and mice succumb to progressive disease. This example shows that the timing of exposure to a particular cytokine may be more important than to which side of the traditional Th1/Th2 spectrum it has been classified. Stat4 is a second messenger important in signaling Th1 cytokine production. The Th2 cytokines, IL-4, and IL-10, suppress Stat4 induction in DCs and macrophages if present during maturation and activation, respectively, thus diminishing IFN-γ production. In contrast, IL-4 has no effect on Stat4 levels in mature DCs and actually augments IFN-γ production by DCs during Ag presentation, indicating that IL-4 acts differently in a spatiotemporal manner.

A phosphorylcholine-containing glycoprotein secreted by the filarial nematode, *Acanthocheilonema viteae*, which generates a Th2 antibody response *in vivo*, has (as would be expected) been found to induce the maturation of DCs with the capacity to induce Th2 responses (increased IL-4 and decreased IFN-γ). The switch to either Th1 or Th2 responses is not affected by differential regulation through CD80 or CD86 and the Th2 response is achieved in the presence of IL-12. This is counter-intuitive, as the traditional model would imply that the presence of IL-12 would dominate and bias a Th1 response.

Surprisingly, and somewhat contrary to what was hypothesized (based on the simplistic view of Th1/Th2), IL-2 knock-out mice develop severe Th1-mediated autoimmune mediated disease (hemolytic anaemia, Crohn's disease). TNF is a prototype proinflammatory Th1 cytokine and yet under some circumstances, TNF can cause immunosuppression. Anti-TNF therapy has been associated with dramatic improvements in many rheumatoid arthritis and Crohn's disease patients. The same treatment however has been associated with worsening multiple sclerosis with enhanced demyelination in some human patients. Under some circumstances and in some animal models, TNF treatment can reduce the severity of some prototypic Th1-type diseases, including diabetes, experimental autoimmune encephalomyelitis, and arthritis. Typically, TNF enhances antigen presentation by APCs. It has, however, been shown that in some circumstances TNF can inhibit the function of mature DCs and might induce their apoptosis and inhibit antigen presentation. It is hypothesized that the timing and duration of TNF production are important in deciding the protective versus harmful effects of TNF.

Both type I and type II interferons are used to treat infectious disease, but in up to 19% of patients treated with these cytokines, autoimmune conditions develop (typical Th1-type diseases). Paradoxically, type I interferons are also used to treat autoimmune diseases like multiple sclerosis (where it is the corner-stone of therapy in some countries). These are cytokines that induce APC maturation and enhance antigen presentation. The mechanisms of IFN-induced immune downregulation are not completely clear, but type I IFNs are known to be able to inhibit IL-12 production and enhance IL-10 production. Both IFN-α and IL-10 can promote the differentiation of CD25+CD4+ regulatory T cells. IFN-γ is regarded as the canonical type I cytokine and consistent with this, it activates APC and promotes Th1 differentiation. Despite large amounts of evidence incriminating it as a promoter of disease, there is also evidence for its protective roles in autoimmunities. The immunosuppressive action of this and other cytokines relates to their ability to induce members of the **s**uppressor **o**f **c**ytokine **s**ignaling (SOCS) family of intracellular messengers that serve as classic feedback inhibitors of cytokine signal transduction. *Socs1* gene knock-out mice show extensive IFN-γ-dependent pathology and show excessive IFN-γ responses. It thus seems possible that low-level exposure to IFN-γ might attenuate subsequent responsiveness to the harmful effects of these cytokines, the net effect thus being inhibitory. Recently, it was shown that the host counterbalanced the excess of inflammatory response production of regulatory cytokines and novel anti-inflammatory eicosanoids, lipoxins; such novel molecules and their stop signaling pathways keep proinflammatory response under control during chronic disease.

MODULATION OF IMMUNE RESPONSE AND PREVENTION OF PATHOLOGY DURING INFECTION; LIPOXINS

LXA_4 or (5S,6S,15S)-5,6,15-trihydroxy-7,9,13-*trans*-11-*cis*-eicosatentraenoic acid is an eicosanoid derived from the lypoxigenase (LO)-metabolism of arachidonic acid (AA). Initial studies found that LXA_4 was secreted by neutrophils and inhibited the activating effects of LTB_4 on platelets [20]. Since then, a growing list of anti-inflammatory effects have been associated with LXA_4 and its analogs [21]. LO-catalyzed oxygenation of unsaturated fatty acids initiates the formation of a variety of compounds with diverse biological activities. The biosynthetic pathways involved in LXA_4 formation are complex, involve the actions of at least two independent LOs, and can occur through transcellular cascades. However, the activity of 5-LO seems to be a common step in LXA_4 synthesis [20]. 5-LO is produced as a pro-peptide that is activated by cleavage. Low levels of active 5-LO are found in different cell types, including macrophages, platelets, DCs, and neutrophils [22]. The expression of a 5-LO-activating protein (FLAP) is a key signal for induction of 5-LO activity. Of note, aspirin acetylation of cyclo-oxygenase (COX) [20] leads to a change in functionality that can drive synthesis of the "aspirin-triggered LXA_4" (ATL): 15-epi-LXA_4. ATL has the same activity, but is more metabolically stable than LXA_4 [20].

LXA_4 are thought to bind to at least two types of receptors [23]: (a) a surface membrane seven-transmembrane G-protein–coupled receptor, FPRL-1 or ALXR [24]; and (b) the nuclear ligand–activated transcription factor, the Aryl hydrocarbon receptor (AhR) [25]. The binding of ligands (i.e., LX) to AhR results in the formation of an active transcription factor that binds to DNA domains – dioxin-responsive elements (DRE) – that activate the expression of a specific panel of genes [26]. In the absence of ligand, AhR is located in the cytosol in association with several molecules, Hsp90, the immunophilin-like protein XAP2, and the co-chaperone p23. Ligand binding allows for release of the AhR, its translocation to the nucleus, and heterodimerization with the AhR nuclear translocator (ARNT; HIF-1β). The AhR/ARNT heterodimer binds to specific DNA sequences (aryl hydrocarbon elements [AHREs], also known as DRE or XRE), where, in concert with the co-activators CBP/p300 (ARNT) and RIPI40 (AhR), they lead to target-gene transcription, leading to expression of target genes [26].

REPRESSION OF IMMUNE RESPONSE AND PREVENTION OF IMMUNOPATHOLOGY

Proinflammatory mediators such as IL-12, IFN-γ, and TNF, which are essential in controlling parasite growth (as discussed earlier), can be unfavorable to the host if produced in excess. For example, uncontrolled immune response during *T. gondii* infection induces cachexia, tissue damage, and early death. In addition, during other inflammatory diseases such as arthritis or Crohns' disease, sustained or uncontrolled type 1 cytokine responses can cause serious damage. To prevent such damage, the proinflammatory effects need to be counterbalanced by the induction of several host regulatory factors and receptors. The presence of such control mechanisms is absolutely essential for the homeostasis of the immune response, and given its efficiency, has been held and used by pathogens to their own benefit to prevent parasite eradication. The IL-10 is an important factor that has been shown to have anti-inflammatory actions in various pathologies associated with increased IFN-γ, IL-1, or TNF-α production [27,28]. Accordingly, neutralization of IL-10 during chronic toxoplasmic encephalitis leads to increased leukocyte infiltration in the CNS, indicating a role for this cytokine in controlling CNS inflammation [29]. Moreover, IL-10-deficient mice have no control over inflammatory responses and succumb to *T. gondii* infection early in the acute phase, with severe leukocyte infiltration and tissue necrosis in the liver and small intestines, mostly due to uncontrolled IFN-γ and TNF-α production [30,31]. Given its immune-modulatory activities, it was hypothesized that IL-10 induction by the pathogen could be used to escape from host immune responses during *T. gondii* infection, therefore contributing to virulence of this parasite. In agreement with this, the IL-10 has been shown as a soluble factor used by pathogens to inhibit immune responses. For example, some viruses, such as Epstein–barr virus, encode a viral homologue of IL-10 that can trigger the same signaling cascade as the mammalian cytokine [32], resulting in the inhibition of antigen processing and presentation by APCs, inactivation of IFN-γ-triggered microbicidal pathways, and inhibition of T-cell cytokine production and cytotoxic activity [33]. Therefore, other mechanisms of immune modulation by the parasite are essential to allow survival of the host long enough for transmission and, ultimately, for parasite survival as a species. Recently, novel molecules and stop signaling pathways were referred to as checkpoint controllers of inflammation [34]. Evidence for another anti-inflammatory mechanism operating during *T. gondii* infection has been provided from observation showing that in vivo injection of STAg triggers the production of endogenous LXA_4 in a 5-lipoxygenase (5-LO)-dependent manner, consequently causing in vivo suppression of IL-12 production and CCR5 expression by DCs, and confers protection to IL-10-deficient mice [35]. Moreover, LXA_4 is an eicosanoid, which has been found to inhibit STAg-induced DC migration and IL-12 production in vivo and in vitro [36]. Lipoxins have been shown to have potent anti-inflammatory properties in a growing list of models, including periodontitis, arthritis, nephritis, inflammatory bowel disease, cystic fibrosis, tuberculosis, and toxoplasmosis [37–42]. They inhibit leukotriene function, natural killer cell function, leukocyte migration, TNF-induced chemokine production, translocation of NF-κB, and expression of chemokine receptor and adhesion molecules [43–47]. Such eicosanoids are thought to bind to two receptors: a seven-transmembrane G-protein–coupled receptor, LXAR/FPRL-1 [48] and a nuclear receptor, AhR [25]. Mice overexpressing human LXAR have shorter and less severe inflammatory responses, indicating that this receptor is important to mediate the anti-inflammatory actions of lipoxins in vivo [49]. Moreover, Machado et al. recently showed that LXA_4 signaling through AhR and LXAR receptors in DCs modulated the innate and acquired immune responses in mice during *T. gondii* infection [50]. Serhan and colleagues initially proposed that Lipoxin biosynthesis arose from arachidonic acid via interaction of the 5-lipoxygenase and 15-lipoxygenase pathways. Since this initial discovery, different routes for cellular and transcellular biosynthesis of endogenous LXA have been described, including the discovery that acetylation of cyclooxygenase-2 (COX-2) by aspirin could lead to transcellular biosynthesis of aspirin-triggered lipoxins (ATL). The biological actions of LXA_4 have been characterized in many cell and tissue types, both in vitro and in vivo. Regarding LXA_4 biosynthesis during *T. gondii* infection, Aliberti et al. have shown that LXA_4 production seen after STAg injection was completely abolished in the absence of 5-lipoxygenase, indicating that the biosynthetic pathways involving this enzyme were crucial in this experimental setting for the production of LXA_4 [36]. 5-lipoxygenase is produced as a propeptide that is activated by cleavage. Low levels of active 5-lipoxygenase are found in different cell types, including macrophages, platelets, DCs, and neutrophils [22]. The expression of a 5-lipoxygenase-activating protein (FLAP) seems to be the key signal for induction of 5-lipoxygenase activity. Although, at the moment, it is not completely clear which cells are the source of lipoxygenase activity in vivo during *T. gondii* infection, it is evident that 5-lipoxygenase is required for biosynthesis of LXA_4. During infection with *T. gondii*, serum levels of LXA_4 increase steadily over the course of the acute phase, and remain at high levels during chronic disease [37]. Such high levels found in the serum of chronically infected animals indicate that this mediator might exert relevant biological functions during this stage of the disease. Consistent with this, 5-lipoxygenase-deficient animals succumbed to *T. gondii* infection at the early onset of chronic disease. Upon further investigation, it became clear that immunity against the parasite was actually increased in the absence of 5-lipoxygenase, with significantly less brain cyst formation than in control animals, indicating that this was not the cause of mortality. By contrast, excessive proinflammatory cytokine production and massive cerebral infiltration was found, including atypical meningitis. The conclusion was that the excessive proinflammatory response in the brain ultimately caused the death of the 5-lipoxygenase–deficient hosts [37]. *T. gondii* infection in 5-lipoxygenase–deficient mice resulted in more extensive tissue pathology, mainly due to lack of LXA_4 production, as treatment of 5-lipoxygenase–deficient mice with lipoxin analogs restored the resistance to tissue pathology with no mortality associated with uncontrolled proinflammatory responses, in a similar manner as for wild-type animals [37]. Interestingly, a recent study showed that suppressor of cytokine signaling-2 (SOCS-2) is a crucial intracellular mediator of the anti-inflammatory actions of lipoxins in vivo [50]. 5-LO and AhR-deficient mice failed to upregulate SOCS-2 expression in vivo after STAg injection, unlike WT mice, indicating that induction of SOCS-2 expression depends on lipoxin-generating enzymes and receptors. Confirming the involvement of SOCS-2 in LXA_4-triggered immune regulatory pathways, SOCS-2-deficient and AhR-deficient mice did not show repression of IL-12 production upon in vivo restimulation with STAg (DCs cells paralysis model) [50]. During infection with *T. gondii*, SOCS-2 deficient mice had significantly higher IL-12, IFN-γ, and TNF-α production; reduced number of brain cysts; higher frequency of TNF-α and iNOS infiltrating cells in the central nervous system; and higher mortality rates, consistent with the results found in *T. gondii*–infected 5-LO deficient mice. In addition, *T. gondii*–infected SOCS-2 deficient mice significantly produce higher levels of chemokine, and increase monocyte and neutrophil migration to the peritoneal cavity compared with WT counterparts [50]. All together, these studies demonstrated that during *T. gondii* infection, the proinflammatory cytokine responses, leukocyte infiltration, T-cell activation is controlled by a lipoxin-regulatory pathway dependent on SOCS-2 in vivo. Accordingly, there is growing evidence of a role for SOCS molecules in the induction of the anti-inflammatory effects seen after lipoxin exposure [51]. Interestingly, SOCS-2 has a role in the intracellular pathways triggered by ATLs [50]. ATLs share the same receptors and mediate the same biological effects as those observed with LXA_4 [45]. Indeed, Machado et al. showed that wild-type mice, but not BOC_2 (peptide antagonist for LXRA receptor)-treated or AhR-deficient mice, upregulated the SOCS-2 expression after treatment with aspirin. Moreover, aspirin-treated SOCS-2-deficient mice did not inhibit neutrophil infiltration into the peritoneal cavity after thioglycollate challenge, and SOCS-2 was also essential for in vivo aspirin-dependent inhibition of lipopolysaccharide-induced secretion of TNF, a hallmark of the anti-inflammatory actions of this drug [50]. Together with the previous observation, Machado et al. also demonstrated that endogenous LXA_4 and ATL, signaling through AhR and LXAR, modulate innate and acquired immune responses through inducing SOCS-2 expression [50].

The SOCS-family proteins, SOCS-1, -2, and -3, are thought to mediate their actions by docking to the intracellular domains of cytokine or hormone receptors, thereby preventing binding and activation of downstream signaling elements [52]. Alternatively, these proteins might facilitate proteasome-dependent degradation of transcription factors through the induction of ubiquitylation [52,53].

THE ROLE OF LIPOXINS IN IMMUNE REGULATION DURING INFECTIOUS DISEASES

The immune-regulatory roles of LX have been described during different infectious diseases. For example, patients with cystic fibrosis fail to generate lipoxins in the lungs and the continuing infiltration of activated neutrophils that ultimately lead to serious tissue damage with organ failure [54]. This contributes the major pathology for the lung form of cystic fibrosis.

Another lung pathogen that causes a chronic disease with enormous public health relevance is *Mycobacterium tuberculosis*. It is estimated that approximately 30% of the world's population is infected with this bacteria, and only 5%–10% of this group develop active disease. The resistance to mycobacterial infection is dependent on Th1 cytokines such as IL-12 and IFN-γ. Breakdown of host resistance, as occurs in HIV-infected patients, leads to reactivation of latent infection by unclear mechanisms involving the failure of tissue granulomas to contain and prevent the spread of the bacteria, particularly in lungs.

In addition to the Th1-type response generated during the course of infection, down-regulatory mediators may be important players in controlling excessive synthesis of proinflammatory cytokines and consequently tissue damage. Nevertheless, the Th2 cytokines IL-4, IL-13, and IL-10 have been described to play no or only limited role in vivo during infection with *M. tuberculosis* [55], for example IL-10-deficient mice, that shows nearly normal control of *M. tuberculosis* infection. On the other hand, Bafica and colleagues have shown that in the absence of endogenously generated LXA_4, mice become more resistant to infection, with longer survival rates, lower bacterial counts, and higher type 1 cell-mediated immunity against the bacilli [38].

It is possible to speculate that pathogens may take advantage of this regulatory pathway to promote host survival, or even allow a less toxic environment in which replication can occur. Accordingly, Bannenberg and colleagues identified an enzymatic activity in tachyzoite forms of *T. gondii* exposed to calcium ionophore in the presence of arachidonic acid in vitro. Moreover, using the proteomics analysis of tachyzoite-derived lysates revealed the presence of peptides homologous to plant-derived type 1-lipoxygenases [56], indicating that the induction of lipoxin biosynthesis by *T. gondii* is the result of a collaboration of the parasite-derived

15-lipoxygenase-like protein, which together with the actions of host-derived 5-lipoxygenase results in high-level lipoxin production. In turn, it dampens the immune responses so that hosts can control parasite proliferation without succumbing to the damaging consequences of excessive inflammation or tissue destruction.

Although the genes that are responsible for the 15-lipoxygenase activity in *T. gondii* have not been formally identified, another intriguing point that contributes to the argument for a role for *T. gondii*-expressed 15-lipoxygenase in immune evasion is the presence of such enzymatic activity in an organism that does not have lipids that could function as substrates for lipoxygenases. Therefore, the substrate must be derived from infected host cells. Consistent with the results found with *T. gondii* was the description of the cloning of a 15-lipoxygenase-like enzyme from the bacterial pathogen *P. aeruginosa* [57]. Although the production of LXAs was not formally shown in this report, injection of exogenous 15-lipoxygenase into naive mice, as shown by Bannenberg et al. [56], was sufficient to induce the production of LXA_4 and to have biological effects, such as inhibition of IL-12 production and inflammatory infiltration. Moreover, *P. aeruginosa* is associated with chronic lung infections in patients with cystic fibrosis, and it is possible that this bacterium may use the 15-lipoxygenase synthetic pathway leading to lipoxin production to promote suppression of inflammation and consequently providing the favorable environment to persist throughout chronic disease in immune-competent individuals. The relevance of pathogen-derived 15-lipoxygenase given the lack of lipoxin generation in the lungs of patients with cystic fibrosis and the severity of disease still remains to be elucidated.

Taking into account the effects of endogenously generated lipoxins during *M. tuberculosis* infection, it is possible to hypothesize that in the *P. aeruginosa* infection in patients with cystic fibrosis, the resulting lung pathology due to the lack of lipoxin generation may indicate that the cystic fibrosis gene defect affects lipoxygenase activity directly, even in the presence of a pathogen-derived 15-lipoxygenase.

Taking the two infection models of *T. gondii* and *M. tuberculosis*, one notices an apparent discrepancy in the outcome of infection of 5-lipoxygenase-deficient animals, indicating a protective versus a host detrimental role for endogenously produced lipoxins, respectively. It is possible to conjecture that from the perspective of *T. gondii*, which is a fast-replicating pathogen, the host must be kept alive so that transmission can occur through predation. Thus, the host needs well-balanced immunity against the parasite, the number of which is kept low but not completely eliminated. To accomplish this, lipoxins are induced to keep immunity present, but not intensified. By contrast, *M. tuberculosis* is a slow growing, silent pathogen, that requires high proliferation rates in lungs of infected hosts for transmission to occur. To achieve this, lipoxins may be generated and may lower ongoing immune responses allowing enough bacilli to expand. Thus, both cases suggest that lipoxin-dependent inhibition of proinflammatory type 1 responses provides a favorable environment for transmission and propagation of the pathogens.

In summary, the emerging role for lipoxins as immune-modulatory mediators, and the potential use of their inhibitory effects for pathogen survival and replication, is still a new and poorly understood field. Important questions such as the nature of the pathogen-derived signal that contributes to lipoxin generation, or whether the anti-inflammatory effects of LXA_4 have a critical role in the balance between type 1, type 2, and regulatory T-cell responses await to be answered. Furthermore, another important key element is whether the "piracy" of lipoxygenases by pathogens constitutes a general trend for immune escape and induction of persistence in vivo. The elucidation of some of those issues may provide support for the development of therapeutic intervention in the 5-lipoxygenase/LXA_4 axis.

REFERENCES

1. Mukhopadhyay, S., Herre, J., Brown, G. D., and Gordon, S. 2004. The potential for Toll-like receptors to collaborate with other innate immune receptors. *Immunology* **112**(4):521–530.
2. Takeda, K., and Akira, S. 2005. Toll-like receptors in innate immunity. *Int Immunol* **17**(1):1–14.
3. Kopp, E., and Medzhitov, R. 2003. Recognition of microbial infection by Toll-like receptors. *Curr Opin Immunol* **15**(4):396–401.
4. Vaidya, S.A., and Cheng, G. 2003. Toll-like receptors and innate antiviral responses. *Curr Opin Immunol* **15**(4):402–407.
5. Weber, J.R., Moreillon, P., and Tuomanen, E.I. 2003. Innate sensors for Gram-positive bacteria. *Curr Opin Immunol* **15**(4):408–415.
6. Carbone, F.R., and Heath, W.R. 2003. The role of dendritic cell subsets in immunity to viruses. *Curr Opin Immunol* **15**(4):416–420.
7. Sher, A., Pearce, E., and Kaye, P. 2003. Shaping the immune response to parasites: role of dendritic cells. *Curr Opin Immunol* **15**(4):421–429.
8. Gazzinelli, R.T., Wysocka, M., Hayashi, S., et al., 1994. Parasite-induced IL-12 stimulates early IFN-gamma synthesis and resistance during acute infection with Toxoplasma gondii. *J Immunol* **153**(6):2533–2543.
9. Denkers, E.Y. 2003. From cells to signaling cascades: manipulation of innate immunity by Toxoplasma gondii. *FEMS Immunol Med Microbiol* **39**(3):193–203.
10. Reise Sousa, C., Hieny, S., Scharton-Kersten, E., et al., 1997. In vivo microbial stimulation induces rapid CD40 ligand-independent production of interleukin 12 by dendritic cells and their redistribution to T cell areas. *J Exp Med* **186**(11):1819–1829.

11. Allen, J.E., and Maizels, R.M. 1997. Th1-Th2: reliable paradigm or dangerous dogma? *Immunol Today* **18**(8):387–392.
12. Cooper, A.M., Roberts, A.D., Rhoades, E.R., Callahan, J.E., Getzy, D.M., and Orme, I.M. 1995. The role of interleukin-12 in acquired immunity to Mycobacterium tuberculosis infection. *Immunology* **84**(3):423–432.
13. de Jong, R., Janson, A., Faber, W., Naafs, B., and Ottenhoff, T.H.M. 1997. IL-2 and IL-12 act in synergy to overcome antigen-specific T cell unresponsiveness in mycobacterial disease. *J Immunol* **159**(2):786–793.
14. Flynn, J.L., and Chan, J. 2001. Immunology of tuberculosis. *Annu Rev Immunol* **19**:93–129.
15. Flynn, J.L., Goldstein, M.M., Triebold, K.J., Sypek, J., Wolf, S., and Bloom, B.R. 1995. IL-12 increases resistance of BALB/c mice to Mycobacterium tuberculosis infection. *J Immunol* **155**(5):2515–2524.
16. Flynn, J.L. 2006. Lessons from experimental Mycobacterium tuberculosis infections. *Microbes Infect* **8**(4):1179–1188.
17. Cooper, A.M., Kipnis, A., Turner, J., Magram, J., Ferrante, J., and Orme, I.M. 2002. Mice lacking bioactive IL-12 can generate protective, antigen-specific cellular responses to mycobacterial infection only if the IL-12 p40 subunit is present. *J Immunol* **168**(3):1322–1327.
18. Altare, F., Durandy, A., Lammas, D., et al., 1998. Impairment of mycobacterial immunity in human interleukin-12 receptor deficiency. *Science* **280**(5368):1432–1435.
19. Jouanguy, E., Altare, F., Lamhamedi, S., et al., 1996. Interferon-gamma-receptor deficiency in an infant with fatal bacille Calmette-Guerin infection. *N Engl J Med* **335**(26):1956–1961.
20. Serhan, C.N., Hamberg, M., and Samuelsson, B. 1984. Lipoxins: novel series of biologically active compounds formed from arachidonic acid in human leukocytes. *Proc Natl Acad Sci USA* **81**(17):5335–5339.
21. Aliberti, J. 2005. Host persistence: exploitation of anti-inflammatory pathways by Toxoplasma gondii. *Nat Rev Immunol* **5**(2):162–170.
22. Funk, C.D., Chen, X.S., Johnson, E.N., and Zhao, L. 2002. Lipoxygenase genes and their targeted disruption. *Prostaglandins Other Lipid Mediat* **68–69**:303–312.
23. Fiore, S., Romano, M., Reardon, E.M., and Serhan, C.N. 1993. Induction of functional lipoxin A4 receptors in HL-60 cells. *Blood* **81**(12):3395–3403.
24. Fiore, S., Maddox, J.F., Perez, H.D., and Serhan, C.N. 1994. Identification of a human cDNA encoding a functional high affinity lipoxin A4 receptor. *J Exp Med* **180**(1):253–260.
25. Schaldach, C.M., Riby, J., and Bjeldanes, L.F. 1999. Lipoxin A4: a new class of ligand for the Ah receptor. *Biochemistry* **38**(23):7594–7600.
26. Mandal, P.K. 2005. Dioxin: a review of its environmental effects and its aryl hydrocarbon receptor biology. *J Comp Physiol [B]* **175**(4):221–230.
27. van de Loo, F.A., and van den Berg, W.B. 2002. Gene therapy for rheumatoid arthritis. Lessons from animal models, including studies on interleukin-4, interleukin-10, and interleukin-1 receptor antagonist as potential disease modulators. *Rheum Dis Clin North Am* **28**(1):127–149.
28. Wille, U., Villegas, E.N., Striepen, B., Roos, D.S., and Hunter, C.A. 2001. Interleukin-10 does not contribute to the pathogenesis of a virulent strain of Toxoplasma gondii. *Parasite Immunol* **23**(6):291–296.
29. Deckert-Schluter, M., Buck, C., Weiner, D., et al. 1997. Interleukin-10 downregulates the intracerebral immune response in chronic Toxoplasma encephalitis. *J Neuroimmunol* **76**(1–2):167–176.
30. Gazzinelli, R.T., Tragoolpua, K., Inoue, N., et al. 1996. In the absence of endogenous IL-10, mice acutely infected with Toxoplasma gondii succumb to a lethal immune response dependent on CD4+ T cells and accompanied by overproduction of IL-12, IFN-gamma and TNF-alpha. *J Immunol* **157**(2):798–805.
31. Suzuki, Y., Sher, A., Yap, G., et al. 2000. IL-10 is required for prevention of necrosis in the small intestine and mortality in both genetically resistant BALB/c and susceptible C57BL/6 mice following peroral infection with Toxoplasma gondii. *J Immunol* **164**(10):5375–5382.
32. Salek-Ardakani, S., Arrand, J.R., and Mackett, M. 2002. Epstein-Barr virus encoded interleukin-10 inhibits HLA-class I, ICAM-1, and B7 expression on human monocytes: implications for immune evasion by EBV. *Virology* **304**(2):342–351.
33. Fiorentino, D.F., Zlotnik, A., Vieira, P., et al. 1991. IL-10 acts on the antigen-presenting cell to inhibit cytokine production by Th1 cells. *J Immunol* **146**(10):3444–3451.
34. Nathan, C. 2002. Points of control in inflammation. *Nature* **420**(6917):846–852.
35. Reis e Sousa, C., Yap, G., Schulz, O., et al. 1999. Paralysis of dendritic cell IL-12 production by microbial products prevents infection-induced immunopathology. *Immunity* **11**(5):637–647.
36. Aliberti, J., Hieny, S., Reis e Sousa, C., Serhan, C.N., and Sher, A. 2002. Lipoxin-mediated inhibition of IL-12 production by DCs: a mechanism for regulation of microbial immunity. *Nat Immunol* **3**(1):76–82.
37. Aliberti, J., Serhan, C., and Sher, A. 2002. Parasite-induced lipoxin A4 is an endogenous regulator of IL-12 production and immunopathology in *Toxoplasma gondii* infection. *J Exp Med* **196**(9):1253–1262.
38. Bafica, A., Scanga, C.A., Serhan, C., et al. 2005. Host control of *Mycobacterium tuberculosis* is regulated by 5-lipoxygenase-dependent lipoxin production. *J Clin Invest* **115**(6):1601–1606.
39. Goh, J., Godson, C., Brady, H.R., and MacMathuna, P. 2003. Lipoxins: pro-resolution lipid mediators in intestinal inflammation. *Gastroenterology* **124**(4):1043–1054.
40. Kieran, N.E., Maderna, P., and Godson, C. 2004. Lipoxins: potential anti-inflammatory, proresolution, and antifibrotic mediators in renal disease. *Kidney Int* **65**(4):1145–1154.
41. Samuelsson, B. 1991. Arachidonic acid metabolism: role in inflammation. *Z Rheumatol* 50(Suppl 1):3–6.
42. Van Dyke, T.E., and Serhan, C.N. 2003. Resolution of inflammation: a new paradigm for the pathogenesis of periodontal diseases. *J Dent Res* **82**(2):82–90.
43. Bandeira-Melo, C., Bozza, P.T., Diaz, B.L., et al. 2000. Cutting edge: lipoxin (LX) A4 and aspirin-triggered 15-epi-LXA4 block allergen-induced eosinophil trafficking. *J Immunol* **164**(5):2267–2271.
44. Clish, C.B., O'Brien, J.A., Gronert, K., Stahl, G.L., Petasis, N.A., and Serhan, C.N. 1999. Local and systemic delivery of a stable aspirin-triggered lipoxin prevents neutrophil recruitment in vivo. *Proc Natl Acad Sci USA* **96**(14):8247–8252.
45. Hachicha, M., Pouliot, M., Petasis, N.A., and Serhan, C.N. 1999. Lipoxin (LX)A4 and aspirin-triggered 15-epi-LXA4 inhibit tumor necrosis factor 1alpha-initiated neutrophil responses and trafficking: regulators of a cytokine-chemokine axis. *J Exp Med* **189**(12):1923–1930.

46. Ohira, T., Bannenberg, G., Arita, M., et al. 2004. A stable aspirin-triggered lipoxin A4 analog blocks phosphorylation of leukocyte-specific protein 1 in human neutrophils. *J Immunol* **173**(3):2091–2098.
47. Ramstedt, U., Ng, J., Wigzell, H., Serhan, CN., and Samuelsson, B. 1985. Action of novel eicosanoids lipoxin A and B on human natural killer cell cytotoxicity: effects on intracellular cAMP and target cell binding. *J Immunol* **135**(5):3434–3438.
48. Maddox, J.F., Hachicha, M., Takano, T., Petasis, N.A., Fokin, V.V., and Serhan, C.N. 1997. Lipoxin A4 stable analogs are potent mimetics that stimulate human monocytes and THP-1 cells via a G-protein-linked lipoxin A4 receptor. *J Biol Chem* **272**(11):6972–6978.
49. Devchand, P.R., Arita, M., Hong, S., et al. 2003. Human ALX receptor regulates neutrophil recruitment in transgenic mice: roles in inflammation and host defense. *FASEB J* **17**(6):652–659.
50. Machado, F.S., Johndrow, J.E., Esper, L., et al. 2006. Anti-inflammatory actions of lipoxin A(4) and aspirin-triggered lipoxin are SOCS-2 dependent. *Nat Med* **12**(3):330–334.
51. Leonard, M.O., Hannan, K., Burne, M.J., et al. 2002. 15-Epi-16-(para-fluorophenoxy)-lipoxin A(4)-methyl ester, a synthetic analogue of 15-epi-lipoxin A(4), is protective in experimental ischemic acute renal failure. *J Am Soc Nephrol* **13**(6):1657–1662.
52. Alexander, W.S., and Hilton, D.J. 2004. The role of suppressors of cytokine signaling (SOCS) proteins in regulation of the immune response. *Annu Rev Immunol* **22**:503–529.
53. Kile, B.T., Schulman, B.A., Alexander, W.S., Nicola, N.A., Martin, H.M., and Hilton, D.J. 2002. The SOCS box: a tale of destruction and degradation. *Trends Biochem Sci* **27**(5):235–241.
54. Karp, C.L., Flick, L.M., Park, K.W., et al. 2004. Defective lipoxin-mediated anti-inflammatory activity in the cystic fibrosis airway. *Nat Immunol* **5**(4):388–392.
55. Jung, Y.J., LaCourse, R., Ryan, L., and North, R.J. 2002. Evidence inconsistent with a negative influence of T helper 2 cells on protection afforded by a dominant T helper 1 response against Mycobacterium tuberculosis lung infection in mice. *Infect Immun* **70**(11): 6436–6443.
56. Bannenberg, G.L., Aliberti, J., Hong, S., Sher, A., and Serhan, C. 2004. Exogenous pathogen and plant 15-lipoxygenase initiate endogenous lipoxin A4 biosynthesis. *J Exp Med* **199**(4):515–523.
57. Vance, R.E., Hong, S., Gronert, K., Serhan, C.N., and Mekalanos, J.J. 2004. The opportunistic pathogen Pseudomonas aeruginosa carries a secretable arachidonate 15-lipoxygenase. *Proc Natl Acad Sci USA* **101**(7):2135–2139.

INDEX